The Art and Science of Growing Crystals

WILEY SERIES ON THE SCIENCE AND TECHNOLOGY OF MATERIALS

Advisory Editors: J. H. Hollomon, J. E. Burke, B. Chalmers, R. L. Sproull, A. V. Tobolsky

THE ART AND SCIENCE OF GROWING CRYSTALS
J. J. Gilman, editor

COLUMBIUM AND TANTALUM
Frank T. Sisco and Edward Epremian, editors

MECHANICAL PROPERTIES OF METALS
D. McLean

THE METALLURGY OF WELDING
D. Séférian—Translated by E. Bishop

THERMODYNAMICS OF SOLIDS
Richard A. Swalin

TRANSMISSION ELECTRON MICROSCOPY OF METALS
Gareth Thomas

PLASTICITY AND CREEP OF METALS
J. D. Lubahn and R. P. Felgar

INTRODUCTION TO CERAMICS
W. D. Kingery

PROPERTIES AND STRUCTURE OF POLYMERS
Arthur V. Tobolsky

MAGNESIUM AND ITS ALLOYS
C. Sheldon Roberts

PHYSICAL METALLURGY
Bruce Chalmers

FERRITES
J. Smit and H. P. J. Wijn

ZONE MELTING
William G. Pfann

THE METALLURGY OF VANADIUM
William Rostoker

INTRODUCTION TO SOLID STATE PHYSICS, SECOND EDITION
Charles Kittel

The Art and Science of Growing Crystals

J. J. Gilman, Editor

DIVISION OF ENGINEERING
BROWN UNIVERSITY

John Wiley & Sons, Inc., New York · London

Library of Congress Catalog Card Number: 63-11432
Printed in the United States of America

Contributors

Dr. K. T. Aust
General Electric Research Laboratory
Schenectady, New York

Dr. A. A. Ballman
Bell Telephone Laboratories
Murray Hill, New Jersey

Dr. W. H. Bauer
Ceramics Department
Rutgers University
New Brunswick, New Jersey

Drs. C. R. Berry, W. West, and F. Moser
Research Laboratories
Eastman Kodak Company
Rochester, New York

Dr. R. S. Bradley
School of Chemistry
University of Leeds
Leeds, England

Dr. S. S. Brenner
E. C. Bain Research Laboratory
United States Steel Corporation
Pittsburgh, Pennsylvania

Professor W. G. Burgers
Technische Hogeschool
Laboratorium voor Fysische Chemie der Vaste Stof
Delft, Holland

Professors N. Cabrera and R. V. Coleman
Department of Physics
University of Virginia
Charlottesville, Virginia

Dr. R. W. Dreyfus
IBM Research Center
Yorktown Heights, New York

Drs. P. H. Egli and L. R. Johnson
Solid State Division
Naval Research Laboratory
Washington, D. C.

Mr. W. G. Field
Air Force Cambridge Research Laboratories
Bedford, Mass.

Dr. R. Glang
IBM Federal Systems Division
Kingston, New York

Dr. A. J. Goss
Marconi's Wireless Telegraph Company, Ltd.
Chelmsford, Essex
England

Dr. R. O. Grisdale
Bell Telephone Laboratories
Murray Hill, New Jersey

Dr. G. T. Kohman
18 Glendale Road Summit, New Jersey
Formerly: Bell Telephone Laboratories

Dr. R. A. Laudise
Bell Telephone Laboratories
Murray Hill, New Jersey

Drs. W. D. Lawson and S. Nielsen
Royal Radar Establishment
Worcestershire, England

Prof. B. J. Mason
Department of Cloud Physics
Imperial College of Science and Technology
London, England

Dr. J. O'Connor
Massachusetts Institute of Technology Lincoln Laboratory
Lexington, Massachusetts

Dr. D. C. Reynolds
Aeronautical Research Laboratory
Wright Air Development Center
Dayton, Ohio

Dr. H. W. Schadler
General Electric Research Laboratory
Schenectady, New York

Dr. W. Tiller
Metallurgy Department
Westinghouse Research Laboratories
Pittsburgh, Pennsylvania

Dr. E. S. Wajda
IBM Federal Systems Division
Kingston, New York

Dr. L. Weisberg
R. C. A. Laboratories
Princeton, New Jersey

Dr. R. H. Wentorf
General Electric Research Laboratory
Schenectady, New York

Preface

Crystals have been admired by man for as long as he has appreciated beauty. The significance of that beauty for a technological society and for the development of scientific knowledge has only begun to be realized, however. The basis of the beauty is now known to be such things as symmetry, structural simplicity, and purity. These characteristics endow crystals with unique physical and chemical properties which have already been used to cause a major transformation of the electronics industry, and systems based on it. Yet to come is the impact of these properties on mechanical technology, on chemical processes, and on the biological sciences.

At one time natural specimens were the only source of large, well-formed crystals. Most of these were gems or museum pieces, and so were not readily available for study or utilitarian purposes. The first step in the transition of crystals from museums to technology was the development of methods for producing large crystals artificially. This provided a source of specimens for intensive scientific study which caused major developments to occur in inorganic chemistry (metallurgy, ceramics, and geophysics), as well as the fast expansion of solid-state physics. Finally, crystal production and fabrication became a major commercial activity when the transistor and its sister electronic devices were invented.

The systematic production of artificial crystals might be viewed as a new "agriculture" that has begun to flourish. It differs from true agriculture in that its products are mostly inorganic at present, but it has many features in common with normal agriculture and promises to have a somewhat comparable effect on society. This new agriculture consists of "growing" solid crystals from a "nutrient" phase (gas, liquid, or solid). To start the growth process, the nutrient is often "seeded" with a small piece of the crystal to be grown, and some workers speak of "reaping the harvest" after a certain length of time. One can judge that growing crystals is already a sizable industry from the large number of references cited in this book and the rate at which patents are issued. There is little doubt that the industry will continue to grow and along with it the scientific knowledge of crystals. In fact, what seems to be taking place is a "materials revolution." We are passing from a period when solid matter was used mainly in unsophisticated forms into a period when increasing emphasis will be placed on its use in the form of highly organized crystals.

Natural crystals have been used for some time both as gems and for a few special devices such as crystal-controlled oscillators. Now, however, the list of uses for artificial crystals is growing almost daily for such things as frequency controlled oscillators (quartz); polarizers (calcite, sodium nitrate); transducers (quartz, Rochelle salts, ammonium dihydrogen phosphate); grinding (diamond); radiation detectors (anthracene, potassium chloride); infrared optics (lithium fluoride); bearings (aluminum oxide); transistors (germanium, silicon); magnetic devices (garnets); strain gages (silicon); ultrasonic amplifiers (cadmium sulfide); masers and lasers (ruby, gallium arsenide, calcium tungstate); lenses (fluorite); melting crucibles (magnesium oxide); and tunnel diodes (gallium arsenide). This list is far from exhaustive, but it shows how several old substances (and some new ones) develop unique properties when they have the form of crystals. It seems almost certain that many other substances will reveal unique behaviors when they have also been prepared as crystals. The entire complex of organic chemicals would seem to be especially fertile in this respect, as suggested by the unusual properties of crystallized high polymers.

It is somewhat ironic that although artificial crystal growth has had so much effect on various sciences, the process itself has retained more of the flavor of art than of science. This is partly due to a lack of really powerful experimental tools for studying the critical steps of the growth process, and partly due to gradual changes in the "definition" of what is meant by thc tcrm crystal. An example of the effect of the lack of experimental tools is the fact that the crucial role of surface diffusion rates has been proved only recently. It has long been known that surface diffusion is one step in the growth process, but there was no demonstration that this is often the critical step.

When artificial crystals were first produced, workers were satisfied with the quality of the product if it had the same symmetry as its natural counterpart. Now they often insist that the crystal have remarkable purity (less than one part per million of impurity), and some crystals have been produced that are not only pure but completely free of such subtle defects as dislocation lines.

This book attempts to present the state of crystal growth as seen by some of the leaders in its development. Both the theory and the art are discussed in great detail. This should help the novice to appreciate the state of sophistication of the subject. It is also hoped that this compilation of numerous aspects of the subject will aid experienced workers to advance further the art and science.

The editorial work on this book was supported in part by the United States Air Force Office of Scientific Research. I am also grateful to Mrs. J. Bradsley for carefully making many revisions.

J. J. Gilman
Division of Engineering
Brown University

March 1963

Contents

VAPOR GROWTH

General Principles

1

Theory of Crystal Growth from the Vapor

N. Cabrera and R. V. Coleman

Crystal growth has been for a long time one of the fascinating observations where the atomic nature of matter clearly shows itself. Since the first attempts to study crystal growth, it has been realized that the subject can be divided into two parts: (1) The study of the equilibrium between the crystal and the surrounding medium and (2) the study of the kinetics of growth.

Gibbs (1878) developed very early a consistent phenomenological (thermodynamical) treatment of the equilibrium problem, which is still essential as an introduction to the study of crystal growth. Part I of this chapter reviews this treatment following the recent extensions by Herring (1953) and Chernov (1961). This very successful phenomenological treatment is made possible through the existence of the variational principles of thermodynamics and does not need the introduction of the atomic structure of matter. Consideration of atomic structure and the use of statistical mechanical methods, however, are necessary in order to develop a complete theory and, most important, to bring out the possible atomic processes responsible for the growth of the crystal. It should be emphasized that much remains to be done to develop the statistical mechanics of crystal surfaces to a satisfactory state. This, however, is not a serious difficulty since the phenomenological treatment is sufficient to give us a basis on which to classify all the very diverse types of surface equilibrium phenomena.

In trying to understand the kinetics of growth, the earlier workers, particularly Curie (1885), attempted unsuccessfully to extend Gibb's phenomenological treatment to the growth problem. For a long time the problem remained very obscure until it was realized that the atomic structure of matter had to be introduced from the beginning. Following an illuminating footnote by Gibbs himself, Kossel (1927) analyzed the atomic inhomogeneity of a crystal surface and emphasized the fundamental importance to the growth process of *molecular steps* and *molecular kinks* along the steps. This was, however, not enough to provide a complete basis for understanding the problem. The final breakthrough came when Frank (1949) showed that *crystal dislocations* were capable of providing the *sources of steps* required for the continuous growth of a crystal. This was followed by great advances in the understanding (at least qualitative) of the many characteristics of crystal growth. There seems to be no doubt at present that all the necessary atomic mechanisms are now available and that only mathematical difficulties prevent a quantitative understanding of crystal growth. Partially successful theories have already been developed, starting with the paper by Burton, Cabrera, and Frank (1951). Much remains to be done, however, particularly in the understanding of the very important role played by impurities in the process of growth. Part II of this chapter attempts to review these theories and emphasizes the difficulties that still remain.

Unfortunately, all these atomic theories are necessarily very complicated; thus it would be helpful if a simpler phenomenological theory could be developed that would be the counterpart of thermodynamics and would allow us to classify the very numerous and apparently contradictory experimental facts. Attempts in this direction have been made by Frank (1958) and by Cabrera and Vermilyea (1958). Part II includes discussion of these efforts, although admittedly they have not been fully successful. It might be that this failure is because it has not been possible as yet to establish a variational principle with sufficient generality to be applicable to the nonlinear kinetic processes involved in crystal growth.

This review on the theory of crystal growth from the vapor appears when several other reviews covering the same material have just come out or are in the process of being published. Among them are the excellent review by Chernov (1961) and the book by Hirth and Pound. Therefore we have not attempted to cover all the literature, but have tried to emphasize viewpoints that might seem controversial.

I. STRUCTURE OF CRYSTAL SURFACES

The structure of crystal surfaces in equilibrium is important in determining the mechanisms and kinetics of crystal growth from the vapor or otherwise. Conclusions about the orientations and structure of crystal bounding surfaces can be drawn by considering the general properties of surface energy and by applying simple nearest neighbor models of molecular cohesion.

Early work on the theory of crystal growth by Gibbs (1878), Curie (1885), and Wulff (1901) made extensive use of the polar diagram of surface free energy as a function of orientation, and Wulff in particular worked out a construction for deriving the equilibrium shape of a crystal from the polar diagram. This approach to equilibrium shapes and the various corollaries to Wulff's theorem, particularly those referring to the equilibrium structure, are useful concepts. Some details are discussed further in the section on surface energy. We should emphasize, however, that the actual growth form of a crystal is not the equilibrium form, but depends largely on the growth and dissolution rates as a function of crystallographic orientation and on the kinetics of step motion on these faces.

As yet there is no rigorous or complete theory for the type of crystal surface to be expected for an arbitrarily shaped crystal under various conditions of temperature, supersaturation, impurity content, etc. However, clear classifications of interfaces between two phases according to their thermodynamic properties, their atomic structure, and their kinetic behavior have been made and definite predictions are possible.

1. CLASSIFICATION OF SURFACES

The following three types of surfaces which have been introduced by Cabrera (1959), Frank (1958a) and others, form a clear basis for the subsequent discussion: (1) singular surfaces, (2) vicinal surfaces, and (3) nonsingular or diffuse surfaces.

From the thermodynamical viewpoint, singular surfaces are those corresponding to a cusped minimum in Wulff's plot giving surface energy as a function of orientation. The orientations immediately around a singular one correspond to vicinal surfaces, and for these the surface energy increases linearly for all orientations away from the singular one. In general, vicinal faces would correspond to a stepped structure in which fairly wide, low index plateaus are separated by monomolecular "risers" or *steps*. All other orientations can be included in the nonsingular type of surfaces, their surface tension being roughly independent of orientation.

From the atomic viewpoint, singular and nonsingular interfaces differ in the number of molecular layers *parallel* to the surface involved in the transition from the crystal to the vapor. For a singular surface the transition occurs in one layer, whereas several layers are involved in a nonsingular interface; hence the name "diffuse interface" used particularly by Cahn and Hilliard (1958). This does not mean, however, that the actual thickness of the transition region is very different, because the thickness of the molecular layer is larger for the singular case.

The differentiation between the two kinds of surfaces is also quite important during the process of growth, as will become evident in Part II. Singular interfaces can only grow by the lateral motion of molecular layers, whereas nonsingular interfaces could be able to grow normal to themselves.

Returning to equilibrium considerations, Landau (1950), and Herring (1953) pointed out that, for an ideal lattice of fixed molecules, there cannot be any nonsingular surfaces. All orientations corresponding to rational Miller indices should correspond to cusps in the Wulff plot and consequently be singular, the remaining orientations being vicinal. Most of these cusps, fortunately, are washed out by the zero point vibrations or by the thermal vibrations at very low temperatures, leaving only the most important ones corresponding to the surfaces with higher symmetry surface-lattices. Even for these there should be, in principle, a critical temperature above which the cusp in the Wulff plot disappears and the transition region increases from one to several molecular layers. The first definition of critical temperature at which a surface becomes nonsingular was proposed by Herring (1953), the second by Burton, Cabrera, and Frank (1951). For a crystal in equilibrium with its vapor all the evidence is that all solids have singular surfaces up to their melting temperature. This might, however, not be true for crystals in the presence of their melt when the latent heat of fusion is small.

2. SURFACE ENERGY AND SURFACE STABILITY

The general problem of the equilibrium shape of a crystal and its relation to surface energy and chemical potential has been formulated most clearly by Herring (1953) and reviewed in detail by Chernov (1961). In the following section we limit our discussion to the problem of macroscopic surface structure, which is particularly valuable in reaching conclusions about the growth of a real crystal.

Let us consider some crystal surface of arbitrary orientation in contact with the vapor phase. In the rest of this chapter, we define this orientation with respect to a reference singular orientation which we take to coincide with the plane $z = 0$. In general, the crystal surface is represented by a function

$$z = z(x, y) \tag{1}$$

and the orientation at the point x, y is given by the two independent components

$$p = -\frac{\partial z}{\partial x} \quad \text{and} \quad q = -\frac{\partial z}{\partial y} \tag{2}$$

It is very useful to realize that if h is the thickness of molecular layers parallel to the singular surface $p = q = 0$, then p/h represents the density of steps in the surface in the orientations p, $q = 0$. In the following discussions we often simplify our problem by considering only the two-dimensional case of surfaces all having $q = 0$, but the generalization to the three-dimensional case is pointed out.

We wish now to examine the stability of a particular infinite plane surface ($z = z_0 - px$) against break up into other surface profiles, and if it is unstable, to determine which new orientations p_1 and p_2 should appear (see Fig. 1).

Let an area on the reference surface be denoted by B and consider an area A of the arbitrary surface such that its projection onto the reference surface is equal to $B = A\sqrt{1 + p^2}$. The surface energy of surface A is given by $\alpha(p)A$, which in terms of the reference surface area is equal to $\alpha(p)\sqrt{1 + p^2}\,B$. In order to examine the stability of the surface A we look at the function $\beta = \alpha\sqrt{1 + p^2}$ as a function of p. The general form of this function is shown in Fig. 2, with the cusp corresponding to the singular reference surface at $p = q = 0$ and

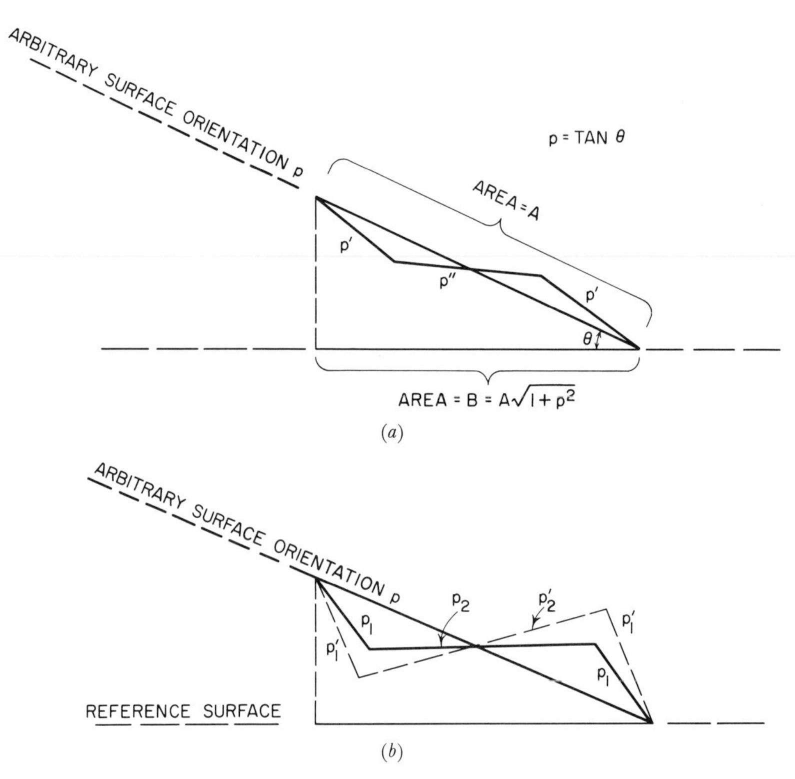

Fig. 1. (*a*) Figure indicates relations between close-packed reference surface and arbitrary surface. Stability is examined by allowing reference surface to form new surfaces of orientations p' and p''. (*b*) Arbitrary surface is unstable and breaks up into surfaces of orientations p_1 and p_2. Stability of new surfaces p_1 and p_2 is examined by allowing new surfaces p_1' and p_2' to form.

another cusp at another singular surface at $p = p_0$, $q = 0$. Consider now a surface A with orientation $p, q = 0$. In order to examine the conditions for its stability imagine that two new surfaces of slightly different orientation designated by p' and p'' are formed such that their total projected areas B' and B'' on the reference surface satisfy the condition

$$B = B' + B'', \qquad p'\,B' + p''\,B'' = pB \quad (3)$$

These two conditions insure that constant volume is maintained during the surface rearrangement. The initial energy of surface A corresponding to the function β and orientations p is given by

$$F = \beta(p)B$$

For the new surface, made up of orientations p' and p'', we have

$$F' = \beta(p')B' + \beta(p'')B''$$

Suppose now that $p' = p + \delta p'$ and $p'' = p + \delta p''$ then the increase in the surface energy of the system is given by

$$\Delta F = F' - F = \tfrac{1}{2}\beta''(p)[B'(\delta p')^2 + B''(\delta p'')^2]$$

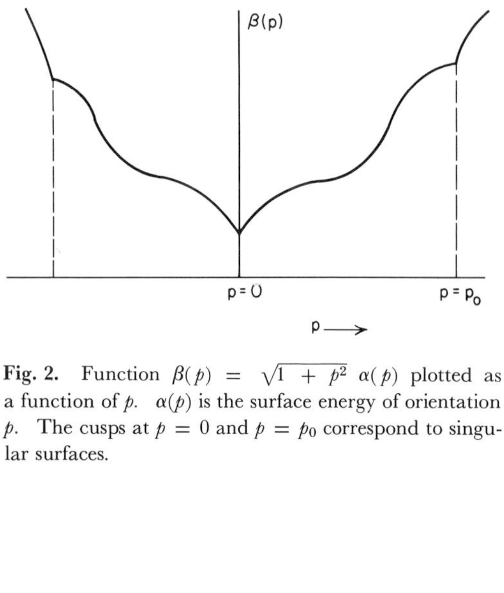

Fig. 2. Function $\beta(p) = \sqrt{1 + p^2}\,\alpha(p)$ plotted as a function of p. $\alpha(p)$ is the surface energy of orientation p. The cusps at $p = 0$ and $p = p_0$ correspond to singular surfaces.

where $\beta' = d\beta/dp$, and $\beta'' = d^2\beta/dp^2$. The last expression is obtained since the zero and first-order terms in $\delta p'$ and $\delta p''$ vanish because of the constant volume condition (3). The condition for stability of surface A is then obtained by requiring the change in energy to be positive, $\Delta F > 0$. This condition is satisfied if

$$\beta'' > 0 \tag{4}$$

which requires the curve of β as a function of p to be concave upward for values of p corresponding to stable orientations.

For orientations on the curve for which $\beta''(p) < 0$, the surface is unstable and new faces form. To determine the new orientations which will develop we look at the conditions for stability of two new faces designated by p_1 and p_2 in Fig. 1. We suppose each of these to be misoriented by a small amount to orientations $p_1 = p_1 + \delta p_1$ and $p_2 = p_2 + \delta p_2$ and again require the change in energy to be positive. The constant volume conditions here are given by

$$B_1 + B_2 = B_1' + B_2' \quad \text{and}$$
$$B_1p_1 + B_2p_2 = B_1' p_1 + B_2' p_2' \tag{5}$$

The change in energy is given by

$$\Delta F = \beta(p_1')B_1' + \beta(p_2')B_2' - [\beta(p_1)B_1 + \beta(p_2'')B_2]$$

which is again expanded in powers of δp_1 and δp_2. The first-order terms do not vanish any longer unless

$$\beta_1' = \frac{\beta_2 - \beta_1}{p_2 - p_1} = \beta_2' \tag{6}$$

which implies also that

$$\Delta F = \tfrac{1}{2}[\beta_1'' \; B_1(\delta p_1)^2 + B_2'' \; B_2(\delta p_2)^2]$$

which will always be positive, provided

$$\beta_1'' \; (p) > 0 \quad \text{and} \quad \beta_2'' > 0 \tag{7}$$

According to these conditions information on the specific orientations which will develop from an arbitrary surface can be obtained by drawing tangents to the curve of β as a function of p. Three possibilities are indicated in Fig. 3. Orientations corresponding to $p = 0$ and p_0 are singular orientations.

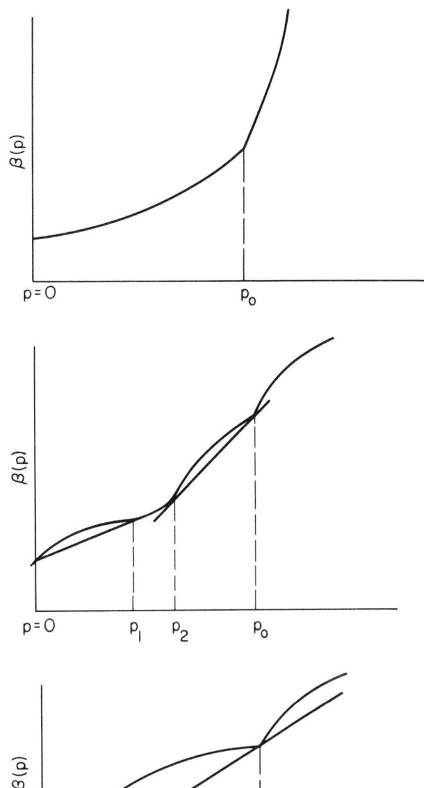

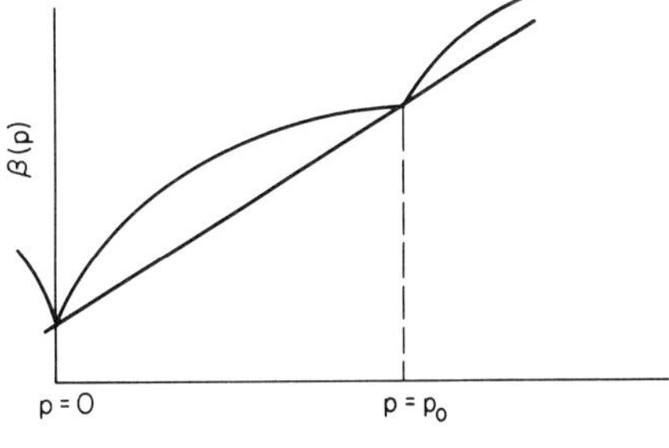

Fig. 3. Case I. All orientations between $p = 0$ and $p = p_0$ are stable. Case II. Orientations between p_1 and p_2 are stable in addition to the singular surfaces $p = 0$ and $p = p_0$. Other orientations are unstable. Case III. Only the orientations $p = 0$ and $p = p_0$ are stable.

The two limiting possibilities are labeled case I and case III. For case I all orientations of the crystal in equilibrium are stable, and for case III only the singular orientations $p = 0$ and p_0 are stable. Case II represents the intermediate possibility where there is a range of orientations between p_1 and p_2, which are also stable in equilibrium, as well as the two singular orientations. Facets of these orientations therefore develop in addition to the singular faces.

Remembering that p is proportional to the density of steps it is useful to expand $\beta(p)$ in terms of p, for small p,

$$\beta = b_0 + b_1p + \tfrac{1}{2}b_2p^2 \tag{8}$$

where $b_2 > 0$ in case I and $b_2 < 0$ in cases II and III. The p^2 term actually represents a free energy of interaction between steps, repulsive in case I and attractive in the other cases. It follows then naturally that in case I there is no tendency for the agglomeration of molecular steps into big macroscopic steps, whereas this occurs in the other two cases. Furthermore from formula (8) we can write the free energy per unit length of step ϵ as

$$\epsilon = (b_1 + \tfrac{1}{2}b_2 p)h \tag{9}$$

which will be useful later.

In three dimensions, the analysis of the problem proceeds along similar lines. The surface orientation is specified by two functions p and q. The surface energy β is plotted as a function of p and q as shown in Fig. 4. The coordinates of a point on the resulting surface then specify p, q, and β for a given surface orientation. The conditions for stability of an arbitrary orientation will again be determined by the derivatives of β with respect to p and q. The surface orientations that will develop for the equilibrium crystal shape can be specified by drawing tangent planes at points on the surface. A plane can be tangent to the surface at one, two, or three points. For a plane with only one tangent point, the orientation corresponding to this point is stable, providing the surface β is everywhere above the tangent plane. If two tangent points exist for a plane, orientations along a line between the two tangent point orientations transform into the two stable orientations specified by the tangent points. When a plane is tangent at three points on the surface, there exist three stable orientations forming a triangle in the p, q plane. Orientations within this triangle transform into these three stable ones.

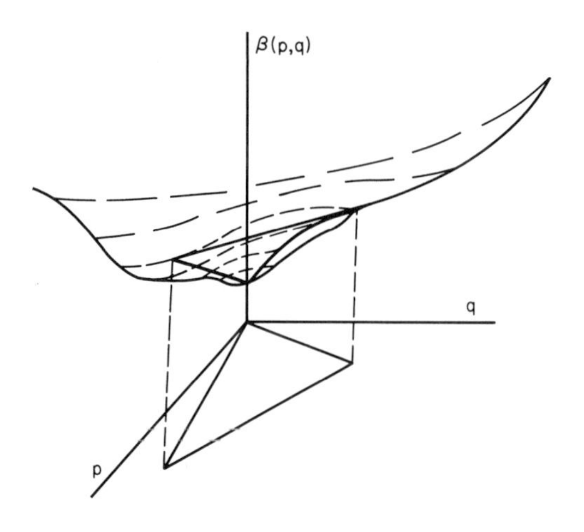

Fig. 4. Three-dimensional plot of $\beta(p, q)$ as a function of p and q. Vertices of triangle represent points of tangency of plane to the general three-dimensional surface.

The relative projected areas of the three orientations that develop are easily calculated by considering the three areas of the triangle found by drawing lines from the initial orientation to the vertices of the triangle.

As just considered, these concepts are strictly applicable only under equilibrium conditions, and as such are extremely valuable for the analysis of the phenomena variably referred to as "thermal etching," striations, surface equilibration, etc. This was first observed by Chalmers, King, and Shuttleworth (1948), who proved that silver in the presence of an inactive gas tended to be smooth at high temperature, whereas striations appeared readily as soon as an active gas (such as oxygen) was introduced. Furthermore, this was proved to be reversible, indicating definitely that it occurred under equilibrium conditions. These experiments have been extended to many other systems by several workers, particularly Benard (1960) and Mykura (1961) and their collaborators. The over-all conclusion of these experiments is that metallic crystal surfaces in the presence of their own vapor or of inactive gases (including apparently hydrogen) appear to be stable for any orientation. This is even more noticeable for face centered cubic crystals than in body centered cubic crystals; hence they correspond to case I. On the contrary, small amounts of active gases (such as oxygen) tend to change $\beta(p)$ by decreasing its value as well as changing its shape from case I to case II and sometimes even case III. Furthermore, the orientations that were singular under pure conditions sometimes become nonsingular under the effect of active impurities; this was suggested by Cabrera (1952) sometime ago. According to Blakely and Mykura (1961), the reverse appears also to happen for nickel. It is still premature to deduce definite conclusions from the experimental data known at the present time. However, from the data collected by Mykura it appears that the magnitudes of the constants b_0, b_1, and b_2 in the development (Eq. 8) of β in powers of p are roughly the same.

As far as crystal growth is concerned, all this work is extremely valuable as will be discussed in Part II. At least it indicates that during growth, crystal surfaces should show no macroscopic steps if they belong to case I, but will do so if they belong to case II or III.

3. SURFACE DEFECTS AND THEIR MOBILITY

It is well known that a crystal lattice in equilibrium contains a certain concentration of vacancies which depends exponentially on temperature according to the following relation

$$n_v = n_0 \exp\left(-\frac{W_v}{kT}\right) \qquad (10)$$

where n_0 ($\simeq 10^{23}\ \text{cm}^{-3}$) is roughly equal to the number of lattice points and W_v is the energy required to form a vacancy. This expression is to be compared to the concentration of molecules in the vapor.

$$n = n_0' \exp\left(\frac{-W}{kT}\right) \qquad (11)$$

where $n_0' \simeq 10^{19}\ \text{cm}^{-3}$ and W is the heat of vaporization. On using a simple nearest-neighbor-bond approximation, W_v and W should be equal; however, because of relaxation effects occurring in the lattice around the vacancy, W is usually much larger than W_v, and so $n_v \gg n$, sometimes by several orders of magnitude.

On exactly the same statistical mechanical basis we can also predict that singular surfaces will develop surface vacancies, with surface concentration $n_v^{(s)}$, and surface adsorbed molecules with surface concentration $n^{(s)}$. Both surface concentrations are given by formulas analogous to Eqs. (10) and (11)

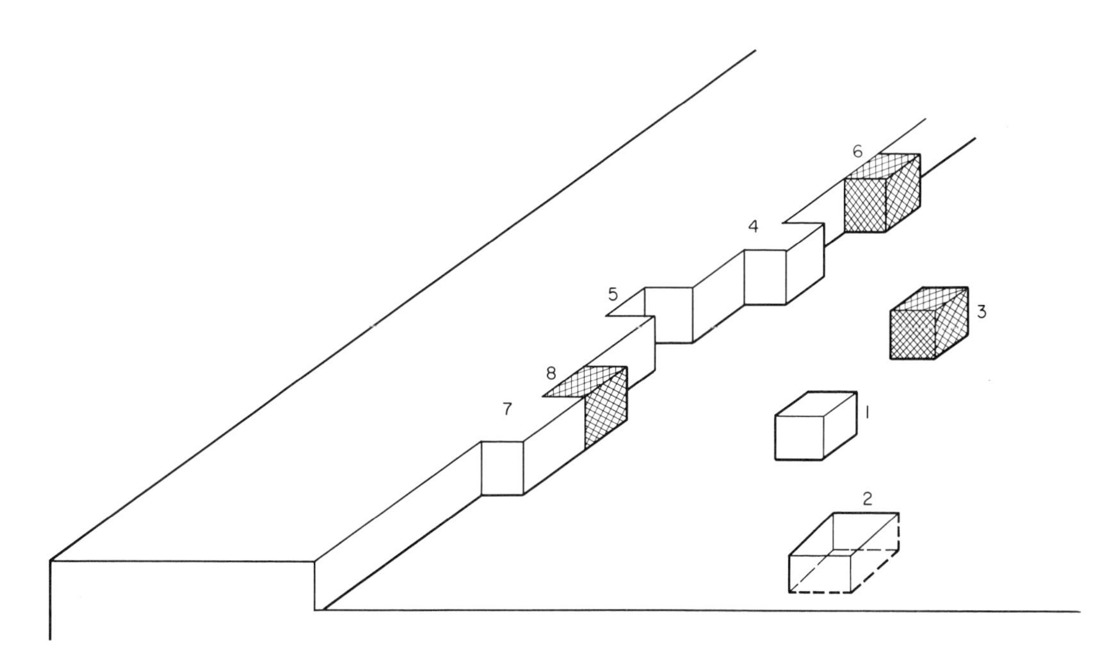

Fig. 5. Point defects on surfaces and steps. 1. Adsorbed surface molecules. 2. Surface vacancies. 3. Adsorbed impurity molecules. 4. Adsorbed edge molecules. 5. Edge vacancies. 6. Adsorbed edge impurity. 7. Pure kinks. 8. Poisoned kinks.

$$n_v^{(s)} = n_0^{(s)} e^{-W_v^{(s)}/kT}, \qquad n^{(s)} = n_0^{(s)} e^{-W^{(s)}/kT} \tag{12}$$

where $n_0^{(s)} \approx 10^{15}$ cm^{-2} is roughly equal to the number of surface-lattice positions. Figure 5 shows the two kinds of point defects (position 1, surface vacancy and position 2, adsorbed molecule) which were pointed out by Volmer (1922) as the intermediate states in the formation of a vacancy in the crystal or a molecule in the vapor. The two relations

$$W_v = W_v^{(s)} + A_v \qquad W = W^{(s)} + A$$

define the adsorption energies A_v and A for the two surface defects, or in other words, the energies required to evaporate them into the corresponding third dimension. For the same reason that $W_v < W$ one would also expect $W_v^{(s)} < W^{(s)}$. However, the relaxation effect around surface defects is likely to be considerably smaller, so that $W_v/W < W_v^{(s)}/W^{(s)} < 1$. It is therefore likely that the concentration of surface vacancies is higher than that of adsorbed molecules. Regarding the absolute values one might expect as a rough estimate $W^{(s)} \sim \frac{1}{2}W$. On this basis the proportion of point defects on the surface will be substantially larger than in the volume. For the lowest energy singular surfaces this proportion will still be small up to the melting point. However, for the higher energy singular surfaces the proportion of surface defects increases rapidly with temperature in a cooperative way up to a critical temperature, above which the surface "melts" and becomes nonsingular.

The importance of point defects on the surface is, of course, due to their role in any transport process. Thermal vibrations occur both parallel and normal to the crystal surface. Fluctuations in the energy of these oscillations cause the defects to either jump to neighboring sites on the surface or leave the surface. The frequency with which a point defect jumps on the crystal surface determines its surface diffusion coefficient which can be expressed

$$D_v^{(s)} \sim a^2\nu \exp\left(\frac{-U_v^{(s)}}{kT}\right), \qquad D^s \sim a^2\nu \exp\left(\frac{-U^{(s)}}{kT}\right) \tag{13}$$

where $\nu \sim 10^{12}$ sec^{-1} represents some average of molecular frequencies and $U_v^{(s)}$ and $U^{(s)}$ are activation energies between neighboring equilibrium positions on the surface distant a from each other.

It is very likely that $U_v^{(s)} > U^{(s)}$, which combined with the result $W_v^{(s)} < W^{(s)}$ does not allow us to give a general answer to the question of which of the two factors, $n_v^{(s)} D_v^{(s)}$, or $n^{(s)} D^{(s)}$ (measuring the magnitude of the surface currents of vacancies and adsorbed molecules) will be largest. To simplify the problem we presume in what follows that the current of adsorbed molecules is more important. However, we point out some cases where this might not be true.

The frequency with which an adsorbed molecule leaves the surface defines its lifetime $\tau^{(s)}$ which can usually be written as

$$\tau^{(s)} = \nu^{-1} \exp\left(\frac{A}{kT}\right) \tag{14}$$

where A has already been defined. Combining $D^{(s)}$ and $\tau^{(s)}$ we define the *average mean displacement of an adsorbed molecule* λ_s by

$$\lambda_s \sim (D^{(s)}\tau^{(s)})^{1/2} \sim a \exp\left(\frac{A - U^{(s)}}{2kT}\right) \tag{15}$$

which plays a very important role in problems of crystal growth from the vapor. In fact, most of the information one deduces from these experiments consists of the determination of the corresponding value for λ_s. From the data collected so far it appears that λ_s can have values between 10^{-6} to 10^{-3} cm. Since $A \gg U^{(s)}$, one expects λ_s to increase very rapidly with decreasing temperature. This is indeed what appears to occur from the data. It is unfortunate, however, that no systematic observation of the temperature effect has been made as yet.

The discussion until now refers only to a singular surface under high purity conditions. As is well known, extremely small amounts of impurities will change things considerably, but we know of no effort to collect systematic information in this direction. We might suspect that in the presence of an adsorbate, the surface vacancies will be predominant in controlling surface transport.

If we now turn to nonsingular surfaces, we expect to find a very different situation. Indeed, a nonsingular surface can be considered to represent a liquid surface; so the concept of point defects loses its meaning. On the same basis, the diffusion coefficient becomes much smaller, corresponding to an activation energy of the order of the energy to move a vacancy in the interior of the crystal. This viewpoint has been considered particularly by Blakeley and Mykura (1961), who have recently measured the diffusion coefficient on nonsingular surfaces.

4. STRUCTURE OF MONOMOLECULAR STEPS

As mentioned in the previous section, vicinal surfaces slightly inclined to singular orientations have stepped structures as shown in Fig. 7. The spacing of these *steps* varies with orientation, the average distance between them being $\lambda = h/p$, where h is their height. At absolute zero, the front of the step is atomically smooth. As the temperature is raised, however, point defects appear at the edge of the steps, consisting of both adsorbed molecules and vacancies like those that appear on the surface itself. Again we might expect the energies of formation to be such that $W_v^{(e)} < W^{(e)}$, but their ratio should definitely be near unity as the possible relaxations are no longer very different.

This is not, however, the whole story as it was for the surface. Indeed, it is easy to see that groups of edge-adsorbed molecules or edge vacancies should form without having to expend very much extra energy, so the significant point defects along a step are the *kinks* formed in the steps as shown in Fig. 5. The two types are designated *positive* and *negative,* corresponding to a jump forward in the step or a drop backward as indicated. During the growth of a crystal the kinks form the active sites for the addition of atoms to the crystal as has already been suggested by Kossel (1927). Indeed, the growth process involves surface adsorption of atoms, diffusion to a step, diffusion along the step to a kink, and incorporation into the crystal at the kink site, so that the equilibrium concentration of kinks on a step becomes a fundamental quantity. To calculate the kink density in a given step, let q be the probability that there is no jump of any kind at a point on the step, and let n_+ and n_- be the probabilities of finding a positive or negative kink at any point along a step of orientation $p = \tan\theta$ with respect to a close-packed orientation. For a long, straight step the result is deduced from the following expressions:

$$n_+ + n_- + q = 1,$$

$$n_+ - n_- = p,$$

$$n_+ n_- = q^2 e^{-2w/kT}$$

where w is the energy required to form a kink. To the approximation used here we neglect the difference between $W_v^{(e)}$ and $W^{(e)}$; in fact, $W_v^{(e)} = W^{(e)} = 2w$. These formulas lead to a mean distance between kinks given by

$$\lambda(p) = \left[1 - \tfrac{1}{2}\left(\frac{\lambda_0}{a}\right)^2 p^2\right]_{\lambda_0}$$

for orientations corresponding to p small, where

$$\lambda_0 = \tfrac{1}{2}a(e^{w/kT} + 2)$$

is the separation of kinks in a close-packed step $p = 0$ with interatomic distance a. From simple considerations it appears that $w \sim W/10$, where W is the heat of vaporization. Then for a typical value $w/kT \sim 2$, $\lambda_0 \sim 4a$, and the separation of kinks even in a close-packed step is so small that it is not necessary to consider the edge diffusion problem at all.

We shall see later that the concentration of kinks in the steps remains practically unchanged as the vapor becomes supersaturated, and the growth problem becomes a simple diffusion problem. Indeed, the step can be regarded as a continuous line sink for surface diffusing atoms.

Furthermore, the high entropy of the structure of the step makes a substantial contribution to the edge energy of the step, so that actually the edge energy instead of having a cusp as it has at $T = 0$ for the close-packed orientations becomes proportional to p^2 for $T > 0$ and small p. This is to be compared to the situation on the surface where the cusp cannot be eliminated until very high T.

We now consider the influence of the presence of a second component (impurity) on the structure of the step. As we have mentioned several times, impurities change the equilibrium structure considerably, particularly for steps, as has been shown recently by Chernov (1961).

We consider for simplicity a step in a close-packed orientation. The various types of sites that must be considered now are q (no kink of any kind), $n_+ = n_-$ (free kink), $m_+ = m_-$ (poisoned kink), m (impurity adsorbed at edge of step). The density of kinks is now determined by the equations

$$q + 2n_+ + 2m_+ + m = 1, \qquad n_+ = qe^{-w/kT},$$

$$m_+ = q\frac{P_i}{P_0}e^{(w_i - w)/kT},$$

$$m = q\frac{P_i}{P_0}e^{w_i{}^e/kT}$$

where w is the energy to form a free kink, w_i and $w_i{}^e$ the adsorption energy of an impurity at a kink and at the edge respectively. P_i is the partial pressure of the impurity, and $P_0{}^2 = (2\pi m v^2/kT)^3$. Solving for n_+ we find the distance between free kinks given by

$$\lambda(P_i) = \lambda(0) + \xi P_i,$$

$$\xi = \frac{a}{2P_0}(e^{(W_i{}^e + w)/kT} + 2e^{w_i/kT})$$

For typical values of the constant ($w_i \sim w_i{}^e + w \sim 0.5$ ev, we obtain $\xi \sim 10^4 a/$mm Hg, so that relatively small values of P_i are sufficient to increase substantially the distance between *free kinks*. On the other hand, solving for m_+ we find the distance between poisoned kinks $\lambda_i(P_i)$ to be given by $\lambda_i(P_i)/\lambda(P_i) = (P_0/P_i)\exp(-w_i/kT)$ so that if $\lambda(0) \ll \xi P_i$ nearly every position in the step becomes a poisoned kink.

The first consequence of this result is that the line free energy of the step will be drastically reduced by the presence of the impurity as both the energy is reduced and the entropy is increased. The situation from the viewpoint of the growth is discussed more fully in Section II. If the incorporation of molecules into the crystal can only occur at the free kinks, as assumed by Chernov (1961), the growth rate should then be drastically reduced at reasonable impurity levels. On the other hand, the diffusion currents to poisoned kink sites, which depend on the relative lifetimes of the various atomic jump processes, might under some circumstances contribute a significant amount to the growth rate.

5. CURVATURE OF STEPS. SURFACE NUCLEATION

Until now we have only considered conditions of stable equilibrium when the vapor is saturated (the chemical potential has the equilibrium value μ_e) and the crystal surface must be atomically flat or composed of pieces of atomically flat surfaces with stable orientations (singular orientations such as $p = 0$ and $p = p_0$ and nonsingular orientations between p_1 and p_2 in Fig. 3).

The situation changes if the vapor is supersaturated (or undersaturated) so that

$$\mu = \mu_e + \Delta\mu, \qquad \Delta\mu = kT\ln\frac{P}{P_e}$$

where P is the actual vapor pressure and P_e the equilibrium vapor pressure. If the surface is not singular, then orientations between B and B' (such that $\beta'' > 0$) become possible, and the actual surface when in (unstable) equilibrium with the vapor shows a curvature $R_c{}^{-1}$ which will be connected to the supersaturation $\Delta\mu$ according to Herring's formula (for two dimensions)

$$\Delta\mu = \Omega(1 + p^2)^{3/2} \frac{\beta''(p)}{R_c}$$

where Ω is the volume per molecule in the crystal. This formula generalizes the familiar Gibbs-Thomson formula for fluids, where $\beta = \alpha\sqrt{1 + p^2}$ with α equal to a constant. The equilibrium is unstable in the sense that the nonsingular surface grows if its curvature $R^{-1} < R_c^{-1}$ and evaporates if $R^{-1} > R_c^{-1}$.

On the other hand, if the surface is a singular one, a metastable equilibrium is established in which the surface cannot advance. If, however, a step happens to be on the crystal surface, this step is also in unstable equilibrium with the vapor if its two-dimensional curvature ρ_c^{-1} is again given by the two-dimensional analogue of Herring's formula

$$\Delta\mu = \frac{\Omega\epsilon}{h\rho_c} \tag{16}$$

where ρ_c is the radius of curvature, ϵ the energy of the step per unit length (assumed independent of orientation), and h its height. The step is in unstable equilibrium, so that for $\rho > \rho_c$ the step will grow, increasing its radius of curvature, and for $\rho < \rho_c$ it will evaporate.

If no step existed on the surface originally, there is still the possibility of creating one by a process of *nucleation,* where critical nucleus with radius ρ_c is created by a fluctuation and then continues to grow to form a new molecular layer.

The number of critical nuclei created per second is proportional to $\exp(-\Delta F^*/kT)$, where ΔF^* is the activation energy for nucleation and is equal to one-half the total edge free energy of the nucleus. Indeed, if we consider a disk-shaped nucleus on the crystal surface (in agreement with the assumption that ϵ is constant), then the free energy of formation is given by

$$\Delta F = 2\pi r\epsilon - \pi r^2 h \frac{\Delta\mu}{\Omega}$$

where $\Delta\mu$ = the change in chemical potential equal to $-kT\ln(P/P_e)$
r = radius of nucleus
h = height of monolayer
ϵ = edge free energy of monolayer
P = vapor pressure
P_e = equilibrium vapor pressure
Ω = volume per molecule in the crystal

The free energy of formation of a critical nucleus is obtained by maximizing ΔF with respect to r and is given by

$$\Delta F^* = \pi\rho_c\epsilon = \frac{\pi\epsilon^2\Omega}{h\,\Delta\mu} \tag{17}$$

and ρ_c is given by the same expression (16).

In general, the nucleation rate is very sensitive to the value of $\Delta\mu$, increasing very rapidly as P/P_e changes in a narrow range. This is usually expressed in terms of a critical supersaturation ratio determined by the condition $\Delta F^* = 50kT$, above which nucleation is appreciable and below which it is very small. This critical value is given by the following expression

$$\frac{P}{P_e} = \exp\left[\frac{\pi\epsilon^2\Omega}{50h(kT)^2}\right] \tag{18}$$

For typical values of the constants involved we obtain $(P/P_e)_c \sim 2$.

On the other hand, real crystals are known to grow from the vapor for values of $P/P_e \sim 1.01$, for which the nucleation rates are negligibly small. The following section briefly describes Frank's dislocation mechanism of crystal growth which solved the preceding fundamental disagreement between theory and experiment.

We then conclude that a crystal should only grow if (1) some of its surfaces are nonsingular surfaces and (2) some of its singular surfaces contain sources of steps, either surface nucleation sources or dislocation sources (Fig. 5).

In the case of evaporation ($\Delta\mu < 0$), we should add to the two conditions above the possibility of the edges of the crystal to act as a source of steps. If the crystal under consideration corresponds to either case I or case II in Fig. 3, the edges of the crystal should be composed of nonsingular surfaces, and therefore one would again refer to case (1) which has been studied by Hirth and Pound (1957). However, if the crystal corresponds to case III in Fig. 3, then only singular surfaces appear in the edges and again a *nucleation* process is required for the evaporation from the edges to take place. This exception appears not to have been considered before and consequently is treated here with some detail.

Let us suppose that a crystal edge consists of two singular surfaces $p = 0$ and $p = p_0$ (Fig. 6). The initiation of the evaporation at the edge consists of the motion of a monomolecular layer from the edge A to A', distant λ from each other. Assuming that the surface energy $\beta(p)$ can be represented by the expression (8) with $b_2 < 0$, then the energy of the step at position A' is given by

$$\frac{\epsilon}{h} = b_1 - |b_2|p = b_1 - |b_2|\left(\frac{h}{\lambda}\right)$$

As the step moves it is submitted to an attractive force (per unit length) towards the edge equal to $|b_2|(h/\lambda)^2$. This force is only compensated by the undersaturation $-|\Delta\mu|$ surrounding the crystal if

$$\tfrac{1}{2}|b_2|\left(\frac{h}{\lambda}\right)^2 = \frac{h|\Delta\mu|}{\Omega}$$

or

$$\left(\frac{\lambda}{h}\right)^2 = \frac{\Omega|b_2|}{h|\Delta\mu|}$$

if $|b_2| \approx b_0 \sim 10^3$ erg/cm^2 one would again require substantial undersaturations ($P_e/P \sim 2$) in order to make λ a few molecular distances so that the nucleation barrier could be overcome. The interesting point here is that this particular effect would depend directly on the derivatives of β rather than on the value itself and would therefore allow us to study these derivatives directly. To our knowledge there has been no direct evidence of this required undersaturation for the evaporation from straight edges other than the experiments of Dittmar and Newmann (1960) on potassium whiskers to which we will refer in Part II.

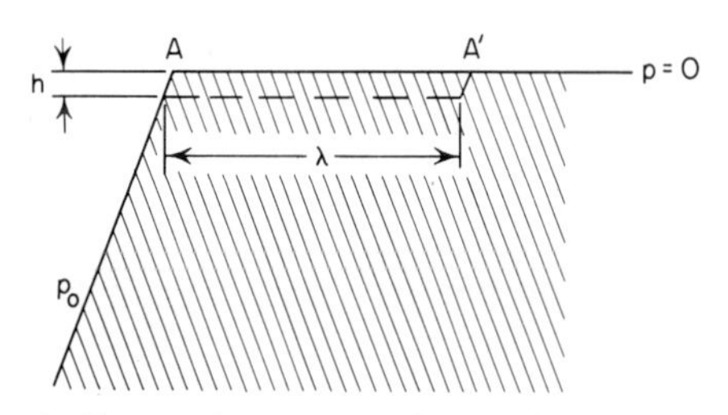

Fig. 6. Evaporation at a crystal corner between the two singular surfaces with orientations $p = 0$ and $p = p_0$.

6. DISLOCATIONS AS SOURCES OF STEPS

The classical theories for the growth of a perfect crystal predicted that they should become bounded by atomically smooth, close-packed faces. These faces should then grow by the spread of successive layers created by a two-dimensional nucleation process on the face as was mentioned in the previous section. The growth rate would then be controlled by the probability of nucleation and usually would not be appreciable until the supersaturation reached the order of a few tens of per cent. Under this condition the subsequent growth of a layer after nucleation would be extremely fast and would not be a limiting factor in controlling the growth rate.

It is observed, however, that most real crystals grow at very low supersaturations, leading to the conclusion that sources of steps are always present which make the nucleation process unnecessary. The entire kinetics of growth is then determined by the motion and interaction of steps on the crystal surface, which is the primary subject of Part II.

The problem of the continuous presence of steps on the crystal surface was resolved by Frank's (1949) screw dislocation mechanism of crystal growth, and we now mention very briefly the main features of this theory. A screw dislocation emergent at a point on the crystal surface provides a step on the surface equal in height to the projection of the Burgers vector of the dislocation on the normal to the surface. During growth, this step forms a spiral growth hill, which then advances the face of the crystal with some characteristic growth rate. Under a given condition of supersaturation the dislocation rapidly winds itself up into a spiral, centered on the dislocation, until the curvature at the center reaches the critical value, at which curvature the rate of advance falls to zero and the whole spiral then rotates steadily with stationary shape. The quantitative theory is considered in Part II.

II. KINETICS OF CRYSTAL GROWTH

The purpose of Part II is to discuss the basic ideas of the theory of crystal growth from the vapor, only referring to the experimental evidence when it is relevant to our point of view. This is admittedly rather unfair, but it is justifiable on two accounts. First, although the basic ideas of how crystals grow are clear enough, we are nevertheless unable to explain in detail most of the experimental facts and should, in our opinion, remain for the present at least vague and general. Second, other chapters in this book take up a more exhaustive discussion of the experimental conditions of growth.

In the first section we attempt a classification of the conditions of growth from the vapor (or for that matter also growth from dilute solutions). This classification is based on the kind of surface (whether singular or diffuse) where the growth promoting source is located. We presume that at least some of the crystal faces are singular (or vicinal), so that we exclude the cases where *all* surfaces are diffuse, as might be true in growth from the melt or concentrated solution.

Most of our efforts are dedicated to what we call "layer growth" where two simultaneous processes are involved: (1) creation of steps at a source somewhere on the growing surface and (2) motion of the steps away from the source.

The importance of this distinction has been recognized long ago, but it is only recently that a clear understanding of their interplay has been achieved. They will be studied in Sections 2, 3, and 4, starting with the motion of steps under steady state conditions on a flat ∞-surface. Section 3 then considers the problem of a growing crystal surface containing a source again under steady state conditions. Finally, Section 4 discusses the transients on the basis of the so-called kinematic theory of crystal growth.

1. CLASSIFICATION OF GROWTH CONDITIONS

Under the conditions of growth from the vapor some of the faces of the growing crystal are singular or vicinal. Thus it is only natural to classify the growth conditions under two headings. (1) The growing surfaces are nonsingular (diffuse) and as such do not require a source of steps. We call this *dendritic growth.* (2) The growing surface is singular (or more precisely, a combination of vicinal surfaces) and requires a source of steps. We call this *layer growth.* The type of growth occurring on a given crystal surface might depend on the rate. A layer type occurring at low rates of growth might become dendritic at sufficiently high rates when the driving force for growth is very large. Cahn (1960) has considered the problem and concluded that for growth from the vapor, layer growth is not likely to transform into dendritic growth.

Dendritic Growth

For a great number of lattices with low symmetry, growth from the vapor can occur easily without the need of sources of steps. Any layered structure, for example, the hexagonal metals zinc and cadmium, will grow under certain circumstances as very thin plates normal to the *c*-axis. The growing surfaces, probably prismatic planes in the case of hexagonal metals, appear to behave as nonsingular surfaces, and as such do not seem to require sources of steps for their growth.

It is quite obvious that this type of growth should occur as long as the other singular surfaces present do not have sources of steps. However, it is not obvious which are the conditions that will still favor dendritic growth when there are sources operating on the singular surfaces. From the experimental point of view dendritic growth appears to be favored by the following conditions: (1) a volume diffusion field operates which favors the nonsingular surfaces and (2) high supersaturation *and* high equilibrium concentration producing a high rate of growth in such a way that the growing surface does not wait for the condensing atoms to diffuse to it but goes forward to catch them.

Unfortunately, no serious theoretical analysis of whether or not these two conditions are sufficient to explain dendritic growth has been done to our knowledge. They certainly must play some role, but the situation appears to be more complicated, and in our opinion needs more careful consideration before the problem could be considered understood. For instance, we believe that surface diffusion might have quite an important place; a large surface current will tend to destroy dendritic growth. Also and probably more important, small amounts of impurities might stop the sources at the singular surfaces from becoming active.

Dendritic growth, in the sense described here, is by no means limited to low symmetry crystals. Cubic crystals also show dendritic growth. Sometimes they show platelet growth, but it is usually because they are twinned crystals, becoming really a layered structure. Much more interesting is the onset of dendritic growth for cubic crystals when dendrites develop in the direction of nonsingular surfaces, as for instance (111) surfaces in NaCl-type ionic crystals. The conditions required for this to happen are not quite clear, although the conditions enumerated for platelet growth seem to be again important. It is our opinion that both surface diffusion and/or impurities must also be important in favoring the growth of the nonsingular surfaces against the operation of the sources that must exist in the singular surfaces.

Layer Growth

Our main discussion refers to "layer growth," in which two simultaneous and, to some extent, interdependent processes are going on: (1) creation of steps at a source and (2) motion of the steps away from the source. These two aspects of the problem are discussed in Sections 2 and 3. Here we would like to point out that layer growth should be divided into two types, according to whether or not every singular surface of the growing crystal has active sources.

In the first case the crystal will develop into a more or less *regular polyhedron.* A perfectly regular polyhedron will only result if the rates of all sources are exactly equal, which is not very likely.

In the second case, particularly if only one of the singular surfaces possesses an active source, the crystal will have an asymmetrical shape. We denote this type of growth as *whisker growth* since it corresponds to many of the conditions of growth of whiskers. Although many similarities exist between dendritic and whisker growth, there are differences that make it worthwhile to distinguish between them.

In Part I we have already introduced and discussed the types of sources of steps which can occur: dislocation sources, nucleation sources, and (in the case of evaporation) crystal edges. The following discussion generally does not stress the particular type of source we have in mind.

Steady State Growth Versus Transient Growth

Here we distinguish between steady state and transient conditions. The distinction between the two are best established on the following simple mathematical basis.

Let the plane $z = 0$ represent the singular surface before layer growth has started on it. Let us assume that a source of steps exists at $x = y = 0$, which promotes the growth of the crystal in such a way that the growing surface is represented by the mathematical surface

$$z = z(x, y, t) \tag{19}$$

The slope of the growing surface at (x, y) and time t is given by the two slopes

$$p(x, y, t) = -\frac{\partial z}{\partial x}, \qquad q(x, y, t) = -\frac{\partial z}{\partial y} \tag{20}$$

Similarly, the rate of growth parallel to the axis z is given by

$$R(x, y, t) = \frac{\partial z}{\partial t} \tag{21}$$

The vector **p** (with components p and q) has a very important place in the following discussions. If h is the thickness of the molecular layers parallel to $z = 0$, then $|\mathbf{p}|/h = \sqrt{p^2 + q^2}/h$ represents the local density of steps normal to the vector **p**, or the number of steps per unit length in the direction of the vector **p**. Similarly, R/h represents the flow of steps or number of steps normal to **p** passing a point on the crystal surface per unit time.

Our final purpose is, of course, to determine $z = z(x, y, t)$ for all times. We will assume in our treatment that after a *transient* the growth of the surface sets itself into a *steady state,* where both the profile of the crystal $z = z_0(x, y)$ and the rate of growth R remain constants; then

$$z = z_0(x, y) + Rt$$

We are, of course, aware that the really steady state configuration will only occur under exceptional conditions. It is, however, the easiest problem to consider and a very useful basis for at least an approximate treatment of the *transient state.*

2. STEADY STATE. MOTION OF STEPS

In this section we consider the motion of steps on an ∞-crystal surface having an average slope given by $p, q = 0$. Where and how these steps are created is the subject of the next section. The steps are all assumed to be parallel to each other and to have the monomolecular height h.

Once the velocity $v(p)$ of the steps in the direction of the vector p is known, the normal rate $R(p)$ is clearly given by

$$R = \frac{hv}{\lambda} = pv \tag{22}$$

where $\lambda = h/p$ is the distance between steps. Figure 7 shows the relation between these quantities.

The computation of the step velocity under given external conditions (temperature and supersaturation) is a very complicated problem. The step moves because of the addition of molecules to the kinks in the steps after these molecules have diffused successively through the volume in the vapor, on the surface between steps, and finally along the edge of the step to the kinks. We call $j(p)$ the current of molecules per second per unit length of step. Once $j(p)$ has been computed,

$$v(p) = \frac{\Omega j}{h} \tag{23}$$

where Ω is the volume per molecule in the crystal.

Before doing any computations it is worthwhile to consider, for given external

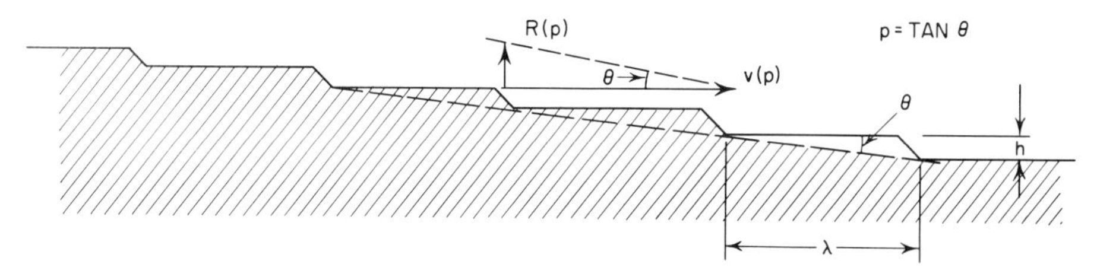

Fig. 7. Profile of a growing crystal surface with orientation p. v is the horizontal velocity of the monomolecular steps of height h and distance $\lambda = h/p$ from each other. R is the rate of growth normal to the orientation $p = 0$.

conditions, the qualitative dependence of $R(p)$ that one can expect. Two types of dependence stand out, which are represented in Fig. 8 by curves I and II. A curve of type I corresponds presumably to the growth of a metallic crystal surface under high purity conditions when the exchange in the neighborhood of the step is so fast that equilibrium is maintained near it. The velocity of a step $v = R/p$ is then a maximum for very small slopes p when the steps are very far apart and the diffusion fields do not interfere with each other. Hence, R should be proportional to p for p small. As p increases, R should approach a flat maximum and thereafter decrease again and become zero for the next singular orientation of the crystal surface. In the upper half of Fig. 8 the crystal profiles corresponding to different slopes are illustrated. The important characteristic of this type I curve is that it always satisfies the condition

$$c < v \qquad \text{where} \quad c = \frac{dR}{dp} \tag{24}$$

which will be shown later to play an important role in the motion of steps.

A curve of type II will probably correspond to the situation where impurities interfere with the motion of steps, particularly when the steps are very far apart. It is conceivable that the velocity might even become zero at $p = 0$ as it is indicated in Fig. 8 ($R \propto p^2$ for small p). The effect of impurities is likely to decrease as p increases, and a maximum must, in any case, be reached. The significant characteristic of a type II curve, as distinct from a type I curve, is that $c \gtrless v$ and there is a particular slope p_c for which v and c are identical.

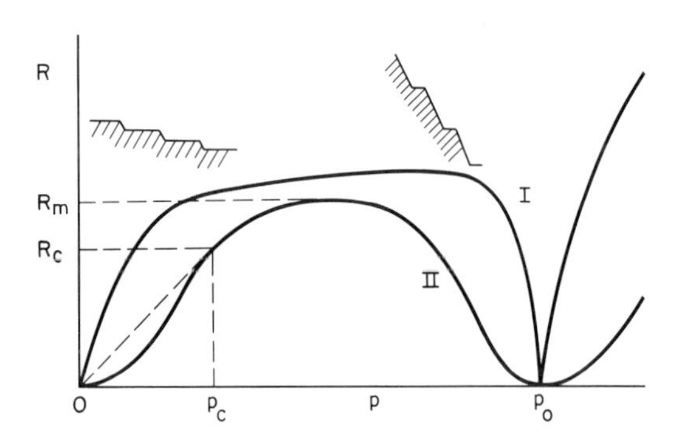

Fig. 8. Schematic rate, slope curves $R(p)$.

The kinetic behavior of a given slope p is to a great extent determined by the behavior of $R(p)$ in the same way that $\beta(p)$ controls the equilibrium properties of the same slope.

These considerations particularly apply simply to a thermodynamically stable orientation ($\beta'' > 0$, see type I) when there is no tendency for the formation of macroscopic steps on the basis of surface energy. If, furthermore, there is no hindrance to the motion of steps, the steady state corresponds to curve I in Fig. 8 and appears to be stable (see Section 4); this particularly simple case is considered in the next subsection, Monomolecular Steps.

Even if the orientation is stable there might be obstacles to the motion of steps (as, for example, slow moving impurities). Then the steady state, described by curve II in Fig. 8, does not appear to have sufficient stability and tends to a new macroscopic steady state configuration with a distribution of macroscopic steps of different heights around an average value. This situation is further complicated if the orientation is thermodynamically unstable, so that the tendency to form macroscopic bunches of steps is already present without the need of any kinetic effects. The way impurities are capable of producing the type of curve II is discussed under Impurities Effects. However, the more interesting problem of the macroscopic steady state is not discussed in any detail since the situation remains still rather confused. There is promise of serious advances in this respect after the very interesting theoretical investigations of Chernov (1957, 1961) and the experimental studies of Howes (1959).

Monomolecular Steps

The simplest case theoretically, and the most difficult to achieve experimentally, is that of a crystal surface for which all slopes are equilibrium orientations. Then there is a free energy repulsion between steps, and

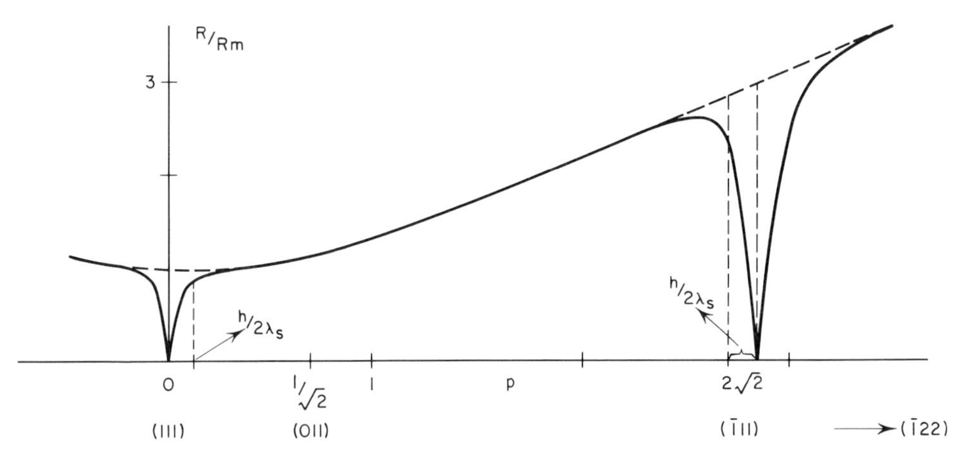

Fig. 9. Rate-slope curve for growth from the vapor under ideally pure conditions of a face centered cubic crystal, assuming $h/2\lambda_s = 0.1$. Only the orientations between (111) $p = 0$, (011) $p = 1/\sqrt{2}$, and $(\bar{1}11)$ $p = 2\sqrt{2}$ are represented.

we know there is no tendency for the formation of macroscopic steps.

Furthermore we know from its equilibrium properties that every step will have a high concentration of kinks; consequently it is safe to assume that for motions of the step which are not exceedingly fast the equilibrium configuration is continuously maintained near the step, and consequently the chemical potential remains equal to the equilibrium value.

Assuming then a set of parallel straight sinks at a distance λ from each other, the diffusion problem can be solved in quite general terms. First one sees easily that the motion of the step can be neglected as its velocity is very small in respect to the mobility of surface defects.

The solution of this problem for the particular case of growth from the vapor was given by Burton, Cabrera, and Frank (1951), assuming that the mean free path in the vapor is so long that one can assume the concentration to be uniform everywhere in the volume. The calculation showed that the current going into every step was due to the molecules condensing from the vapor in a strip of width λ_s at both sides of the step. The characteristic length λ_s has already been introduced (see type I) and represents the average distance $\sqrt{D_s\tau_s}$ that an adsorbed atom moves on the surface between condensation and evaporation.

Defining the supersaturation of the vapor by the ratio $\sigma = (P - P_e)/P_e$ it is easy to show that the current j is given by

$$j = \frac{2\lambda_s P_e \sigma}{\sqrt{2\pi mkT}}$$

and consequently

$$v_\infty = \frac{2\lambda_s}{h} \frac{P_e \Omega \sigma}{\sqrt{2\pi mkT}} \tag{25}$$

Assuming $\lambda_s/h \sim 10^2 - 10^5$ and taking characteristic values for a typical metal, we find $v_\infty \approx (10^{-2}\sigma - 10\sigma)$ cm/sec. This is, of course, the maximum velocity that a step can have when it is isolated from any other. If the distance between steps $\lambda = h/p$ becomes of the order of λ_s or smaller, the steps compete with each other so that the current going into each of them diminishes and consequently their velocity. Then v turns out to be

$$v(\lambda) = \frac{2\lambda_s}{h} \frac{P_e \Omega \sigma}{\sqrt{2\pi mkT}} \tanh\left(\frac{\lambda}{2\lambda_s}\right) \tag{26}$$

From v we deduce the following expression for the rate-slope relationship $R(p)$

$$R(p) = R_m \frac{2\lambda_s p}{h} \tanh \frac{h}{2\lambda_s p} \tag{27}$$

where $R_m = P_e \Omega \sigma / \sqrt{2\pi mkT}$ is the maximum rate of growth of any crystal surface normal to itself. This expression is valid only for $P < 1$. Figure 9 shows how R

varies for a face centered cubic crystal, taking a (111) surface as reference surface. We see that it is only in very small ranges ($\Delta p \sim h/\lambda_s \ll 1$) of slopes that R is different from the maximum value $R_m\sqrt{1 + p^2}$. In any case, the curve $R(p)$ corresponds to type I in Fig. 8 as $c < v$ always.

Impurity Effects

The problem of the effect of impurities on the motion of steps is extremely complex and very poorly understood. Efforts have been made recently, but much remains to be done both from experimental and theoretical points of view to understand this subject.

We attempt here to point out several aspects of the problem and indicate the important features that remain to be solved. Discussion can be divided naturally into two limiting cases which will require quite different analysis: (1) immobile surface adsorbed impurities and (2) completely mobile surface adsorbed impurities. For immobile impurities we assume that impurity molecules stick and remain at the points where they strike the crystal surface. As first pointed out by Price, Vermilyea, and Webb (1958), a step moving along the crystal surface will then be required to squeeze through between the impurities, and in so doing incorporate them into the crystal. If a pair of impurities is less than $2\rho_c$ apart, the step cannot squeeze through and motion is effectively blocked. If a certain constant current of impurity atoms J_i reaches the crystal surface, we assume that the concentration of surface impurities at a particular point builds up until a step passes this point and effectively adsorbs all of the impurities into the crystal. For large impurity current, of course, the buildup may be sufficiently rapid to stop all growth. In cases where growth continues, steps will squeeze through the two-dimensional array of impurity molecules and will assume various curvatures as they advance, and hence will move with an average velocity considerably less than the velocity v_∞ of a long, straight step moving under pure conditions and given by (25). Then $v_\infty \propto (\mu_{\text{vap}} - \mu_e) = \Delta\mu$, where $\mu_{\text{vap}} = \mu_e + \Delta\mu$ is the chemical potential of the vapor and μ_e is the equilibrium chemical potential at the long, straight step. For a curved step with curvature ρ

$$v_\rho \propto (\mu_{\text{vap}} - \mu_\rho) = [\mu_{\text{vap}} - \mu_e - (\mu_\rho - \mu_e)]$$
$$= \Delta\mu\left(1 - \frac{\mu_\rho - \mu_e}{\mu_{\text{vap}} - \mu_e}\right) = \Delta\mu\left(1 - \frac{\rho_c}{\rho}\right)$$

The average velocity of the steps moving through the impurity distribution could then be obtained by calculating an average μ_ρ existing in the neighborhood of the step. A proper treatment would be a statistical one which would give a nonzero but very rapidly decreasing velocity for $d < 2\rho_c$ if impurities are assumed to be an average distance d from each other. The following expression is a rough estimate given by Cabrera and Vermilyea (1958):

$$v = v_\infty\left(1 - \frac{2\rho_c}{d}\right)^{1/2}$$

In the presence of an impurity current equal to J_i, a certain critical supersaturation will be required before step motion and growth occur. This critical supersaturation is equal to the supersaturation, which in the absence of impurities produces a rate $R(0)$ given by

$$R(0) = 14h^2\rho_c^2 J_i$$

If $\rho_c \sim \sigma^{-1}$ and $R(0) \sim \sigma^m$ growth in the presence of impurities take place only at supersaturations $\sigma > \sigma^* = \text{constant } J_i^{1/(2+m)}$, the relation $R(p)$ is then of type II in Fig. 8. Another important point with immobile impurities is that the source strength or rate of step generation is not changed by the presence of impurities, provided the critical supersaturation is exceeded. The source strength will, of course, be affected by any change in crystal profile affecting the diffusion field, which we discuss further in Section 3.

For completely mobile adsorbed impurities, the concentration of impurities along the step is in equilibrium with the impurity concentration in the vapor. The principal effect on step motion is caused by impurity adsorption at kink sites which may appreciably lower the velocity of step motion. As mentioned previously in Part I a calculation of the change in step velocity requires a rather exact analysis of the lifetimes involved in the various atomic jump processes occurring at the kink sites along the step. In general, the step velocity will probably be lowered either by direct poisoning of kink sites or by a second order process involving the exchange of impurity atoms at the kink sites. Chernov (1961) has calculated growth rates in the presence of impurities, assuming that impurities poison kink sites and that only free kinks contribute to growth.

Under those conditions the chemical potential in the neighborhood of the step μ_e is no longer maintained equal to the equilibrium value because the rate of exchange with the crystal is not fast enough; in fact, at the limit of very slow exchange the surface diffusion gradient is eliminated so that $\mu(0) = \mu_{\text{vap}}$ and the exchange at the step becomes rate controlling. Then the velocity becomes independent of the distance between steps and $R \propto p$.

For mobile impurities a second important effect is the lowering of the edge energy ϵ by impurity adsorption in the step. This effectively increases the strength of the source, both with a dislocation or a nucleation source.

The intermediate case between the extremes of immobile and completely mobile impurities is at the present time very difficult to analyze. The impurity concentrations at moving steps become more difficult to calculate since the velocity of steps relative to the mobility of impurities must be analyzed carefully. Here the step height will probably play a fairly important role in the impurity effect.

3. STEADY STATE. SOURCE STRENGTH

In this section we consider the question of which sources are capable of producing the steady state slopes that we considered in the previous one. The possible types of sources available have already been described in Part I: dislocation source, nucleation (mostly inhomogeneous) source, in (with evaporation) crystal edges.

As we noticed in the previous section under given external conditions of temperature and supersaturation, an arbitrary constant slope corresponds to a possible steady state condition. As long as we do not consider the source producing the steps, the problem of rate of growth under steady state conditions is undetermined. It is the source that fixes this rate. A great simplification would be achieved if the actual steady state could be determined by a variational principle, such as for instance "maximum rate of growth" under given external conditions. Several authors have proposed such an extra condition to solve similar problems such as the rate of dendritic growth from the melt. Unfortunately, there has been as yet no proof that the atomic mechanisms which constitute the known sources of steps satisfy any such variational principle, and therefore it is necessary to consider every source on its own.

Once the mechanisms at the source are known, the problem of determining the actual steady state rate of growth is a self-consistent problem involving diffusion fields which has really only been properly considered in the simplest possible case, that of a dislocation source operating at very low supersaturation, which is considered in some detail in the next subsection.

For a given kind of source under given external conditions there will be only one strength of source R_0 which will correspond to a certain slope p_0 of the crystal surface.

Dislocation Source. "Back-Stress Effect"

As already described in Part I, a dislocation ending at a fixed point on a crystal surface will have a step connected with it, which will wind itself into a spiral around the fixed point. Let us first consider the case of high purity described on p. 19. We use the expression for the velocity of a step having a radius of curvature ρ and growing under an external supersaturation $\Delta\mu =$

$kT \ln (P/P_e) \approx kT\sigma$, we obtain

$$v = v_\infty\left(1 - \frac{\rho_c}{\rho}\right)$$

where $\rho_c = \Omega\epsilon/h\Delta\mu$ and v_∞ is given by (25), and does not depend on the curvature. On this basis, which is a good approximation for not too large a curvature, one can solve completely the steady state problem and determine both the rate of growth or source strength R_0 and the slope p_0 or the equivalent distance between steps $\lambda_0 = h/p_0$. This was first done by Burton, Cabrera, and Frank (1951) and more accurately by Cabrera and Levine (1956) and the answer is

$$R_0 = \frac{hv_\infty}{\lambda_0}; \qquad \lambda_0 = 19\rho_c \tag{28}$$

For very low supersaturations, when every turn of the step moves independently of each other, one finds $R_0 \propto \sigma^2$, so that a parabolic law should be observed.

An interesting problem is what happens if the supersaturation increases (ρ_c decreases) to the point where the diffusion fields of successive turns overlap. The problem then becomes very complicated and no accurate treatment of it has been published to our knowledge. However, one can predict that the solution (28) will remain correct to a good approximation, provided two modifications are introduced into it: first the expression for v becomes the general one (26) with the competing effect of the diffusion fields of the turns of the step at a distance λ_0 from each other. This effect was already considered by Burton, Cabrera, and Frank (1951). Furthermore, the center of the spiral will no longer see the same supersaturation $\Delta\mu$ in the vapor because of the diffusion field due to the first turn of the spiral with average radius $\sim \lambda_0$. An approximate estimate of this "back-stress effect" can be made if we replace $\rho_c = \Omega\epsilon/h\Delta\mu$ in the expression for λ_0 by $\rho_0 = \Omega\epsilon/h\Delta\mu_0$ where $\Delta\mu_0 < \Delta\mu$ has to be determined and represents the supersaturation existing at the center of the spiral.

In the particular case of growth from the vapor, with only surface diffusion operating, $\Delta\mu_0$ is given approximately by

$$\frac{\Delta\mu_0}{\Delta\mu} = 1 - \mathrm{I}_0^{-1}\left(\frac{\lambda_0}{\lambda_s}\right)$$

where $I_0(x)$ is the Bessel function of order zero and imaginary argument. When we write $\Delta\mu_0$ in terms of λ_0, the following equation determines the ratio λ_0/λ_s

$$\frac{\Delta\mu_1}{\Delta\mu} = \frac{\lambda_0}{\lambda_s}\left[1 - \mathrm{I}_0^{-1}\left(\frac{\lambda_0}{\lambda_s}\right)\right]$$

with $\Delta\mu_1 = 19\Omega\epsilon/h\lambda_s$. Therefore the ratio λ_0/λ_s is given by the following limits:

$$\begin{aligned} \Delta\mu < \Delta\mu_1: \quad & \frac{\lambda_0}{2\lambda_s} = \frac{\Delta\mu_1}{2\Delta\mu} \\ \Delta\mu > \Delta\mu_1: \quad & \left(\frac{\lambda_0}{2\lambda_s}\right)^3 = \frac{\Delta\mu_1}{2\Delta\mu} \end{aligned} \tag{29}$$

This shows that the slope of a growing crystal (under conditions of very high purity) will not go far beyond $p_0 = h/\lambda_s$, which for the typical values $\lambda_s \sim 10^2 - 10^5 h$ is extremely small. This is, of course, because the steps have to move rapidly away from the dislocation source for it to able to send new turns.

The rate of growth R_0 is still given by (27)

$$R_0 = R_m \frac{2\lambda_s}{\lambda_0} \tanh \frac{\lambda_0}{2\lambda_s}$$

so that for large supersaturations

$$R_0 = R_m\left[1 - \tfrac{1}{3}\left(\frac{\Delta\mu_1}{2\Delta\mu}\right)^{2/3}\right] \tag{30}$$

which approaches the maximum linear rate R_m very slowly.

The conclusion from this discussion is that under the conditions of very high purity, growing surfaces should always be practically flat, there should not be any whisker type of growth out of a large close-packed surface, nor should there be any observable etch pits. This will be more noticeable the larger the value of λ_s. Furthermore, the rate of growth R_0 will only be significantly different from the classical linear law R_m if the characteristic value for the supersaturation $\sigma_1 = \Delta\mu_1/kT$ can be attained.

Whisker Growth

Crystals growing in the form of whiskers have been the subject of many experimental and theoretical papers (for example, Nabarro and Jackson, 1958). Many theories have been proposed to explain their growth under a great variety of conditions. For growth from the vapor, the theory of Burton, Cabrera, and Frank (1951) is generally applicable, the only important additional point remaining to be answered is why the sides of the whisker do not grow at an appreciable rate. With very anisotropic substances this can, perhaps, be explained by the existence of an easy growth direction and sides of the whisker on which nucleation is very difficult. However, many cubic metals and substances of all types grow in whisker form and require further mechanisms for explanation. The most popular theory of whisker growth from the vapor has required three steps consisting of (1) adsorption of atoms on perfect sidewalls of the whisker, (2) surface diffusion to the whisker tip, and (3) incorporation at a step in the tip associated with a single axial screw dislocation. Experiments by Sears (1953) and Gomer (1957) on mercury whiskers have shown that steps (1) and (2) are indeed correct, but they were unable to verify step (3).

The three steps of this theory undoubtedly form the general basis for whisker growth from the vapor, but the exact reasons for the whiskerlike profile are not yet completely understood. Modification of growth by impurities is very likely and provides possible mechanisms for producing whisker profiles.

For immobile adsorbed impurities it was previously pointed out that if the critical supersaturation is exceeded, the presence of immobile impurities will not have much effect on the source strength. On the other hand, they will slow considerably the velocity of steps moving away from the source. This effect will produce a much steeper growth hill, eliminating the "back-stress effect" described in (1) and enhancing considerably the probability of whisker formation. For some types of impurities it is also possible that the impurity current J_i can be such that, during the time required for one revolution of the spiral, surface impurities will build up to the critical value necessary for blockage of step motion. Under this condition, the first turn of radius ρ_c will continue to move since it will now be moving over new surfaces laid down by the previous turn. However, steps at a greater distance from the source will still be moving on surfaces where impurities have built up beyond the critical value and step motion will be blocked. This will result in a very rapid generation of steps at the source, causing the center of the crystal to grow out into a whisker as impinging atoms all diffuse to the central portion of the spiral. A careful analysis of the flow of steps from the source should enable one to predict the diameter of the whisker for a given impurity concentration.

For mobile surface impurities, the effects on step motion and generation will be completely different. They may, however, also give rise to steeper growth profiles which favor whisker growth as they again eliminate the "back-stress effect" described in (1). Impurities that establish an equilibrium concentration along the steps may considerably lower the edge free energy with a consequent decrease in ρ_c. This will increase the source strength, whereas the velocity of step motion away from the source is probably decreased by the impurities. The net result is again a steeper growth profile than would exist under pure conditions.

The validity of these possible impurity effects in whisker growth will require much further experimental study.

4. TRANSIENTS

The previous two sections have attempted to present the status of the theory of steady state rates of growth. The problem now is to study the ways this steady state is created and whether or not it has sufficient stability to resist small perturbations. The second part that we call the transient state has been analyzed only recently, and the theory is in such a preliminary state that it is rather doubtful whether a review of it is of much use. There is, however, a general result which is becoming apparent and it is that most of the experimental conditions correspond to transient conditions rather than steady state conditions. The time required for the establishment of really steady state conditions is usually quite long. The steady state theory might then be considered useless if it were not for the fact that it turns out to be of fundamental importance for at least an approximate treatment of the transient state, the *kinematic theory of crystal growth* first considered by Frank (1958) and others.

This theory is the basis of the next subsection, and its application to two particular transients, the establishment of the steady state and its stability against perturbations, is treated in the following two subsections.

Kinematic Theory of Crystal Growth

In the introduction to Part II we gave the general relation between the crystal surface

$$z = z(x, y, t) \tag{31}$$

the slopes p, q and the rate of growth R at x, y and t.

$$p = -\frac{\partial z}{\partial x}, \qquad q = -\frac{\partial z}{\partial y},$$

$$R = \frac{\partial z}{\partial t} \tag{32}$$

From these relations it follows that

$$\frac{\partial p}{\partial t} + \frac{\partial R}{\partial x} = \frac{\partial q}{\partial t} + \frac{\partial R}{\partial y} = \frac{\partial p}{\partial y} - \frac{\partial q}{\partial x} = 0 \tag{33}$$

which will be the basis for the solution of the transient problem. Derived in this form these equations are a consequence of the existence of (31) and are not limited to the conditions of layer growth to which we are going to apply them; it is, however, interesting to point out that the equations (33) are the result of the law of conservation of monomolecular steps. This is particularly clear if we consider the two-dimensional cases ($q = 0$) and realize that p/h is the density of steps and R/h is the flow of steps crossing a point on the surface per unit time.

In order to use formulas (33) we must make some extra assumptions relating R, p and q. The kinematic theory of crystal growth makes the following assumptions concerning the strength of the source R_0 and the relations between the rate of growth R, p and q away from the source.

1. For constant external conditions (temperature, supersaturation, etc.) the source strength $R_0(t)$ is only a function of the time.
2. Again for constant external conditions it is assumed that away from the source the rate R only depends explicitly on the local slope p, q, and furthermore that the relation $R(p, q)$ is the same as we obtained for the steady state. For the sake of continuity we also assume that the slopes at the source itself are such that $R(p_0, q_0) = R_0$.

It is clear that these assumptions can only be considered an approximation; indeed the assumption (2), assuming that $R(p, q)$ remains the same for all times and positions, neglects the changes in diffusion fields that are bound to occur in any transient. Nevertheless, the conclusions one can deduce are quite valuable, at least from a qualitative point of view.

Let us now consider the method of solution. Once a relation

$$R = R(p, q)$$

has been assumed we can eliminate R from Eq. (33), so that

$$\frac{\partial R}{\partial x} = \frac{\partial R}{\partial p}\frac{\partial p}{\partial x} + \frac{\partial R}{\partial q}\frac{\partial q}{\partial x}$$

$$= \frac{\partial R}{\partial p}\frac{\partial p}{\partial x} + \frac{\partial R}{\partial q}\frac{\partial p}{\partial y}$$

and consequently,

$$\frac{\partial p}{\partial t} + \frac{\partial R}{\partial p}\frac{\partial p}{\partial x} + \frac{\partial R}{\partial q}\frac{\partial p}{\partial y} = 0 \tag{34}$$

and similarly,

$$\frac{\partial q}{\partial t} + \frac{\partial R}{\partial p}\frac{\partial q}{\partial x} + \frac{\partial R}{\partial q}\frac{\partial q}{\partial y} = 0$$

which give two first order, nonlinear differential equations that can be solved for $p(x, y, t)$ and $q(x, y, t)$; hence, R and finally the actual crystal surface $z = z(x, y, t)$.

These first-order equations are now solved by the method of characteristics. The method is applicable in three dimensions, as was shown by Frank (1958), but for simplicity let us consider the two-dimensional problem when only one slope p has to be considered.

Then Eq. (34) becomes

$$\frac{\partial p}{\partial t} + c\frac{\partial p}{\partial x} = 0 \tag{35}$$

where

$$c = \frac{dR}{dp} \tag{36}$$

The solution of Eq. (35) is obtained in terms of the straight lines

$$\frac{dx}{dt} = c(p) \tag{37}$$

representing the motion of "kinematic waves" which are regions on the crystal surface having constant slope p (as well as constant rate R). These waves, of course, do not contain the same monomolecular steps all the time, as the step velocity $v = R/p$ can be larger or smaller than c. If, finally, we know from the boundary conditions the function $p(x, t = 0)$ along the axis x from $x = 0$ (source) to $x = \infty$, and also $R_0(t) = R(x = 0, t)$ along the axis t for $t > 0$, one can build up the solution $p(x, t)$ for all values of x, t and then by integration determine $z(x, t)$.

Another very important property of the solution, pointed out by Frank (1958) follows if one considers lines of equal slope in the plane z, x giving the profile of the crystal. As

$$dz = \frac{\partial z}{\partial t}dt + \frac{\partial z}{\partial x}dx = \left(R\frac{dt}{dx} - p\right)dx$$

it is clear that these lines are given by

$$\frac{dz}{dx} = \frac{p}{c}\left(\frac{R}{p} - c\right) = \frac{p}{c}(v - c) \tag{38}$$

As a consequence $dz/dx \gtrless 0$, according to whether $v \gtrless c$.

Furthermore, it is readily shown that whenever two kinematic waves with slopes p_1 and p_2 (and rates R_1 and R_2) meet a discontinuity in slope must exist which will follow in the x, t plane a trajectory

$$\frac{dx}{dt} = \frac{R_2 - R_1}{p_2 - p_1} \tag{39}$$

These are called "shock waves" to distinguish them from the "kinematic waves." The shock waves will not move with constant velocity nor will they be straight in the x, t plane.

Establishment of Steady State

Let us now consider the application of the theory to the establishment of the final steady state shape of a crystal surface.

The problem we have in mind would be either the formation of an etch pit, starting from a close-packed singular surface, or the initiation of a whisker, etc. This problem was considered by Cabrera (1960) as applied particularly to etching with the following results.

Two cases have to be distinguished, depending on the type of curve in Fig. 8.

If the $R(p)$ relation is of type I, any constant source strength R_0 will be able to produce a growth hill (or an etch pit) with smooth slope going continuously from $p = 0$ to the ultimate steady state p_0 given by $R_0 = R(p_0)$. Figure 10 shows the successive crystal profiles and the lines of equal slopes $(0 < p < p_0)$ for this case. The actual value of R_0 under given external conditions has to be determined in the manner discussed in section 3, and we know that probably in all cases when $R(p)$ is of type I it will lead to a very small final slope p_0.

On the contrary, if $R(p)$ is of type II, only source strengths larger than a certain critical value R_c shown in Fig. 8 will be able to initiate and maintain a growth hill. The reason for this is that all lines of equal slope starting from the origin must have a positive slope dz/dx which, according to

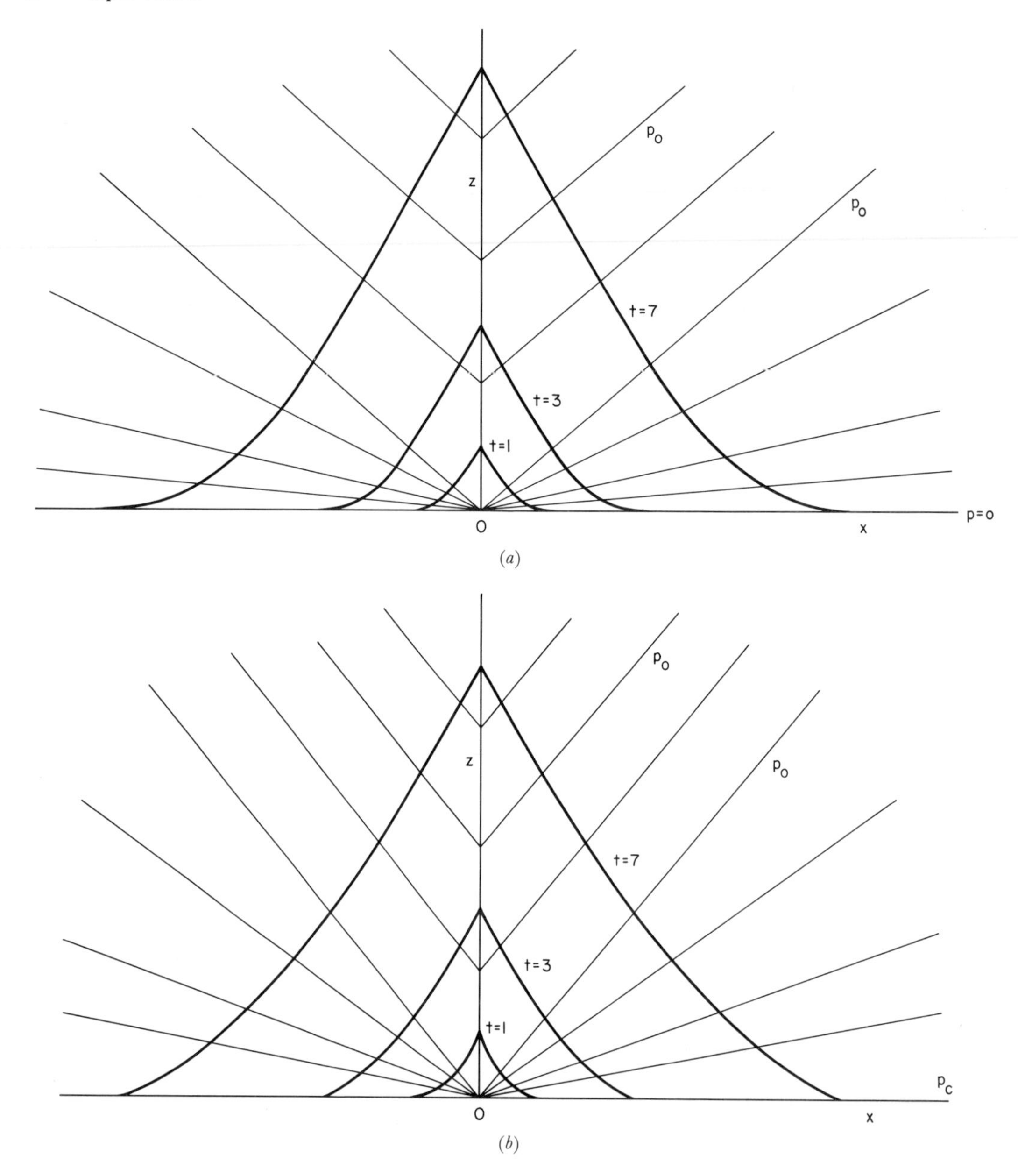

Fig. 10. (*a*) Crystal profiles for successive times $t = 1, 3, 7$, and lines of equal slopes from $p = 0$ to $p = p_0$, for a type I rate-slope curve. (*b*) Crystal profiles for successive times $t = 1, 3, 7$, and lines of equal slopes from $p = p_c$ to $p = p_0$ for a type II rate-slope curve.

(37) is only possible if $v > c$, which excludes all the slopes smaller than p_c (see Fig. 8). Furthermore if $R_0 > R_c$ then the successive profiles will appear as in Fig. 10b, with only slopes between p_c and p_0 appearing in the profile and showing a sharp discontinuity (shock wave) at the foot of the growth hill.

It appears therefore that when $R(p)$ is of type II the flow of the steps on the crystal surface puts a lower limit R_c to the strength of the source and an upper limit corresponding to the maximum R_m of $R(p)$. In between these two limits the actual atomic mechanism at the source will fix the actual value of R_0, but its order of magnitude is already determined. As we already discussed in Section 3 the same mechanisms that are likely to produce type II curves (for example, impurities) will also admit rather large slopes p_0 because of the elimination of the "back-stress effect."

The assumptions (1) and (2) made by the kinematic theory clearly represent an oversimplification of the facts. Some improvement of the theory can be achieved by allowing a dependence of R_0 on the time, still remaining within the kinematic theory. This dependence will be particularly noticeable when a volume diffusion field is partially controlling the rate R_0; then it is quite clear that the microscopic diffusion field established around the source will change noticeably as the height of the growth hill increases. In fact, it is to be expected that R_0 will increase in the case of a growth hill and will decrease in the case of an etch pit.

In the first case, if R_0 is initially already sufficiently near the maximum value R_m, it might be that the R_0 will become larger than R_m, in which case we have the situation corresponding to whisker growth. On the other hand, in the case of an etch pit, R_0 might become in time smaller than R_c, and then lines of equal slope with $dz/dx < 0$ will appear which are now possible because they do not originate at $z = x = 0$. Detailed studies on the profile of etch pits performed by Ives and Hirth (1960) show the existence of lines of equal slope with negative orientation, but they also show definitely that these lines are curved and not straight as the theory predicts. This is clearly due to the assumption that the rate R only depends on the local slope p. Curved lines of equal slope can be obtained by introducing in R either an explicit dependence on x and t or, alternatively, a dependence on z. Unfortunately, in both cases the method of solution breaks down.

Stability of Steady State

Another problem that can be treated on the basis of the kinematic theory of crystal growth is the question of steps which were considered in Section 2. Suppose, for instance, that the strength R_0 of the source producing a steady state with slope p_0 changes during a short period, producing near the source slopes p which differ from p_0 and are equivalent to a bunch of monomolecular steps. The question is what is going to happen to this bunch as the source continues to function with the strength R_0. If the steady state is unstable, the bunch should tend to grow with slopes more and more different from p_0.

First, the kinematic theory cannot show an instability in the sense stated because of the assumptions introduced regarding the dependence of R. In order to obtain slopes which are outside the range introduced at the source, we have to assume a dependence of R on the local p and on $\partial p/\partial x$.

On the other hand, the kinematic theory can discuss the way in which the slopes other than p_0 disappear by the formation of a "shock wave" with a decreasing discontinuity in p. This problem has been considered in detail by Chernov (1960, 1961) with the following interesting result. For a sufficiently small initial change in p, the time required for the disappearance of the bunch is inversely proportional to the square of (d^2R/dp^2) for $p = p_0$. Furthermore, the magnitude of this time is typically of the order of minutes so that should be easily observable.

The fact that the stability of the steady state decreases as d^2R/dp^2 decreases shows again a difference between the two types of curves I and II in Fig. 8. A curve of type II will lead to a very weak stability of the steady state for slopes in the neighborhood of the inflection point.

ACKNOWLEDGMENT

This work has been supported by the United States Office of Naval Research contract No. NONR 474(05) Project No. NRO 17-615.

References

Benard, J. (1960), *Fortschr. Miner.* **38,** 22.

Blakely, J. M., and H. Mykura (1961), *Acta Met.* **9,** 23; *Acta Met.* **9,** 595.

Cabrera, N. (1952), *Verlag Chemie* **56,** 294; (1959), *Discussions Faraday Soc.* No. 28, 16; (1960), *Reactivity of solids,* editor, J. H. de Boer, Elsevier, Amsterdam, p. 345.

Burton, Cabrera, and Frank (1951), *Phil. Trans. Roy. Soc.* **243,** 209.

Cabrera, N., and M. M. Levine (1956), *Phil. Mag.* **1,** 450.

Cabrera, N., and D. A. Vermilyea (1958), *Growth and Perfection of Crystals,* John Wiley and Sons, New York, p. 411.

Cahn, J. W. (1960), *Acta Met.* **8,** 554.

Cahn, J. W., and Hilliard (1958), *J. Chem. Physics,* **28.**

Chalmers, B., R. King, and R. Shuttleworth (1948), *Proc. Roy. Soc.* **A193,** 465.

Chernov, A. A. (1960), *Soviet Physics—crystallography* **5,** 423; (1961), *Soviet Physics Uspekhi* **4,** 116.

Curie, P. (1885), *Bull Soc. Min. de France* **8,** 145.

Dittmar, W., and K. Newmann (1960), *Z. Elektrochem., Ber. Bunsenges. physik. Chem.* **64,** 297.

Frank, F. C. (1949), *Discussions Faraday Soc.* **5,** 48; (1958a), *Growth and Perfection of Crystals,* John Wiley and Sons, p. 304; (1958b), p. 393.

Gomer, R. (1957), *J. Chem. Phys.* **26,** 1333; (1958), **28,** 457.

Herring, C. 1953, *Structure and Properties of Solid Surfaces,* editors, R. Gomer and C. S. Smith, University of Chicago, p. 5.

Gibbs, J. W. (1878), *On the Equilibrium of Heterogeneous Substances, Collected Works,* Longmans, Green and Co., New York, 1928.

Hirth, J. P., and G. M. Pound (1951), *J. Chem. Phys.* **26,** 1216.

Howes, V. R. (1959), *Proc. Phys. Soc.* **74,** 616.

Ives, M. B., and J. P. Hirth (1960), *J. Chem. Phys.* **33,** 517.

Kossel, W. (1927), *Nach. Ges. Wiss. Gotlingen,* **135.**

Landau, L. D. (1950), *Symposium in Commemoration of 70th Birthday of A. F. Ioffe,* Acad. Sci. Press, Moscow, p. 44.

Mykura, H. (1961), *Acta Met.* **9,** 570.

Nabarro, F. R. N., and P. J. Jackson (1958), *Growth and Perfection of Crystals,* John Wiley and Sons, New York, p. 13.

Price, P. B., D. A. Vermilyea, and M. B. Webb (1958), *Acta Met.* **6,** 524.

Volmer, M. (1922), *Z. Physik Chem.* **102,** 267.

Wulff, G. (1901), *Z. Krist.* **34,** 449.

Specific Substances

2

Metals

S. S. Brenner

Crystals can be grown from the vapor phase by two general methods: condensation of a supersaturated vapor and chemical reactions. In the condensation process, the vapor atoms or molecules impinge on the substrate surface; they become adsorbed, releasing part of their latent heat of condensation, and migrate across the crystal surface. If during their mean life of adsorption they encounter a growth site, they become incorporated into the crystal lattice, releasing the remaining heat of condensation; otherwise they reevaporate.

In the chemical process, the vapor atoms or molecules are chemically different from those in the growing crystal. For metals the vapor species may be halides, carbonyls, or metal-organic compounds. The vapor molecules adsorb on the surface of the substrate where they either are thermally decomposed or are chemically reduced to release the metal atoms. The mechanism of growth is very complex and the atomistic details are almost completely unknown.

Vapor-phase growth has commercial application in the production of luminescent crystals (zinc sulfide, cadmium sulfide) and semiconducting crystals (silicon, germanium, II–VI compounds). The growth of metal crystals from the vapor is technologically of little importance, however, because large, pure metal crystals can be obtained more easily by solidification. Nevertheless, there is an increasing demand for vapor grown crystals because these crystals sometimes possess unusual properties. The crystals often have perfectly flat external surfaces that cannot be duplicated by other means, and they often contain less substructure and imperfections than melt-grown crystals. Perhaps the greatest impetus in recent years for the vapor growth of metal crystals has come from the discovery that filamentary crystals (or whiskers) grown from the vapor have remarkable mechanical, magnetic, and surface properties. There is also an increasing interest in thin monocrystalline films grown epitaxially from the vapor.

1. GROWTH BY CONDENSATION

Crystals grown from their own vapor are small, their volume rarely exceeding a few cubic millimeters. The general procedure is to seal the metal into a tube that is either evacuated or filled with an inert gas; the tube is then heated in a temperature gradient with the metal stock at the highest temperature T_1. The metal vapor effuses throughout the tube and condenses on the colder walls. The vapor pressure within the tube is assumed to be the equilibrium vapor pressure of the bulk metal at T_1, so the supersaturation of the vapor in contact with the crystal growing at T_2 is assumed to be given by

$$\sigma = \frac{p_1 - p_2}{p_2} \tag{1}$$

In practice, a large number of crystals nucleate on the walls of the growth chamber and compete for the vapor. Thus the actual supersaturation near any crystal is lower than indicated by Eq. (1). It depends on the proximity and the number of surrounding crystals.

Because of their high vapor pressures (see Table 1), Zn, Cd, and Mg have been studied most frequently. Straumanis (1931), studying the growth process of each of these metals, used the experimental arrangement shown in Fig. 1*a*. The metal stock (sealed in an evacuated tube) was held in a temperature gradient of about 10°C/cm. The temperature of the Zn and the Cd vapor source was about 20°C above the melting point, whereas Mg (mp 651°C) was vaporized at 530°C. The Zn and Cd crystals nucleated primarily on the walls of the central Pyrex thermocouple tube; the Mg was condensed onto iron sheets. A large variety of crystal shapes were observed, depending on the temperature of condensation. Individual, discrete crystals predominated at condensation temperatures just below the melting points. The Zn and Cd crystals first formed as thin hexagonal plates with their *c*-axes perpendicular to the glass walls. New layers grew out from the centers of the platelets, with the edges of each succeeding layer remaining some distance from the edges of the underlying layer. For Cd, Straumanis found the height of each layer to be 0.8 to ±0.1 μ, or multiples thereof. McNutt and Mehl (1958), on the other hand, found no characteristic step height, only a range of heights from 300 to 3000 A. The Mg crystals were more regular in shape and considerably larger than the Cd or Zn crystals. The Mg crystals were

TABLE 1

VAPOR PRESSURE AT THE TRIPLE POINT OF SOME METALS CRYSTALLIZED FROM THE VAPOR (DUSHMAN, 1949)

Metal	Vapor Pressure (Torr)
Cr	63.5
Mg	2.2
Zn	0.16
Cd	0.10
Ti	8.4×10^{-2}
Fe	3.7×10^{-2}
Ni	4.4×10^{-3}
Ag	1.8×10^{-3}
V	6.5×10^{-4}
Cu	3.1×10^{-4}
Pt	1.6×10^{-4}
Au	6.0×10^{-6}
K	9.8×10^{-7}

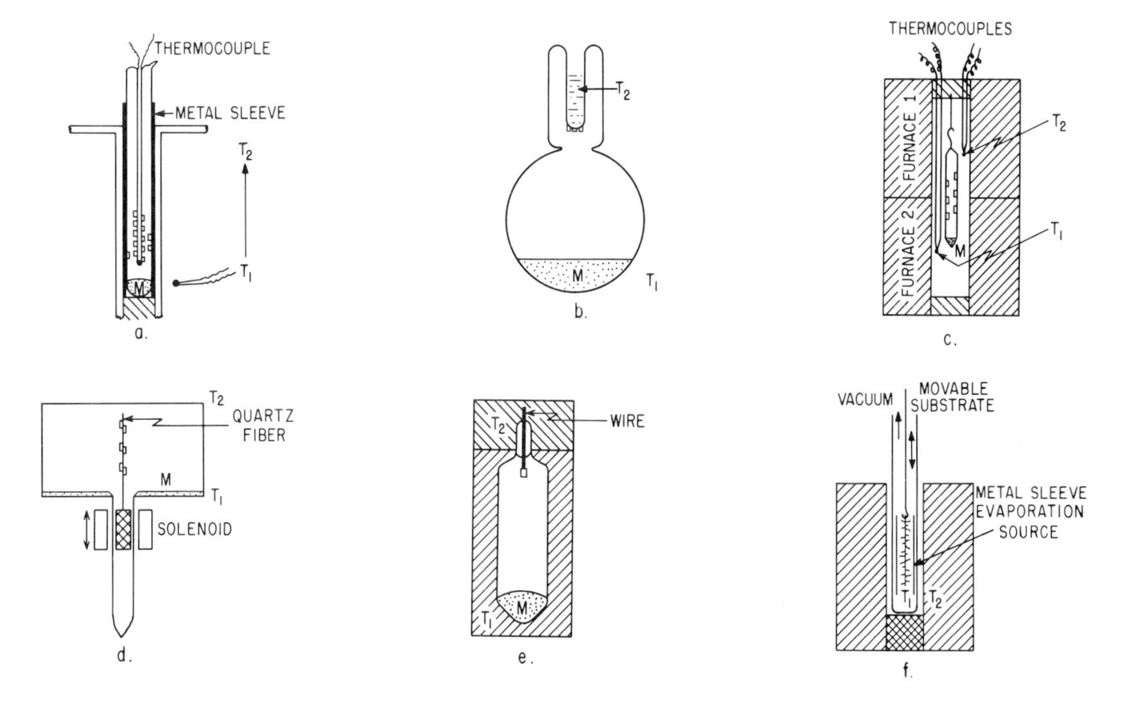

Fig. 1. Growth cells for the condensation of metal vapors.

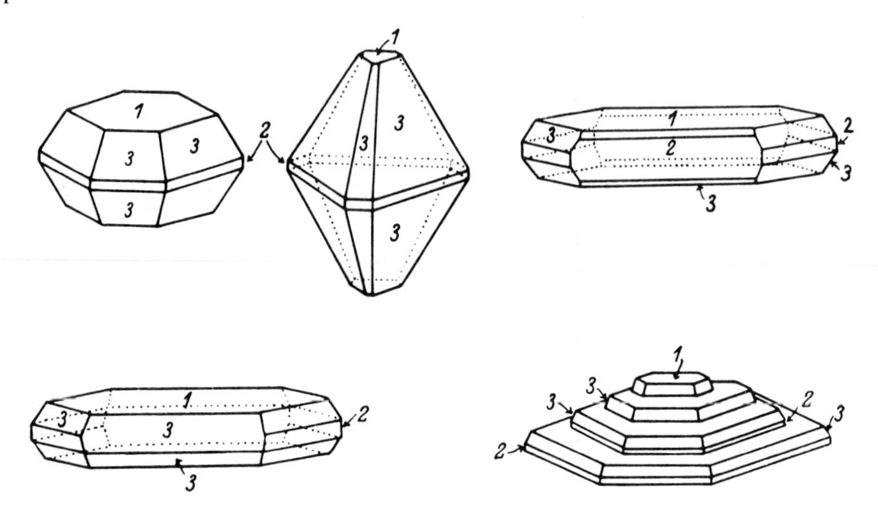

Fig. 2. Shapes of magnesium crystals grown from the vapor (Straumanis, 1931). 1. Basal plane. 2. Prism plane. 3. Pyramidal plane.

bounded by fewer and better developed surfaces, primarily of the (0001), (10$\bar{1}$0), and (10$\bar{1}$1) types. Some of the forms are shown in Fig. 2.

As the condensation temperature was decreased and the supersaturation increased, dendritic crystals and polycrystalline aggregates were observed. Assuming no vapor depletion near the surface of the growing crystals, it can be estimated from Straumanis' data that the larger Zn and Cd crystals grew at a supersaturation of about 3 to 4. Individual Zn crystals were still observed at supersaturations of about 10. Above this, however, the metal condensed to form a powder.

The presence of a foreign gas had a significant effect on the shapes of the crystals. Zn grew as regularly shaped crystals when hydrogen was present at pressures below 4 Torr*, but grew in a layered fashion at higher hydrogen pressures. Cd, growing in an atmosphere of hydrogen, formed "hair-crystals" and thin platelets as well as the layered crystals observed in a nominal vacuum. The effect of hydrogen on the growth of Mg crystals was even more striking: Sakui (1938) found that in the presence of hydrogen, Mg condensed to form crystals, whereas in a high vacuum a thin film was deposited. He suggested that this is because the gas helps to dissipate the heat of condensation. This seems unlikely, however, because monocrystal formation is enhanced by higher substrate temperatures. It is more likely that the inert gas reduces the supersaturation near the crystal surfaces, and thus lowers the nucleation rate of new crystals. Keepin (1950) made use of an inert gas atmosphere to grow "large" crystals of Cd and Zn. The metal was vaporized in a Mo crucible, and was collected on independently heated metal plates suspended above the crucible. The crucible and collector plates were in a bell jar containing purified nitrogen. The optimum conditions for the growth are shown in Table 2 and are very similar to those of Straumanis. Dendritic crystals were observed when either the collector temperature or the nitrogen pressure was decreased from its optimum value.

* 1 Torr = 1 mm Hg pressure.

Growth of Whiskers and Thin Platelets

The anisotropic growth of metal crystals was first studied by Volmer and Estermann (1921). They observed that thin platelets of mercury nucleating on a glass surface at −63.5°C, grow at least 10,000 times faster in width than in thickness. From this they concluded that there is an intermediate stage of condensation in which the atoms are in a mobile, adsorbed state. They postulated that the mercury atoms impinging on the basal surfaces of the crystal are unable to form new layers and that they migrate to the platelet edges where they are incorporated into the lattice. Sears (1955a) repeated this work thirty years later and found whiskers in addition to platelets. In a growth cell similar to that used by Volmer and Estermann (Fig. 1*b*), Sears studied the kinetics of whisker and platelet growth. He proposed that the edges of the platelets and the tips of the whiskers are preferred growth sites because they contain one or more emerging screw dislocations, whereas the basal surfaces of the platelets and the lateral surfaces of the whiskers are free of dislocations and thus do not contain permanent growth sites.

The dendritic crystals observed by Straumanis, Keepin, and others probably would be described as whiskers today. Once their significance in terms of dislocation theory was recognized, study of them was greatly intensified.

Sears (1955b) found that Zn and Cd whiskers can be grown in large numbers from the triply distilled metals in vacuum of 10^{-6} Torr. The apparatus is shown in Fig. 1*c*, and the operating conditions (Table 3) are similar to those of Keepin for dendritic crystals.

Coleman and Sears (1957) found that Zn whiskers grow in greater profusion and to a much greater length in the presence of pure inert gases than in vacuum. The largest number of long whiskers grew at substrate temperatures between 340 and 420°C (melting point of Zn) in a hydrogen or helium atmosphere at pressures in excess of 10 Torr. Whiskers as long as 17 mm and 1 to 3 μ diameter were observed. As the pressure of helium or hydrogen was increased from 10 to 600 Torr, the thickness of the whiskers decreased to a few tenths of a micron.

Using the technique of Coleman and Sears, Cabrera and co-workers (Cabrera and Price, 1958; Coleman and Cabrera, 1957) grew a profusion of Zn whiskers in an atmosphere of helium containing 10^{-3} mole per cent nitrogen as the major impurity. The Zn was vaporized at 490 to 500°C in a temperature gradient of 20 to 40°C/in. The impurities in the whiskers were estimated from residual resistivity measurements to be less than 10^{-2} atomic per cent. In addition to whiskers, hexagonal platelets, thin ribbons, rhomboids, and triangular

TABLE 2

GROWTH OF Zn, Cd, AND Mg CRYSTALS

	Temperature °C					
Metal	Vapor Source	Substrate	Substrate	Environment	Shape	Refs.
Zn	439	419 and lower ~10°/cm gradient	Glass	Vacuum	Regular	(1)
				$H_2 < 4$ Torr	Layered	(2)
	409–419	410 ±3	Cu, Ni	$N_2 = 1 - 3 \times 10^{-2}$ Torr	Regular	(3)
Cd	341	321 and lower ~10°/cm gradient	Glass	Vacuum, H_2	Layered	(4)
	310–321	305 ±3	Ag	$N_2 = 1 - 5 \times 10^{-2}$ Torr	Regular	(5)
Mg	530–800	Gradient	Fe	$H_2 = 10^{-5} - 380$ Torr		(6), (7)

(1) and (2) Straumanis, 1931. (3) Keepin, 1950. (4) Straumanis, 1931. (5) Keepin, 1950. (6) Straumanis, 1931. (7) Sakui, 1938.

TABLE 3

GROWTH OF METAL WHISKERS BY CONDENSATION

Metal	Temperature °C		Environment	l_{max}(mm)	Ref.
	Vapor Source	Substrate			
Hg	−50	−63.5	Vacuum	~2.5	(1)
Zn	375	350–375	Vacuum	~1	(2)
	421	380–420	H_2 = 10–600 Torr	17	(3)
	490–500	20–40°C/in. gradient	He = 760 Torr	Several millimeters	(4), (5)
Cd	330	250–321	Vacuum	~1	(6)
	320	~290	A = 760 Torr		(7)
Ag	940	850–940	Vacuum	~.1	(8)
K	54–57	58–62	Vacuum	0.2 to 0.7	(9), (10)

(1) Sears, 1953, 1955a. (2) Sears, 1955b. (3) Coleman and Sears, 1957. (4) Cabrera and Price, 1958. (5) Coleman and Cabrera, 1957. (6) Sears, 1955b. (7) Price, 1960. (8) Sears, 1953, 1955a. (9) Hock and Neumann, 1954. (10) Dittmar and Neumann, 1958.

plates were observed. The whiskers were hexagonal in cross section and their axes usually formed an angle of 30° with the *c*-directions. In this orientation two sides of the whiskers are prismatic $(10\bar{1}0)$ planes and the other four are $(10\bar{1}1)$ planes. Whiskers with growth directions 42° and 90° from the *c*-direction were also observed. Growth along the *c*-direction, however, was never observed. The thin ribbons are rectangular in cross section, two sides lying in the basal plane and two in the prismatic plane; the growth direction in the *a*-direction.

The Cd and Zn whiskers were found (Coleman, Price, and Cabrera, 1957) to be 100 to 1000 times stronger than bulk crystals of the metal. Some of the platelets were found to be sufficiently thin ($<$2000 A) for electron transmission and have been used by Price (1961) for the study of dislocation formation and motion.

The growth of Cd crystals was studied by Price (1960) using the growth cell shown in Fig. 1*d*. A temperature and supersaturation gradient was established between two flat plates. Then a thin, preheated quartz fiber was inserted to serve as a substrate for condensation. The cell contained an atmosphere of purified argon, and as long as the number of crystals on the quartz fiber remained small, the supersaturation of the environment could be assessed. This is a considerably better technique than earlier ones where only the upper limit of the supersaturation can be evaluated.

Price found that the number and the form of the crystals varied greatly with supersaturation. For $\sigma < 2$, whiskers and thin platelets formed predominantly (Fig. 3). For $\sigma > 2$, the density of crystals increased, and the whiskers and thin platelets thickened rapidly to yield thick hexagonal plates like those of Straumanis. Thus whisker growth (at least for metals) is not a unique phenomenon but is a transition stage during crystal growth at moderate supersaturations.

At supersaturations less than 0.4, Price did not observe the nucleation of crystals. This contrasts with the work of Parker and Kushner (1961) who found that Zn crystals, under carefully controlled temperature and vacuum conditions, nucleate on tungsten surfaces at supersaturations as low as 0.009. The discrepancy may result either from the much longer periods (up to 150 hr) used by Parker and Kushner or from the difference in substrates.

Condensation of Other Metal Crystals

POTASSIUM. The growth of crystal needles of potassium has been described by

Fig. 3. Growth of cadmium whiskers and platelets. (Price, 1960.)

Neumann and co-workers (Hock and Neumann, 1954; Dittmar and Neumann, 1958) and Parker and Kushner (1961). Their experimental arrangement is shown in Fig. 1*e*. Potassium, sealed in a thermostatically controlled glass tube, is condensed on a silver wire or on a crystal of potassium formed by solidification. The cell is maintained at 55 to 60°C and the growth substrate a few degrees lower, where the vapor pressure of potassium is about 5×10^{-7} Torr. The supersaturation of potassium was varied between 0.1 and 9.0. At lower temperatures (lowest for growth being -78°C), it was possible to use supersaturations greater than 10^3. After an induction period of as much as several hours, numerous potassium needles grew from the surfaces of the solidified potassium drop or silver wire. The needles were hexagonal in cross section with the [111] direction parallel to their axes and were bounded by six (110) planes; they formed on the (111) surfaces of the substrate and thus had its orientation. The needles, grown at a rate of 2–3 μ/min, were about 1000 A thick and up to 2 mm long. Thicker needles ($\sim$5 μ diameter) were reported in the paper of Hock and Neumann (1954).

SILVER. Howey (1939) condensed silver vapor on solidified spherical drops of silver to form crystal needles up to 3 mm long and $\sim$0.5 mm thick. The needles had [110] growth directions.

Sears grew short silver whiskers in an evacuated quartz tube with evaporation at 940°C and condensation at $\sim$850°C.

Growth of Submicron Whiskers from the Vapor

The whisker-growth techniques of Sears (1953, 1955b) and Coleman and Sears (1957) have been applied by Melmed and Gomer (1961) to grow thin crystals of numerous metals for use in field-emission microscopes. This method (Fig. 4) permits the use of metals which had previously caused difficulties. Whiskers grown in high vacuum ($p < 10^{-9}$ Torr) have clean sur-

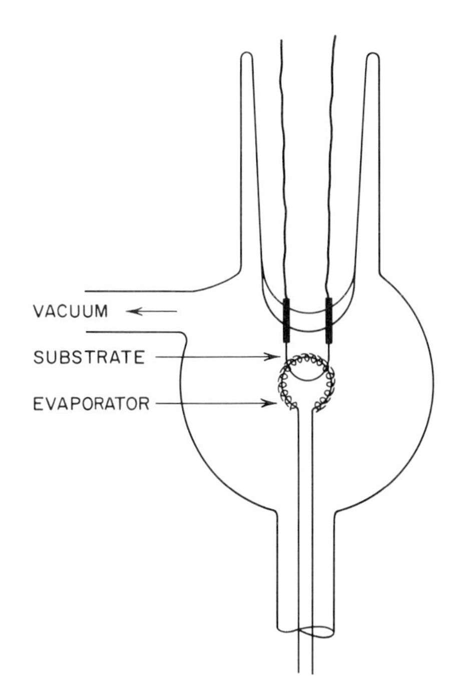

Fig. 4. Growth of whiskers for field emission studies. (Melmed and Gomer, 1961.)

faces and the necessary strength to withstand the high stresses caused by the electrostatic field.

The source material is wrapped around a 1 cm-diameter tungsten loop, and lies 1 to 3 mm from the cathode substrate which must be "rough." The oxide coating on electropolished tungsten suffices as long as it is not heated above 1000°C; however, a stranded substrate, consisting of forty twisted 0.0005 in. tungsten wires, is more satisfactory. Whiskers of various metals were grown in 5 to 10 min when the operating conditions given in Table 4 were used. In general, the substrate temperatures listed are 0.5 to 0.7 T_{melting} (°K), and the source temperatures correspond to equilibrium vapor pressures of 10^{-5} to 10^{-6} Torr. The crystals were 50 to 200 A thick and about 15 μ in length; their axes were parallel to low index directions, with the [110] direction predominant in fcc, and the [100] direction in bcc crystals.

Using Gomer's technique, Parker and Hardy (1961) grew potassium field emitters; so the technique is also applicable to alkali metals. The whiskers were grown with the source vapor pressure between 10^{-6} and 10^{-8} Torr, and substrate temperatures as low as 77°K.

The results of Melmed and Gomer and of Parker and Hardy show that at low temperatures whiskers can grow at very high supersaturations. For example, gold whiskers grew at 600°C at an apparent supersaturation of about 6×10^5. The rate of nucleation of two-dimensional nuclei, $\dot{N}$, is usually given as:

$$\dot{N} = B \exp - \left[\frac{\pi a \gamma^2 M}{\rho k R T^2 \ln (\sigma + 1)}\right] \text{cm}^{-2}\,\text{sec}^{-1} \tag{2}$$

where a = distance between lattice planes
γ = surface energy of material
ρ = density of crystal
B = frequency factor (10^{20} cm^{-2} sec^{-1})
M = molecular weight of vapor
k = Boltzmann constant

TABLE 4

GROWTH OF SUBMICRON WHISKERS

Metal	Temperature °C Source	Substrate	
Ni	1049–1271*	(700)†–900	Melmed and Gomer, 1961
Cu	>900	(600–700)	
Fe	1198	661–753	
Pt	1616	750–(960)	
Ti	1247	765	
Au	1092–1111	557–630	
Al	—	500–630	
Ba	—	(550–600)	
Ge	>900	(750–800)	
V	—	(800–850)	
Ag	>800	(600–650)	
Cr	1050	$\Delta T <$ 150**	Morelock, 1961
Ni	1200	"	
Au	875–1050	"	
Cu	930–1000	"	
Fe	1060–1150	$\Delta T \sim$ 17°C	

Residual gas pressure during growth $\leq 10^{-6}$ mm Hg.
ΔT = temperature difference between source and substrate.
* Growth times = 10 min.
** Growth times usually 16 hr.
† Values in parentheses estimated.

If 1600 ergs/cm^2 is used for the surface energy of gold (Buttner, Udin, and Wulff, 1951), N should be about 10^{11} nuclei/cm^2/sec. The whiskers should therefore thicken at a considerably faster rate than was observed.

This discrepancy may be due to an error in the value of the frequency factor B or to a thermal accommodation factor of less than unity as suggested by Sears and Cahn (1960) for a related case. The latter would imply that the temperature of the adsorbed layer is significantly higher than the temperature of the substrate.

Submicron whiskers of a variety of metals have also been grown by Morelock (1961) at higher temperatures and lower supersaturations. The metals were vaporized in a radial temperature gradient in a vacuum of about 10^{-6} Torr residual pressure. Substrates of the same metal or molybdenum screens were used at the temperatures given in Table 4. The experimental arrangement is shown in Fig. 1*f*.

Morelock found that the metal vapor, in addition to being the source of material for the crystals, acts as a "getter" for the residual oxygen and water vapor. When the pressure of the metal vapor was less than 10^{-6} Torr, the gettering was incomplete and whiskers of the reactive metals did not form. However, gold, which does not form a stable oxide at elevated temperatures, could be grown at much lower pressures.

The whiskers grown by Morelock ranged in diameter from 400 to 5000 A and were up to 300 μ long. In general, shorter whiskers were obtained at the lower temperatures. The whiskers grew to their terminal lengths in the first few hours of the experiment and would then remain constant in length. The cause for the cessation of growth was not understood.

2. EPITAXIAL GROWTH OF SINGLE CRYSTAL FILMS

In a sense all crystals form by epitaxial growth, each crystal layer depositing with the same orientation as the layer beneath it. However, the term *epitaxy,* first coined by Royer (1928), is usually reserved for oriented crystallization on foreign substrates. The subject has recently been reviewed at length by Seifert (1953), Pashley (1956), and Bassett, Menter, and Pashley (1959) and is discussed only briefly here.

The formation of thin films occurs by the nucleation and growth of many individual crystallites as shown in Fig. 5, taken from Bassett, Menter, and Pashley (1959). The crystallites are usually three-dimensional even when the film thickness, as calculated from the volume of evaporated metal, is less than a monolayer. However, in some cases, such as the deposition of Ag, Pb, and Tl on Ag substrates, two-dimensional nuclei have been reported (Newmann 1957). As deposition proceeds, the nuclei grow until a continuous film results. The thickness at which the porosity disappears depends on various factors such as deposition rate, temperature, and nature of substrate. For thin continuous films, the nuclei must be able to spread more readily than they thicken. A small misfit between nuclei and substrate promotes this lateral spread (Pashley, 1956; Engel, 1953).

Gold deposited on cleaved NaCl at 270°C is still discontinuous at 500 A nominal thickness. Only at 800 A does the network of voids disappear. Even then a variation of as much as 25% in thickness is reported (Bassett, Menter, and Pashley, 1959). Gold films deposited onto Ag become continuous at about 100 A. Heavens et al. (1961) report that Ni and Ni—Fe alloys deposited on Cu are continuous at 100 A thickness, but are still porous at that thickness when deposited on NaCl.

Crystals prepared by epitaxial growth usually are highly imperfect (Basset, Menter, and Pashley, 1959). They frequently contain twins, stacking faults, low-angle boundaries, and 10^{10} to 10^{11} dislocations/cm^2. The large dislocation density is not surprising since it is improbable that the crystallites are separated by exact multiples of their lattice dimensions, so that defects must be generated when the crystallites join. The twins and stacking faults are low energy defects; thus the probability of nucleating crystallites with these defects is

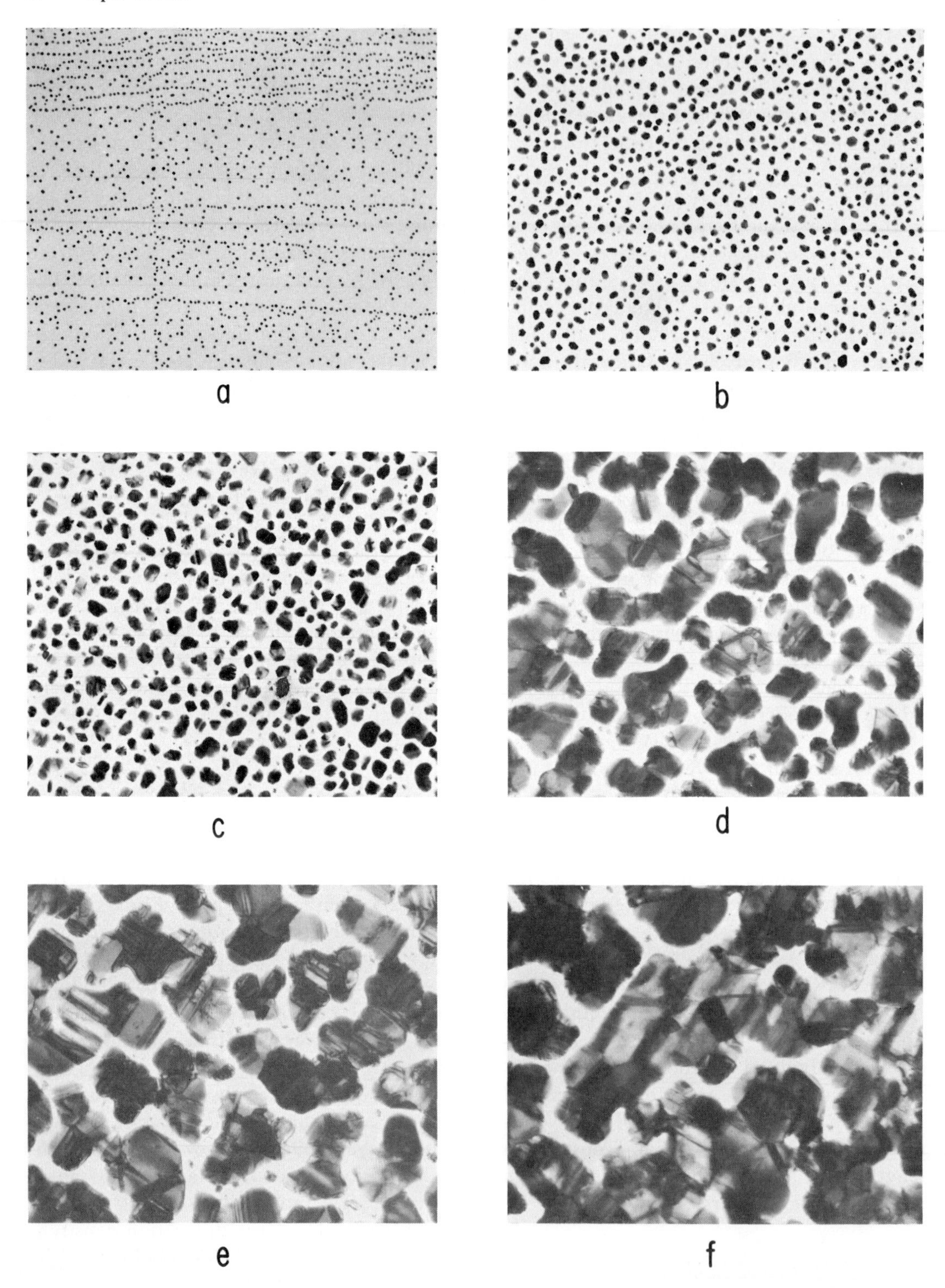

Fig. 5. Various stages of growth of gold deposited onto a NaCl cleavage surface at 270°C, (*a*) 10 A, (*b*) 30 A, (*c*) 100 A, (*d*) 200 A, (*e*) 300 A, and (*f*) 500 A. (68,000 X). (Bassett, Menter and Pashley, 1959.)

nearly the same as without them. It is frequently difficult to judge whether a thin film is, in fact, a continuous crystal or not. The structure is usually determined by electron diffraction from small areas. This fails to detect porosity, and does not distinguish between large-grained films and single-crystal films.

The main factors controlling epitaxial growth are the following.

1. Nature of Substrate. Monocrystalline substrates are required and fewer defects form when the surface is smooth. Cleaved surfaces of NaCl and mica are frequently used, but even these contain irregular cleavage steps. When the height of these steps is small, they present no problem. Bassett (1958) has shown that even monatomic ones are preferred nucleation sites, but they do not produce misorientations in the nuclei. When the cleavage steps are large, correspondingly large steps appear in the films (Goche and Wilman, 1939).

For some purposes metal substrates are desirable because films deposited on them become continuous at lower thicknesses and exhibit less twinning than do films grown on ionic substrates. Sometimes their use also facilitates film removal. Metal substrates can be prepared by epitaxial growth on NaCl or mica or by careful cutting and electropolishing of bulk crystals. With the latter technique any orientation can be obtained, but the preparation of atomically flat surfaces is difficult. Pashley (1959) has used Ag substrates grown epitaxially on mica to prepare smooth Au films which were then removed by dissolving the Ag in nitric acid. Similarly, Heavens et al. (1961) made films of Ni and Ni-Fe by deposition on Cu-coated NaCl, and then detached them by immersion in chloride. Electropolished Ag surfaces were used by Brame and Evans (1958) to make films of Au, Au-Pd, Pt, and Rh. The films were detached with 35% nitric acid.

When metallic substrates are used, the surfaces must be free from any oxide layer that might interfere with growth. For instance, Haase (1956) reports that completely oriented films of Cu, Ag, Au, Pd, Fe, and Zn were obtained at room temperature on (111) surfaces of Cu crystals that were first electropolished and then bombarded with argon ions. If the bombardment step was omitted, only partially oriented films or polycrystalline films were obtained. The effect was especially striking with Zn deposits.

Another difficulty with metallic substrates is alloying at the interface. This can be minimized by using a low substrate temperature.

2. Temperature of Substrate. This has three major effects. It affects the critical size and rate of formation of nuclei, the mobility of the adsorbed atoms, and the annealing of defects in the condensed films. At very low temperatures the critical size of the nuclei is very small and the rate of nucleation is large. Also the lateral growth rate of the nuclei approaches zero because of the low mobility of the adsorbed atoms, and only the atoms that impinge near the nuclei can reach them. Furthermore, the driving force for condensation is very large, so misoriented crystallites are easily formed. The result of these effects is highly imperfect or polycrystalline films.

At higher temperatures, as surface mobility increases and the rate of nucleation decreases, the film forms from fewer nuclei, so fewer closure defects will be incorporated. However, if the temperature is too high, the crystallites will grow three-dimensionally as rods or whiskers or as large, faceted crystals, and a uniform continuous film will not be formed. There is therefore, as was first pointed out by Brück (1936), an optimum range of growth temperatures. These "epitaxial temperatures" as determined by Brück are given in Table 5 together with the optimum conditions reported by various other investigators. The "epitaxial temperature" is not an intrinsic property of the substrate, but depends on various factors, including the rate of deposition (Yelon and Hoffman, 1960). Thus Bassett obtained oriented films of Au on NaCl at 270°C, although the "epitaxial temperature" is usually considered to be 400°C.

TABLE 5

GROWTH OF ORIENTED METAL FILMS

Metal	Plane of Film	Substrate and Temperature*
Ag	(100)	NaCl (150)[1]; CaF_2 (>510)[2]
	(111)	Mica (250–300)[3]; $CaCO_3$ (470)[4] Cu (250)[5]
Au	(100)	NaCl (400)[6]
	(110)	$CaCO_3$ (510)[7]
	(111)	Mica (450)[8]; $CaCO_3$ (360)[9]; CaF_2 (380)[10] Cu (25)[11]; Ag (270)[12], [13]
Cu	(100)	NaCl (300)[14]
		Ag (270)[15]
Pt	Variable	Ag (350)[16]
Pd	(100)	NaCl (250)[17]
	(110)	$CaCO_3$ (490)[18]
	(111)	Mica (470)[19]; $CaCO_3$ (400)[20]; CaF_2 (400)[21] Cu (25)[22]
Au–Pd	Variable	Ag (250)[23]
Rh	Variable	Ag (400)[24]
Pt	Variable	Ag (350)[25]
Ni	(100)	NaCl (370)[26]
	(100), (110), (111)	Polished NaCl (330)[27]; Cu (330)[28]
Ni–Fe	Same as Ni (Heavens et al., 1961)	
Co	(100)	NaCl (≥540)[29]
	(100), (110), (111)	Polished NaCl (500)[30]
Fe	(100)	NaCl (≥540)[31]
	(100), (110), (111)	Polished NaCl (470)[32]
	(111)	Cu (25)[33]
Al	(100)	NaCl (440)[34]
Cr	(100)	NaCl (>540)[35]
Zn	(0001)	Cu (25)[36]
Sn	(200)	Polished, annealed NaCl (−196)[37]

* Nonmetallic substrates cleaved unless indicated.

Conditions for epitaxial growth of metals on MoS_2, ZnS, PbS, and FeS_2 can be found in Uyeda (1942), Miyake and Kubo (1947), and Kainuma (1951).

References: (1) Brück, 1936. (2) Rudiger, 1937. (3) Pashley, 1959. (4) Rudiger, 1937. (5) Haase, 1956. (6) Brück, 1936. (7), (8), (9), and (10) Rudiger, 1937. (11) Haase, 1956. (12) Newman, 1957. (13) Brame and Evans, 1958. (14) Brück; 1936. (15) Kehoe et al., 1956. (16) Brame and Evans, 1958. (17) Brück, 1936. (18), (19), (20), and (21) Rudiger, 1937. (22) Haase, 1956. (23), (24), and (25), Brame and Evans, 1958. (26) Brück, 1936. (27) and (28), Heavens et al., 1961. (29) Brück, 1936. (30) Heavens et al., 1961. (31) Brück, 1936. (32) Heavens et al., 1961. (33) Haase, 1956. (34) and (35) Brück, 1936. (36) Heavens et al., 1961. (37) Vook, 1961.

The substrate temperature can also affect annealing of defects, recrystallization, and grain growth. Although no direct observations of defect annealing have been made during deposition, annealing after deposition can markedly improve the orientation of a film (Bassett et al., 1959; Vook, 1961).

Because of the heat of condensation and the radiation of heat from the vapor source, the temperature of the adsorbate and the condensed film are not necessarily the same as that of the substrate. Wilman (1955) estimated that the temperature at the surface of a deposit growing at 900 to 18,000 A/min can be several hundred degrees higher than that of the substrate. Such a large temperature rise has been disputed by others (Hoffman et al., 1954; Kehoe, 1957).

3. Rate of Deposition. Although epitaxial growth occurs by the nucleation and growth of crystallites, almost no systematic work has been done to establish the effect of beam pressure or deposition rate on the nucleation process. As the deposition rate increases, the films become less perfect. However, whether this is due to an increased rate of nucleation or to more misoriented nuclei is not known.

Under normal deposition conditions one would expect very small nucleation barriers. For instance, when gold is deposited at 400°C at a rate of 20 A/sec the apparent supersaturation, given by the ratio of the rates of arrival and evaporation, is of the order of 10^{15} if the absorbate temperature equals the substrate temperature. The critical radius of a disk-shaped nucleus of height a, equals

$$r^* = \frac{a\gamma_1}{a\dfrac{\rho RT}{M}\ln(\sigma + 1) - (\gamma_1 + \gamma_{12} - \gamma_2)} \quad (3)$$

where γ_1, γ_2, and γ_{12} are the surface energies of the condensed phase, the substrate, and the interface, respectively. Using 1600 ergs/cm^2 for γ_1, 3×10^{-8} cm for a, and assuming $\gamma_{12} \simeq \gamma_2$, the diameter of the critical nucleus is less than the atomic diameter of gold. Thus every atom that strikes the surface can attach itself to the substrate, and discrete nuclei should not be observed. This is contradictory to experimental observation. The explanation is probably that the condensed gold atoms reevaporate at a faster rate than normally. Hence the true supersaturation is less, and the nucleation barrier is greater. Sears and Cahn (1960) have suggested that the energy exchange between the adsorbed atoms and the substrate requires a finite time compared to the half-life of the adsorbed atom; thus the adsorbate temperature is greater than the substrate temperature. Also one should consider the increase of vapor pressure with decreasing radius of curvature. A nucleus with $r = 3$ A at 400°C has a vapor pressure $\sim 10^8$ times greater than for a flat surface, according to the Kelvin relation. The idea that the adsorbate is hotter than the substrate agrees with observations that condensed atoms at low temperatures have long migration paths (Kehoe et al., 1956). One would expect the beam temperature to affect the nucleation rate, but independent variations of the intensity and temperature of the atomic beam have not been studied.

4. Residual Gas Pressure during Deposition. Gas impurities may have major effects on the perfection of an overgrowth; yet no detailed studies have been reported. Improvements in vacuum technology may remedy this situation since most studies have been done in vacua of 10^{-4} to 10^{-6} Torr. Under these conditions, the surface is bombarded with as many residual gas atoms as condensate atoms, and the process variables must be affected, especially by gases that strongly adsorb, such as oxygen, nitrogen, and water. Shlier and Farnsworth (1958), for instance, report that a monolayer of adsorbed oxygen on the surface of titanium inhibits the formation of an oriented copper deposit.

3. GROWTH OF METAL CRYSTALS BY HALIDE REDUCTION

Metal deposition by halide reduction gained prominence during the early days of the lamp industry. de Lodyguine (1893) obtained a patent for depositing tungsten onto hot carbon filaments in a mixture of hydrogen and tungsten hexachloride by the reaction

$$WCl_6 + 3H_2 = 6HCl + W \quad (4)$$

Halide reduction was subsequently used by Koref (1922) to grow single crystal deposits of W, Zr, Ta, Ti, etc., on tungsten wire crystals. The method consists of heating a tungsten wire crystal at about 1000°C in a stream of hydrogen saturated with WCl_6 at 120°C, and a total pressure of 12 Torr. A continuous, coherent deposit is formed as shown in Fig. 6. If the partial pressure of the WCl_6 is too high, a polycrystalline deposit forms. Deposition also occurs upon heating in the halide vapor without hydrogen (van Arkel, 1923), if the wire is hot enough to cause rapid thermal dissociation of the halide; for WCl_6 it is about 1600°C. Very pure, ductile deposits can be obtained, and this process is now the basis for the metallurgical production of high purity Zr, Hf, and Ti. Crystals obtained by this method are usually no better than the substrate on which they are grown. To obtain crystals free of substructure and with smooth crystallographic planes, it is necessary to start with highly perfect substrates.

Whiskers and Dendritic Crystals

The formation of fibrous crystals during the reduction of metal halides has been casually observed for many years. Sometimes, as during vapor plating, this is highly undesirable since a poor coating results. In recent years interest in filamentary crystals has greatly increased because of their unusual mechanical properties and perfection. Whiskers grown by the halide process have received particular attention because relatively large fibers can be grown with very simple equipment. In addition to their perfection, whiskers have better defined orientations than crystals prepared by other means, and are often bounded by extremely smooth, low index crystallographic-planes. Their main disadvantage is small size, necessitating careful handling and manipulation. Whiskers of high strength are usually (Brenner, 1956a) less than 10 μ in diameter and are up to 1 cm in length. Copper and iron whiskers 10 to 20 cm long and 1 mm or more wide can be grown by increasing the scale of operation (Brenner, 1957; Wayman, 1961). These large whiskers are not as perfect as thinner ones.

The apparatus (Brenner, 1956b) for whisker growth is very simple (Fig. 7). A boat filled with the metal halide is placed

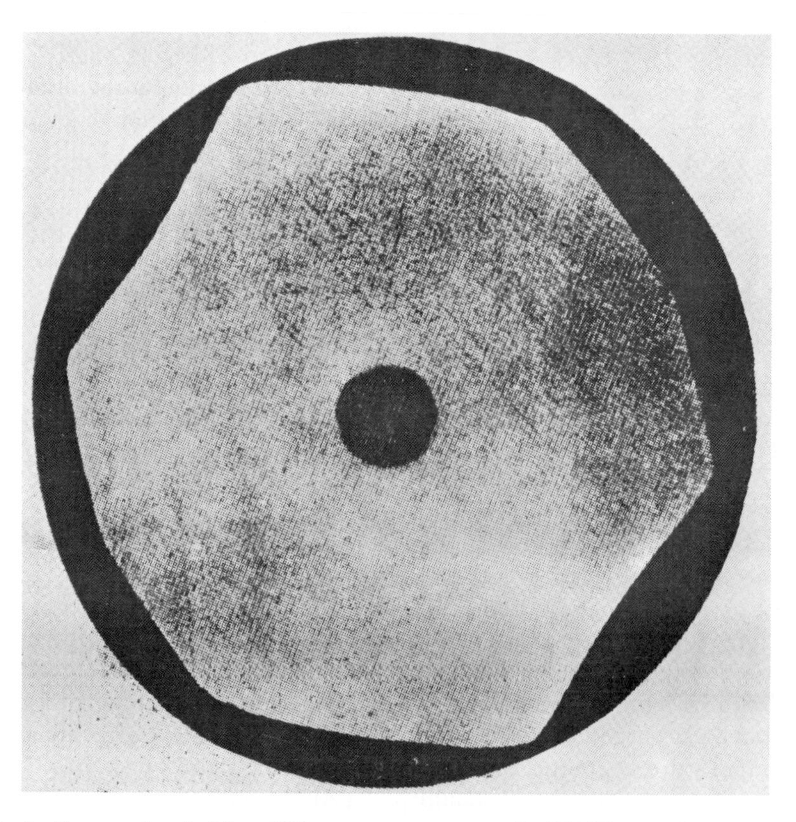

Fig. 6. Tungsten deposited from WCl_6 onto a tungsten crystal. (150 X). (van Liempt 1931).

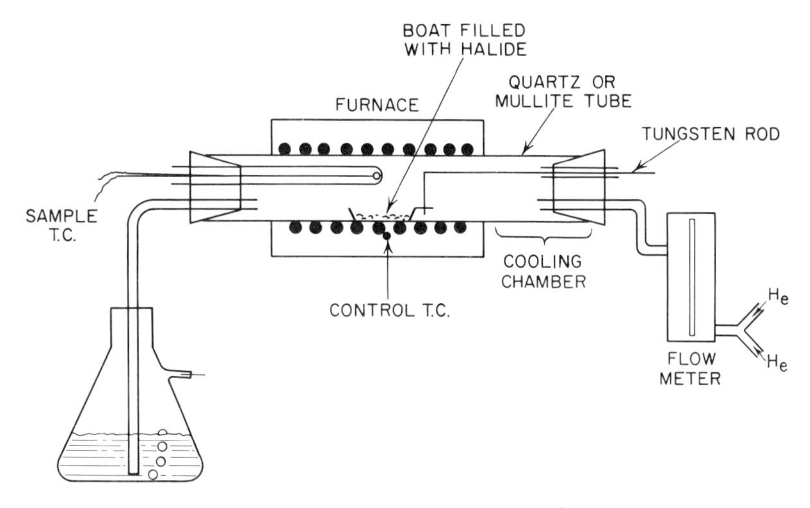

Fig. 7. Apparatus for growth of whiskers from halides. (Brenner, 1956b.)

into a hot furnace through which hydrogen or some other reducing gas is flowing. The furnace should have sufficient power to bring the halide rapidly to the growth temperature; otherwise much reduction may occur at a temperature that is too low.

The mechanism of whisker growth is not well understood. In some cases it is even difficult to determine whether it occurs by vapor-phase transport or by diffusion through the substrate or by capillary flow of the liquid halide along the whisker surfaces. Observations suggest that the whiskers are nucleated on small metal or oxide crystals. For the more noble metals, such as Cu and Ag, small crystals form directly from the halide and the whiskers grow on these. However, for the more reactive metals, such as Ni and Co, whiskers are believed to form on metal particles formed by the reduction of oxide crystals. Cech (1959) has shown the oxide-metal-whisker nucleation sequence particularly clearly for the growth of Ni whiskers from NiBr in atmospheres containing water vapor. For Fe, Gorsuch (1961) has shown that the whiskers can nucleate both on metal and oxide particles, the nucleating agent determining the orientation.

The growth forms of whiskers imply that their tips are preferred growth sites, especially when the whiskers are bounded by equivalent crystallographic planes. Sears has suggested that this is because only the tips of whiskers contain emerging screw dislocations. An alternative explanation for the large growth anisotropy is given by Price, Vermilyea, and Webb (1958) that does not require any particular dislocation density or configuration. They propose that the tip of the whisker remains clean while the sides adsorb impurities that hinder their thickening. The impurities reaching the tip become buried, implying that the whiskers may contain varying concentrations of impurities, depending on their axial growth rates.

To account for the high axial growth rates of whiskers, a large flux of the condensing species must exist. If the condensing species is the metal vapor, a supersaturation of the order of 10^4 to 10^6 must exist. It has been shown thermodynamically by Gorsuch (1961), Webb and Riebling (1958), and Wiedersich (1959) that the actual supersaturations are ~ 1 to 100 rather than 10^4 to 10^6. The vapor transport must therefore occur by means of the halides (Coleman and Sears, 1957; Brenner, 1957; Morelock and Sears, 1961) which have sufficiently high vapor pressures. It has been found by the author (Brenner, 1959) that the growth rate of Cu whiskers, from CuI, is proportional to the CuI pressure in the temperature range where vapor transport appears to dominate. Metal halides strongly adsorb on metal sur-

TABLE 6

GROWTH OF WHISKERS BY THE REDUCTION OF METAL HALIDES

Metal	Halide	Temperature °C	Remarks	Refs.
Cu	CuCl	430–800	[100] growth direction predominant	(1), (2)
	CuBr	600	[100] growth direction predominant	(3), (4)
	CuI	600–650	[100], [110] and [111] whiskers	(5), (6)
	0.8 CuBr+ 0.2 CuI	600	Large yield of whiskers with diameter $\leq 10\ \mu$ and lengths of up to 10 mm; zircon refractory boat.	(7)
Fe	$FeBr_2$	700	Salt partially decomposed containing α Fe_2O_3 and β Fe·OH. Diameter of whiskers decreases with decreasing partial pressure of hydrogen. [100] growth direction predominant.	(8), (9), (10)
	$FeCl_2$	400–900	Additions of Fe_2O_3 and use of wet hydrogen increases yield of whiskers. [100] type whiskers predominant.	(11) (12), (13) (14), (15)
	$FeCl_2 \cdot 4H_2O$	700–850	Diameter = 100–500 μ. [111] growth direction.	(16)
	Fe + NH_4Cl + H_2O	600–650	Short [100] whiskers	(17)
		700–800	[111] whiskers −50–500 μ diameter.	
Ag	AgCl	700–900	Growth does not occur from vapor	(18)
	AgI	600	Growth rate of whiskers is small	(19)
Ni	$NiBr_2$ H_2O^+	700–900	Wet hydrogen beneficial. Reproducibility of growing large whiskers is poor	(20), (21)
Co	$CoBr_2$ + H_2O	550–850	Wet hydrogen and use of covered Al_2O_3 boats is beneficial. Whiskers usually mixtures of fcc and hcp phases	(22), (23)
βMn	$MnCl_2$	940	Al_2O_3 boats. Salt first distilled. Whiskers ~0.25 mm length	(24)
Mn_5Si_3	$MnCl_2$ + SiO_2	940	Length of whiskers ~0.25 mm	(25)
Pd	$PdCl_2$	960	Thermal reduction in argon. Whiskers 1–10 mm long, ~2 μ diameter	(26)
Pt	$PtCl_4$	800	Thermal reduction. Many of the whiskers kinked	(27), (28)
Au	$AuCl_x$	550	Thermal reduction. Most whiskers less than 1 mm long	(29), (30)

(1), (3), and (5) Brenner, 1957. (2), (4), and (6) Brenner, 1956a. (7) Brenner, unpublished work. (8) Brenner, 1957. (9) Brenner, 1956b. (10) Gorsuch, 1961. (11) Brenner, 1956b. (12) Gorsuch, 1961. (13) Wiedersich, 1959. (14) Wan et al., 1959. (15) Gatti and Fullman, unpublished work. (16) and (17) Gorsuch, 1961. (18) Brenner, 1959. (19) Brenner, 1957. (20) Brenner, 1957. (21) Brenner, 1956b. (22) Brenner, 1957. (23) Brenner, 1956b. (24), (25), and (26) Rubling and Webb, 1957. (27) Brenner, 1957. (28) Brenner, 1956b. (29) Brenner, 1957. (30) Brenner, 1956b.

faces, so whisker surfaces are covered with halide during growth. If only the tip contains dislocations as Sears has proposed, and if they promote chemical reactions, the whisker can grow much faster in length than in width. Specific conditions for whisker growth are described in the following sections, and a summary is given in Table 6.

Copper

Crystals can be grown from CuCl, CuBr, and CuI. Optimum conditions vary appreciably from one batch of raw materials to the next, and with the size and geometry of the reduction chamber. For maximum growth velocity, it is generally desirable to grow the whiskers from the liquid halides. The axial growth rate of whiskers from CuI (Brenner, 1959) increases from 15 μ/sec at 580°C to 65 μ/sec at 620°C, the melting point of CuI being 586°C.

Whiskers formed from CuCl and CuBr tend to have square or rectangular cross sections with [100] axial orientations and (100) lateral surfaces. CuI produces (in addition to [100] whiskers) some [110] and many [111] whiskers. The latter are hexagonal in cross section bounded by six (110) planes.

The size of whiskers grown from CuI depends on the dimensions of the growth boat and the quantity of salt it contains. Initially, the whiskers are extremely thin and grow axially at a rate as great as 1 cm/min. Then, for some unknown reason, the axial growth rate abruptly decreases and thickening begins. If the process is stopped, very fine whiskers (1 to 10 μ diameter) are obtained. If reduction is continued, the final dimensions are governed by the size of the growth chamber. To produce a large number of whiskers several inches in length and up to 1 mm in diameter (Fig. 8), the author found it most convenient to heat a copper boat about one-fourth filled with CuI in a 3 in. hydrogen furnace for 2 to 3 days. These large crystals often show lamellar growth markings as shown in Fig. 9 and are not as strong as thinner ones.

Whiskers grown from CuCl and CuBr are more dendritic in nature than those grown from CuI. An increase in growth temperature or the amount of starting material usually results in a mass of interlocked, branched crystals. Thus smooth, defect-free whiskers with diameter > 10 μ are rarely observed. Some improvement in size and perfection is obtained by reducing mixtures of CuBr and CuI in a porous alundum boat at 600°C.

Wan et al. (1959) report that additions of the chlorides of Fe, Co, and Ag to CuCl markedly affect the growth of Cu whiskers. Mixtures of CuCl—$FeCl_2$ held in Ni boats and reduced in an Fe tube at 750 to 800°C produced whiskers 7 to 8 cm in length and 50 μ in diameter in a period of 40 to 70 min.

The quantity, quality, and properties of the whiskers can vary widely, depending on the source of the halide salts. This is particularly true of the mechanical properties. Whiskers grown under seemingly identical conditions from different batches of materials may have average yield strengths that vary as much as 100:1.

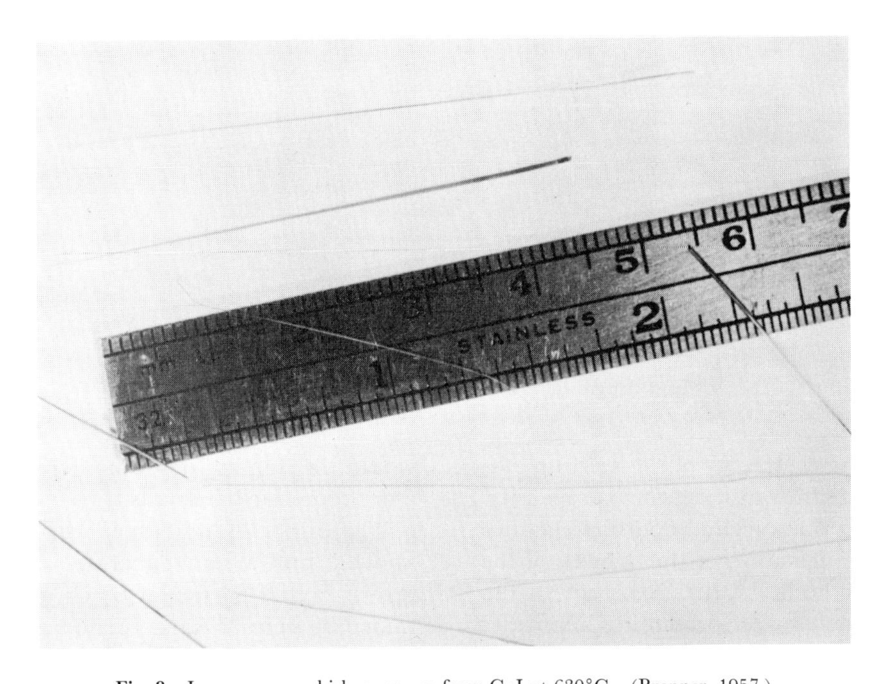

Fig. 8. Long copper whiskers grown from CuI at 630°C. (Brenner, 1957.)

Fig. 9. Lamellar growth marking on copper whiskers. (Brenner, 1957.)

Increasing the purity of the starting material reduces the whisker-growth propensity. This implies that certain impurities enhance either the nucleation or the growth of the whiskers or both. A detailed study of the nature of the whisker-promoting impurities would be highly desirable for obtaining more reproducible results.

The purity of individual whiskers is not known, but a conglomerate of whiskers grown from a particular batch of CuI contained 30 ppm of Ag as the major impurity

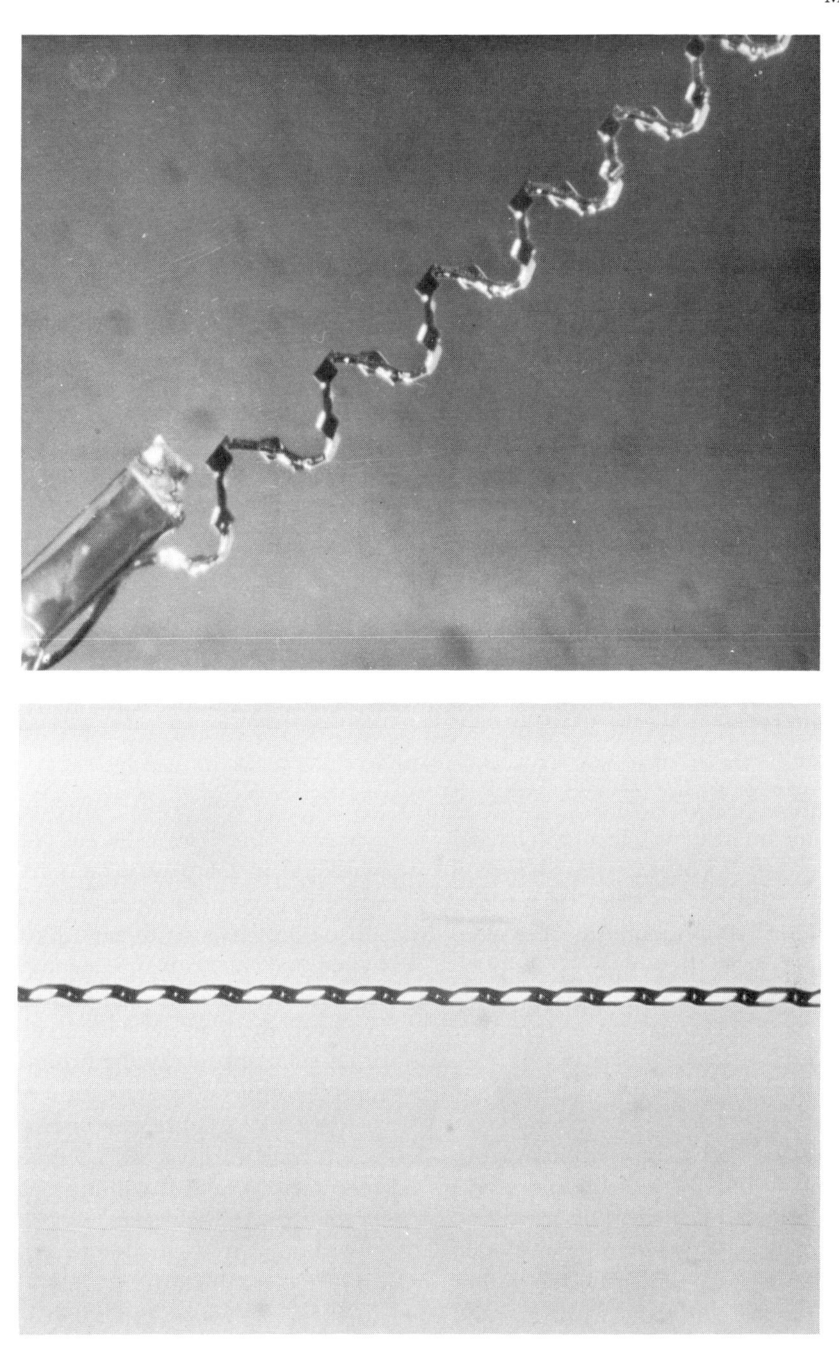

Fig. 10. Helical copper whiskers. (Brenner, 1957.)

(Brenner 1957). There is also good evidence from activation analyses (Cabrera, private communication) that Cu whiskers contain halogens. These affect the low-temperature elasto-resistivity behavior (Cabrera, private communication) and may also affect the mechanical properties (Brenner, unpublished work). Whiskers grown from the iodide contain more trapped halogen than those grown from the bromide.

Crystals of many shapes other than whiskers are formed during halide reduction.

Fig. 11. Iron whiskers grown from $FeBr_2$ at 700°C. (8 X). (Brenner, 1957.)

Of particular interest are faceted, helical crystals (Fig. 10). The external shapes of the helices usually reflect their crystal orientation. Thus [100]-oriented helices often have flat facets at 90° intervals. The pitch of the helix can vary from a few turns to several thousand turns per centimeter.

Iron

Filamentary Fe crystals were first grown by Gatti and Fullman (unpublished work) when they reduced a filtrate of ferric and aluminum hydroxide containing occluded chloride. It was later shown by the author (Brenner, 1956b) and by Cochardt and Wiedersich (1955) that Fe whiskers can be grown directly by hydrogen reduction of $FeCl_2$ and $FeBr_2$. The arrangement is the same as for Cu except that Fe boats are used.

Specific impurities are necessary for the nucleation and probably also for the growth of Fe whiskers. Gorsuch (1961), studying the growth in detail, concludes that whiskers with [100]-oriented axes nucleate on iron oxide particles, whereas [111]-oriented whiskers nucleate on iron particles. The oxide particles can be added to the halide salt or they can be formed by heating the halide in an oxidizing atmosphere for a short period prior to reduction. Usually, however, the commercially available anhydrous salts (especially the bromide) are partially hydrolized, containing α—Fe_2O_3 and γ—$FeO \cdot OH$. To retain the oxide for a sufficient time at the growth temperature, it is necessary to heat the charge rapidly, preferably in wet hydrogen.

Although it is possible to grow filamentary Fe crystals under widely varying conditions as shown in Table 6, it is difficult to grow them with reproducible surfaces and structural perfections. A critical relation exists among the state and composition of the halide, water vapor content of the hydrogen, size and geometry of growth chamber, reduction temperature, etc., which makes it necessary to optimize conditions every time a variable is changed.

For the growth of mechanically strong whiskers, the author has found the following procedure most suitable. Nominally anhydrous ferrous-bromide is used as the starting material. As received, it is usually in various states of decomposition, containing some water. A small amount of the bromide in an iron boat is pushed by means of a rod inside a one-inch quartz tube into the middle of a furnace held at a temperature of 700 to 750°C. A mixture of nitrogen and 20 to 50% hydrogen is passed through the quartz tube at approximately 250 cc/min. The gas mixture may be saturated with water vapor at room temperature. The boat should remain in the furnace at least several hours and preferably longer to reduce the last traces of halide. After this period, the boat is pulled into the cool end of the quartz tube where it is allowed to cool. Some of the resulting high strength whiskers are shown in Fig. 11.

The success of this technique depends on the state of the starting material. Under ideal conditions the crystals are uniform in diameter and have highly reflecting surfaces without pits or other observable defects. Frequently, however, a particular batch of material yields tapered, pitted, and mechanically weak whiskers. More reproducible results may be attainable by partially decomposing pure $FeBr_2$ in oxygen at low temperatures, as suggested by Gorsuch (1961).

As with Cu, Fe whiskers can be grown to lengths greater than 10 cm and diameters of 1 mm by increasing the scale of operation. Whiskers up to several hundred microns in width have been used (Coleman, 1958; Laukonis and Coleman, 1961; DeBlois and Bean, 1958; Coleman and Scott, 1958) for various experimental purposes.

Whether Fe whiskers grow by vapor transport has not been established. Below the melting points of the halides, Gorsuch (1961) and Cochardt and Wiedersich (1955) claim that growth occurs by the solid state reduction of the salts, similar to the well-known growth of silver hair from silver sulfide (Kohlschutter, 1932). However, Wiedersich's (1959) photograph of Fe whiskers growing at 658°C on a wire stretched across a boat containing $FeCl_2$ ($FeCl_2$ melts at 670°C) seems to contradict this contention unless the $FeCl_2$ vaporized and condensed on the wire prior to whisker formation. At temperatures above the melting points of the halides, the whiskers grow at their tips, but it is uncertain whether material is transported through the vapor or by capillary flow along the sides of the whiskers.

Silver

Filaments of Ag can be grown easily from molten AgCl heated rapidly to 460 to 900°C in a stream of hydrogen. The whiskers, 2 to 15 μ in diameter and up to 1 cm in length, are round in cross section. Both [100]- and [110]-oriented fibers have been observed. The whiskers grow from their bases according to visual observations (Brenner, 1959) and do not grow by vapor transport. Many of the whiskers exhibit low angle boundaries and have relatively low mechanical strength.

Silver whiskers can also be grown from AgI at 600°C at a considerably lower growth rate (Brenner, 1947). The whiskers are polygonal in cross section, and their general growth characteristics are similar to those of Cu, suggesting that they, like the Cu whiskers, grow by vapor transport.

Nickel

Whiskers up to 3 mm in length and 5 μ in diameter have been grown from $NiBr_2$ in Ni boats at temperatures between 700 to 900°C.

As Cech (1959) has shown, the environment must first be oxidizing to produce nucleation sites. This condition can be obtained by slightly wetting the anhydrous bromide and by using hydrogen diluted with nitrogen. During the initial stages of heating, the boat containing the NiBr is covered with a blanket of steam which promotes the formation of oxide particles. Later, when the water vapor has effused away, conditions become reducing and metallic Ni forms. Both [100]- and [110]-oriented whiskers have been reported. It is very difficult to obtain reproducible conditions for the growth of whiskers longer than 0.2 to 0.3 mm. Whiskers have also been grown by Cech (1959) on epitaxial NiO films.

Cobalt

The conditions (Brenner, 1956b) for the growth of Co whiskers are similar to those for Ni. Moist $CoBr_2$ is reduced with diluted hydrogen at temperatures ranging from 550 to 850°C. Covered alundum (Brenner, unpublished work) boats contain the cobalt bromide. The reproducibility of growing long whiskers (few millimeters) is greater than with Ni but still is poor. Many of the whiskers consist of a mixture of the hexagonal and face centered cubic phases, and some are polycrystalline. Growth markings, which may correspond to stacking faults have been observed. The whisker axes are parallel to the [110] direction.

Tungsten, Zirconium

Crystals of tungsten and zirconium have been grown by Moliere and Wagner (1957) by the Van Arkel process (Fig. 12). WBr_5 or ZrI_4, sublimed into one end of the apparatus, is pumped over a hot tungsten wire to which short "barbs" of 40 μ diameter tungsten are attached transversely. The tips of the tungsten barbs, deformed from the wire cutting, serve as nucleating sites for the crystals. The "barbs" must be recrystallized at 2000°C prior to the crystal growth.

For the growth of tungsten crystals, the WBr_5 is maintained at 300°C and the tungsten wire at 1800 to 2000°C. During the 8- to 10-hr operation, the tungsten wires increased in thickness from 1 mm to about 4 mm, whereas crystals bounded by flat (110) planes and up to 3 mm in length were observed at the ends of the "barbs."

When the growth chamber was sealed off from the pump, Molière and Wagner observed tungsten dendrites 3 to 4 mm in length perpendicular to the heated tungsten wire. Both [100] and [111] dendrites were observed as shown in Fig. 13.

Zirconium crystals were grown in a sealed system with the halide at 250°C and the tungsten wire at 1150 to 1200°C. The planes of the crystals were of the (110) and (100) type.

4. THE STATE OF THE ART

Growth by Condensation

Although large crystals can be grown by vapor condensation, this has not been exploited for metals. In most condensation studies many crystals are nucleated and the crystals remain small. To grow large crystals, it is necessary that only one, or at most, a few crystals nucleate. The formation of nuclei can be controlled by using a localized substrate, the temperature of which can be adjusted independently of the temperature of the growth chamber. Initially the substrate is maintained at a temperature for a predetermined period of time during which only one crystal nucleates. The temperature difference between vapor and substrate is then reduced to inhibit any further nucleation. This method was employed by Honigmann (1954) to grow crystals of hexamethlyene-tetramine on cooled substrates.

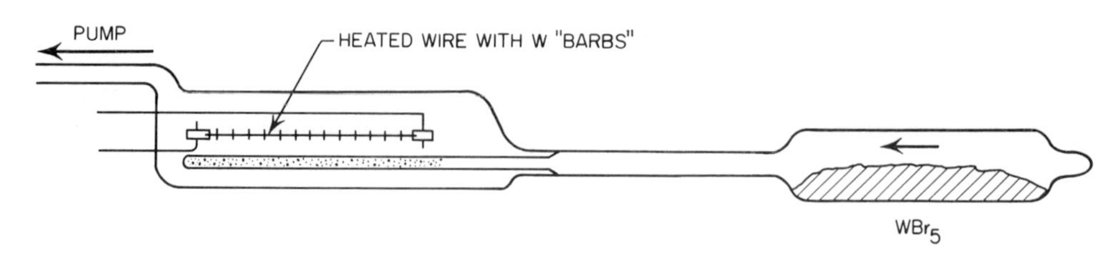

Fig. 12. Apparatus for the growth of tungsten dendrites. (Moliére and Wagner, 1957.)

Fig. 13. Tungsten dendrites: (*a*) [100] growth, (*b*) [111] growth. (Moliére and Wagner, 1957.)

They were 1 cm^2 in cross section and 3 mm thick. The final thickness was limited by the low thermal conductivity of the crystal which limited the cooling effect of the substrate. The high conductivity of metals would reduce this problem.

Another technique that may be applicable to the growth of metals is one described by Piper and Polich (1961) for the growth of large CdS crystals (see Chapter 4). This would only be applicable to metals that have appreciable vapor pressure. The growth rate of an infinitely extended crystal surface can be estimated from gas kinetics

$$\dot{l} = \frac{\sigma p_0}{\rho} \sqrt{M/2\pi RT} \text{ cm/sec} \qquad (5)$$

where σ is the supersaturation of the vapor and p_0 is the equilibrium vapor pressure of the solid. For the highest growth rates $p_0/\sqrt{T}$ and σ must be maximized. $p_0/\sqrt{T}$ is greatest near the melting point of the metal because p_0 increases exponentially with temperature. It is obvious from Eq. (5) that $\dot{l}$ increases linearly with σ. However, if σ becomes too great, the rate of nucleation of noncoherent layers becomes significant. As an approximation, the upper limit of σ is assumed to be the value at which the rate of coherent two-dimensional nucleation given by Eq. (2) is 1 sec^{-1} cm^{-2}. The calculated maximum velocities of growth of some metals near their melting points are tabulated in Table 7, where it may be seen that a large number of metals could be grown at rates in excess of 1 mm/day.

TABLE 7

CALCULATED GROWTH RATES OF METAL CRYSTALS FROM THE VAPOR USING EQ. (5)

Metal	T (°K)	$\gamma\left(\frac{\text{erg}}{\text{cm}^2}\right)$	σ^*	$l\left(\frac{\text{mm}}{\text{min}}\right)$
Cr	2173	1420	0.60	28.8
Mg	924	665	3.2	22.8
Cd	594	630	13.8	2.3
Hg	234	500	3×10^5	1.7
Zn	692	750	4.9	1.2
Mn	1517	1050	0.63	0.51
Ti	2085	1325	1.2	0.11
Zr	2003	1710	1.7	9.1×10^{-2}
Fe	1808	1835	2.1	6.1×10^{-2}
Be	1553	1100	0.4	2.9×10^{-2}
Mo	2883	2240	1.6	2.2×10^{-2}
W	3653	2680	1.4	9.8×10^{-3}
Os	2973	2450	1.6	8.4×10^{-3}
Ni	1728	1850	2.2	8.3×10^{-3}
Pd	1823	1280	1.1	6.8×10^{-3}
Ag	1234	1140	3.8	6.7×10^{-3}
Ta	3250	2330	1.7	4.2×10^{-3}
Ir	2116	2310	1.8	2.7×10^{-3}
Co	1768	1936	2.4	1.3×10^{-3}
Cu	1356	1650	4.3	1.1×10^{-3}
V	2003	1710	1.7	1.1×10^{-3}
Nb	2760	2030	1.8	8.1×10^{-4}
Rh	2239	1130	0.44	2.6×10^{-4}
Pt	2047	1819	2.44	1.95×10^{-4}
K	337	95	1.85	2.5×10^{-5}
Au	1336	1350	5.5	2.3×10^{-5}
Al	933	900	4.2	1.1×10^{-5}

Note: T = Melting temperature
γ = Surface energy
σ^* = Supersaturation at which rate of two-dimensional nucleation is 1 $\text{cm}^{-2}\ \text{sec}^{-1}$ according to Eq. (2).
To calculate σ^*, the height a of the nucleus was assumed to be equal to the distance of closest atomic approach.

Epitaxial Growth of Single Crystal Films

The phenomenology of epitaxial growth has been considerably clarified by electron microscopy. The films form by a process of crystallite nucleation and growth. However, a theoretical definition of the factors that control the formation of nonporous crystals is still lacking. The "best fit" criterion, and its mathematical interpretation by Frank and van der Merwe (1949) does not satisfactorily explain many of the observed orientation relationships. Also no theoretical treatment predicts optimum substrate temperatures, deposition rates, vapor temperatures, etc.

Epitaxial metal films can be useful in many areas of research where well-defined crystallographic surfaces are required. Thus they may find increased application in studies of clean surfaces since they can be produced in situ in ultra-high vacua. At present most epitaxial films are highly imperfect, and the art will be significantly advanced when it is learned how these defects arise and how they can be minimized. It may be that the presence of impurities during deposition plays a major role.

Growth by Chemical Processes

The growth of crystals from vaporized metal compounds has several advantages. Relatively high rates can be achieved at temperatures well below the crystal's melting point and the purity of the crystals can be exceptionally high. However, the technique is limited to metals that form volatile and easily reduced compounds. The growth of alloy crystals using mixtures of metal compounds could be an interesting application of this technique. The key to obtaining greater reproducibility with this method seems to be in the better control of the impurities. However, the mechanism of growth will only be completely understood when the details of the adsorptive, catalytic, and desorptive processes that occur on crystal surfaces become known.

ACKNOWLEDGMENTS

The author is greatly indebted to Dr. R. P. Frankenthal of the Edgar C. Bain Laboratory for Fundamental Research, U. S. Steel Corporation, for his editorial assistance in preparing the manuscript. He also kindly thanks Dr. P. B. Price of the General Electric Research Laboratory, Dr. D. W. Pashley of the Tube Investments Laboratory, Mr. G. A. Bassett of the University of Bristol, and Professor K. Molière of the Fritz Haber Institute for supplying him with some of the photographs.

References

van Arkel, A. E. (1923), *Physica* **3,** 76.

Bassett, G. A. (1958), *Phil Mag.* **3,** 72.

Bassett, G. A., J. W. Menter, and D. W. Pashley (1959), *Structure and Properties of Thin Films,* edited by C. Neugebauer et al., John Wiley and Sons, New York, p. 11.

Brame, D. R., and T. Evans (1958), *Phil. Mag.* **3,** 971.

Brenner, S. S. (1956a), *J. Appl. Phys.* **27,** 1484; (1956b), *Acta Met.* **4,** 62; (1957), Ph.D. Thesis, *Rensllaer Polytechnic Institute;* (1959a), *Acta Met.* **7,** 519; (1959b), *Acta Met.* **7,** 677; Unpublished Work.

Bruck, L. (1936), *Ann. Physik.* **26,** 283.

Buttner, F. H., H. Udin, and J. Wulff (1951), *Trans. AIME* **191,** 1209.

Cabrera, N., Private Communication.

Cabrera, N., and P. B. Price (1958), *Growth and Perfection of Crystals,* edited by Doremus et al., John Wiley and Sons, New York, p. 204.

Cech, R. E. (1959), *Acta Met.* **7,** 787; (1961), **9,** 459.

Cochardt, A. W., and H. Wiedersich (1955), *Naturwissenschaften* **42,** 342.

Coleman, R. V. (1958), *Growth and Perfection of Crystals,* edited by R. H. Doremus et al., John Wiley and Sons, New York, p. 239.

Coleman, R. V., and N. Cabrera (1957), *J. Appl. Phys.* **28,** 1360.

Coleman, R. V., P. B. Price, and N. Cabrera, *J. Appl. Phys.* **28,** 1360.

Coleman, R. V., and G. G. Scott (1958), *J. Appl. Phys.* **29,** 526.

Coleman, R. V., and G. W. Sears (1957), *Acta Met.* **5,** 131.

DeBlois, R. W., and C. P. Bean (1958), *Growth and Perfection of Crystals,* edited by R. H. Doremus et al., John Wiley and Sons, New York, p. 253.

DeBlois, R. W., and C. D. Graham, Jr. (1958), *J. Appl. Phys.* **29,** 528 (1958); 29, 931 (1958).

Dittmar, W., and K. Neumann (1958), *Growth and Perfections of Crystals,* edited by R. H. Doremus et al., John Wiley and Sons, New York, p. 121.

Dushman, S. (1949), *Scientific Foundations of Vacuum Technique,* John Wiley and Sons, New York, p. 746.

Engel, O. G. (1953), *J. Res. NBS* **50,** 249.

Frank, F. C., and J. H. van der Merwe (1959), *Proc. Roy. Soc.* **A198,** 205.

Gatti, A., and R. L. Fullman, Unpublished Work.

Goche, O., and H. Wilman (1939), *Proc. Phys. Soc.* **51,** 625.

Gorsuch, P. D. (1961), *Int. Symp. Phys. Chem. Proc. Metallurgy,* Vol. 2, edited by G. St. Pierre, Interscience, p. 771.

Haase, O., Z. f. *Naturforschung* **11a,** 862.

Heavens, O. S., R. F. Miller, G. L. Moss, and J. C. Anderson (1961), *Proc. Phys. Soc.* **78,** 33.

Hock, F., and K. Neumann (1954), *Z. Physik. Chem. N. F.* **2,** 241.

Hoffman, R. W., R. D. Daniels, and E. C. Crittendon (1954), *Proc. Phys. Soc.* **B67,** 497.

Honigmann, B. (1954), *Z. Elektrochem.* **58,** 322.

Howey, J. H. (1939), *Phys. Rev.* **55,** 1936.

Kainuma, Y. (1951), *J. Phys. Soc. Japan* **6,** 135.

Keepin. G. R., Jr. (1950), *J.A.P.* **21,** 260.

Kehoe, R. B. (1957), *Phil. Mag.* **2,** 455.

Kehoe, R. B., R. C. Newman, and D. W. Pashley (1956), *Phil. Mag.* **1,** 783.

Koref, F. (1922), *Z. Elektrochem.* 28, 511; *Z. Metallkunde* **17,** 213 (1925).

Kohlschutter (1932), *Z. Elektrochem.* **38,** 345.

Laukonis, J. V., and C. V. Coleman (1961), *J. Appl. Phys.* **32,** 242.

van Liempt, (1931), "Die Afscheidung von Wolfram uit Gasformige Verbindingen hare Toepassing," Purmevend.

de Lodyguine, (1893), U. S. Patent 575,000.

Melmed, A. J., and R. Gomer (1961), *J. Phys. Chem.* **34,** 1802.

McNutt, J. E., and R. F. Mehl (1958), *Trans. ASM* **50,** 1006.

Miyake, S., and M. Kubo (1947), *J. Phys. Soc. Japan* **2;** 2, 20 (1947).

Moliere, K., and D. Wagner (1959), *Z. Elektrochem.* **61,** 65.

Morelock, C. R. (1961), *G. E. Report* No. 61-RL-(2756M).

Morelock, C. R., and G. W. Sears (1961), *J. Chem. Phys.* **34,** 1008.

Newman, R. C. (1957), *Phil. Mag.* **2,** 750.

Parker, R. L., and S. C. Hardy (August 1961), *Eighth Field Emission Symposium.*

Parker, R. L., and L. M. Kushner (1961), *J. Chem. Phys.* **35,** 1345.

Pashley, D. W. (1956), *Advances in Physics* **5,** 173; (1959), *Phil. Mag.* **4,** 324.

Piper, W., and S. J. Polich (1961), *J. Appl. Phys.* **32,** 1278.

Price, P. B. (1960), *Phil. Mag.* **5,** 473; (1961), *J. Appl. Phys.* **32,** 1746.

Price, P. B., D. A. Vermilyea, and M. B. Webb (1958), *Acta Met.* **6,** 524.

Riebling, E. F., and W. W. Webb (1957), *Science* **126,** 309.

Royer, M. L. (1928), *Bull. Soc. Franc. Mineral.* **51,** 7.

Rudiger, O. (1937), *Ann Phys.* **30,** 505.

Sakui (1938), *Scient. Papers, Inst. Phys. and Chem. Res.* **34,** 1131.

Sears, G. W. (1959a), *Acta Met.* **1,** 457; 3, 361; (1959b), *Acta Met.* **3,** 367.

Sears, G. W., and J. W. Cahn (1960), *J. Chem. Phys.* **32,** 1317.

Seifert, H. (1953), *Structure and Properties of Solid Surfaces,* edited by R. Gomer and C. S. Smith, University of Chicago Press, p. 318.

Shlier, R. E., and H. E. Farnsworth (1958), *J. Phys. Chem. Solids,* **6,** 271.

Straumanis, M. (1931), *Z. f. Physik. Chem.* **B13,** 316; (1932), **B19,** 63; (1934), **B26,** 246.

Uyeda, R. (1942), *Proc. Phys. Math. Soc., Japan* **24,** 809; (1938) 20, 656; (1940), **22,** 1023.

Volmer, M., and I. Estermann (1921), *Z. Physik.* **7,** 13.

Vook, R. W. (1961), *J. Appl. Phys.* 1557.

Wayman, C. M. (1961), *J. Appl. Phys.* **32,** 1844.

Wilman, H. (1955), *Proc. Phys. Soc.* **B68,** 474.

Webb, W. W., and E. F. Riebling (1958), *J. Chem. Phys.* **28,** 1242.

Wiedersich (1959), *J. Electrochem. Soc.* **106,** 810.

Yao-Kuang Wan, Shih-Sheng Chang, and Ting-Suiko (1959), *Ko Hsuch Tung Pao,* No. 24, 830.

Yelon, H., and R. W. Hoffman (1960), *J. Appl. Phys.* **31,** 1672.

3

Organic Compounds

R. S. Bradley

In recent years it has been realized that organic crystals provide a potential source of solids with interesting electrical properties; for example, guanidine aluminum-sulfate-hexahydrate is a ferroelectric, sorbital hexa-acetate is piezoelectric, and many organic compounds with π electrons are semiconducting. For these studies it is essential that the crystal should be pure and of good quality structurally. Of the various existing processes for growing large single organic crystals, growth from the vapor is probably best for the production of very pure crystals. Crystals grown from solution contain trapped solvent, and a suitable solvent is not always available as, for example, in the case of copper phthalocyanine. For many crystals the melting point is so high that decomposition occurs prior to melting, whereas crystals may be grown from the vapor at much lower temperatures.

There is a limit imposed by molecular weight, however, for the growth of organic crystals from the vapor, since it is necessary to heat the compound to a temperature at which the vapor pressure gives a reasonable rate of crystal growth. With compounds of high molecular weight this temperature may be above the decomposition point even if the compound is heated *in vacuo* so that oxidation is excluded. According to Carothers and co-workers (1930) no organic compound with a molecular weight greater than that of heptacontane, n—$C_{70}H_{142}$, can be distilled under any experimental conditions, however favorable. In fact, n—$C_{80}H_{162}$, octacontane, decomposes on distillation. It is interesting that the latent heat of vaporization of the latter, 81,000 calories mole^{-1}, is near to the heat of dissociation of a C—C bond, 83,100 calories (Pauling, 1960). It appears likely that the limit to the use of vapor techniques occurs when the latent heat of vaporization becomes comparable with the heat of disruption of a C—C bond, although it is unlikely that the whole latent heat would become concentrated in one bond. Actually, for most organic compounds the limit will correspond to a molecular weight much less than 983 (corresponding to $C_{70}H_{142}$), since the normal paraffins have exceptionally low cohesive energies per C atom. Also many other organic compounds of exceptional thermal instability have low molecular weights.

In theory it would be possible to raise the vapor pressure of a compound of high molecular weight by applying pressure by using an inert gas. Such high pressures would be required, however, that the inert gas would become in effect liquidlike, and the process of evaporation would then correspond to that of solution in ordinary solvents.

For vapor growth of crystals, nucleation is first necessary unless a seed crystal is introduced. Nucleation is followed by growth under controlled supersaturation. The rate of growth depends on the degree of supersaturation, the absolute value of the vapour pressure at the growing surface, the temperature at the surface, the presence of an inert foreign gas, the removal of heat from the growing crystal, and the degree of perfection of the nucleus or seed.

Extensive nucleation is quite normal during purification by sublimation *in vacuo,* but must be avoided if good single crystals are to be grown from the vapor. Owing to the low surface energy of organic compounds (usually less than 30 ergs cm^{-2} for organic liquids and expected to be only slightly greater for the solid phases), nucleation, especially on the boundary wall, is much easier than for ionic or metallic solids. It is difficult to control the rate of nucleation; it can be done only by working with sufficiently low supersaturations.

It is usual to control the supersaturation by maintaining the condensing surface at a slightly lower temperature T_2, than that of the evaporating surface at temperature T_1. The rate of vapor transfer is greatest if the system is evacuated and is given approximately, for the simplest case, by $A\alpha(p_1 - p_2)(M/2\pi RT_2)^{1/2}$ gram sec^{-1}, where A is the area of the growing surface, α the condensation coefficient, p_1 and p_2 the saturation vapor pressures at temperatures T_1 and T_2, respectively and M the molecular weight. This expression is based on the assumption that there is no impedance arising from the resistance of the vessel to the molecular flow and that the whole of the growing surface is available for the capture of molecules. For molecular crystals α is usually in the range 0.1 to 1 (Bradley, 1951).

In the presence of an inert foreign gas the rate of transfer is reduced and the process may become diffusion controlled. Moreover, there may be a tendency to produce "snow" nuclei in the vapor space, since the thermal conductivity of the inert gas allows abstraction of heat from the bulk of the vapor.

It is desirable to reduce $T_1 - T_2$ to as small a value as possible in order to grow crystals free from visible flaws. Moreover, the value of T_2 at the condensing surface should not be too low since molecules which impinge on the cooled surface must reorient themselves in order that well-organized crystal growth occurs.

It should be realized that the crystal produced by condensation of the vapor is not necessarily the most stable form at the temperature of the condensate. Most organic compounds exist in a number of polymorphic varieties, and confusion may result if this is not born in mind, for example, anhydrous α-oxalic acid on heating to 100° *in vacuo* condenses as needles of the β-form on the cold neck of the flask; this form is unstable relative to the α-form at room temperature (Bradley and Cotson, 1953). A difference in habit should not be confused with a difference in crystal structure; in the former the angles between faces are the same in the different habits, whereas with different polymorphs the angles differ.

It must be admitted that the growth of good crystals of organic compounds from the vapor seems to be governed by factors not yet understood, perhaps because little exact work has been done in this field. It is easy to produce good crystals of many compounds from the vapor, for example, menthol, anthracene, benzoic acid, quinone, naphthalene, carbon tetrabromide, but with the majority of compounds a crystal aggregate is produced. The suggestion which advises leaving a bottle of the compound in the dark for several years seems often to work. The growth of good crystals from the vapor is an extension of techniques of sublimation. Many organic compounds which might be thought to be unstable, such as glucose, may be sublimed by vacuum distillation (Hill, 1932) and presumably could be grown as single crystals under the correct conditions. A list of compounds which may be sublimed *in vacuo* without decomposition is given by Weissberger (1951) and includes caffeine (anhydrous), quinine (anhydrous), saccharin, coumarin, urea, vanillin, alizarin, isatin, cinchonine, cocaine, and atropine.

It should be remembered that crystals that appear perfect optically are in general imperfect and may have a complex structure of dislocations or faults. Organic crystals are not likely, however, to have Frenkel and Schottky defects, owing to the size of the molecules, although there may be "partial" holes less than one molecule in size. The study of the imperfections of organic crystals appears to be in its infancy, perhaps because of the lack of interest in the mechanical properties or organic compounds other than polymers.

EXPERIMENTAL

Before pure crystals of an organic compound can be grown it is essential to purify it vigorously, and for this purpose sublimation is ideal. Figure 1 shows a single apparatus for the sublimation of phthalocyanine and of copper phthalocyanine as used by the author and co-workers. Copper phthalocyanine has extraordinary thermal stability and may be heated *in vacuo* to 580°C without decomposition (Robertson, 1953). The apparatus contains no ground joints before the trap in order to avoid contamination by grease vapor. The aluminum block is heated to 400°C with a high vacuum in the system and liquid air round the trap. After some hours the condensate forms a number of zones, and after cooling and admitting air the tube may be cut off and the zones removed. The middle fraction is resublimed with the same treatment and the process can be repeated several times. If ground joints are not taboo, the vacuum sublimation apparatus may resemble a molecular still, as in Fig. 2.

One of the most promising methods for growing single crystals of organic com-

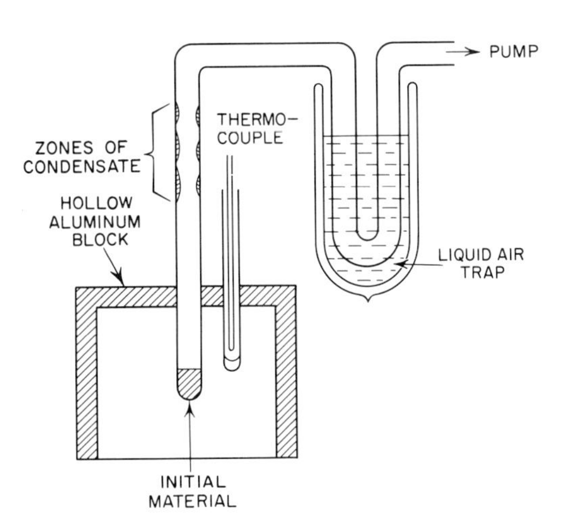

Fig. 1. Vacuum sublimation apparatus with zoning.

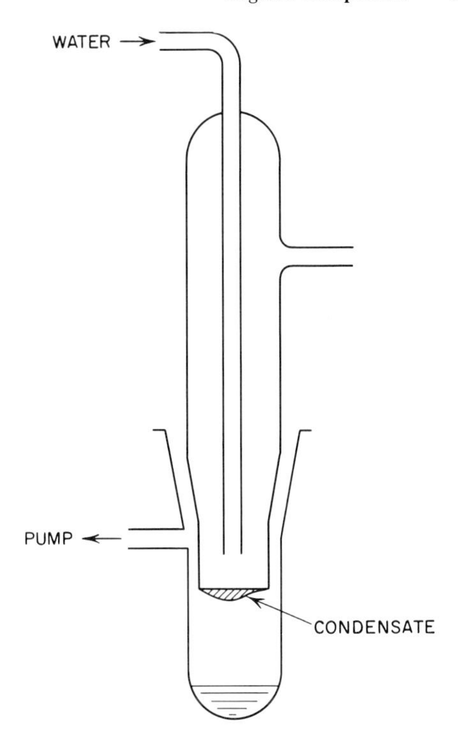

Fig. 2. Vacuum sublimation apparatus of molecular still type.

pounds from the vapor was introduced by Zwerdling and Halford (1955) who grew oriented disks of crystalline benzene 0.1 mm thick and 1 cm in radius. Their method is designed to produce an oriented single crystal in a shape suitable for optical study and is of very general application, including the production of crystals of low melting point. The essential part of the apparatus, the growth chamber, is sketched in Fig. 3. It consists of a nickel ring closed on both faces by greased coverplates of potassium bromide, so that there is a gap of 0.1 mm between the interior faces of the coverplates. One pole of the ring can be cooled by a solid copper rod from which heat is drawn by a cooling mixture. The opposite pole, which can be electrically heated, has a hollow copper tube for the entry of vapor. A temperature difference of about 25°C is maintained between the two sides of the chamber.

The benzene vapor is allowed to leak slowly into the evacuated chamber. Nucleation occurs at a visible rate when the

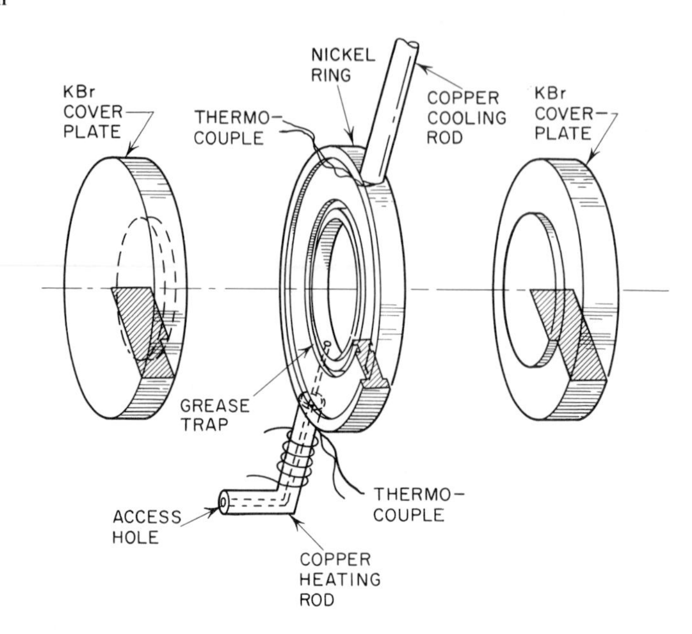

Fig. 3. Growth chamber for production of benzene crystals. (After Zwerdling and Halford.)

ambient pressure in the chamber reaches a critical value, and will, of course, first occur on the cold side of the chamber. If the rate of leakage is sufficiently low, the formation of nuclei will relieve the supersaturation, which will favor the growth of the nuclei rather than the formation of fresh nuclei. The nuclei are formed on the walls of the chamber on which the heat of condensation is readily dissipated. With anisotropic crystals, such as benzene, some nuclei will grow faster than others if suitably oriented, so that only a few suitably oriented seeds will attain a visible size. As a result of further growth a single crystal plate with a definite orientation will be formed, the axis of maximum thermal conductivity lying nearly parallel to the line joining the poles of temperature. The orientation of the remaining principal axes of heat conduction depends on the pattern of isothermal lines on the wall of the chamber near the cold plate.

The supply of vapor is regulated by transferring the benzene from its solid carbon dioxide refrigerant to water at 0°C, which is allowed to warm up slowly to room temperature. In one experiment a visible seed appeared at the coldest part of the chamber after 15 min, although no particular significance is attached to this time interval. The seed appeared as a bright point of light under observation with crossed polaroids and grew to a small crescent several millimeters in arc. The lower edge then moved over the chamber, giving a crystal plate with a cross section parallel to the {010} plane. Examination with parallel oblique light showed that the crystal was in intimate contact with nearly the whole of the salt surface, in a form ideal for study of absorption spectra. The sectional area was 300 mm^2 and the thickness 0.1 mm. The method is clearly capable of extension to most organic compounds which sublime easily without decomposition. The vacuum jacketing is somewhat complex and could probably be simplified for crystals of higher melting point; thicker disks could persumably be grown if required.

J. F. Scott and co-workers (1948) have sublimed films of amino acids, purines, and pyrimidines without decomposition onto quartz slides for the study of absorption spectra, but these films were presumably aggregates. The possibility remains, however, of sublimation onto an ordered surface, such as rock salt, on which many organic compounds form oriented arrangements

(Willems, 1948), so that in theory the aggregates might grow to one large crystal, as in the work of Zwerdling and Halford.

The simplest way to control crystal growth from the vapor is to maintain the growing crystal and the source of vapor at temperatures which differ very slightly, as in the work of Bradley and Drury (1959) on CBr_4. Although this apparatus was designed to study the mechanism of growth and to test Frank's theory, it could with little modification be used as a preparation method. The apparatus (Fig. 4) consists of an evacuated inverted U tube approximately 2 cm in diameter with one limb containing crystal aggregate and the other the growing crystals. The former is maintained at a slightly higher temperature than the latter by means of two constant-temperature Dewar flasks controlled to 0.001°C by thyratron relays. The temperature difference ranged from 0.01 to 1°C. The connecting tube was heated a few degrees higher than the vertical limbs by metal tubing through which thermostated water was pumped. The crystals grew on an optical window and could be observed by an arrangement of prisms. Carbon tetrabromide is cubic above 47°C and monoclinic below (and in addition there are many high pressure polymorphs). The cubic form nucleated on glass at the lowest supersaturations used, but

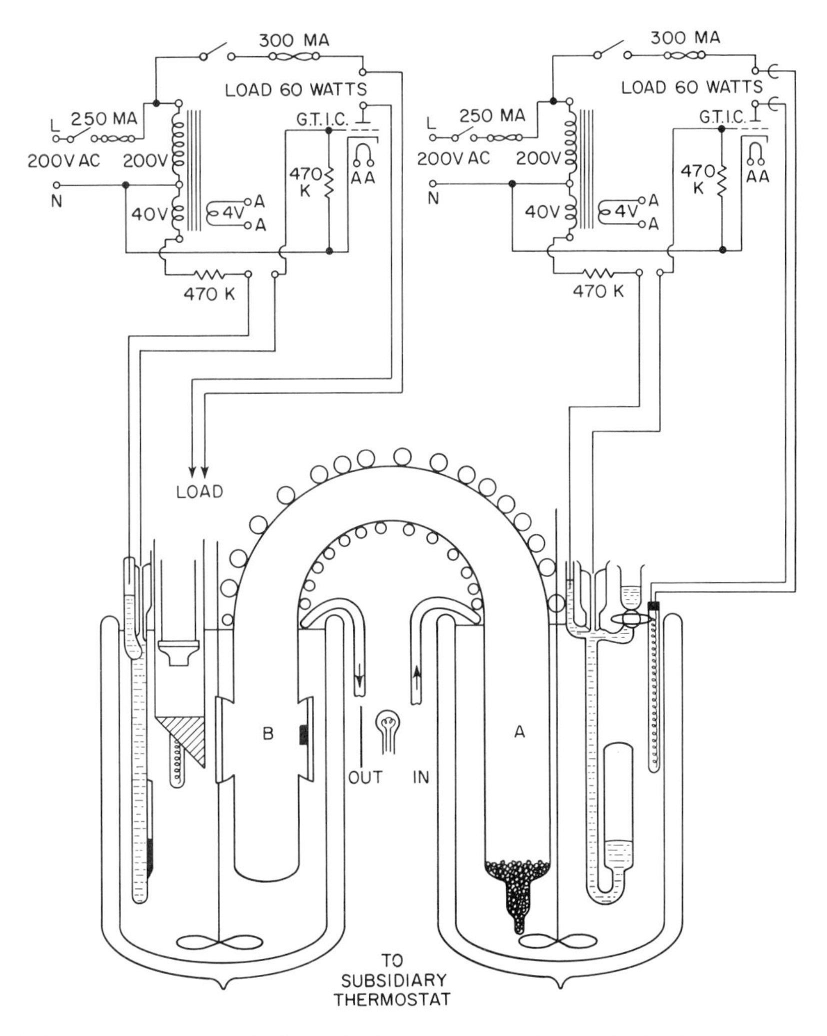

Fig. 4. Apparatus for growth of crystals of carbon tetrabromide. (After Bradley and Drury.)

the monoclinic form required a touch of solid carbon dioxide on the window. The two forms behaved differently with respect to growth at equal supersaturations, the monoclinic growing the faster, even though the temperature was lower. The cubic form appeared as pyramids with {110} faces, sometimes with {111} flat tops, and the square root of the linear rate of growth (R) was proportioned to $(p_1 - p_2)/p_2$. The monoclinic form appeared as flat hexagonal plates and R was a linear function of $(p_1 - p_2)/p_2$. The authors discussed these results in the light of Frank's theory. According to Dunning (1961) the growth step for monoclinic CBr_4 is about ten times that for the cubic form. Screw dislocations of large Burgers vector are considered by Dunning to be more stable in the monoclinic form. In the cubic form mobility of vacancies could cause a large dislocation to split into a number of small dislocations.

With compounds of much lower vapor pressure, such as copper phthalocyanine, it is, of course, necessary to work at much higher temperatures. Fielding and Gutman (1957) prepared single crystals of this compound by sublimation in pure nitrogen (5 to 7 mm Hg) by heating in a tube at 415 to 440°C and growing at 320 to 350°C. Crystals 3 cm × 2 mm × 1 mm were obtained. An evacuated tube gave similar results.

In the apparatus of Honigmann and Heyer (1955) shown in Fig. 5, which is used for the preparation of good crystals of hexamethylene tetramine, the tube (15 cm × 2.5 to 3 cm) was evacuated, sealed off, and heated in a metal-block thermostat to 70°C. Growth occurred below the copper exit tube which acted as a controlled heat leak, the temperature difference $T_1 - T_2$ being 2°C and corresponding to $(p_1 - p_2)/p_2 = 0.16$. The copper tube formed one element of a copper-constantan thermocouple. A small heater above the furnace opening allowed the degree of supersaturation to be controlled. If more than one nucleus appeared, the nuclei were sublimed off by heating, and the growth was repeated until only one nucleus was formed. Excellent crystals were produced with {011} faces perpendicular to the wall.

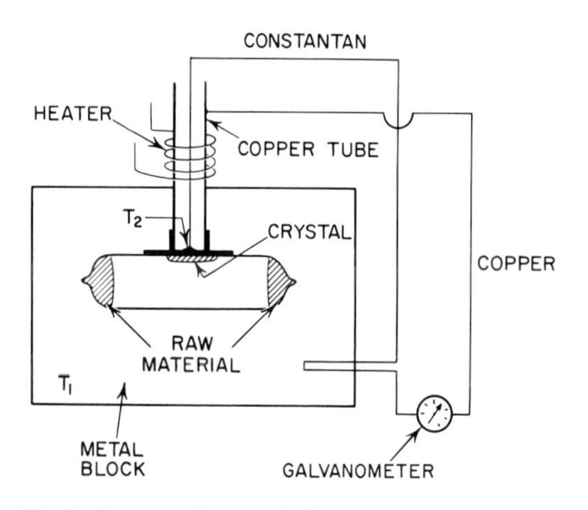

Fig. 5. Apparatus for growth of crystals of hexamethylene tetramine. (After Honigmann & Heyer.)

Some of the more perfect organic crystals have been large in one dimension only and assumed the habit of whiskers, many centimeters long but only a few microns in thickness. Theory suggests that such whiskers grow on a single screw dislocation, and that the side faces are free from imperfection and do not nucleate. The writer showed that whiskers of phthalic anhydride could be formed by subliming the solid in a flask and was able to demonstrate their structural perfection (Bradley, 1958).

In conclusion, the best method of crystal growth from the vapor appears to be that of Zwerdling and Halford, since it provides the greatest degree of control and is most likely to produce single crystals. Provided the experimenter is willing to construct the somewhat complicated apparatus, all crystals that can be sublimed without decomposition should be capable of growth. It is to be hoped that the production of good quality crystals will increase our knowledge of the electrical and mechanical properties of molecular solids, since it is still in a rudimentary state.

References

Bradley, R. S. (1951), *Proc. Roy. Soc.*, **205A,** 553.

Bradley, R. S., and S. Cotson (1953), *J. Chem. Soc.*, 1684.

Bradley, R. S. (1958), in *Growth and Perfection of Crystals*, R. H. Doremus, B. W. Roberts, and D. J. Turnbull, Editors, John Wiley and Sons, p. 133.

Bradley, R. S., and T. Drury (1959), *Trans. Faraday Soc.,* **55,** 1848.

Carothers, W. A., J. W. Hill, J. E. Kirby, and R. A. Jacobsen, (1930), *J. Amer. Chem. Soc.,* **52,** 5279.

Dunning, W. J. (1961), *Phys. Chem. Solids,* **18,** 21.

Fielding, P. E., and F. Gutman (1957), *J. Chem. Phys.,* **26,** 411.

Hill, J. W. (1932), *Science,* **76,** 218.

Honigmann, B., and H. Heyer (1955), *Zeit. Krist.,* **106,** 199.

Pauling, L. (1960), *Nature of Chemical Bond,* 3rd Edition, Cornell University Press, p. 85.

Robertson, J. M. (1955), *Organic Crystals and Molecules,* Cornell University Press, p. 262.

Scott, J. F., R. L. Sinsheimer, and J. R. Loofbourow (1948), *Science,* **107,** 302.

Weissberger (1951), *Technique of Organic Chemistry,* Vol. IV Interscience, p. 611.

Willems, J. (1948), *Naturwiss.,* **35,** 375.

Zwerdling, S., and R. S. Halford (1955), *J. Chem. Phys.,* **25,** 2215.

4

Sulfides

D. C. Reynolds

Early interest in sulfide crystals was generated chiefly by their luminescent and photoconductive properties. Some of the earliest investigations were carried out by Gudden and Pohl (1920). They investigated commercial phosphors and such natural crystals as zinc-blende, greenockite, and cinnabar. These natural crystals often contained a sufficiently high concentration of various impurities so that the intrinsic properties of the material were completely masked (Merriman, 1912). However, these investigators were successful in describing some of the main regularities of the electric conductivity connected with phosphorescence. This provided a stimulus for later investigators to search for new techniques that would permit synthetic growth of sulfides. Early experiments on growth from the vapor phase were reported by Lorenz (1891) and from solution by Allen and Crenshaw (1912).

From the time of the early work of Gudden and Pohl until the successful growth of synthetic cadmium sulfide crystals by Frerichs (1946), very little effort was devoted either to the investigation of the properties of sulfide crystals or to methods for growing them synthetically. Since the time of Frerichs' success, research on the growth and properties of sulfide crystals has been continually expanding.

1. GROWTH METHODS

The initial attempts at synthetic growth of the sulfides were devoted largely to CdS and ZnS. The ways these crystals may be grown are: from solution (Allen and Crenshaw, 1912); by sublimation*, and from the melt (Medcalf and Fahrig, 1958; Addamiano and Aven, 1960).

The sublimation method first yielded crystals of usable size and reasonably good quality. Although several growth procedures have been reported in the literature, they all are modifications of two basic methods, dynamic and static. The dynamic method of Frerichs (1946) was the first that succeeded in growing usable crystals.

Dynamic Method

Frerichs and co-workers produced a chemically purified grade of CdS which, when fired in a suitable atmosphere, exhibited remarkable photoconductive properties. In order to stabilize the photoconductive properties, single crystals were required. Frerichs used a modification of the method of Lorenz (1891). The furnace arrangement is shown in Fig. 1. Cadmium metal is heated in a small porcelain boat inside a quartz tube to a temperature of about 800 to 1000°C. The cadmium vapor formed in the boat is carried by a slow stream of hydrogen into contact with a stream of H_2S that flows into the furnace through a quartz tube. A reaction occurs forming CdS. As the furnace is slowly cooled, with the gases still flowing, crystals of very pure CdS grow. This method has two advantages: (1) since purification by

* See the article "Photoconductivity of the Sulfide, Selenide, and Telluride of Zinc or Cadmium," by R. H. Bube, *Proceedings of the IRE* **43,** 1836 (1955), for bibliography on growth of single crystals of CdS and ZnS.

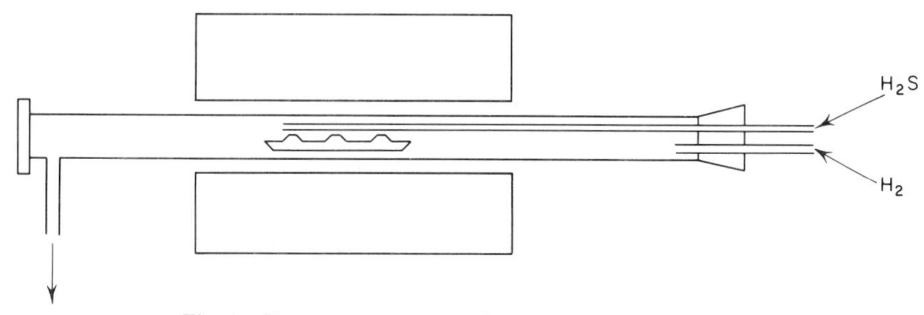

Fig. 1. Furnace arrangement for Frerichs' growth method.

fractional sublimation occurs, it is not necessary to use exceptionally pure cadmium as a starting material; (2) as the furnace is cooled in a steady flow of hydrogen sulfide any tendency toward dissociation is avoided. Crystals produced by this method are generally of the plate and ribbon type. The size of the crystals depends on the size of the reaction tube, since they tend to grow radially from the walls toward the center of the tube. Crystals 1 to 2 cm in length can be obtained in tubes of 10 cm diameter. The crystals vary in thickness from 0.01 to 1.2 mm. In every case the "*C*"-axis lies in the plane of the plate. Many of the plates contain striations on the surface; these striations lie parallel to the *C*-axis. The crystals always have the wurtzite structure.

The method of Frerichs was later improved by Bishop and Liebson (1953) by using argon instead of hydrogen as the carrier gas. The improvement resulted from the greater density of sulfur in the reaction area. These authors further modified the method by arranging the apparatus as shown in Fig. 2. Here cadmium and sulfur are heated separately and the vapors are carried to the reaction zone by argon. The reaction volume inside the flask is maintained at approximately 1000°C, causing crystals to grow up to several cm^2 in area and 1 mm thick. Their quality is comparable to those grown by Frerichs' method.

These growth techniques described have employed a chemical reaction to form the CdS. Stanley (1956) has instead sublimed reagent grade CdS in an oxygen-free atmosphere. His experiments were conducted in fused quartz, vycor, and ceramic tubes. The CdS powder was sublimed by heat and the vapor was carried by nitrogen or argon to a cooler region of the furnace for condensation. Temperatures ranging from 900 to 1040°C for periods of 16 to 48 hr were used. Both prism and platelike crystals are grown by this technique, the largest plates being 22 mm × 25 mm × 1 mm.

Static Growth

In the preceding discussion the methods employed flowing gas during the growth process. They are therefore named *dynamic growth methods.* A "static method" for growing sulfide was devised by Reynolds and Czyzak (1950) and was first applied to ZnS. Chemically pure ZnS powder was distributed over a length of approximately $2\frac{1}{2}$ in. in a quartz tube. This charge was placed in the center of a horizontal resistance furnace. The quartz tube was then filled with a H_2S atmosphere and sealed. A slightly positive pressure was maintained at the growth temperature of 1150°C for periods of 48 to 96 hrs. Crystals formed directly on the powder charge in the shape

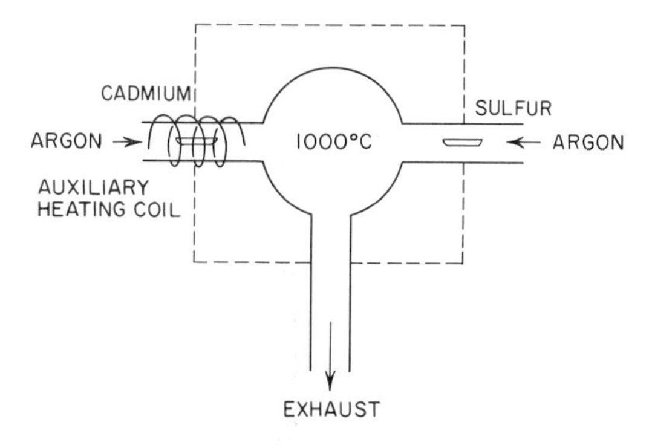

Fig. 2. Apparatus used by Bishop and Liebson.

of hexagonal prisms, the largest having dimensions of 2 × 2 × 10 mm.

This method was later applied to CdS (Czyzak et al., 1952). Now the growth temperature was 1000° C for periods of 48 to 72 hrs. As with ZnS, the crystals formed directly on the powder charge in the form of hexagonal prisms. The crystals were bounded by the two planes $(11\bar{2}0)$ and $(10\bar{1}0)$ with the *C*-axis in the direction of the dihedral angle between these two planes. It was possible to grow crystals approximately 1 cm^3 in size as shown in Fig. 3.

A modification of this method (Greene et al., 1958) has made it possible to grow large sulfide crystals weighing as much as 300 grams. Since the CdS and ZnS crystals are grown from commercially available luminescent-grade powder, the purity of the powder varies from one source to another. One method for obtaining purer powder is to use distilled sulfur and zone-refined cadmium. The two purified elements are then reacted to form CdS powder. This reduces metallic impurities to levels that cannot be detected spectrographically. However, the reaction is carried out in quartz containers, where approximately 5 ppm of silicon are picked up.

Another technique is that of fractional distillation. Two advantages to this technique are (1) the distillation (performed near the sublimation temperature) not only removes impurities but also degases the

Fig. 3. Prism-type CdS crystal approximately 1 cm^3 in size.

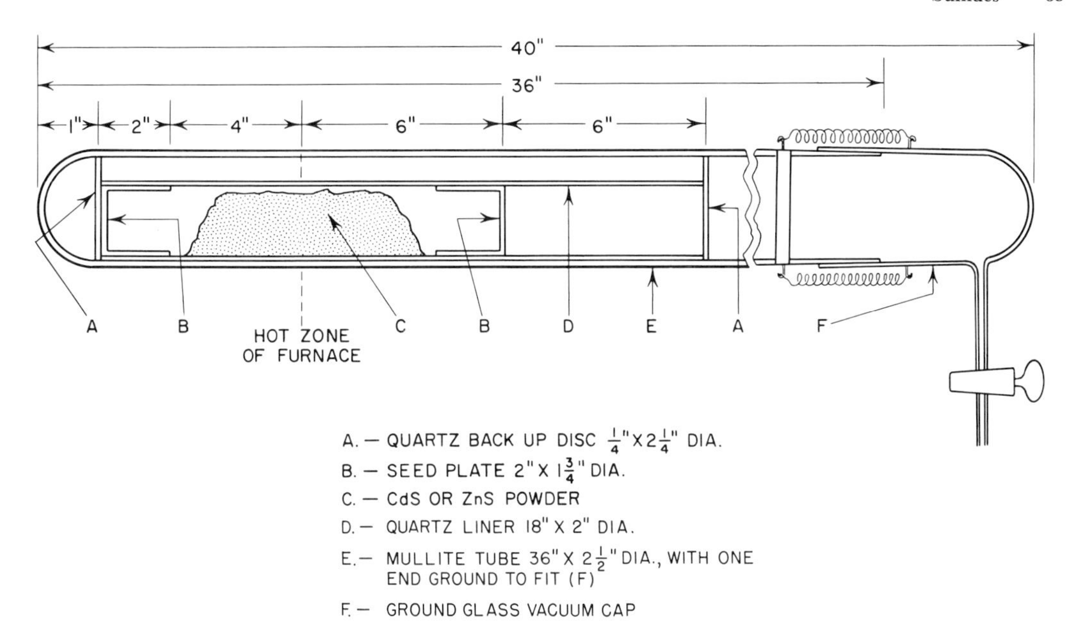

Fig. 4. Arrangement of sublimation chamber used by Greene et al.

powder, and (2) as a consequence of the fractional distillation the powder is thoroughly dried. The distillation is first performed in an H_2S atmosphere at a pressure of ~760 mm and at a temperature of 930 to 950°C for CdS, and 1150 to 1170°C for ZnS, for periods of 48 to 72 hr. For best results the powder is spread loosely over a distance of 6 to 8 in. inside the center section of a quartz tube 30 in. long and 2 in. in diameter. The H_2S gas is introduced and then the tube is sealed. The center section of the tube lies in the hot zone of the furnace. After sublimation has occurred and the contents of the tube have cooled, the H_2S atmosphere is removed and the distillation is performed again, but this time in a vacuum (same temperature and same time). Spectroscopic analysis afterwards shows that the powder has a total impurity content of less than 0.0001%. The following impurity elements are usually found in the CdS and ZnS powders: B, Si, Mg, Pb, Cu, and Fe. To obtain maximum purity, it is advisable to repeat the fractional distillation process two or three times, although this depends on the original purity of the powder.

The dried and purified CdS or ZnS powder is placed in a quartz liner with a clear quartz seed plate sealed to one end as shown in Fig. 4. The powder is positioned in the quartz liner so that it is approximately 2 in. from each end of the liner. The liner and contents are then placed inside a mullite tube. Another clear quartz seed plate is next placed against the open end of the liner. The mullite tube is positioned in a glowbar furnace with the charge centered in the hot zone of the furnace (Fig. 5). After the system is filled with argon to a pressure that equals ~1 atmosphere at the growth temperature, it is sealed. Alternative gases such as H_2S have been used successfully. The best temperatures for growing large crystals are given in Table 1.

DISCUSSION OF RESULTS. Figure 6 shows three crystals of CdS, each weighing ~250

TABLE 1

TEMPERATURES FOR GROWING LARGE CRYSTALS IN APPARATUS OF FIG. 5

Crystals	Center Zone	Growing Zone
ZnS	1550–1600°C	1475–1500°C

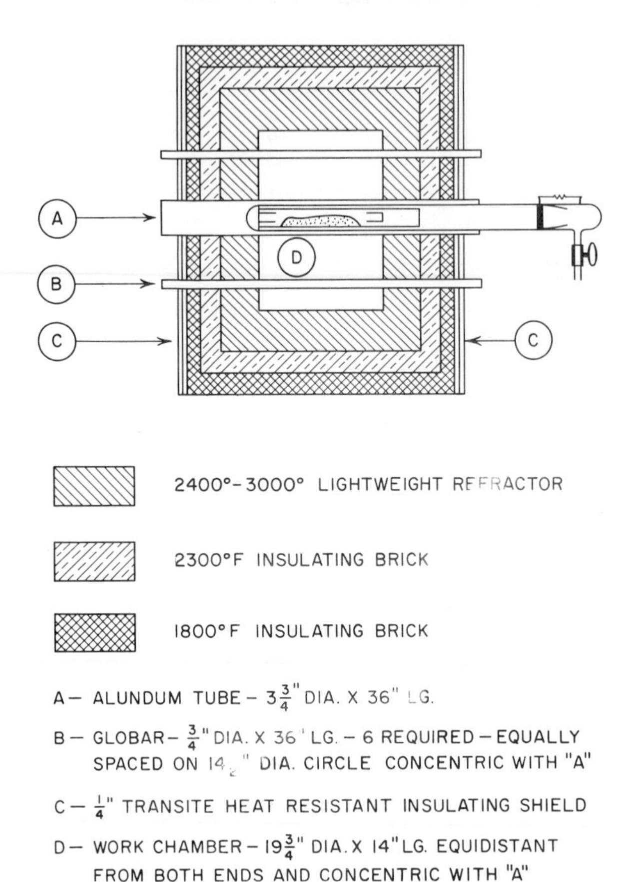

Fig. 5. Cross section of Globar furnace with sublimation chamber in place.

grams. These crystals are representative of runs where only one crystal grew on the seed plate; usually 3 to 4 crystals grow on a seed plate. Specimens cut and polished from these crystals are also shown in Fig. 6. Similar crystals of ZnS can be grown by the same technique.

Spectroscopic analysis of crystals grown by this method shows that a total impurity content as low as ~0.0001% for CdS and ~0.0005% for ZnS can be obtained. However, typically, the impurity content is somewhat higher and is shown in Table 2 for each type of crystal. It should be remembered that this analysis does not account for any of the anions such as chlorine, oxygen, etc., that may be present.

TABLE 2

Element	ZnS Crystal	CdS Crystal
Cu	<0.00001	<0.0001
Fe	<0.0001	<0.0001
Mg	<0.00001	<0.00001
Pb	not detected	<0.00001
B	<0.0001	<0.0001
Si	<0.0001	<0.0005

Crystals grown by the seed plate technique have different growth habits than those grown by other techniques. The growth direction is normal to the seed plate; however, the *C*-axes of the crystals are tilted with respect to the normal by angles as great as 14°. In the other growth methods the growing surfaces are crystallographic planes.

X-ray diffraction and optical analysis showed that the CdS crystals were completely hexagonal. The ZnS crystals invariably contain stacking disorders. Allan and Crenshaw (1912), who first proved that the wurtzite-sphalerite reaction was reversible, found that to convert sphalerite to wurtzite, the crystal had to be cooled from the inversion temperature, that is, 1020 to 100°C in a period to 2 hr. To reverse this process they required 66 hr at 800 to 900°C. Greene et al. (1958) were successful in converting to the wurtzite phase but were unable to get complete conversion to the sphalerite phase.

Recently, Samelson and Brophy (1961) have grown structurally pure cubic and hexagonal crystals of ZnS. Using a static system and starting with pure ZnS powder, they were able to grow structurally pure wurtzite crystals. Starting with ZnS powder doped with KCl they were able to grow structurally pure zinc-blende crystals.

Their furnace arrangement and a typical temperature gradient are shown in Fig. 7.

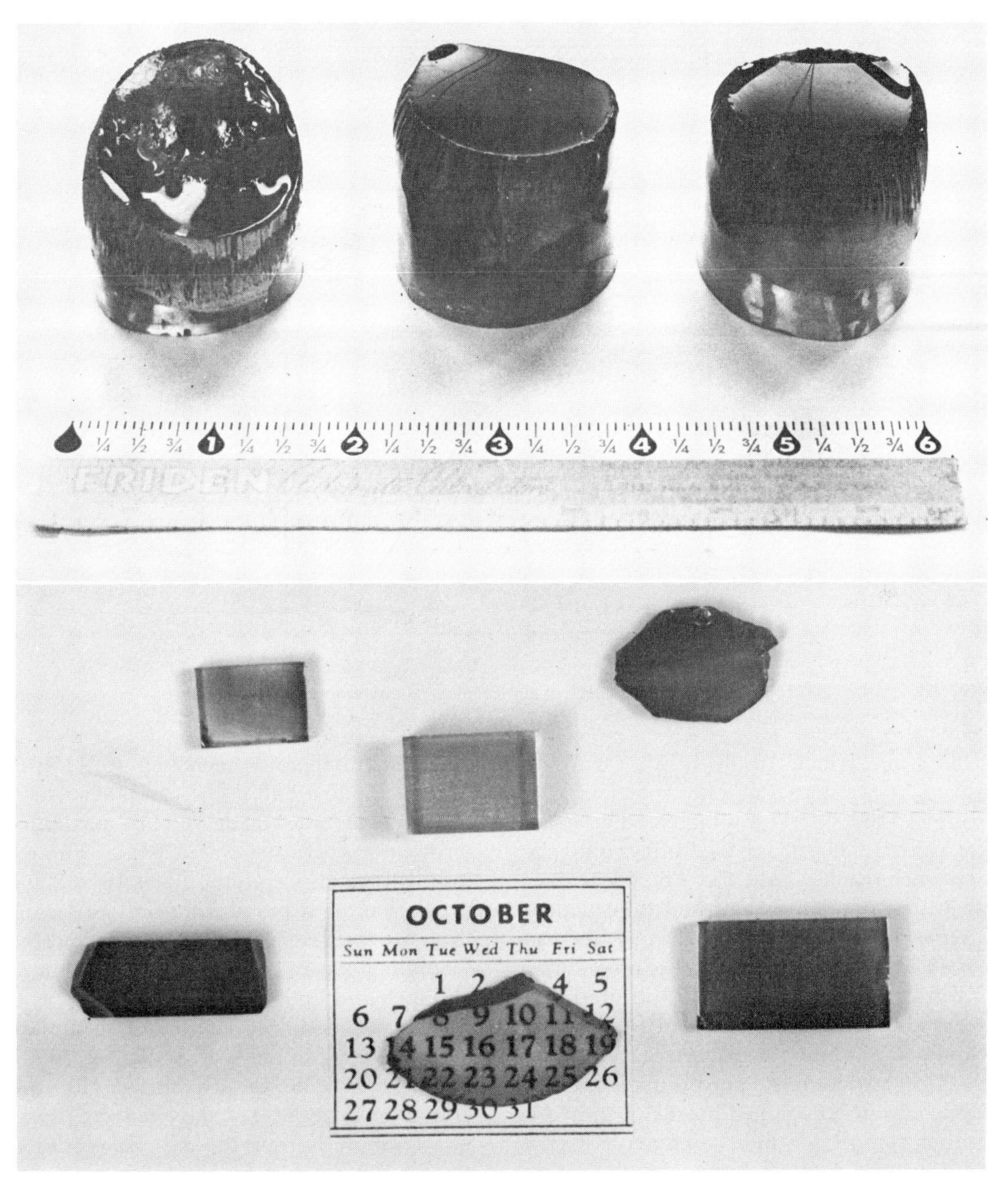

Fig. 6. As-grown crystals of CdS from a seed plate and some cut and polished specimens.

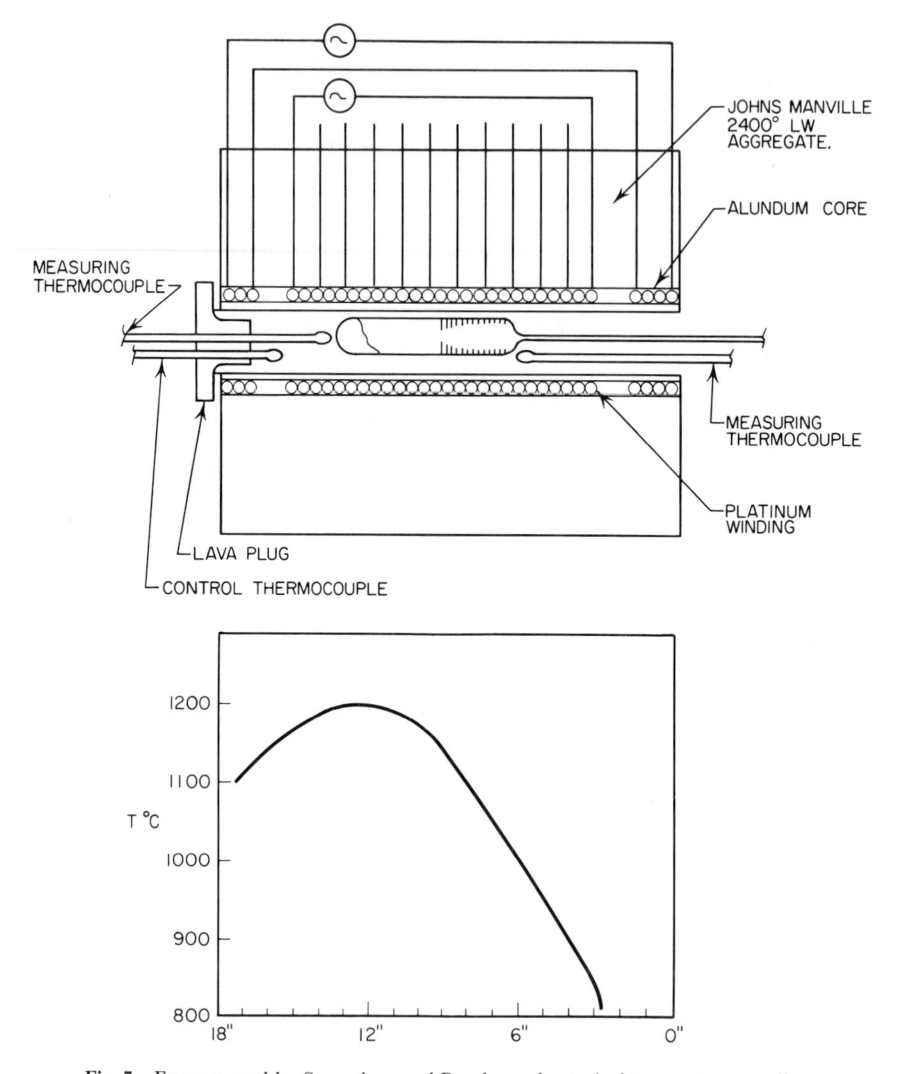

Fig. 7. Furnace used by Samuelson and Brophy and a typical temperature gradient.

The runs which result in the wurtzite structure are started at a hot zone temperature of 1100°C. This is increased at a rate of about 10°C per day for 14 days until the hot zone reaches 1230°C. The linear temperature gradient is adjusted to provide a temperature which varies from 990 to 1020°C in the cold end of the tube. Growth occurred at 1090°C and produced crystals of the rod type 0.5 to 1.0 cm long and about 0.1 mm in diameter.

The structurally pure cubic crystals grown from KCl-doped powder required a much steeper gradient. The runs had a duration of 7 to 10 days, with the initial temperature of the hot zone set at 1100°C, and then uniformly increased to 1150°C during the first half of the run, after which no change was made. The maximum growth temperature was 800°C. The resulting crystals were thin plates 10 to 200 μ thick, having areas of from 0.01 to 1 cm^2. The flat faces were the (111) planes and the bounding edges (110) planes.

Piper and Polich (1961) have made two significant modifications of the seed-plate technique: (1) in place of the flat seed plate they use a cone-shaped surface for the initial nucleation; (2) the seed cone was slowly moved from the hot zone into a cooler region of the furnace during growth. A cross-sectional view of their furnace and crucible is shown in Fig. 8 along with the temperature profile used for growing CdS crystals. The crucible was made from clear quartz tubing tapered to a blunt conical tip. A quartz rod, sealed to the tip, aided in removing the heat of sublimation by conduction through the quartz rod.

The powder charge was prepared by being heated in a vacuum at 500 to 700°C for ~1 hr, then in a stream of H_2S at 900 to 1000°C for 10 hr. This purified and sintered the charge, thereby preventing large volume changes during crystal growth. The prepared charge was placed in the crucible and then a close fitting, closed quartz tube was inserted behind the charge (Fig. 8). Condensing vapors formed a seal with the crucible, thereby isolating the charge. The assembly was placed inside a gas-tight mullite tube. This tube was evacuated and slowly heated to 500°C, where the system was held for 1 hr to remove volatile impurities. A slow stream of argon at one atmosphere of pressure was then introduced into the system, and the furnace temperature was increased to the growth temperature.

At the start of each run the tip of the crucible is nearly at the maximum temperature of the furnace. The mullite tube is then mechanically pushed into the cooler region of the furnace at a rate of 0.3 to 1.5 mm/hr. As the tube moves, the supersaturation at the crucible tip increases until nucleation occurs. Usually one seed dominates and a single crystal grows. A typical crystal is shown in Fig. 9.

This technique has been used successfully

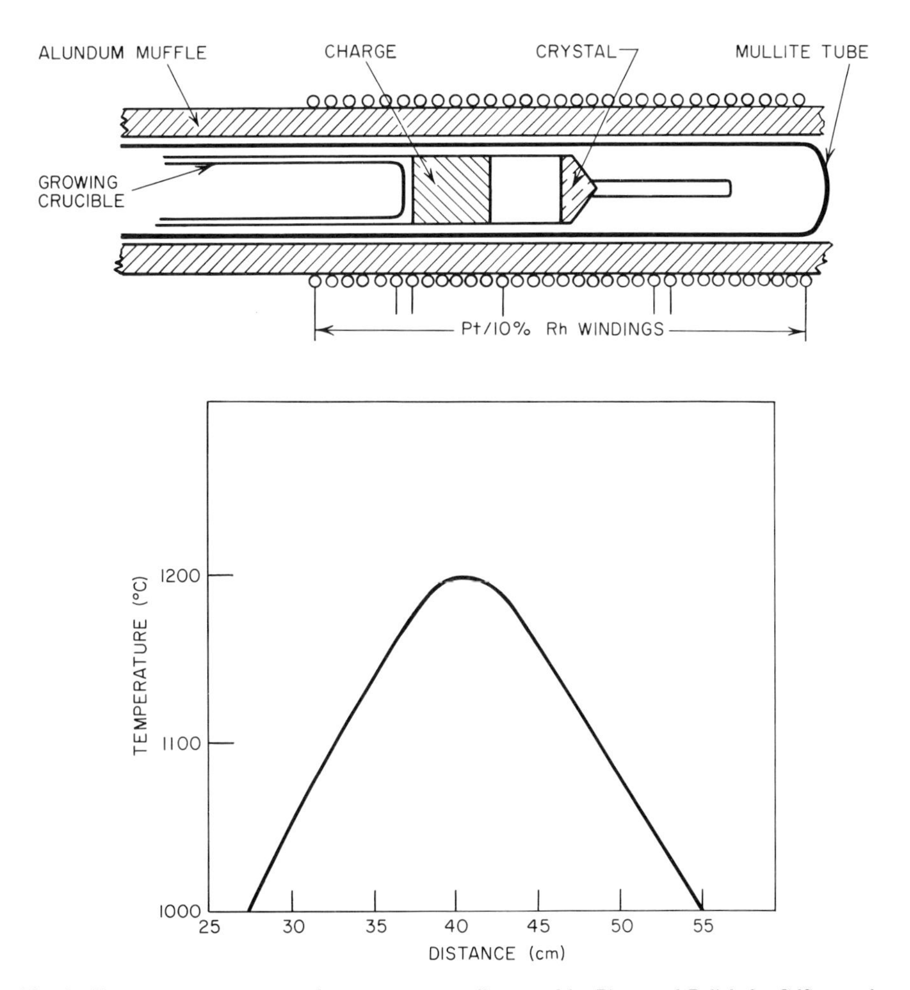

Fig. 8. Furnace arrangement and temperature gradient used by Piper and Polich for CdS crystals.

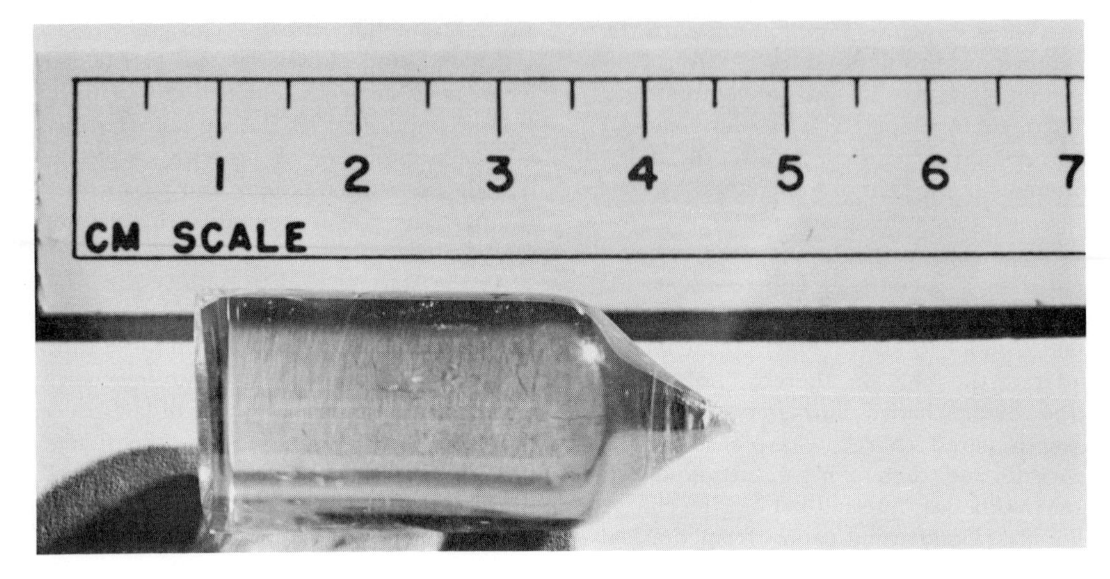

Fig. 9. Crystal of CdS produced by method shown in Fig. 8.

to grow most of the II–VI compounds and many solid solutions of these compounds.

An additional technique for growing CdS crystals, different from any of those previously mentioned, was used by Miller and Bachman (1958). This technique employs streams of evaporated cadmium and sulfur directed in a vacuum to a common point on a temperature-controlled substrate. Several types of crystals were produced that the authors described as globules, spiroidals, ribbons, and needles. With the substrate temperature between 180 and 390°C the growth was predominantly of the globule and spiroidal type. As the substrate temperature was increased from 550 to 650°C, the ribbon and needle types of growth became evident. This technique allows the possibility of introducing a third element that might serve as a phosphorescence activator.

2. CRYSTAL GROWTH MECHANISM

The crystal growth theory of Volmer and Weber (1925), Becker and Doring (1935), and others depends upon the formation of a stable two-dimensional nucleus on a crystal surface or edge for the generation of new layers. The theory of Frank (1952) and others postulates the generation of new layers with the aid of screw dislocations.

Growth patterns on CdS crystals show that both screw dislocations and edge nucleation occur. The influence of screw dislocations is shown in the growth pattern of Fig. 10*a*. A well-developed spiral with interlaced spirals is illustrated. The spiral occurs in the (0001) plane and a step travels across a diameter of the spiral, which may indicate a collapse of the lattice after growth was complete. This would result from slip parallel to the *C*-axis (this is the observed direction of slip in CdS).

Theory predicts that growth of a stepped crystal surface results from three processes: (1) a transport of molecules from the vapor to the surface; (2) diffusion of adsorbed molecules to the steps; and (3) diffusion of molecules along the edge of a step to a kink. The rate of advance of a straight step is

$$V = 2(\sigma - 1)X_s v e^{-w/kT}$$

where $(\sigma - 1)$ is the supersaturation, X_s is the mean distance a molecule will wander over the surface of the crystal between the time it strikes and the time it evaporates, v is an atomic vibration frequency-factor, and w the evaporation energy. When the mean distance between kinks X_0 is small and the

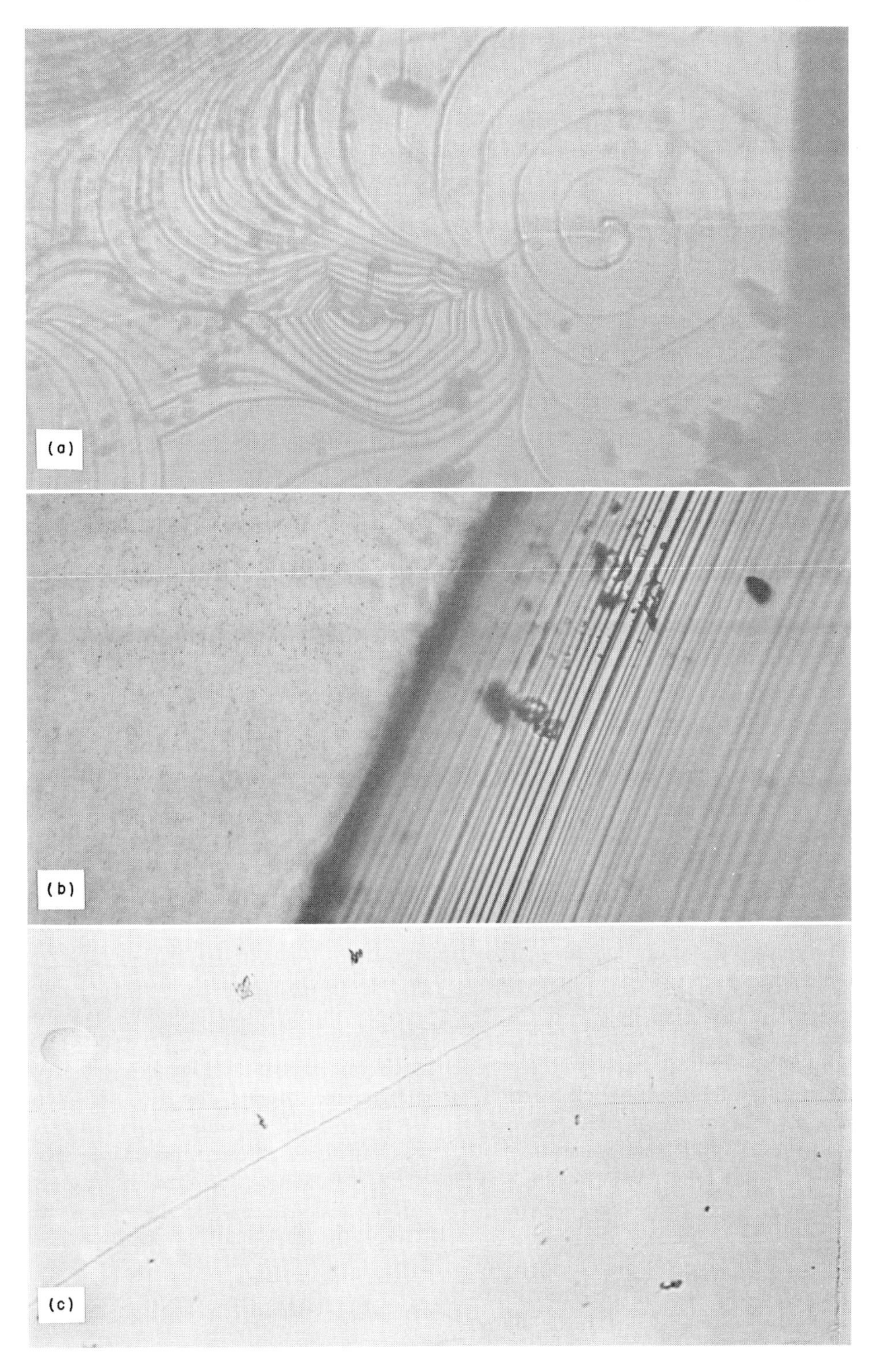

Fig. 10. Growth features on CdS crystals. (*a*) Spiral growth pattern on crystal of the prism type (144 X). (*b*) Growth steps along the side of a CdS pyramid and a portion of the plateau where the growth terminated (224 X). (*c*) Oriented growth layer on the surface of the plateau shown in (*b*) (160 X).

mean diffusion length of adsorbed molecules X_s is large $X_s \gg X_0$, the molecules will have a high probability of adhering to the step if adsorbed near it, regardless of the orientation of the step. Therefore the rate of advance of the step is independent of crystal orientation and rounded spirals result as shown in Fig. 10*a*.

If the rate of advance of a step is dependent of its orientation on the crystal face, a directional dependence of macroscopic growth will be observed. The rate of advance of a step is a minimum when it lies parallel to a close-packed direction. The kink-energy is highest for this orientation, so the step has few kinks. If the relation $X_s \gg X_0$ is no longer true, directional growth results, the edges of the steps becoming straight and perpendicular to the slow growth direction. Examples of this type of growth pattern are shown in Figs. 10*b* and 10*c*. This growth occurred after the furnace was shut off so the supersaturation decreased, making the exponential term in the velocity equation dominant and accounting for the oriented growth.

Figure 10*b* shows steps along the side of a pyramid. The step heights are very large, ~50,000 A. Smaller steps on the surface of a plateau are shown in Fig. 10*c*. The pattern consists of a series of layers spreading out from a region at or near the crystal edge. The exact mechanism for the generation of the steps is not known; however, Newkirk (1956) suggests that many screw dislocation sources may be present in a particular area of the crystal, each source generating a step too small to be seen by itself. Elbaum and Chalmers (1955) propose another mechanism for the generation of closed loops when the growth plane is parallel to the growth surface. Growth could take place by surface nucleation on the plateau, with layers being formed by edge growth on the edges of the critical nucleus. It is believed that layers of many atomic steps in height result from the coalescence of monatomic steps generated at or near the crystal edge. Slowing down of a monatomic step caused by impurities or imperfections would allow succeeding layers to catch up with it to form a polyatomic step. The slowing down of closely spaced steps could also result from evolution of latent heat at the steps.

Studies by Woods (1959) showed that the screw dislocation mechanism can operate when CdS crystals are condensed from vapor. He was not able to observe monatomic spirals however. His evidence suggests that although growth from screw dislocations does occur on some CdS crystals, this does not form the principal growth mechanism. The principal growth process at the higher supersaturations, where the major part of growth occurs, appears to be two-dimensional nucleation near the centers of plateaus, leading to the formation of hexagonal-terraced hills. No evidence was found, even with phase-contrast microscopy, to indicate that the closed hexagonal steps are generated from a pair of screw dislocations of opposite sign.

One might expect that if screw dislocations emerged on the growing surface of a crystal, at low supersaturations, growth peaks would be observed at these sites. Also, if the supersaturation is low, the growth peaks should consist of oriented growth. This has been observed by Reynolds and Czyzak (1960) and is shown in Fig. 11*a*. The growth may result from the spiral step being blocked by an absorbed impurity as suggested by Amelinckx (1957) in explaining dendritic growth. Upon etching, the growth peaks should be the sites of dislocation etch pits. The results of a 2-sec etch are shown in Fig. 11*b*. The growth peaks are indeed the sites of etch pits. Votava (1958) has also observed the influence of screw dislocations on the growth of CdS. Evidence for growth of ZnS from screw dislocations has been observed by Greene et al. (1958).

3. PREFERRED GROWING SURFACE

Lavine, Rosenberg, and Gatos (1958) have observed that the two faces of indium antimonide crystals resulting from a cut normal to the ⟨111⟩ direction show different etching characteristics. They point out

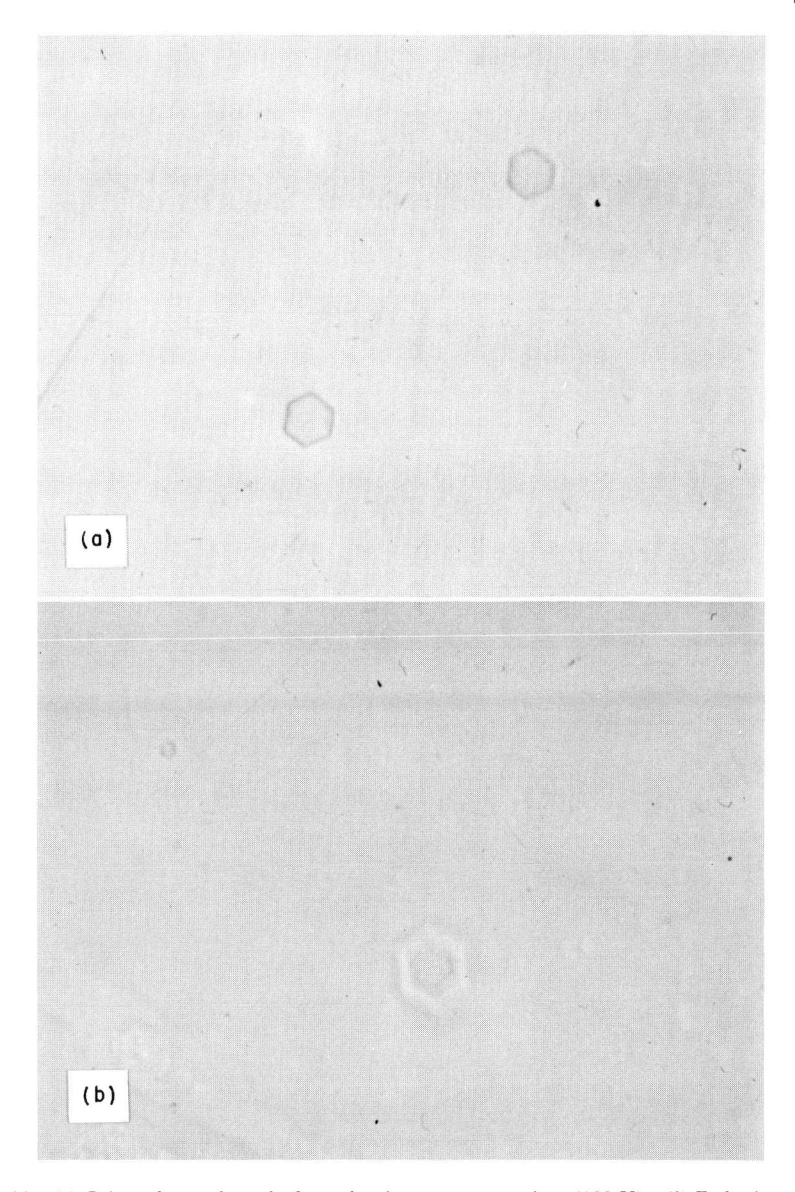

Fig. 11. (*a*) Oriented growth peaks formed at low supersaturations (132 X). (*b*) Etch-pits resulting from a 2 sec etch of the growth peaks (200 X).

that the two faces arbitrarily designated as (111) and $(\bar{1}\bar{1}\bar{1})$ are geometrically similar, but chemical differences would be expected because In and Sb atoms have different arrangements of neighbors at the two surfaces. The structure along the *C*-axis in the wurtzite structure is very similar to that along the ⟨111⟩ direction in the zincblende structure. The wurtzite structure is shown in Fig. 12. Vertical bonds between silver and white spheres appear in the photograph. If a horizontal cut were made through a set of these bonds, it may be seen that silver spheres would lie nearest the surface on top of the cut and white spheres nearest the bottom surface of the cut. For this structure, as for the InSb structure, differences in the chemical etching of these two surfaces are observed. Using a HCl etch on CdS crystals, Reynolds and Czyzak (1960) showed that hexagonally shaped etch pits form on one face,

Fig. 12. Model of wurzite crystal structure.

whereas a velvetlike appearance develops on the opposite face. Using a chromic acid etch, Woods (1960) has also shown etching differences for the two basal planes of CdS.

The difference between the basal planes affects the growth of crystals. The growth plane is (0001) for crystals that grow in the form of hexagonal prisms. The growing face is always the one on which hexagonally shaped etch-pits form. X-ray analysis by Warekois (1961) has shown that this face is the one on which sulfur is exposed. For the seed-plate technique, preferred growth on one basal plane is not observed. This further indicates a difference in the mechanism of these two growth methods.

4. GROWTH OF WHISKERS AND PLATELETS

Galt and Herring (1952) have shown that small "whiskers" of Sn are relatively perfect single crystals in the sense that they may be elastically bent to a strain of 1% without plastic deformation. Piper and Roth (1953) grew single crystals of ZnS with similar properties. These crystals were grown by sublimation in an atmosphere of hydrogen. They crystallized as needles to about 5×10^{-3} cm in diameter and several millimeters long. It was believed that these crystals grew from single screw dislocations with their Burgers vectors parallel to the needle axes. X-ray analysis showed that the needles had the wurtzite structure. Whereas most vapor grown ZnS crystals possess a high density of stacking faults, these crystals appeared free from such defects since diffuse streaking in the X-ray photographs was absent. The needles were quite flexible and exhibited considerably greater strength than ordinary ZnS crystals. Their Young's moduli were estimated to be about 7×10^{11} dynes/cm^2. Elastic strains up to $1\frac{1}{2}\%$ could be induced by bending the crystals.

Whiskers and platelets of CdS were grown by Reynolds and Greene (1958) by a technique similar to that used by Sears (1955). The furnace arrangement was the same as that shown in Fig. 5. The seed plates were removed from the ends of the sleeve; thus the whiskers form further down the tube where it becomes cooler. The center of the charge is heated to approximately 1000°C and the whiskers form at 800 to 900°C. The system is filled with about one atmosphere of H_2S or argon and is sealed off. Whiskers up to a few microns in cross section and 3 cm long have been grown that appear to be very perfect. In Fig. 13, a hexagonal whisker $\sim$10 μ in cross section is shown.

CdS whiskers exhibit unusual elastic properties. A 2.7 μ whisker has been bent enough to produce a maximum elastic strain in it (as deduced from its curvature) of 2.4%.

In the same region of the furnace where the whiskers grow, blades and platelets also grow. The platelets vary from a fraction of a micron to several microns in thickness and are as large as a square centimeter in area. They have elastic properties comparable to whiskers, and measurements of their electrical and optical properties show them to be superior to bulk crystals.

Platelets of CdS and ZnS have been grown by Fochs (1960). He used a separately controlled two-element furnace, one winding used to heat the charge, the other used to heat the crystal. For platelet growth, flowing gas was used with a steep gradient between the two windings. The temperature of the charge was usually set between 1000 to 1100°C. Close control over fluctuations in both temperature and gas flow was necessary.

5. GROWTH OF OTHER MATERIALS FROM THE VAPOR

This section describes the growth of some materials other than CdS and ZnS from their vapor. The discussion is not intended to be exhaustive but rather to point out instances where vapor phase growth has been successful. Several of the Group II–VI compounds such as ZnSe, ZnTe, CdSe, and a number of solid solutions of these compounds have been grown by the techniques previously described. The vapor phase method is easily applicable to all the

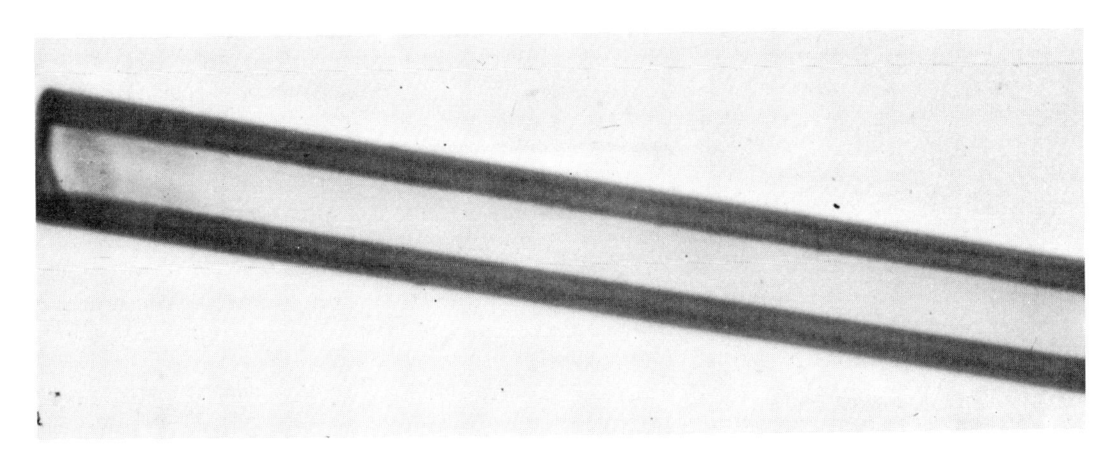

Fig. 13. Hexagonal CdS whisker (thickness of whisker 10 μ).

II–VI compounds mentioned because these compounds sublime at ordinary pressures.

PbSe and PbS can be grown readily from either liquid or vapor. Prior (1961) grew PbSe crystals from vapor under conditions that allowed almost the whole charge (a few grams) to be converted into one crystal. The growth technique was designed to gain composition control by controlling the selenium vapor pressure as the crystal grew. Prior's furnace arrangement is shown in Fig. 14. The charge consisting of pieces of melt-grown PbSe crystals was contained at point *A* in a fused silica crucible. This material was sublimed in a small temperature gradient (approximately 1°C at 775°C) to the point *B*. Excess pure selenium was contained at point *D*. The bore of capillary *C* was small enough to minimize the loss of PbSe during growth but large enough to allow the control of selenium. Impurity gases were pumped out through capillary *E,* which is of a size that prevents reduction of the selenium pressure. The duration of a run was 2 to 3 days, resulting in crystals with volumes of $\sim$0.2 cm^3. Lattice defect concentrations were minimized by using low temperatures; thus the standard growth temperature of 775°C was chosen to compromise between minimizing lattice defects and achieving a reasonable growth rate.

The size and quality of the crystals depend on small, well-controlled temperature gradients. These were obtained by means of an auxiliary winding on the inner furnace, permitting control to better than 0.1°C/cm. Precautions were taken in preparing and charging crucibles to limit the number of nucleation centers. A small number of large crystals was desired, and with proper seeding and control the entire charge could be converted into one crystal. Crystals grown by this method are shown in

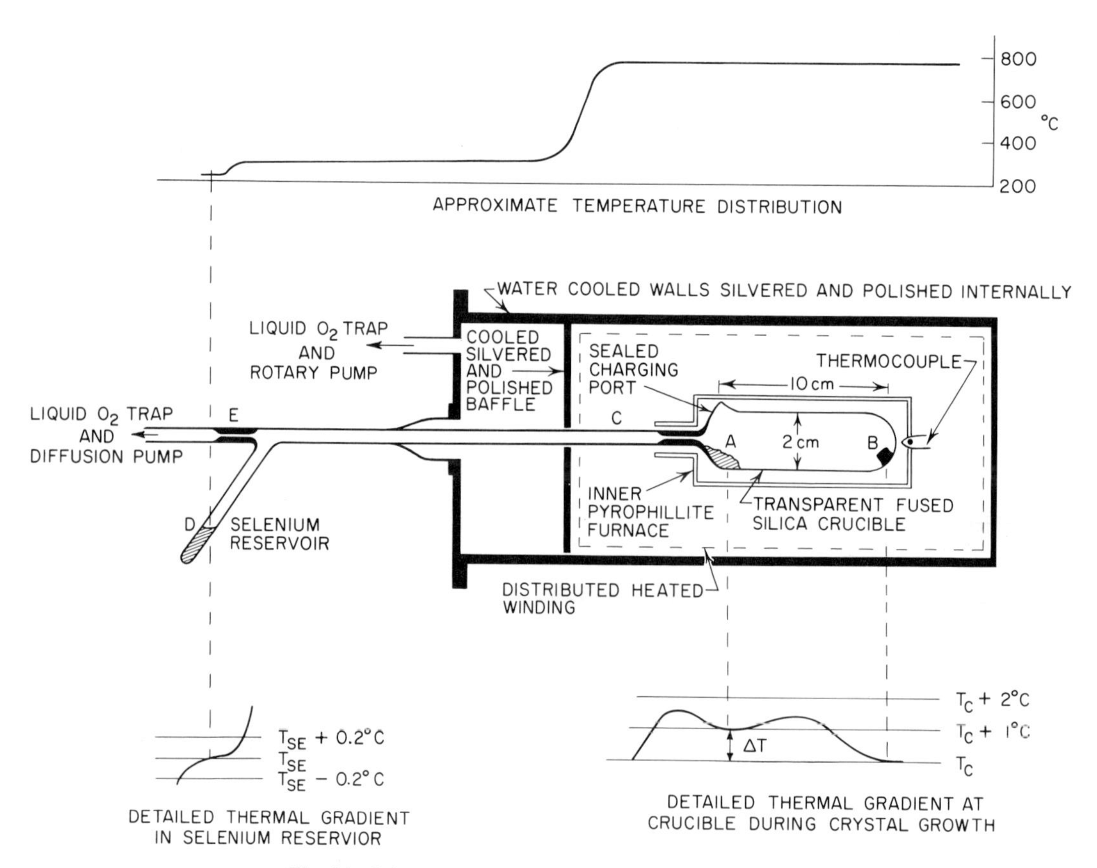

Fig. 14. Schematic diagram of arrangement used by Prior.

Fig. 15. These crystals have electrical carrier concentrations as low as $3 \times 10^{16}/cm^3$. This is lower than has been achieved in melt-grown PbSe crystals.

Crystals of PbS were grown from the vapor by Pizzarello (1954) using a Bridgman furnace with an inverted temperature gradient. This temperature distribution was obtained by placing a thick-walled Inconel tube in the core of the vertical furnace. The Inconel tube (two-thirds as long as the furnace core) was placed in the bottom of the furnace. The sample, which was contained in a quartz tube just large enough to fit into the Inconel tube, was prepared by reacting lead with sulfur. The two elements were placed in a sealed quartz tube with a pressure of 10^{-6} mm Hg. The reaction was initiated at ~500°C and completed at 800°C to insure homogeneity. After this process, the sample was drawn through the temperature gradient at a rate of $\frac{1}{16}$ in./hr, causing the PbS vapor to condense at the top of the sample chamber in the form of one or a few large crystals. To increase the probability of growing one crystal, the sample chamber was pointed at the top.

Seeds formed at 950°C, and the source temperature was maintained at 1100°C. The rate of growth was calculated from a mass flow calculation and from the difference in the vapor pressures at these two temperatures. This growth rate was $\sim 10^5$ times the observed rate. It was believed that foreign gases were liberated from the quartz tube and the reacting elements, making the growth rate dependent on diffusion of PbS through the foreign gases. This mechanism is consistent with the experimentally observed growth rate.

InAs, InP, GaAs, and GaP are examples of compounds whose crystals are generally grown from the melt. However, Antell and Effer (1959) have successfully grown them by a vapor phase reaction. The monochloride of indium can sometimes be reacted to form the trichloride plus a compound of indium. Antell and Effer first tried this using arsenic vapors to form InAs. This material was selected for the initial experi-

Fig. 15. Crystals of PbSe. (*a*) A particularly well-formed specimen. (*b*) Typical quality.

ments because $AsCl_3$ is unstable at the reaction temperature and it had been determined that chlorine had little effect on the electrical properties of InAs.

The crystals were grown in evacuated silica reaction tubes approximately 20 cm long and ~50 cm³ in volume. Two methods were used. In the first method, the monochloride of indium was heated uniformly, in the presence of arsenic, to a temperature above the reaction temperature. The following reaction then occurred:

$$3\ InCl + \tfrac{1}{2}\ As_4 \rightleftharpoons InCl_3 + 2\ InAs$$

and the temperature was gradually lowered until crystal growth occurred. Similar results were obtained if the chlorine was replaced by iodine or bromine.

In the second method, small amounts of the chloride or iodide of indium, or only

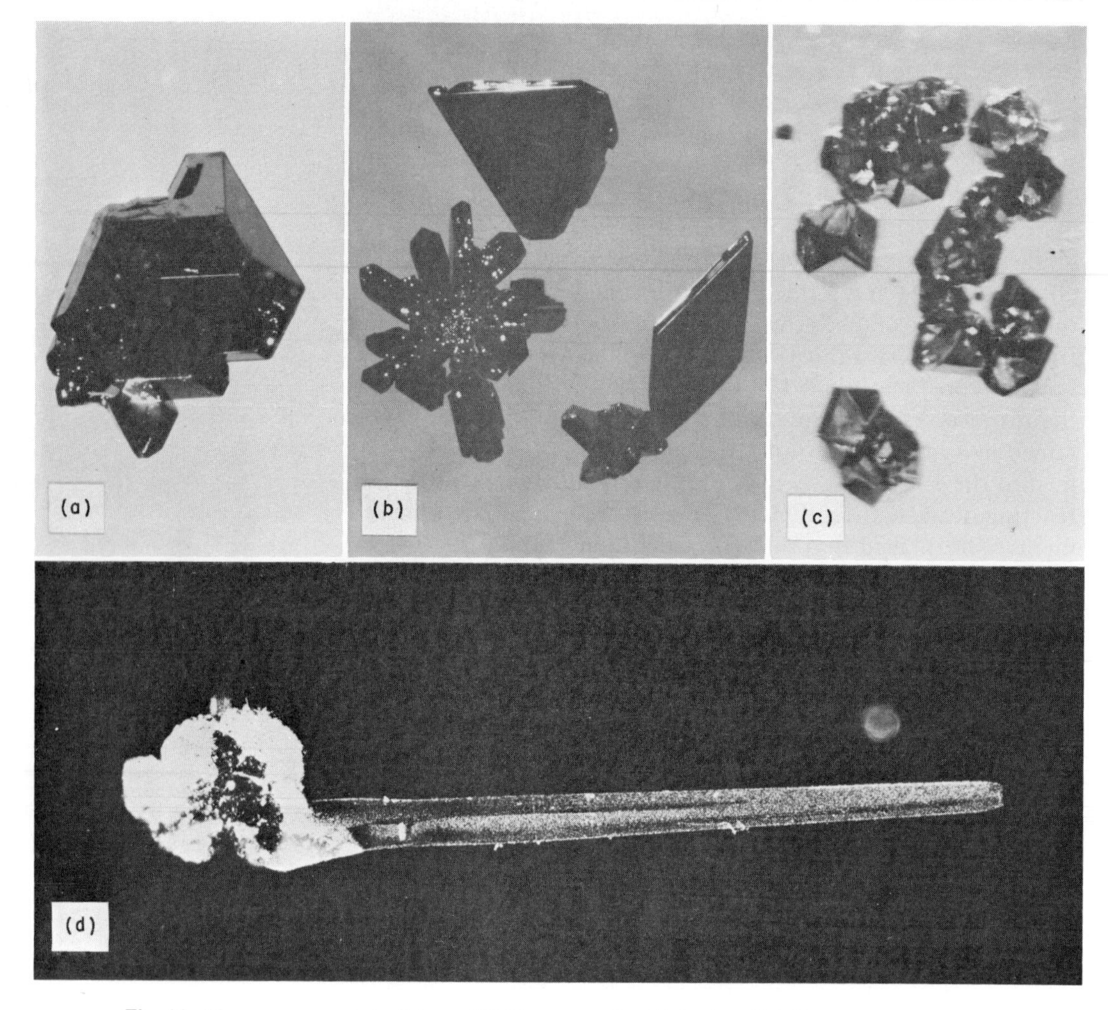

Fig. 16. Vapor grown crystals of Group III–V compounds. (*a*) GaAs. (*b*) InP. (*c*) GaP. (*d*) InAs.

iodine itself, together with a sample of InAs, were placed in a furnace with a temperature gradient of 40 to 70°C over the length of the tube, the compound being at the hotter end. As the temperature of the furnace was lowered, crystals formed in the cool end. The furnace was then held at this temperature and it was possible to transport the entire sample of InAs to the cool end of the tube. This follows from the reversibility of the reaction shown.

The same methods were used successfully to grow crystals of GaAs, InP, and GaP. At that time, the solubility of chlorine or idoine in these compounds was not known. Examples of crystals grown by these two methods are shown in Fig. 16. X-ray measurements of the lattic parameters of these crystals were in very good agreement with previous measurements made on the same compounds.

References

Allen, E. J., and J. L. Crenshaw (1912), *Am. J. Sci.* **34,** 310.

Addamiano, A., and M. Aven (1960), *J. Appl. Phys.* **31,** 36.

Amelinckx, S. (1957), *Conference on Dislocations and Mechanical Properties of Crystals,* John Wiley and Sons, New York.

Antell, G. R., and D. Effer (1959), *J. Electrochem. Soc.* **106,** 509.

Bishop, M. E., and S. H. Liebson (1953), *J. Appl. Phys.* **24,** 5, 660.

Becker, R., and W. Doring (1953), *Ann. Phys. Paris* **24,** 719.

Czyzak, S. J., D. G. Craig, C. E. McCain, and D. C. Reynolds (1952), *J. Appl. Phys.* **23,** 932.

Elbaum, C., and B. Chalmers (1955), *Can. J. Phys.* **33,** 196.
Fochs, P. D. (1960), *J. Appl. Phys.* **31,** 1733.
Frank, F. C. (1952), *Advances in Physics* **1,** 91.
Frerichs, R. (1946), *Naturewissenschaften* **33,** 387; (1947), *Phys. Rev.* **72,** 594.
Galt, J. K., and C. Herring (1952), *Phys. Rev.* **85,** 1060.
Greene, L. C., D. C. Reynolds, S. J. Czyzak and W. M. Baker (1958), *J. Chem. Phys.* **29,** 6, 1375.
Gudden, B., and R. Pohl (1920), *Z. Physik* **2,** 181, 361.
Lang, A. R., *J. Appl. Phys.* **28,** 497 (1957).
Lavine, M. C., A. J. Rosenberg, and H. C. Gatos (1958), *J. Appl. Phys.* **29,** 1131.
Lorenz, R. (1891), *Chem. Ber.* **24,** 1509.
Medcalf, W. E., and R. H. Fahrig (1958), *J. Electrochem. Soc.* **105,** 719.
Merriman, H. E. (1912), *Am. J. Sci.* **34,** 341.
Miller, R. J., and C. H. Bachman (1958), *J. Appl. Phys.* **29,** 1277.
Newkirk, J. B. (1956), *Acta Met.* **4,** 3.
Piper, W. W., and S. J. Polich (1961), *J. Appl. Phys.* **32,** 1278.
Piper, W. W., and W. L. Roth (1953), *Phys. Rev.* **92,** 503.
Pizzarello, F. P. (1954), *J. Appl. Phys.* **25,** 804.
Prior, A. C. (1961), *J. Electrochem. Soc.* **108,** 82.
Reynolds, D. C., and S. J. Czyzak (1950), *Phys. Rev.* **79,** 543; (1960), *J. Appl. Phys.* **31,** 94.
Reynolds, D. C., and L. C. Greene (1960), *International Conference on Solid State Physics,* Academic Press, New York.
Samelson, H. (1961), *J. Appl. Phys.* **32,** 309.
Samelson, H., and V. A. Brophy (1961), *J. Electrochem. Soc.* **108,** 150.
Sears, G. W. (1955), *Acta Met.* **3,** 4.
Stanley, J. M. (1956), *J. Chem. Phys.* **24,** 1279.
Verma, A. R. (1953), *Crystal Growth and Dislocations,* Butterworth Scientific Publications.
Volmer, M., and A. Weber (1925), *Z. Phys. Chem.* **119,** 277.
Votava, V. E. (1958), *Z. Naturforschg.* **13a,** 542.
Warekois, E. P. (1961), Private Communication.
Woods, J. (1959), *British J. Appl. Phys.* **10,** 529; (1960), **11,** 296.

5

Silicon

R. Glang and E. S. Wajda

The preparation of crystals of silicon by techniques other than solidification has until recently remained relatively unexplored. This chapter presents various aspects of an alternative method of single crystal formation in which silicon to be added to the seed crystal is supplied by a chemical reaction between the solid and a gas. This process is called *epitaxial growth,* or *vapor growth.*

The term *epitaxial* refers to a layer of solid material whose crystal structure and orientation are determined by the lattice of the substrate or seed crystal on which it lies. An epitaxial layer is not necessarily the same material as the parent seed. Epitaxial films of copper on sodium chloride (Mayer, 1955), of germanium on gallium arsenide, or gallium arsenide on germanium (Anderson, 1960) have been grown repeatedly. These examples, however, are more accurately described as hetero-epitaxial. In this chapter, only the iso-epitaxial case of silicon growing onto silicon substrate are discussed.

1. GENERAL CONSIDERATIONS

Many reactions for depositing high purity polycrystalline silicon from a gas have been investigated. Table 1 lists three categories of processes. Some that yield solid reaction products other than silicon, such as the reduction of $SiCl_4$ by zinc or sodium vapors, are not included. All the chemical processes have certain principal steps in common. First, a stable silicon compound is formed at a temperature T_1 and if necessary purified in a separate operation. Second, a certain vapor pressure of this compound is maintained by evaporating the compound from a source at a suitable temperature. The silicon-compound vapors are then carried into a deposition zone of temperature T_2. This temperature is chosen to make the silicon-compound undergo chemical reaction which deposits elemental silicon on surfaces exposed to the vapors.

The rates and yields of these deposition reactions depend on a number of process variables, such as residence time, supersaturation of the gaseous compound in the reaction zone, and deposition temperature.

In principle, all the reactions listed in Table 1 should be suitable for growing monocrystalline layers of silicon. Additional requirements are necessary for epitaxial growth, however, that refer not to the chemical reactions but to the properties of the substrate surface. The crystal structures, atomic dimensions, and thermal expansions of the substrate and deposition lattices must be compatible. These conditions are automatically satisfied for iso-epitaxial growth. Furthermore, the initial substrate surface must be atomically clean and without serious surface damage to insure undisturbed growth. To prevent random nucleation of crystallites, the deposition temperature (which determines the surface mobility of the depositing atoms) must be compatible with the deposition rate.

Although most of the chemical reactions listed in Table 1 have produced monocrystalline epitaxial deposits, experience has shown that some do not permit the close control necessary for manufacturing semiconducting devices. Control is needed of thickness, surface topography, resistivity,

TABLE 1

SILICON DEPOSITION PROCESSES

Chemical Reaction	Initial Pressure of Silicon Compound (mm Hg)	Deposition Temperature (°C)	Flow Rate Of Carrier Gas (liter/min.)	Yield (%)	Deposition Rate (gram Si/hr)	References
Pyrolytic Decomposition						
$SiH_4 \rightarrow Si + 2H_2$	3–18	770–1335	—	33–85	3–7	(1)
$SiI_4 \rightarrow Si + 4I$	$10^{-3} - 2 \times 10^{-2}$	1000	—	50–80	1–10	(2), (3), (4), (5)
Reduction						
$SiCl_4 + 2H_2 \rightarrow Si + 4HCl$	—	1100	—	(Small)	2.5	(6)
$SiBr_4 + 2H_2 \rightarrow Si + 4HBr$	6	1275	1.5	40	0.36	(7)
$SiI_4 + 2H_2 \rightarrow Si + 4HI$	100	1000	0.5	53	2.6	(8)
$SiHCl_3 + H_2 \rightarrow Si + 3HCl$	15–20	1050	13	45–50	10	(9), (10)
Disproportionation						
$2SiCl_2 \rightarrow Si + SiCl_4$	~240	800–1220	—	—	0.1–0.2	(11), (12)
$2SiI_2 \rightarrow Si + SiI_4$	15–2200	950	—	—	—	(13)

(1) Mayer, 1953. (2) McCarty, 1959. (3) Herrick and Krieble, 1960. (5) Litton and Anderson, 1954. (6) Rubin et al., 1957. (8) Szekely, 1957. (9) Glang and Wajda, 1961. (10) van der Linden and DeJonge, 1959. (11) Kempter and Alvarez-Tostado, 1958. (12) Schafer and Nickl, 1953. (13) Glang and Kippenhan, 1960.

and the conductivity type of the deposited layer.

To illustrate the importance of the surface topography, two monocrystalline deposits having the same (111) orientation, but grown under different conditions, are shown in Fig. 1. The growth with the chevron surface pattern is not adaptable to device fabrication.

Silicon vapor reactions (Table 1) have

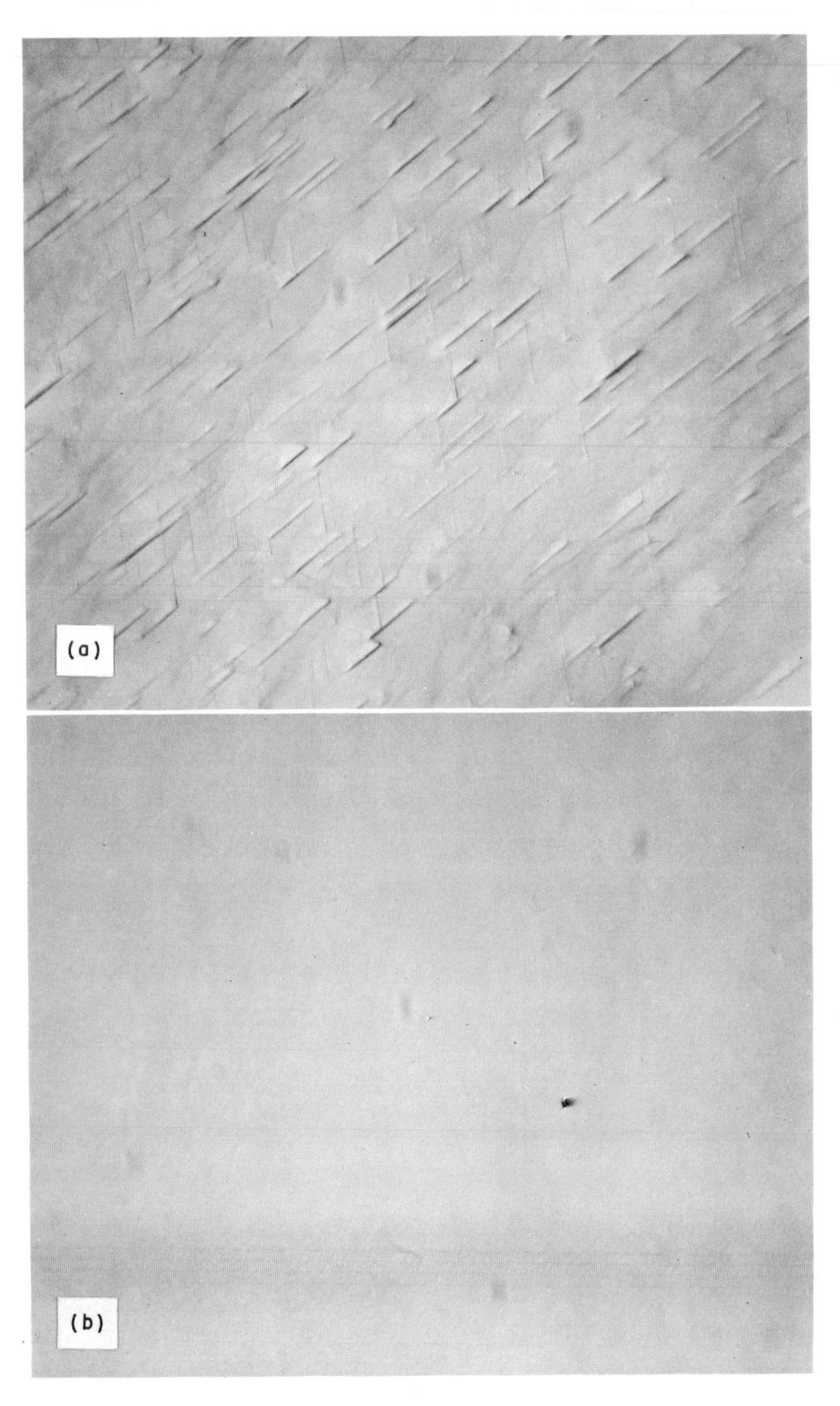

Fig. 1. Surface topography of (111) silicon deposits grown by the hydrogen reduction of trichlorosilane. (500 X)

been conducted under widely different conditions of deposition. Since very few detailed studies have been made of the factors affecting crystal growth, an empirical approach is usually used to produce the desired results.

In the following sections, typical deposition systems in each of the reaction categories listed in Table 1 are discussed.

2. PYROLYTIC DECOMPOSITION

Of the variety of silicon compounds that can be used for silicon deposition by thermal decomposition, the halides seem to be most easily synthesized, purified, and decomposed into elemental silicon. The silanes, siloxanes, and silicates have been studied (Hirshon) but present more difficulties than the halides. Of the four tetrahalides, SiF_4 is the most stable, only decomposing under extremes of temperature and pressure. Furthermore, it is a gas under normal conditions and so is difficult to handle and purify. Since $SiBr_4$ and $SiCl_4$ do not decompose at temperatures below the melting point of silicon, most process studies have been limited to SiI_4.

The pyrolytic breakdown of SiI_4

$$SiI_4(g) \rightarrow Si(s) + 4\,I(g)$$

has been investigated by several authors (McCarty, 1959; Herrick and Krieble, 1960; Litton and Anderson, 1954; Rubin et al., 1957) for the purpose of preparing transistor-grade polycrystalline silicon. Epitaxial growth could also be achieved if the additional conditions discussed in the previous section were fulfilled.

The chemistry of silicon-iodine compounds is rather complex because there are at least two iodides, SiI_2 and SiI_4, existing in the gaseous state. The feasibility of a deposition process depends on the thermodynamic equilibria between these iodides and their constituents in the gas. These equilibria have been investigated (Wajda and Glang, August 1960) in order to establish the pressure and temperature conditions that will allow silicon deposition. An indication of how much silicon the gas phase will hold at constant volume in the presence of a given amount of iodine and excess solid silicon, is shown in Fig. 2. Here the combined normalized pressure of SiI_2 and SiI_4 is plotted as a function of the equilibrium temperature and iodine concentration. For iodine pressures smaller than 30 mm Hg, the silicon content in the gas phase decreases with rising temperatures, whereas beyond pressures of 100 mm Hg the silicon concentration increases with temperature. It is this behavior that determines the type of reaction and experimental conditions for a deposition process.

The SiI_4, prepared in a separate operation, is purified either by recrystallization, distillation, or zone refining (Rubin et al., 1957). Pure SiI_4 vapors are then admitted into a reaction chamber at temperatures greater than 800°C. A calculation (Wajda and Glang) of the yield of silicon under equilibrium conditions (based on the amount of SiI_4 fed into the reaction chamber) shows that for high SiI_4 pressures no silicon deposition occurs, since the oncoming SiI_4 vapors are not fully saturated with silicon. Therefore they react with the silicon substrate to form SiI_2.

$$SiI_4(g) + Si(s) \rightarrow 2\,SiI_2(g)$$

Substantial yields can only be obtained for tetraiodide pressures well below 1 mm Hg. This conclusion is confirmed by the experimental work of McCarty (1959), Herrick and Krieble (1960) and Litton and Anderson (1954), all of whom did their work in the 1 to 10 μ pressure range.

Reaction at low pressures yields slow growth because the amount of silicon per unit volume of gas is rather small. This can be compensated by having a very high flow rate through the reaction chamber, but a high flow rate shortens the residence time of the gas in the deposition zone; therefore the optimum yield is not obtained. Because of this, pyrolytic decomposition of SiI_4 is not attractive as a practical method for epitaxial silicon growth.

3. DISPROPORTIONATION PROCESS

The disproportionating reaction of silicon dihalides

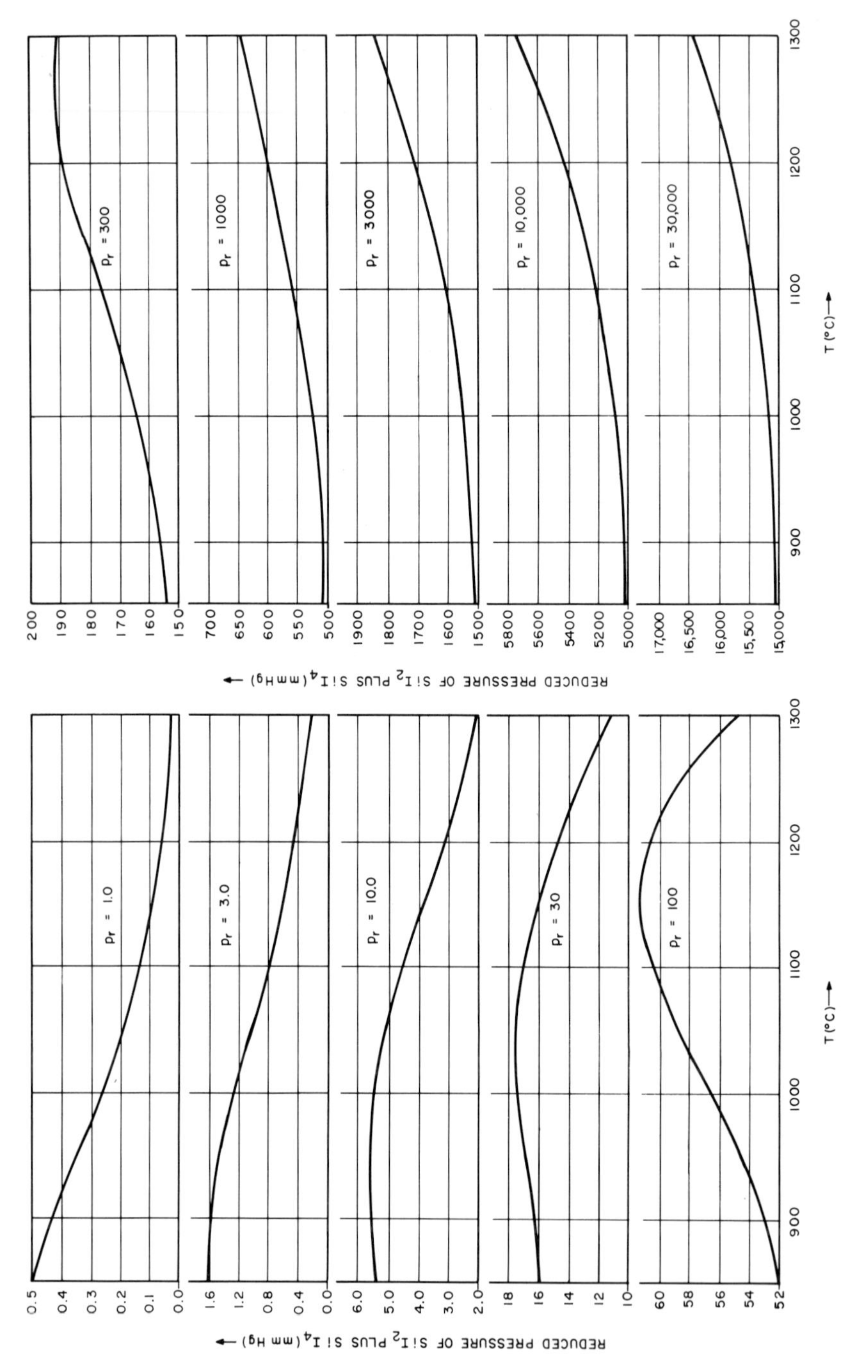

Fig. 2. Iodine in thermodynamic equilibrium with an excess of silicon; combined equilibrium pressures of SiI_4 and SiI_2, normalized to 850°C, plotted versus temperature for various iodine reference pressures. (All pressures in mm Hg.)

$$2\ SiX_2 \rightleftharpoons SiX_4 + Si$$

has been used for transporting silicon through a temperature gradient (Wajda and Glang, 1960; Kempter and Alvarez-Tostado, 1958; Schaefer and Nickl, 1953; Wajda et al., 1960). The iodine system is discussed here, but similar considerations apply for the other halide systems.

Since SiI_2 cannot be isolated as a pure compound but exists only in equilibrium with SiI_4, the reaction must be conducted so that SiI_2 is generated in one part of the reaction system and disproportionated in another zone of the same apparatus. The reaction can be carried out in a sealed, evacuated quartz tube, placed in a two-zone furnace as shown in Fig. 3. In the high temperature zone, silicon reacts with iodine to form SiI_4, which then reacts with more silicon to form SiI_2. This SiI_2 is transported to the cooler substrate zone

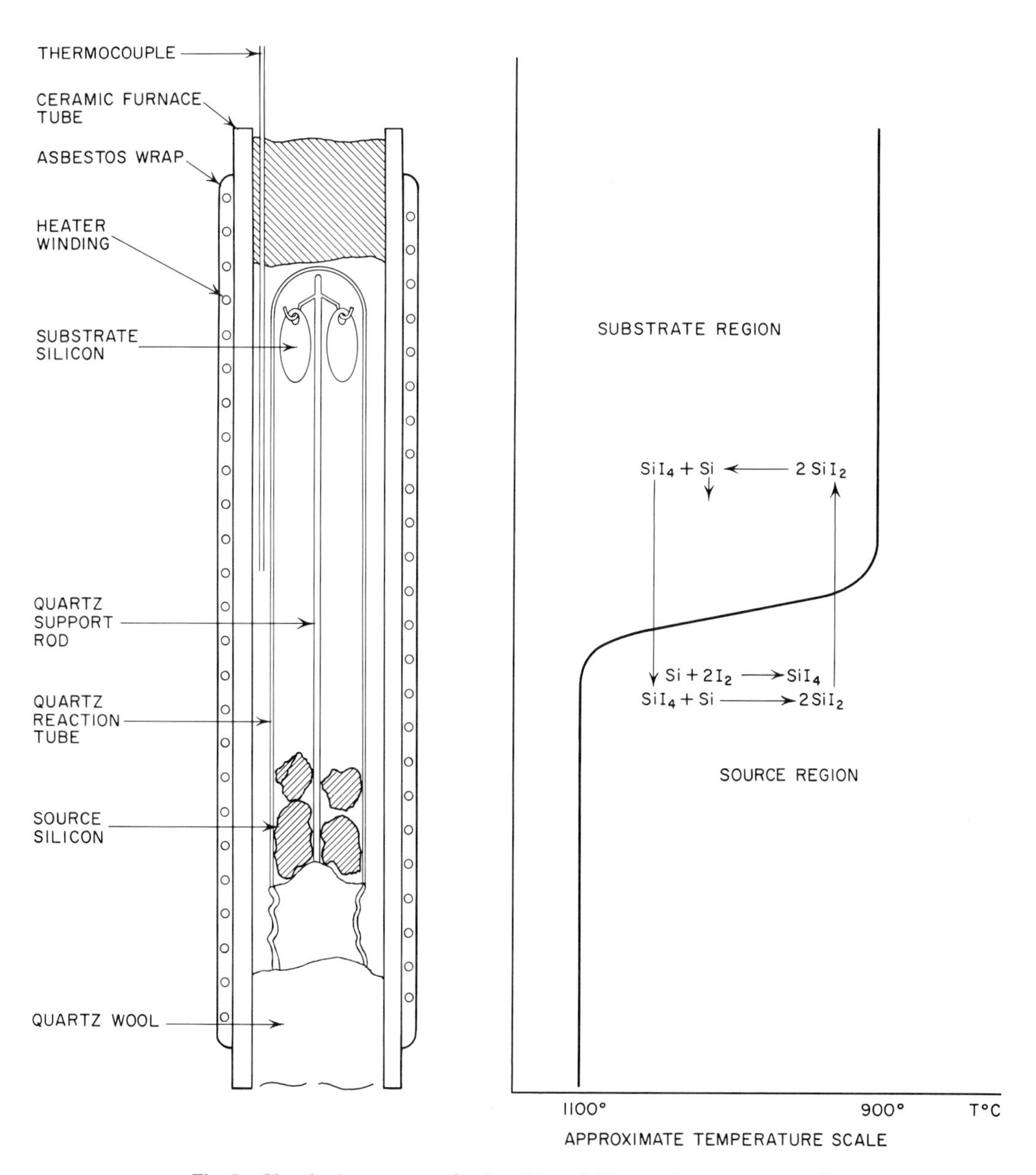

Fig. 3. Closed tube apparatus for deposition of silicon and temperature profile.

where it disproportionates into elemental silicon and SiI_4. The silicon deposits on the substrate, and the SiI_4 vapor returns to the source zone to repeat the cycle. Thermodynamic calculations (Wajda and Glang) have shown that the transportation of silicon through the reversible disproportionating reaction involves only a modest fraction of the total iodine in the system. The main portion of the vapor consists of SiI_4, forming a nonuseful cycle of transport as shown by the schematic flow chart in Fig. 4.

Suitable quantities of SiI_2 can be generated at source temperatures above 1000°C. Disproportionation and hence silicon deposition occur below the generating temperature. To obtain monocrystalline growth, a minimum substrate temperature of 780°C is required. Since silicon transport from the source to the substrate is controlled by diffusion and convection, the reaction tube geometry greatly influences the rate of transport and the macroscopic uniformity of the deposit. A large diameter tube, mounted in a vertical position, gives a uniform concentration of SiI_2 in the substrate region.

Uniform nucleation of crystal layers is seriously impaired by oxides of silicon on the substrate surface. Since these oxides are extremely stable and form readily in the presence of oxygen or water vapor at elevated temperatures, special cleaning procedures (Wajda and Glang, 1960) are necessary in the preparation of the substrate surfaces and in the final sealing of the reaction tube.

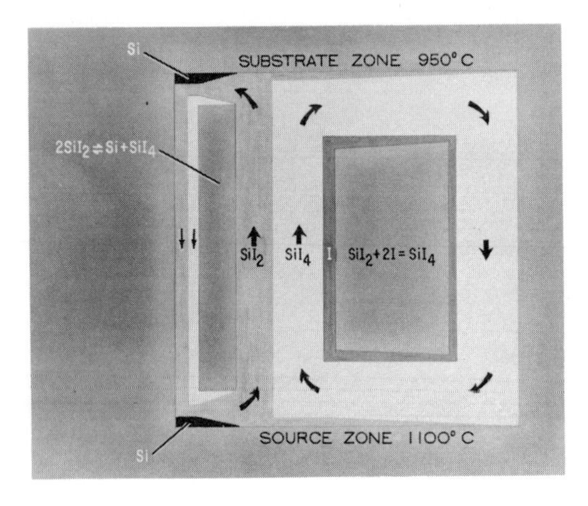

Fig. 4. Schematic representation of flow and concentrations of the chemical reactants during a silicon diiodide disproportionation cycle.

Typical epitaxial deposits on (111) and (100) silicon substrates (grown under 5 atmospheres of iodine at 1150°C source and 950°C substrate temperature) are shown in Fig. 5.

The growth rate, defined as the average deposit thickness divided by the total time of deposition, ranges from 0.5 to 20 μ/hr, depending on the prevailing growth conditions. The growth rate increases with increasing temperature difference between source and substrate and with an increase in iodine pressure. Under identical deposition conditions, the growth rate also varies with the crystallographic orientation, being largest for the $\langle 211 \rangle$ direction and decreasing in the following order: $\langle 110 \rangle$, $\langle 111 \rangle$, and $\langle 100 \rangle$.

The surface topography of the deposit is dependent on the orientation, perfection, and initial surface preparation of the substrate, as well as on the deposition rate and the temperature. Slower growth rates and higher deposition temperatures yield smoother surfaces than those shown in Fig. 5.

Controlled introduction of impurities for obtaining desired resistivity and conductivity type in the deposit can be achieved by using silicon alloys with known impurity concentrations as the source for the disproportionating reaction. Boron, phosphorus, arsenic, and antimony enter the silicon deposit in the same concentration as that in the source alloy. With these impurities, doping concentrations from 5×10^{16} to 2×10^{20}/cc have been achieved (Glang and Kippenhan, 1960) as shown in Fig. 6. Elements such as indium, gallium, and aluminum produce deposits with impurity concentrations orders of magnitude smaller than those in the source alloy. This results because the free energies of AlI_3, InI_3 and GaI_3 are significantly larger on the negative scale than those of BI_3, PI_3, AsI_3, and SbI_3.

Several methods have been described (Wajda and Glang, 1960) for changing the

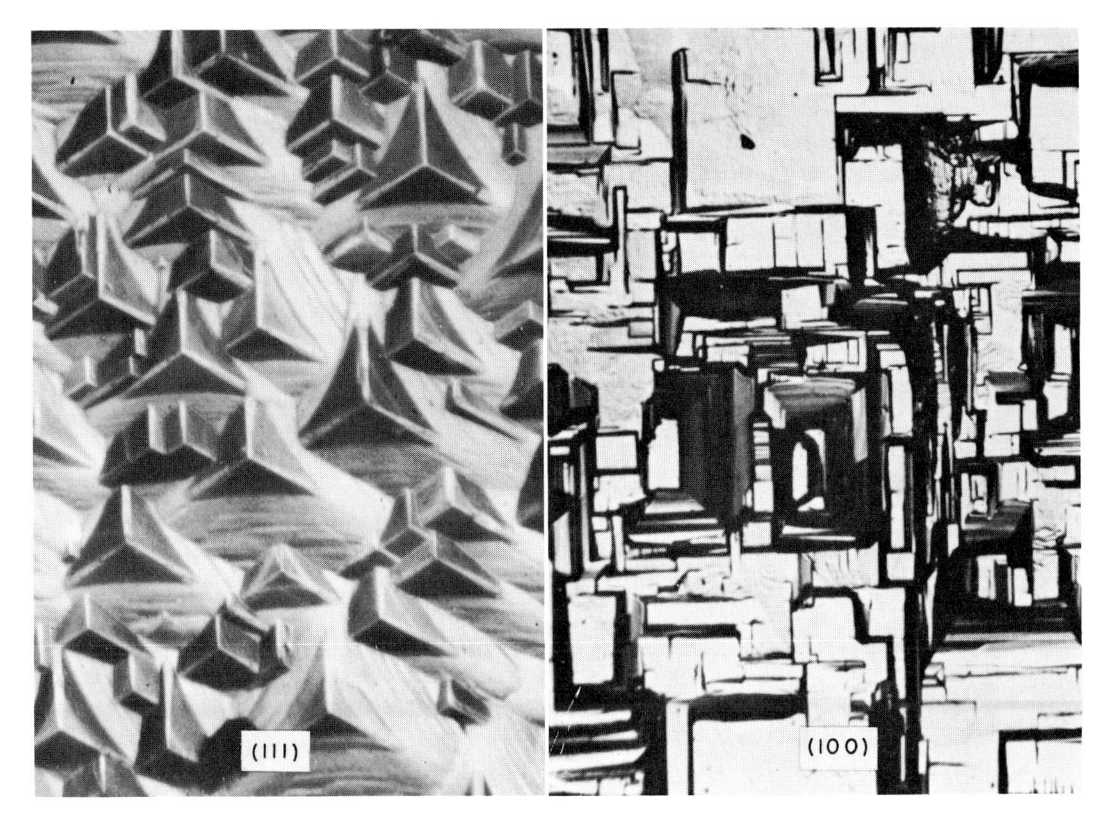

Fig. 5. Typical epitaxial growths on (111) and (100) silicon substrates (X 175 before publication reduction).

conductivity type during the deposition process in order to grow *p-n* junctions. This requires changing the source and reaction gases and is experimentally difficult. The lack of a purging capability causes compensation effects in the deposits and restricts the type of junctions that can be formed.

It is concluded that a practical process based on the silicon disproportionation reaction is unattractive because of the necessity of a closed tube, difficulties in preparing the substrate surface, and limitations for junction formation.

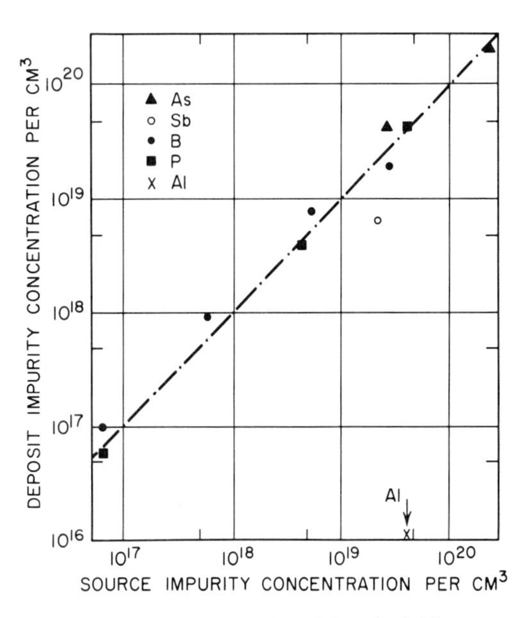

Fig. 6. Impurity concentration of deposited silicon versus impurity concentration of master alloys used as a source.

The three experimental points representing concentrations smaller than 10^{18}/cc have been obtained from "diluted" master alloy sources.

4. HYDROGEN REDUCTION PROCESS

The hydrogen reduction of trichlorosilane has found wide application in the commercial fabrication of transistor structures. The equation

$$SiHCl_3(g) + H_2(g) \rightleftharpoons Si(s) + 3\,HCl(g)$$

describes the part of the reaction that results

in silicon deposition. The full reaction mechanism and the rate determining step are not known. The kinetics of the reaction involves intermediate products. Some of these polymerize because a viscous oil that hydrolyses easily is sometimes observed as a byproduct.

In contrast with the disproportionating process previously discussed, the gases of the hydrogen reduction process are not recycled. They are swept through the deposition zone in a continuous flow. Therefore this is frequently referred to as an "open tube process." An advantage is that the deposition chamber can be purged.

Apparatus for the reduction process varies widely in design and construction materials. Quartz, glass, Teflon*, Viton*, graphite, and corrosion-resistant metals like molybdenum show little or no corrosion when constantly exposed to the chemically aggressive silicon-halides. Other plastics,

* Trademark of E. I. DuPont de Nemours & Co., Inc.

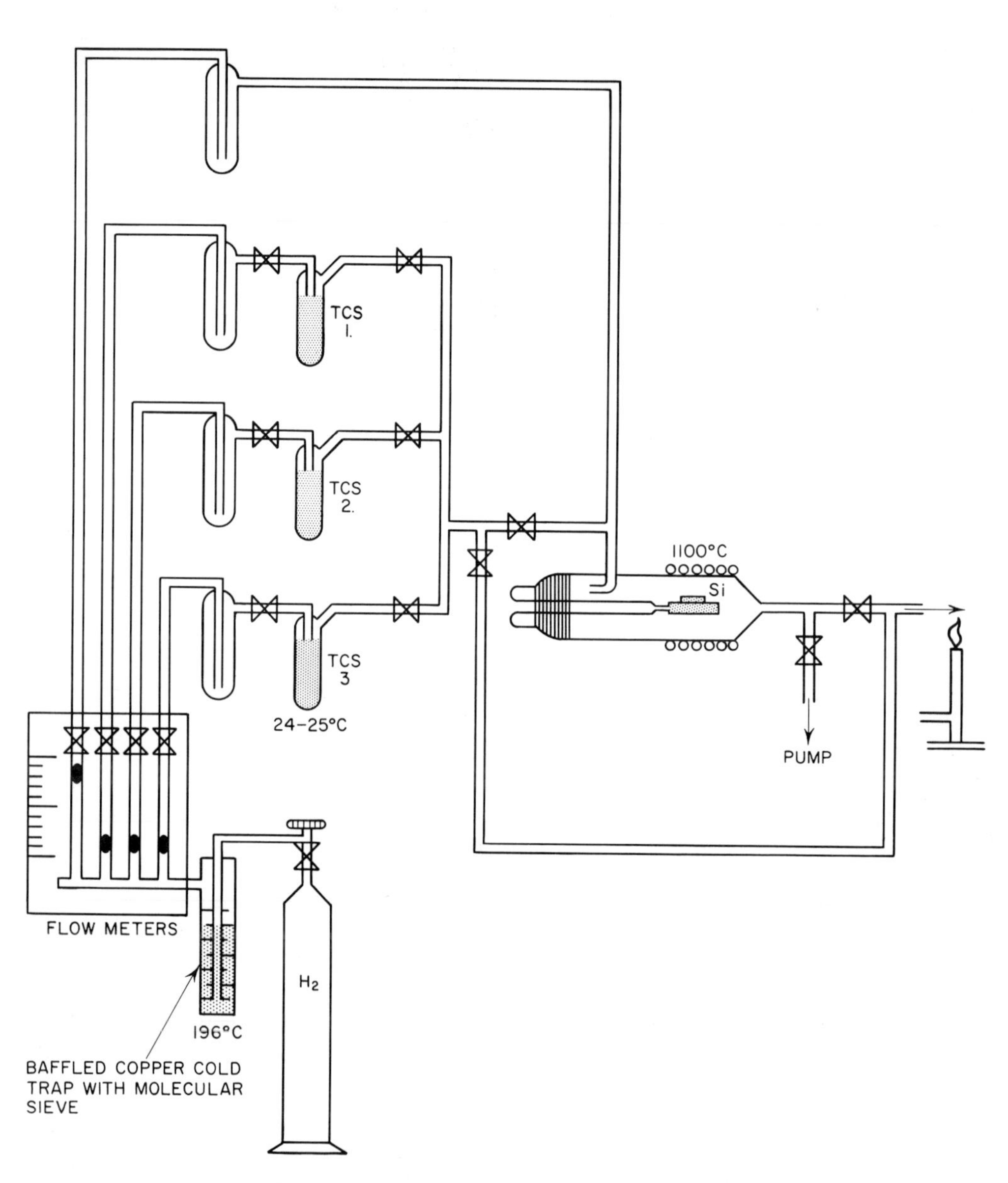

Fig. 7. Radio frequency-heated silicon deposition system.

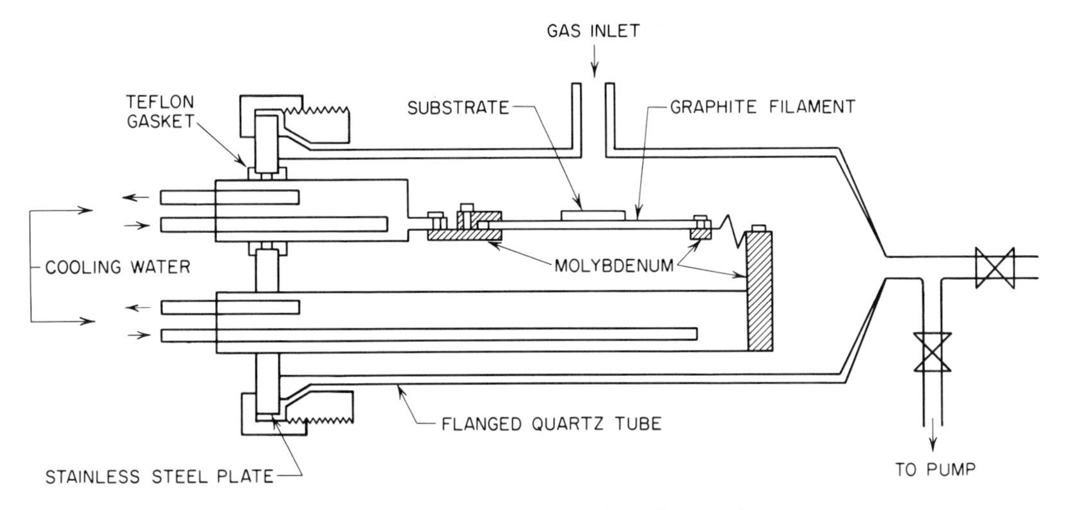

Fig. 8. Resistance-heated silicon deposition chamber.

elastomers, and especially all types of stopcock grease slowly degenerate under the influence of trichlorosilane vapors.

The deposition apparatus consists of three stages: a hydrogen source and purification train; a glass system to inject and direct the flow of trichlorosilane and impurity halides; and the reaction chamber.

Prepurified hydrogen is preferable as a source because traces of oxidizing gases and water vapor must be less than about 10 ppm if crystal perfection comparable to melt-grown crystals is desired. Since gas purity is so important, the entire system must be leaktight.

The second part of the deposition apparatus, where the halides are mixed with the hydrogen stream, can be constructed in many ways. The example shown in Fig. 7 has three different trichlorosilane bubblers in order to allow production of multiple *p-n* junctions. The amount of compound injected into the hydrogen depends on the flow rate and the temperature of the trichlorosilane. Instead of bubbling the hydrogen through it, the trichlorosilane may also be injected by evaporating it continuously from the opening of a capillary. However, the required flow of liquid is very small and difficult to control.

In designing the reaction chamber, primary consideration must be given to the method of heating the seed (usually a monocrystalline silicon wafer). Externally heated tube furnaces do not produce satisfactory results because the inner walls of the tube are hot and offer a large area for deposition. Therefore heat should be generated inside the reaction tube and transferred to the seed by direct contact. In the chamber shown in Fig. 7, the seed wafer rests on an induction-heated graphite block. Induction-heated silicon pedestals in vertical reaction tubes have also been used with good success (Theurer, 1961). An alternative heating method is shown in Fig. 8, where the silicon seed lies on a thin graphite strip. Regardless of the heating method chosen, several precautions need to be observed. Materials used in the hot deposition zone must be very pure, chemically resistant, and must not give off volatile contaminants. The heating element should be designed to heat the seed uniformly because temperature gradients cause dislocations in the growing layer. Finally, the surface area of the heating element should not be large compared to the seed surface in order to avoid major concentration changes of the reaction gases.

In most practical cases, it is desirable to deposit layers of uniform thickness. The distribution of deposited material is strongly influenced by the gas flow pattern. If the gas flow is perpendicular to the seed surface, curved deposit profiles result. Gas flow parallel to the seed surface leads to wedge-shaped deposit profiles. Then, the thickness variation can be reduced by increasing the flow rate. This shortens the residence time of gases and prevents major concentration changes as the gas stream passes over the substrate. By proper adjustment of the flow rate, it has been possible to obtain deposits several inches long with only 1 to 5% over-all variation in thickness (Corrigan, 1961). However, the process efficiency is low; yields of 5 to 10% being typical.

To produce epitaxial layers with specific physical properties, the deposition temperature, reaction time, trichlorosilane concentration, and gas flow rate must be adjusted and controlled. Optimum operational data depend on the mode of operation as well as on the design and size of the deposition apparatus. Therefore these data are usually specific for an individual system. Unfortunately, most of the process variables affect the quality of deposits in several ways. Table 2 summarizes these influences.

At deposition temperatures from 1150 to 1300°C epitaxial deposits have been grown with dislocation densities of $\sim 10^4/cm^2$. At lower temperatures, polycrystalline growth frequently occurs. Rapid diffusion of impurities between the seed and the deposit occurs at higher temperatures. In addition, the chance of contamination from the heater increases sharply as the temperature increases.

For uniform deposition onto a single wafer in a small laboratory system (tube diameter of about 1.5 in.), a hydrogen flow rate of ~ 1 to 5 liters/min is required (Glang and Wajda, 1961).

The deposition rate is determined by the amount of trichlorosilane injected into the hydrogen stream. Growth rates of ~ 1 to 4 μ/min are possible in the temperature range mentioned. Higher growth rates cause random nucleation and polycrystalline growth. Typical trichlorosilane concentrations are 0.25 to 2 mole per cent, depending on the hydrogen flow rate used. In the range below 1 mole per cent the growth rate is a linear function of the trichlorosilane concentration.

With a fixed set of process variables, the thickness of an epitaxial crystal is a func-

TABLE 2

DEPENDENCY OF DEPOSIT PROPERTIES ON VARIOUS PROCESS PARAMETERS

	Deposit Properties / Process Parameters	Crystal Perfection (*a*) Mirror surface (*b*) Dislocations	Deposit Thickness (*a*) Controllable (*b*) Uniform	Deposition Rate (*a*) Controllable (*b*) Constant	Deposit Resistivity (*a*) Variable (*b*) Reproducible
Apparatus and Experimental Techniques	Flow pattern, tube, and heater geometry		√	√	
	Purity of system: (*a*) Trichlorosilane (*b*) Heater material (*c*) Hydrogen carrier gas (*d*) Stop cocks and lines	√			√
	Substrate preparation	√			
	Temperature gradient in substrate	√			
Process Variables	Deposition temperature	√		√	√
	Deposition time		√		
	Trichlorosilane feed rate	√		√	√
	Hydrogen flow rate		√	√	
	Impurity halide concentration in trichlorosilane reservoir	?			√

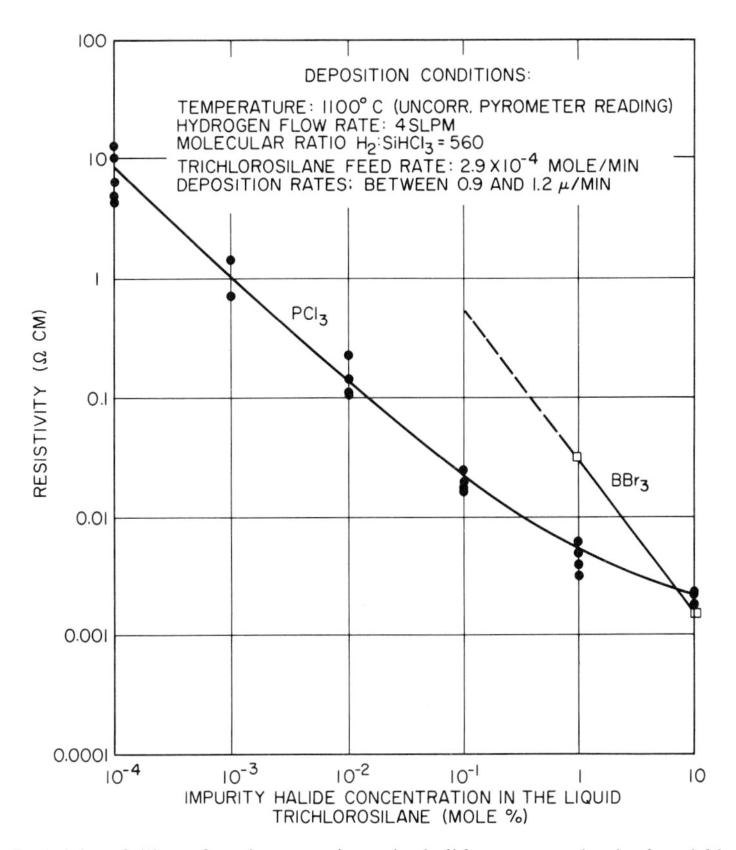

Fig. 9. Resistivity of silicon deposits versus impurity halide concentration in the trichlorosilane solution.

tion of time. Deposits as thin as 1 μ and as thick as ~200 μ have been grown with reasonable control.

By adding halides of group III or V elements to the reaction gases, impurities can be deposited simultaneously with the silicon. This can be done by evaporating the doping compound into a separate stream of hydrogen or by flowing hydrogen through a mixture of trichlorosilane and impurity halide. The latter method requires less experimental effort and is indicated in Fig. 7. To minimize fractionating effects during the evaporation of the solution, impurity halides with nearly the same vapor pressure as trichlorosilane are desired. Therefore PCl_3 and BBr_3 are often used. However, both compounds gradually accumulate in the liquid, so the composition of the gas changes constantly. This is reflected in resistivity variations in the deposit, so that only 20 to 50% of a doped solution is usable, depending on the required tolerances.

Both n and p-type deposits with resistivities from 0.001 to about 100 ohm-cm can be produced. Usually, the resistivity is more sensitive to changes in concentration of BBr_3 than that of PCl_3. A typical relationship between concentration and resistivity, valid only for the system shown in Fig. 7, is plotted in Fig. 9.

Silicon halides other than $SiHCl_3$ can also be used for hydrogen reduction processes and epitaxial deposition. Similar system design considerations apply for the tetrahalides and the operating conditions are not very different.

H. C. Theurer (1961) investigated the hydrogen reduction of $SiCl_4$, and a comparison of his results with data concerning the $SiHCl_3$ process shows that both compounds can be used in the same way and with equivalent results. The reduction of $SiBr_4$ has been studied by Sangster et al. (1957). Although oriented deposits have been obtained, a complete epitaxial process has not been reported. The hydrogen reduction of SiI_4 has also been attempted (Szekely, 1957), and epitaxial silicon films have been produced (Beattly et al.).

References

Anderson, R. L. (1960), *IBM Journal of Research and Development* **4,** 283.

Beatty, H. J., et al., U. S. Signal Corps, Contract No. DA-36-039-sc-87395.

Corrigan, W. J. (August 1961), *Proc. AIME Conf. Semiconductor Materials,* Los Angeles, California.

Glang, R., and B. W. Kippenhan (1960), *IBM Journal of Research and Development* **4,** 299.

Glang, R., and E. S. Wajda (August 1961), *Proc. AIME Conf. Semiconductor Materials,* Los Angeles, California.

Herrick, C. S., and J. G. Krieble (1960), *J. Electrochem. Soc.* **107,** 111.

Hirshon, J. M., U. S. Signal Corps, Contract No. DA-36-039-ac-72768.

Kempter, C. P., and C. Alvarez-Tostado (1958), U. S. Patent No. 2,840,489.

van der Linden, P. C., and J. DeJonge (1959), *Rec. des Travaux Chimiques des* Pay-Bas, **T78,** 962.

Litton, F. B., and H. C. Anderson (1954), *J. Electrochem. Soc.* **101,** 287.

Mayer, H. (1955), *Physik duenner Schichten,* Vol. II, 105, Wissenschaftliche Verlagsgesellschaft m.b.H, Stuttgart.

McCarthy, L. V. (1959), *J. Electrochem. Soc.* **106,** 1036.

Rubin, B., G. H. Moates, and J. R. Weiner (1957), *J. Electrochem. Soc.* **104,** 656.

Sangster, R. C., E. F. Maverick, and M. L. Groutch (1957), *J. Electrochem. Soc.* **104,** 317.

Schaefer, H., and J. Nickl (1953), *Z. Anorg. Allgem. Chemie* **274,** 250.

Szekely, G. (1957), *J. Electrochem. Soc.* **104,** 663.

Theurer, H. C. (1961), *J. Electrochem. Soc.* **108,** 649.

Wajda, E. S. and R. Glang (August 1960), *Proc. AIME Conf. Semiconductor Materials,* Boston, Mass.

Wajda, E. S., B. W. Kippenhan, and W. H. White (1960), *IBM Journal of Research and Development* **4,** 288.

6

Silicon Carbide

J. R. O'Connor

The nature of the silicon-carbon binary system makes five crystal growth methods possible: (1) sublimation, (2) epitaxial growth, (3) gaseous cracking, (4) growth from solution, and (5) growth from the melt. All these methods except the last have produced single crystals. Sublimation, the most successful method, is emphasized in this discussion, and the experimental procedures followed in each of the other methods are also discussed in detail. Recent interest in silicon carbide crystals stems from the fact that they are semiconducting with a large band gap. Various aspects of this material as a semiconductor were the subject of a recent conference. See O'Connor (1959) and O'Connor and Smiltens (1960).

Silicon carbide can crystallize in two basic forms, cubic and hexagonal. The cubic form, called β-SiC, may be formed by reacting silicon and carbon at temperatures below 2000°C. The hexagonal form has at least six variations of its basic structure. These types, designated α-I, II, III, IV, V, VI, are formed in the temperature range from 2400 to 2600°C. At one time it was believed that an α-to-β phase transformation took place in the solid at 2000°C. However, Slack (1960) has found that the transformation takes place with reasonable velocity only via the vapor phase.

Hexagonal silicon carbide has been produced commercially for many years in open electric furnaces charged with a mixture of coal, sand, and minor additives. Two kinds are common, "green" and "black." Green silicon carbide, the purest, contains iron, aluminum, nitrogen, calcium, and nickel as impurities. Black silicon carbide is often referred to as "electrical grade" and differs from green by containing a higher percentage of aluminum and iron.

Occasionally, cavities are formed in the furnace charges as a result of burning and settling, and large flat crystals may extend from the cavity walls. These crystals have one smooth, flat side with a hexagonal shape and one stepped side. The flat hexagonal faces tend to lie parallel to each other and perpendicular to the wall.

1. THE SILICON CARBIDE BINARY SYSTEM

Probably the simplest view that we may have of a substance is given by a knowledge of its phase-equilibrium relationships. This view eliminates details concerning the structure and physical properties of the substance. It is subject to precise definition and determination, and thus provides a sturdy framework upon which to build a more detailed understanding. Although SiC is one of the oldest of man-made materials, the exact nature of its binary phase diagram has proved to be quite elusive. However, in recent years progress has been made in its study.

Before one can synthesize a compound efficiently, the effects of pressure and temperature on its composition must be determined. The temperature-composition diagram for SiC has been derived by Nowotny (1954) and is shown in Fig. 1. The only

Most of these data were compiled while the author was a member of the Air Force Cambridge Research Laboratories, Bedford, Massachusetts.

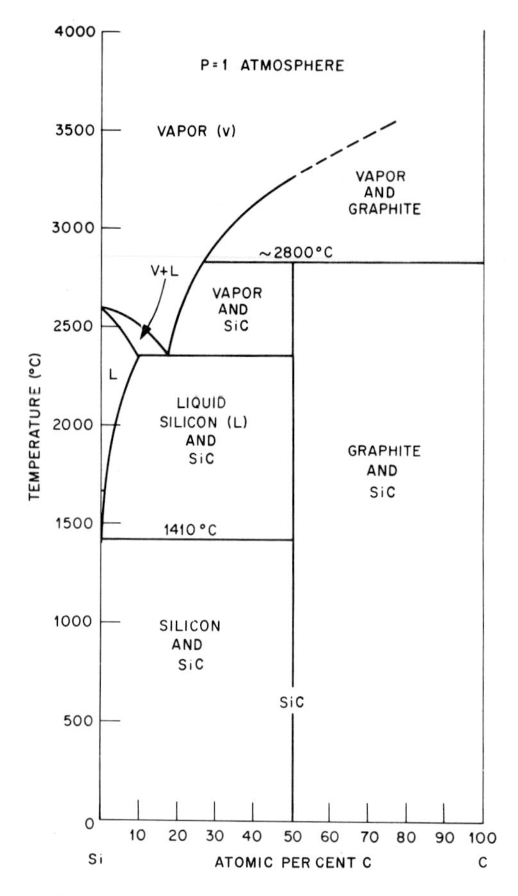

Fig. 1. Temperature-composition phase diagram after Nowotny (1954).

condensed phases that occur in the binary system are silicon, silicon carbide, and graphite. According to Hall (1958) and Scace and Slack (1959) the solubility of carbon in liquid silicon is very low at the melting point of silicon but increases progressively with temperature and reaches one atomic per cent at 2150°C. According to Lely (1955), SiC (crystals grown at 2500 to 2600°C) is stoichiometric; if there is a deviation from stoichiometry at all, it should be less than 10^{-5} atomic per cent. The graphite phase is quite pure since silicon is insoluble in graphite.

Pressure-Composition Phase Diagram

Above 2800°C a peritectic transformation occurs in which SiC decomposes into graphite and a silicon-rich vapor. In studying the phase relationships in this system, especially those involving the gaseous phase, the pressure parameter must also be considered. The phase diagram should be visualized as a prismatic body with composition-, temperature-, and pressure-coordinate axes as its edges.

By applying both statistical mechanics and van't Hoff equilibrium methods, Smiltens (1960) has been able to study various phase equilibria in the SiC system as a function of pressure. One isothermal section of the phase diagram is shown in Fig. 2. At some elevated pressure the distance *A–B* vanishes and a theoretical quadruple point is reached. At this point, vapor, liquid silicon, silicon carbide, and graphite are in equilibrium.

Mass Spectrometer Studies at High Temperature

Drowart et al. (1958, 1960) have studied the vapor phases of graphite + SiC and

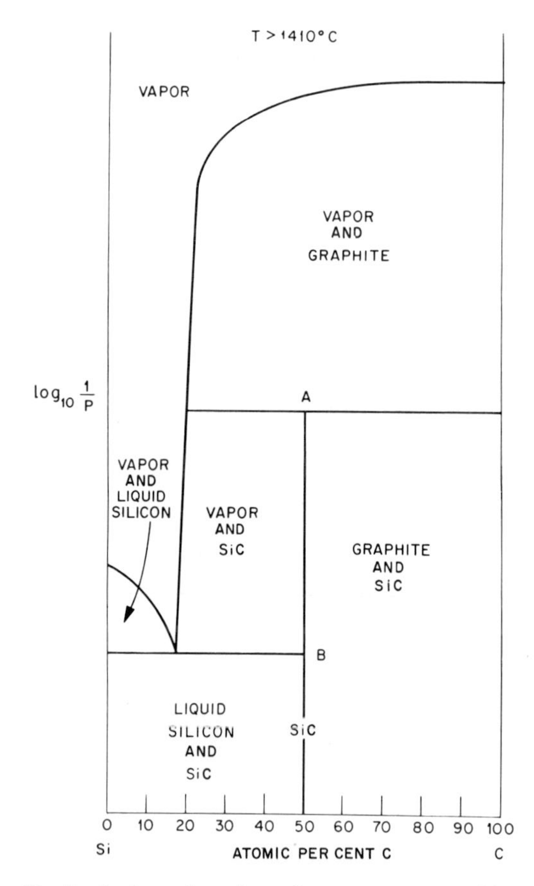

Fig. 2. Isothermal section of a pressure-composition phase diagram after Smiltens (1960).

TABLE 1

PARTIAL PRESSURES (ATMOSPHERES) IN THE SYSTEM SiC GRAPHITE

T	Si	SiC	SiC_2	Si_2	Si_2C	Si_2C_2	Si_3	Si_2C_3	Si_3C
2149	2.1×10^{-5}		1.9×10^{-6}	3.8×10^{-8}	1.4×10^{-6}				
2168	2.7×10^{-5}		2.5×10^{-6}	4.8×10^{-8}	1.9×10^{-6}				
2181	3.3×10^{-5}	2.2×10^{-9}	4.2×10^{-6}	6.7×10^{-8}	2.6×10^{-6}				
2196	4.1×10^{-5}		4.4×10^{-6}	1.1×10^{-7}	3.9×10^{-6}	8.5×10^{-9}			1.5×10^{-8}
2230	6.5×10^{-5}		6.5×10^{-6}	1.6×10^{-7}	5.1×10^{-6}	1.6×10^{-8}	3.2×10^{-9}	3.6×10^{-9}	1.8×10^{-8}
2247	8.3×10^{-5}	6.3×10^{-9}	1.1×10^{-5}	2.1×10^{-7}	8.1×10^{-6}				
2316	2.0×10^{-4}	1.9×10^{-8}	3.1×10^{-5}	7.0×10^{-7}	2.2×10^{-5}	7.5×10^{-8}	1.6×10^{-8}	1.7×10^{-8}	8.5×10^{-8}

TABLE 2

PARTIAL PRESSURES (ATMOSPHERES) IN THE SYSTEM Si + SiC

T	Si	Si_2	Si_3	Si_4	SiC_2	Si_2C	Si_3C
1703	8.1×10^{-7}	8.1×10^{-9}	4×10^{-9}	1×10^{-9}			
1765	2.1×10^{-6}	2.4×10^{-8}	8×10^{-9}	3.5×10^{-9}		4.0×10^{-8}	
1825	5.2×10^{-6}	6.8×10^{-8}	2×10^{-8}	1×10^{-8}		5.0×10^{-8}	1.4×10^{-9}
1890	1.4×10^{-5}	1.9×10^{-7}	5×10^{-8}	3×10^{-8}		1.3×10^{-7}	9.0×10^{-9}
1935	2.3×10^{-5}	3.6×10^{-7}			1×10^{-9}	3.5×10^{-7}	
1960	3.1×10^{-5}	5.2×10^{-7}			1.5×10^{-9}	5.0×10^{-7}	
2000	5.0×10^{-5}	8.8×10^{-7}				1.2×10^{-6}	
2100	1.5×10^{-4}	3.0×10^{-6}			3×10^{-8}	6.0×10^{-5}	
2160	2.9×10^{-4}	6.4×10^{-6}			1.5×10^{-7}	1.3×10^{-5}	

Si + SiC at elevated temperatures by using a Knudsen cell and mass spectrometer. The cell was heated by electron bombardment. Its temperature was measured with an optical pyrometer seen through a quartz window. An electrically grounded plate containing a movable slit was included between the Knudsen cell and the ionization chamber of the mass spectrometer. This beam-defining slit provided a method for distinguishing between ions formed from the vapor effusing from the cell and those formed from residual gases in the apparatus. In addition, this slit made it possible to distinguish between vapor effusing through the orifice of the cell and vapor evaporating from the surface of the cell.

The Knudsen cell was made of high purity graphite inserted in a tantalum container. Both cells had a knife-edge orifice 1 mm in diameter. In order to insure thermodynamic equilibrium, the number of small SiC crystals used was such that the ratio of the orifice area to the geometrical area of the crystals was less than 0.001.

The measured pressures of Si, Si_2, Si_2C, SiC_2 and other less important gaseous molecules, summarized in Table 1, are believed to be accurate within a factor of two.

At this point it is interesting to note that the partial pressure of the molecule SiC is three orders of magnitude lower than that of either SiC_2 or Si_2C.

Silicon was inserted in graphite cells previously heated in the presence of SiC crystals up to 2500°K. The silver, used to calibrate the pressure, was contained in a small graphite thimble to prevent direct contact with silicon. Pressures for a number of Si_x and Si_xC_y molecules are given in Table 2. When the increase in temperature is taken into account, the partial pressures of Si_2, Si_3, and Si_4 agree with the pressures calculated from the relative ion intensities determined by Honig (1954).

Here again it is interesting to note that there is no appreciable number of SiC molecules in equilibrium with either liquid Si or condensed SiC.

Solubility Studies at High Temperature

To investigate the behavior of SiC at elevated temperatures and pressures, the furnace shown in Fig. 3 was built by Scace and Slack (1959). A water-cooled pressure vessel surrounded the graphite furnace assembly. The vessel held a pressure of as

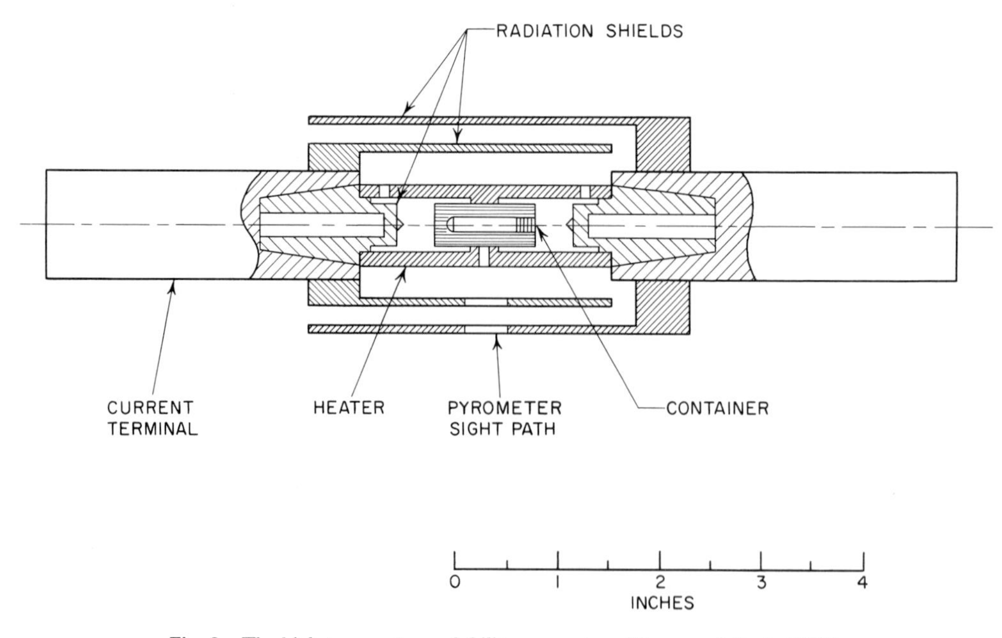

Fig. 3. The high temperature solubility apparatus of Scace and Slack (1959).

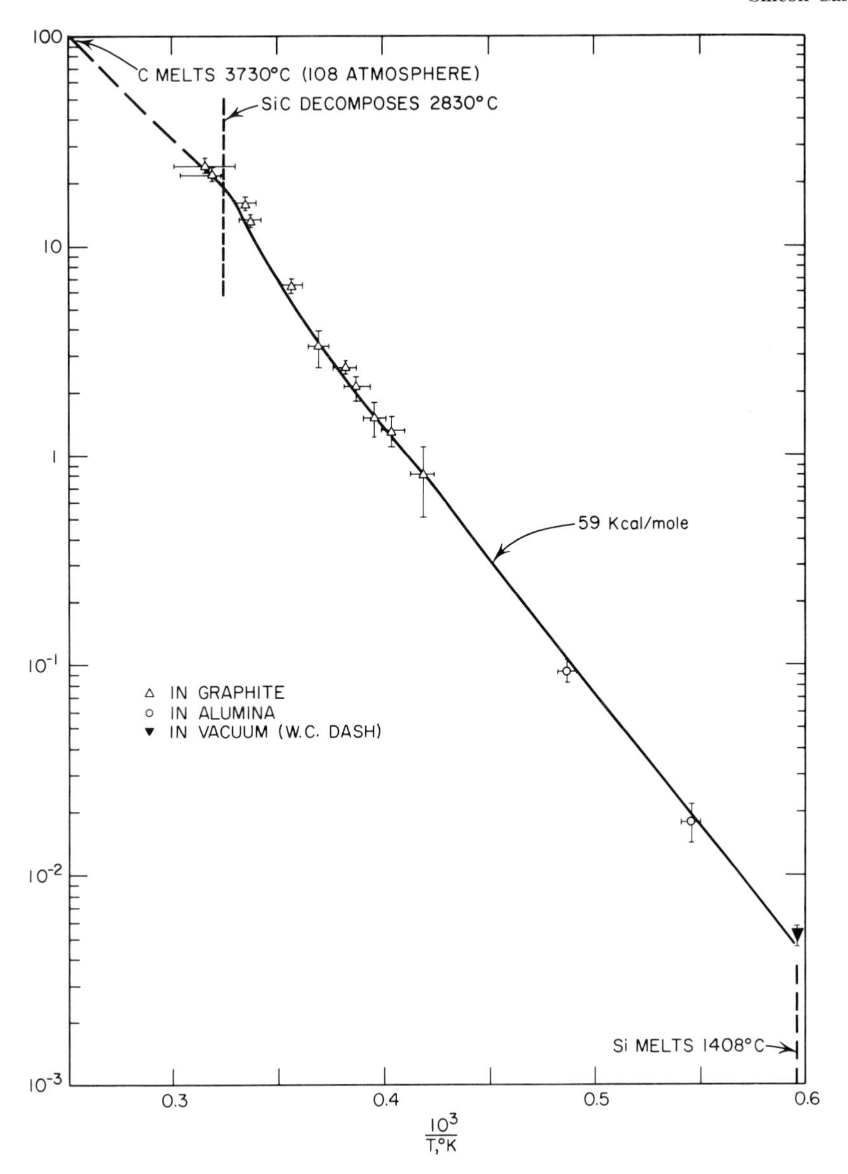

Fig. 4. The solubility (atomic per cent) of carbon in liquid silicon as a function of reciprocal temperature.

much as 1000 lb/in.² This dense atmosphere retarded the evaporation of silicon, which otherwise would be quite troublesome. All parts were made of graphite. The two current terminals supported a tubular heater, two radiation shields, and a container that was used in the experiments on the solubility of carbon in silicon. The container was supported in such a way that it carried no current. About 50 kw was available to heat this assembly, permitting temperatures of ~3500°C to be reached.

Optical pyrometry is the only reasonably accurate way of measuring temperature above 1900°C, which is the limit for platinum-alloy thermocouples. A sight tube and window were provided in the side of the pressure vessel, and openings were left in the radiation shields and heater tube to allow measurement of the internal temperature of the furnace.

In order to determine the solubility of carbon in Si, the inner container was filled with Si, sealed with a tight-fitting plug,

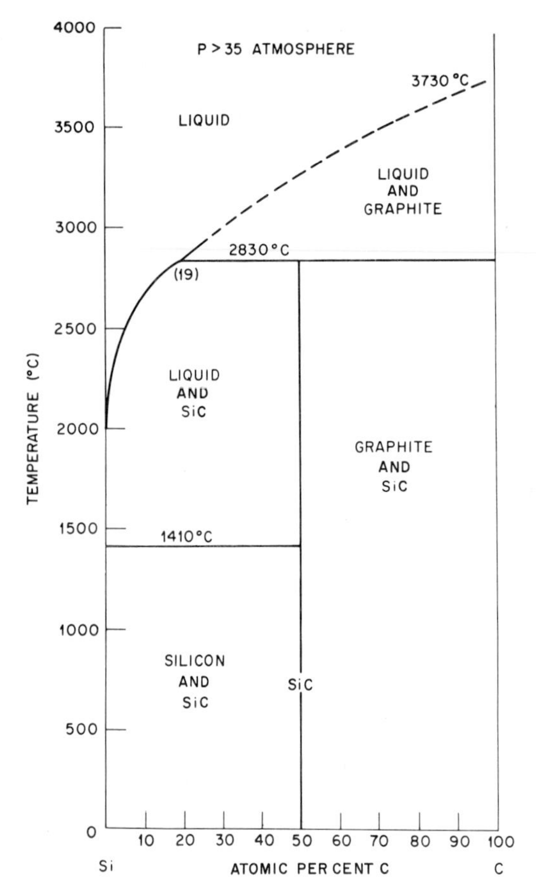

Fig. 5. Temperature-composition phase diagram of Scace and Slack (1959).

placed in the furnace, and heated to a temperature between 2100 and 2900°C. Thermal equilibrium was achieved quite rapidly, and the sample was held at a fixed temperature for 1 or 2 min, after which the power was shut off. Then the furnace temperature decreased rapidly through radiation cooling and the Si froze.

The solubilities obtained are summarized in Fig. 4. Of particular interest are the solubilities of carbon in liquid Si at the Si melting point, 5×10^{-3} atomic per cent (5×10^{17} C/cm^{3}Si), and at the decomposition temperature of 2830°C. This is the maximum temperature at which SiC could be grown from an Si solution, and the carbon content of the melt here is 19 atomic per cent. This high solubility indicates that growth from solution is feasible. The vapor pressure of Si over liquid Si and over SiC at this temperature is of the order of one atmosphere and rising rapidly; therefore the need for a pressurized atmosphere to maintain stable conditions for any length of time is apparent.

Another method for plotting the data obtained by Scace and Slack is on a temperature versus composition phase diagram, Fig. 5. Here the behavior of SiC at the decomposition temperature may be clearly seen.

2. METHODS FOR GROWING SILICON-CARBIDE CRYSTALS

The occasional growth of large SiC crystals in commercial furnaces has already been mentioned. These crystals grow from the vapor created by the decomposition of SiC at 2800°C. The reaction, which takes place at atmospheric pressure, is shown in Fig. 1. Methods using this phase transformation are described in a later section, *Growth by Sublimation.*

A low pressure furnace can be used to grow SiC (~20 mm Hg) from the vapor phase onto an oriented crystal substrate. This process, called *epitaxial growth,* is also described.

A somewhat artificial manner of growing SiC from the vapor consists of bringing Si- and C-bearing vapors together at high temperatures. In this method, usually called *growth by gaseous cracking,* a gas such as trimethychlorosilane can be decomposed at elevated temperatures, or a mixed gas such as silicon tetrachloride and toluene can be reduced by hydrogen.

Figures 4 and 5, which show the solubility of carbon in liquid silicon as a function of temperature, indicate that crystal growth from solution is feasible. The data of Scace and Slack (1960) show that, at pressures greater than 35 atmosphere, liquid silicon containing 19 atomic per cent carbon is in equilibrium with SiC at 2830°C. Under these conditions a maximum growth velocity would be obtained. Experiments using these methods are described in the section, *Growth from Solution.*

At greater pressures (perhaps in the order

of 100,000 atmospheres), the solubility of carbon in liquid silicon is increased until a concentration of 50 atomic per cent is reached. At this point solid and liquid phases of SiC are in equilibrium and direct crystallization from the melt can occur. Field (1961) has indicated that experiments of this kind are being conducted in an apparatus originally designed for the growing of diamond from metallic solutions. This method is not discussed.

3. CRYSTAL GROWTH MECHANISMS

The general theories of crystal growth distinguish between two major mechanisms: (1) the surface advances by the lateral motion of steps or (2) the growth interface advances normal to itself without steps (Burton et al., 1950). The actual mechanism that is operative depends on the nature of the surface. For a diffuse surface (Jackson, 1958), that is, a surface in which the change from one phase to the other is gradual, crystal growth occurs normal to the interface. In other words, the interface advances normal to itself without the need of steps. For a singular surface (Cabrera, 1959), that is, an interface for which the surface tension as a function of orientation has a sharp minimum, the interface requires steps in order to advance normal to itself. Herring (1951) has shown that surfaces which are neither diffuse nor singular will lower their free energy by forming facets.

The Growth of α-SiC

In considering the growth of a single crystal of SiC, one must take the following factors into account. The crystals of hexagonal types (the α forms) are usually in the form of thin platelets, the faces of which are singular (0001) planes. The edges of the plates are composed of $\{21\bar{3}0\}$ planes, probably diffused. In addition, the (0001) planes are smooth, but the edges must be atomically rough because the stacking faults act as low-energy twin boundaries. Each twin plane will introduce one re-entrant corner on these faces. Therefore with α-SiC both types of crystal growth are operative; on the smooth (0001) planes, steps are necessary for crystal growth, whereas the edges of these crystals can readily advance normal to themselves into regions of supersaturated vapors.

The nature of steps on the (0001) faces is a matter of some concern. The steps can form by classical two-dimensional nucleation if $\pi d^2\sigma < 50kT$, where d is the step height and σ the surface free energy. If the minimum values of $d = 10$ A and $\sigma = 3000$ ergs/cm^2 are used, it is obvious that the inequality is not true for temperatures as high as 2800°. However, Amelinckx (1960) has reported that steps can be observed on all sizable α-SiC crystals and that these steps are the results of operative screw dislocations during crystal growth. The conclusion is that growth normal to the (0001) faces proceeds by screw dislocations.

The Growth of β-Silicon Carbide

The growth of cubic SiC (the β form) should be similar to that of other zinc-blende semiconductors such as Ge and Si. In recent studies of Ge grown from the melt (O'Connor, 1961), it appears that {211} planes are diffuse, whereas {111} planes are singular. Planes with other orientations have faces and are partially diffuse. Steps are usually required for growth on {111} planes. With Ge, however, the surface free energy is substantially reduced by the liquid (and by impurities in the melt at the interface) to approximately 100 ergs/cm^2, so that two-dimensional nucleations are quite feasible. These facts probably account for the growth of β-SiC from solution. Growth from the vapor is more complicated.

Mass Transport Phenomena

In addition to complexities of growth, it has already been mentioned that there is very little SiC in equilibrium with either liquid Si or solid SiC. Silicon carbide does not seem to exist as a monomer in the vapor in any appreciable concentration. Therefore the composition of the vapor species arriving at the growing interface must somehow disproportionate before solid SiC can be formed. Since the vapor is silicon-rich, excess silicon is present and must be removed. We are therefore confronted with a mass transport problem that would not exist if we were able to grow from a system where the composition of the vapor was the same as that of the growing crystal. Since the concentration of the various vapor species is extremely sensitive to temperature, rather accurate temperature control as well as temperature uniformity are required during growth.

With growth from a solution containing an excess of silicon, we do not know what the solution species are. However, we do know that the composition of the solution is different from that of the desired SiC crystal.

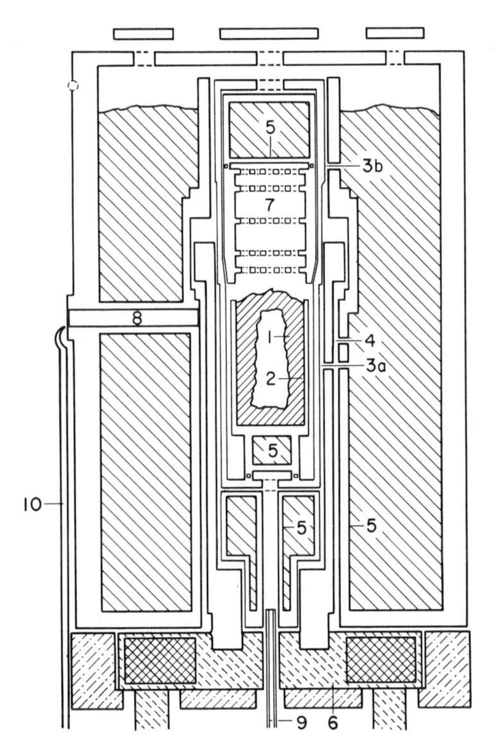

Fig. 6. Diagram of Lely furnace. (1) SiC for sublimation; (2) graphite crucible; (3) graphite crucible made of two parts for protection of SiC; (4) heating element, a graphite tube split three-fourths of its length; (5) insulating body made with graphite lampblack; (6) water-cooled copper electrodes; (7) graphite network; (8) tube for pyrometer; (9) tube for introducing protective gas; (10) protective gas inlet. The furnace is surrounded by a water-cooled vacuum-type envelope.

According to Scace and Slack, the maximum concentration of carbon that can be dissolved in silicon below the decomposition temperature of SiC is approximately 19 atomic per cent. Therefore we again have a mass transport problem. Since mass transport is generally slower in the condensed phase of a liquid than in a vapor, we might expect growth from solution to be slower than growth from the vapor phase. However, the fact that the solution is much more dense than the vapor helps alleviate the mass transport problem. Indeed, it appears that growth from silicon solutions is at least as fast as, and perhaps somewhat faster than, growth by sublimation (or by gaseous cracking).

The complex problems outlined in this section make it impossible to describe in any detail either the exact mechanism of crystal growth or the kinetics of the various processes. Therefore the remainder of the discussion is confined to methods of growing single crystals.

4. GROWTH BY SUBLIMATION

It has been pointed out that the large single crystals of SiC sometimes occur in cavities accidentally formed when the more volatile constituents of the charge are vaporized. Patrick (1957) has made an exhaustive study of these crystals, and of the *p-n* junctions that are occasionally found in them.

Lely (1955), modifying equipment described by Kroll (1946), constructed the furnace shown in Fig. 6 to grow single crystals from the vapor phase using commercial SiC as a starting material. The crystals he obtained were similar to those found in commercial charges, having one nearly flat hexagonal face and one face resembling a fine staircase. Adaptations of Lely's process made by Chang and Kroko (1957) and Hamilton (1958) have led to the growth of high purity single crystals with parallel faces, modifications that will be discussed in some detail.

In Lely's furnace, as in commercial furnaces, crystals are grown inside cavities formed in bulk SiC but with much greater precision and control. The crystals are grown from SiC vapor, which sublimes and supersaturates the interior of the cavity. Nucleation occurs on the inner wall. Vapor pressure in the cavity is maintained by virtue of the constant decomposition of charge material at a temperature greater than that of the cavity; an equilibrium vapor pressure of the sort normally assumed in closed systems does not exist.

In such a system, the maintenance of vapor pressure can be improved by making the charge material as dense as possible, but there are very serious practical limitations to this approach. For the most part little can be done beyond careful preparation of the charge to assure maximum density and homogeneity. Since crystal growth depends on the ability of the crystal to dissipate the heat of vaporization of incoming molecules or atoms, temperature and, in particular, temperature gradients are very important. In fact, they constitute the only controllable parameters in the growth process. Moreover, each crystal can achieve maximum size and perfection only if it has sufficient unrestricted room for growth. Therefore it is desirable to achieve some control of the number and spacing of nucleations.

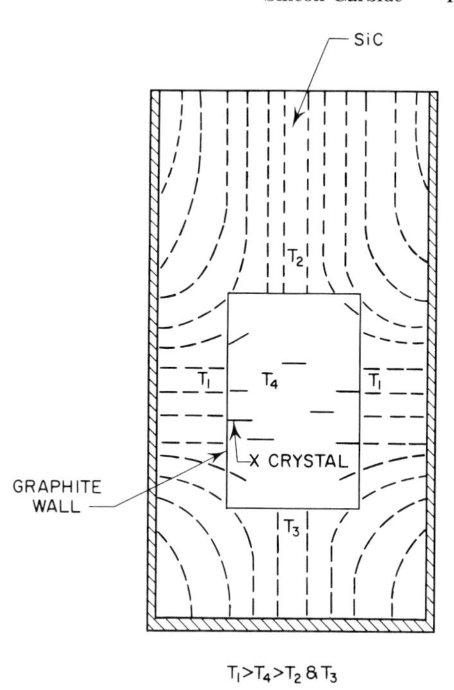

Fig. 8. Silicon carbide growth cavity.

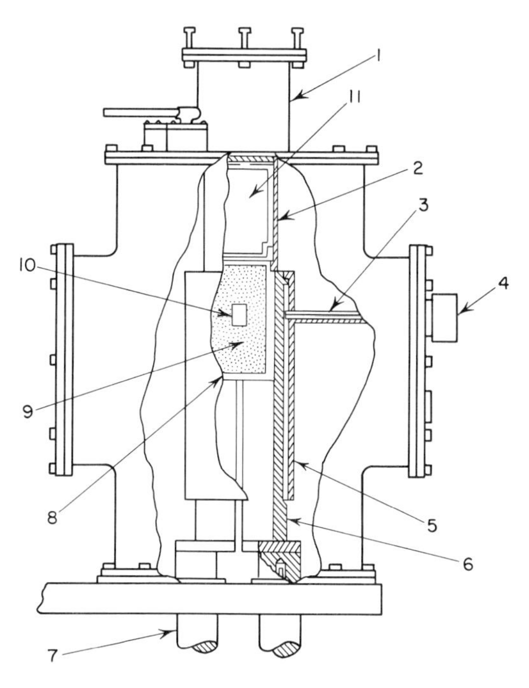

Fig. 7. High temperature furnace designed by Chang and Kroko (1957) and described in the text.

Growth in a Closed Cavity

Single crystals of SiC usually grow from a substrate of SiC crystallites that presents to the vapor a number of nucleation centers of many orientations and varied sizes. Study of a cross section parallel to the growth direction reveals that the material crystallizes along definite paths or lines, resulting in elongated crystals or needles. The rodlike growth is suggestive of the heat flow lines that one might expect.

Figure 7 shows the general details of the Chang-Kroko system and the location of the crucible in the furnace. Figure 8 is a schematic cross section of a typical furnace charge consisting of a circular graphite crucible filled with SiC in which there is a central cavity. The crucible is heated so that the sides are hot and the ends are cool; hence heat tends to flow along the dotted lines in the illustration. The striking re-

Fig. 9. Planar α-SiC crystals as grown in a closed cavity by Chang and Kroko at the Westinghouse Electric Corporation (1959).

semblance between these schematic heat flow lines and the growth pattern of the crystals is shown in Fig. 9.

Well-shaped single crystal platelets can be grown only if a limited number are present. Therefore it is necessary to make use of a catalytic substrate for control of nucleation. Because of the temperatures involved and the high reactivity of silicon vapor, graphite is considered the most likely prospect. Graphite offers other advantages for control of the growth process; differences in thermal conductivity permit control of the temperature gradient within the growth region, and the rigid form of the cavity permits the use of finely powdered SiC and/or mixtures of silicon and carbon for the charge material. Experiments with graphite substrates show that a wall thickness of 0.003 to 0.007 in. is satisfactory. Thicker tubes produce no growth, whereas thinner ones vaporize. It has been observed that crystals generally grow from the graphite surface with no connection to the bulk SiC in the charge. Between crystals, the substrate is often pure graphite devoid of crystals. In the remaining instances, the graphite substrate is transformed to SiC.

When the growth temperature is increased slightly for each of a series of runs, the substrate undergoes a sequence of changes in surface appearance. At the minimum temperature for growth, myriad nucleations of microscopic size cover the substrate between crystals, whereas at the highest growth temperatures no nucleations other than those of full grown crystals appear. In an effort to control the number of nucleations by controlling vapor pressure within the growth cavity, a variety of graphite tubes with certain sections thinned to allow free passage of SiC vapor from the base material have been tested. Generally, no crystals grow on the surface of the thicker sections of graphite. One interesting exception consisted of a reasonably thick tube with very thin slits oriented horizontally and vertically. At normal growth temperatures, no crystals were produced in such tubes; however, when the temperature was increased some crystals were observed.

A graphite nucleation substrate with slits or thin sections is used to restrict nucleation. Modifications intended to encourage nucleation have also been tested. The first of these is that of placing a single crystal seed inside the growth cavity. Rather than being a place for preferred growth, however, the crystal actually becomes a point for new nucleation, and many single crystals grow in more or less random directions from this seed. A more successful experiment involved seeding of crystals by vapor deposition of silicon strips on the surface of a graphite substrate. This deposition supplies microscopic nuclei of SiC. Crystal growth is indeed promoted by such strips.

Since the growth of crystals is completely dependent on the thermal conditions within the charge, careful consideration must be given to the isothermal lines. In Fig. 8, a schematic of the growth cavity, appropriate heat flow lines are shown as dashed lines. The flow pattern in the central portion of the charge is significantly modified by the presence of a graphite tube whose thermal conductivity is significantly higher than that of bulk SiC. Heat flow lines, which might normally be curved in the central region if there were no cavity, are constrained by the pseudo-isothermal surface to be perpendicular to the graphite wall of the growth chamber. As a result, the thin graphite nucleation surface extends into colder sections of the furnace, and there is some tendency for that surface to assume a temperature lower than that of the hottest section. The degree to which this surface becomes isothermal is readily determined by observing the angle at which the growth lines intersect it. In this instance, it will be assumed that they intersect at nearly 90°. We may now conclude that this surface is nearly an isothermal surface, which may be either cool enough to allow condensation of SiC or hot enough to prohibit the formation of any critical nuclei during the growth period.

Assuming that a crystal has grown to some significant size, it is possible to determine the mechanism by which the heat of vaporization of incoming molecules is dissipated. There are three possibilities: radiation from the faces of the crystal; conduction to the substrate from which the crystal grows; conduction and convection to the gas inside the growth cavity. The nucleation substrate can and often does support growth at a temperature where new nucleations cannot form. Therefore it is reasonable to assume that it is hotter than the crystal proper and can conduct no heat away from the crystal. If a significant amount of the heat of condensation were conducted to the gas in the growth cavity, one would expect to find equal numbers of crystals oriented in all directions in all charges since no single orientation would be capable of dissipating heat to the gas any better than any other. Experience, however, has shown that usually most crystals, and certainly the largest ones, grow with plane faces parallel to the hot-zone normals (that is, parallel to the top and bottom of the crucible as shown in Fig. 8). When a large number of crystals is nucleated and grown in a cavity, it is observed that there is an apparent contradiction to this rule. However, such is not so, for it is entirely possible for any one crystal to radiate energy to several others rather than to the cold object directly. Computations have shown that, if certain reasonable simplifying assumptions are made, the amount of energy transferred at temperatures in the vicinity of 2400°C considerably exceeds that removed by conduction or convection.

Since the energy radiated is a function of the difference of the fourth powers of the temperatures of the high and low temperature regions, and is proportional to the area of the radiating and receiving source, one would expect crystals whose broadest areas are aligned parallel to the cold sources to be capable of radiating more energy and hence to grow more in a fixed period of time. Practically all charges have confirmed this analysis. Using the foregoing considerations, the following description of crystal growth may be postulated (Westinghouse, 1957).

A crystal is shown in Fig. 10 growing from a substrate towards the center of the cavity. Molecules arriving at *a* give up heat h_1, which flows into the crystal and is radiated from the surfaces *b* and *c* to the

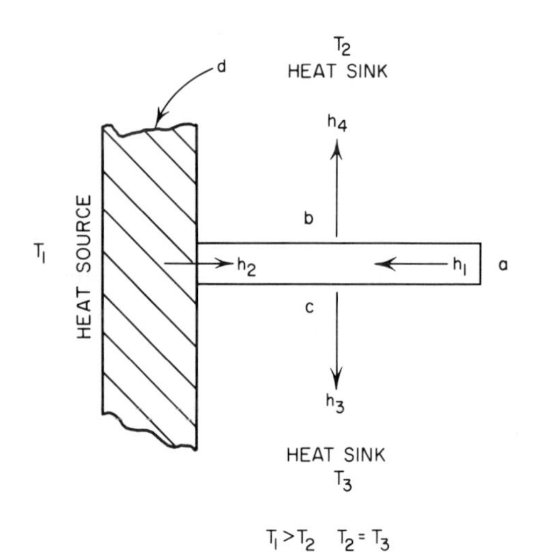

Fig. 10. Heat flow during crystal growth.

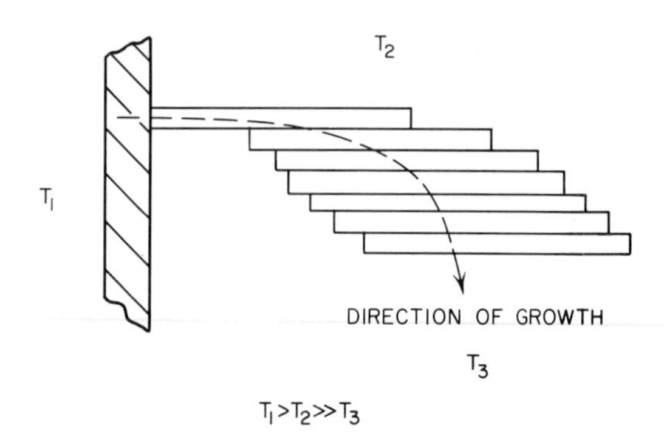

Fig. 11. Schematic drawing of crystal growth due to an unsymmetrical temperature distribution.

cold ends. Heat h_2 from *d,* the hot substrate, also flows along the crystal and is radiated from *b* and *c*. Molecules arriving at *b* and *c* may join the lattice, but these areas are not preferred growth regions and it may be assumed, as in accepted growth theory, that most molecules migrate to *a* and join the lattice there. In any case, when $T_2 = T_3$, temperature and vapor pressure at *b* and *c* are equal and hence growth rates are equal. If T_2 is greater than T_3 or vice versa, the vapor on the side nearest the cold region is more supersaturated, and hence growth proceeds more rapidly on that side, giving rise to a crystal like that in Fig. 11. When T_2 and T_3 are equal but much lower than T_1, growth of both sides will proceed at equal rates, producing a thick crystal. Thin planar crystals generally have a ratio of preferred to perpendicular thickness of the order of 20 to 1. The shape of the crystal in Fig. 11 approximates that of crystals found in commercial charges.

The Nucleation of New Crystals

The heterogeneous nucleation of SiC crystals on graphite or on other substrates is presumed to occur by the formation of small critical nuclei of the order of 10 to 100 atoms, whose size is increased by the migration of atoms or molecules along the surface of the substrate. An atom or molecule joining a nucleus must give up energy and, on such a small atomic scale, it is unreasonable to expect this energy to be taken away by radiation, especially when one considers the extreme roughness of the graphite surface. Initially then it can be assumed that the heat of vaporization of the critical nucleus and the heat of vaporization during the initial stages of growth are absorbed by the substrate. Assuming this to be so, one would conclude further that nuclei can form only during the period when the substrate is below the vaporization temperature of SiC. This rather unusual conclusion has been substantiated by the following experiment.

Normally, SiC crystals are blue or green, depending upon the nature of the impurity present in the vapor. When aluminum or boron is present, the crystals are blue. When nitrogen is present, the crystals are green. When the impurity present in the gas is changed during the growth of a crystal, the crystal color changes abruptly at the point of impurity change. It is possible to allow an impurity, nitrogen for example, to be admitted for short periods of time, thereby marking the growth process. In one typical run, nitrogen was admitted for periods of approximately 5 min each half-hour of the growth period. Upon inspection, it was found that a great number of crystals were present. Some were very small, of almost microscopic dimensions, whereas others grew as large as $\frac{1}{4}$ in. In every crystal, however, all nitrogen marks were observed. Because the first mark was made within the first 10 min at the growth temperature, it can be seen that all crystals that grew to any appreciable size were nucleated within a period of approximately 10 min at the very beginning of the run. This gives credence to the supposition that the heat of vaporization for the formation of the critical nucleus is absorbed by the heat capacity of the substrate before it reaches the growth temperature, which may be above the decomposition temperature of SiC.

Crystal Purity

Despite careful preparation of both starting materials and equipment, crystal purity has been improved only by about a factor of ten over the best available commercial stock, except for one or two cases that are not well understood. Although it is believed that nitrogen is the dominant impurity in the purest samples, analysis techniques are not adequate to prove this. Other significant impurities are aluminum, iron, titanium, calcium, magnesium, copper, and vanadium.

The most obvious approach to pure crystal production is that of utilizing purer starting materials, and to this end use has been made of semiconductor grade silicon and spectrographic carbon. Although crystals produced from these materials are somewhat better than commercial green crystals, the beginning purity level has been markedly reduced by impurities in the furnace. Figure 7 shows many furnace parts that may contribute to contamination.

In an attempt to clean the furnace, the heavy metal impurities have been exposed to chlorine at temperatures in excess of 2700°C for several days before a growth run with noticeable but insufficient effects. Charges fired after such a cleaning treatment produce bright green crystals of purity comparable to commercial green SiC.

Removal of nitrogen has been attempted by putting titanium inside the furnace as a getter and as a denitrogenizer in the inlet gas flow. Although no absolute analysis has been made, variations in color intensity and in resistivity measurements indicate purity no more than ten to one better than commercial stock.

Furnace Design and Operation

The high temperatures (2400 to 2700°C) required for the growth of SiC crystals are not obtained in ordinary laboratory furnaces. Because the reactivity of silcon vapor precludes the use of metallic resistance heaters, graphite heating elements have been developed. The furnace shown in Fig. 7 is described as an example.

The heater (6) is a hollow carbon tube slotted into halves that are joined at one end. Current can be passed in one side through the electrodes (7) and out the other. The hot zone of the furnace is located near the free end of the heater where the current path reverses direction. A heater made in this manner requires a minimum of power to produce a given temperature. Insulation is provided by finely powdered graphite loosely placed in the surrounding container. To keep the heater from shorting across the slot, a graphite baffle (5) is hung on the outside of the heater from a ridge at the top. The inside of the heater is protected from silicon vapor by a thin graphite liner (2), which also serves to hold the crucible (8) at the proper position in the hot zone. The crucible is removed through a hole in the top of the furnace (1), sealed during operation by a graphite container filled with powdered graphite. A carbon sighting tube (3), which opens near the heater, is constantly supplied with argon to keep the sighting channel clear.

The connection between the graphite heater and the copper electrode must be cooled to at least 300°C to prevent electrical deterioration of the contact. At the same time, it is important for safety reasons that the cooling medium be prohibited from entering the furnace in the event of failure of the cooling supply and subsequent melting of part of the electrodes. The only insulation found to be satisfactory for this purpose is a form of finely powdered graphite, which is used wherever the temperature is too high to allow the use of ordinary ceramic bricks.

Access to the furnace is through the top. During operation it is necessary to close this passage with a good insulator. The temperature above the hot zone gradually drops, and because the density of SiC vapor in this configuration is less than that of argon, there is a tendency for vapor to rise in the furnace and condense. Unless sufficient clearance is provided, the condensed material may freeze and join the insulator to the lining. An insulator containing a solid top section filled with powdered graphite and a lower section made up of stacked $\frac{1}{8}$ in. disks has been found to be the most suitable.

To obtain a representative temperature reading, the graphite sighting tube allows visual observation of the outside of the heating element. It is not practical to view the crucible itself because of the high density of vapor near it and consequent clogging of the sight tube by condensation. The sight tube itself must be kept very thin in cross section in order to minimize the amount of heat conducted away from the hot zone. A sight tube with a hole $\frac{3}{8}$ in. in diameter and walls $\frac{1}{8}$ in. thick produces excessive conduction.

The heater is constructed so that the hot zone extends for a distance of approximately 3 in. below the uppermost extremity of the slot. Under normal conditions, a heater can be operated for at least 300 hr at 2500°C in an inert atmosphere. Failure usually occurs from deterioration of the current-carrying sections. Good electrical contact surface to the heater is made through a plated copper coating. Two gaskets of copper cloth are used between the heater and the electrodes to compensate for misalignments.

The graphite baffles used to insulate the heater from the silicon carbide vapor have a pronounced effect on the life of the heater. They should be designed so that no part of the heater is directly exposed to the crucible. This is accomplished by using overlapping joints. Erosion of baffles, particularly the liner containing the crucible, is rather rapid, allowing only 50 to 60 hr of operation before replacement is required.

The power consumption of these furnaces varies somewhat, depending on the quality of the electrical connection and the thermal conductivity of the insulation. For low thermal conductivity the powdered graphite must be very loosely packed. Optimum power consumption is about 10 kw for a temperature of 2400°C in the hot zone. Under poor conditions it has been as high as 15 kw. Current input varies between 1200 and 2200 amperes. During the initial heating period of about $\frac{1}{2}$ hr, the power input is about 25 kw, and is then decreased slowly. Under normal operating conditions, the electrodes reach a temperature of approximately 350°C, the adjoining graphite reaches about 750°C, and the thermal drop in the insulation is approximately 1600°C.

Mass Transport in a Static Gas

Mass transport in a static gas occurs by interdiffusion. It may be accelerated by reducing the gas density through a reduction of pressure or through an increase in temperature. The rate of diffusion also depends on the ambient gas. For example, aluminum diffuses through hydrogen about four times faster than through argon. Crystals grown in hydrogen might consequently be expected to contain less aluminum than crystals grown under identical circumstances in argon.

In practice, diffusion takes place in an ambient of many components with strong temperature and concentration gradients. Under these circumstances, consideration must be given to thermal diffusion and, if convection currents are present, to the cascade effect. Because of thermal diffusion, heavy molecules diffuse to cooler regions of the system, leaving only the lightest molecules or atoms in the hottest section. This tendency is counteracted by the resulting imbalance in concentrations. Because of the cascade effect, this reverse diffusion is opposed by convection currents, permitting more marked separation of the components. The cascade effect is most pronounced in systems with geometry that favors selective convection currents.

When the graphite furnace operates in argon, thermal diffusion will not be pronounced, since the masses and volumes of the atomic or molecular species are similar. In hydrogen, the species are very different. It is believed therefore that thermal diffusion may affect the contamination of crystals grown in hydrogen. Evidence of this was obtained by comparing the resistivity and optical absorption of crystals prepared in argon and hydrogen when both gases contained about the same amount of nitrogen. The samples prepared in hydrogen contain considerably less nitrogen than those prepared in argon. Furthermore, the amount of nitrogen incorporated into crystals prepared in hydrogen depends on position in the well, suggesting the existence of nitrogen concentration gradients. Finally, the difference between the crystals decreases if lids and baffles are used to reduce convection currents. No such effects are observed in argon. It seems then that the two ambients differ greatly in the way impurities find their way into the growing crystals.

A temperature-dependent solubility phenomenon has also been observed. The amount of nitrogen incorporated into SiC crystals depends markedly on the growth temperature. The effect is too large to be ascribed to the diffusion effects in the ambient. The resistivities of a number of crystals that were prepared in argon at various temperatures were measured at room temperature. Preliminary data (Fig. 12) show that the mean resistivity increases rapidly with increasing temperature of preparation. It is believed that the data in Fig. 12 reflect a very large and genuine change in the nitrogen donor density. This claim is strengthened by observations of marked differences in optical absorption associated with nitrogen centers and by subtle variations in the photoluminescence of the various crystals. The changes are too great to be explained by diffusion effects in the ambient or by growth rate changes or segregation. Similar effects have been observed for hydrogen ambients.

The method described here produces crystals in the form of planar platelets of $\frac{1}{4}$ in. maximum dimension and 0.01 to 0.03 in. thickness. Larger crystals may be ob-

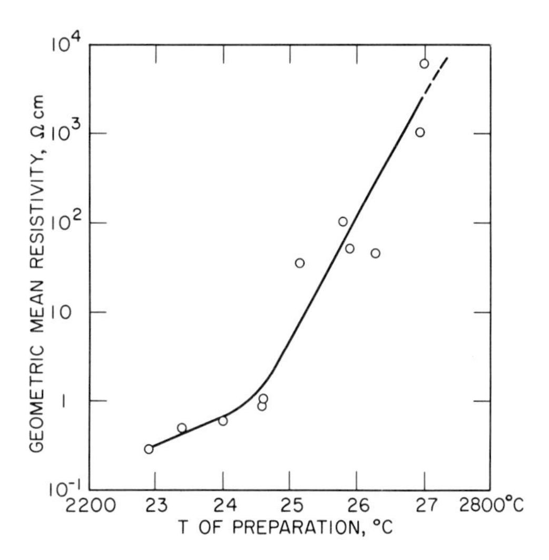

Fig. 12. The mean resistivity of several lots of *n*-type SiC prepared in argon at several temperatures.

Fig. 13. Pure α-SiC photographed on the abstract of Hamilton (1958).

tained by minor modifications of the growth cavity. Measurements indicate that the crystals are of a quality suitable for semiconductor work and good rectifying junctions can be made in them. A photograph of representative crystals is shown against a portion of the text of Hamilton (1958) in Fig. 13.

5. EPITAXIAL GROWTH

The preceding section has dealt with the growth of SiC by sublimation at atmospheric pressure. Hergenrother et al. (1960) have shown that SiC can also be grown from the vapor onto a seed crystal at low pressures. This method is commonly called *epitaxial growth.*

For a seed crystal to grow from the vapor phase, it is necessary for molecules to be adsorbed onto the crystal surface and then to diffusc to surface steps. High supersaturation leads to spurious growth nuclei on the seed's surface which destroys any chance of epitaxial crystal growth. If, however, the supersaturation of the vapor is just enough to allow growth, but not high enough to allow nucleation, the seed crystal will grow by adding epitaxial layers. In practice, a furnace that is to be used to grow epitaxial crystal layers must be designed so that SiC vapor does not reach the seed until the seed's temperature exceeds 2000°C. Then the vapor is only moderately supersaturated at the seed's surface.

Since high growth rates are desirable, the seed should be near the SiC charge so the vapor does not need to diffuse far. Therefore, the charge may not be heated appreciably faster than the seed at temperatures below 2000°C, and this requires a furnace with a low thermal inertia (Fig. 14).

Low-Pressure Furnace

The heater in the low-pressure furnace is a slotted graphite tube with a resistance of about $\frac{1}{3}$ ohm. It is heated by d-c current through two water-cooled electrodes and is insulated solely by molybdenum shields above, below, and around the heater. A graphite crucible containing the charge and seed is placed inside the heater on a graphite stand. All this is covered by a water-cooled vacuum enclosure.

In designing the crucible it should be re-

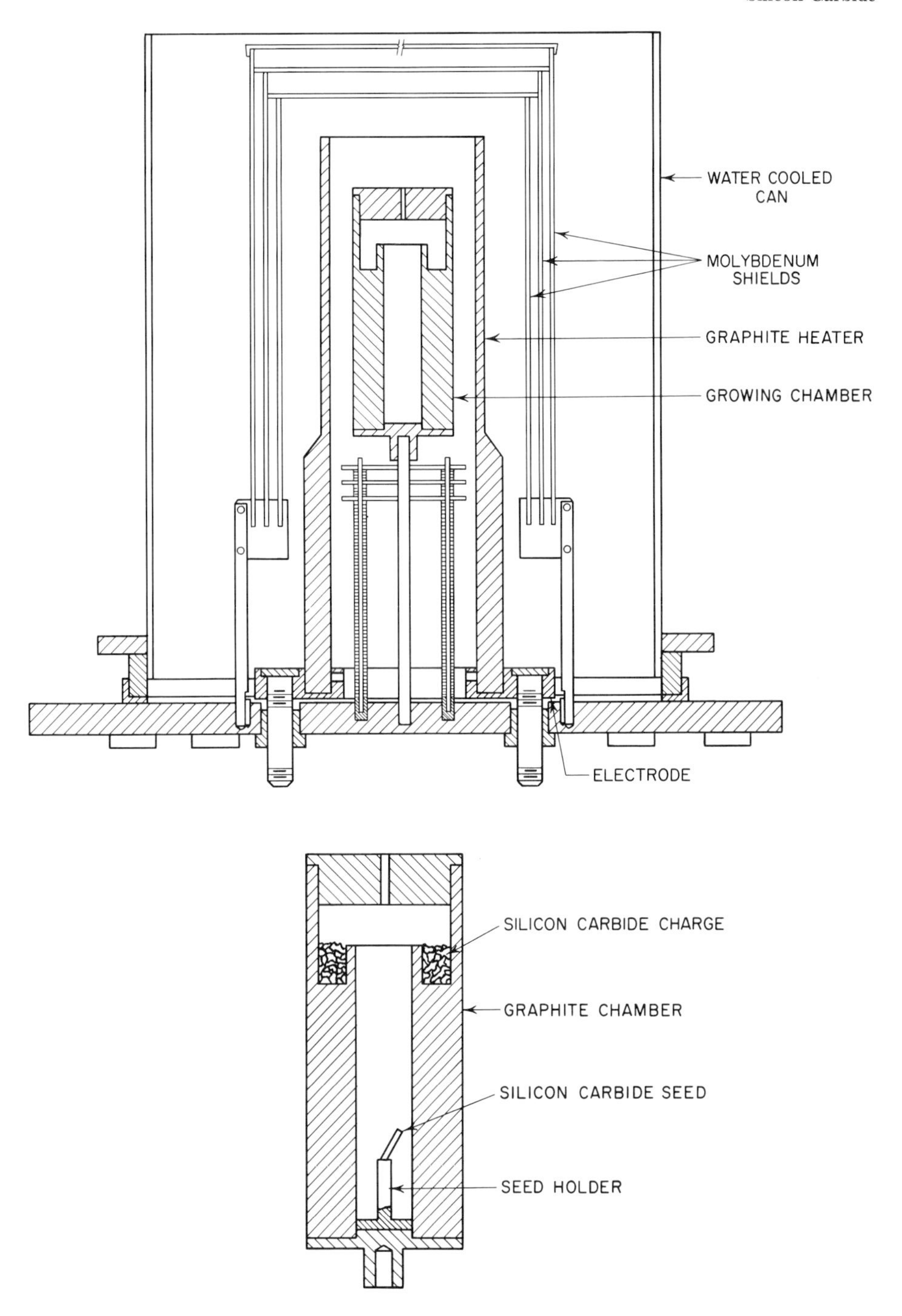

Fig. 14. Vacuum furnace and seed holder designed by Hergenrother et al. (1960).

membered that the major heat transfer mode is radiation. Therefore the temperatures of various parts of the crucible depend on the orientation of their surfaces toward the heater and its unheated ends rather than the temperatures of adjacent parts. This circumstance is quite useful because the temperature of the seed can be controlled by changing its orientation relative to the heater.

This design has low thermal inertia and can be raised to operating temperatures in

3 to 6 min. Therefore the seed reaches 2000°C before enough SiC diffuses to cause random nucleation.

A typical operating procedure is as follows. A seed is etched in a fused alkali-metal hydroxide to remove surface contamination. The seed is then oriented in a graphite seed holder at 5 to 45° with respect to the long axis of the heater. A charge of about 15 grams of SiC is packed into the crucible. After the shields are in place, the pressure is reduced to 5 μ. Then the system is backfilled with purified helium and the pressure lowered to 20 mm. The heater current is raised to 270 amperes at 80 volts, and maintained for $1\frac{1}{2}$ hr. When the power is shut off, the furnace requires $\sim\frac{1}{2}$ hr to cool.

The growth rate depends on the crucible geometry and position, the time, the temperature, and the orientation of the seed. The average growth rate is about 5 mils/hr. Since an average run lasts 10 hr, a typical uniform film 8 mils thick forms. If the seed axis lies parallel to the long axis of the crucible, a thin film, only $\frac{1}{2}$ mil thick forms, except for an 8 mil rim. This last happens because the edges of the seed radiate to the cool ends of the crucible, whereas the flat faces radiate to the hotter side walls. Most of the seed's surface is in a state of only slight supersaturation and therefore grows very slowly.

This dislocation density in epitaxial layers formed on tilted seeds is $\sim 10^6/\text{cm}^2$ compared to $\sim 100/\text{cm}^2$ for the seed crystal. The dislocation density approaches that of the seed crystal when the layer is grown with the seed lying parallel to the crucible axis (slow growth).

Experimental Results

Epitaxial layers vary from dark green through colorless to dark yellow. The resistivity of the colorless and near-colorless films is in excess of 1000 ohms cm. Some films make point contact diodes with peak inverse voltages of 1000 volts.

The high resistivity films just described were grown using commercial green SiC. The resistivity of epitaxial films seems to be insensitive to the purity of the charge material. Fractional distillation of the SiC purifies the deposited material, first because the segregation constant of nitrogen and other impurities is probably less than 1, and second because all the impurities are diluted by the ambient gas. At least two distillations occur since the top plug of the crucible collects a layer of SiC. The impurities in the charge material can be reduced by using SiC made by decomposing pure organo-silanes. When the temperature of the top plug is kept below 2050°C, the collected crystals are cubic, and they grow up to several cubic millimeters in volume.

6. GROWTH BY GASEOUS CRACKING

Kendall (1953) has shown that SiC crystals can be grown by gaseous cracking. Volatile compounds of silicon or carbon are heated to cause thermal decomposition to silicon or carbon. One method for manufacturing semiconductor-grade silicon uses the decomposition of trichlorosilane.

Growth on a Graphite Filament

By flowing mixtures of silicon tetrachloride, toluene, and hydrogen past a hot (2000°C) graphite filament, β-SiC crystals can be grown in the apparatus shown in Fig. 15.

A number of compounds have been tried, but the most success has been obtained with mixtures of silicon tetrachloride and toluene. A considerable excess of silicon tetrachloride appears to be necessary, a ratio of about 20 to 1.

The diameter of the hot filament was only 10 mils, but crystals up to 25 mils long could be produced. Still larger crystals could be obtained if the diameter of the filament were increased and if it were surrounded by a reflecting radiation screen. Excessive nucleation on a hot filament is difficult to avoid. However, this drawback would be minimized by conducting the experiment on a larger scale.

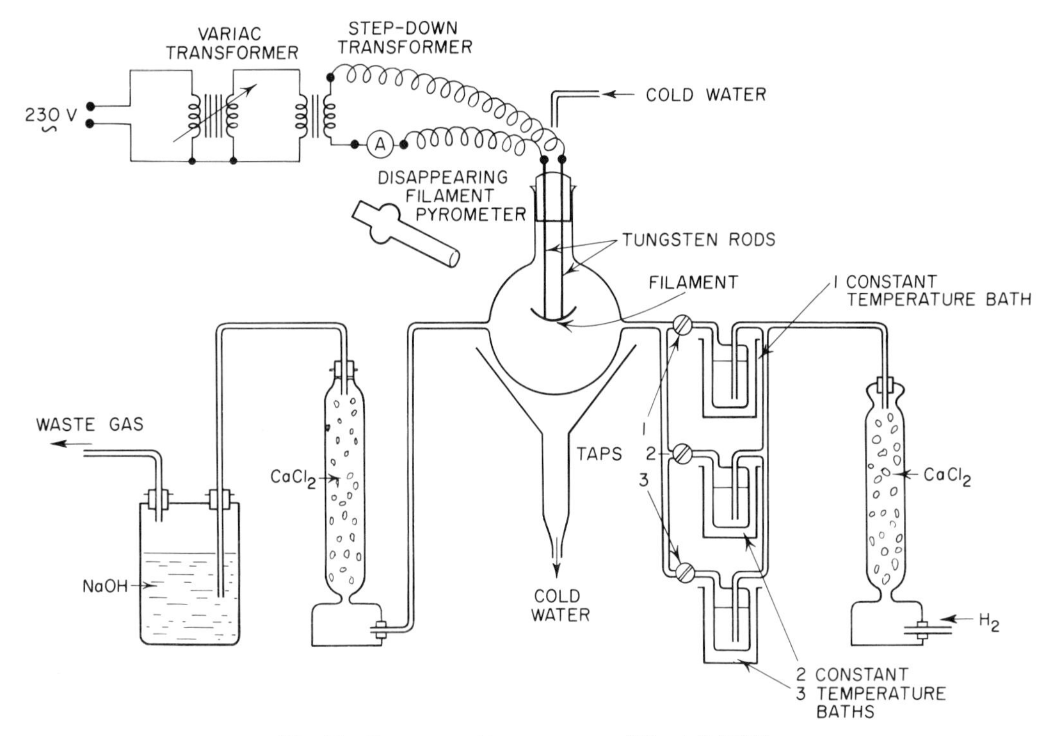

Fig. 15. Gaseous cracking apparatus of Kendall (1953).

Growth in an Induction-Heated Graphite Tube

Experiments have been performed by Merz (1960) using $SiCl_4 + C_7H_8 + H_2$, and $CH_3SiCl_3 + H_2$ in an inductively heated tube-furnace (Fig. 16). A 30-in. graphite tube served as both reaction container and susceptor. In early experiments, moldstock graphite was used, but was found to be permeable to HCl and $SiCl_4$; therefore, a dense graphite tube was substituted. No loss of HCl or $SiCl_4$ from this tube could be detected with ammonia fumes at temperatures up to 2000°C. However, it is probable that small amounts of O_2 and N_2 could diffuse into the tube and that H_2 could diffuse in the reverse direction. The reaction tube was insulated with $1\frac{3}{4}$ in. of 1000-mesh SiC for temperatures below 2100°C. A quartz container held the insulation in place and supported the induction coil. The ends of the quartz cylinder were closed by end plates held in place by water-cooled copper bands bolted to the reaction tube. Water-cooled copper

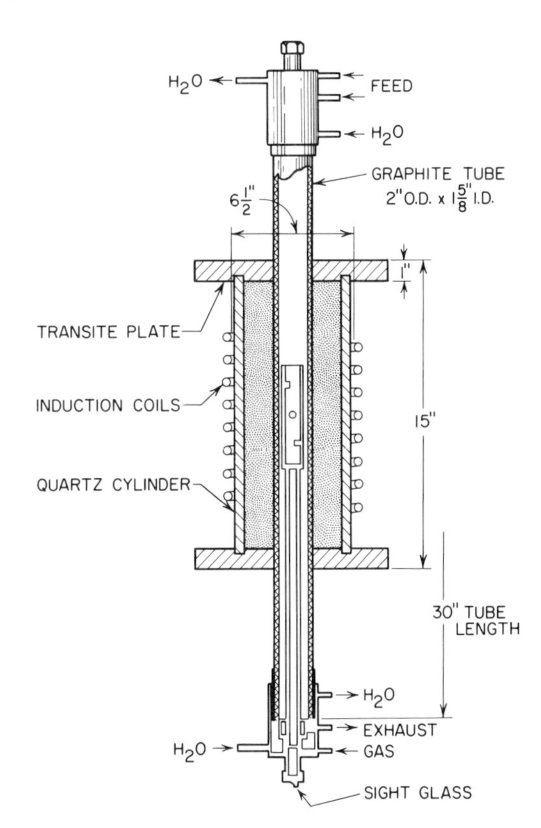

Fig. 16. Graphite tube furnace constructed by Merz (1960).

caps were sealed to the end of the reaction tube with Gooch crucible tubing. The caps were provided with two inlet tubes and a sight glass that allowed temperature measurement and control. A graphite tube supported the crucible and provided a sight path for the radiation pyrometer. A purging gas was fed into this tube to keep the sight path free of smoke. The crucible was 5 in. long and contained three evenly spaced graphite sight rods that projected into the gas stream to provide means for temperature measurements at various positions along the crucible. The reaction gases were fed into the open top of the crucible and passed out through the holes in the crucible support plate, which also served as a cap for the crucible support tube. To remove HCl and $SiCl_4$, the reaction and purge gases were exhausted into gas-washing bottles containing aqueous NaOH.

A typical temperature profile measured from top to bottom of the crucible was 1400°, 1600°, 1730°, and 1550°C at four equally spaced positions. The temperature was controlled by a proportioning control system coupled to phase-shifting rectifiers in a 400-kc induction generator. The sensing element was a radiation pyrometer. The control equipment maintained the temperature constant within ±10°C at 1750°C.

The carrier gas H_2 bubbled through the liquid SiCl and toluene at measured rates that were varied to provide definite Si/C ratios. The bubbling bottles were weighed before and after each run to determine consumption.

For experiments in the temperature range 2000 to 2500°C, a carbon resistance furnace with a 3-in. diameter tube was used. A dense graphite tube of 2-in. diameter and 7 ft long was placed inside the furnace tube with protruding ends. This reaction tube was capped as in Fig. 16.

Experiments below 1500°C were done in a $1\frac{1}{4}$-in. alumina tube furnace heated by SiC elements. The ends of the alumina tube were closed with rubber stoppers provided with inlet-outlet tubes and sight glasses. The chlorinated reagents used in this investigation, such as $SiCl_4$ and CCl_4, were found to react with the alumina tube at temperatures >1300°C. Hence both aluminum and oxygen must be considered part of the reaction system in these experiments.

Attempts were made to grow β-SiC crystals at 1450 to 2500°C from a mixture of $SiCl_4$, C_7H_8, and H_2. Crystal growth did occur at 1600 to 2250°C, but below 1750°C the growth rate was quite slow. At 1750°C, the fastest growth was observed with a nominal Si/C ratio of 1 to 4 and 90 to 98.7 mole per cent H_2. The true Si/C ratio at the point of reaction was unknown since the H_2 reacted somewhat with the furnace interior. The gas velocity varied from 40 to 280 cm/min (NTP).

Figure 17 shows some typical β-SiC crystals found growing from a $\frac{1}{4}$-in. graphite sight rod that was held at 1910°C. The crystals are green, dendritic, and grow in irregular clumps. The largest crystals shown is about $1 \times 1 \times 2$ mm.

In order to determine the effect of oxygen, a series of experiments was made in which controlled amounts of compounds such as H_2O, CO_2H_5OH were added. Crystals were grown successfully with Si/O ratios as low as 1:1 at 1760°C. The habit of the crystals was somewhat altered. They tended to grow in large isolated clumps and to be lamellar in shape. It was concluded that small amounts of oxygen had no beneficial effects on SiC growth.

Crystals grown at 2000°C and above showed some anisotropic behavior under the polarizing microscope, indicating the presence of α-SiC. As the growth temperature increased, larger amounts of α-SiC were found. The α-SiC was intimately intergrown with β-SiC, some crystals having lamellae of α-SiC running through a matrix of β-SiC.

At temperatures above 1850°C, it was not necessary to introduce toluene since the hydrogen reacted with the reaction tube to supply the required carbon. The carbon of the crucible was badly eroded and pitted in the hot zone. Above 2250°C, no crystal growth was observed. The α-SiC crystals grown near 2250°C grew quite rapidly,

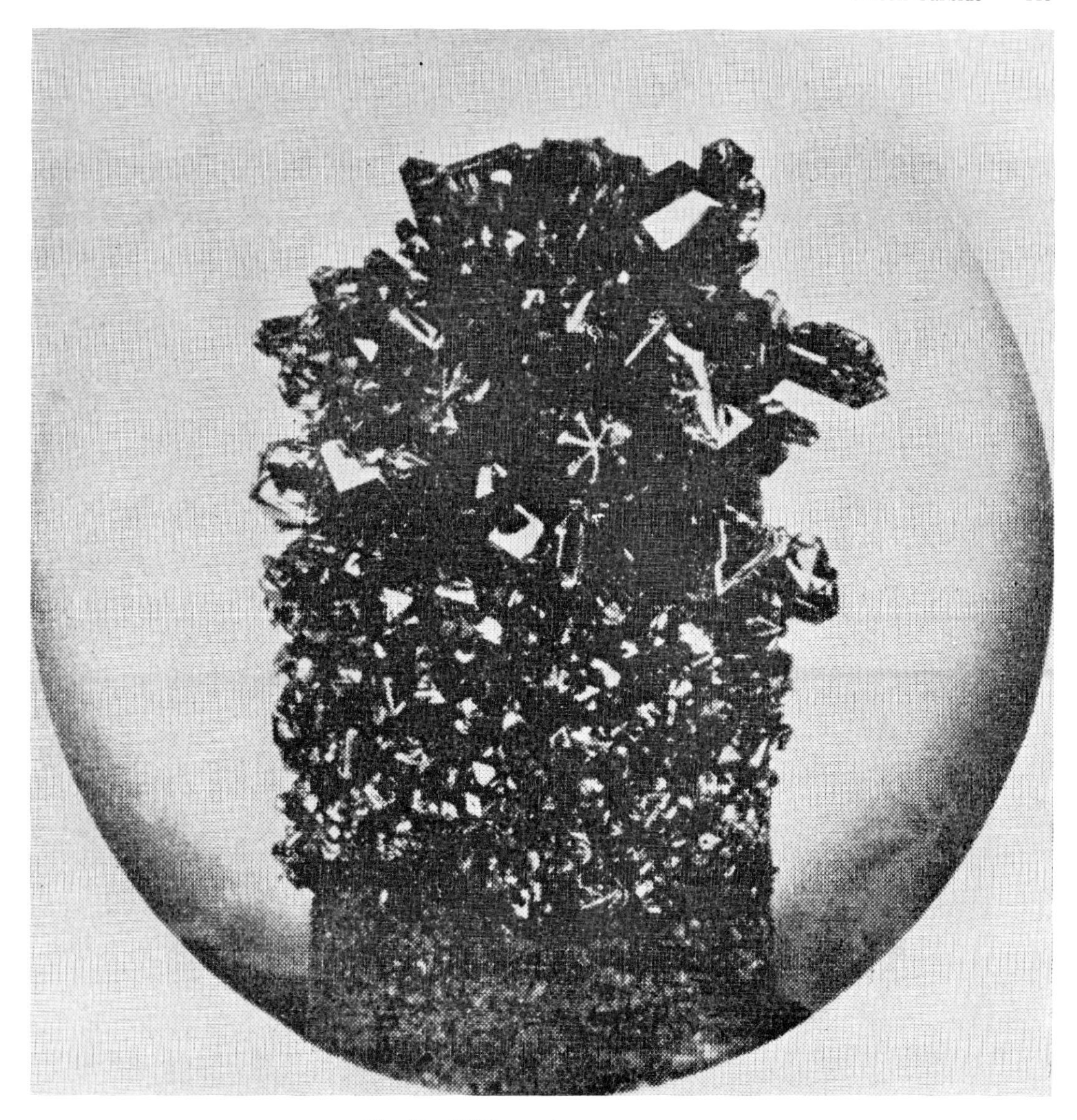

Fig. 17. β-SiC crystals on a graphite rod.

attaining a length of ≥ 1 cm in 4 hr. These crystals were generally intertwined masses of dark green, dendritic needles.

In some experiments, continuous layers of microcrystalline SiC were deposited on the crucible wall. A layer 3 mm thick formed in 6 hr from a gas stream of $SiCl_4$ + C_7H_8 + H_2 or $SiCl_4$ + CCl_4 + H_2. Material grown at 1400°C was glassy black and broke with a conchoidal fracture. Further along the tube, where the temperature was 1520°C, the material was dull in fracture, thinner in cross section, and the surface was raised into hillocks. A concentration of 10 mole per cent $SiCl_4$ and a gas velocity of about 200 cm/min was used to grow this form of SiC. Chemical analysis indicated that the material contained 98.28% SiC. The density was found to be 3.20 grams/cm^3, which is very close to the value 3.217 found by single crystal and X-ray determinations. The material was insoluble in HF-HNO_3 mixtures, indicating that there was no free Si metal present.

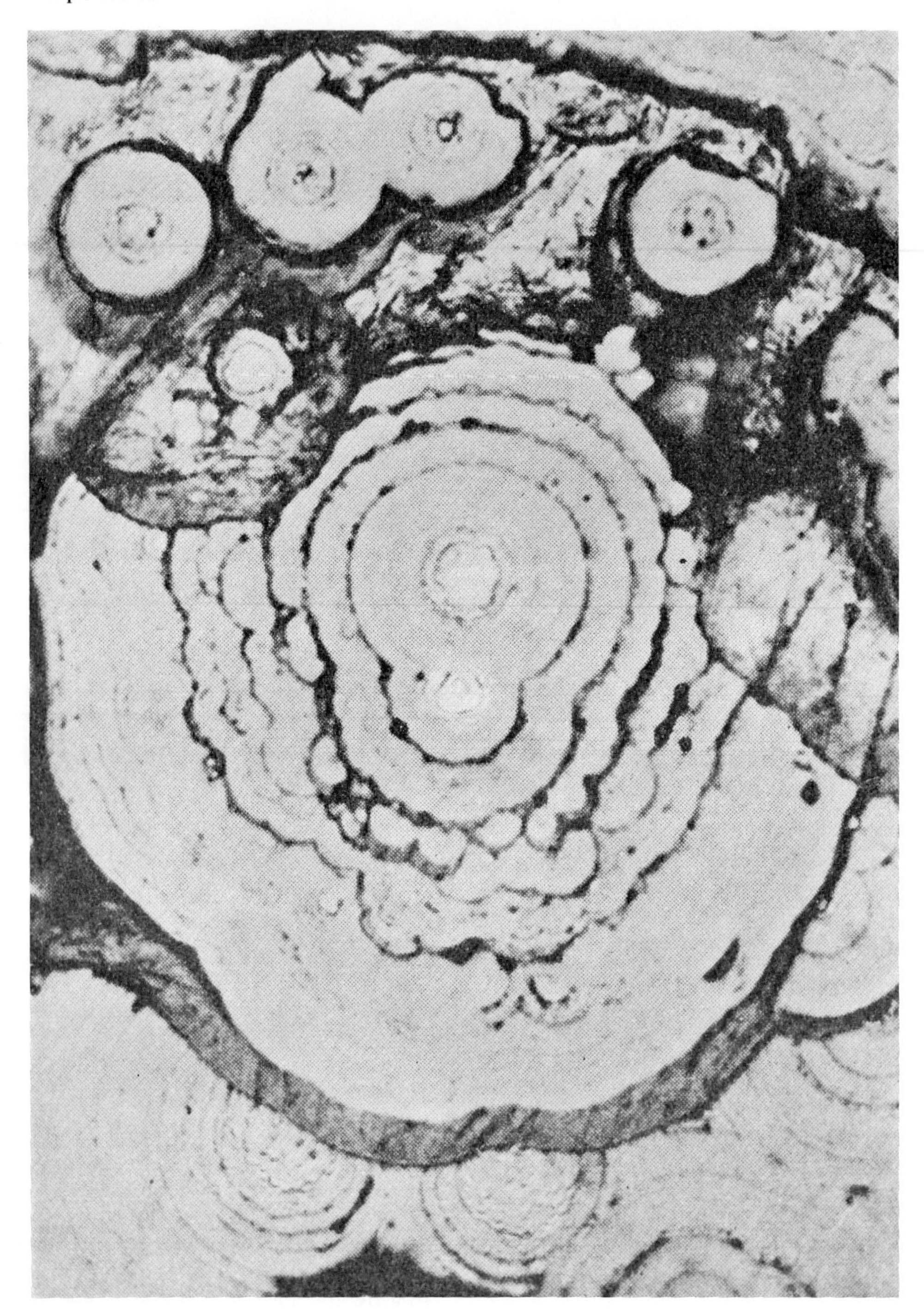

Fig. 18. Microcrystalline β-SiC from Merz (1960).

Figure 18 is a photomicrograph of a polished section of microcrystalline β-SiC. The direction of the gas stream was in the plane of the figure. It can be seen that the material is laid down in layers which form small hillocks. Periodically the layer formation is disturbed by what appear to be inhomogeneous open-structured regions. These might be due to fluctuations in either temperature or the flow of the gas stream.

7. GROWTH FROM SOLUTION

The expectation that pure SiC crystals might be grown from liquid silicon solutions is based on the fact that SiC bonds are considerably stronger than the average of silicon-silicon and carbon-carbon bonds. Therefore a solution saturated with SiC may be very unsaturated with respect to graphite. Furthermore, Shockley (1957) has pointed out that near its melting point silicon tends to reject metallic impurities by a factor 10^3 to 10^5. At lower temperatures the rejection ratio increases. It seems plausible that the rejection ratio at the melting point might be higher for SiC than for silicon. Furthermore, 1500°C is only about half of the melting temperature, so the rejection ratio might be squared. Thus the alloy used for dissolving SiC might be rejected by a factor as high as 10^{10}, thereby yielding material of adequate purity for many purposes.

Several investigators have observed the precipitation of β-silicon carbide from metal alloys. Baumann (1952) reported growth of tiny crystals in a silicon-nickel eutectic held one hour at 1000°C, and in a silicon-aluminum-zinc alloy held for three hours at 525°C. Chipman (1954) reports the formation of SiC in iron-silicon alloys and Hall (1958) has attempted to grow SiC from saturated silicon melts.

Carbon Solubility Studies

An ideal solvent for growing SiC crystals would readily dissolve SiC and have considerable temperature dependence of the solubility. Moreover, the dissolution process would be reversible.

Figure 19 summarizes data on the solubility of carbon in molten silicon. Hall (1958) measured the weight loss of SiC crystals immersed in silicon. However, oxygen from the crucibles was present in the Hall's melts causing excess solubility. This defect has been corrected by Scace and Slack (1959), who contained the silicon in graphite. Errors in their analytical technique would result in slightly low values of carbon solubility. The data of Stevenson and Halden were obtained by sampling molten silicon held in SiC-coated graphite crucibles. Analytical errors in this technique would also lead to low values of carbon solubility.

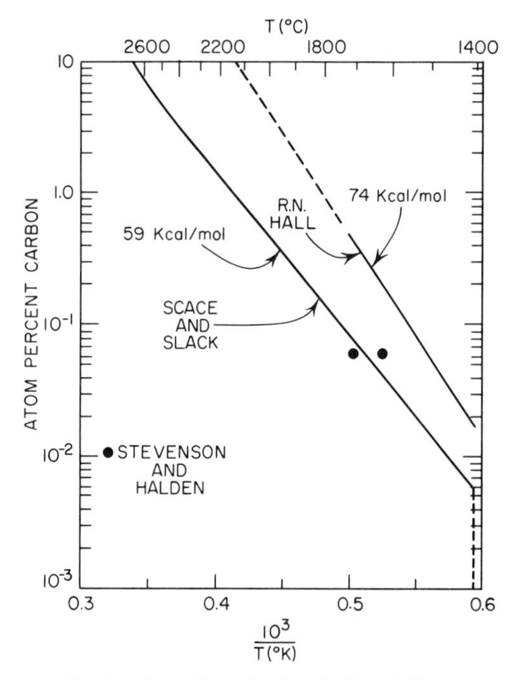

Fig. 19. Solubility of carbon in liquid silicon.

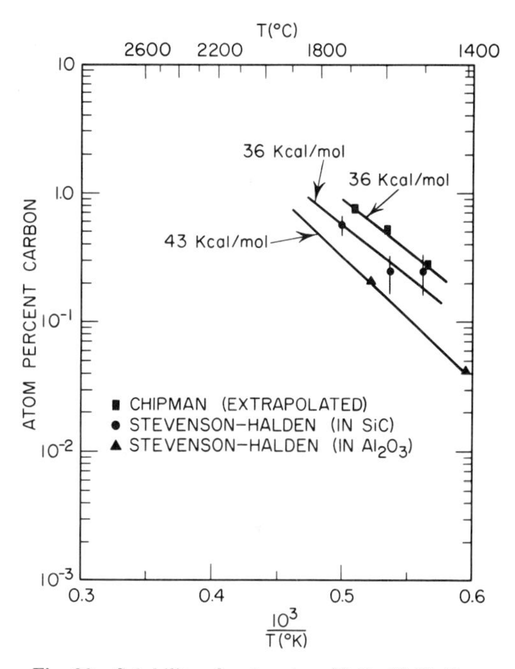

Fig. 20. Solubility of carbon in a 70 Fe–30 Si alloy.

Figure 20 shows approximate solubility

data for carbon in a 70 Fe-30% Si alloy. It appears that the solubility of carbon in the iron-silicon alloy is eight to ten times that in pure silicon at 1500 to 1700°C, but the dependence of solubility on temperature is somewhat less.

These data suggest that it might be possible to dissolve carbon from a graphite crucible and crystallize SiC on a seed placed at the coolest portion of the melt. It will be necessary to determine whether a sufficient temperature gradient can be established to permit growth at reasonable rates without spontaneous nucleation.

Experimental Procedure

Halden (1960) has used a modified Czochralski apparatus. A graphite resistance heater, powered by a 21-kw variac transformer combination, is manually controlled. The hot zone is surrounded by molybdenum radiation shields and a water-cooled shell. The seed is rotated and raised by motors at variable speeds from 0.1 μ/sec to 50 μ/sec. Rapid adjustments are accomplished by another motor connected through a friction clutch. The crucible pedestal is also rotated by a variable-speed motor, and can be raised and lowered through the hot zone. The heater provides a uniform temperature zone approximately 2 in. long, with a thermal gradient above it of approximately 60°C/in.

The furnace is normally heated to the melting point of the solution under a vacuum of 10^{-4} mm. Then a purified inert atmosphere in introduced, and the temperature of the solution is raised to the operating temperature. After about 30 min, the crucible is raised into the thermal gradient region and the melt is seeded.

Studies with various pulling rates revealed that growth occurred primarily at the surface of the melt and where the vertical temperature gradient was excessive. This was corrected by placing a graphite plate over the melt, with only a small hole for the seed crystal and a thin slit for temperature measurement. Using a *KT*-silicon carbide point as a seed, a large decrease in growth rate was noted, but little improvement of crystalline perfection. The temperature gradient was provided primarily by heat conducted through the seed. Figure 21 shows a crystal grown during a 48 hr run.

One reason for poor quality in the crystals was too high a melt temperature; however, when it was reduced so the bottom of the melt was at 1525°C and the top at about 1475°C, significant improvement was noted in the crystals. First, a polycrystalline material was observed to grow out at the surface of the melt to about $\frac{1}{2}$ in. diameter. Then, individual crystallites began to grow down into the melt. These crystallites were in the form of needles and platelets. The crystals were very thin, only 5 to 50 μ thick. Stirring of the melt was extremely important. Without crucible or seed rotation, no crystals grew into the melt. A crucible rotation rate of ~5 rev/min was required for growth down into the melt.

A few crystals have been grown from a 65 Fe–35% Si melt. These crystals grew at the crucible-alloy surface interface, with the melt at ~1650°C. The crystals formed as thin, hexagonal platelets varying from transparent yellow to green or blue.

8. DISCUSSION

The most successful method for growing SiC to date is the sublimation process. Material from this process is good enough for the fabrication of unipolar transistor devices.

Several problems remain unresolved. One is the determination of the factors that control crystal modification. In general, β-SiC is formed at temperatures less than 2000°C and α-SiC forms at higher temperatures. However, cubic material sometimes forms at a temperature as high as 2700°C and α-SiC at a temperature as low as 1500°C. Another problem is that while the supersaturation at the growing interface must be controlled, so that a reasonable growth rate is obtained while excessive nucleation is avoided, the exact conditions of supersaturation for SiC are not known.

Fig. 21. β-SiC crystal grown from a silicon melt at 1750°C by Halden (1960).

In the gaseous cracking methods, it is difficult to introduce the reagent gases into the decomposition chamber in a controlled manner and to transport the active gases to the growth interface. Still another difficulty is the maintenance of stoichiometry at the growing interface.

Growth from solution is simpler perhaps than growth by gaseous cracking; nevertheless it has its own complexities. With growth from a solution containing excess silicon, we do not know what the species in solution are. Also there are some important apparatus difficulties. One of these is the crucible problem. The most successful crystal growth has used molten silicon which is very reactive. Two attempts to solve this problem have been mentioned. The first is the use of a siliconized graphite crucible, and the second uses a pyrolytically deposited graphite film that is dense enough to reduce the attack by molten silicon. To achieve appreciable solubility of carbon in silicon, a temperature of about 1700°C is needed where evaporation of silicon becomes rapid and should be suppressed. It is quite a problem to grow a crystal while the solvent is rapidly evaporating and the crucible is being dissolved! A pure SiC crucible or a floating zone arrangement might solve the problem.

Growth from solution, using solvents other than silicon, is possible and should not be abandoned. Perhaps a solvent could be found that is either insoluble in solid SiC or not electrically active, which would permit SiC to be dissolved at low temperatures.

The question of growth habit is another

problem that is unsolved. In a given crystal growth experiment, several habits are often obtained, such as plates, rods, and needles. To achieve the growth of large crystals, this action cannot be tolerated. This leads to another practical consideration, how large should one attempt to grow SiC crystals. Although control of purity is certainly essential for practical applications, shape and size control is also essential. It may be that small crystals about 1 mm square and with a uniform thickness of a few microns would be useful. Therefore information concerning natural growth habits and how they may be influenced is important.

The problem of impurity control cannot be separated from crystal growth because, for example, temperature appears to affect the solubility of nitrogen in SiC. The distribution coefficients of most impurities are expected to be temperature dependent. Data describing these dependences, together with the corollary information mentioned, are required before we can hope to really control crystal growth.

ACKNOWLEDGMENT

The author gratefully acknowledges helpful discussions with several of his former associates at the Air Force Cambridge Research Laboratories, where most of these data were compiled. Thanks are also due Mrs. Josephine M. Vigneault and Mrs. Dolores A. Gunnerson for the preparation of the manuscript.

References

Amelinckx, S. (1960), "Silicon Carbide," p. 201.* Pergamon Press, New York.

Baumann, H. M. (1952), *J. Electrochem. Soc.* **99,** 109.

Burton, W. K., N. Cabrera, and F. C. Frank (1950), *Phil Trans. Roy. Soc.* **A243,** 299.

Cabrera, N. (1953), *J. Chem. Phys.,* **21,** 1111.

Chang, H. C., and L. J. Kroko (1957), *A.I.E.E. Conference Paper* **57,** 1131.

Chipman, J., J. C. Fulton, N. Gokcen, and G. R. Caskey (1954), *Acta. Met.* **2,** 439.

Drowart, J., and G. DeMaria (1960), "Silicon Carbide," p. 16.*

Drowart, J., G. DeMaria, and M. G. Inghram (1958), *J. Chem. Phys.* **29,** 1015.

Field, W. G. (1961), Private Communication.

Halden, F. A. (1960), "Silicon Carbide," p. 115.*

Hall, R. N. (1958), *J. Appl. Phys.* **29,** 914.

Hamilton, D. R. (1958), *J. Electrochem. Soc.* **105,** 735.

Hergenrother, K. M., S. E. Mayer, and A. I. Mlavsky (1960), "Silicon Carbide," p. 60.*

Herring, C. (1951), *Phys. Rev.* **82,** 87.

Honig, R. E. (1954), *J. Chem. Phys.* **22,** 1611.

Jackson, K. A. (1958), *Growth and Perfection of Crystals,* R. H. Doremus, B. W. Roberts, and D. Turnbull, Editors, John Wiley and Sons, New York, p. 319.

Kendall, J. T. (1953), *J. Chem. Phys.* **21,** 831.

Kroll, W. J., A. W. Schlechten, and L. A. Yerkos, (1946), *Trans. Electrochem. Soc.* **89,** 317.

Lely, J. A. (1955), *Ber. deut. heram. Ges.* **32,** 229.

Merz, K. M. (1960), "Silicon Carbide," p. 73.*

Nowotny, N. (1954), *Monatsh. Chem.* **85,** 225.

O'Connor, J. R. (1959) *Physics Today* **12,** 14; *J. App. Phys.,* To be published.

O'Connor, J. R., and J. Smiltens, Editors (1960), *Silicon Carbide—A High Temperature Semiconductor,* Pergamon Press, New York.

Patrick, L. A. (1957), *J. App. Phys.* **28,** 765.

Scace, R. I., and G. A. Slack (1959), *J. Chem. Phys.* **30,** 1551.

Shockley, W. (December 1957), *Bureau of Ships Contract No.* Nobar-72706, p. 4.

Slack, G. A. (1960), "Silicon Carbide," p. 30.*

Smiltens, J. (1960), "Silicon Carbide," p. 4.*

Westinghouse Electric Corporation (1957), *Research Reports on Contract* AF 19 (604)-217.

* See J. R. O'Connor and J. Smiltens (1960).

7

Ice

B. J. Mason

Research on the formation of ice crystals from the vapor has been concerned largely with the growth of snow crystals in both the atmosphere and the laboratory. These studies in cloud physics are important because the higher atmospheric clouds are composed almost entirely of ice and most of the rain, at least in temperate latitudes, results from the growth, aggregation, and melting of snow crystals. Also the remarkable variety of shape and pattern of snow crystals poses some very interesting and perplexing problems in crystal physics.

The collection, photography, and classification of snow crystals have long been an attractive pastime, the books by Bentley and Humphreys (1931) and by Nakaya (1954) being particularly famous for their thousands of beautiful photomicrographs. Furthermore, valuable data have been obtained on the masses, dimensions, and terminal velocities of the different crystal forms. However, it is only recently that any real progress has been made in relating the shape, structure, and growth rate of the crystals to the environmental conditions in which they are formed. These studies, which have revealed that the growth mechanisms of ice crystals are very complicated, and with many puzzling features that are apparently peculiar to ice, are now reviewed.

1. THE OCCURRENCE OF ICE CRYSTALS IN NATURAL CLOUDS

Although extensive observations have been made of snow crystals reaching the ground, and attempts have been made to correlate the relative frequencies of the various crystal forms with the temperature at the place of observation, very few consistent results have emerged. This is not, perhaps, surprising since only the conditions prevailing during the growth of the crystal are likely to be of major importance in determining its shape, and it is only in recent years that crystals have been collected from different types of cloud having widely different conditions of temperature, water vapor concentration, and supersaturation relative to ice. The most comprehensive collection of ice crystals from natural clouds has been obtained by Weickmann (1947) at different heights in the troposphere up to cirrus levels and with the temperature of the sampling region measured. His observations, summarized in Table 1, are not as detailed as one might wish, but the sampling of crystals from aircraft presents considerable difficulties, particularly at high speeds when it becomes difficult to avoid the shattering of the crystals during their collection.

Weickmann found that, at temperatures below $-25\,^{\circ}\mathrm{C}$, the dominant crystal form was the hexagonal, prismatic column which develops by preferential growth along the principal *c*-axis, that is, normal to the basal plane. These crystals are typically $\frac{1}{2}$ mm long, the ratio of length to breadth varying from one to five, and appear as single crystals, as basal twins (Fig. 1) and in clusters originating from a central frozen droplet (Fig. 2). If grown under conditions of high supersaturation, the crystals usually contain pronounced funnel-shaped cavities, whereas in clouds that are only slightly supersatu-

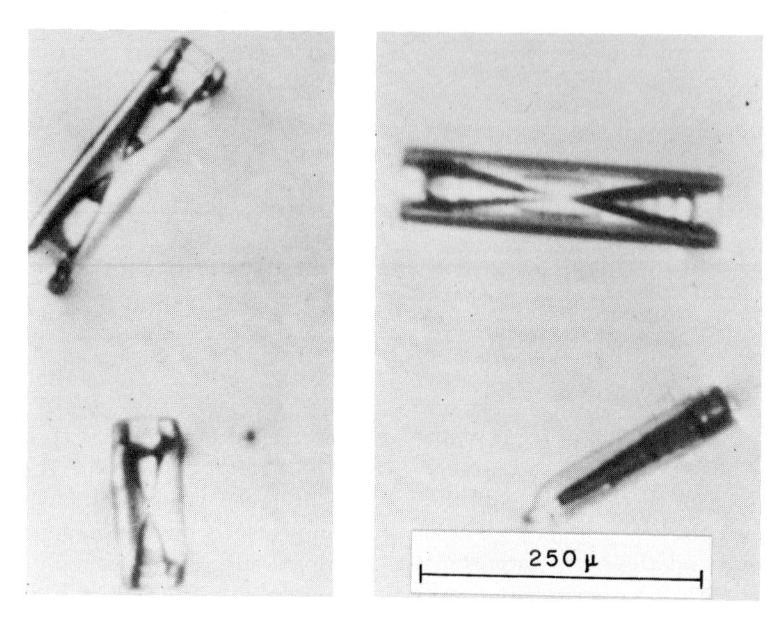

Fig. 1. Individual hollow prismatic ice columns collected from cirrus clouds at $-40°C$. (Photograph by H. Weickmann.)

rated with respect to ice, occasional prismatic columns with a single pyramidal end (Fig. 3) are found.

In proceeding from high- through medium- to low-level clouds, and therefore to higher temperatures, Weickmann reported that there was a gradual transition from prisms through thick plates to thin, hexagonal plates, the latter resulting from preferential growth of the prism face. The thin, hexagonal plates (Fig. 4), which occur at temperatures usually above $-20°C$, are generally less than 500 μ in diameter and less than 50 μ in thickness.

At temperatures near $-15°C$ the striking, star-shaped crystals are dominant. They usually possess six arms, which may grow rapidly and develop side branches to assume the fine dendritic structure illustrated in Fig. 6, and attain diameters of several millimeters. Although the development of all six arms is often superficially very simi-

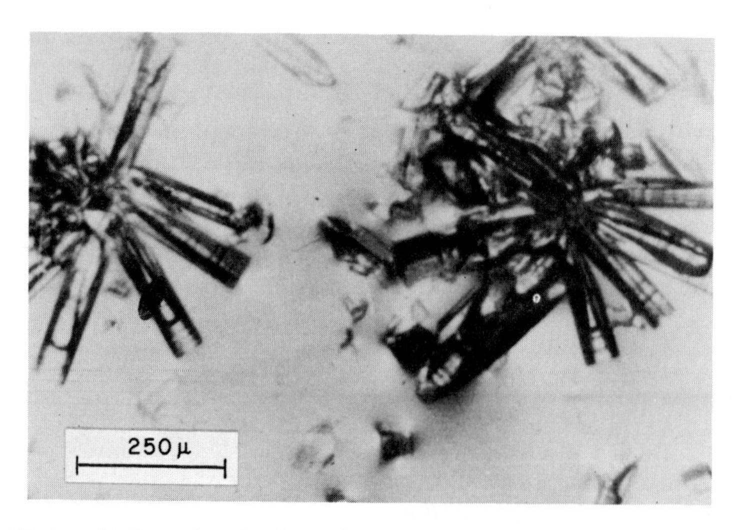

Fig. 2. Clusters of hollow prismatic columns from cirrus clouds. (Photgraph by H. Weickmann.)

TABLE 1

WEICKMANN'S OBSERVATIONS OF PREDOMINANT CRYSTAL FORMS IN DIFFERENT CLOUD TYPES

Level of Observation	Temperature Range	Cloud Types	Crystal Forms
Lower troposphere	0°C to −15°C	Nimbostratus	Thin, hexagonal plates
		Stratocumulus	Star-shaped crystals showing dendritic structures
		Stratus	
Middle troposphere	−15°C to −30°C	Altostratus Altocumulus	Thick hexagonal plates. Prismatic columns—single prisms and twins
Upper troposphere	< −30°C	Isolated cirrus	Clusters of prismatic columns containing cavities. Some single hollow prisms
		Cirrostratus	Individual complete prisms

lar, the crystals rarely show exact hexagonal symmetry in their detailed fine structure. A stellar crystal often develops from a hexagonal plate, which at a certain stage of its development begins to sprout at the corners, but some photographs suggest that the crystal may originate from a frozen droplet that has developed crystal faces.

Needle-shaped crystals, such as are illustrated in Fig. 7, are not represented in Weickmann's classification but are often observed at the ground when the temperature is only slightly below 0°C, and there are other observations to show that these originate from clouds with temperatures between about −3°C and −8°C. The needles, which often occur in clusters, are a skeletal form of hollow prism. They arise as the result of rapid growth along the principal axis; not only are the basal faces incomplete

Fig. 3. Solid prismatic columns from cirrostratus cloud at −26°C. (Photograph by H. Weickmann.)

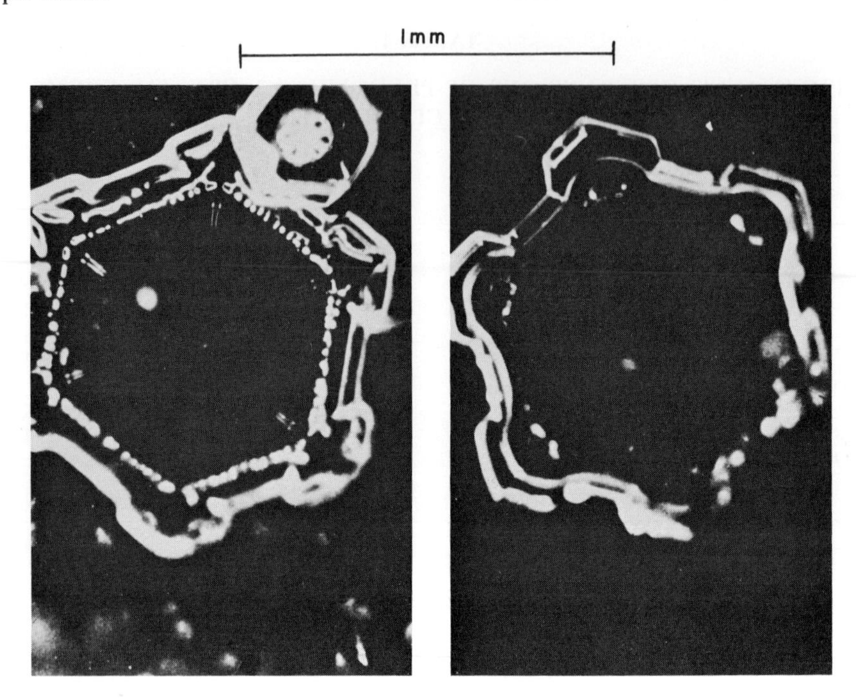

Fig. 4. Thin hexagonal ice plates. (Photograph by H. Weickmann.)

but growth may occur on only some of the edges of the basal plane or, in extreme cases, only from one or more corners.

The data obtained by Weickmann from an aircraft have been supplemented by a comprehensive series of observations made by Wall (1947) at the Friedrichshafen Observatory during the winter of 1940–1941 when the crystals fell from thin, well-defined cloud sheets with no higher clouds; thus the temperatures at which the crystals originated could be determined within rather narrow limits. Similar observations have been made in Canada by Gold and Power (1952), on the Hohenpeissenberg in Germany by Weickmann (1957), in New England by Kuettner and Boucher (1958), and in Japan by Murai (1956) and Magono (1960). The temperature ranges in which the different

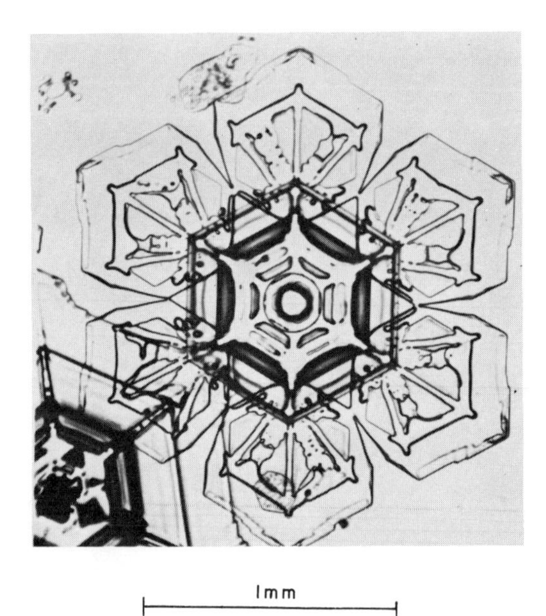

Fig. 5. A sector-plate ice crystal photographed in transmitted light. (Photograph by Urs Beyeler.)

TABLE 2

TEMPERATURE RANGES FOR THE ORIGIN OF SNOW CRYSTALS IN THE ATMOSPHERE

$-3°C$ to $-8°C$	Needles
$-8°C$ to $-25°C$	Hexagonal plates, sector plates
$-10°C$ to $-20°C$	Stellar dendrites
$< -20°C$	Prismatic columns, single crystals and twins
$< -30°C$	Clusters of hollow prisms

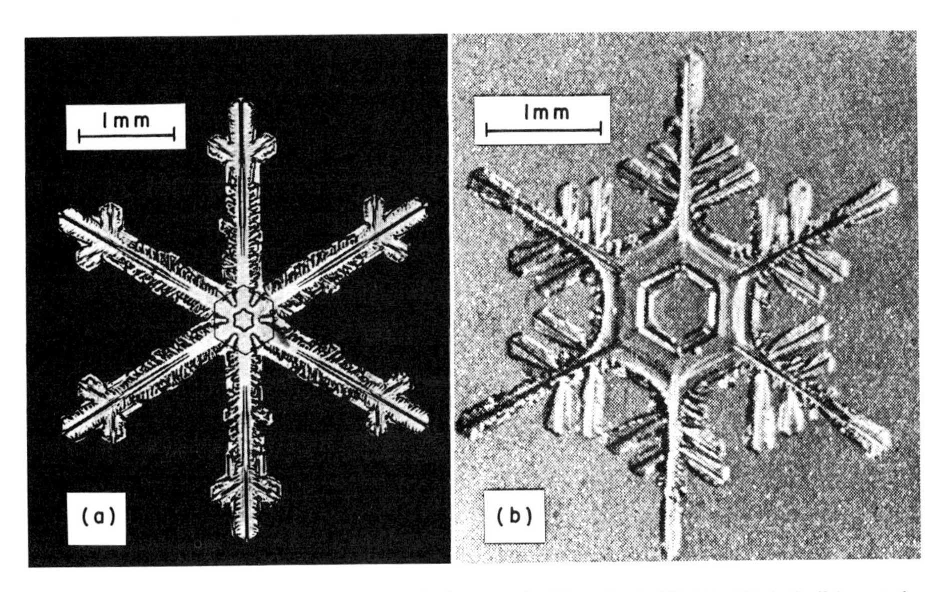

Fig. 6. (*a*) A dendritic stellar snow crystal photographed in reflected light. (*b*) A similar crystal photographed in oblique illumination. (Photograph by Professor U. Nakaya.)

basic crystals forms were believed to have originated are very similar in all these series of observations and agree, in general, with Weickmann's aircraft data. The results of all the observations are summarized in Table 2.

Crystals which are a combination of two or more of the basic types (prismatic column, plate, star) are not uncommon; the prismatic column with end plates and the combination of plate and dendritic star shown in Figs. 8 and 6 are good examples.

Snowflakes are agglomerates of individual crystals in which the star-shaped dendrites are generally prominent but in which the needle and plate forms may also be found. Adhesion is effected partly by the interlocking of the arms of stellar crystals, partly by the sublimation of water vapor and the freezing of collected supercooled droplets acting as a cement, and, perhaps also, by the fusion at points of contact through the action of localized plastic flow and diffusion. Large snowflakes are not observed in very cold weather, but when the temperature is very near 0°C, aggregates of as many as a hundred crystals may be seen. Aggregation of crystals probably does not readily occur at temperatures much below −10°C.

2. THE FIXATION AND PHOTOGRAPHY OF SNOW CRYSTALS

The direct photography of snow and ice crystals presents difficulties apart from the obvious one of their melting; the crystals have to be photographed in subfreezing conditions when trouble may arise from condensation on the crystal surfaces and from frosting of lenses, etc. Most of the difficulties were removed when Schaefer

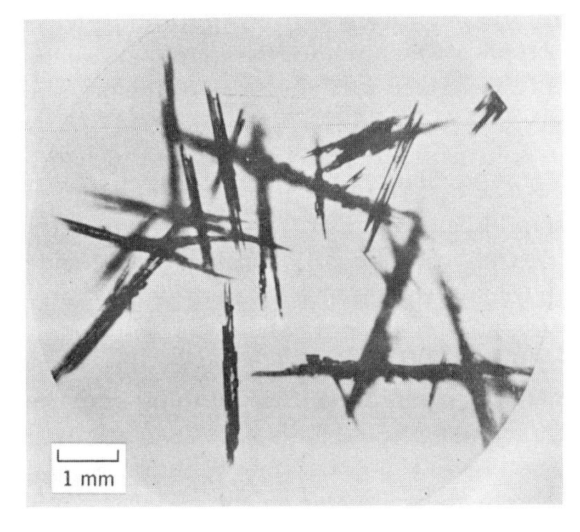

Fig. 7. Thin ice needles. (Photograph by Professor U. Nakaya.)

Fig. 8. A hexagonal column capped with hexagonal end plates. (From *Snow Crystals* by W. A. Bentley and W. J. Humphreys, McGraw-Hill, 1931.)

(1946) invented his technique of making permanent plastic replicas of the crystals. A solution of formvar (polyvinyl acetal resin) in ethylene dichloride is kept at a temperature of about −5°C. A clean glass microscope slide is immersed in the solution for about 30 sec and then exposed horizontally to the falling crystals. A captured crystal becomes submerged in the solution, after which the slide should be kept at a subfreezing temperature for a few minutes until the solvent evaporates, leaving the crystal encased in a thin, but tough, plastic shell. The slide may now be exposed to room temperature when the crystals will melt, the water diffusing out of the plastic membrane as it evaporates to leave a replica which retains in microscopic detail all the surface configurations of the original crystal.

A little practice and care are required to get the best results. The formvar solution should be dehydrated by shaking it up with calcium chloride or phosphorus-pentoxide to remove the dissolved water, which otherwise will come out when the solution is chilled and form spurious ice crystals. It is particularly important to use solution of the right strength. If it is too viscous, small crystals will not become submerged and merely make a crater on the surface; if it is too thin, it will run off the slide and not cover a large crystal. Good replicas of natural snow crystals may be obtained with a 1 to 3% solution but, for small crystals less than 0.1 mm diameter, such as may be produced in a laboratory experiment, a 0.1% solution should be used.

The replicas or the crystals themselves may be photographed through a microscope by either reflected or transmitted light. The use of transmitted light produces a clear picture of the boundary and internal structure of the crystal (Fig. 5). Reflected light increases the beauty of the photograph by producing a white image on a black background (Fig. 6*a*) and reveals something of the surface topography. Perhaps the best results are obtained with oblique illumination which combines the advantages of both transmitted and reflected light and reveals both the internal and surface structure (Fig. 6*b*). A very fine collection of several hundreds of photographs taken by this type of illumination has been published by Nakaya (1954) in his book *Snow Crystals.*

3. THE NUCLEATION OF ICE CRYSTALS AND EPITAXIAL GROWTH

In both the atmosphere and the laboratory, ice crystals growing from the vapor usually originate either from a frozen water droplet or on a small foreign nucleus (seed) which promotes the direct sublimation of ice or the adsorption and subsequent freezing of a film or liquid water.

It is comparatively easy to supercool micron-size droplets of pure water down to −40°C. In the absence of foreign surfaces, nucleation of the ice phase may occur only by the chance orientation of localized groups of water molecules into an icelike configuration. Such molecular aggregates continually arise and disappear as a result of thermal agitation, but the lower the temperature the greater is their size and frequency of formation until eventually they obtain a critical size above which they survive and continue to grow, forming nuclei for the ice phase. At −40°C, a nucleus of critical size is formed spontaneously in this manner within about 1 sec in a droplet of 1 μ diameter.

At appreciably higher temperatures, such a droplet has a low probability of freezing unless it is infected by a suitable foreign particle which may cause water molecules to become "locked" in an icelike configuration under the influence of its surface force field. The formation of a stable ice nucleus is largely determined by the degree of supercooling of the liquid and the configuration of the surface force field of the substrate, which, in general, is not uniform but contains some specially favored sites for nucleation.

The supercooling and freezing of water, and the nature, origin, and properties of the ice nuclei have been intensively studied by cloud physicists in recent years; the work has been reviewed by Mason (1957a, 1958, 1961).

The ice-nucleating ability of a wide variety of both natural and artificial substances has been tested, usually by dispersing them as fine dusts, smokes, or sprays into a cloud of supercooled water droplets, and determining the highest temperature at which about 1 in 10^4 of the particles produced an ice crystal.

In an attempt to discover the nature and origin of the atmospheric ice nuclei responsible for initiating natural snow, Mason and Maybank (1958) and Mason (1960) tested the nucleating properties of thirty-five different types of soil and mineral-dust particles. Twenty-one of these, mainly silicate minerals of the clay and mica groups, were found to produce ice crystals in supercooled clouds at temperatures of −15°C, or above, and of these, 10 were active above −10°C (see Table 3). The most abundant of these is kaolinite with a threshold temperature of −9°C; the kaolin minerals together with the illites and halloysite are considered to be the most important natural sources of efficient ice nuclei. Additional support for

TABLE 3

SUBSTANCES ACTIVE AS ICE NUCLEI

Natural Nuclei			Artificial Nuclei		
Substance	Crystal Symmetry	Threshold Temperature (°C)	Substance	Crystal Symmetry	Threshold Temperature (°C)
Covellite	Hexagonal	−5	Silver iodide	Hexagonal	−4
Vaterite	Hexagonal	−7	Lead iodide	Hexagonal	−6
β-tridymite	Hexagonal	−7	Cupric sulfide	Hexagonal	−6
Magnetite		−8	Mercuric iodide	Tetragonal	−8
Kaolinite	Triclinic	−9	Silver sulfide	Monoclinic	−8
Anauxite }	Monoclinic	−9	Silver oxide	Cubic	−11
Illite }	Monoclinic	−9	Ammonium fluoride	Hexagonal	−9
Metabentonite }	Monoclinic	−9	Cadmium iodide	Hexagonal	−12
Glacial debris	Hexagonal	−10	Vanadium pentoxide	Orthorhombic	−14
Hematite	Hexagonal	−10	Iodine	Orthorhombic	−14
Brucite		−11			
Gibbsite	Monoclinic	−11			
Dickite	Monoclinic	−12			
Halloysite		−12			
Volcanic ash	Hexagonal	−13			
Dolomite		−14			
Biotite	Monoclinic	−14			
Attapulgite	Monoclinic	−14			
Vermiculite		−15			
Phlogophite	Monoclinic	−15			
Nontronite		−15			

this view is provided by Isono (1959) and his colleagues who have used the electron microscope and electron-diffraction techniques to examine the nuclei at the centers of natural snow crystals. About 90% of the particles were identified as soil particles, 75% of these being clay particles, mostly kaolinite.

The possibility of inducing snow and rain by introducing into clouds artificial nuclei has stimulated many investigations of the ice-nucleating ability of a wide variety of chemical compounds, but there has been little agreement in the results published by different workers. Careful tests in the author's laboratory indicate that many of the published results are spurious because of the presence, in the air or in the chemicals, of small traces of silver or free iodine, leading to the formation of silver iodide, which is the most effective of all substances studied so far. If all such trace impurities are removed, many of the suggested substances are found to be quite ineffective. There remain those substances listed in Table 3; of these, the first six, which are only slightly soluble, are active to the extent of about one particle in 10^4 producing an ice crystal at the indicated threshold temperature when introduced into a supercooled cloud formed in a cloud chamber. NH_4F, CdI_2, and I_2, being soluble in water, are inactive in an atmosphere saturated relative to liquid water, but produce ice crystals in an environment supersaturated relative to ice but subsaturated relative to water, at the temperatures indicated. More detailed accounts of these experiments are given by Mason and Hallett (1956, 1957) and by Mason and van den Heuvel (1959).

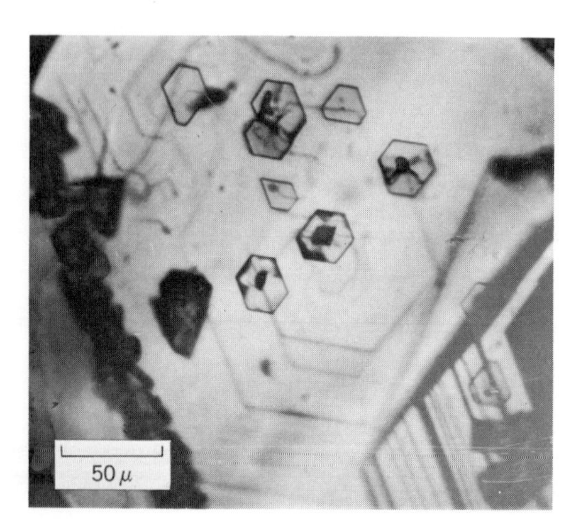

Fig. 9. Ice crystals growing on a single crystal of cadmium iodide. They form preferentially at the edges of growth steps which emanate in a spiral terrace from the cadmium iodide surfaces.

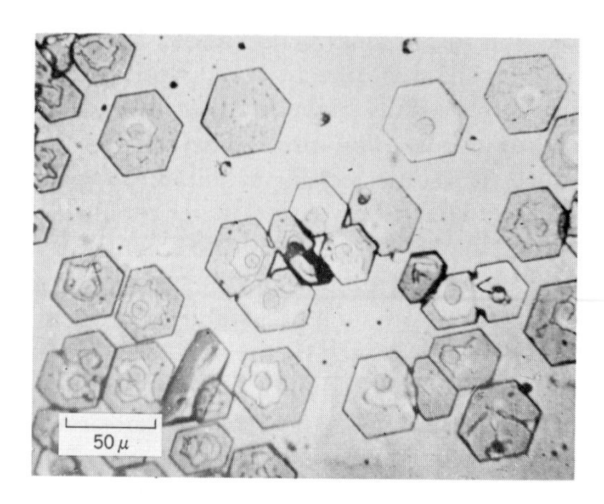

Fig. 10. A deposit of hexagonal ice crystals oriented parallel to one another on the basal surface of a single crystal of silver iodide.

Although there is a tendency for the more effective nucleators to be hexagonally symmetrical crystals in which the atomic arrangement is reasonably similar to that of ice, Table 3 shows that there are a number of exceptions; but all of the substances which are active above $-15°C$ possess a low index crystal face in which the degree of misfit between the ice and substrate lattices is less than 15%. However, there is not, in general, a high correlation between the threshold nucleation temperature and the degree of misfit, indicating that nucleating ability is only partly determined by geometrical factors.

In an attempt to investigate the nucleating mechanism in more detail, Bryant, Hallett, and Mason (1959) have studied the growth of ice on well-defined faces of single crystals of various nucleating agents under carefully determined conditions of temperature and supersaturation of the vapor. Epitaxial deposits of ice crystals have been observed on hexagonal crystals of silver iodide, lead iodide, cupric sulfide, cadmium iodide, and brucite, and also on freshly cleaved muscovite mica, on mercuric iodide, iodine, calcite, and iodoform. This study has revealed the great influence of the surface structure and topography of the host crystal. The ice crystals show a marked tendency to form at special sites on the surface, especially at the edges of growth or cleavage steps and on etch pits and other imperfections. This is illustrated in Figs. 9 and 10. Crystals appear at these preferred locations for supersaturations of about 10%, but much higher supersaturations, exceeding perhaps 100%, are required for nucleation on the very flat, nearly perfect areas of the substrate surface.

4. VARIATION OF ICE CRYSTAL HABIT WITH TEMPERATURE AND SUPERSATURATION

The existence of at least three basic forms of natural snow crystals, and of an almost infinite number of variations on each of these main themes, suggests that their growth and development are complicated matters which may best be studied in the laboratory, where the whole life history of a crystal can be observed under controlled conditions.

The growth of ice crystals in supercooled water clouds, the latter being produced in room size cold chambers whose temperature could be controlled, has been studied by Aufm Kampe et al. (1951), and by Mason (1953). The clouds were produced by the introduction of steam into the chamber. Within a few seconds, the fog cooled to the temperature of the chamber and was then seeded with dry ice or silver iodide if insufficient natural ice nuclei were present. Mason projected a small pellet of dry ice to produce a horizontal seeded track near the top of the chamber, and caught the crystals on plastic-coated slides near the floor after they had fallen through about 3 meters of supercooled cloud. The plastic replicas of the crystals were subsequently studied and photographed under the microscope.

The observed changes of crystal habit with temperature were very similar in both sets of experiments and are summarized in Table 4.

The classification of these laboratory-produced crystals according to their temperature of formation bears a marked similarity to that of natural snow crystals (Table 2), showing that it is possible to simulate quite well the early stages of growth of snow crystals in the laboratory and, at the same time, determine the transition temperatures for the different crystal forms more precisely than can be done in the atmosphere. The most striking feature of Table 2 is the remarkable sequence of habit, plates → prisms → plates (and stars) → prisms, which occurs as the temperature is lowered from 0 to −25°C.

A series of experiments to study the growth of individual ice crystals in different environmental conditions has been de-

TABLE 4

CHANGES OF CRYSTAL HABIT WITH TEMPERATURE IN ARTIFICIALLY PRODUCED WATER CLOUDS (AFTER AUFM KAMPE ET AL. AND MASON)

Temperature Range	Crystal Habit
0°C to −5°C	Simple, clear hexagonal plates with no surface markings; some trigonal shapes
−4°C to −9°C	Prisms, some showing marked cavities and similarity to needles
−10°C to −25°C	Hexagonal plates showing ribs, surface markings, and tendency to sprout at corners
	Sector stars
	Dendritic stars, most prominent around −15°C
−25°C to −40°C	Single prisms, twins, and hollow prisms
	Aggregates of prisms and irregular crystals (Aufm Kampe et al.)

scribed by Nakaya (1951, 1954). The crystals were grown on a fine rabbit hair stretched on a frame and suspended in an air stream whose temperature and water vapor content could be varied. The crystal development was followed over a period of 30 min or so by time-lapse photography. The apparatus, shown in Fig. 11, consisted of two concentric glass cylinders, the warm water vapor from an electrically heated reservoir being convected upwards inside the inner tube, cooled on its way up, and returned through the annular space. The whole apparatus was placed in a thermostat and located in a cold chamber maintained at about $-30°C$. The degree of supersaturation in the experimental space was varied by altering the temperature T_w of the water in the reservoir R. The temperature T_a of the air in the immediate neighborhood of the growing crystal was a function of both T_w and the temperature T_t of the thermostat. For a given value T_w, the air temperature was regulated by adjusting T_t. Crystals were produced at various combinations of T_a and T_w, with a view to studying the relationship between the crystal form and the external conditions.

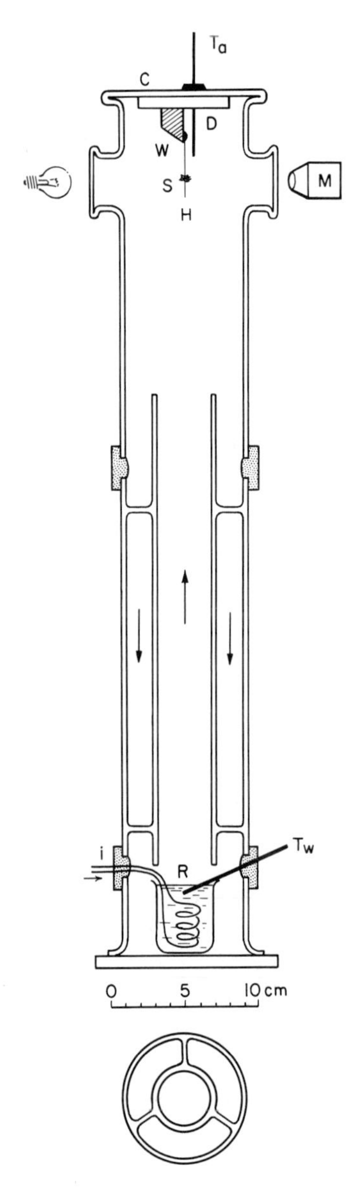

Fig. 11. Nakaya's ice-crystal growth apparatus. (From "The Compendium of Meteorology," *Amer. Met. Soc.* 1951, p. 207.)

Unfortunately, the conditions in Nakaya's apparatus were not steady nor well defined because the strong convection gave rise to large fluctuations in both temperature and supersaturation. The average temperature of the air in the neighborhood of the crystal was measured with an alcohol-in-glass thermometer. However, only rather crude relative estimates of the supersaturation were made by sucking the air through a filter containing P_2O_5 and measuring the increase in weight caused by absorption of both liquid and vapor. Nevertheless, it emerged that T_a and T_w were the two main parameters controlling the growth forms of the crystals; and Nakaya's classification of habit in relation to the air temperature T_a, shown in Figs. 12 and 13, is in broad agreement with the data of Table 4. Nakaya was of the opinion that the temperature rather than the supersaturation of the environment was the main factor controlling the crystal shape, except for dendritic growth, which occurred only at relatively high supersaturations and at air temperatures between -14 and $-17°C$. However, Marshall and Langleben (1954) interpreted Nakaya's results as showing that the habit is principally determined by the excess of the ambient vapor density over that which would be in equilibrium with the surface of the ice crystal, that is, by a quantity closely related to the supersatura-

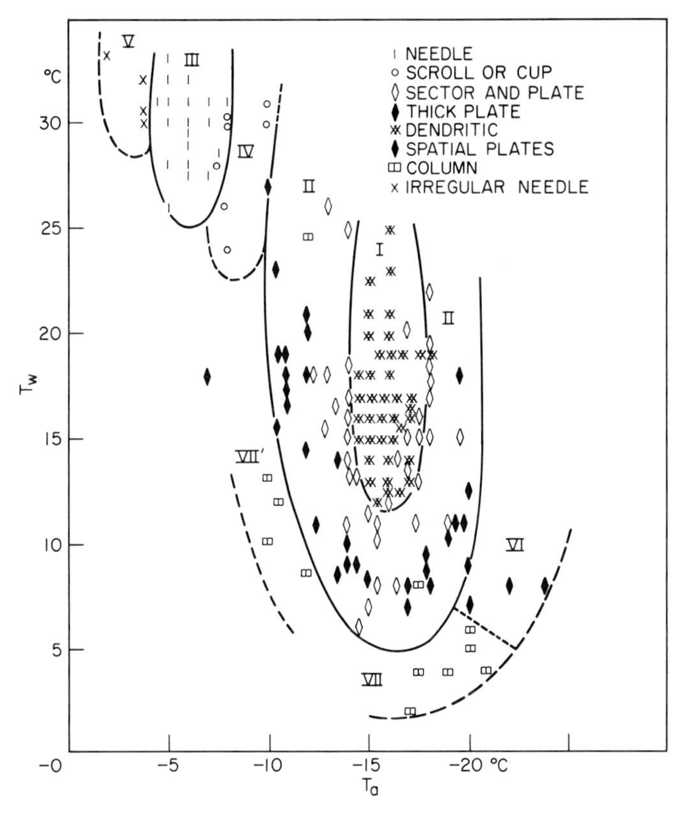

Fig. 12. Relation between Nakaya's crystal forms and the temperatures T_a and T_w. (From "The Compendium of Meteorology," op. cit.)

tion and proportional to the flux of vapor directed towards the crystal. A similar suggestion was made earlier by Weickmann (1950). Marshall and Langleben hypothesize that resistance to growth will be greatest at the crystal corners and greater on the prism faces than on the basal faces, so that growth of the corners and the prism faces will occur only when the excess vapor density $\Delta\rho$ becomes sufficiently large to overcome these inhibitions. Thus, prismatic columns would be expected to develop at relatively low values of $\Delta\rho$, plates only when $\Delta\rho$ is relatively large, and the corners (dendritic growth) only when $\Delta\rho$ achieves very high values.

Strong experimental evidence against such an interpretation was obtained by Shaw and Mason (1955), who studied the growth of individual ice crystals growing on a metal surface under conditions such that the temperature and supersaturation of the surrounding air could be controlled independently. The clean, smooth metal surface was located in the center of a cylindrical metal chamber, (Fig. 14), the hollow walls of which were cooled to the desired temperature by circulating chilled petrol through the annular space. The inner walls and base of the chamber were coated with ice, so that the air in the experimental space, which was stirred by a small fan, was saturated with ice at a temperature T_1, indicated by a thermocouple embedded in the surface of the ice layer in the bottom of the chamber. The metal plate on which the crystals grew was supported on a copper rod which was insulated from the chamber and dipped into a Dewar flask of liquid air. The plate was then cooled to a temperature T_2, lower than that of the surrounding air, which could be controlled by varying the current through a heating coil. The temperature T_2 of the crystals was recorded

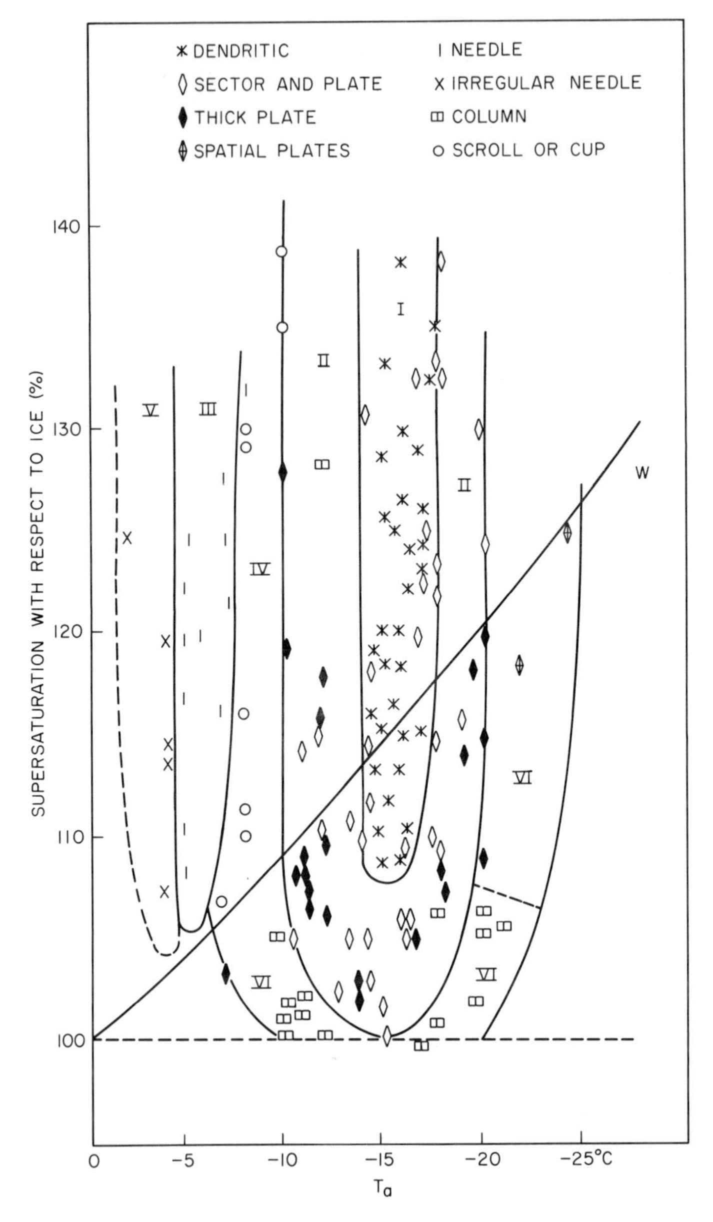

Fig. 13. Relation between the crystal type, air temperature T_a and the supersaturation relative to a plane ice surface, as observed by Nakaya.

by a thermocouple placed immediately below the metal surface, and the saturation ratio of the air in the immediate vicinity of the crystals was given by the ratio of the equilibrium vapor pressure over ice at the temperatures T_1 and T_2 respectively. The crystals were viewed through a metallurgical microscope and photographed at 1-min intervals. The growth rates of individual crystal faces were determined from measurements made on the negatives with a micrometer eyepiece.

In order to determine whether the various crystal forms appearing in natural and laboratory-produced clouds could be produced at the same temperatures on the metal plate (the plate temperature was held constant at various temperatures between -5 and $-40°C$), the crystals were grown under conditions of saturation relative to liquid water. The crystal habit was found to vary as follows:

$-5°C$ to $-9°C$	Prisms
$-9°C$ to $-25°C$	Plates
Below $-25°C$	Prisms

in a manner very similar to that observed in the cloud experiments. Stars were absent, but a new crystal form appeared quite often on the plate at temperatures between -4 and $-8°C$ and also below $-22°C$. This took the form of a prism terminated at one end only by a pyramid. The appearance of such hemimorphic forms, which have been found only very occasionally in natural clouds (usually cirro-stratus), but not at all in small-scale laboratory clouds, is of considerable interest in connection with the possibility that ice possesses a polar lattice.

When they were maintained at constant temperature and supersaturation, the crystals grew towards a limiting habit defined by

$$\Gamma = \left|\frac{c}{a}\right|_{\text{lim}} = \sqrt{\left(\frac{dc^2}{dt}\right)\Big/\left(\frac{da^2}{dt}\right)}$$

where c and a are respectively the principal and secondary crystal axes. Shaw and Mason established that this habit, and therefore whether a crystal developed as a prism or a plate, was determined very largely by the temperature, the supersaturation, although varied over wide limits, having no systematic effect. It was also established that a minimum critical value of

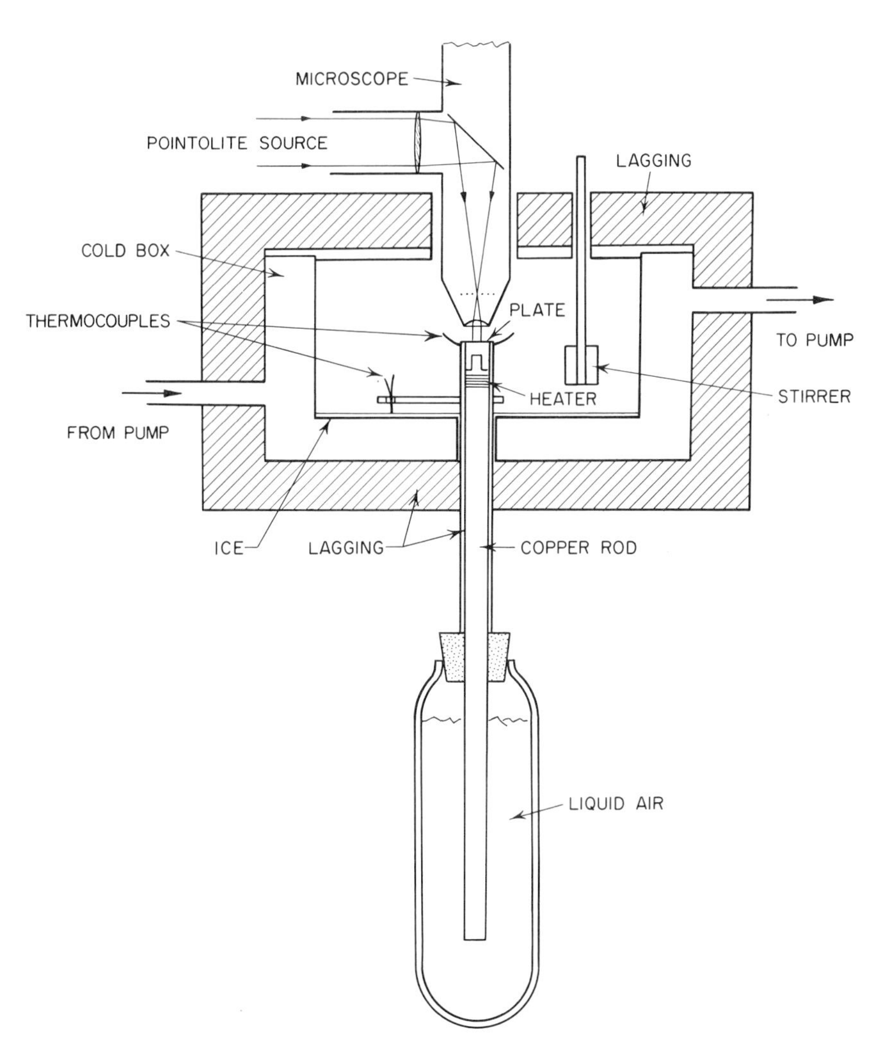

Fig. 14. The apparatus of Shaw and Mason for the growth of ice crystals from the vapor.

the supersaturation was required to start growth on any crystal face, but this varied in an apparently random manner from face to face and from crystal to crystal; there was no systematic difference between the critical supersaturations required for growth on the basal and prism faces as required by the hypotheses of Marshall and Langleben.

It must be admitted, however, that conditions for crystal growth on the metal surface may not have simulated fairly those occurring in the free air. To meet this point and to study the whole problem in more detail, a new series of experiments were begun by Mason and Hallett in 1955 (see Mason 1957b, Hallett and Mason 1958a).

The crystals are grown on a thin nylon or glass fiber running vertically through the center of a water vapor diffusion chamber, 50 cm high and 30 cm diameter, constructed of Perspex and resting on a solid aluminum block maintained at about $-60°C$ by dry ice. The apparatus is shown diagramatically in Fig. 15. Cooled from below, with its top maintained either at or above room temperature, the chamber encloses a thermally stratified, convectively stable atmosphere in which steady state conditions are readily achieved. Water vapor, evaporated from an extended source, diffuses downwards through the chamber towards the low-temperature sink, the supersaturation regime being largely determined by the temperatures of the source and sink, whose vertical separation can be varied. In such a chamber containing room air, condensation of water vapor upon the aerosol particles produces a dense cloud of tiny water droplets which freezes spontaneously on falling beneath the $-40°C$ level. In the presence of a persistent droplet cloud, the air may be regarded as saturated relative to liquid water, the supersaturation relative to ice at any level being determined solely by the temperature at the level. The variation of temperature with height is measured with a thermocouple, the separation of the 0 and $-40°C$ levels being about 10 cm.

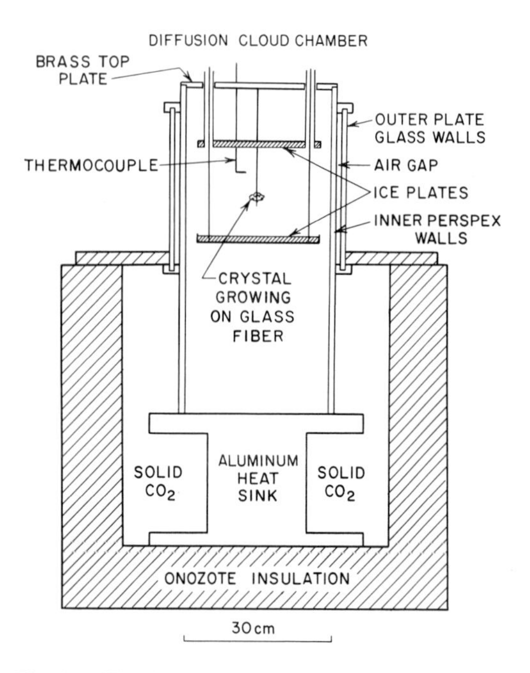

Fig. 15. The diffusion cloud chamber used by Mason and Hallett for growing ice crystals at various temperatures and supersaturations.

If the chamber is sealed and left for some hours, the condensation nuclei are progressively removed by sedimentation, leaving clean, highly supersaturated air in which condensation tracks produced by cosmic rays may be seen.

In order to simulate conditions in natural clouds, where ice crystals may grow at humidities below water saturation and therefore at supersaturations of only a few per cent relative to ice, the chamber was modified so that the crystals are grown on a fiber suspended centrally between two plane parallel sheets of ice. The vertical profiles of both temperature and supersaturation between the ice plates, the upper of which serves as the vapor source, is determined by their positions in the chamber, which are adjustable; additional control is provided by electrical heating of the top plate. A thermocouple passing through a small hole in this plate measures the temperature profile between the plates, and this, together with the solution of the diffusion equation for the profile of water vapor concentration, allows the supersaturation at each level to be calculated.

With these facilities, Mason and Hallett were able to grow crystals over the temperature range 0 to $-50°C$ and under super-

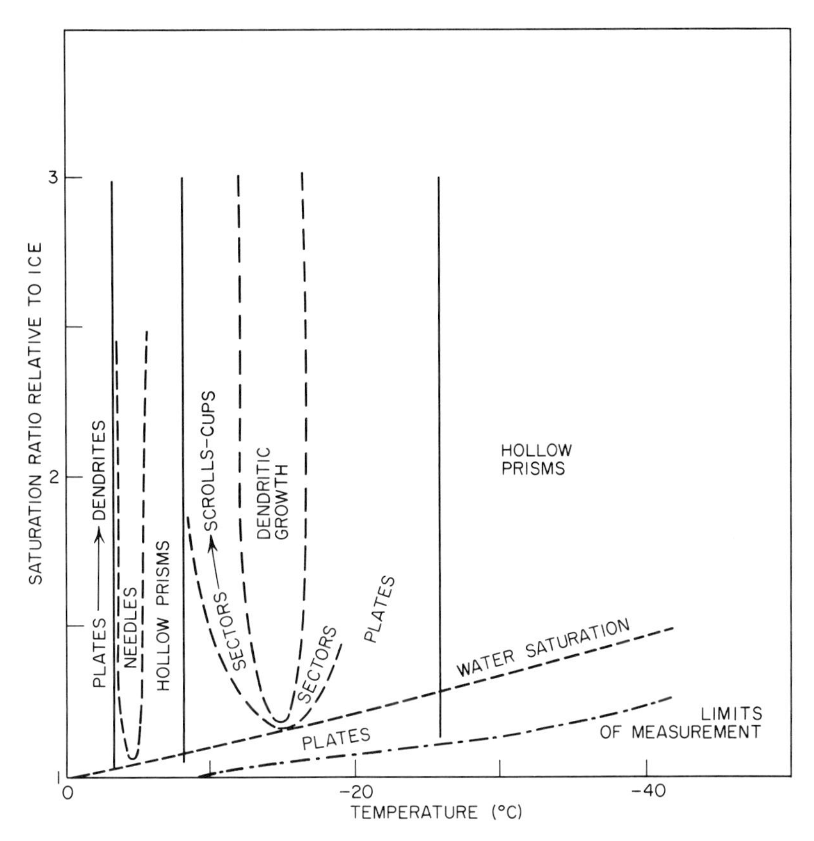

Fig. 16. The growth habits of ice crystals in relation to the temperature and supersaturation of the environment as observed by Mason and Hallett.

saturations ranging from a few per cent to about 400%. The results of many experiments are summarized in Fig. 16. Consistently, the crystal habit varied along the length of the fiber as shown in Table 5.

This scheme is very similar to that of Table 4, but the simultaneous growth of all the crystal forms on the same fiber brought to light the sharpness of the boundaries between one habit and another. For example, the transition between the plates and needles at $-3°C$ and that between hollow prisms and plates at $-8°C$ occurred within temperature intervals of less than one degree. A photograph showing the variation of crystal habit along the length of the fiber is reproduced in Fig. 17.

Figure 16 is similar, in many respects, to the diagram published by Nakaya (1954), but it covers much wider ranges of temperature and supersaturation. It differs from his in that it shows the existence of plates between 0 and $-3°C$, of hollow prisms rather than needles between -5 and $-8°C$, and does not show prisms to occur between -10 and $-15°C$ at low supersaturations.

Crystals having an almost identical variation of habit with temperature have also been grown by Mason and Hallett from the

TABLE 5

VARIATION OF CRYSTAL HABIT WITH TEMPERATURE (HALLETT AND MASON, 1958)

0°C to −3°C	Thin hexagonal plates
−3°C to −5°C	Needles
−5°C to −8°C	Hollow prisms
−8°C to −12°C	Hexagonal plates
−12°C to −16°C	Dendritic crystals
−16°C to −25°C	Plates
−25°C to −50°C	Hollow Prisms

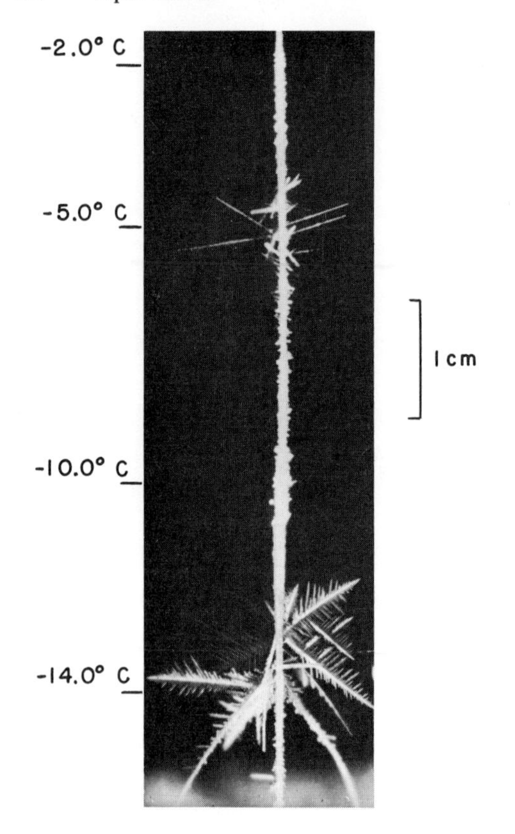

Fig. 17. The variation of crystal habit along a part of the fiber suspended in the diffusion cloud chamber. The temperature varied from -2 to $-16°C$ and the saturation ratio everywhere exceeded 2.0. The sequence, from top to bottom, is plates → needles → hollow prisms → plates → dendrites. (From J. Hallett and B. J. Mason, *Proc. Roy. Soc.*, **A247**, 440, 1958.)

vapor of heavy water (99.75% pure) but with the transition temperatures all shifted upwards by nearly 4°, in conformity with the difference between the melting points of H_2O and D_2O.

These experiments appear conclusive in showing that very large variations of supersaturation do not change the basic crystal habit as between prism and platelike growth although, of course, the growth rates are profoundly affected. On the other hand, the supersaturation appears to govern the development of various secondary features such as the needlelike extensions of hollow prisms, the growth of spikes and sectors at the corners of hexagonal plates, and the fernlike development of the star-shaped crystals, all of which occur only if the supersaturation exceeds values which, in these experiments, correspond roughly to saturation relative to liquid water.

The effect of suddenly changing the temperature and supersaturation on the growth form of a particular crystal could be observed simply by raising or lowering the fiber in the chamber. Whenever a crystal was thus transferred into a new environment, the continued growth assumed a new habit characteristic of the new conditions. Thus when needles grown at temperatures between -3 and $-5°C$ were suddenly moved up in the chamber, to about $-2°C$, plates developed on their ends (Fig. 18), and when similar needles were lowered to about $-14°C$, they gave way to star-shaped crystals as in Fig. 19. These are only some examples of metamorphoses which were observed when crystals were transferred to a new environment; in fact, combination forms at all the basic crystal types shown in Fig. 16 were readily produced this way. Such radical changes in crystal shape could not be produced by varying the supersaturation at constant temperature, but in some cases were produced by only a degree or two change in temperature at constant supersaturation.

Kobayashi (1957) has also used a diffusion cloud chamber to grow ice crystals on a rabbit hair at temperatures ranging from 0 to $-30°C$ and supersaturations up to about 90% relative to ice. He also concluded that the temperature was the main factor controlling the crystal habit and obtained a very similar scheme to that of Table 5. Kobayashi's technique of estimating the supersaturation in his chamber by observing the rate of growth of an ice sphere does not appear to be satisfactory, but he also finds that sector plates and dendrites do not often occur unless the air is supersaturated relative to liquid water, that is, supersaturated by at least 12% relative to ice at $-12°C$.

It is hardly possible to produce and measure very small supersaturations of only a few per cent in the diffusion chamber. Kobayashi (1960) attempted to achieve this in a convective-mixing chamber of the

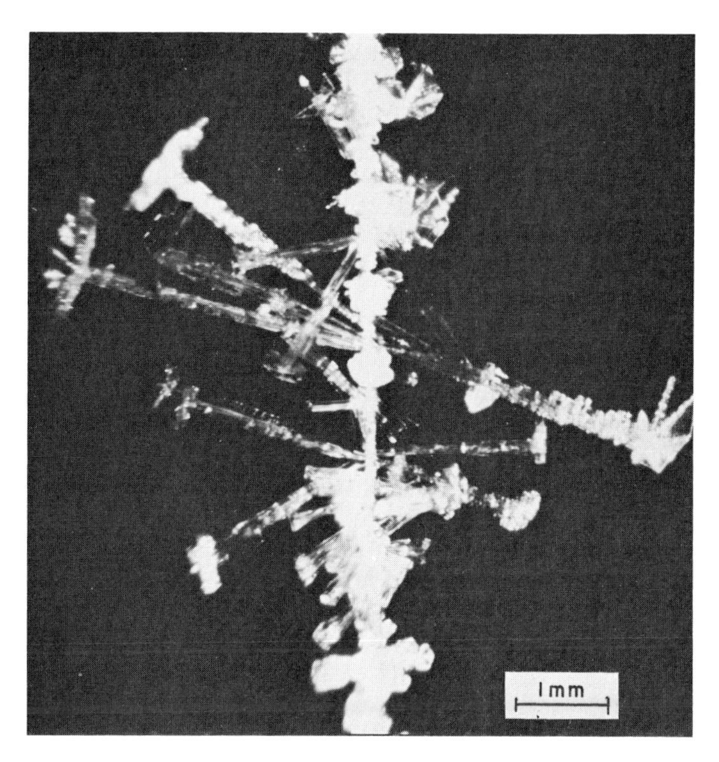

Fig. 18. Crystal hybrids showing how the crystal habit is dictated by temperature. Needles grown at $-5°C$ developed plates on their ends when shifted to a temperature of $-2°C$.

Fig. 19. Needles grown at $-5°C$ developed stars on their ends when shifted to $-14°C$. (From J. Hallett and B. J. Mason, op. cit.)

Nakaya type, the supersaturation being measured by sucking a known volume (about 2 liters) of air from the chamber through dried methanol and determining the *total* water content by titration with Karl Fischer reagent. Although an accuracy of ±5% was claimed for this method, it has the serious disadvantages of disturbing conditions in the chamber and that it can give only a mean value for the whole chamber and not that prevailing in the neighborhood of the growing crystal. The errors involved in determining the supersaturation are indicated by the fact that crystals continued to grow although the measurements indicated a *sub*saturation of 10%. Nevertheless, at the higher supersaturations, Kobayashi obtained a very similar variation of crystal habit with temperature to that given in Table 5, except that he found dendrites only between −12 and −14°C. At very low supersaturations, he claimed that only short, solid columns occurred at all temperatures between −8 and −23°C, the implication being that the temperature-dependent variation of habit shown in Table 5 no longer held under these conditions.

This apparent contradiction has recently been resolved by Kobayashi (1961) working in the author's laboratory with an improved technique in which two air streams saturated with respect to ice at different temperatures are mixed to produce a supersaturated mixture at an intermediate temperature. The ice crystals are grown on a fiber suspended in the mixing air currents, the temperature of which is measured by a thermistor and the supersaturation by a frost-point hygrometer. Careful measurement of time-lapse photographs of the crystals, taken over periods of about one hour, reveal that at very low supersaturations of a few per cent the crystals tend to grow as nearly isometric prisms, but in the temperature ranges at which plates usually appear (Table 5) the crystals grow as very thick plates and approach a limiting c/a ratio of 0.8, whereas at temperatures normally associated with prisms they approach a limiting c/a ratio of 1.4. The results of all our laboratory experiments on the variation of ice crystal habit with temperature and supersaturation or excess vapor density are consolidated in Fig. 20. It is clear that the principal factor controlling the *basic habit* as

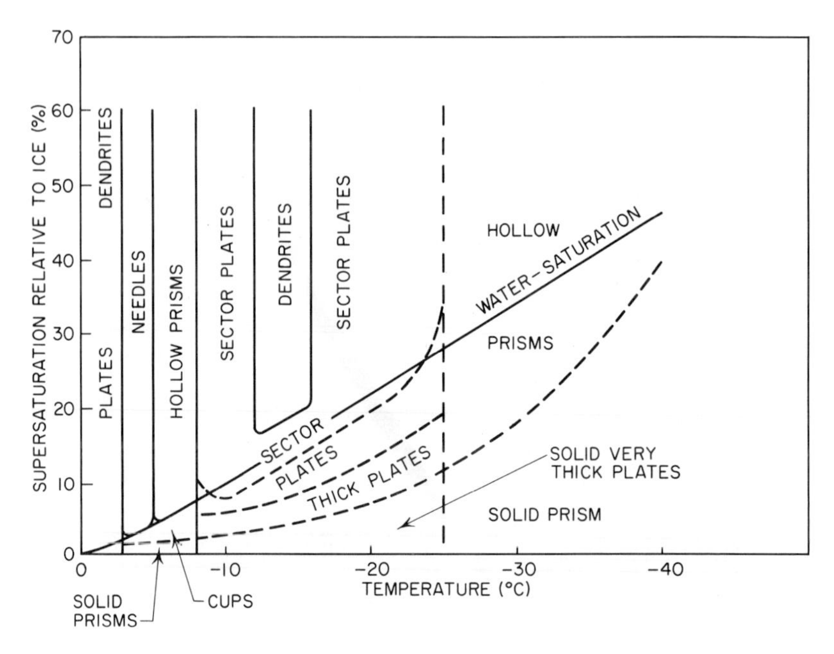

Fig. 20. The variation of crystal habit with temperature and supersaturation (according to experiments in the author's laboratory) including the results obtained at low supersaturations.

determined by the relative growth rates along the *c*- and *a*-axes is the temperature. But as the diagrams show equally clearly, the secondary growth features are determined by the supersaturation or flux of vapor. Increasing supersaturation causes transitions from very thick plate → thick plate → sector plate → dendrite, or from solid prismatic column → hollow prism → needle.

If, as Marshall and Langleben suggested, the basic habit was determined by the flux of vapor towards the growing crystal, that is, by $D\Delta\rho$, where D is the diffusion coefficient of water vapor and $\Delta\rho$ the excess of the ambient vapor density over that in equilibrium at the crystal surface, it might be expected to change if the crystals were grown in other gases or in air at reduced pressure. In fact, Isono et al. (1957, 1958) reported that when crystals were grown in hydrogen and in air at low pressure, they observed nearly isometric growth ($c/a \simeq 1$) throughout the temperature range -7 to $-16°C$, a region in which plates and dendrites are formed in air at normal pressure. These authors claim that the environment was kept saturated relative to liquid water and attribute the observed habit changes to the more rapid diffusion of vapor in the hydrogen and low-pressure atmospheres. On the other hand, van den Heuvel and Mason (1959) found that the variation of habit with temperature, involving plates, needles, hollow prisms, and dendrites, was unaffected by reducing the air pressure to 20 mm Hg or by replacing the air by carbon dioxide, hydrogen, or helium; only the *growth rate* of the crystals was affected as was anticipated in view of the differences in the diffusion coefficients and thermal conductivities of the various gases.

Kobayashi (1958) has also investigated the effect of lowering the air pressure on the growth habit. At pressures greater than 300 mm Hg, the normal variation of habit with temperature was observed; at pressures between 300 and 70 mm Hg, hollow prisms were observed in the temperature range -10 to $-22°C$, but below 70 mm Hg these gave way to solid prisms. These latter crystals grew very slowly, suggesting that the increased value of D was more than offset by a reduction in $\Delta\rho$. Kobayashi has concluded that his experimental arrangement was such that a large reduction in the air pressure caused the supply of water vapor to the crystals to be much reduced, something which did not occur in the experiments of van den Heuvel and Mason. It seems therefore that the latter were correct in stating that the basic crystal habit is not modified by the nature and pressure of the carrier gas.

5. THE INFLUENCE OF IMPURITIES ON ICE CRYSTAL HABIT

It has been reported by Nakaya (1955) that the growth forms of ice crystals may be modified by the presence of atmospheric aerosols. For example, he stated that when the air was filtered to remove the suspended particles, the plate and dendritic crystals which normally occur in the temperature range -10 to $-20°C$ were replaced by hollow prisms. Hallett and Mason (1958) have been unable to confirm this result. Crystals growing in their diffusion chamber show exactly the same variation of habit with respect to temperature irrespective of whether they grow in the presence of a water cloud formed by condensation on atmospheric aerosol particles, or in very clean air from which all the particles have been removed by sedimentation.

Later, Nakaya, Hanajima, and Mugurama (1958) reported that the changes of habit had nothing to do with the removal of aerosol but were caused by a trace of silicone vapor leaking into the chamber. This recalls the earlier findings of Schaefer (1949) that the habits of ice crystals produced by seeding a supercooled cloud were changed by the presence of vapors of ketones, fatty acids, silicones, aldehydes and alcohols.

Hallett and Mason (1958b) also found that traces of organic vapors profoundly influenced the habit of ice crystals grown in their diffusion chamber. For example, needlelike crystals appeared at all temperatures between 0 and $-40°C$ when a small

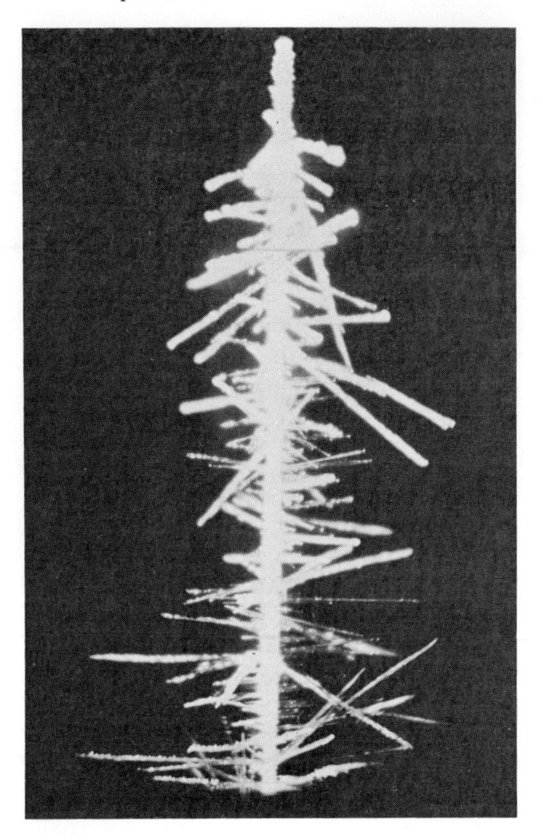

Fig. 21. The effect of introducing a small quantity of camphor vapor into the apparatus; the crystals now assume a needlelike form at all temperatures.

trace of camphor vapor was introduced into the chamber (see Fig. 21). Even more remarkable changes followed the introduction of isobutyl alcohol. In the temperature range -12 to $-16°C$, when the partial pressure of alcohol exceeded about 10^{-3} mb, the normal dendritic growth was suppressed and replaced by platelike crystals. A further rather small increase in alcohol concentration caused these to be replaced by either hollow prisms or needles which persisted until the concentration was raised beyond 0.1 mb, above which there was a further transition to plates, and finally at higher concentrations, a reversion to a rather malformed type of dendritic growth.

An earlier observation by Vonnegut (1948) that a cloud of tiny hexagonal platelike crystals growing at $-20°C$ was transformed into prismatic columns by the addition of butyl alcohol at about 10^{-2} mb partial pressure was confirmed by Hallett and Mason, with a reversion to plates if the concentration exceeds about 0.2 mb. Similar transitions are also observed in the temperature range -8 to $-12°C$.

6. THE MECHANISM OF HABIT CHANGE

It can be seen from Table 5 that the habit of ice crystals changes from being essentially platelike to being columnlike four times over the narrow temperature range 0 to $-25°C$; these changes, which reflect changes in the relative growth rates of the basal and prism faces of the crystals, are governed by the temperature and may be induced when this changes by as little as one degree. There has been no convincing explanation of these habit changes, which appear peculiar to ice, because present theories of crystal growth do not contain parameters that are likely to be strongly face-dependent and also sensitive to temperature changes of only a few degrees. The author has long been of the opinion that the variations of habit are to be attributed to a *surface* property of ice (a view which receives some support from the fact that they may also be induced by the adsorption of traces of certain impurities) and that the surface diffusion of molecules on the growing faces is most likely to be the responsible parameter.

The first direct experimental evidence that water molecules arriving on the surface of a growing ice crystal may travel considerable distances before being built into the crystal structure was obtained by Bryant, Hallett, and Mason (1959) when studying the epitaxial growth of ice crystals on the basal faces of crystals of natural cupric sulfide (covellite). The ice crystals, being only a few thousand angstroms high, exhibited interference colors when viewed in reflected white light. The colors gave a measure of the crystal thickness, where changes could be measured to within an accuracy of about 150 A. Under low supersaturations, some crystals grew considerably in diameter with no discernible change of thickness. This suggested that molecules arriving on the upper basal surface were not being assimilated but were migrating over the surface and being built into the prism faces. This was also suggested by the fact that the lateral growth rate of two neighboring platelike crystals of constant, nearly equal thickness did not decrease as they approached each other, even when the intervening gap narrowed to about 1 μ. Had the crystals been growing mainly by direct deposition of vapor molecules on their edges they should have slowed down as they approached and shadowed each other, but this was not to be expected if growth occurred mainly by migration of molecules from the top surface.

Frequently, a small platelike crystal did not thicken until it collided with a cleavage step on the substrate or a neighboring crystal; colored growth layers, often originating from the point of contact (see Fig. 22), then spread across the crystal surface. Hallett (1961) has established that at constant temperature and supersaturation the layers travel with a velocity inversely proportional to their thickness, which indicates that they advance as the result of molecules diffusing across the surface and becoming incorporated at the edges of the growth fronts. If x_s is the average distance which a molecule travels on the ice surface before reevaporating, all molecules arriving within a distance x_s on either side of the growing step will contribute to its growth. If F(gram cm^{-2} sec^{-1}) is the flux of vapor reaching the surface, the rate of direct deposition on unit width of a step of height h will be Fh(gram sec^{-1}), the rate of deposition by surface diffusion $2x_s/h\ Fh$, and the total deposition rate $(1 + 2x_s/h)Fh$. The growth front therefore advances at a velocity $v = (1 + 2x_s/h)F/\rho_i$, where ρ_i is the density of ice. Then, if growth occurs mainly by surface diffusion, $v \propto 1/h$.

If the separation of neighboring growth fronts exceeds $2x_s$, the fronts will move independently, but if their separation is less than $2x_s$, their collecting zones will overlap, and they will compete for the available material and therefore advance more slowly than more widely separated layers. By measuring the velocity of layers of compara-

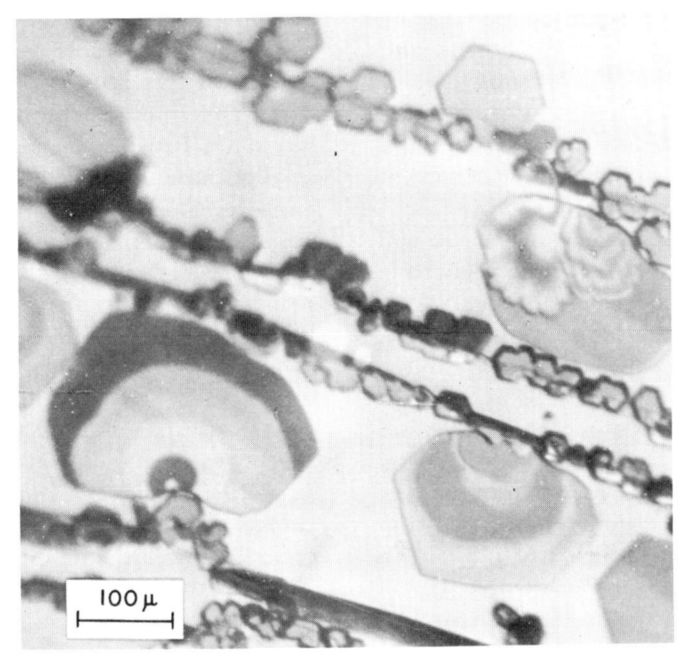

Fig. 22. Growth layers spreading across ice crystals growing epitaxially upon a basal surface of covellite. The layers often spread out from the point of contact with another crystal. The ice crystals form preferentially along cleavage steps on the substrate surface.

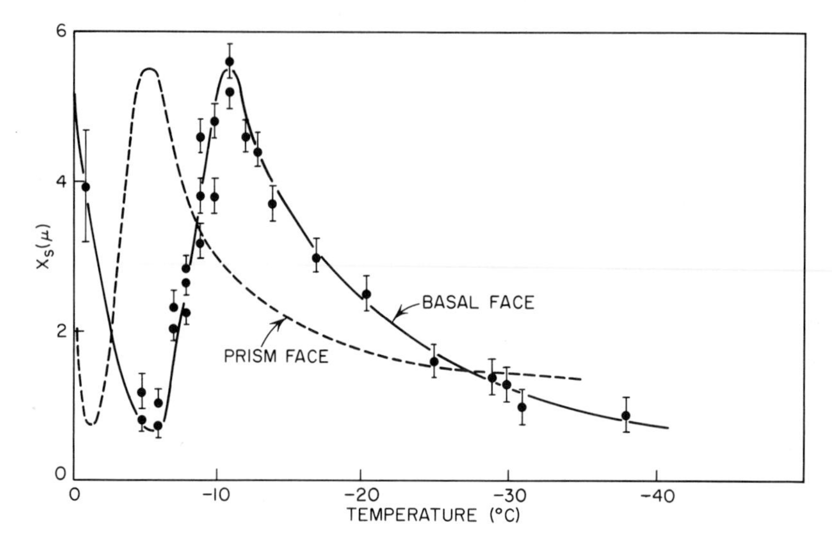

Fig. 23. The mean surface migration distance x_s as a function of temperature.

———— Basal face ----- Suggested curve for prism face.

ble but unequal thickness as a function of their separation, Hallett (1961) attempted to determine the critical separation at which they began to slow down, and hence a value for $2x_s$. He obtained two values of $x_s = 5\ \mu$ at $-10°$C and $2.5\ \mu$ at $-17.5°$C. A much more complete and accurate set of measurements has recently been completed by Mason and Bryant (to be published), and the values of x_s obtained at different temperatures are shown in Fig. 23. The variation of x_s with temperature is remarkable and suggests an explanation for the variation of crystal habit with temperature.

Since molecules arriving on the crystal surface have a surface migration length of several microns, they may reach an adjacent face before becoming built into the lattice. The initial habit development from the embryo stage may therefore be determined by the relative values of x_s for the basal and prism faces. If x_s is greater for the basal face, there will be a net transport of material by surface diffusion to the prism faces and the crystal will start to develop a platelike habit. The reverse will be true if x_s is greater for the prism faces. Figure 23 shows that x_s for the basal face varies markedly with temperature.

Unfortunately, we do not yet have corresponding measurements for prism faces because it is difficult to grow prisms thin enough for use with the interference technique. However, it is not unreasonable to suppose that the curve of x_s versus temperature for the prism faces will be of the same general shape as that for the basal faces but displaced a little along the temperature axis as indicated by the broken curve of Fig. 23. The two curves may then intersect at three points, giving four temperature ranges in which the ratio $x_s^{(B)}/x_s^{(P)}$ alternates between being larger than and smaller than unity. When $x_s^{(B)} > x_s^{(P)}$, the early development will be platelike, whereas prismatic columns will develop at temperatures for which $x_s^{(P)} > x_s^{(B)}$. To explain the sequence of habit changes shown in Table 5, the two curves would have to intersect similarly to that shown in Fig. 23; the intersections at $-3°$C and $-8°$C suggest that transitions of habit would occur very sharply at these temperatures, whereas that at $-25°$C suggests a much more gradual transition from plates to prisms. This is entirely in accord with observation.

Once the crystal dimensions become large compared with x_s, surface diffusion will have little further effect on habit development, which will now be mainly controlled by the three-dimensional diffusion field. However, once a habit has been established in the early stages of growth, the diffusion field around the crystal will orient itself to conform to the crystal geometry and tend to maintain it. For example, the lines of diffusive flux will tend to concentrate towards the edges and corners of a hexagonal plate and accentuate its development. The fact that, at moderate supersaturations, crystals continue to develop as polyhedra suggests that the excess material arriving at the edges and corners is redistributed over the surface by surface diffusion. However, as the supersaturation of the vapor phase is increased, surface diffusion becomes unable to cope with the nonuniform deposition, so the corners and edges begin to sprout to form sector plates, dendrites, hopper crystals, and other skeletal forms.

Accordingly, we regard the crystal habit to be determined by the interaction between its surface properties (surface migration) and the diffusion field. The relative rates of surface diffusion on the basal and prism faces are responsible for determining the habit in the very early stages of growth, and this is later maintained and accentuated by diffusion of vapor to, and conduction of latent heat of crystallization away from, the crystal surface.

7. THE SURFACE STRUCTURE OF ICE CRYSTALS

Crystals grown under moderate and high supersaturations, both plates and prismatic columns, show hopper development with terraced faces as shown in Figs. 24 and 25. When the thin, hexagonal plates are viewed in transmitted light, this development is barely discernable, but when seen in reflected light the surfaces are covered with transverse and radial ridges as shown in Fig. 24. The formation of these terraces has been studied by time-lapse photography and their contours examined by making formvar replicas of the surface, which were then stripped, silvered, and examined under the microscope. The ridges shown in Figs. 24*a*,*b* are several

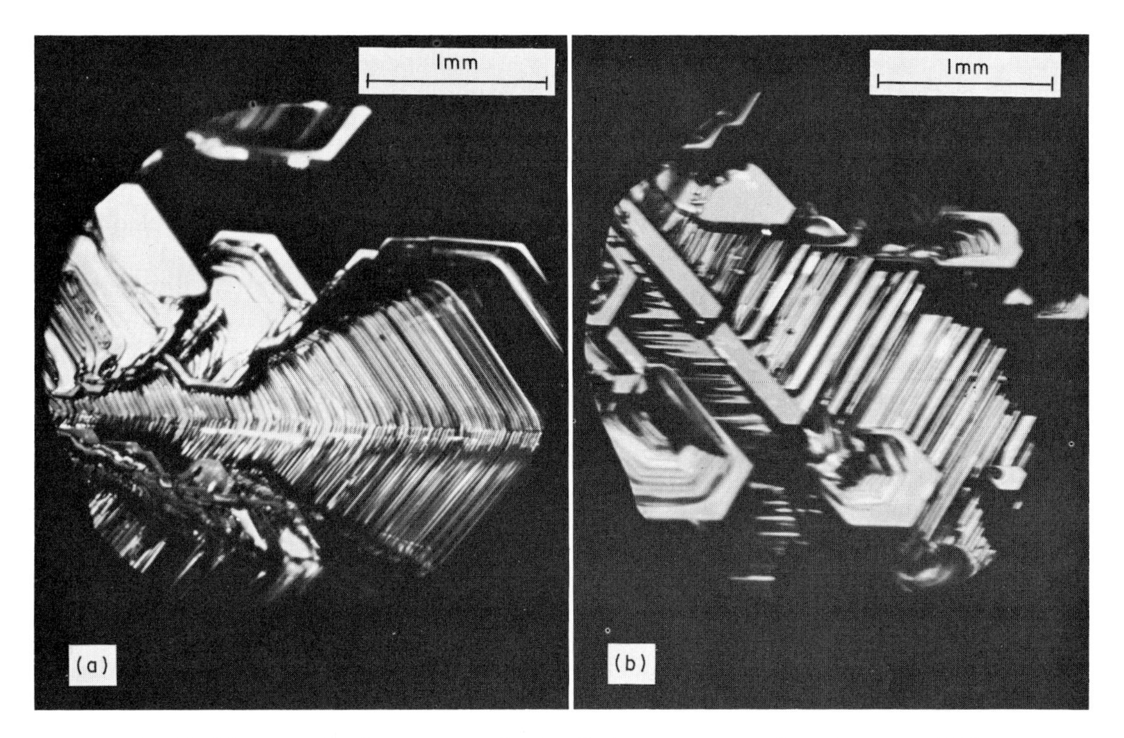

Fig. 24. The stepped-surface structure of a platelike ice crystal grown from the vapor. (*a*) The most rapidly growing face with steps about 5 μ high and 25 to 50 μ across. (*b*) This face was partly shielded and grew more slowly, the steps now being at least 10 μ high and usually 50 to 150 μ across.

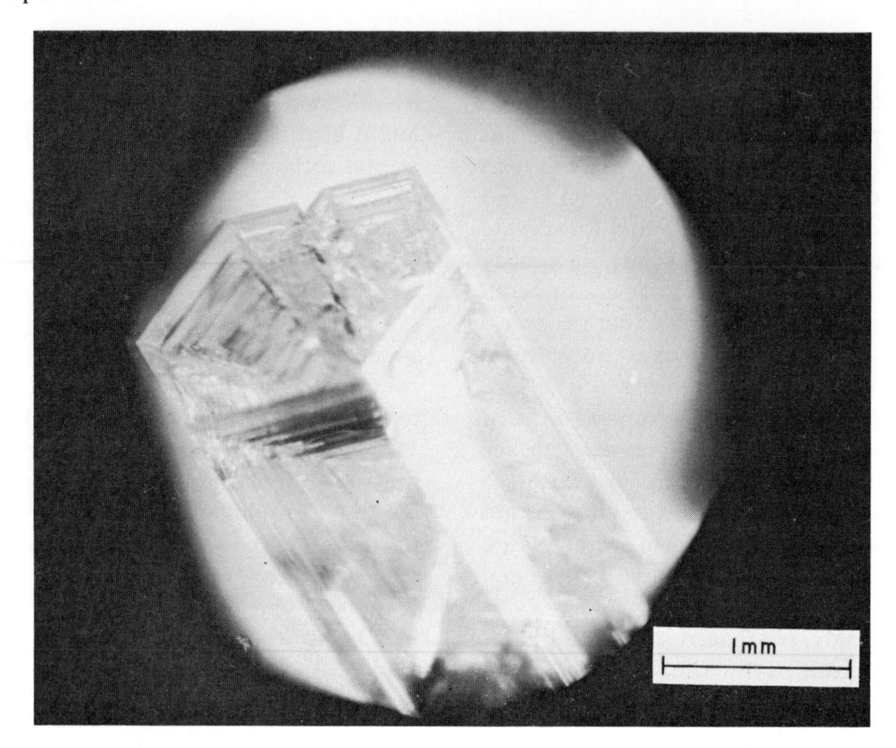

Fig. 25. A hopper prismatic column grown from the vapor showing the incomplete and stepped basal and prism faces.

microns high; visible steps appear only when the crystal diameter exceeds a few hundreds of microns. The author believes that they result from the bunching of much thinner growth layers spreading out from the leading edge of that crystal where they have been nucleated under the high local supersaturation. The growth and amalgamation of layers, 150 to 5000 A high, have been directly observed on ice crystals by the interference technique described in Section 7.

It may be shown (Mason, unpublished) that if, for example, a random fluctuation in supersaturation causes two successive members of a family of growing steps to be nucleated with separations less than $2x_s$, then they may close up, amalgamate, and form a double step traveling at half the speed. This, in turn, may be caught by a third layer and result in the formation of a triple step traveling at one-third the speed of a unit step. This bunching process will continue to produce progressively thicker and slower moving steps against which thinner layers will pile up as in a traffic jam.

It may be shown that the time taken for two layers of comparable but unequal thickness to close, having reached a separation of $2x_s$, is given by $t = (4x_s \ln 2)/v$, where v is the uniform rate-of-change of separation beforehand. The distance traveled by an m step in growing to $(m + 1)$ units is $2x_s/(m - 1)$. The horizontal displacement of a step while growing to N units is $2x_s \sum^{N=2} \frac{1}{n}$, and the time taken is

$$\frac{4x_s \ln 2}{v_0} \sum_{1}^{N=2} \frac{1}{1 - \frac{1}{n}} \simeq N \frac{4x_s \ln 2}{v_0},$$

if N is large, v_0 being the velocity of a unit step. The minimum distance from the edge of a crystal at which a step many units high may appcar is about $20x_s$.

These calculations are found to be in good agreement with measurements made from ciné-films of growing ice crystals.

If the crystal is practically stationary relative to the surrounding vapor, layers will usually be nucleated near the edge of the crystal where the supersaturation is highest and will then spread more slowly as they approach the center of face where the supersaturation is lower. This gradient of supersaturation, together with the onset of bunching, allows new growth layers to form at the edge before earlier ones have completed their travel to the face center. This leads to a preferential thickening at the periphery, more pronounced starvation of the face center, and thus to hopper development, either in the form of shallow hexagonal dishes (plate regime) or as hollow prisms (see Fig. 26). Simultaneous hopper development of basal and prism faces is shown in Fig. 25, where the incomplete development of basal planes leads to the formation of hollow funnel-shaped cavities which are also in evidence in the natural ice crystals of Fig. 1.

If, on the other hand, the crystal grows in a well-stirred medium, the supersaturation tends to be more uniform over the crystal faces, and hopper development is less marked.

We have seen that both the growth rate and growth habit of ice crystals are governed partly by the environmental conditions—the temperature, supersaturation, disposition of concentration and thermal gradients; and partly by conditions at the crystal surface—the surface concentration of material and its migration across the surface to the growth sites where it becomes built into the crystal lattice. Growth involves the initiation and spreading of new layers. It is now a generally recognized feature of crystal growth that new layers cannot be formed at an observable rate by two-dimensional nucleation on a molecularly flat surface except at very high supersaturations. At modest supersaturations, observable growth rates can be achieved only if the free-energy requirement for successive two-dimensional nucleation can be avoided. A way out was

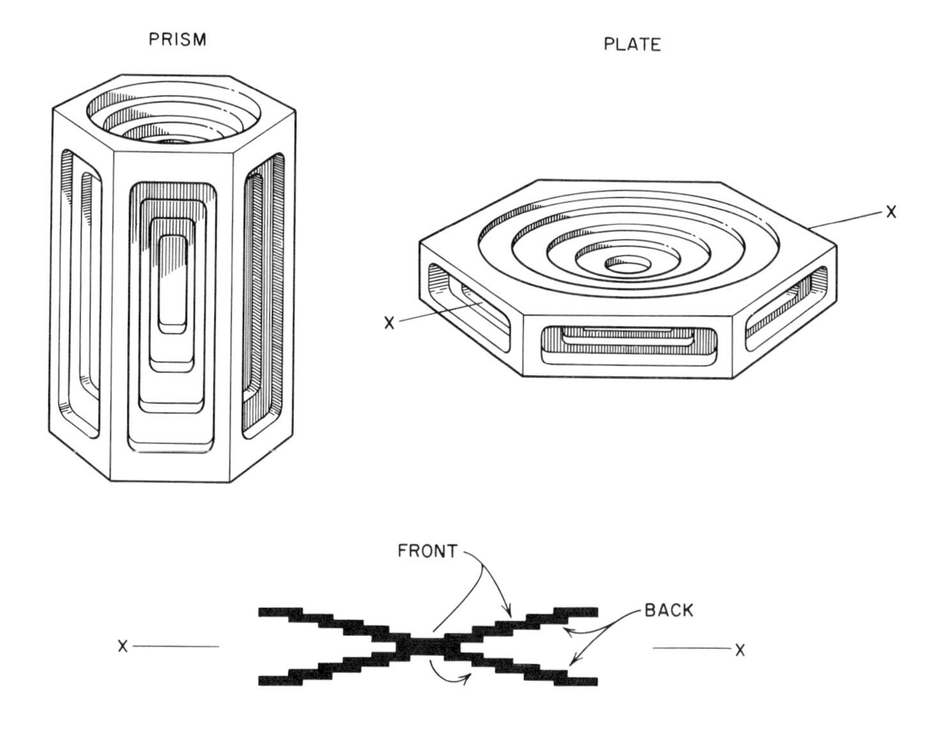

Fig. 26. A schematic diagram illustrating the hopper development of prismatic columns and hexagonal plates.

proposed by Frank (1949), who suggested that screw dislocations might terminate in a crystal face to produce a spiral terrace of steps, each of which would be capable of acting as a two-dimensional nucleus and which would not disappear during the deposition and evaporation of molecules. Growth fronts spreading out from the dislocation would form a growth pyramid.

Spiral growth fronts have since been identified on crystals of many different substances, whereas the dislocations themselves have been revealed by decoration techniques; but not with ice. However, it has recently been demonstrated by several workers that etch pits mark the location of emergent dislocations in many other substances; this method has now been used successfully with ice. The appearance of etch pits on the surfaces of ice crystals was first reported by Truby (1955). Hexagonal pyramidal features of base diameter $\frac{1}{2}$ to 20 μ and depth $\frac{1}{4}$ to $\frac{1}{2}$ μ were observed and sometimes exhibited a stepped-layer concentric with the c-axis. Higuchi (1958) and Mugurama and Higuchi (1959) discovered that etch pits could be produced on the surfaces of ice and snow crystals by etching with a formvar solution. Hexagonal pits appeared on the basal surface of a snow crystal in concentrations up to 10^5/cm^2. After prolonged etching the bottom surfaces of the pits revealed the layer-type structure noted by Truby, the height of the individual steps varying between 3 and 16 μ.

A detailed investigation of etch pits on ice crystals and an attempt to relate them to the dislocation structure has been made by Bryant and Mason (1960). The surfaces of crystals grown from both the vapor and the melt were etched with a 1% solution of formvar in ethylene dichloride. Replicas of etch pits were studied at successive stages of development by peeling off the plastic film with adhesive tape, evaporating a thin layer of silver onto the film, and viewing it

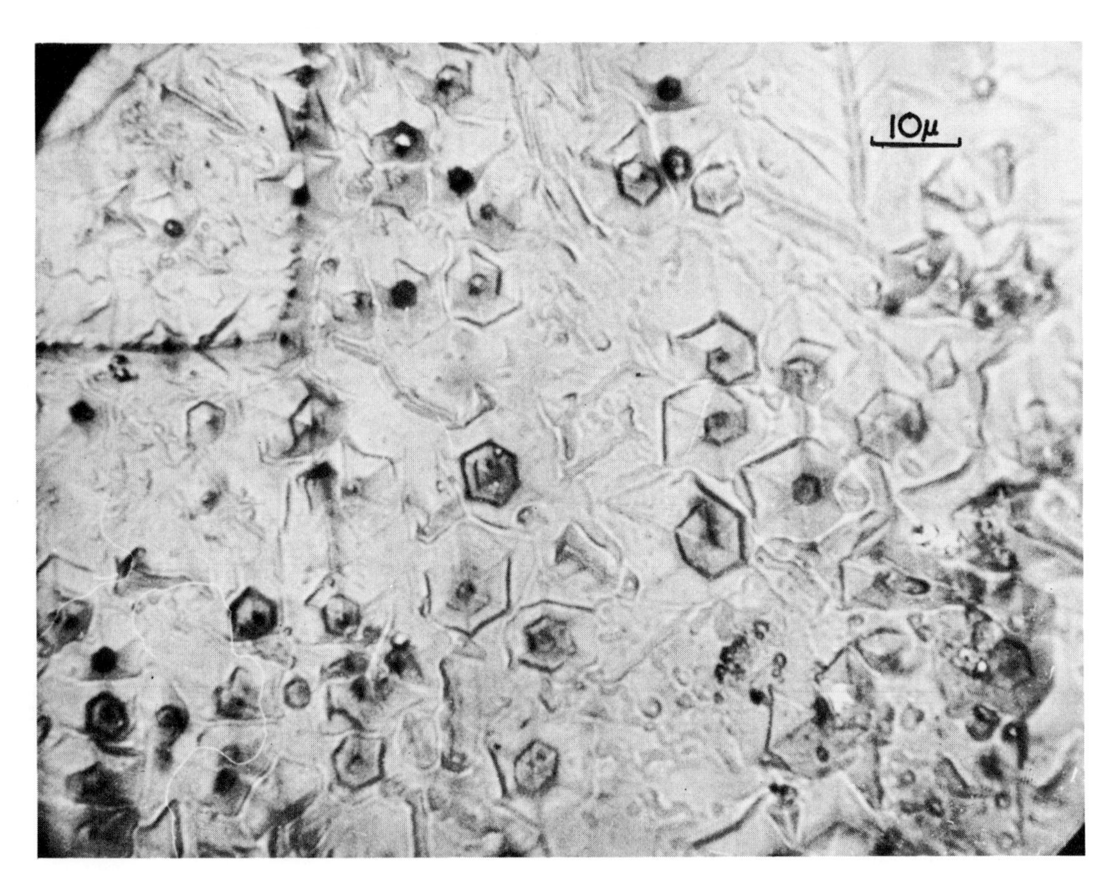

Fig. 27. Replica of a basal face of ice showing spiral steps on etch pits

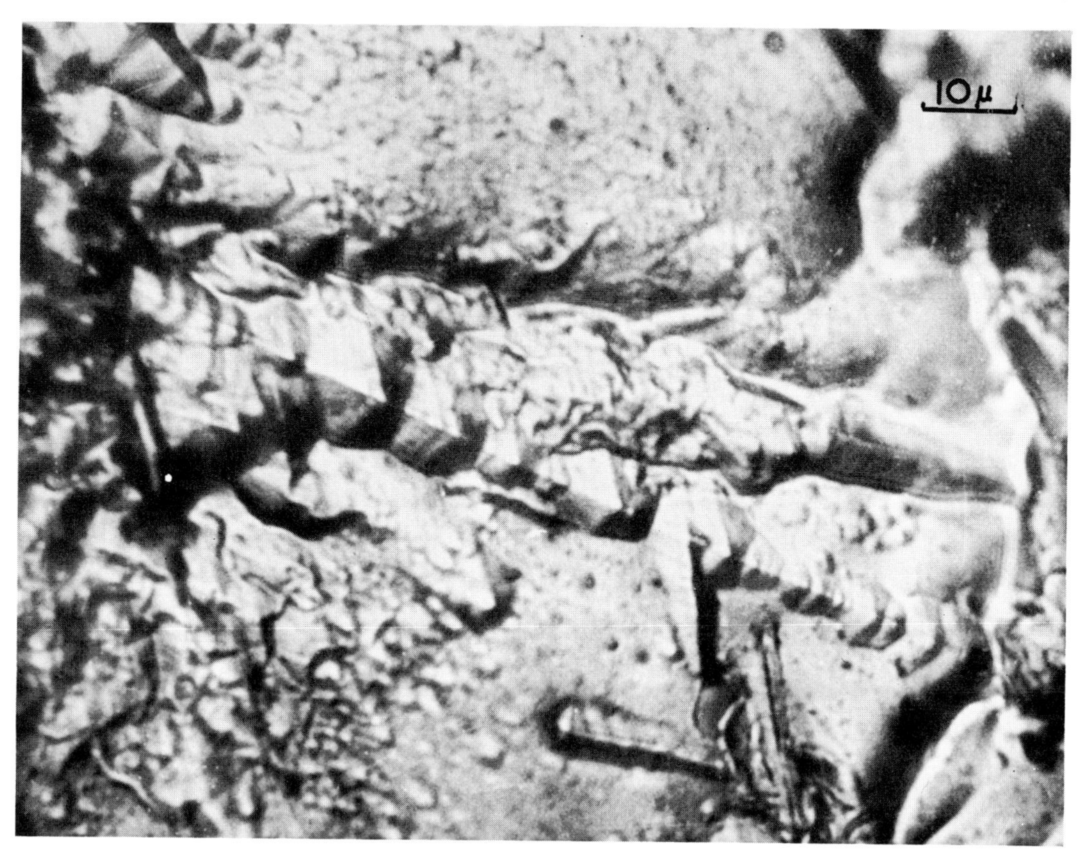

Fig. 28. A line of pyramidal etch pits along a low-angle grain boundary.

in reflected light with magnifications of $\times 400$.

Replicas of ice crystals grown from the melt showed many small pyramidal etch pits, 5 to 10 μ diameter, occurring in concentrations up to $10^5/\text{cm}^2$. They were mostly hexagonal in cross section, their sides being built up of several concentric steps each a few tenths of a micron in height. On many of the larger pyramids ($d > 10\ \mu$), these were accompanied by one, and sometimes two, large concentric steps of height up to 1 μ. A few per cent of these pits exhibited hexagonal spirals of similar step height as shown in Fig. 27. A silvered replica showing an (inverted) line of pyramidal pits formed at a grain boundary is shown in Fig. 28.

Ice crystals were grown from the vapor in the diffusion cloud chamber as described in Section 4. Replicas of previously unetched basal faces of platelike crystals revealed, instead of etch pits, raised hexagonal pyramids, usually less than 10 μ in diameter. Under high magnification these exhibited concentric rings and spirals similar to those observed in etch pits. These hillocks, shown in Fig. 29, were eroded by subsequent etching; they may have been formed by growth rather than by dissolution at preferred sites on the crystal surface.

Etching with formvar solution produced hexagonal etch pits on the basal surfaces of platelike crystals in concentrations of order $10^4/\text{cm}^2$, the edges of the pits being parallel to the crystallographic $\langle 11\bar{2}0 \rangle$ directions (see Fig. 30). Pits formed on the prism faces of ice needles were rectangular in cross sections and mostly flat at the bottom as shown in Fig. 31. Occasionally, a star-shaped etch channel, such as that in Fig. 32, was observed. The channels run parallel to the $\langle 11\bar{2}0 \rangle$ axes and are marked along their length by very small pits of $d < 1\ \mu$; they may have been originated on helical dislocations spreading out from a central

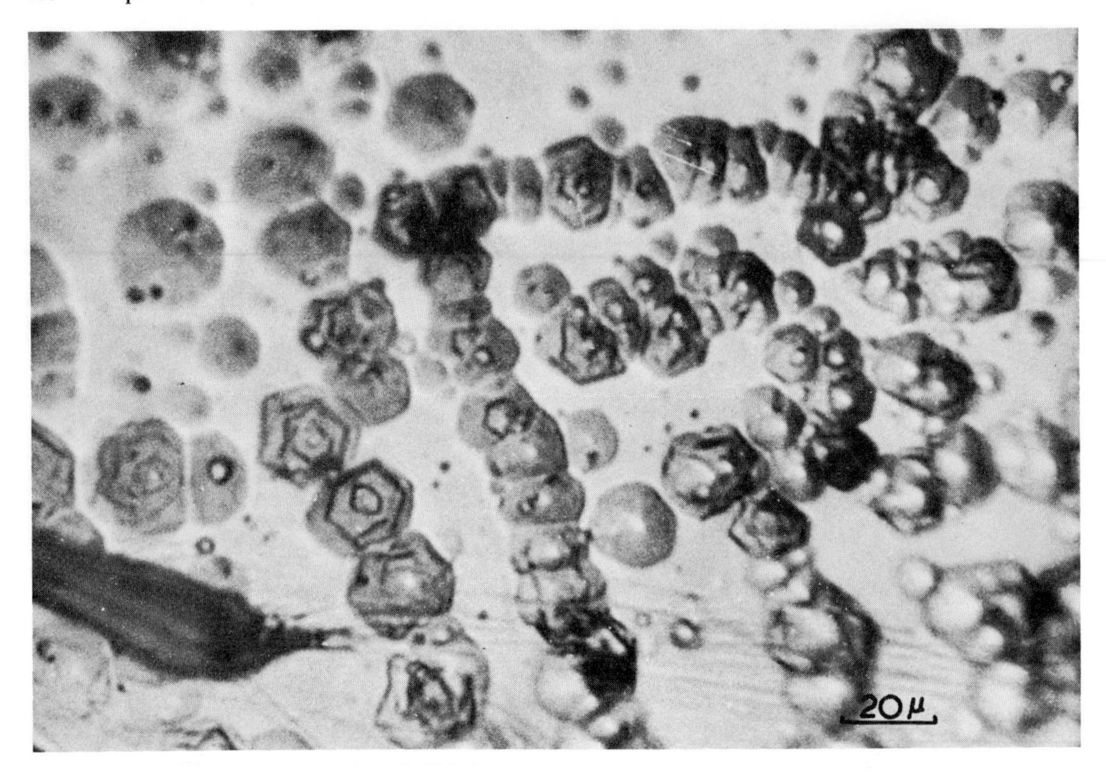

Fig. 29. Pyramidal hillocks showing concentric and spiral steps on the basal face of a crystal grown from the vapor.

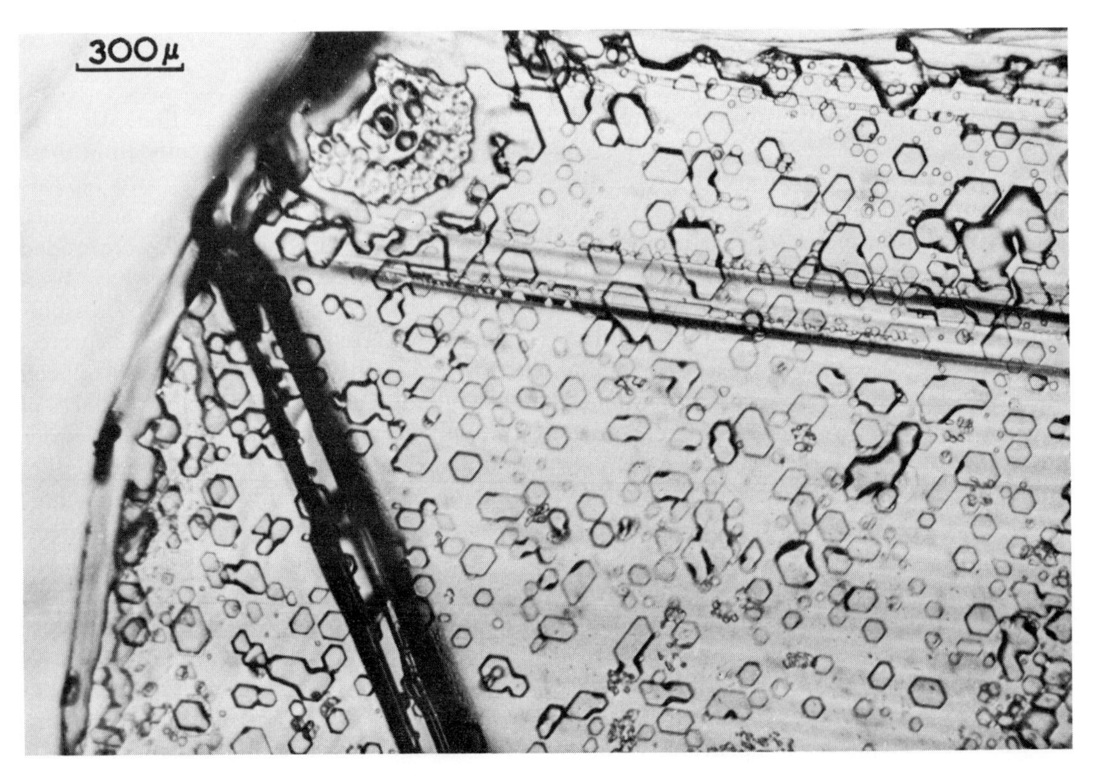

Fig. 30. A high concentration of hexagonal etch pits on the basal surface of a platelike crystal grown from the vapor.

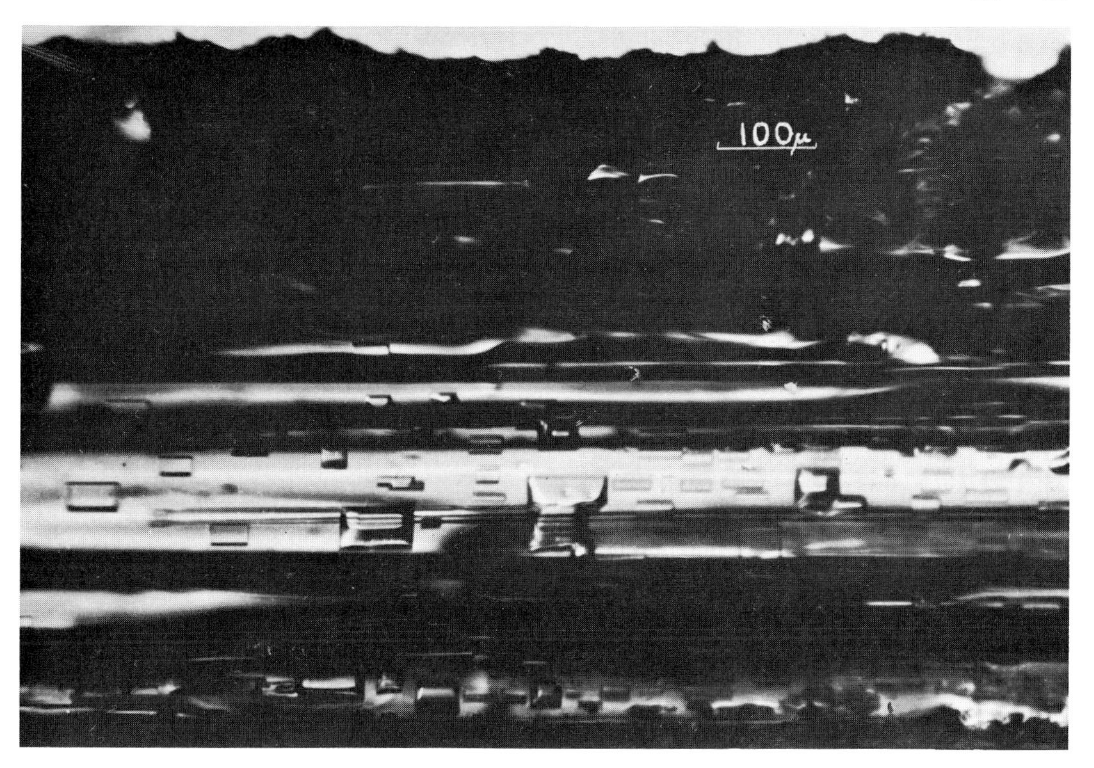

Fig. 31. Rectangular etch pits on the prism faces of an ice needle.

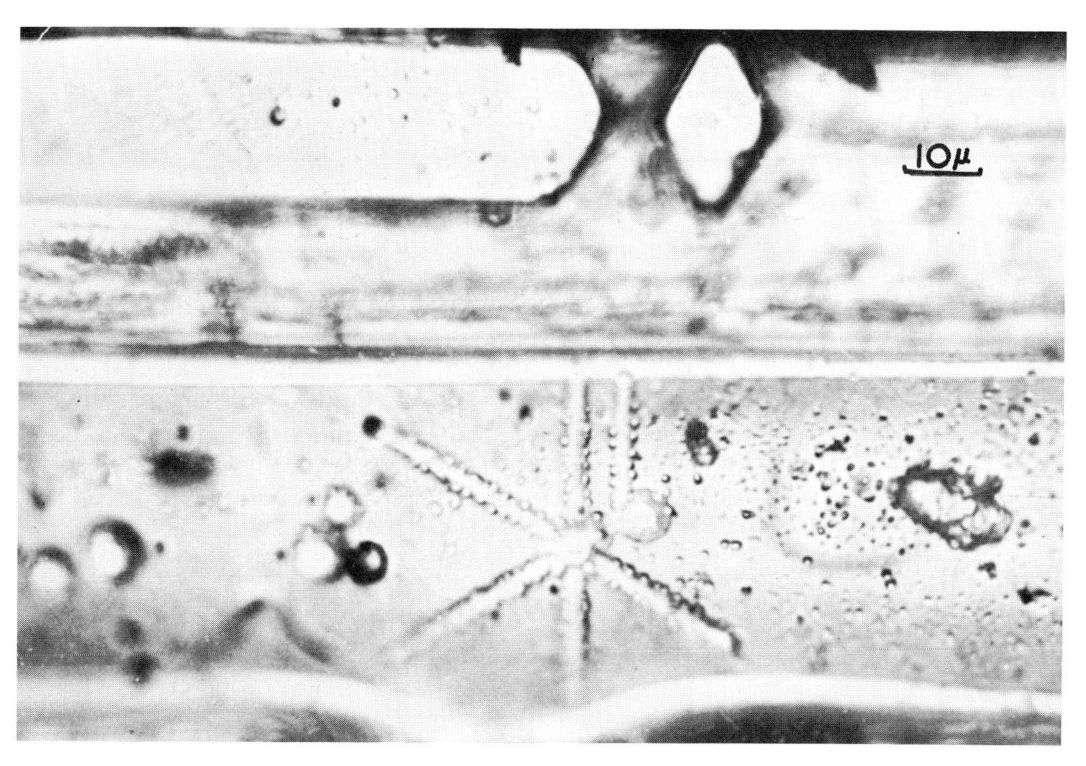

Fig. 32. Etch pits radiating from an impurity speck in a platelike crystal. Each channel consists of a double row of small etch pits which probably mark the intersection of dislocation loops with the surfaces. (Figures 27 to 32 from Bryant and Mason, *Phil. Mag., Ser. 8.* **5**, 1221, 1960.)

disturbance caused by an included dust particle and traveling with their axes parallel to the surface.

The fact that the surface densities of the smaller etch pits on ice lie in the range 10^5 to $10^6/cm^2$ and are comparable with the observed densities of dislocations in other materials suggests that they probably arose by the etching of emergent dislocations. A close association between dislocations and etch pits is also suggested by the latter tending to group preferentially along grain boundaries and slip lines. Bryant and Mason did not find it possible to say whether etch pits were generated about edge or screw dislocations, but were best able to interpret some of their etching patterns in terms of the etching of helical dislocations.

8. THE GROWTH RATES OF ICE CRYSTALS

The rate at which the mass of an ice crystal will increase when suspended in an atmosphere of given supersaturation and temperature will be governed by the rate of diffusion of water vapor to, and of latent heat away from, the crystal surface. If it is assumed that the crystal is at rest relative to the air and that the diffusion field has achieved a steady state the governing equations are

$$\frac{dm}{dt} = 4\pi CD\{\rho - \rho(s)\} \qquad (1)$$

and

$$L_s\frac{dm}{dt} = 4\pi CK(T_s - T) \qquad (2)$$

where dm/dt is the rate of increase of mass, C is a shape factor for the crystal equivalent to its electrostatic capacity, D the coefficient of diffusion of water vapor in air, K the thermal conductivity of air, $\rho(s)$, ρ, are respectively the vapor densities in the immediate vicinity of the crystal surface and at infinity, T_s, T the temperatures of the crystal surface and of the undisturbed environment, and L_s the latent heat of sublimation. Combining these equations with the Clausius-Clapeyron equation, $(1/p_e)dp_e/dT = L_sM/RT^2$, where p_e is the equilibrium vapor pressure of water vapor at temperature T, M the molecular weight of water, and R the universal gas constant, Mason (1953) obtains the equation

$$\frac{dm}{dt} = \left(\frac{4\pi C\sigma}{\dfrac{L_s^2M}{KRT^2} + \dfrac{RT}{DMp_e}}\right) \qquad (3)$$

where $\sigma = p/p_e - 1$ is the supersaturation of the environment relative to a plane ice surface. At constant air pressure, the term in parentheses is a function of temperature only.

Equation (3) can be applied to calculate the growth rate of ice crystals whose shapes approximate those of conductors of known electrostatic capacity. To a first approximation, one may follow Houghton (1950) and treat the thin, hexagonal plate as a circular disk, and a hexagonal prismatic column as a prolate spheroid of large eccentricity. The following substitutions then apply:

Crystal Shape:	Sphere	Circular Disk	Prolate Spheroid
Capacity C:	r	$2r/\pi$	$2ae/\ln[(1 + e)/(1 - e)]$

where the eccentricity $e = (1 - b^2/a^2)^{1/2}$ and a and b are the major and minor semi-axes respectively.

When an ice crystal in the atmosphere grows to such a size that it attains an appreciable fall speed, its growth becomes complicated because it can no longer be considered stationary with respect to the air and the diffusion field. The crystal will always be moving into new surroundings and the diffusion field will be limited to a boundary layer around the crystal in which the vapor concentration and temperature gradients will be larger than those directed towards a stationary crystal. For a spherical crystal of radius r, Mason (1953) shows the concentration gradient at the crystal surface at a time t after being introduced into a new environment to be

$$\frac{\partial \rho}{\partial r} = [\rho - \rho(s)]\left[\frac{1}{r} + \frac{1}{(\pi D t)}^{1/2}\right] \quad (4)$$

where the symbols have the same meaning as in Eq. (1). The second term in the square brackets is a measure of the enchancement of the concentration gradient at the crystal surface relative to the steady state value, and the effect will be marked if $(\pi Dt)^{1/2} \ll r$. It implies that the mass growth rate of the falling crystal will be increased by a factor $1 + (vr/2\pi D)^{1/2} = (1 + 0.22Re^{1/2})$, where Re is the Reynolds number. $(\pi Dt)^{1/2}$ has the dimensions of a length and may be defined as the thickness of the diffusion boundary layer.

Experiments designed to measure the rate of mass increase of crystals growing in a supercooled water cloud have been described by Reynolds (1952) and by Mason (1953). In both cases a supercooled cloud was produced in a large cold chamber and seeded with dry ice, the crystals being collected near the floor at successive time intervals, measured under the microscope, and the masses calculated from their linear dimensions. Mason made measurements on simple, regular, hexagonal plates grown at $-2.5°C$ and on prisms grown at $-5°C$ for growth periods up to 3 min., during which time the crystals achieved maximum linear dimensions of order 100 μ. The experimentally determined values of dm/dt agreed with those predicted by Eq. (3) to within about 10%, which was as good as could have been expected in view of the experimental conditions and the difficulty of making very accurate measurements on such small crystals. Reynolds used much the same procedure, but made his measurements on plane dendritic stellar-crystals grown at $-18°C$, the masses of which are more difficult to determine from their dimensions because of their irregular outline and nonuniform thickness. Because of this and also because the temperature and supersaturation in his chamber varied considerably during the course of the experiment, Reynolds obtained rather poorer agreement between his measurements and the theory.

The growth rates of ice crystals produced by seeding an outdoor supercooled fog have been measured by Yamamoto et al. (1952), Okita and Kimura (1954), and by Isono et al. (1956). Crystals carried away from the seeding site by the wind were collected on slides at various distances downwind corresponding to various growth times in the cloud. In the first mentioned papers, the authors claim that the growth rates obtained from their measurements were in good agreement with those calculated from Eqs. (1) and (2). However, their plotted data showed a good deal of scatter which may be attributed to spatial variations in the cloud of supersaturation, wind velocity, and the fact that the larger crystals would tend to settle out preferentially. Isono et al., who assumed that the *largest* crystals caught on their slides had grown under conditions of water saturation, reported that the mass growth rate of prisms growing at $-4°C$ was nearly twice that given by Eq. (3). However, since the crystals were typically 300 to 400 μ in length, falling at, say 20 cm sec^{-1}, the correction term $(1 + 0.22\ Re^{1/2})$ would largely account for the discrepancy.

Although Eq. (3) may predict quite well the rate of mass increase of stationary ice crystals and of small crystals falling only very slowly through a supersaturated environment, it cannot predict how the mass will be distributed, and hence the detailed shapes of the crystals. Measurements on the growth rates of individual faces were made by Shaw and Mason (1955) for crystals growing on their metal plate. Under conditions of constant temperature and supersaturation, the square of the linear dimensions of a crystal increased linearly with time, this relationship being in accord with theory for a crystal growing slowly in a steady state diffusion field. However, the growth rates of both prism and basal faces were often different in different crystals, whereas crystallographically similar faces of the same crystal sometimes grew at different rates. Furthermore, the growth rate of a particular face sometimes changed abruptly, even though the external conditions remained constant.

Measurements have also been made in this laboratory on needles, plates, and dendrites growing on a fine fiber in the diffusion cloud chamber. Under fairly constant conditions of temperature and supersaturation, the lengths of needles and of the arms of dendrites increased linearly and uniformly with time at rates of a few microns per second over periods up to half an hour. Very small plates, of diameter $<100\ \mu$, growing at constant thickness on the surface of covellite also increased in diameter at a uniform rate, but plates larger than $\frac{1}{2}$ mm across, growing in the diffusion chamber, did not show a linear or uniform increase in diameter. Plots of the distance of a growing edge from the crystal center against time revealed pauses in growth at intervals of usually between 2 and 4 min, pauses coinciding with the appearance of visible ridges on the surface such as are shown in Fig. 24 and discussed in Section 7. These results indicate that the growth of crystal faces is not determined solely by the flux of material to, and heat away, from the crystal, but is partly influenced by certain individual characteristics of the faces themselves which control the rate at which the material can be built into the crystal structure.

References

Aufm Kampe, H. J., H. K. Weickmann, and J. J. Kelly (1951), *J. Meteor.* **8,** 168.

Bentley, W. A. and W. J. Humphreys (1931), *Snow Crystals,* McGraw-Hill, New York.

Bryant, G. W., J. Hallett, and B. J. Mason (1959), *J. Phys. Chem. Solids* **12,** 189.

Bryant, G. W. and B. J. Mason (1960), *Phil. Mag.* 8, **5,** 1221.

Frank, F. C. (1949), *Faraday Soc. Discussion No. 5 on Crystal Growth,* p. 48.

Gold, L. W. and B. A. Power (1952), *J. Meteor.* **9,** 447.

Hallett, J. (1961), *Phil. Mag.* **6,** 1073.

Hallett, J. and B. J. Mason (1958a), *Proc. Roy. Soc.* **A247,** 440. (1958b), *Nature, London* **181,** 467.

Higuchi, K. (1958), *Chamonix Symp. Ins. Ass. Sci. Hydrol. Pub.* **47,** 249.

Houghton, H. G. (1950), *J. Meteor.* **7,** 363.

Isono, K., M. Kombayasi, and A. Ono (1957), *J. Met. Soc. Japan* **35,** 327.

Isono, K. (1958), *Nature, London* **182,** 1221.

Isono, K., M. Komabayasi, Y. Yamanaka, and H. Fujita (1956), *J. Met. Soc. Japan,* **34,** 158.

Kobayashi, T. (1957), *J. Met. Soc. Japan,* 75th Anniversary Volume, p. 38; (1958), **36,** 193; (1960), **38,** 231; (1961), *Phil. Mag.* **6,** 1363.

Kuettner, J. and R. J. Boucher (1958), *Final Rept. to U.S. Army Signal Corps.,* Eng. Lab. Contract No. DA-36-039.

Magono, C. (1960), "Physics of Precipitation," *Geophys. Monogr.* No. 5, Amer. Geophys. Union, p. 142.

Marshall, J. S. and M. P. Langleben (1954), *J. Meteor.* **11,** 104.

Mason, B. J. (1953), *Quart. J. Roy. Met. Soc.* **79,** 104; (1957a), *The Physics of Clouds,* Clarendon Press, Oxford, Chapter IV; (1957b), *Nature London* **180,** 573; (1958), *Advances in Physics (Supp. Phil. Mag.)* **7,** 235; (1960), *Quart. J. Roy. Met. Soc.* **86,** 552; (1961), *Discussion Faraday Soc.* No. 30, p. 20.

Mason, B. J. and J. Hallett (1956), *Nature, London* **177,** 681; (1957), **179,** 357.

Mason, B. J. and J. Maybank (1958), *Quart. J. Roy. Met. Soc.* **84,** 235.

Mason, B. J. and A. P. van den Heuvel (1959), *Proc. Phys. Soc.* **74,** 744.

Muguruma, J. and K. Higuchi (1959), *J. Met. Soc. Japan* **37,** 71.

Murai, G. (1956), *Low Temp. Science (Japanese)* **A15,** 13.

Nakaya, U. (1951), *Compendium of Meteorology,* Amer. Met. Soc., Boston, p. 207; (1954), *Snow Crystals,* Harvard University Press, Cambridge, Massachusetts; (1955), *J. Fac. Sci. Hokkaido Univ., Ser. 2,* **4,** 341.

Nakaya, U., M. Hanajima, and J. Muguruma (1958), *J. Fac. Sci. Hokkaido Univ., Ser. 2,* **5,** 87.

Okita, T. and K. Kimura (1954), *J. Met. Soc. Japan* **32,** 129.

Reynolds, S. E. (1952) *J. Meteor.* **9,** 36.

Schaefer, V. J. (1946), *Science* **104,** 457; (1949), *Chem. Rev.* **44,** 291.

Shaw, D. and B. J. Mason (1955), *Phil. Mag.* 7, **46,** 249.

Truby, F. R. (1955), *J. App. Phys.,* **26,** 1416.

van den Heuvel, A. P. and B. J. Mason (1959), *Nature, London* **184,** 519.

Vonnegut, B. (1948), *G.E. Res. Lab. Project Cirrus. Occas.* Rept. No. 5.

Wall, E. (1947), "Wiss. Arbeit deutsch," *Dienst. franz. Zone* **1,** 151.

Weickmann, H. (1947), "Die Eisphase in der Atmosphäre," *Reports and Translations* **716,** Völkenrode Ministry of Supply, London; (1950), *Die Umschau,* **50,** 116; (1957), *Artificial Production of Rain,* Pergamon Press, London, p. 315.

Yamamoto, G. et al. (1952), *Sci. Rep. Tohoku Univ. Ser. 5. Geophys.* **5,** 141.

PRECIPITATION FROM LIQUID

General Principles

8

Precipitation of Crystals from Solution

G. T. Kohman

Any discussion of precipitation from liquids necessarily relates in many ways to precipitation from vapors, from melts, and from solids because the basic problem in each case is control of solubility. It is appropriate to review solubility relations in this section not only because of their basic importance but also because solid-liquid equilibria have been investigated for a variety of systems. No doubt this is why Buckley (1951) introduced his very informative and detailed treatment of crystal growth with a discussion of "solution, solubility, and supersolubility." Also for these reasons many of the factors discussed by Egli and Zerfoss (1949) apply to crystal growth from solution as well as from melts.

There are two reasons why precipitation from a liquid is favorable for growth of a wide variety of crystals. First, the concentration of many compounds in liquid solution can be high, and second, they often have high mobilities in solution. Both factors are favorable for crystal growth. It is natural therefore that Holden and Singer (1960) should stress growth of crystals from liquid in their very stimulating book, *Crystals and Crystal Growing.* They explain that many interesting crystals can be grown with little or no understanding of principles simply by evaporation of solvent or temperature change; but when well-formed crystals of complex composition are required, it is helpful, and sometimes essential, to have a knowledge of solubility relations and the properties of solutions.

The purpose of this introduction is to review general principles primarily from the viewpoint of the experimentalist who faces the problem of growing a large crystal or a crystal that has not yet been grown artificially. A complete review of the extensive literature on the theory and mechanism of crystal growth is not included because rates and perfection of growth are more often influenced by careful control of the medium in which a crystal grows than on detailed knowledge of mechanism of growth. It is hoped that this situation will change in the future. References to recent theories and growth mechanisms are included because of the increasing evidence that growth mechanisms do influence certain crystal properties.

1. THEORY AND MECHANISM OF GROWTH

According to Bunn (1949), the growth of a crystal is similar to the formation of polymer chains. This process is initiated by activated molecules or radicals and terminates when these high energy species become deactivated. Kossel (1927) has estimated the energy change that occurs when an ion or atom is removed from solution and fits itself into active sites in various types of crystal lattices. The rate of growth of the crystal depends on the probability that an ion or atom finds a low energy site in a plane, corner, or edge of the solid. According to the theories of Stranski (1928) Volmer (1932), and Burton, Cabrera, and Frank (1951), thc growth rate should not be a simple or linear function of supersaturation. Different mechanisms possessing various activation energies appear as the supersaturation is increased. It is difficult to demonstrate these transitions in mechanism experimentally apparently because more than one mechanism operates at a given supersaturation. Sears (1957) demonstrated the reverse process by growing platelets of p-toluidine free of screw dislocations and then observing the rate of evaporation at various partial pressures of p-toluidine vapor. Evaporation did not occur until the partial pressure was less than 48% of the equilibrium pressure. Howard (1939) was able to demonstrate a sharp discontinuity in the growth of mercuric iodide crystals as the supersaturation of the vapor was increased.

Bunn and Emmett (1949) attempted to relate rates of growth of particular faces to the supersaturation of the solution in contact with the face. After many experiments they concluded that local energy changes, in accordance with the theories of Kossel, Stranski, and Volmer, control the behavior and make it difficult to relate the rate of growth of a face to the supersaturation of its environment. Experiments with sodium chloride showed that layer growth begins from points roughly in the centers of faces. The thickness increases as the layers spread towards the edges of a face.

The growth of crystals at very low supersaturations can be explained according to Frank (1958) by the presence of Burgers' screw dislocations (1940). The dislocations provide a continuous transition in growth from one layer to the next, making it impossible to complete a full layer.

For a more detailed discussion of theory and mechanisms the reader is referred to Verma (1952), to Frank (1952), and to Read (1953). Much of the literature prior to 1951 is reviewed in detail by Buckley (1951, p. 169).

2. ADVANTAGES OF GROWTH FROM LIQUID

Probably the most important advantage of crystal growth from solution is the control that it provides over the temperature of growth (Tanenbaum, 1961). This makes it possible to grow crystals that are unstable at their melting points or that exist in several crystal forms depending on the temperature. A second advantage is the control of viscosity, thus permitting crystals that tend to form glasses when cooled from their melts to be grown. Low viscosity solvents such as water greatly accelerate the devitrification of silica glass as shown by Walker (1950). Presumably this occurs by a combination of solvent and hydrogen bonding actions, thus providing the increased molecular mobility required. In general, rates of growth from melts are orders of magnitude greater than those from solution. Crystals grown from solution usually have well-defined faces as compared with those grown from melts.

3. CHOICE OF SOLVENT—ADDITIVES

Although crystal growth from the melt and from solution is basically the same process, in the latter case the choice of solvent provides some control over crystal habit as explained by Remika (1954), and discussed in greater detail by Buckley (1951, p. 339). This effect depends on the interaction of the surface of the crystal as it grows and the solvent molecules. Sometimes this is sufficient to result in the precipitation of a new crystalline phase. The effect is not unrelated to the influence of impurities or additives upon habit as discussed by Buckley (p. 339).

A solvent is preferred in which the solute is soluble to the extent of 10 to 60%. If the solvent is volatile, precautions must be taken to prevent volatilization which promotes spurious nucleation due to temperature and concentration changes. The experiments of Bunn and Emmett (1949) show that electrostatic forces between crystal and solution are involved in crystal growth, and therefore the polarity of the solvent must be considered. These investigators observed that as the dielectric constant of the solvent increased, it became more difficult to detect layer growth, possibly because of a decrease in thickness of the layers. The author has observed that crystals of nonpolar organic compounds grow from nonpolar organic solvents at rates comparable to the solidification of melts, suggesting that the principal difference between the two processes is the interaction of solute and solvent molecules. Probably no other solvent is as generally useful for growing crystals as is water. Some properties that account for this are its high solvent action, which is related to its high dielectric constant, its stability, its low viscosity, its low toxicity, its availability, and the fact that convenient means for control of supersaturation exist by change of temperature or removal of solvent.

The wide variety of additives that influences crystal growth morphology and the differences in their effects suggest that the molecular size of the additive is only one of several important factors. For example, Sears (1957) states that additives may increase growth rates by affecting the activation energy for nucleation of layers. According to Bunn (1933), it is not necessary for the adsorbed impurity to be present as a complete molecular layer to alter relative growth rates. Isolated molecules on a crystal face can result in the formation of an unstable mixed crystal. In effect this reduces the degree of supersaturation over the face because of the greater solubility of the mixed crystal. According to Stranski (1949), foreign adsorbed molecules affect the growth and dissolution processes in two ways. They favor or catalyze the elementary process by decreasing the amount of energy required, and they can also retard diffusion in the adsorption layer. The latter effect may outweigh the former. To explain the large effects of small impurity concentrations on habit, Buckley (1951, p. 169) favors a model of the Kossel or Stranski type in which an occasional impurity ion can attach itself to a high energy site, thereby interrupting a chainlike sequence upon which normal growth depends.

The effect of impurities can often be employed to practical advantage for altering crystal morphology. Walker (1950) has reported that the amount of taper in the growth of EDT and ADP crystals can be altered by control of *pH* and by the concentration of chromium and tin in the solution. Similar effects have been reported by Jaffe and Kjellgren (1949). Egli (1958) reports that properly selected additives aid in the production of flawless crystals, probably by suppressing spontaneous nucleation, thus permitting growth at higher supersaturations. Ammonium phosphate containing one mole per cent of iron can be grown flawlessly from solution more than ten times as fast as from pure solution. The impurity is not necessarily grown into the crystal. According to Egli (1958), the maximum rate at which flawless crystals can be grown is inversely proportional to the area of the crystal face. He interprets this to be an indication that growth generally proceeds by a classical process such as layer growth rather than by one of the several defect mechanisms.

With ionic crystals, it is frequently desirable to put ionic impurities into a lattice in order to control its electronic properties. According to Ballman (1961), two things that affect this procedure are the relative ionic diameters of the host and impurity ions and the balance of electric charges on the ions in the crystal.

4. SOLUBILITY RELATIONS

It is helpful to recognize that dissolution and melting are similar processes so that many considerations that apply to growth from solution also apply to crystallization from melts. The phase diagram in Fig. 1 (solubility relations) for an ideal two-com-

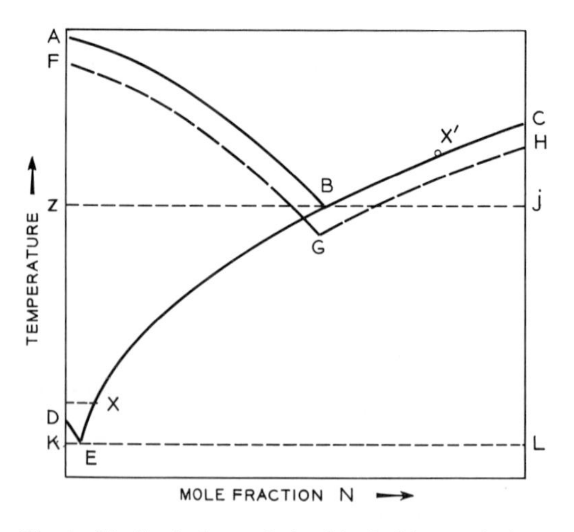

Fig. 1 Idealized phase relationship for binary solutions.

ponent system applies to a system of fused salts such as sodium fluoride and sodium chloride as well as to a water solution of sodium chloride. Ideally, if the appropriate units are employed in plotting the curves, the common branches of the diagram are the same, and their slopes are determined by the heat of fusion or the heat of solution of the solute. Solubility is often expressed as grams of salt per 100 cc of water at some temperature, say 25°C. For the system sodium chloride and water, C represents the melting point of sodium chloride, and D the melting point of ice. The solubility of sodium chloride in water at 25°C is the point X on the CE branch of the phase diagram. This is analogous to point X' on the phase diagram for a melt of sodium chloride and sodium fluoride, representing the solubility of sodium chloride in sodium fluoride. This diagram indicates the general rule that the temperature coefficient of solubility of a compound in a given solvent decreases as the melting point of the solute increases. This rule accounts in part for the low temperature coefficient of solubility of sodium chloride in water and the difficulty of growing crystals from this solution by lowering the temperature of the saturated solution. In a single component system, melting occurs at constant temperature; the process can be regarded as the saturation of the melt with a solid phase of the same composition. A discussion of the wide variety of solubility diagrams is given by Findley (1951), and their use in determining the conditions for crystal growth is described by Buckley (1951). See also the following section on molten salt solutions by Laudise.

It is often helpful to recognize that solubility in a given solvent depends on the number of solvent molecules per unit volume. This applies to above the critical temperature as well as to below. It has been incorrectly reported that precipitation of solute occurs as the solution is carried through the critical temperature. Examination of the experimental technique discloses that when this is observed, it is probably the result of a density decrease as the critical temperature is approached. As will be shown later, the critical temperature has no special significance with respect to solubility. Therefore many of the considerations that apply to crystallization from solution apply also to vapor as well as to melt crystallization. Also, it is to be expected that a sublimation process might be improved by the addition of a second gaseous phase, thereby increasing the concentration of the compound being sublimed without increasing the temperature.

If one is concerned with crystallization from solution at an elevated temperature, where the vapor pressure of the solution is appreciable as in the hydrothermal process, the distribution of solute between the liquid and vapor phases becomes important. It is therefore necessary to consider the effect of temperature on the concentration of solvent molecules in both the liquid and the gaseous phases. The normal effect of temperature on solubility is given by the relationship

$$\frac{d \ln N_A}{dT} = -\frac{H}{RT^2} \tag{1}$$

For the ideal case this reduces to

$$\log N_A = -\frac{H_A}{4.579T} + \frac{H_A}{4.579T_A} \tag{2}$$

where $\ln N_A$ is the natural logarithm of N_A, the mol fraction in the solution of the substance crystallizing, H_A is its molal heat of melting, R the gas constant, T the absolute temperature of the solution, and T_A the melting temperature of the crystal. Equation (2) is in the form

$$\log N_A = -\frac{a}{T} + b \tag{3}$$

On the other hand, at constant pressure

$$D = a' - b'T \tag{4}$$

where D is the volume density of solvent molecules which, according to Eq. (4), decreases with temperature. Because of these two opposing effects of temperature, the solubility may pass through a maximum if the temperature range is sufficiently large as is illustrated in Fig. 2. Figure 2 represents the amount of solute S which can be retained in solution for three degrees of fill,

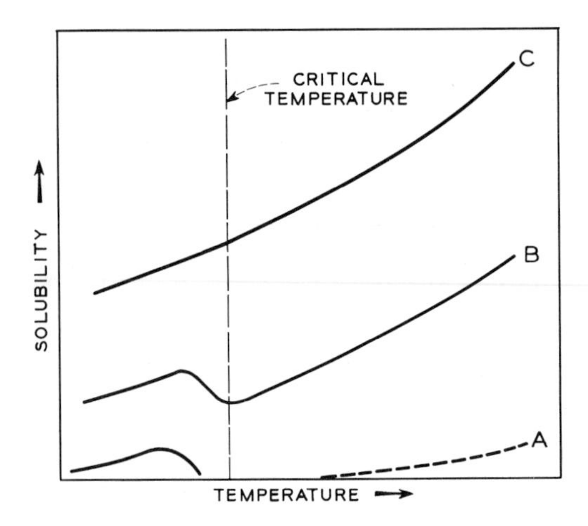

Fig. 2. Solubility changes in the vicinity of the critical temperature. *A*. Below the critical fill. *B*. Above the critical fill. *C*. One hundred per cent fill-constant density.

A, *B*, and *C*, as the temperature of a constant volume pressure vessel is varied over a wide range. Curve *A* represents a low degree of fill where the liquid phase is lost before the critical temperature is reached and solute is precipitated because of the low density of the vapor phase. The dotted portion of the curve above the critical temperature indicates a limited solubility in the gas phase. Curve *B* represents an intermediate degree of fill showing the retrograde solubility effect that is commonly experienced in the hydrothermal process upon raising and lowering the temperature or when the pressure drops as the result of a leak. Curve *C* represents a fill of 100% and therefore no change in density as the temperature is increased. Under this condition, no precipitation of solute occurs at the critical temperature and no effect of the critical temperature on solubility is observed.

Since a knowledge of solubility relations is essential in the controlled growth of crystals, some of the shortcuts in obtaining such information are reviewed. A considerable amount of information can be obtained from time-temperature curves as described by Johnston (1925) and Andrews and Kohman (1925). Figure 3 represents a time-temperature curve where *AB* represents the temperature of a heat sink separated from a sample by a suitable thermal insulator, for example, the walls of a small Dewar vessel containing the sample. The temperature of the sample is then determined by a differential thermoelement and

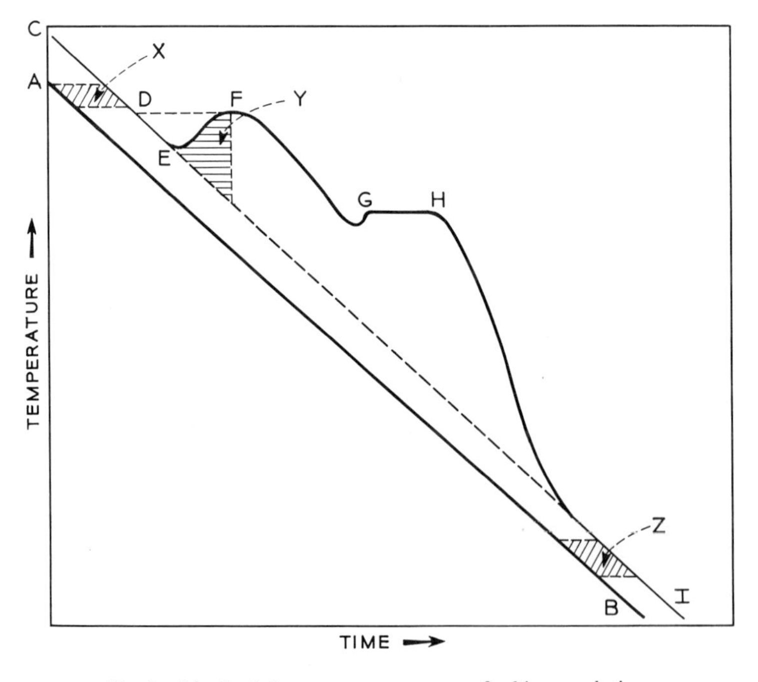

Fig. 3. Idealized time-temperature curve for binary solutions.

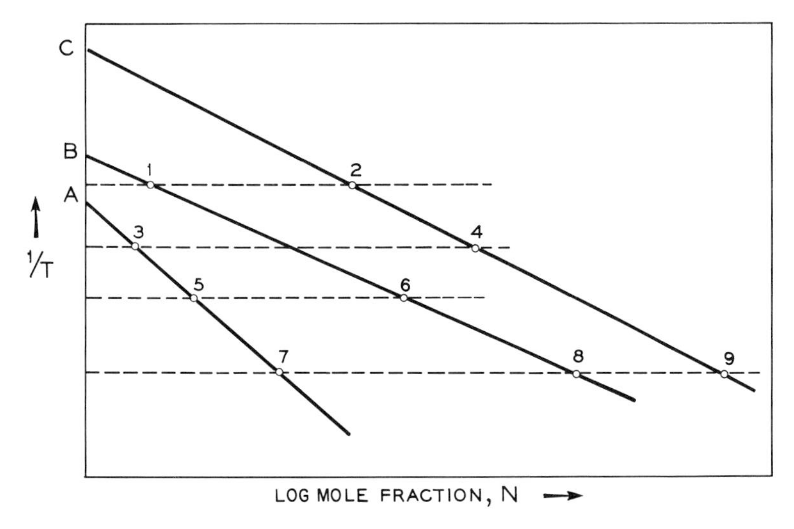

Fig. 4. Linear Solubility—Temperature plots for location of eutectic temperatures and compositions.

is shown by curve *C D E F G H I*. It is possible from a few such curves to construct the complete solubility diagram for a two- or three-component system. In the ideal case, the area *X* between *AB* and *CE* is proportional to the specific heat of the melt over this temperature range, and the specific heat of the solid can be obtained similarly from the area *Z* after solidification is complete. The temperature *F* corrected for undercooling, according to the method of Andrews (1923), is the temperature at which the solute will begin to crystallize if nucleated. This correction requires the experimental determination of at least two cooling curves where the degree of supercooling differs.

The corresponding areas *Y* are then plotted against the maximum temperature *F*, and a straight line is drawn through the points which these areas represent. This line, extrapolated to zero area, intersects the ordinate at the highest temperature at which solid can exist in equilibrium with the melt, which is the true freezing point of the solution. Each point on the curve *FG* represents an equilibrium between solid and liquid and therefore is a point on the binary phase diagram. The composition of the melt corresponding to each temperature can be approximately established by the appropriate areas between the two curves. These areas are approximately proportional to the amount of solid phase that has separated. Temperature *G* represents the appearance of a second phase and establishes the eutectic temperature. Crystals of the solute can be grown between temperatures *F* and *G*. The temperature difference between *E* and *F* is related to the metastable supersaturation range in which supersaturation must be controlled (known as the "Ostwald-Miers range").

As required by Eq. (2), a plot of log N versus $1/T$ is linear. For binary and ternary ideal solutions a family of curves as shown in Fig. 4 can be constructed from the freezing points of the components and a knowledge of their heats of fusion or solution. The latter determines the slopes of the curves. A single freezing point for each binary solution can also be used. A very limited amount of experimental data thus enables the complete binary or ternary phase diagram to be approximately reproduced. The binary eutectic temperatures and compositions can be determined by locating the horizontal line which will intersect any pair of lines *A, B,* and *C* at points corresponding to values of *N,* the sum of which equals unity. Similarly, the ternary eutectic temperature and composition can be determined by locating the horizontal line so that the sum of three points of intersection corresponding to a value of $N_A + N_B + N_C$ equals unity. The slopes of the straight lines together with the position of the horizontal lines thus determine the amount of the solute which can be crystallized over a given range of temperature.

5. NUCLEATION

Experience indicates that the best nucleus for the growth of a monocrystal is probably a small "seed" crystal of the same material. However, there is much evidence that other materials of similar lattice constants and perhaps even replicas of a crystal face will nucleate crystal growth. The most common procedure for developing the nucleus is to cool a solution beyond the "Ostwald-Miers" metastable range causing spontaneous nucleation and then to select a well-formed crystal for use as a seed crystal under more carefully controlled conditions. It is usually advantageous to select a seed crystal and growing conditions that result in growth in only one or two directions. To accomplish this, a seed crystal of the desired cross-sectional area is produced. Plates cut from this crystal at the appropriate orientation are then employed as seeds in a manner described by Walker (1950). An increase in cross-sectional area can sometimes be obtained by forcing the crystal to grow at near maximum rate, by the use of additives which are selectively absorbed on certain faces, or by control of *pH*. Experience shows that crystals containing macroscopic flaws as a result of forced growth or inclusions grow abnormally rapidly, probably because such flaws alter the growth mechanism. This effect can be employed in the production of seed crystals of the desired geometry. According to Bunn (1949) the outstanding generalization of crystal morphology is the universal tendency for the bounding surfaces to be faces of low index. Faces of high index have higher surface energies and tend to break up into large steps which generate low indice faces. The small faces that survive on a crystal are the fastest growing faces. Seed plates cut parallel to such faces tend to grow at the fastest rates. A very useful procedure in locating this face is to use a spherical crystal as a seed and observe the growth rates of new faces on the sphere. This is described by Artemiev (1910). All possible orientations are present on the sphere and the corresponding growth rates can be established. Seed plates are then cut parallel to the face that grows fastest.

Significant increases in growth rates of quartz have been achieved at Clevite and Bell Telephone Laboratories by using seed plates whose surfaces are not parallel to natural faces. It had been observed that growth on such surfaces (ADP crystals) was very rapid, but disorganized, until the natural faces were restored. This process was referred to by Walker (1950) as "capping" and was used in the preparation of seed crystals for the growth of large ADP crystals. Experience with the growth of quartz suggests that if the supersaturation is properly controlled, it might be possible to increase the rate of crystal growth generally by growth on surfaces that are not natural faces. The appearance of the new growth on quartz suggests that such surfaces are composed of a large number of high energy sites that can nucleate crystal growth. The geometry of these sites is such that it promotes simultaneous growth of elements so orientated that they will eventually join to form a single crystal. In this manner it may be possible to decrease the time interval between the growth pulses suggested by Kossel's analysis.

The production of large crystals in quantity is often limited by a lack of large area seed plates. In addition to the procedure described by Walker (1950), attempts have been made to cause carefully oriented sections of small crystals to join together and form a single crystal. These attempts have been only partially successful, partly because of the difficulty of obtaining exactly oriented sections and also because the layer growth process will normally not cross a crack in a single crystal, but each portion of a cracked face will grow separately at somewhat different rates. On the other hand, Gwathmey and Dyer (1959) have demonstrated that two carefully oriented spheres cut from copper crystals join to form a single crystal when heated to 400 to 750°C.

Control of nucleation, to avoid "spurious seed formation," is one of the most important requirements in the growth of large single crystals. Therefore the saturated growth solution must be carefully prepared by filtration and heating to remove (or deactivate) nuclei in the form of foreign particles. Dorsey (1940) has found that the presence of such particles in water taken from various natural sources has a great effect on the temperature to which the water can be supercooled. Van Hook and Bruno (1949) report that the crystallization of supersaturated solutions of sucrose can be inhibited by heating the solution at 20°C above the saturation temperature for at least 20 min. Egli and Zerfoss (1949) state that the incidence of nucleation depends on the previous history of the system. According to these authors, it has not been possible to detect nuclei above the saturation temperature by thermal or Tyndall effects or by small angle X-ray scattering. However, increasing the time of superheating decreases nucleation during subsequent cooling.

6. TRANSPORT OF MATERIAL

There are many theories regarding the mechanisms of transport of material, and considerable disagreement as to their relative importances. The fluid mechanics of crystal growth from solution is discussed by Carlson (1958). He states that it has been commonly believed that crystal growth is controlled by diffusion rates through a stagnant layer about 10^5 A thick which exists at the surface of a growing crystal. Carlson concludes that this layer probably does not exist, so that diffusion is not the rate controlling step. Experimentally, it is found that as the rate of circulation of solution is increased, a point is soon reached beyond which no effect on growth rate is observed. For ADP crystals, a flow rate of 10 to 15 cm/sec is sufficient to prevent a high incidence of veiling. In a crystallizer of the Holden type, a rate of rotation of the seed crystals of 20 rpm with reversal of direction approximately twice per minute was found adequate according to Walker (1950). In the hydrothermal process, according to Walker the temperature differential in the autoclave is usually more than adequate to provide material transport. Although there is reason to expect that transport of material might be a maximum at the critical temperature, recent results reported by Laudise* suggest that this may not be so. If the rate of heat transfer is taken as a measure of the rate of material transport, the maximum occurs when one of the phases, liquid or gas, just disappears. Since solubility is also a factor, the hydrothermal process would be expected to be most efficient when the fill is such that the miniscus just disappears at the top or bottom of the vessel.

An attractive feature of the hydrothermal process is the efficient and uniform transport of material to the growing surfaces that is automatically provided by the temperature differential. If the process is operated near the critical temperature, the temperature differentials as measured outside the autoclave are 20 to 60°C. This probably results in considerable turbulence. Observations made in glass capillaries and of suspended particles are consistent with this. For a variety of vessel shapes, no variation in growth rates due to variations in transport rates has been observed. The rate controlling step in the hydrothermal growth of quartz probably lies at the crystal solution interface (Laudise, 1959). Maximum growth rates are obtained when a baffle is placed between the growth and solution zones, but this results from a change in the temperature distribution rather than from a change in the rate of material transport.

7. CONTROL OF SUPERSATURATION

The most critical factor in the crystal growth process is probably supersaturation. It can be controlled by changes in temperature, by changes in composition through removal of solvent, by chemical action, or by

* R. A. Laudise (1959), *Proc. 13th Annual Symp. on Fire Control,* U. S. Army Signal Corps, p. 17.

an unstable form. The solubility diagram provides the information required to relate supersaturation to temperature changes. The following three general methods are used:

1. The temperature of the saturated solution is raised or lowered depending on the sign of the temperature coefficient of solubility.
2. A temperature gradient is established in the solution, resulting in continuous dissolution of excess nutrient in one zone and continuous extraction by crystal growth in a second zone.
3. A solution contained in one vessel at one constant temperature is continuously saturated and then pumped into a second vessel that contains seed crystals at another temperature.

Control of supersaturation by chemical action was employed by Holmes (1917) in the growth of crystals of barium thiosulfate. In general, it is difficult to control supersaturation by chemical action because most chemical reactions have very different rates from crystal growth rates.

The use of unstable forms for controlling supersaturation is well illustrated by the work of Nacken (February 28, 1946). He grew quartz at constant temperature by using vitreous silica as the unstable form to supply the nutrient material. The solubility relations for two forms differing in stability are shown schematically in Fig. 5, where AB represents the solubility of substance A which is unstable below the temperature O. The solubility of substance C, a form of A which is stable below temperature O and unstable above this temperature, is represented by CD. At temperature T_1, A dissolves and recrystallizes on seeds of the stable form C, whereas at T_2 the reverse reaction occurs. As T_1 and T_2 are removed from temperature O where both forms are equally soluble, the supersaturation of the stable form increases. With quartz, this proved to be a serious limitation of the process because the temperature range over which quartz is the stable form is several hundred degrees lower than the point O. Here point O represents the temperature at which molten cristobalite, fused silica and water exist in equilibrium. The spread between curves AO and CO in the temperature range where quartz is the stable form is so large that spontaneous nucleation cannot be suppressed.

A similar situation was encountered in the growth of EDT crystals (Kohman, 1950), where AB represents the solubility of EDT and CD the solubility of EDT $\cdot$ 1 H_2O. Here it is possible to grow EDT crystals at constant temperature by supplying excess hydrate to the solution at a temperature above point O. Because of its higher solubility, the hydrate must go into solution and then recrystallize onto the less-soluble anhydrous crystals. There is some evidence that crystals of the monohydrate have a tendency to nucleate anhydrous crystals. Thus it was necessary to operate the process at a temperature only slightly above point O where supersaturation is low, resulting in a low rate of growth.

8. CRYSTAL GROWTH FROM COLLOIDAL SOLUTION

A few attempts to grow crystals from colloidal solutions have been made. Some indications of success and numerous widely scattered observations have been reported which suggest that this possibility cannot be ignored. The most positive results are those reported by Hauser and LeBeau (1938), who investigated the properties of films formed by the gradual extraction of water from bentonite sols of very small and uniform particle size. They observed that a concentration was reached at which Brownian motion ceased and the particles tended to align themselves. On further removal of water the particles appeared to snap into position and to form "crystallites" of highly anisometric shape. The films that form are chemically similar to mica and possess some of its properties including its structural characteristics as revealed by X-ray diffraction. This phenomenon gives new insight about crystal growth from colloidal dispersions and its connection with mineralogical and geological processes.

Unpublished results of attempts to detect

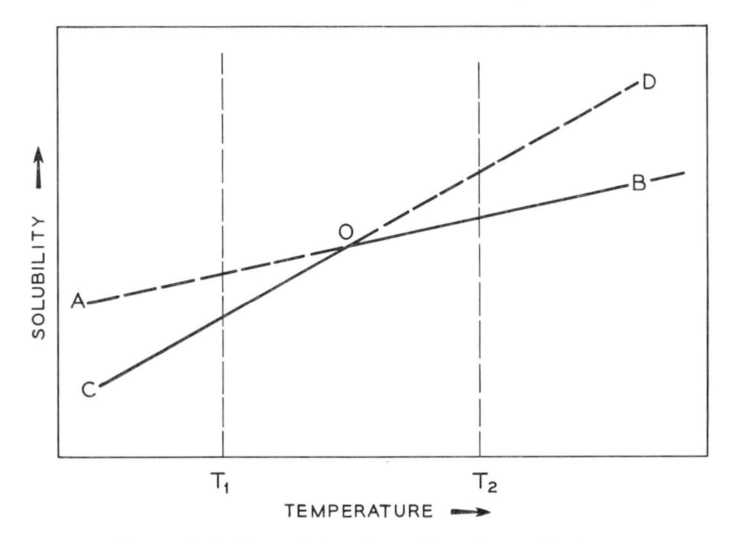

Fig. 5. Solubility relations for stable and unstable forms.

growth of quartz crystals suspended in silica sols have been brought to the author's attention (Philadelphia Quartz Co.). The experiment extended over a period of several years, but no evidence of growth was observed.

In the growth of ferrite crystals, Grisdale has made many observations suggesting that these crystals grow by the accretion of molecular complexes. He describes this in detail in the following chapter.

It has recently been shown by Schulman and Montagne (December 1960) that it is possible to prepare colloidal solutions with negative interfacial tensions. This allows the preparation of dispersions of extremely small particle size which are stable because of the low surface energy of the dispersed phase. Since particle sizes of the order of 100 A can be achieved, it is difficult to distinguish these dispersions from true solutions. In view of Buckley's observation (1951, p. 25) that small crystals can become oriented and joined together by surface tension forces, crystal growth from such systems appears possible.

References

Andrews, D. H. (1923), *Ph. D. Thesis*, Yale University.

Andrews, D. H., and G. T. Kohman (1925), *J. Phys. Chem.* **29,** 1317.

Artemiev, D. (1910), *Z. Krist.* **48,** 415.

Ballman, A. A. (1961), *Am. Mineralogist,* p. 439.

Buckley, H. E. (1951), *Crystal Growth,* John Wiley and Sons, New York.

Bunn, C. W. (1933), *Proc. Roy. Soc. London* **A141,** 567; (1949), *Discussions Faraday Soc.* **5,** 132.

Bunn, C. W., and H. Emmett (1949), *Discussions Faraday Soc.* **5,** 119.

Burgers, J. M. (1940), *Proc. Phys. Soc.* **52,** 23.

Burton, W. K., N. Cabrera, and F. C. Frank (1951), *Phil. Trans. Roy. Soc.* **A243,** 299.

Carlson, A. (1958), *Growth and Perfection of Crystals,* John Wiley and Sons, New York, p. 421.

Dorsey, N. E., *Properties of Ordinary Water Substances* (1940), Reinhold Publishing Corp., New York.

Egli, P. H., *Growth and Perfection of Crystals* (1958), John Wiley and Sons, New York, p. 408.

Egli, P. H., and J. Zerfoss (1949), *Discussions Faraday Soc.* **5,** 61.

Findley, A. (1951), *Phase Rule,* Dover Publications, New York.

Frank, F. C. (1958), *Growth and Perfection of Crystals,* John Wiley and Sons, New York, p. 5.

Frank, F. C. (1952), "Advances in Physics," *Phil. Mag. Supp.* **1,** 91.

Gwathmey, A. T., and L. D. Dyer (1959), *Friction and Wear,* Elsvier Publishing Co., New York, p. 165.

Hauser, E. A., and D. S. LeBeau (1938), *J. Phys. Chem.* **42,** 961.

Holden and Singer (1960), *Crystals and Crystal Growing,* Doubleday and Co., Garden City, New York.

Holmes, H. H. (1917), *J. Franklin Inst.* **184,** 743.

Howard, R. N. (1939), *Trans. Faraday Soc.* **35,** 1401.

Jaffe, H., and B. R. F. Kjellgren (1949), *Discussions Faraday Soc.* **5,** 319.

Kohman, G. T. (1950), *Bell Laboratories Record* **28,** 13.

Johnston, (1925), *J. Phys. Chem.* **29,** 882.

Kossel, W. (1927), *Nachr. Ges. Wiss.,* Gottingen, p. 135.

Laudise, R. A. (1959), *J. Amer. Chem. Soc.* **81,** 562.

Nacken, R. (February 22, 1946), *Captured German Report RDRDC*/13/18.
Philadelphia Quartz Company, reported with permission.
Read, W. T., *Dislocations in Crystals* (1953), McGraw-Hill, New York.
Remika, P. (1954), *J. Amer. Chem. Soc.* **76,** 940.
Schulman, J. H., and J. B. Montagne (December 1960), *Ann. N.Y. Acad. Sci.*
Sears, G. W. (1957), *J. Chem. Phys.* **27,** 1308.
Stranski, I. N. (1928), *Z. Physik. Chem.* **136,** 259.
Stranski, I. N. (1949), *Discussions Faraday Soc.* **5,** 74.
Tanenbaum, M. (1961), *Methods of Experimental Physics,* Vol. 6, Academic Press, New York.
Verma, A. R., *Crystal Growth and Dislocations* (1953), Academic Press, New York.
Volmer, M. (1932), *Trans. Faraday Soc.* **28,** 359.
Walker, A. C. (1950), *J. Franklin Inst.* **250,** 481.
Van Hook, A., and A. J. Bruno (1949), *Discussions Faraday Soc.* **5,** 112.

9

Growth from Molecular Complexes

R. O. Grisdale

Crystals of many substances are grown in nature and in laboratories by a variety of processes; among them are vapor phase reactions between metallic halides and water. Reactions of this type have much in common with the process of "mineralization" in which "crystallization catalysts" are important and are the starting point for the following discussion.

The synthesis of crystals is largely an empirical "art," and growth by one process has not always been integrated with another. Nevertheless various processes must have more in common than is immediately obvious, for crystals grown from aqueous or organic solutions, from high temperature flux solutions, from pure melts, by vapor phase reaction, by sublimation, and by electrolysis all exhibit similar growth structures. Ferrite crystals grown by vapor phase hydrolysis appear to offer morphological and chemical evidence of growth from a complex fluid phase intermediate between the crystalline solid and the molecular vapor. Since other crystals grown by other processes are morphologically similar, it appears that they too may grow from phases intermediate between crystals and truly molecular fluids. This is not a new viewpoint (Volmer, 1939; Devaux, 1953; Rae and Robinson, 1954; Bradley, 1951; Ubbelohde, 1950), but its generality and comparison with the more usual theories of crystal growth have not heretofore been discussed.

Crystals of the ferrites of iron, nickel, manganese, and zinc were grown by reaction between gaseous metal halides and water vapor at temperatures ranging from 900 to 1250°C. Halide vapors produced by reaction between chlorine and metals or ferrites were reacted with water vapor generated by combustion of hydrogen. Laminar flow systems were employed, concentric streams of the reactants in nitrogen flowing with equal velocities through a furnace. Crystals grew at the entrance to the reaction zone on the end of the tube separating the reactants, invariably extending at an angle into the halide zone. Water diffuses more rapidly than the halides, and growth apparently followed an interface defined by the vector sum of its diffusion and flow velocities. The largest, cleanest, and most perfect crystals grew in regions of diffusional mixing when the halide/water ratio was high. If this ratio was smaller than stoichiometric or if turbulent mixing occurred, the crystals were smaller and far more numerous. Crystals did not grow if the ratio were too large, and if, after growth had occurred, the ratio was increased to large values, the crystals disappeared or were often reduced to their metallic pseudomorphs.

The reaction between metal halides and water at elevated temperatures thus may be more complex than is often assumed, for the reaction of a divalent metal M, is usually written as

$$MCl_2 + H_2O \rightleftharpoons MO + 2\,HCl$$

However, reactions such as

$$\begin{aligned}
&MCl_2 + H_2O \rightleftharpoons MO_xH_yCl_z\\
&MO + MCl_2 \rightleftharpoons MO_xCl_z\\
&MO_xH_yCl_z \rightleftharpoons MO + HCl\\
&MO_xH_yCl_z + MCl_2 \rightleftharpoons M + H_yO_xCl_z
\end{aligned}$$

and others must also be concerned, with many of the complex oxy- or hydroxyhalide "compounds" being volatile. Consequently, there is the possibility that oxyhalide or hydroxyhalide complexes, some of low melting point and high volatility, may be involved as intermediates in growth of the crystalline ferrites.

Studies of lower temperatures and in solutions have established that such reactions are of common occurrence. The structures, compositions, disproportionation, methods of formation, and reactions of similar complexes have been widely studied (Feitknecht, 1953). They are important in many materials and technical processes, and their existence has been confirmed in many high temperature reactions. Indeed, the occurrence of such complexes is widespread; and, although their possible function as intermediates in the growth of crystals by vapor phase reaction apparently has not previously been suggested explicitly, their significance in mineralization is well recognized (Buerger, 1948).

1. CRYSTAL MORPHOLOGY AND GROWTH MECHANISM

Evidence for fluid intermediate complexes also came from a study of crystal morphology, for the products of every growth experiment suggested the production of crystals by solidification of liquid or molten material. Often the crystal surfaces were rounded, as though they had been formed by solidification of a viscous fluid; and on flat faces of some crystals circular flat-topped "islands" occurred, as if a droplet of fluid had spread over the surface and then solidified (Pandeya and Tolansky, 1961; Graf, 1951). More frequently these islands exhibited a geometric symmetry determined by the structure of the underlying crystal face, all being preferentially oriented but of different thicknesses and sizes; an example is shown in Fig. 1. Throughout there was an unmistakable impression of the growth of crystalline material by the accretion, merging, and solidification of fluid droplets or lamellae.

Discrete globules of ferrite were frequently observed, some being nearly spherical and composed of a large number of small crystals oriented at random, whereas others were monocrystals with well-developed faces. Often these globular and polygonal artifacts were attached to crystal or ceramic substrates by thin necks as though they had been formed from larger fluid droplets that

Fig. 1. Ferrite crystals showing at the right, polygonal "islands" on a (111) face and, at the left the concentration of these "islands" at the edge of a face.

Fig. 2. Polycrystalline ferrite globules near the end of the ceramic tube through which water vapor is admitted to the reaction zone. Through the annular space between this tube and the surrounding coaxial furnace core the halide vapor is admitted, and then mixes largely by diffusion with the water vapor in the reaction zone. Note the roughly spherical shape of the globules and their attachment by thin necks to the surface.

Fig. 3. Similar to Fig. 2 at higher magnification. Note the polygonization of the globules. Some of these polygons are single crystals, whereas others consist of two or more crystals. The small circular craterlike marks are spots from which globules have broken off in handling.

wet the substrate only at the point of attachment. The appearance of these globular artifacts is shown in Figs. 2 and 3.

These observations can be explained by postulating that a complex of the form $MO_xH_yCl_z$ is produced. Through a process of multiple collisions and condensation in the gas phase, small molten droplets of the complex form that are quasistable under the growth conditions (Grisdale, 1953). With high hydrogen and chlorine content the complex is probably fluid; but, as these decrease, its fluidity and vapor pressure must decrease until the end member, the solid oxide or ferrite, is reached. Beyond a possibly critical oxygen content, solidification occurs and crystallization of the oxide may take place by disproportionation.

The complex is not stable at the growth temperature. A droplet of it, suspended in the furnace atmosphere, changes in composition with time; and at some point in its lifetime the droplet will crystallize if left undisturbed. Solidification as a mono- or polycrystal is probably determined by the uniformity of its composition and by external influences. If, when the droplet has not crystallized in its own way, it comes in contact with a solid, crystallization may proceed from the point of contact. Epitaxial crystallization may result, and extensive volumes of growth are frequently obtained that consist of large numbers of connected and similarly oriented crystals. Indeed, growth seems to proceed on the basis of an oriented nucleus mechanism, the nuclei being the droplets of fluid complex.

Solidification of the fluid complex depends also on the nature of the droplet. If its composition is close to the presumed critical hydrochloride concentration, crystallization may occur shortly after the moment of contact, with the solid product being roughly related in shape to that of the fluid globule, as in Figs. 2 and 3. Also the droplets may supercool or have, while still fluid, an oxygen concentration more than sufficient to produce normal solidification. Then, the product consists of a mass of connected polycrystalline globules. On the other hand, if the hydrochloride content is large, with the material quite fluid and far removed from the solidification composition, the droplet may wet and flow over the surface, with solidification and crystallization being postponed in time. Under these conditions, the droplets may spread over the solid surface to give the circular or polygonal "islands," or, as often noted, a "growth layer" with a relatively large thickness and a peripheral contour not fundamentally related to the structure of the underlying crystal (Pandeya and Tolansky, 1961; Graf, 1951). At any moment some droplets are probably quite fluid, whereas others have already solidified, with the result that the products of any one run invariably exhibit all these characteristic morphologies. Electron microscopy shows that these also occur on the submicroscopic scale (Dillon, 1954).

As noted, oxyhalide complexes generated by reaction between the desired ferrite and chlorine and reacted with water vapor yielded the largest and cleanest crystals. Examples are shown in Fig. 4.

2. THE MORPHOLOGY OF FERRITE CRYSTALS

As shown in Figs. 4 and 5, ferrite crystals grow dendritically along three intersecting ⟨100⟩ directions. Following this, but not always keeping up with the corners, the edges grow along ⟨110⟩ directions. Finally, but growing initially more slowly, the octahedral {111} faces develop. The result is that the crystals grow at one stage as "hopper-faced" octahedra. The octahedral faces subsequently extend laterally in a series of separated sheets with growth originating on the ⟨110⟩ directions in plates of varied thickness that grow with constant thickness in the {111} plane (Fig. 4). Often the octahedral plates growing across the hopper openings advance simultaneously from two or three of the ⟨110⟩ edges. In the former instance, shown in Fig. 4, these plates sometimes meet, whereas in the latter case the otherwise complete face may be pierced by a small triangular hole. Not infrequently, octahedral plates grow across the hopper opening from ⟨110⟩ edges at

Fig. 4. Crystals of nickel ferrite growing from the end of the water vapor tube into the halide vapor, showing the octahedral plates across the hopper opening. Background in millimeters.

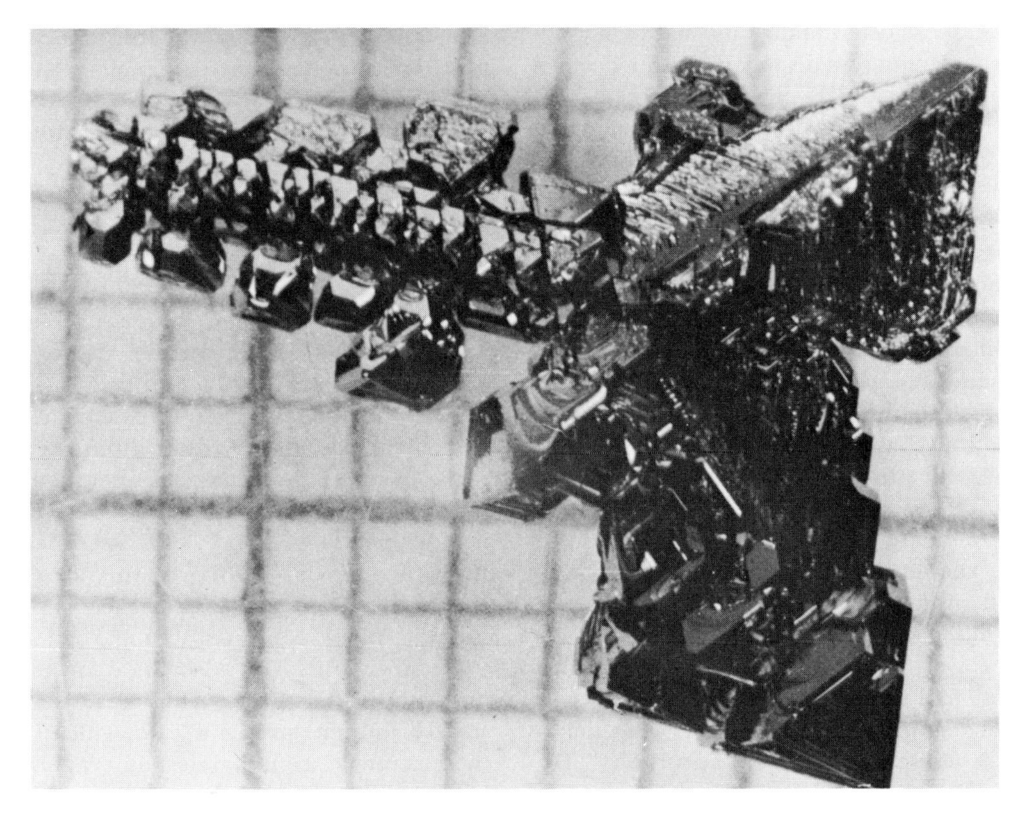

Fig. 5. Development of a nucleus showing to the left growth along ⟨100⟩; to the upper right the development of the hopper structure by growth along ⟨100⟩ from the ⟨100⟩ limbs; and to the bottom right development of the "cap-on-cap" structure. Background in millimeters.

different levels until they overlap, resulting in an interleaving of separated plates in the hopper face.

A very generally observed growth feature is shown in Fig. 4. Growth advances along a ⟨110⟩ direction and then halts, with further growth being along a second ⟨110⟩ direction. The result is growth inwards along a three-sided spiral.

A very common type of growth, shown in Figs. 4 and 5, consists of a dendritic "cap-on-cap" structure. Here an octahedron grows to a given size followed by nucleation and growth of a new octahedron at an apex, and this repeats itself many times over. The individual octahedra of these cap-on-cap structures may be solid or have spiraled hopper faces.

Small crystals tend to have filled-in octahedral faces, those of intermediate size are most frequently hopper-faced, and the largest ones (Fig. 4) tend to have filled-over octahedral faces. This is so general that the size of the crystal itself must be important in determining its morphology. Indeed, ferrite crystals follow a sequence of shapes as their sizes increase: (1) sphere, (2) rounded octahedron, (3) octahedron, (4) hopper-faced octahedron, and (5) hoppered octahedron with filled-over faces. The sizes of the crystals in the first three stages in this series may be related to sizes of the droplets of the complex from which they are produced.

Differences in adsorption potential can account for variations in the probability of trapping building blocks, but adsorption anisotropy has not usually been considered to influence their surface diffusion. However, studies (Drechsler, 1954) have shown that there are preferred directions for diffusion, and pronounced "directed diffusion" has been shown to occur for instances where the adsorbed film constitutes less than a monolayer. It is also expected to occur for diffusion of thicker films or of spatial nuclei with some internal "structure." This may be the cause of rapid growth in the ⟨100⟩ and ⟨110⟩ directions. (See also Chapter 7.)

Although crystal anisotropy influences the course of crystal growth, whether from molecular units or from spatial nuclei, growth is additionally dependent on the relative "concentrations" of corners, edges, and faces. As a crystal grows, the number of corners remains constant, but the length of edges increases linearly and the area of faces rises quadratically as its linear dimension. Thus, with increasing size, the probability that a building block will arrive at corners, edges, and faces increases in the order that the adsorption potential at these locations decreases—the two influences tend to cancel each other and the over-all growth anisotropy can be expected to decrease. The evolution of a growing crystal from octahedron to hoppered crystal to hoppered crystal with filled-over faces may be explicable on this basis.

3. THEORIES OF CRYSTAL GROWTH

The morphological suggestion of fluidity at an intermediate stage in vapor phase growth of ferrite crystals is difficult to reconcile with the usual theories of crystal growth. However, since similar morphologies are quite generally observed, their origin may be fundamentally significant in crystal growth.

Theories of crystal growth usually start from Kossel's "repeatable step," involving the repetitious incorporation into a growing crystal surface of atomic or molecular building blocks (Buckley, 1961; Read, 1953; Verma, 1953). An atom or molecule adsorbed on a crystal face diffuses over it before incorporation or reevaporation. When the rate of arrival of atoms at the surface equals the rate of evaporation or solution, the crystal is in equilibrium with its surroundings, and characteristically most theories deal with growth very near equilibrium. On a close-packed face a building block, conveniently regarded as a unit cube, is held by adsorptive forces on only one of its six faces; it diffuses over the surface until it finds a site to which it is more strongly bound unless it evaporates beforehand. The reentrant corner between a close-packed face and an incomplete "growth layer" on it provides such a site; and, depending on whether the edge of the layer is straight or kinked, the cube can fit into a site at the edge where it is bound to either two or three neighbors. It requires little elaboration of this picture to appreciate that close-packed faces should grow laterally to the crystal edges and that the resultant crystals should be bounded by these faces and not by stepped faces of higher index.

One feature of this mechanism requires close scrutiny, namely, that once a close-packed face has grown to its full extent, a new two-dimensional island with a step at its periphery must be formed before growth can proceed. The probability that a stable nucleus will form thus depends on local fluctuations in vapor pressure or concentration; and to form one larger than the critical size below which evaporation occurs requires a high degree of apparent supersaturation. Theory and experiment differ widely in the required degree of apparent supersaturation. Crystals actually grow at supersaturations below 1%, whereas theory predicts that values of 25 to 50% are necessary. One way of surmounting this difficulty postulates a spiral dislocation in the crystal that continuously provides the surface step needed for growth at low supersaturation (Read, 1953). Beautiful examples of the growth spirals expected from this mechanism have been observed, indicating that the process indeed occurs in crystal growth (Verma, 1953). Although the occurrence of such dislocations is clearly established in certain instances, it may be, however, only one of many mechanisms operative in the growth of the great majority of crystals.

Crystal growth by the accretion of droplets of a complex intermediate may provide a mechanism by which the large local supersaturation required for surface nucleation can be provided. Indeed, the droplet may represent an effective degree of supersaturation so large and the diffusion of "spatial nuclei" subject to such influences that the applicability of these theories is questionable unless they are modified.

The droplets of complex apparently possess fluid properties together with a structural organization intermediate between the liquid and the solid states. The concept of such an intermediate metastable phase is not new since it occurs frequently in discussions of nucleation and crystal growth (Volmer, 1939; Bradley, 1951; Ubbelohde, 1950; Devaux, 1953). The suggestion is that there may be vitreous or semifluid transition phases between the truly crystalline and the liquid or "monomolecular" states in which "association" or a degree of organization between the two occurs. This state has been described by such terms as subnuclei, submicrons, cybotactic groups, swarms, metastable phases, ion hydration, solvent fixation, clusters, micelles, complexes, cooperative effects, or association. The very widespread use of such terms is itself suggestive of some frequently occurring state of aggregation.

In most theories, this state, if considered at all, is regarded as significant only in nucleation of the crystalline phase. However, it may be fundamentally important in all stages of crystal growth, since the building blocks in growth may be volume elements of this metastable phase of appreciable size (Rae and Robinson, 1954). They provide a mechanism for surface nucleation and crystal growth analogous to that observed for the droplets of complex intermediates in the growth of ferrite crystals. The usual concepts of supersaturation can thus be revised and crystal growth can, as it does, proceed at low values of apparent supersaturation without need for dislocations to induce surface nucleation. Thus rather than occurring only in the vapor phase growth of ferrite crystals by reason of the existence of low-melting chemical complexes, the growth of many kinds of crystals by widely different processes may proceed through the accretion of complexes of the material in an intermediate state of aggregation.

4. MEANING OF SUPERSATURATION

Maintenance of crystal growth is limited both by the difficulty of nucleating low index faces and by slow diffusion of building blocks to the surface of the growing crystal or, in other words, by insufficient supersaturation at the crystal face. This gives rise to the theoretical requirement for larger supersaturations than are experimentally necessary and, in turn, leads to postulation of nucleation by spiral dislocations. Since this latter mechanism seems to lack the necessary generality in explaining crystal growth and morphology, some other concept of surface nucleation is desirable. One means of developing such a concept starts with inquiry into the meaning of "supersaturation."

Supersaturation is always defined in terms of a particular phase, and it is usually assumed that the solution is supersaturated with respect to a molecular dissolution of the crystalline solid. However, if there exists a metastable phase intermediate between the true molecular solution and the crystalline solid, this is incorrect. [It has been repeatedly stated that crystalline precipitates are always preceded by formation of amorphous aggregates (Volmer, 1939; Devaux, 1953).] It seems reasonable therefore to postulate that a "supersaturated" solution is not really supersaturated with respect to molecular dissolution of the ultimate crystal phase. Rather, excess solute over the saturation limit may exist as a separate "phase" in the form of cybotactic groups or clusters. In other words, a supersaturated solution may be regarded as composed of randomly oriented molecular complexes or cybotactic groups dispersed in an essentially homogeneous phase of the saturated molecular solution (Frenkel, 1946). These clusters may then be the building blocks concerned in crystal growth. It has, in fact, already been suggested that supersaturated solutions must contain clusters which, if they are not capable of independent growth, can deposit as single units on crystal faces, producing thick growth layers that resemble those of molecular dimensions (Rac and Robinson, 1954).

Thus the nature of the equilibrium in crystal growth may require more careful definition. Instead of being a simple equilibrium between a crystal and its isolated atomic or molecular elements, the "equilibrium" may be twofold: first, between the crystal embryos and the true molecular solution and second, between these embryos and the crystalline solid. The evidence is that the equilibrium between the saturated solution and the crystal embryos can be stable for all practical purposes but that the "equilibrium" between the crystal and its embryos is less stable in that once the macroscopic crystalline phase appears it will expand in volume until the remaining solution decreases to its usual saturation limit.

Complex aggregates or clusters also occur in undersaturated solutions and in pure liquids (Frenkel, 1946; Volmer, 1939; Pryor and Roscoe, 1954; Green, 1952), and therefore it cannot be the mere presence of molecular aggregates which characterizes a supersaturated or supercooled state. Considerations similar to those governing the formation of two-dimensional surface nuclei are also applicable here in that there exists a critical degree of supersaturation below which the embryonic crystal or cybotactic group is too small to be stable. Of course, on a fluctuation basis, groups larger than this critical size will always be present, but the probability of their occurrence may be vanishingly small. This probability rises exponentially with supersaturation, the increase being so extremely rapid in some cases that there appears to exist a sharply defined metastable limit, which in reality may be simply the point at which nucleation becomes experimentally observable in a finite time.

Although these cooperative and fluctuation phenomena in liquids need further study, it is sufficient to recognize that there is abundant evidence for the existence in solutions, melts, and vapors of molecular complexes or clusters which are probably fundamentally important in all stages of crystal growth. The characteristics of these aggregates—their critical sizes, the dependence of these on supersaturation, and the degree of internal latticelike organization which they possess—are determined, differently in each case, by statistical fluctuations and by the various intermolecular forces responsible for "association" (Green, 1952).

These kinetic units, large in comparison with atomic or molecular size, together with the media in which they occur, constitute typical colloidal systems. Colloidal particles are usually regarded as being prevented from coagulating by a potential barrier between them originating in their electric charges. However, since the system as a whole is electrically neutral, the dispersion medium contains an equivalent charge of opposite sign; the result is that each particle immersed in the dispersion medium is surrounded by an electric double layer. Small amounts of ionic material, many times no more than are adventitiously present, suffice to create the double layers essential to long time stability of colloidal systems (and also have profound influences on crystal growth). Until relatively recently it was commonly believed that the short-range repulsive forces acting between particles as a consequence of interactions between these layers are the only forces important in colloidal systems. It has become increasingly apparent, however, that long-range attractive forces are also operative and that a more correct picture of the interaction of colloidal particles should be based on a combination of the short-range repulsive double layer interaction with these long-range attractive forces (Verwey and Overbeek, 1948; Kruyt, 1952; Alexander, 1950).

The long-range attractive potential between the particles of a colloidal system is due to van der Waals-London forces which exist among all atoms, molecules, or ions. As a result of their additive character, these forces cooperatively yield an attraction between two colloid particles acting over an appreciable distance. The attraction decays as an inverse power of interparticle distance, but for particles of colloidal size the forces can be effective over a distance of some hundreds or thousands of angstroms (Verwey and Overbeek, 1948; Alexander, 1950; Kruyt, 1952; Casimir and Polder, 1948). Thus these forces may provide a "driving potential" in the process of crystal growth, for, rather than reaching a crystal surface through operation of random diffusional processes, the molecular clusters may be attracted to the growing crystal.

The adsorption potential important in theories of crystal growth also arises from van der Waals-London forces; but, since these theories deal with growth from atomic or molecular elements, the force attracting a single molecular building block to the surface is of short range. It is only for interactions between two aggregates of appreciable size or between these and a flat surface that the full effects of the cooperative nature of these forces are realized in range and magnitude. However, even with molecular units the forces are significant, and the surface of a growing crystal should probably not be regarded as a geometrical plane separating the solid from its melt, solution, or vapor. It may be more correct to think of the "surface zone" of a growing crystal in which the degree of latticelike organization changes over a finite distance from that of the crystal to that of the medium in which it grows. With growth occurring, as it appears to, from media containing colloidal clusters this surface zone may be of appreciable thickness. In this viscous zone random diffusional processes may be inhibited; but the long-range forces operative between clusters or between the crystal and the clusters may be sufficient to overcome viscous resistance.

These long-range attractive forces are always active, depending in magnitude and range on the sizes of the particles; under their influence alone coagulation of the aggregates would be expected to occur. That it does not is due in colloidal systems to the repulsive interaction of the electrical double layers at small interparticle distances. Superposition of the attractive and repulsive potentials gives rise to a steep potential maximum at small separation and to a shallow potential minimum at a larger particle separation (Verwey and Overbeek, 1948; Kruyt, 1952). The potential minimum, of course, represents an equilibrium separation of the particles which may be concerned in the formation of gels from colloidal suspensions or of vitreous phases from supercooled media. Furthermore, because of differences in attractive potentials, the maximum in the potential curve which prevents coagulation is smaller between a crystal surface and a particle than it is between two particles themselves (Kruyt, 1952, pp. 260 and 325). This may explain why a macroscopic crystal will immediately start to grow in a supersaturated solution.

The possibilities for forming chemical complexes are legion, and they are probably already of wide, although little recognized importance in crystal growth. For example, substances commonly effective as mineralizers such as metal or hydrogen halides are prone to form complex addition compounds of relatively low melting point and high fluidity (Buerger, 1948). Furthermore, the solvent influences the habit and growth rate of many crystals, which may be tracable to the formation of complexes (Wells, 1949; Egli and Zerfoss, 1949). It has also been stated that growth occurs most regularly and rapidly from solutions where the "degree of association of the solute" is greatest and that for a given solute this varies from one solvent to another (Egli and Zerfoss, 1949). Adventitious impurities as well as those intentionally added in small quantities to melts or solutions are profoundly important in crystal growth and are presumed to function by influencing the degree of association in the fluid phase.

The art of crystal growing would therefore appear to consist of fostering the formation of this labile metastable state of aggregation. A reexamination of crystal growing processes from the viewpoint of the structural chemistry of fluids and solutions may thus be profitable.

The concept of crystal growth from such an intermediate phase is, of course, conjectural and its validity must be tested. One possible way of doing this may be to examine phenomena arising during growth itself. There is, for example, a "phase difference potential" between a crystal and its growth medium; and incorporation of a building block into a crystal must be accompanied by a potential fluctuation, whose magnitude is dependent on the size of the block (or its charge). Thus it might be possible electronically to "hear" or "see" these fluctuations and from their magnitudes to determine the sizes of the building blocks. Furthermore, careful observation of the influences of electric field on nucleation and growth should be significant (Hirano, 1954). Nuclear or electron paramagnetic resonance techniques might also be applicable to studies of clustering in supersaturation and crystal growth in suitable systems. In addition, of course, optical scattering studies at and near the growing crystal face should yield valuable information on clustering. Finally, study of the mosaic structure of crystals and how it is influenced by growth parameters might be important, for it is conceivable that this universally observed feature of crystals is a natural consequence of their growth from colloidal complexes (Wells, 1949; Stranski, 1948).

References

Alexander, J. (1950), *Colloid Chemistry,* Vol. VIII, Reinhold Publishing Corp., New York, p. 47.

Bradley, R. S. (1951), "Nucleation in Phase Changes," *Chem. Soc. London, Quart. Reviews* **5,** 315.

Buckley, H. E. (1951), *Crystal Growth,* John Wiley and Sons, New York.

Buerger, M. J. (1948), *Am. Mineralogist* **33,** 744.

Casimir, H. B. G. and D. Polder (1948), *Phys. Rev.* **73,** 360.

Devaux, H. (1953), *Compt. rend.* **237,** 287. See also *Compt. rend.* **196,** 1091 (1933) for crystal growth from droplets of such a phase.

Dillon, J. F. Jr. (1954), Private Communication.

Drechsler, M. (1954), *Z. f. Elektrochemie* **58,** 334.

Egli, P. H., and S. Zerfoss (1949), *Discussion Faraday Soc.* **5,** 61.

Feitknecht, W. (1953), *Forschr. chem. Forschung* **2,** 670. See also A. F. Wells, *Structural Inorganic Chemistry,* Second Edition, Oxford University Press, London.

Frenkel, J. (1946), *Kinetic Theory of Liquids,* Oxford University Press, London.

Graf, L. (1951), *Z. Metallkunde* **42,** 336, 401.

Green, H. S. (1952), *Molecular Theory of Fluids,* Interscience, New York.

Grisdale, R. O. (1953), *J. Appl. Phys.* **24,** 1082. This process is similar in some respects to that observed in the pyrolysis of hydrocarbons.

Hirano, K. (1954), *Nature* **174,** 268.
Kruyt, H. R. (1952), *Colloid Science,* Elsevier, Amsterdam.
Pandeya, D. C., and S. Tolanski (1961), *Proc. Phys. Soc. London* **78,** 12.
Pryor, A. W., and R. Roscoe (1954), *Proc. Phys. Soc.* **B67,** 70.
Rae, H., and A. E. Robinson (1954), *Proc. Roy. Soc.* **A222,** 558.
Read, W. T. (1953), *Dislocations in Crystals,* McGraw-Hill, New York.
Stranski, I. N. (1948), *Optik* **3,** 17.
Ubbelohde, A. R. (1950), *Chem. Soc. London, Quart. Reviews* **4,** 356.
Verma, A. R. (1953), *Crystal Growth and Dislocations,* Butterworths Scientific Publications, England.
Verwey, E. J. W. and J. Th. G. Overbeek (1948), *Theory of the Stability of Lyophobic Colloids,* Elsevier Publishing Co., Amsterdam.
Volmer, M. (1939), *Kinetik der Phasenbildung,* Dresden and Leipzig.
Wells, A. F. (1949), *Discussion Faraday Soc.* **5,** 197.

Specific Substances

10

Elements

R. H. Wentorf, Jr.

The elements discussed in this chapter are those that may exist in two or more allotropic forms, such as tin, iron, sulfur, boron, phosphorus, or carbon.

Usually crystals of the chemical elements are grown directly from their vapors or their own melts because such methods are physically and chemically simple. However, with elements that may exist in more than one crystal form or allotrope, the stability of a particular allotrope depends upon the pressure and the temperature. It may be desired to prepare an allotrope that is not thermodynamically stable at the pressures and temperatures concomitant with ordinary liquid or vapor phases of the element. Again, for various reasons, kinetic or otherwise, the desired allotrope may not form rapidly enough (or at all) from the liquid or vapor under ordinary conditions. Under such circumstances the allotropic form sought may often be prepared by precipitation from a solution held at the appropriate conditions of pressure and temperature.

As usual, the best crystals are obtained by maintaining a relatively low supersaturation for a long time. This may be accomplished by the slow removal of solvent from the system, by a slow change of temperature or pressure, or by the transport of solute from a dissolving zone to a crystal growing zone. The selection of the solvent may depend on its chemical and physical affinity for the allotrope in question, on its chemical stability against decomposition or reaction with the container, and sometimes on its "catalytic powers" for producing the desired allotrope and not another.

The interplay of these various factors is illustrated in the following discussions of the preparation of crystalline elements from solution, beginning with sulfur which behaves most simply, and concluding with diamond, whose formation appears to be the most complex.

1. LOW TEMPERATURE PROCESSES

In this section are discussed the preparations of crystals of sulfur, gray tin, and fine particle iron crystals for magnets. All these crystals are prepared at temperatures below about 200°C.

Sulfur

Almost any textbook of inorganic chemistry discusses the various forms of sulfur. (See, for example, Sidgwick, 1951; or Deming, 1940.) The growth of sulfur crystals from various solutions illustrates in a direct, simple way the factors involved in crystallizing the allotropes of an element from solution. Below about 96°C, rhombic sulfur is the stable phase at one atmosphere. Between 96 and 119°C monoclinic sulfur is stable at one atmosphere. When crystals of one phase are subjected to conditions appropriate for the other phase, the change of phase eventually takes place; rate studies have been made by Elias, Hartshorne, and James (1940).

In both forms of sulfur the S atoms are joined in rings of 8, and the slight differences between the two forms are merely the result of slight differences in arrangement of the S_8 rings with respect to one another; no marked differences in chemical bonding are involved. It is not surprising then that the precipitation from solution of one form or the other can be fairly closely controlled simply by selecting the temperature of crystallization, so as to produce the thermodynamically stable phase. This is conveniently done by permitting the solvent to evaporate at the desired temperature. Useful solvents include carbon bisulfide, benzene, turpentine, and halogenated hydrocarbons.

It is also found that at those temperatures where the rate of conversion of one form of sulfur to the other is slow, crystals of the nominally unstable phase may be produced by providing seed crystals of that phase. Thus one may produce monoclinic sulfur crystals from cool carbon disulfide or benzene solutions if monoclinic seed crystals are used, even though rhombic sulfur is the stable phase under these conditions.

The change of solubility with temperature provides the basis for growing the largest crystals. Sulfur is dissolved in a warm solution and crystallizes when the solution is cooled. By recirculating the solvent, or by permitting the sulfur to diffuse from a warm place to a cooler one in undisturbed solvent, simple, straightforward techniques will produce satisfactory crystals (Elias et al., 1940).

Gray Tin

In contrast to the forms of sulfur, the gray form of tin is rather difficult to prepare in large crystals. Thc cquilibrium measurements of Becker (1958) showed that gray tin should be the stable form for temperatures below about 20°C at 1 atmosphere, and that the pressure dependence of the equilibrium line was -2×10^{-2} °C/atmosphere in accordance with the 27% increase in volume on going from white to gray tin. Although the change from white to gray occurs easily enough in the metal, so that many tiny crystals are produced, the gray form is surprisingly difficult to prepare by other routes. Vapor deposition, electrolytic deposition, reduction of stannons ions in solution, all produced only white tin (Becker, 1958; Wentorf, 1952). The reasons for this behavior are not clearly understood.

The increasing interest in semiconducting materials and the resulting desire to study large gray tin crystals led Ewald and Tufte (1958) to study the tin-mercury system. They found that relatively large (1 cm) crystals of gray tin could be grown from mercury solution at temperatures of about −30°C.

A sketch of their apparatus is shown in Fig. 1. The entire apparatus is placed in a refrigerator at −55°C, but the charge of tin and mercury in the lower part of the vessel is held at −20°C with a heating coil. Thermal convection, as well as the reduced density associated with dissolved tin, carry amalgam upward in the central tube. As it flows downward along the outer walls it becomes cooled to about −30°C and supersaturated with tin. Gray tin crystals grow on the vertical walls beneath the mercury surface. Every 24 hr the white tin charge

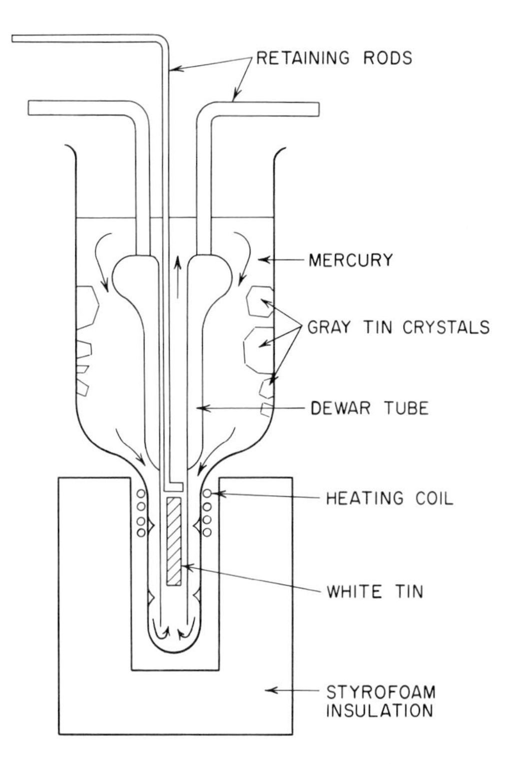

Fig. 1. Apparatus used by Ewald and Tufte (1958) for growing gray tin crystals.

is replenished with a rod 4 mm in diameter and 2.5 cm long.

It was found that although gray tin crystals were not the first to form, seeding was unnecessary. [If seeds are desired, they may be made by placing white tin, freshly electroplated at high current density from stannous chloride solution, in a refrigerator at $-15°C$ (Wentorf, 1952).] The first crystals to form were hexagonal plates, presumably $HgSn_{12}$. (Such crystals were always formed if the minimum mercury temperature were raised to $-10°C$.) If, however, after a week's growth at $-30°C$ the hexagonal crystals were removed, gray tin crystallites formed in great abundance. By regular weekly removal of unwanted nuclei, the deposition of gray tin could be concentrated on a few crystals. After a month of growing, crystals such as shown in Fig. 2 could be produced. Some of the most perfect crystals are shown in Fig. 3.

Almost all the crystals have one imperfect face that exhibits a ringlike growth structure as shown in Fig. 4. This face lay against the wall of the apparatus. Many crystals also contained internal pockets of occluded mercury which were exposed

Fig. 2. Large gray tin crystals. (Ewald and Tufte, 1958.)

Fig. 3. Gray tin crystals with best faces uppermost. (Ewald and Tufte, 1958.)

by cutting. However, the solubility of mercury in gray tin is evidently extremely low because the ostensibly mercury-free portions of a crystal, when converted to the white form and tested for residual resistance at low temperatures, indicated a maximum mercury content of 0.001 atomic per cent or lower.

Gray tin is brittle and has a high luster similar to germanium. Both *p*- and *n*-types of semiconducting crystals have been grown; the growth of *n*-type crystals requires careful attention to purity. Measurements on an *n*-type crystal yielded a value for the energy gap of 0.08 ev.

A crystal of gray tin free of mercury usually transforms to white tin in a few hours at room temperature, although a few crystals have remained unchanged for several weeks. The change to white tin is greatly accelerated by the presence of mercury. At about 1°C, mercury adhering to a gray tin crystal suddenly goes into solid solution, and the region around a mercury droplet changes from gray to white. The volume of tin transformed is approximately equal to that of the mercury.

Fig. 4. Gray tin crystals showing growth rings. (Ewald and Tufte, 1958.)

Fine Particle Magnet Iron

Usually growers of crystals focus their attention on the production of large crystals. The recent development, however, of fine particle magnets has emphasized the factors involved in the growth of small crystals —a few thousand angstroms or less in size. This concern with small particles has arisen as a result of a greater understanding of magnetic domains. If a ferromagnetic crystal is small enough, about 150 A, it may consist of a single magnetic domain whose direction of magnetization may be changed only with difficulty (Kittel and Galt, 1956). When large numbers of such small particles are suitably aligned and bound together, they form a permanent magnet having a high coercive force many powers of ten greater than that of a single large crystal. The coercive force may be increased by using particles which have a high anisotropy of shape and/or crystal structure.

Small crystals of iron or cobalt-iron having suitable sizes and shapes for use in fine particle magnets have been prepared during the past several years (Luborsky, Paine, and Mendelsohn, 1959; Luborsky, 1961). The method used by these workers was the electrodeposition of iron or iron-cobalt into a pool of mercury under controlled conditions, followed by suitable heat and surface treatments of the particles to produce the desired rodlike shapes.

A schematic drawing of the electrodeposition cell is shown in Fig. 5. The cell was mechanically isolated to reduce vibration or disturbance of the mercury cathode

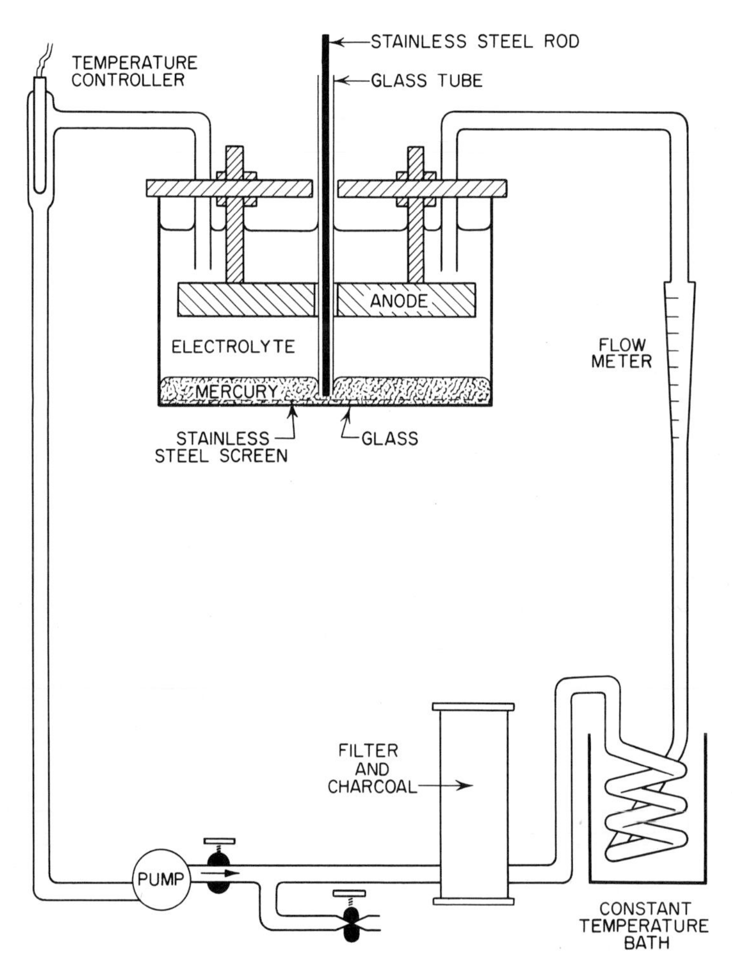

Fig. 5. Cell for electrodeposition of small iron crystals. (Luborsky, 1961.)

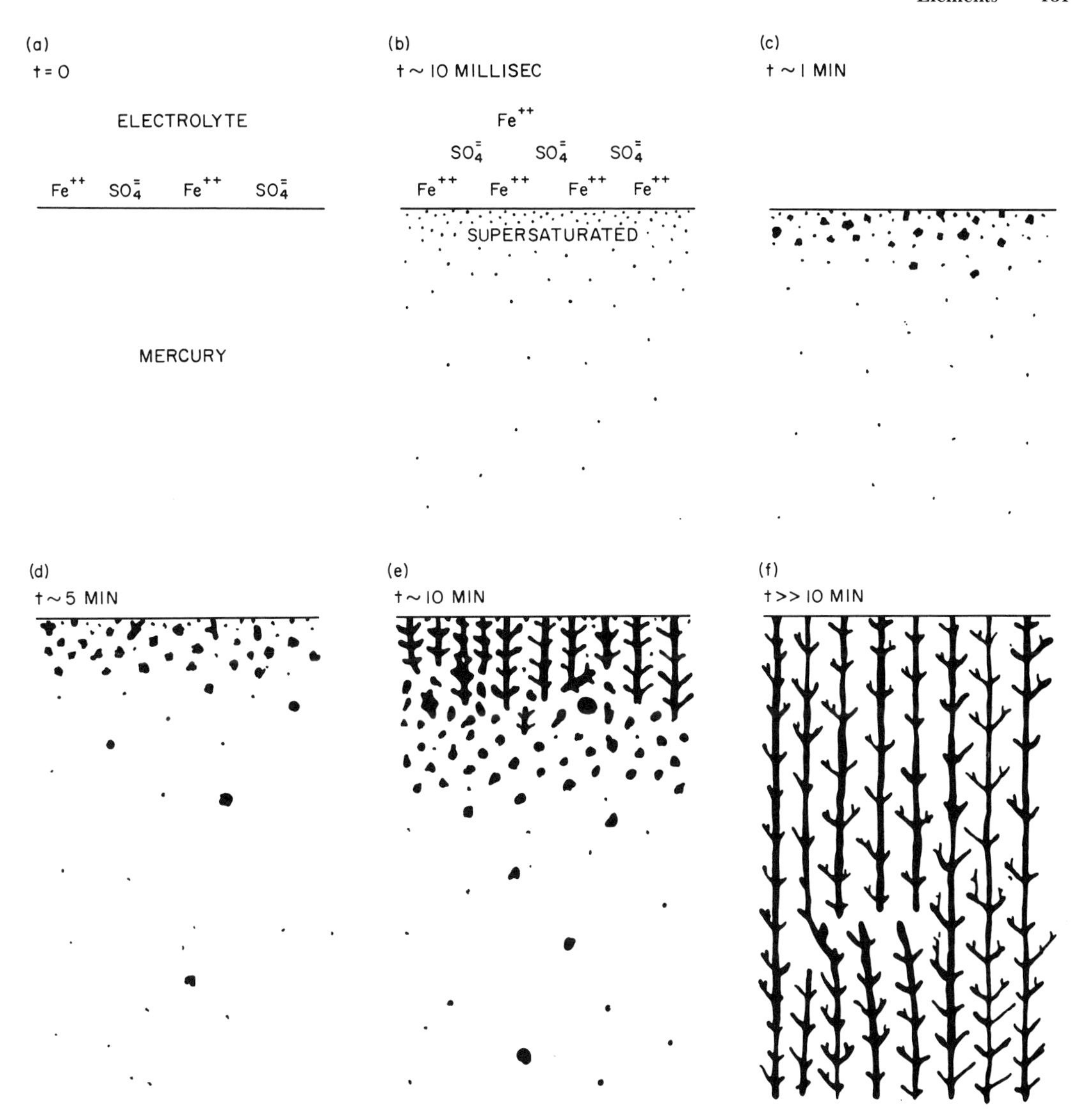

Fig. 6. Schematic illustration of steps in the growth of iron crystals in mercury. (Luborsky, 1961.)

surface. A stainless steel screen welded to a stainless steel rod made contact with the mercury. The electrolyte, usually acidified ferrous sulfate solution, was continuously filtered through charcoal in order to control the impurities which might produce hydrogen evolution at the mercury surface and thereby interfere with the uniform deposition of iron into the mercury. Current densities from about 0.001 to 0.1 ampere/cm^2 were used. The process was monitored using magnetic analyses and electron microscopy.

The sequence of events that occurs during electrodeposition is illustrated schematically in Fig. 6 for a current density of 0.01 ampere/cm^2. During the first second of current flow the region near the mercury surface is supersaturated with respect to iron. The arrival rate of iron atoms greatly exceeds their diffusion rate in mercury, and the solubility of iron in mercury at 25°C is about 10^{-6}. Thus many iron metal nuclei form. The nuclei grow rapidly upward toward the mercury surface, which is the source of iron, and compete for the iron that is available. The higher the current density, the more closely the crystals are spaced. Within a wide range of current densities the crystals are spaced closely enough to form a viscous sort of gel in the mercury surface, and if the surface is not disturbed by vibration, etc., the crystals continue to grow at their top ends without the nucleation of new crystals. Thus the mercury-iron "gel" region extends downward from the surface as the crystals become longer.

The iron-mercury gels were solid and easily handled. They contained from 7 to 15% iron by volume, and could be broken to reveal the expected fibrous structure. A study of the structures of gels prepared in different ways gave insight into the mechanisms of crystal formation.

It was found that after the crystals became 0.2 to 0.4 cm long, the buoyant forces of the iron began to overcome the interfacial tension of the wetted iron-mercury, and the layer of mercury above the growing iron crystal became very thin. (Ultimately the iron crystals may protrude into the electrolyte, so that iron is deposited upon them directly and they then become too large to be useful in magnets.) As a result of the thinning of the mercury layer, the crystals changed their appearance as the deposition time increased, as shown in the electron micrographs of Fig. 7. It is seen that the particles are very long, dendritic fibers with average diameters about 200 A.

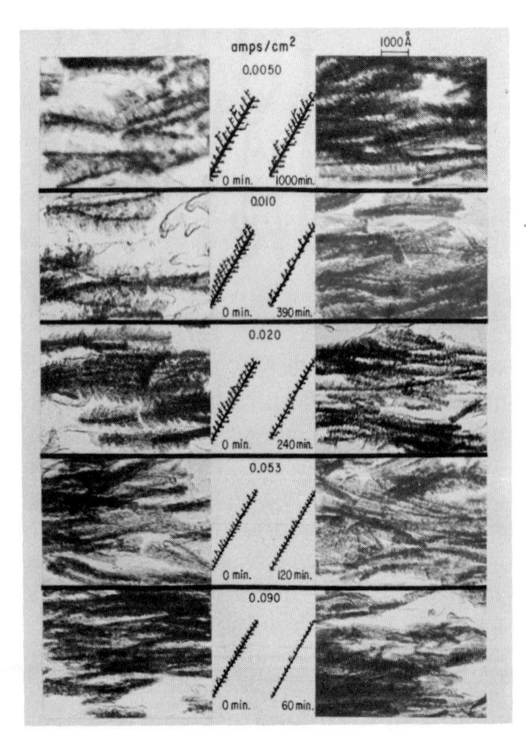

Fig. 7. Electron micrographs of small iron crystals showing the variation with current density and growing time. (Luborsky, 1961.)

Such micrographs were made by oxidizing a freshly broken surface in moist air and casting a cellulose acetate film on it. When the film was removed, it carried some iron particles with it. Carbon was deposited on this side of the film, and then the acetate was dissolved. The remaining carbon film bearing the iron particles was examined by electron transmission microscopy. (See also Schuster and Fullam, 1946.)

As the temperature of the deposition bath was increased above 25°C, the crystals lost more of their feathery dendrites, whereas their central stems and main branches became thicker. At deposition temperatures of 175 to 225°C, the dendritic branches became too thick to be removed by the recrystallization process in the following description in which the dendrite-free, rod-shaped crystals useful for magnets are formed.

Dendritic particles are not as suitable for use in magnets as are elongated or ellipsoidal particles. However, the dendritic particles are relatively easily recrystallized to the desired shapes by heating them to temperatures of 150 to 200°C for times of the order of an hour. The mechanism of growth of spherical particles has been elucidated by Luborsky (1957, 1958); the growth rate is limited by diffusion of iron atoms through the mercury, and is driven by the greater solubility of surfaces of smaller radii. The experimental growth equation for spherical iron particles in mercury is given by

$$|\log|\frac{r}{r_0} = 0.225|\log|t + 3.34 - \frac{1430}{T} \tag{1}$$

where r is the particle radius at any time t

in seconds, and T is the absolute temperature.

After the dentritic branches, with their smaller radii, have dissolved and the particles have become rodlike or ellipsoidal, continued growth will produce spherical particles. Spherical particles are not so desirable for magnets because of their lower coercive force. The conditions of growth are accordingly carefully regulated in order to produce particles with a length/diameter ratio greater than ten; this is a satisfactory shape.

When such particles are compacted into a magnet, less than the ideal coercive force is obtained from the magnet because of the magnetic domain interactions arising from physical contacts between the particles. However if the particles are coated with a nonmagnetic dispersing agent, the performance of the finished magnet is improved. A suitable dispersing agent is tin. It is added by mixing tin amalgam with the slurry of iron and mercury. The tin is adsorbed upon the iron surfaces and helps isolate the particles magnetically.

To make magnets, the particles of tin-treated iron are first aligned in a strong magnetic field and the mercury is distilled off, leaving a porous stable bar. (Bars suitable for testing purposes may be made in the laboratory by simply squeezing out about half the mercury.) The bar is pulverized and the coarse powder is mixed with a plastic or metal binder, aligned with a magnetic field and pressed or sintered into the final desired shapes.

Such magnets are found to have a B_r of about 9000 gauss and a $BH_{\max}$ of 3.5 to 6.5×10^6 gauss-oersteds. They compare very favorably with the well-known alnico alloys in these respects, and are much easier to make in desirable shapes.

2. HIGH TEMPERATURE PROCESSES

In principle, the growth of crystals from solutions at high temperatures, over about 300°C, differs little from the growth at low temperatures. In practice, higher temperatures place restrictions upon the solvents and containers that may be used; the processes involve considerable "art" and may partly rely upon special chemical or physical phenomena.

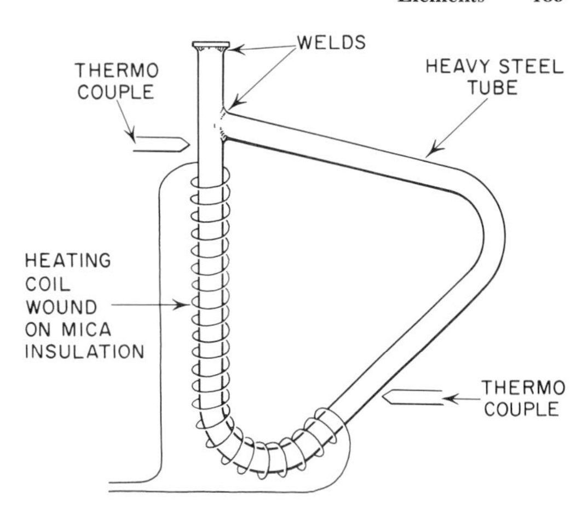

Fig. 8. "Harp" or "thermal induction loop" used to study the solvent properties of mercury at high temperatures.

Iron from Mercury

In the early 1930's extensive studies were made of the use of mercury as a high temperature working fluid in a power plant cycle. Higher efficiencies at lower pressures were possible with a combined mercury-steam cycle than with steam alone (Nerad, 1932). One of the problems faced in the use of mercury was its tendency to dissolve iron once the iron oxide film had been penetrated and the iron could be wetted by the mercury. About one ppm of iron was dissolved in the mercury at about 500°C; the solubility fell with temperature. In this way iron was dissolved in hot portions of the machinery, and precipitated in cold portions. Certain metals, such as titanium or zirconium, when added to the mercury in small amounts of a few ppm, inhibited the transport of iron without affecting the wetting necessary for good heat transfer.

Many studies on the solvent powers of mercury were carried out in laboratory devices called "harps" or "thermal induction loops." One is sketched in Fig. 8 (Nerad, 1933).

A thick-walled steel tube about 4 ft long was bent, and one end was welded into a hole near the other end so as to form a

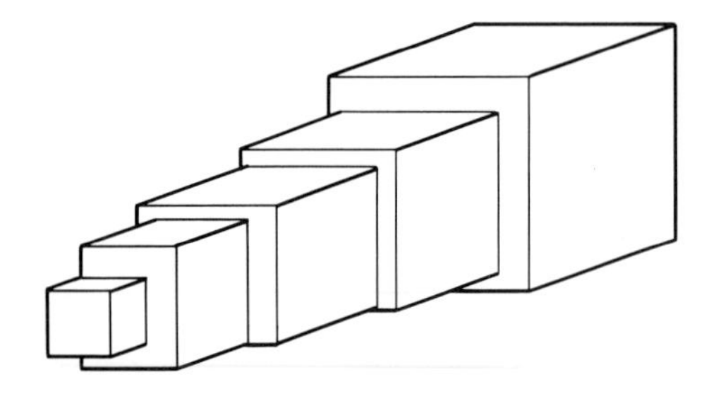

Fig. 9. Sketch of iron crystals produced at high temperature from mercury solution.

complete internal circuit. Various test specimens could be lowered into the straight part through the open end. After it had been filled with mercury and inert gas, the end was welded shut. Resistance heaters wound on mica tapes heated about one-half the tube. The resulting temperature differences in various portions of the tube created convection currents which circulated the mercury around the loop. Average temperatures of 400 to over 700°C were maintained for periods as long as 10,000 hr.

Figure 9 is a sketch of some iron crystals grown from a hot mercury solution. They had the form of cubes or rectangular parallelepipeds, often stacked end to end to form "whiskers." The cubes were typically 0.02 or 0.04 mm after about 100 hr of growth. Growth spirals were often observed on the faces of such iron crystals. Mercury-free crystals were usually quite active chemically and rapidly acquired an oxide film in the atmosphere.

Boron from Platinum

Three well-defined allotropes of boron have been identified (Hoard and Newkirk, 1960). Above about 1500°C the common form is beta-rhombohedral. Crystals of this form may be conveniently prepared by freezing a boron melt. Between 1500 and 1200°C several forms of boron have been reported; the best defined of these is the tetragonal form.

Below about 1200°C boron crystallizes in a red alpha-rhombohedral form. This allotrope was first obtained as one of the products of the reaction of boron halides with hydrogen near a tantalum filament maintained below about 1200°C (McCarty et al., 1958). At temperatures above 1200°C this form of boron is thermally unstable; above 1500°C it transforms into the beta-rhombohedral form.

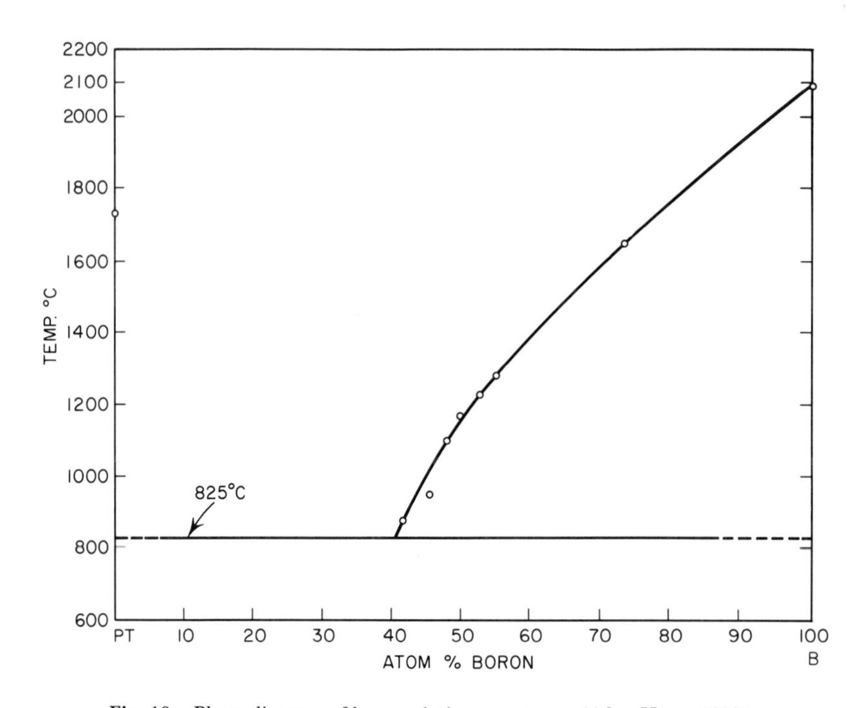

Fig. 10. Phase diagram of boron-platinum system. (After Horn, 1959.)

Horn (1959) has reported on the crystallization of the alpha-rhombohedral form of boron from platinum-boron melts. Figure 10 shows a phase diagram of the platinum-boron system as determined by Horn. It is evident that a melt containing about 50 atomic per cent boron would begin to deposit boron at about 1200°C, so that this boron should crystallize in the alpha-rhombohedral form.

A method for putting these findings into practice is to melt a 50 atomic per cent boron-platinum alloy in a hot-pressed boron-nitride crucible which has previously been fired in vacuum to remove traces of volatile impurities. The melting is conveniently done in vacuum or inert atmosphere by induction heating. The method and rate of cooling affect the boron crystal size. With rapid cooling the crystals are only a few tenths of a millimeter in average diameter.

There is no foreseeable reason why larger crystals may not be grown by simply reducing the over-all cooling rate while maintaining the entire melt in a mild temperature gradient. The precipitated boron, being less dense than the alloy, would tend to rise to the top of the melt. An alternative method might make use of this difference of density between the two phases and let a supply of boron float at the warmer upper portions of the melt while boron was deposited on seed crystals mechanically held near the cooler bottom portions of the melt. Thus with a constant temperature difference and a constant average melt temperature, boron would diffuse from the supply zone to the crystallizing zone at a rate slow enough to favor the growth of good crystals.

A difficulty appears when it is desired to separate the crystals from the solidified eutectic. Unfortunately, the boron dissolves in aqua regia at about the same rate as platinum. A more discriminating though slower acting solvent for platinum, perhaps potassium cyanide, might be used. The platinum-boron eutectic is very brittle and may easily be crushed away from the boron crystals if a simple mechanical separation is desired.

If pure starting materials are used and

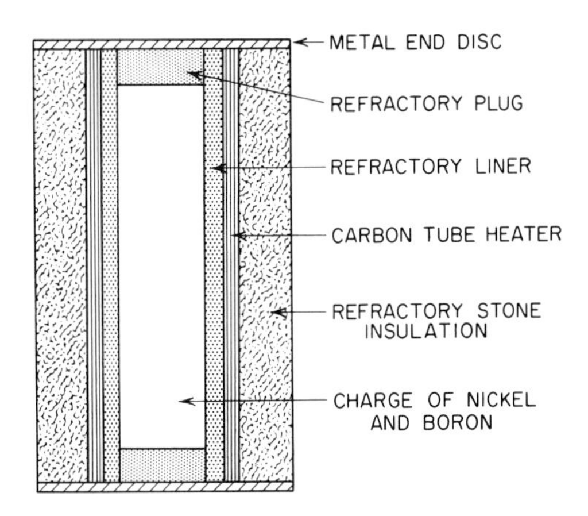

Fig. 11. High pressure, high temperature reaction cell for crystallization of boron.

contamination of the melt is avoided, most of the boron appears as translucent red crystals, indicating that they are fairly pure. They give the characteristic X-ray diffraction pattern of the alpha-rhombohedral boron. Sometimes the red crystals are accompanied by black material which appears to be one of the many complex and possibly impure intermediate forms of boron.

Attempts have been made to grow boron crystals from molten nickel-boron mixtures at high pressures. A mixture of nickel and boron, about 50 atomic per cent of each, was loaded into a refractory sample holder as shown in Fig. 11. This was placed in a "belt" high pressure apparatus (Hall, 1960; see section about diamond growth), and exposed to pressures of about 60 kilobars (kb) while being melted and cooled. Red crystals of boron were obtained; their density, about 2.60 grams/cc, and X-ray diffraction patterns (Decker and Kasper, 1960) indicated the presence of about 4% of nickel atoms in the lattice, so that the crystals were not pure boron.

Black Phosphorus

The black modification of phosphorus was first prepared at high pressures by Bridgman (1914, 1916). Because of its low vapor pressure he concluded that it was the most stable form of phosphorus, although its preparation at one atmosphere was not accomplished until 1955 (Krebs et al.,

1955) by heating red phosphorus with finely divided mercury. The mercury acts as a surface catalyst for the substantial changes in chemical bonding that take place when the graphitelike, closely knit, "giant molecule" lattice of black phosphorus is formed from the more open red phosphorus structure. (Red phosphorus may be regarded as intermediate between the collection of P_4 molecules present in white phosphorus and the giant molecules represented by crystals of black phosphorus. In a sense, black phosphorus may be regarded as a three-dimensional crystalline polymer of phosphorus, whereas white phosphorus may be regarded as the tetramer of phosphorus.)

Although the prepartion of crystals of black phosphorus does not usually involve the use of a solvent, it is an allotrope that must be prepared under special conditions and is thus included in this discussion of the crystallization of the elements.

The most convenient way to prepare large amounts of black phosphorus is to freeze molten phosphorus under pressures high enough to ensure the stability of the black phase. Preparations of the author and co-workers have used pressures of about 22 kb. Russian workers (Butuzov and Boksha, 1956) have employed pressures of about 15 kb; they report the melting temperature of black phosphorus to be 1000°C at 20 kb. If the pressure is too high, a high pressure modification of phosphorus forms instead of the usual black phase; upon release of pressure the change to the black form disrupts the crystals.

The apparatus used by Butuzov and Boksha is illustrated in Fig. 12. A refractory crucible containing phosphorus was exposed to high pressure argon or carbon dioxide gas inside a heavy-walled, steel pressure vessel. The crucible was electrically heated and its temperature was monitored by a thermocouple. The molten phosphorus was allowed to cool slowly so as to produce a mass of crystals.

In the author's preparation, white phosphorus was cast under water into titanium tubes. Just before they were put into the pressure apparatus, the tubes were wiped dry and quickly slipped into a refractory sample holder as shown in Fig. 13. This entire sample holder was confined in a "belt"-type pressure apparatus (Hall, 1960), as shown in Fig. 14. (See also the follow-

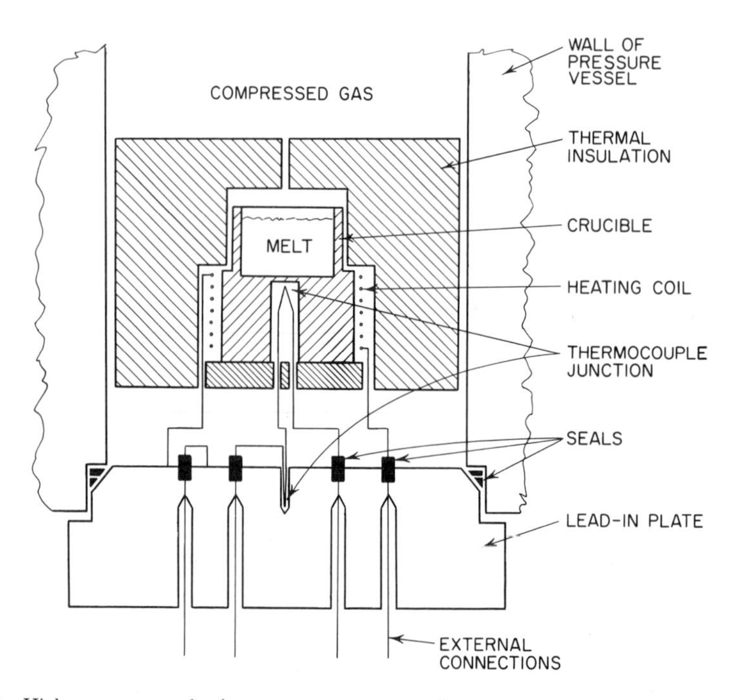

Fig. 12. High pressure vessel using gaseous pressure medium. (After Butuzov and Boksha, 1956.)

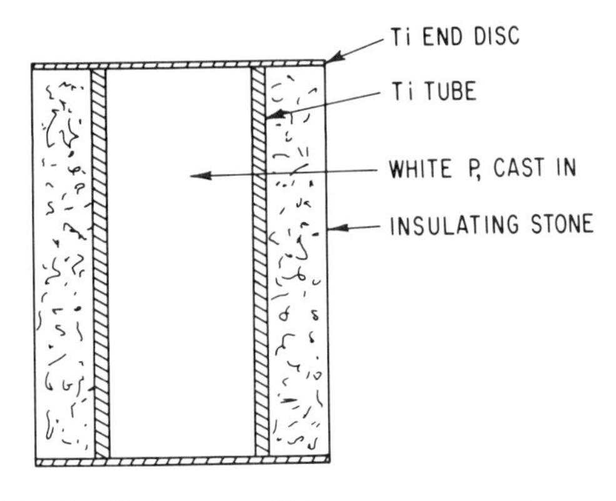

Fig. 13. High pressure, high temperature reaction vessel for making black phosphorus ingots.

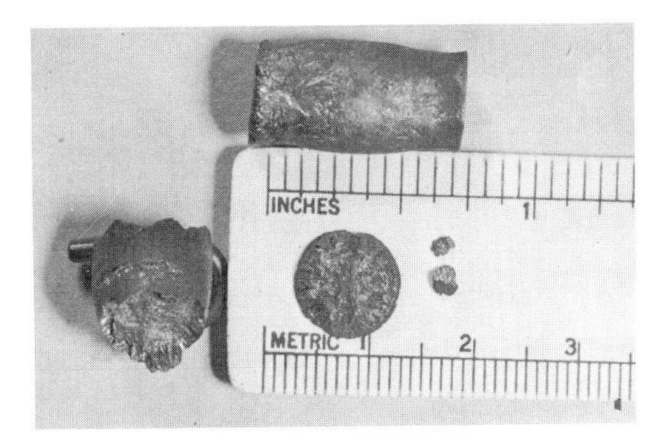

Fig. 15. Black phosphorus ingot showing columnar masses of crystals.

ing section on diamond.) After the pressure on the sample had been increased to about 22 kb, the capsule was heated to about 1200°C by passing a heavy current through the metal tube. The heating current was then slowly decreased, so that the phosphorus gradually solidified. The heat loss from the tube was greatest near the ends, and the black phosphorus froze there first.

After the sample had cooled, the pressure was reduced and the capsule removed from the apparatus. The metal tube surrounding the phosphorus helped prevent extensive shearing or tearing of the ingot, and was removed by dissolution in hydrofluoric acid. A columnar mass of black phosphorus crystals remained as shown in Fig. 15. Coarse, imperfect crystals of black phosphorus could be teased out of a partially crushed ingot. Attempts to limit the number of crystals produced in a single ingot have not yet been successful.

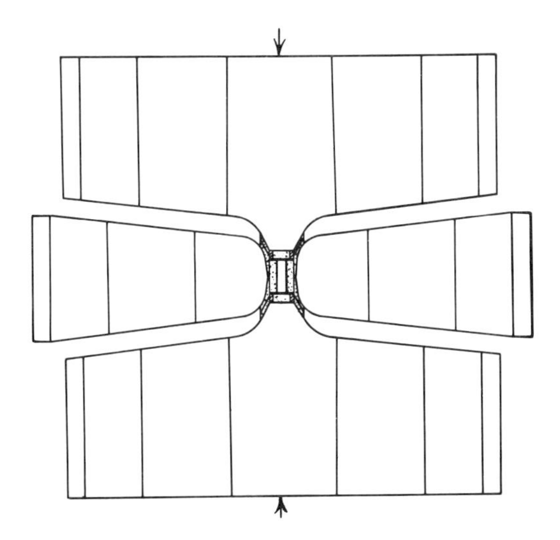

Fig. 14. Cross section of "Belt" high pressure, high temperature apparatus. (Hall, 1960.)

Diamond

Perhaps no crystal has received so much study as has diamond. Although many details about the constitution of natural crystals have been gathered, the manner in which the diamonds were formed is not yet known with certainty. Many attempts have been made to synthesize diamond from other forms of carbon or from various carbon compounds. However, reproducible attainment of this goal was not achieved until sufficiently high operating pressures and temperatures appropriate for the thermodynamic stability of diamond had been reached and the proper chemical systems found (Bovenkerk et al., 1959; Liander and Lundblad, 1960).

In brief, the methods found to be effective for forming diamonds are to expose mixtures of graphite and catalyst metals, such as manganese, iron, nickel, platinum, etc., to pressures of 45 kb or more while the mixture is hot enough for a molten metal phase to form. The transformation from graphite to diamond proceeds via a thin film of the molten catalyst alloy, and usually occurs in a few minutes.

Many of the principles involved in the

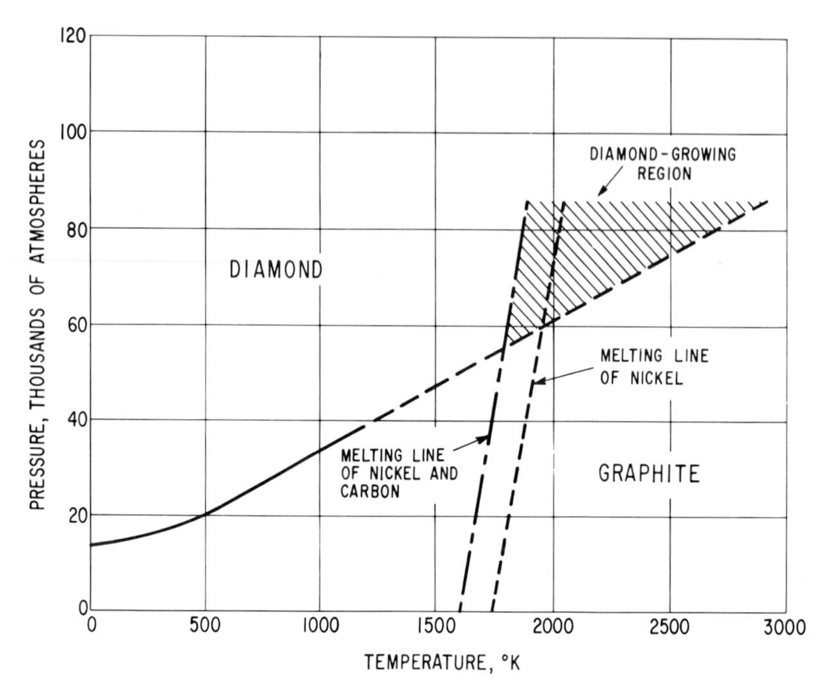

Fig. 16. Pressure-temperature diagram showing a portion of the diamond-graphite equilibrium line and the melting line for nickel and the nickel-carbon eutectic.

transformation are illustrated in the P-T diagram of Fig. 16, which shows the approximate position of the diamond-graphite equilibrium line as well as melting curves for nickel and the nickel-carbon eutectic. Diamond formation occurs when the pressure is high enough for the diamond to be stable and the temperature is high enough to melt the nickel-carbon eutectic. These conditions are satisfied in the shaded region on the diagram. The rate at which the diamond forms is found to be only mildly dependent on temperature but rather sensitive to pressure. The more the pressure exceeds that required for equilibrium, the greater the rate of diamond nucleation and growth.

The crystal habit is affected by the pressure during growth; low pressures favor the cubic habit, and higher pressures favor the octahedral habit. The color of the crystals is governed by the temperature and the chemical environment. At the lowest feasible growth temperatures, the diamonds tend to be black or dark green; higher operating temperatures produce light green, yellow, or almost colorless crystals. The addition of small amounts, less than 1%, of boron to the synthesis mixture produces blue-colored crystals which are also *p*-type semiconductors (Wentorf and Bovenkerk, 1961).

An apparatus in which the required conditions of temperature and pressure may be provided is shown in Fig. 14. This apparatus was named the "Belt" by its designer, H. T. Hall (1960). Two tapered pistons compress a sample confined in a short, tapered cylinder. The pistons and cylinder are made of cemented tungsten carbide and are supported by strong tapered steel rings. A "sandwich" gasket made of pyrophyllite stone and steel provides a compressible seal between the piston flanks and the cylinder. The sample proper is contained in insulating stone such as pyrophyllite and is heated by electrical resistance; heating current passes in through one piston and out through the other.

When it is properly constructed and loaded, a "belt" apparatus is capable of maintaining pressures up to 100 kb and temperatures up to 3000°C for periods of several hours.

The temperature of the sample may be

estimated by thermocouples passed through the compressible gasket; but such measurements are not precise because of errors caused by the effects of pressure on the thermocouples. These errors are probably less than 5% at 1500°C and 60 kb. Usually the sample temperature is roughly proportional to the heating power.

Errors also prevent precise determination of the pressure in the sample. One source of error is the finite shear strength of the thermal insulation; pressure gradients may exist in it so the sample pressure is not uniform. Another source of error exists in the determination of the average chamber pressure. This pressure may not be estimated from the ratio of force/area because friction modifies the true force, and the compressible gasket does not clearly define the area. Therefore the apparatus is calibrated for pressure by using reference standards.

Well-marked changes in volume or electrical resistance of certain compounds or metals that are agreed to occur at certain pressures, for example, bismuth at 24.7 kb or barium at 59 kb, can be used as standards. By observing the external forces required to produce the various reference phenomena in the pressure chamber, one obtains a relationship between external force and chamber pressure. This relationship is valid at 25°C, but the apparatus is used at much higher temperatures for which no well-determined reference phenomena exist. At higher temperatures the physical properties of the chamber contents, such as density and shear strength, may change and these changes may alter in an unknown way the calibration relationships of 25°C. However, one can argue effectively that the errors are probably no more than 10%.

For a more complete discussion of the problems encountered in achieving and measuring combined high pressures and temperatures, the reader is referred to *Modern Very High Pressure Techniques,* Butterworths, London, 1961.

Despite the uncertainties in the absolute measurement of pressure and temperature it is often possible to obtain relatively reproducible conditions of pressure and temperature from one experiment to the next. During the course of a particular experiment there may be slow changes in local temperatures and pressures because of various physical and chemical changes, a catalog of which is too lengthy to be given here. In view of the difficulty of making measurements on the reaction chamber during a run, the trends of these slow changes are best determined from the examinations of runs of varying lengths. In this way empirical rules may be established for producing various desired effects; such rules naturally vary from one kind of sample to another or from one type of apparatus to another.

A typical reaction chamber arrangement for forming diamond is shown in Fig. 17. Here a central graphite or carbon cylinder is capped at each end by catalyst metal cylinders. When such a system is heated at high pressure, the metal ncxt to the graphite usually forms an alloy with the carbon and eventually melts. The catalytic powers of the metal now act, and a thin layer of the graphite next to the metal is converted into diamond. As time passes, more and more of the graphite is changed to diamond and a film of molten catalyst sweeps through the graphite, leaving diamond in its wake. Eventually the entire mass may be changed

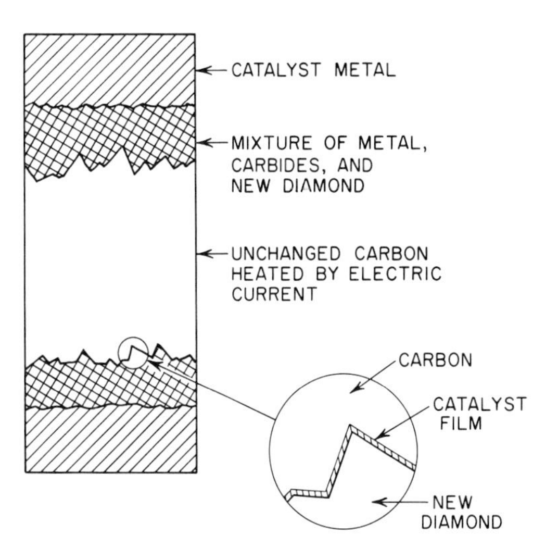

Fig. 17. Sketch of reaction chamber for making diamonds. (After Bovenkerk et al., 1959.)

Fig. 18. Freshly made diamonds covered with their thin catalytic film.

to a mixture of diamond and catalyst alloy. Figure 18 shows a mass of diamonds covered with a film of nickel catalyst obtained by interrupting the diamond-forming process by quenching the sample. The adhering graphite has been removed in a hot mixture of sulfuric and nitric acids. Figure 19 shows the typical appearance presented by the diamonds after the metal has been removed with acid.

In Fig. 20 one may see a freshly grown diamond cube face on which a coarse "spiral" growth pattern appears. This growth pattern, featuring coarse, curved terraces, is usually observed when the synthesis pressure and temperature are quite low. Such diamonds have not been reported to occur in nature and are probably the product of a special growth mechanism, the details of which are not yet thoroughly understood but which are believed to be related to graphite formation.

Figure 21 shows a cube face on a diamond grown with nickel catalyst at higher pressures and temperatures than the crystal of Fig. 20. This face is marked by many tiny ridges and valleys in the diamond. They appear as the result of uneven freezing of the nickel catalyst film on the diamond surface. Because the nickel lattice is a fairly good fit for the diamond lattice, the diamond face influences the crystallization of the nickel. Dendrites of nickel freeze epitaxially upon the diamond first. Where

Fig. 19. Mass of diamonds after acid cleaning.

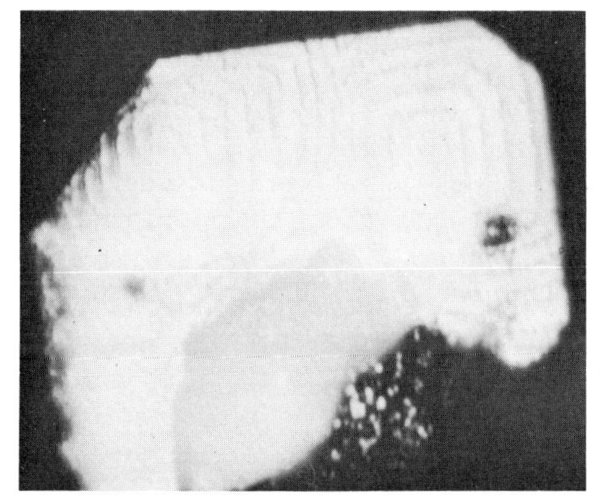

Fig. 20. Spiral growth pattern on cube face of synthetic diamond.

Fig. 21. Freshly grown cube face of diamond showing dendritic freezing patterns.

Fig. 22. Freshly grown octahedral face of diamond showing triangular growth markings.

the nickel has frozen, diamond formation ceases; where the nickel is still molten, ridges of diamond are deposited until this nickel also freezes. Thus the observed pattern is the result of the last stages of the diamond growth process.

Figure 22 shows a freshly grown diamond octahedral face bearing many of the "trigons" (hollow triangular depressions) often found on natural diamond faces. Such trigons, oriented antiparallel to the octahedral face edges, are usually regarded as products of the growth process.

Growth patterns of a different sort are displayed on the octahedral face shown in Fig. 23. Instead of the comparatively

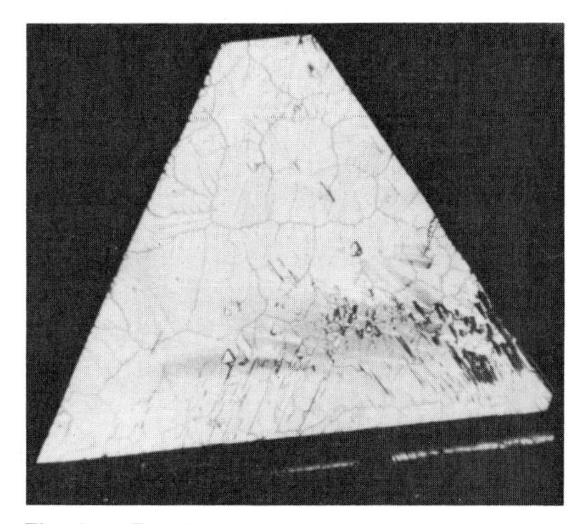

Fig. 23. Freshly grown octahedral face of diamond showing irregular growth ridges.

large, regularly arranged ridges so often found on cube faces, this face shows many tiny crooked ridges meandering about.

Such well-developed diamond faces as the foregoing are usually obtained by slow growth close to equilibrium conditions. When the pressure exceeds that of equilibrium by a few kilobars, the nucleation and growth rates become very high, limited at most by diffusion of carbon and catalyst. Small crystals form such as those shown in Fig. 24.

Crystal twinning is rather frequent, and usually occurs on the octahedral plane. Sometimes fivefold cyclic, octahedral twins form a star as in Fig. 25. Five octahedral apex angles equal 352°40′; the missing 7°20′ is filled in with polycrystalline diamond. The relatively symmetrical growth of these crystals suggests that they grew from a fivefold twin nucleus which might have been present in the original graphite.

Since the diamond crystals often grow relatively rapidly, they are likely to contain inclusions. Nickel is especially prone to be included in diamond, probably because the crystal lattice of nickel has almost the same unit cell size as diamond. Thus very little strain is required. At high pressures a small increase in volume due to strain can correspond to a significant amount of energy. (The energy associated with a volume change of 1 cm^3 at 40 kb is 1 kcal.) It is not surprising then to find that x-ray diffraction patterns of nickel-bearing diamonds show that the nickel crystals are aligned parallel with the host diamond (Lonsdale and Grenville-Wells, 1958). Even relatively transparent nickel-bearing diamonds are ferromagnetic with the same Curie temperature as nickel.

Graphite is also frequently included in

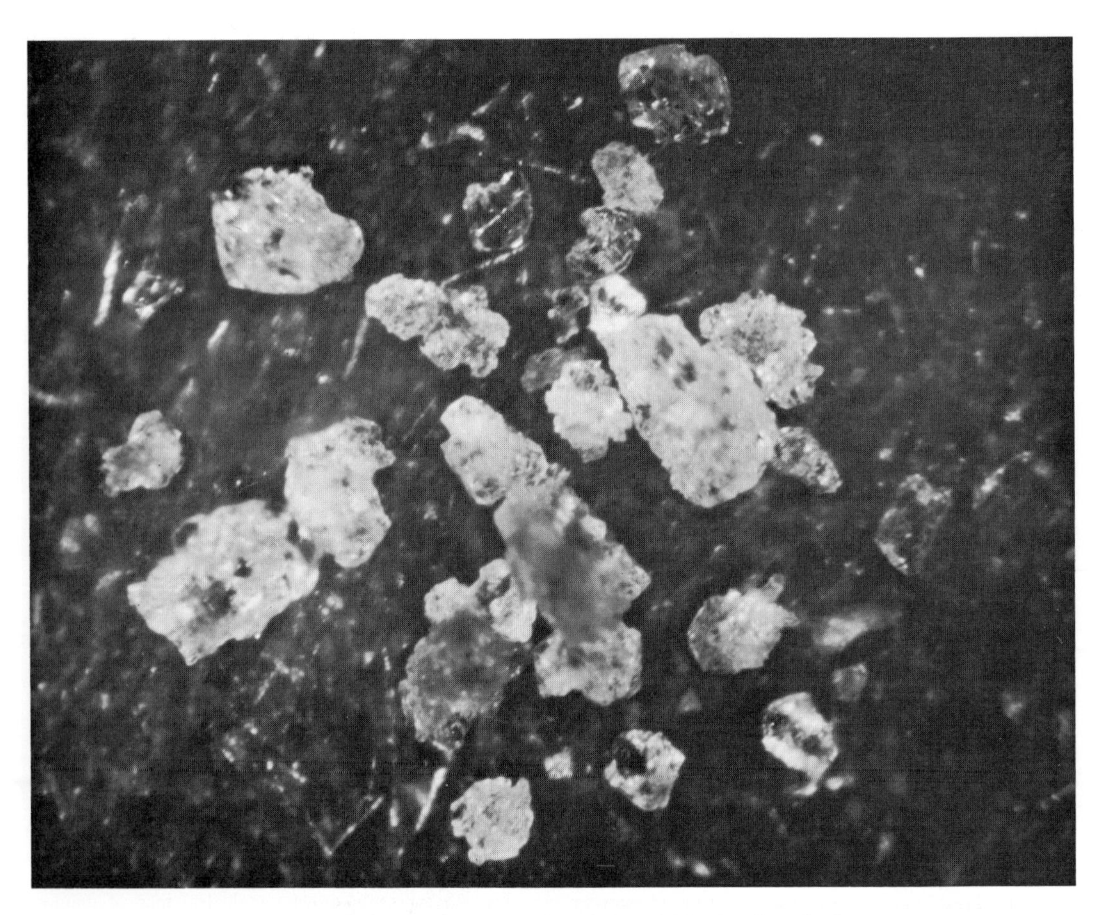

Fig. 24. Sharp, polycrystalline diamonds useful in abrasive wheels.

Fig. 25. Cyclic, fivefold, octahedral twin of diamond.

both natural and man-made diamonds, probably because graphite flakes laid against octahedral diamond faces would not introduce excessive strain. The flat hexagonal rings of graphite match the slightly puckered hexagonal rings of diamond fairly closely; the puckering helps to compensate for the slightly greater bond length in diamond.

The formation of diamond does not appear to consist simply of precipitation from solution, but seems to require catalytic effects, and one finds that small traces of certain elements may poison or accelerate the nucleation and growth of diamond. Small amounts of lithium or iodine, for example, frequently halt the formation of diamond even though the solvent power of the catalyst for carbon appears to be unaffected. In contrast, small amounts of boron greatly accelerate the formation of diamonds in addition to making the crystals *p*-type semiconductors.

Today a detailed explanation for some of the phenomena observed in diamond formation is lacking. It is hoped that future work will provide a complete picture of these interesting processes.

References

Becker, J. H. (1958), *J. Appl. Phys.* **29,** 1110.

Bovenkerk, H. P., F. P. Bundy, H. T. Hall, H. M. Strong, and R. H. Wentorf, Jr. (1959), *Nature* **184,** 1094.

Bridgman, P. W. (1914), *J. Am. Chem. Soc.* **36,** 1344.

Bridgman, P. W. (1916), *J. Am. Chem. Soc.* **38,** 609.

Butuzov, V. P., and S. S. Boksha (1956), *Rost Kristallov,* p. 245. *Conference on Crystal Growth,* March 5–10, 1956, Consultants Bureau Translation, New York.

Decker, B. F., and J. S. Kasper (1960), *Acta Cryst.* **13,** 1030.

Deming, H. G. (1940), *Fundamental Chemistry,* John Wiley and Sons, New York.

Elias, P. G., N. H. Hartshorne, and J. E. D. James (1940), *Chemical Soc.* p. 588.

Ewald, A. W., and O. N. Tufte (1958), *J. Appl. Phys.* **29,** 1007.

Hall, H. T. (1960), *Rev. Sci. Instr.* **31,** 125.

Hoard, J. C. (1960), "Structure and Polymorphism in Elemental Boron," in *From Borax to Boranes,* Advances in Chemistry Series, Washington, Am. Chem. Soc.

Hoard, J. C., and A. E. Newkirk (1960), *J. Am. Chem. Soc.* **82,** 70. See also *Boron: Synthesis, Structure, and Properties,* Plenum Press, New York.

Horn, F. H. (1959), *J. Electrochem. Soc.* **106,** 905.

Kittel, C., and J. K. Galt (1956), *Solid State Physics,* edited by F. Seitz and D. Turnbull, Vol. III, p. 439, Academic Press, New York; Academic Books, London.

Krebs, H., H. Weitz, and K. H. Worms (1955), *Z. Anorg. u. Allg. Chem.* **280,** 119.

Liander, H., and E. Lundbald (1960), *Arkiv för Kemi,* Band 16, nr. 9, p. 139.

Lonsdale, K., and H. J. Grenville-Wells (1958), *Nature* **181,** 758.

Luborsky, F. E. (1957), *J. Phys. Chem.* **61,** 1336; (1958), *J. Phys. Chem.* **62,** 1131; (1961), *J. Electrochem. Soc.* in press.

Luborsky, F. E., T. O. Paine, and L. I. Mendelsohn (1959), *Powder Metallurgy,* No. 4, p. 57.

McCarty, L. V., J. S. Kasper, F. H. Horn, B. F. Decker, and A. E. Newkirk (1958), *J. Am. Chem. Soc.* **80,** 2592.

Nerad, A. J. (1932), *Trans. Am. Inst. Chem. Eng.* **28,** 12.

Nerad, A. J. (1933), Unpublished Work.

Schuster, M. C., and E. F. Fullman (1946), *Indust. and Eng. Chem., Analyt. Edn.* **18,** 653.

Sidgwick, N. V. (1951), *The Chemical Elements and Their Compounds,* Oxford University Press, London.

Wentorf, R. H., Jr. (1952), Unpublished Work.

Wentorf, R. H., Jr., and H. P. Bovenkerk (1962), *J. Chem. Phys.* **36,** 1987.

11

Ionic Salts

P. H. Egli and L. R. Johnson

When the chemist approaches the task of growing a particular crystal from solution, he is confronted by the literature with nearly as many procedures as there are crystals and investigators (*Crystal Growth,* 1949). No one of these systems is a cure-all; and the best combination of equipment and growth methods varies considerably, depending on the crystal and with the scale of the operation. This discussion reviews the factors that determine how best to grow a crystal of a particular compound. It summarizes the experiences of the Naval Research Laboratory with several hundred compounds.

It is necessary first to discuss the factors that control the growth process. Success depends more on attention to the principles of growth and on taking advantage of factors that influence chemical reaction rates than on details of equipment design and procedures. There are four general headings.

1. Character of the Solution. This involves choice of the solvent or alteration of the solution, usually to obtain maximum solubility.

2. Effect of Additives and Purity of Reagents. Purity of the chemicals is frequently important, but the most influential variable is the presence of several types of additives.

3. Operating Variables. For any chemical system, success depends on selecting the optimum temperature range, providing the proper degree of agitation, and controlling the degree of supersaturation.

4. Seed Crystals. Size, orientation, and quality of seeds are all important factors.

1. GROWTH FACTORS

The Character of the Solution

The dominating factor during growth is the character of the solution. The choice of solvent, the purity, *pH,* and the presence of several types of additives are very important, more so than the apparatus design. So little is known about the chemistry of concentrated solutions, however, that the crystal grower must arrive at the optimum composition by trial and error, assisted only by broad generalizations. It is logical to suppose that the ease of crystal growth depends on the similarity between the state or association of a compound in solution and in the crystal. This is demonstrated in a general way by the fact that hydrates like the alums, etc., grow with ease, and in these solutions the solvation of the ions must be similar to the water coordination in the solid crystal. The same applies to H-bonded crystals: $NH_4H_2PO_4$, HIO_3. In the latter case the concentrated solution is known to contain dimers and trimers. Thus the condition of preorganization in solution is an important one for ease of growth. To this end, the importance of choosing the solvent and adjusting the *pH* is obvious. Growth of large, clear crystals is virtually impossible unless a solvent is found in which the solute is appreciably soluble. Beyond that, a correlation has been found between the ease of growth and the degree of supersaturation a solution will support without spontaneous nucleation. This factor (the extent of the supersaturation) can frequently be controlled by chemical additives.

The importance of the choice of solvent is strikingly illustrated by the difficulty of growing benzil from alcohol and a number of other solvents, even though it is appreciably soluble; but it grows very readily from benzene solution. Obviously, the similarity of the chemical structure rather than the amount of solubility dominates this situation.

Materials that are relatively insoluble in water can be grown more readily by the addition of an agent that increases the solubility. For example, the growth of $KBrO_3$ is improved, although it remains difficult, by the addition of 4 *M* solution of $NaNO_3$. Frequently the addition of an acid or base results in a usable composition; for example, the addition of 10% NH_4OH to a water solution makes it possible to grow a series of silver salts,* Ag_2SO_4, $AgIO_3$, AgCl, AgCN, AgI, and $AgNO_2$. $AgIO_3$ grows more readily from pyridine, and both AgI and AgCN grow somewhat better from a $Hg(NO_3)_2$ solution. HgI_2 can be obtained from KI solution.

The effect of *pH* is illustrated by the growth of $Li_2SO_4 \cdot H_2O$. Growth of this compound is moderately difficult under the best conditions, and precise adjustment of *pH* is one of the critical factors. The optimum value is 6.5 to 6.7, and a variation of more than ± 0.25 makes strain-free growth almost impossible. In other instances, the *pH* is less critical. With $NH_4H_2PO_4$, for example, the *pH* is adjusted to about 3.9 ± 0.2 in order to control the solubility of certain additives, but the quality of growth is not particularly sensitive to small changes in *pH* as such. Good growth can be obtained, although with more difficulty, from compositions containing as much as 5% excess of either NH_3 or H_3PO_4. The *pH* may also influence habit as well as the ease of growth. In $NH_4H_2PO_4$, for example, excess NH_3 induces considerable growth on the prism faces where none ordinarily occurs. Conversely, in HIO_3, the addition of 10% H_2SO_4 causes only a minor change in the relative growth rates of various faces, although higher concentrations suppress sidewise growth.

* Precautions must be observed because of the explosive nature of NH_4, Ag compounds.

Effect of Additives and Purity of Reagents

Very rarely is even the purest reagent-grade material of any use for crystal growth. Even when crystals can be grown that are visually perfect, they frequently have physical properties so altered by trace impurities that they are useless for the purpose intended. The effect of impurities on the growth tendency of crystals is very complex. In some instances, certain additives must be present to achieve any sort of sizable growth; conversely, excessive amounts of these same additives or traces of certain other ions may completely block the growth process. Certain physical properties are even more sensitive to impurities than is the growth process. For example, the electrical resistivity of $NH_4H_2PO_4$ changes by a factor of ten with variations of a few ppm of absorbed SO_4^{--}. Instances are common where properties are affected by impurities present in amounts below the usual analytical level. For example, there is no commercial source for such a common substance as NaCl that will produce satisfactory crystals for many physical measurements. Certain brands of table salt have proved to be better than the best material from chemical reagent companies, but all of it contains enough lead and various other ions to have a measurable effect on impurity-sensitive properties, such as luminescence. This point can hardly be overemphasized, since the literature is undoubtedly full of incorrect physical concepts because the data on which they are based were controlled by unsuspected impurities.

Purity of the solution and the effect of additives are thus the most powerful variables and the most likely source of difficulty in the entire problem of crystal growth. No adequate explanations for various aspects of these phenomena are available, but even traces of foreign ions frequently may:

1. Alter the habit or endform of the crystal.
2. Change the degree of supersaturation that the system will support.

3. Change the rate of growth that can be maintained without formation of flaws.
4. Control the texture of the crystal, both with respect to large flaws and small-scale structural defects.
5. Control the impurity sensitive and structure sensitive physical properties.

In view of the uncertainty concerning the effect of many specific impurities, it is generally desirable to produce the purest possible chemical and to introduce any desirable additives to this material in known amounts. The preparation of pure chemicals is beyond the scope of this discussion (Egli et al., 1959).

There are several classes of additives that may produce remarkable improvements in crystals whose growth from water solution is otherwise very difficult. Highly polarizable ions constitute the most important group, and a virtually neglected one. This is a powerful tool, and, properly used, it can make the difference between a growth process being very easy or very difficult. For example, NaCl is nearly impossible to grow from pure solution, but is readily grown in the presence of 0.01 mole per cent lead.

This sort of phenomenon has not been thoroughly explored, and it is not yet possible to predict what additive will be most effective for a particular compound. To illustrate the highly specific nature of these additives consider the scope of Pb^{++} and of Mn^{++}, the ions that give the most aid in growing NaCl. In an investigation of eleven compounds by Yamamoto (1939), the results in Table 1 were obtained.

The list in Table 2 gives instances that have been reported by various investigators in the literature; but these do not represent thorough investigations in every case. The ions are listed (so far as known) in the order of effectiveness; and, for the most part, only the first one or two listed for each compound are strongly beneficial.

On the basis of experience concerning the degree of effectiveness of each ion, some generalizations can be drawn from Table 2. The compounds that respond most readily to this type of additive are the ammonium and alkali-metal salts. The most powerful additives are a series of metal ions recognized to have the common characteristic of basigenic properties. For the alkali metals, Pb^{++} and Mn^{++} are usually most effective; for ammonium salts, Al, Cr, and Fe^{+3} are generally best, although in both cases there are exceptions.

Another frequently helpful type of additive is a solution modifier that presumably changes such properties of the system as viscosity and surface tension. The mechanism for these is even more obscure than for trace concentrations of metal ions, but it has been clearly established that the addition of glycerin, sugar, or various gelatins to saturated aqueous solutions of certain salts results in better crystal growth. Cases that have been conclusively demonstrated are listed in Table 3.

Both the metal ions and the solution-modifying agents usually alter the crystal habit, usually to make the crystal more nearly equidimensional. A vast array of other types of additives is known that lead to the more dramatic habit modifications involving radical changes in the sizes and types of faces on the finished crystal. This

TABLE 1

	Aid Formation		No Effect		
Pb^{++}	NaCl	KBr	CsCl	K_2SO_4	
Pb^{++}	KCl	KI	$LiCl \cdot H_2O$	$KClO_3$	
Pb^{++}	RbCl	KNO_3	NH_4Cl	———	
Mn^{++}	NaCl	K_2SO_4	KCl	KBr	$KClO_3$
Mn^{++}	$LiCl \cdot H_2O$		RbCl	KI	
Mn^{++}	NH_4Cl		CsCl	KNO_3	

TABLE 2

Compound	Additives Helpful in Small Concentrations to Crystal Growth
$AgBrO_3$	Fe^{+3}, Al, Zr, W^{+6}, Th
$AgIO_3$	Exhaustive study—no helpful ion (except NH_4^+)
$Ba(NO_2)_2$	(10% alcohol solution) Mg, Te^{+4}
CsCl	La, Ce^{+3}, Nb^{+3} (all of slight effect)
$KBrO_3$	Pb, Th, Te^{+4}, V^{+5}
KCl	Pb, Bi, Sn^{+2}, Ti, Zr, Th, Cd, Hg, Fe
$KClO_3$	Nothing helpful found
KI	Pb, Ti, Sn, Bi, Fe
KIO_3	Te—slight effect
KNO_2	Fe—slight effect
KNO_3	Pb, Th, Bi
K_2SO_4	UO_2^{+2}, VO_2^+, Cd, Mn, Fe, Ce, Cu, Al, Mg, Bi
$LiCl \cdot H_2O$	Cd, Mn^{+2}, Sn^{+2}, Sn^{+4}, Co, Ni, Fe^{+3}, Ti, Cr, Th
NaCl	Pb, Mn^{+2}, Bi, Sn^{+2}, Ti, Cd, Fe, Hg
$NaNO_2$	Ca
$(NH_4)_2C_2O_4 \cdot H_2O$	Cu, Ca, Fe, Th
NH_4Cl	Zr, Cd, Mn, Fe^{+2}, Cu, Co, Ni, Fe^{+3}, Cr
NH_4F	(In alcohol saturated with NH_3) Ca
$NH_4H_2PO_4$	Fe^{+3}, Cr, Al, Sn
NH_4IO_3	Cd, Ca
RbCl	Pb, Sn, Zr, Ti

subject is outside the scope of this discussion, however, because these latter additives are frequently absorbed and are beneficial to the growth process only in special circumstances. In general, the effect of such additives is to retard the growth of certain faces (Whetstone, 1949; Butchart and Whetstone, 1949). For example, needlelike growth should be taken as indication of the presence of an undesirable habit-modifying impurity. Several excellent reviews of habit modifications are available (Buckley, 1951).

We conclude that it is important to recognize that the composition of the solvent is a controllable variable that is more important to success than details of the growth technique and apparatus. Whether a solution is satisfactory can usually be determined by observing spontaneous crystallization on very slow cooling or evaporation from a beaker at room temperature. It is usually useless to attempt the growth of large crystals if normal precipitation results in fine powders or very thin needles; and time is well spent in altering the composition of the solvent until sizable particles are obtained by simple precipitation.

Operating Variables

Before selecting a growth method and apparatus it is necessary to consider the implications of the operating variables: (1) degree of supersaturation, (2) efficiency of agitation, and (3) temperature range.

DEGREE OF SATURATION. Growth rate increases directly with degree of supersaturation; thus for an efficient process supersaturation must be maintained just as high as possible but below the point that induces flaws or spontaneous nucleation. This

TABLE 3

Crystal	Growth Aid
$Ba(HCOO)_2$	10% glycerin
$Ba(NO_3)_2$	10% glycerin
$Ca(HCOO)_2$	10% glycerin
$(NH_4)_2C_2O_4 \cdot H_2O$	Glycerin
Pentaerythritol	Sucrose
$MgSO_4 \cdot 7H_2O$	Borax (50 grams/liter)
$ZnSO_4 \cdot 7H_2O$	Borax
KH_2PO_4	Borax

requires that the growth apparatus provide very sensitive control of the means for inducing supersaturation. The critical growth rate (beyond which flaws are formed) varies tremendously among different crystals, and a generally useful growth apparatus must provide, in addition to sensitive control, a means of inducing supersaturation over a wide range.

Of the several methods for inducing supersaturation, evaporation is usually the most practical on a laboratory scale. The principal advantage is that the rate is very easy to control simply by adjusting the size and number of holes in the lid of the container. For greater precision, the vapor can be condensed and withdrawn at a very accurate rate independent of humidity and barometric pressure changes. Evaporation at constant temperature avoids the danger of cracks induced by thermal strains, and the single temperature is easy to control with inexpensive equipment. This method works equally well for solutions with any sort of temperature-solubility characteristics, and on a laboratory scale is about the only possibility for systems with flat temperature-solubility curves. On a large scale, vapor removal becomes more difficult to control.

For salts with normal solubility curves, temperature lowering is a practical method of inducing supersaturation for large-scale operations, but the necessary temperature controllers are expensive for laboratory use. This method does have the important advantage that the system is cool when the growth run is completed, so that the crystals can be easily removed without damage from thermal shock.

Recirculation is another method for inducing supersaturation that is efficient on a large scale, but often impractical in the laboratory (Eitel, 1926). This scheme uses a solution make-up tank (at a temperature slightly higher than the growing tank) that provides a continuous source of new material. The solution is filtered through a holding tank to insure complete solution of nuclei. The system provides the advantages of constant temperature and precise control, flexible enough for any sort of solubility characteristics. Moreover, the growing tank is never depleted, so crystals can be removed and replanted in a semicontinuous manner. In the laboratory, however, control is difficult and temperature changes in small tubing and pumps cause difficulties with spontaneous nucleation. A common laboratory version of this scheme is convection current transfer between the two vertical members of an H-cell. Diffusion of new material through the horizontal connector is necessarily slow since the connector is full and the amount of supersaturation developed low. This too is difficult to control at an efficient rate, but is a good scheme for relatively insoluble materials, or any other that must be grown very slowly and gently. Bartlett's apparatus is very promising and simple (Bartlett, 1961).

The fourth general method of inducing supersaturation is by introducing additional components to the system. These salting-out procedures have the common difficulty that the point at which the additional component is introduced is highly supersaturated, and spontaneous nuclei are likely to form before the additive is evenly dispersed.

Various other schemes such as refluxing through a Soxhlet extractor (Eitel, 1926), convection currents from temperature gradients in a single container, and diffusion currents from particles ground fine enough to be supersoluble have been used successfully; but for the most part these represent variations of the general schemes already discussed and are necessary only in unique situations.

EFFICIENCY OF AGITATION. In most crystal-growing systems, the rate of growth depends directly on the efficiency of the agitation up to the point where the process is no longer diffusion controlled, and this factor more than any other is important in the design of apparatus. Agitation sweeps off the depleted solution in contact with the crystal surface, restoring supersaturated solution that would otherwise have a long diffusion path to traverse. It is a frequent experience that a slight improvement in agitation permits much more rapid growth.

The optimum amount of agitation is undoubtedly far greater than the amount provided by most of the proposed apparatus. The limiting factor is cavitation, or sufficient turbulence at a point in the system to induce spontaneous nuclei. Thus small impeller blades as frequently used for laboratory stirring can induce nuclei at the blade tips without providing nearly enough movement throughout the vessel.

The importance of effective agitation is a consideration in mounting crystal seeds. If the growing crystal is near some obstruction, so that solution is channeled through a narrow passage between the crystal surface and the container wall or the temperature control element or an adjacent crystal, the resulting turbulence may induce flaws or nuclei while agitation is still not optimum in other regions of the container. Agitation also determines the direction seeds should be mounted. Normally they should be mounted so that maximum agitation occurs across the faces with the highest critical growth rate. Usually these are the most rapid-growing faces and are the smallest of the end form.

Conversely, agitation can be used to alter the natural habit of the crystal by making most of the solution sweep across faces that are extended little or none under uniform conditions. For example, $NH_4H_2PO_4$ and its isomorphs usually add new material only on the pyramidal faces so that there is no increase in cross section. However, when the crystals are mounted close together or near some obstruction, considerable growth is obtained on the nearest prism face.

The need for considerable agitation can also influence the composition of the solution. With sufficient agitation, diffusion is usually not the rate controlling process in systems with normal viscosity; but in thick solutions, such as form with highly soluble HIO_3, turbulence becomes serious even at low rates of stirring. Thus it may be desirable to add a thinning agent, such as H_2SO_4 in the case of HIO_3, to make stirring easier, even if it has some undesirable effect such as lowering the metastable range of supersaturation.

TEMPERATURE RANGE. The dominating fact is that the critical growth rate increases appreciably with temperature. Therefore, regardless of the solubility characteristics of the solution, good crystals can be grown more rapidly at high temperatures. Very little data are available on the change of the metastable field of supersaturation with temperature, but limited experience has indicated that trouble with spontaneous nuclei is not markedly increased at high temperatures. Other operating difficulties are magnified, however. One problem is that large pieces of crystal are very sensitive to thermal shock. Crystals with highly anisotropic thermal expansions are likely to crack in the slightest thermal gradients. Crystals an inch or two in size are likely to crack, for example, when touched with bare hands in an air-conditioned laboratory, or when moved from one room to another with a temperature difference of a few degrees. This is a particularly difficult problem at the start and finish of growth above room temperature. At the start, the problem is usually less acute because the seeds are usually rather small. They can be heated to within a few degrees of the temperature of the solution before they are introduced. The problem of removing completed crystals from a hot solution without getting cracks is difficult, however, and no satisfactory technique has been developed.

Other disadvantages of hot solutions are the higher vapor pressures and more rapid evaporation which cause handling problems. It is difficult at best to make a precisely saturated solution at a given temperature, and it is virtually impossible at near-boiling temperatures. Moreover, sealing the growth vessel against small vapor leaks around the stirring mechanism, etc., becomes more difficult. Finally, at high temperatures, temporary power failures are invariably disastrous, whereas near room temperature there may be time for repairs before the temperature of the solution changes enough to induce flaws. Because of these difficulties, most crystals are grown in the authors' laboratory at a few degrees above room temperature but it is important to recognize that difficult growth near room temperature may be easy at 80 to 90°C (Tuttle and Twenhofel, 1946). One instance is $Li_2SO_4 \cdot H_2O$ which was originally difficult to grow even up to 50°C. The first success was obtained at 90°C. At this temperature crystals of a usable size were obtained in two weeks, whereas a comparable size and quality could not be obtained in two months at 40°C. Later, as techniques and the quality of seeds improved, the temperature was gradually lowered to about 50°C. Special circumstances may require growth below room temperature. A technique has been described by Wolfe (1946).

Seed Crystals

Providing the proper *seed* is another important factor. The classical method is by evaporation or slow cooling of a saturated solution; but it is frequently more practical to use a piece from a natural or a previously grown crystal, or to produce oriented overgrowth on a seed of similar material. By proper planning, it is possible to grow in the most efficient direction to minimize flaws and to a certain extent control the habit.

Any fragment of the desired crystal can serve as a seed. A piece bounded by the natural faces is not always the most desirable. The most efficient seed is a plate whose faces are normal to the most rapid growth direction. Usually, this means the seed should be cut parallel to the smallest face of the finished crystal. An important factor in selecting a seed is the difficulty of starting the growth process. The first layers of growth are the most critical, and it often appears to be virtually impossible to avoid small flaws at this point. If a flaw is large, it apparently induces strain in subsequent layers and the flaw is likely to recur periodically throughout the growth process. It is thus important to select a seed on which material deposits most easily. This was extensively studied in connection with $NH_4H_2PO_4$. The various seeds used are illustrated in Fig. 1.

First, two cap ends were glued back to back, making all the natural pyramidal faces available. Slight mismatches at the glue joint induced flaws that persisted throughout the length of the prism faces. Subsequently, it was found that the pyramidal faces would cap over from a flat face normal to the *Z*-axis so that a single cap end was sufficient (seed *B*). The uneven length and the frequent failures of the flat face to cap symetrically led to the use of a *Z*-cut plate. This was capped over in a

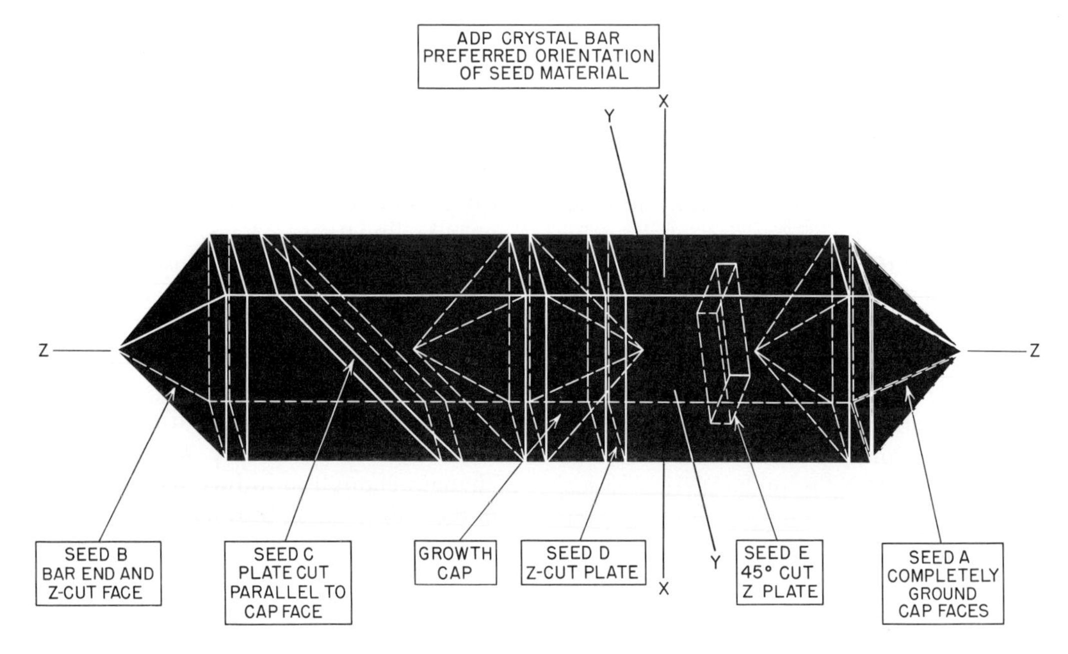

Fig. 1. $NH_4H_2PO_4$ seeds.

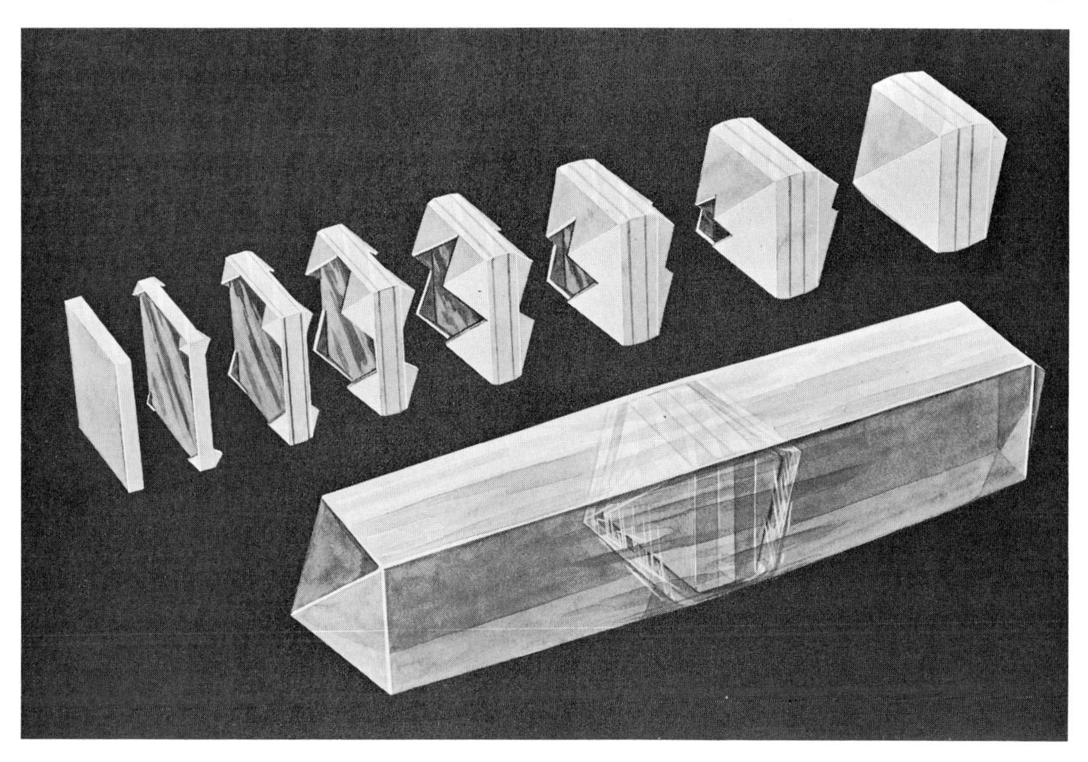

Fig. 2. Capping $NH_4H_2PO_4$ crystal.

separate growth operation. Growth occurs only at the very perimeter of the plate and proceeds to build a thin skin of material with a hollow center as shown in Fig. 2. Since growth originates only at the edges, the center section of the plate could be badly flawed or completely cut away without affecting the process. The completed caps were replanted as seeds for the growth of long bars. The natural pyramid faces of these seeds were never entirely satisfactory because it was impossible to start growth without flaws outlining the planes of the original cap. These flaws were large and recurring.

The next seed that was tried was a plate cut parallel to a single cap face (seed *C*). New material could be joined smoothly to this, so the seed could not be detected in the finished crystal. The missing pyramid faces developed slowly as growth proceeded, so the finished bar attained the normal habit with four equivalent pyramid faces on each end. It was eventually found that the best seed was a section with all the natural faces but with roughly ground surfaces (seed *A*). The rough surface apparently adds new material very smoothly so the seed interface is barely detectable. Figure 3 shows bars grown under equivalent conditions in a large-scale operation. This was operated at a rate that formed a harmless flaw on the ground surfaces but induced a series of large flaws from smooth, natural surfaces.

Seed *E* (Fig. 1) was used in an unsuccessful attempt to grow bars of only the portion of the crystal intended for use. The normal sides filled in very quickly and usually were so badly flawed that subsequent lengthwise growth was ruined. Such a scheme is successful in other instances, however.

In some crystals, certain zones persist in developing flaws while adjacent zones are growing perfectly at lower rates than necessary. In such instances, by properly orienting the seed, the flawed areas can be reduced to a minimum. For example, the pyramid zone in $KLiSO_4$ and the tetrahedral zone in $NaClO_3$ persistently formed closely spaced veils, whereas the adjacent zones

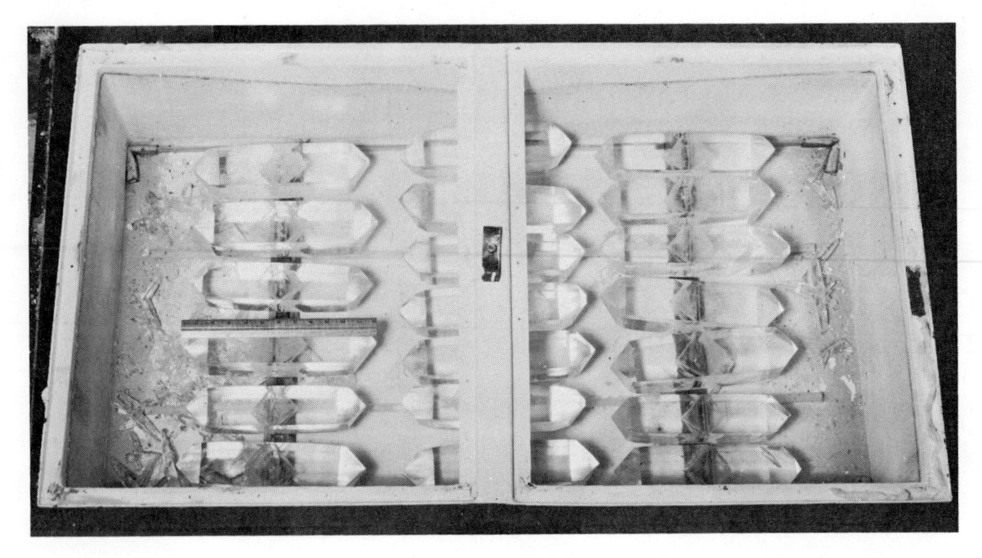

Fig. 3. Clear growth of $NH_4H_2PO_4$ on a large scale.

were clear. By successive growth on plates cut parallel to the clear faces, crystals nearly 2 in. in size were made that were perfect except for a very small zone at each end.

In other instances, a capping operation such as described for $NH_4H_2PO_4$ can be used to stop the propagation of persistent flaws. With seeds cut at an appreciable angle from a direction of rapid growth, a skin of strain-free material may cap over that does not propagate the flaws of the body of the seed. Whether capping is feasible depends on the relative growth rates of the various crystallographic planes. For example, HIO_3 is roughly similar to $NH_4H_2PO_4$ in that almost all growth occurs on cap faces when the solution is diluted with H_2SO_4. Hence the crystal lengthens appreciably with little increase in cross section. A surface cut normal to the length direction of $NH_4H_2PO_4$ usually caps over regardless of the growth technique, because the plane perpendicular to the Z-axis normally does not support growth. In HIO_3, however, a similarly cut seed develops the cap faces by successive growth of layers across the plate. This is very rapid compared to the subsequent increase in length by deposit on the normal cap faces. It indicates that there is a basal growth plane normal to the Z-axis that is so rapid that the resulting face is normally completely eliminated. Occasionally such a basal face is observed on a well-developed HIO_3 crystal, whereas none occurs on $NH_4H_2PO_4$ except when polyphosphates are added to the solution (Carlson, private communication).

The quality of the seed crystal is always important. In this connection it is important to distinguish between growth flaws and mechanical flaws. Mechanical flaws are not usually harmful, and may even be helpful, but growth flaws usually propagate and multiply. Common growth flaws include veils, lineage, and mosaic structure, vicinal faces and twins, splinter growth, cracks, holes, and included insoluble particles. Any one of them present near the surface of a seed is likely to propagate. A tiny blemish that appears innocuous may be the origin of a network of flaws. Such blemishes can sometimes be detected by examination of the plate in polarized light. The presence of a flaw is revealed by a blotchiness or nonuniformity of interference colors. There is a strong suspicion among crystal growers that the texture of seeds continues to improve with successive generations of growth even beyond the state where flaws cannot be observed or detected by physical measurements. Lithium sulfate, for example, is moderately difficult to grow; and in the early stages of development, the yield of flawless material increased with each successive run with no change in technique and no obvious change in the quality of material selected as seeds. The improved texture of the material was indicated by a greater ease (with each succeeding generation) in sawing sections without inducing cracks.

Mechanical flaws, such as those induced by sawing, are usually readily healed by subsequent growth. Cracks from thermal shock are bridged over, and corners or edges chipped by handling are filled in the first few minutes of growth. Mechanical flaws may contribute to the ease of starting growth as previously described for $NH_4H_2PO_4$. Thus seeds should be carefully inspected for growth flaws, and sawing out or breaking off faulty portions is likely to benefit the growth process.

In the absence of seed material, it is sometimes possible to get oriented overgrowth on a different crystal. The requirements of such seeds are severely limiting, however, and very few successful results have been reported. The crystallographic requirements have been the subject of numerous investigations. The necessary degree of structural similarity varies considerably with the similarity in chemical composition and with the type of growth process; that is, it varies with rate, temperature, and whether the crystal is deposited from vapor, melt, or solution. For growth from a melt rather faint similarities exert orienting influences, and a single face may serve as a seed. Thus a wide variety of compounds have been oriented by growth in contact with a plate of mica. For growth from solution, however, overgrowth does not occur so readily unless the compounds are chemically similar. On mica, for example, a few compounds deposit small, oriented crystallites; but no case has been reported where a continuous crystal layer grew over a mica plate. Moreover, the growing crystal usually encloses all sides of the seed, so all faces of the seed must be similar to corresponding faces on the growing crystal. In addition, the seed must be relatively insoluble in the saturated solution. This usually limits seeds for oriented overgrowth to less soluble isomorphs of the desired compound. When such a seed can be found, however, it saves considerable time in developing large crystals of the desired product. For example, perfect bars of KH_2PO_4 (2 in. in cross section) were grown on the first attempt using $NH_4H_2PO_4$ plates as seeds. A most striking case of overgrowth results if one deposits successive layers of several of the alums to form a rainbow crystal.

2. EXPERIMENTAL PROCEDURES

Equipment

The best growth apparatus depends on a combination of the growth factors previously discussed. For most situations, some variation of the basic apparatus shown in Fig. 4 can be used effectively (Imber). The most critical elements are the means for controlling temperature, evaporation, and agitation. In addition, there are certain refinements, such as minimizing creep of the solution up the container walls and protecting the growing crystals from floating nuclei. The following generalizations should aid in the choice of a system for a particular crystal.

SIZE OF CONTAINER. For laboratory operation, an 18-liter glass jar, 12 in. in diameter and 12 in. high, is optimum. With a smaller diameter it becomes difficult to mount seeds sufficiently distant from obstructions to prevent turbulent channeling. Temperature measuring and control elements, the agitator, and the container wall should be $1\frac{1}{2}$ in. removed from the nearest face of the finished crystal. For larger containers it becomes inconvenient to prepare solutions.

TEMPERATURE CONTROL. *Heat Source.* External heat sources are generally preferable to any kind of immersion heaters because of the elimination of corrosion problems and because of the more even temperature distribution. For salts with normal solubility characteristics, it is often sufficient to heat from the bottom only. For better temperature control, however, there should be two heat sources, one continuous, supplying about 75% of the required heat, and the

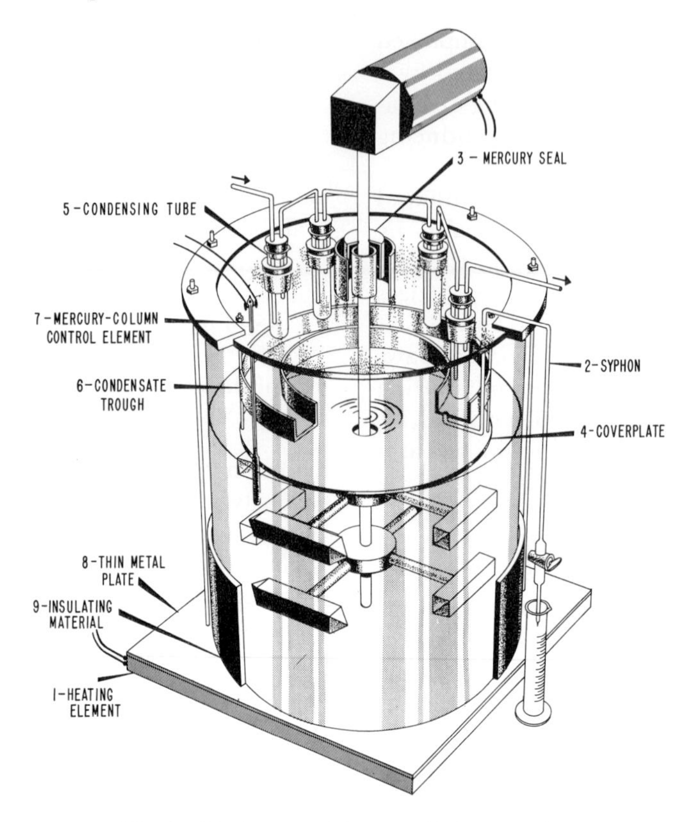

Fig. 4. Solution growth by controlled evaporation.

other controlled. The continuous heater is beneath the jar, and its power is supplied by a variable transformer. The controlled heater is a radiant heater, such as an infrared heat lamp or reflector-spot lamp, with a rating of 150 to 250 watts at 110 volts. An incandescent lamp gives good temperature control, since little time is lost when the controller turns the heat on and little overshoot occurs when the control point is reached. The bottom heater has the important advantage that extraneous nuclei tend to drop to the bottom and dissolve. Extraneous nuclei are difficult to avoid; frequently they become so numerous that they present more growth area than the planted seeds. Thus for salts with large temperature-solubility characteristics, the bottom heater is an important refinement. For those with flat solubility curves, it is obviously less helpful, and with inverse solubility curves, it is definitely undesirable.

For systems with flat or inverse solubility characteristics, an even temperature distribution, such as given by a well-stirred external water-bath, is best. Water, covered with a thin film of oil or paraffin, is preferable to an all-oil bath in case a container breaks. When a mass of crystals becomes tightly packed in the bottom of the growth container, a break may be caused by slight temperature fluctuation or perhaps by the growth forces.

Temperature Control. Temperature control is the most critical and most serious problem in the growth apparatus. The growth rate that can be maintained is directly dependent on the sensitivity of the controller. A high growth rate requires a high degree of supersaturation, but supersaturation must not, even temporarily, become so high that flaws or spontaneous seeds are formed. Growth from solution is slow at best, and therefore it is usually desirable to operate as close to the maximum supersaturation as the sensitivity of the controller, and thus the temperature fluctuations, permit. A system with a maximum supersaturation tolerance of 0.5°C could not be operated very close to the critical rate with a controller sensitivity of ±0.2°C.

Dependability of temperature control apparatus is extremely important from the operational viewpoint. Most growth operations run continuously for a period of months, and very few of the inexpensive instruments that are available commercially are sufficiently reliable for this work. Control elements based on bimetal strips or thermistors are inherently unstable, so that the control point slowly drifts as much as several tenths of a degree per day. Many of the conventional mercury column devices drift slightly or the control point does not come back to precisely the same setting after an appreciable temperature change. Moreover, all types appear to be subject to burnt points or stuck relays if used continuously for long periods.

For growth at constant temperature, the device described in Fig. 5 has proved the most reliable of a long series tried at the Naval Research Laboratory. The mercury relay does not stick, and the mercury-

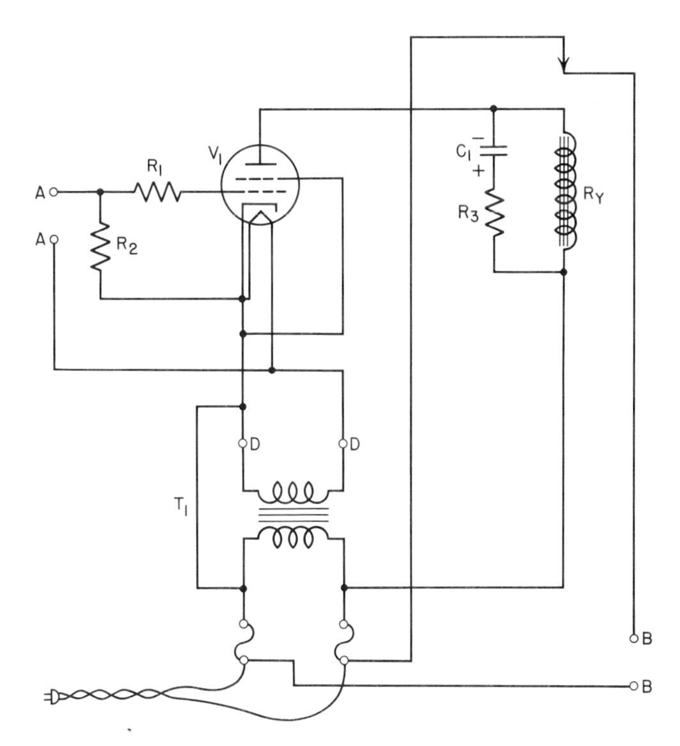

Fig. 5. Schematic diagram of thyratron relay.

R_1, R_2	100,000 ohms, 1 watt
R_3	100 ohms, 10 watts
C_1	8 microfarad, 400 volts
V_1	Tube type 2050 or 2D21
T_1	Filament transformer, secondary: 6.3 volts, 1 ampere
R_y	Relay, mercury, single pole, normally open, Contact rating: 35 amperes at 115 volts ac Coil rating: 0.1 ampere, 115 volts ac (Ebert type EM-1)
A, *A*	Connections to thermoregulator
B, *B*	Connections to heater

If the relay does not open when terminals *A*, *A* are shorted, reverse the leads of the transformer to terminals *D*, *D*.

column control element (Fig. 4) does not drift and rarely changes setting while out of use. The simple thyratron circuit limits the current through the contacts so that there is no appreciable burning over a period of several years. The system is thoroughly reliable so long as the thyratron tube is replaced every four to six months. For crystal growth by temperature lowering, only one type of inexpensive control element approaches the necessary sensitivity, reliability, and stability.* This is a mercury-in-glass differential thermoregulator whose fine adjustment (about 0.3°C per turn) may be altered to be driven by a geared-down clock motor. Variable cooling rates are easily obtained by the use of a proportional timer, which will turn the clock motor on for 5, 10, 15, etc., sec. every minute.

If a program controller is to be used, the choice should be strongly influenced by rugged construction and dependability rather than extreme sensitivity. Considerable care in selecting the temperature controller is a profitable use of time, because the early progress reports from nearly every laboratory involved in crystal growth consist primarily of a list of runs ruined by temperature control failures.

EVAPORATION CONTROL. Control of evaporation is very simple at moderate temperatures. A series of holes in the top of the container is often sufficient; the rate can be adjusted by stoppers in some of the holes. Four or five half-in. holes in the top of a 12 in.-diameter container operated at about 35°C permit a reasonable growth rate for many compounds. In systems that require more precise control, the vapor can be condensed and the rate of evaporation regulated by the amount of condensate withdrawn. A syphon arrangement for this purpose is shown in Fig. 4(2).

At high temperatures an arrangement of this kind is essential, and the problem of unintentional leaks is considerable. The lid should be bolted to the container using well-fitted rubber gaskets. Even a mercury-seal device for the stirrer or crystal support will pass an intolerable amount of vapor if constructed of glass. This can be minimized by making the upper part of the seal out of metal or some other material that is wetted by mercury. Several commercial varieties of stirring glands are available that fit into standard taper joints, which may in turn be cemented into the lid.

* Fisher Scientific Co., No. 15-180-15; Scientific Glass Apparatus Co., T5760; Chicago Apparatus Co., No. 55227.

MINIMIZING CREEP AND SURFACE GROWTH. Spurious crystals forming at the solution surface (or growing up the sides of the container) are a serious problem with some compounds, particularly during growth by evaporation. When these spurious particles attain sufficient size to fall, they frequently lodge on the growing crystal and start growth in the wrong direction. One means of protection is a coverplate as shown in Fig. 4(4). Since most of the agitation is in a horizontal plane, this is only a partial solution. Figure 4(5) and (6) attempt to minimize the trouble at the source. The condensate of a coil or thimble condenser is collected in a flat-bottomed trough about 3 in. wide, mounted to make the overflow even from both the inner and the outer circumference. This distributes water droplets rather evenly to form a diluted surface layer in which crystals do not tend to form. In several instances this has been a highly successful refinement.

A simple expedient that often prevents creeping up the walls of the jar is the use of a tall jar (18 or 24 in. high) filled only about half full of solution. The solvent that condenses on the upper part of the jar keeps the walls washed down.

SUPPORT FOR SEEDS. The objective in mounting seeds is to position them firmly with the least possible contact with any part of the apparatus. For the best distribution, it is usually desirable to mount the crystals at the end of the rods that radiate from a central shaft. The best attachment method depends on the quality and size of the seeds and on the amount of growth expected in various directions. If a crystal grows so as to enclose the supporting rod, the strains induced are very likely to initiate a series of flaws. The same is true to a lesser extent when even one side of a crystal is in contact with the apparatus over an appreciable area. For example, crystals glued to a plate or even lying on the bottom of the container are likely to develop flaws originating from the surface in contact with the apparatus. Changing the material is of little help. Various materials including glass, a series of plastics, several metals, and rubbers of various hardnesses have been tried to find which causes the least difficulty. A highly polished surface where there is no chemical adhesion may reduce but not eliminate this trouble.

Accordingly, the crystal should be mounted, if possible, on a surface to which no new material deposits. For example, no growth takes place on the prism faces of $NH_4H_2PO_4$ as it is normally grown. Thus a support attached to the prism face makes no obstacle to new growth. Various methods of mounting seeds are illustrated in Fig. 6. The support rods are usually lucite or bakelite and are threaded into the stem to permit easy removal of the rod without disturbing the crystal. In strongly acid or oxidizing solutions, glass must be used. At *A*, the roughly ground seed is glued to a flat end of the support rod. Duco cement or lucite dissolved in chloroform has been used. Holes are ground in the seeds *B, C,* and *D* using a No. 60 drill operated at medium high speed. This drilling should

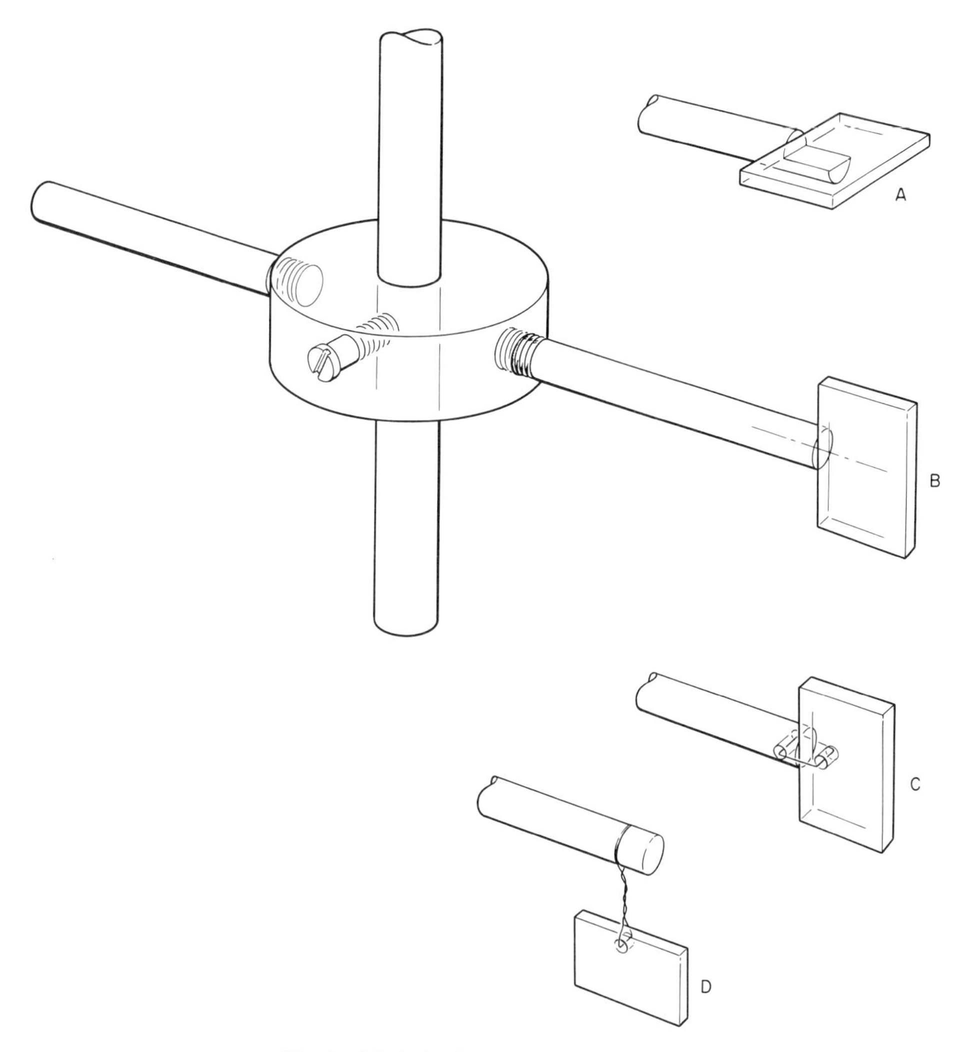

Fig. 6. Methods of mounting seed crystals.

be done under water to prevent cracking. Seed *B* is mounted with a stainless steel or platinum pin glued into the seed and rod. For seed *C*, a gold or platinum wire holds the seed onto the rod.

Frequently, the crystal grows around the support rod. If this must be avoided (as with HIO_3), seed mount *D* can be used. Then only the wire will be enclosed during growth. A variety of crystals has been grown so as to enclose long wires without cracking. Gold or platinum wire is most frequently used, but stainless steel, nickel, or copper may be substituted if the solution is chemically compatible. Single filament saran or nylon has also been used. This relatively loose method of mounting tends to suppress the transfer of vibrations to the seed. Small seeds may also be mounted by pressing them into rubber or plastic tubing on the ends of the support rods (Bartlett, 1961).

AGITATOR. If the seeds are mounted from rods radiating from a central shaft, agitation is most easily furnished by rotating the shaft. In a 12-in. diameter container with a water solution of average viscosity, 20 to 30 rpm is usually satisfactory. In very viscous solutions this may cause excessive turbulence, and should be reduced to 5 to 15 rpm. Occasionally, crystal forms are encountered for which the lee side grows less rapidly than the stoss side. This can usually be minimized by increasing the rate of agitation; otherwise, it is necessary to reverse the direction of rotation periodically. The latter should be done at rather short intervals, 1 to 5 min in each direction, and the reversals should be gradual to avoid causing excessive turbulence.

Bartlett (1961) has introduced a simple centrifugal system for solution agitation. The "spider" or "tree" on which the seeds are mounted is hollow, with an impeller at the top, and its rotation draws solution up from the bottom of the jar and discharges it at the top. In addition to the regular temperature control system, there is a small independent heater at the center of the bottom of the jar. This heater may be used to control the rate at which a supply of solute dissolves, and thus is made available for crystal growth, or it may alternatively be used to dissolve spontaneous seeds. Bartlett also suggests the use of auxiliary paddles for use when the initial seeds are small.

If the seeds are mounted on stationary racks, the best agitator is one with radial arms such as shown for the seed mounts, with paddles in place of the crystals. It should operate at about the speeds previously described so as to provide uniform movement of the solution with a minimum tip turbulence.

Other schemes of providing agitation, such as rocking or rolling of the whole container, have not proved as efficient.

Operating Techniques

PREPARING THE SOLUTION. Making up the solution is the most time-consuming step. There appear to be no short cuts for obtaining a solution precisely equilibrated at a desired temperature, but it may be helpful to mention some common pitfalls. A precisely saturated solution can never be made simply by combining the necessary amount of water and salts as determined by solubility curves, first, because an astonishingly large amount of published solubility data is not accurate, and second, because evaporation during heating to complete dissolution introduces gross errors. Vigorous heating beyond the time when dissolution appears complete is recommended. It provides insurance against spurious nuclei that dissolve increasingly slowly as they become smaller. The amount of evaporation during a given heating period is difficult to predict, but the resulting change in saturation temperature may be very large. For example, an open 12-liter container of highly soluble salt solution heated one hour at 80°C can change saturation temperature as much as 30°C. Accordingly, it is necessary to make up the solution with excess water and evaporate to the correct volume. Since the necessary data are seldom available, and

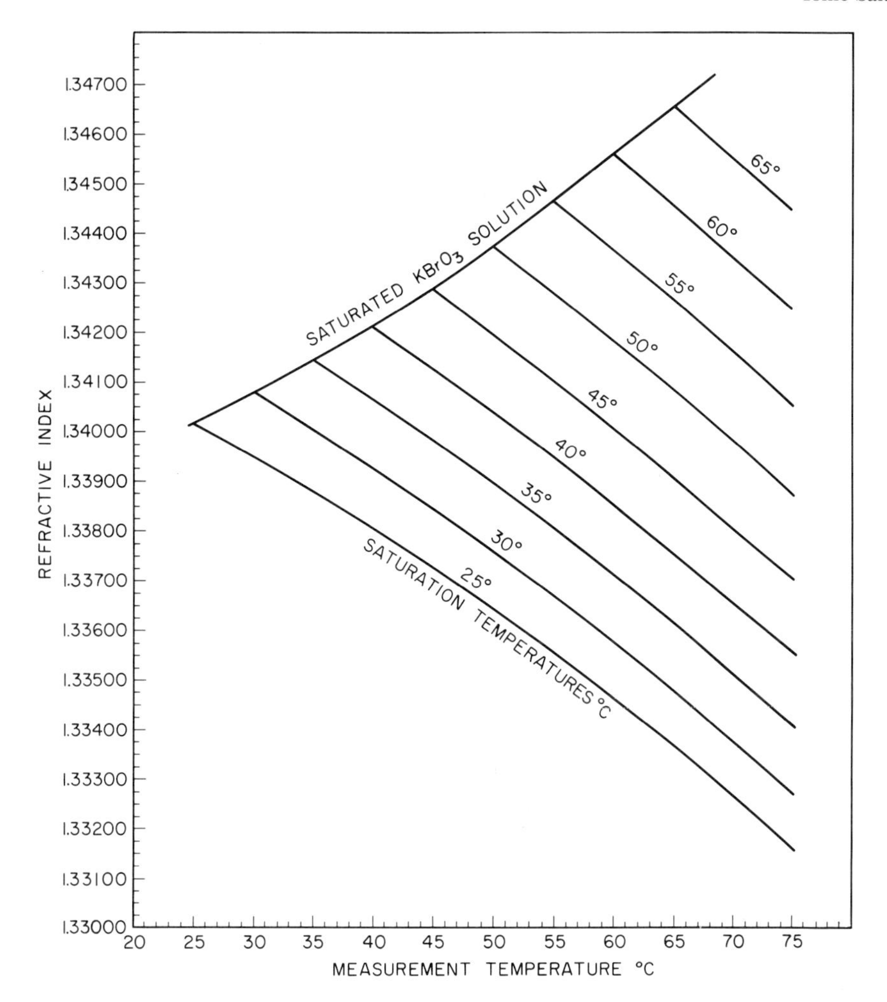

Fig. 7. Variation of refractive index with temperature and concentration of potassium bromate solutions. (After Parkerson, 1951.)

since volume measurement in large containers is not very precise, it is necessary finally to measure the saturation temperature before the crystal seeds are introduced. This can be done by measuring the specific gravity or better the refractive index; but again sufficiently reliable data are rarely available. If a number of runs is to be made with the same material, it is worthwhile to obtain such data in tabular form (Fig. 7). A method for refractive index measurement of solutions at elevated temperatures is described by Parkerson (1951).

When previous data are not available, it is necessary to introduce a test crystal. If the solution is not agitated, it is often possible to observe Schlieren streams, rising, if the solution is supersaturated (since the solution is being depleted of solute as the crystal grows), or falling, if the solution is undersaturated (since the test crystal is dissolving, locally raising the density of the solution). For the Schlieren streams to be easily observable, the solution must be fairly far from saturation. If the solution is close to saturation, the extremely sensitive method of Kovalevskii (1957) may be used. Unfortunately, this method requires a good optical system, a darkened room, etc. Thus it is better suited to industrial or pilot plant operations than to laboratory use.

A simple method is periodic observation

of a particular feature of the test seed, such as a bored hole or a small notch. Undersaturation is fairly easily detected in the course of a day, or less, by the rounding of corners or the enlargement and change of shape of the bored hole. Slight supersaturation will often produce partial filling of the hole overnight. This is much easier to observe than the onset of growth on a perfect seed. When the test seed remains essentially unchanged for 18 hr or so, it is safe to introduce the seeds for the growth run.

SEED PREPARATION. An essential factor in processing seed material is the use of proper tools. For sawing water soluble crystals, circular saws with diamond or carborundum blades are the most efficient, with a few exceptions. For large crystals, the type of saw commonly used for quartz is satisfactory. Blades with 100 grit diamond in brass rims, or 120 grit carborundum bonded in rubber are best for most crystals. A 50:50 mixture of propylene (or ethylene) glycol and water is a suitable coolant. This mixture is good for most inorganic compounds, but the fraction of water must be reduced for very soluble crystals. Saturated, or nearly saturated, water solutions of the crystals being cut have been tried, but are rarely used because of the difficulty of keeping the equipment clean. Both the crystal and the coolant must be at the same temperature as the room to avoid dissolution and thermal shock. The coolant should be evenly distributed on both sides of the front radius of the wheel. This minimizes the chief source of difficulty, crystal particles adhering to the wheel and masking the cutting grit. Very soft materials such as Rochelle salt and many organic crystals tend to load the wheel. They may be cut more efficiently by a rougher blade or by a specially sharpened band saw. Small crystals are more conveniently cut with a small, very high speed (18 to 35,000 rpm) radial saw. The blades are rubber-bonded carborundum grit, 4 in. in diameter and 0.010 in. thick.

Many varieties can be cut most easily on a string saw similar to that described by Maddin and Asher (1950). This saw uses a reciprocating thread wet with water or acid. Modifications permitting horizontal and vertical displacements of the specimen relative to the string have proved very convenient.

Drilling of holes for pegs or wires is best done with a small, high speed bench drill. The bit should be sharpened with an edge angle of 30° (rather than the conventional 59°) to facilitate rapid removal of the drilling chips. Drilling should be done in a water-glycol mixture to reduce cracking. Fairly neat holes may be formed by dissolution using a miniature version of the apparatus of R. Vaska (1959).

Grinding can be done on inexpensive belt-sanders or vertical plate grinders. The carborundum wheel recommended for a saw blade makes an excellent grinding surface. It can be cemented to the plate of a disk sander with Vulcalock, and dressed periodically with a diamond-tipped cutter.

After machining operations, seeds should be thoroughly scrubbed in alcohol to remove loose particles that might serve as spurious nuclei. Seeds should then be inspected in polarized light to locate growth flaws that might propagate in subsequent growth.

STARTING A RUN. The most critical stage in the growth process is the deposition of the first few layers onto a seed. Seeds and the supporting mechanism should be heated to the same temperature as the solution, or slightly higher, before they are introduced. It is also preferable that the solution be very slightly undersaturated at the time of planting, so the seeds dissolve slightly at first. This is most easily done by raising the temperature 0.1 or 0.2°C just before planting. The exact amount depends upon the slope of the solubility-temperature curve, the size of the seeds, and the amount of solvent that is lost while the jar is open. The temperature should be returned to the original control point slowly, over a period of one or two days, so that growth is started very gently. After a good layer is visible, the growth rate can be safely increased.

Some crystal growers, instead of raising

the temperature just before planting, prefer to add a few milliliters of solvent. This is very easy to overdo, however, and it is very difficult to correct.

REMOVING FINISHED CRYSTALS. If the system is above room temperature at the end of a run, the crystals are very likely to crack if they are removed into cool air. Even rapid transfer into an oven usually produces a few cracks. The safest procedure is to slowly displace the solution with an immiscible liquid at the same temperature. Once the crystals are enclosed in the inert liquid, the temperature of the system can be gradually lowered.

Eliminating Flaws

By way of summary, is may be desirable to list types of flaws frequently encountered and probable sources of difficulty.

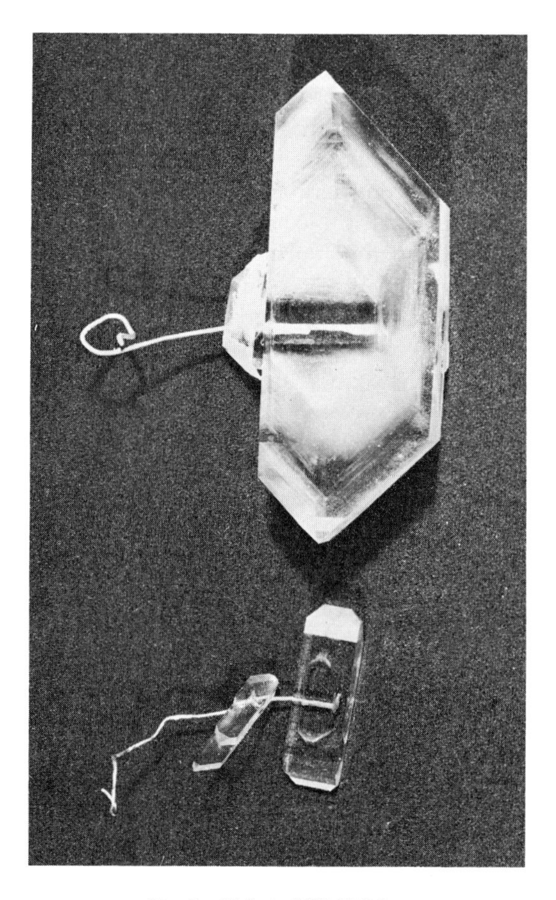

Fig. 8. Veils in $NH_4H_2PO_4$.

Fig. 9. $NH_4H_2PO_4$ crystal with curved faces.

VEILS. Veils (Fig. 8) are the most common type of flaw in solution-grown crystals. They form when the growth rate is temporarily greater than the crystal can tolerate. Veils that occur periodically have often been traced to daily or weekly changes in temperature, line voltage, or humidity. (In many laboratories there is a pronounced rise in line voltage as the end of the working day approaches and the demand for power becomes smaller. Thus the heater power may rise, temporarily halting growth, which resumes with a rush when the voltage drops to normal.) Improved agitation may increase the permissible growth rate. Sometimes the seeds may be cut with a different orientation, one that minimizes veil formation. Otherwise, it is necessary to grow more slowly or to obtain better temperature control.

CURVED SURFACES. Curved surfaces (Fig. 9) may occur for two reasons. First, trace impurities may hinder growth in a particular direction, causing a type of habit modification. The responsible impurity may be undetectable in the finished crystal or even in the solution. A second source of curved surfaces is nonuniform agitation. The flow of solution may bring more material to deposit at one end of a crystal. This can usually be cured by increasing the over-all rate of agitation.

TWINS. True twinning (Fig. 10) as distinguished from spurious growth in random directions is not often encountered in well-controlled growth. Depending on its symmetry, a crystal may twin on a plane, on an axis, or in an interpenetrating fashion. These may be simple as shown by $NH_4H_2PO_4$, or multiple (polysynthetic) as shown by $BaCl_2 \cdot 2\ H_2O$.

Most often a twin can be traced to a twinned seed, but spontaneous twinning is observed and is sometimes difficult to eliminate. In one instance, twinning was eliminated by addition of a "thickening agent" to the system, but in other cases the problem is unsolved.

Spurious growth, and sometimes twins, can be caused by stray nuclei that lodge on a growing seed, furnishing loci for rapid growth in new directions. This is, of course, best eliminated by avoiding stray nuclei formation, especially during planting. The condenser system for evaporation control, is usually effective in preventing undesired nucleation. Plates covering the seeds are an alternative preventative.

Fig. 10. Twins of $NH_4H_2PO_4$.

HOLES. Holes may be produced by inclusion of macroscopic bubbles. Sometimes, a bubble remains on the surface as growth proceeds, leaving behind a cylindrical hole. Bubbles that adhere too tightly to be scrubbed off by solution agitation must be scraped off manually.

Care in transferring the solution, and in introducing seed crystals, is usually sufficient to avoid bubbles. Recurrent trouble with bubbles may require the addition of a wetting agent to the solution. Bubbles on lithium sulfate hydrate were eliminated by adding aerosol.

SURFACE GROWTH FLAWS. Small bumps and fairly symmetrical steps of vicinal faces frequently occur on slowly growing surfaces. The pattern may remain fairly constant or may shift considerably as growth proceeds. They result from exaggerated growth arising from small surface flaws that grow rapidly at their edges. They can usually be minimized by increased agitation and by adjustment of the composition of the solution so that flaws have less advantage in the growth process.

Another type of surface flaw is a needle-like splinter growth on a prism face parallel to the long axis of the crystal. This arises from flaws on the prism face of the seed or from spontaneous nuclei that lodge on the surface.

With the alkali halides, the surface pattern may consist of 0.1 to 5.0 mm blocks in almost parallel array, but slightly tilted with respect to the original seed axis. This kind of growth usually results from (1) poor seeds or (2) improper agitation (insufficient diffusion). For the alkali halides it can be minimized by adding an impurity (for example, Pb^{++} as previously described).

References

Bartlett, B. M. (1961), *J. Sci. Instr.* **38,** 54.

Buckley, H. E. (1951), *Crystal Growth,* John Wiley and Sons, New York, pp. 386–387.

Butchart, A., and J. Whetstone (1949), *Discussions Faraday Soc.* **5,** 254.

Carlson, Alan, Private Communication.

Crystal Growth (1949), *Discussions Faraday Soc.* **5,** 261.

Egli, P., L. Johnson, and W. Zimmerman (1959), "Purification," p. 29, *Experimental Physics,* Vol. 6A, edited by K. Lark-Horowitz and Vivian A. Johnson, Academic Press, New York.

Eitel, W. (1926), "Krystallzüchtung," *Handbuch der Arbeitsmethoden in der anorganischen Chemie,* Bd. IV, Gruyter and Co., Berlin and Leipzig, pp. 448–463.

Imber, O., U. S. Patent 2,647,043.

Kovalevskii, A. N. (1957) (original Russian text: *Rost Kritallov* Vol. I, Acad. Sciences USSR Press, Moscow), "Growth of Crystals," *Report at the First Conference on Crystal Growth,* March 5–10, 1956, Moscow. (English Translation, 1958, New York, p. 264.)

Maddin, R., and W. R. Asher (1950), *Rev. Sci. Instr.* **21,** 881.

Parkerson, C. R. (1951), *Analytical Chemistry* **23,** 610.

Tuttle, O. F., and W. S. Twenhofel (1946), *Am. Mineral.* **31,** 569–573.

Vaska, R. (1959), *Kristallografiya* **4,** No. 2, 260 (March–April, 1959); *Soviet Physics—Crystallography* **4,** No. 2, 239 (February 1960).

Whetstone, J. (1949), *Discussions Faraday Soc.* **5,** 261.

Wolfe, C. W. (1946), *Am. Mineral.* **45,** 1211.

Yamamoto, T. (1939), *Sci. Pap. Inst. Phys. Chem. Res.* **35,** 228.

12

Silver Halides

C. Berry, W. West, and F. Moser

Three classes of processes are discussed: (1), growth from aqueous solution, (2), growth of thin sheets from the melt, and (3), growth of large bulk crystals by slow solidification.*

Crystals as large as a few millimeters or smaller than a micron may be produced from aqueous solution. The smallest crystals are frequently precipitated into a gelatin solution to prevent coagulation of the solid particles. Small crystals grown from solution have as special features: well-determined crystallographic surfaces of low Miller indices and larger surface areas relative to their volumes.

Thin crystals, varying in thickness from a few microns to 100 to 200 μ and of large areas, are made from the melt. Crystalline sheets of AgCl and AgBr, made by compressing a globule of the melt between glass or quartz plates, have been used to study the photoconductivity and the absorption spectra of these substances. Boissonas (1949, 1950, 1951) showed that photographic images composed of reduced silver could be formed in such crystals by treatment of the exposed sheets with normal photographic developers. At about this time, Mitchell proposed that such crystals could serve as model systems in the study of fundamental problems in photographic sensitivity, provided the crystals were free of impurities which produce spontaneous reduction to metallic silver (fog) on immersion in a photographic developer. In the following years, he and co-workers devised methods for the purification of AgCl and AgBr and for making very pure, thin crystals. These methods have been widely utilized. A detailed account of the purification methods is given by Clark and Mitchell (1956).

Large crystals of AgCl or AgBr with dimensions of several centimeters are quite readily grown from the melt. Both the pulling method (Czochralski, 1918; Kyropoulos, 1926) and the method of moving a confined melt through an appropriate temperature gradient (Bridgman, 1925; Stockbarger, 1936) have been used. Most workers have preferred the latter method because it makes control of atmosphere, crystal shape, and growth rate easier. Such crystals have been used in a variety of studies including optical absorption, light-induced coloration, electron and hole mobility and lifetime, and mechanical properties. Recent review articles (Brown and Seitz, 1958; Mitchell, 1959; Meiklyar, 1957) provide complete bibliographies of these studies.

Most experiments with silver halides have been done using AgCl and AgBr. Silver fluoride and AgI have not received as much attention. A complication with AgI is that it exists in more than one crystal structure. At room temperature, it consists of both hexagonal and cubic crystals and at 146°C there is a transition to a high-temperature structure (Strock, 1934). Silver fluoride differs from the other silver halides by having high solubility in water.

PRECIPITATION AND GROWTH FROM AQUEOUS SOLUTION

At room temperature, the solubility in water of AgCl is 1.9×10^{-3} gram/liter and

* Because of photodecomposition in the silver halides, all the methods of crystal growth described here need to be carried out in red light.

that of AgBr is 1.5×10^{-4} gram/liter. Therefore they precipitate rapidly when small concentrations of silver and halide ions are brought together. The silver ions are usually provided as an aqueous solution of $AgNO_3$. The halide ions are in aqueous solutions of salts such as KCl, or of acids such as HCl. The two kinds of ions may be brought together simply by running the solutionof $AgNO_3$ into the solution of halide ions.

A detailed description has been given of the precipitation of pure AgCl for use as the starting material in growing large crystals from the melt (Nail, et al., 1957). The $AgNO_3$, selected on the basis of spectrographic analyses, is available as Special Product X-491 from Eastman Organic Chemical Sales Division. Reagent-grade HCl was fractionally distilled from a constant-boiling mixture using a quartz condenser. To produce about 1 kg of precipitate, 8 moles of $AgNO_3$ was dissolved in 5 liters of redistilled water and 8.25 moles of distilled acid was diluted to 5 liters. The solutions, heated to 70°C, were run simultaneously through glass nozzles into 5 liters of heated redistilled water. The flow of acid was started a few seconds before the $AgNO_3$ solution. The precipitation was completed in 15 min with constant stirring by a Pyrex glass paddle. The supernatant liquid was decanted from the precipitate which was washed many times with redistilled water until the *p*Ag of the wash water reached a constant value. Complete drying of the precipitate was accomplished by heating it to not more than 40°C in vacuum.

Many impurities are rejected strongly by precipitating silver halides. In particular, there is no evidence of the presence in the silver halide of nitrate or hydrogen ions. However, small quantities of these impurities are more difficult to detect than many metallic impurities. On the other hand, some impurities do appear to be incorporated in the crystals. For example, Pb^{++} is incorporated in AgBr during precipitation (Wakabayashi and Kobayashi, 1957). This was proved by noting changes in the conductivity of thin sheets prepared from the melt when various precipitates were used as the starting materials. The possibility that the Pb^{++} was merely adsorbed onto the silver halide was disproved by showing that a precipitate behaved like pure AgBr when made in the absence of Pb^{++}, but dispersed in a $PbBr_2$ solution before washing and fabricating.

Another example of the incorporation of an impurity during the growth of AgCl was shown in crystals grown from a solution containing some dyes (Reinders, 1911). These crystals were tinted throughout their volume, as compared with pure AgCl which is colorless.

There appears to be a slight change in the stoichiometry of the silver halide crystals, depending on whether silver or halide ions are present in excess during precipitation. Most of the excess of ions is adsorbed on the surfaces of the crystals. Evidence for the difference in composition is found in photographic emulsions, where it is observed that emulsions precipitated in an excess of silver ion have a high level of fog on being developed. Additional evidence for the adsorption of ions is given by potentiometric measurements (Boyer et al., 1959); Herz and Helling, 1961). It was found that, in the presence of 0.001 *M* or more of bromide ions, microcrystals of AgBr have an excess of bromide ions corresponding to about 10% of the surface sites. A slight excess of halide ions is customarily used to suppress the fog.

To put large amounts of silver halide into solution in water, a variety of silver-complexing agents may be used. The two most commonly used agents are ammonia and halide ions, which form complexes such as $Ag^+(NH_3)_2$, $AgBr_2^-$, and $AgBr_3^=$. An advantage of dissolving AgBr in ammonia water or in HBr is that these solvents have considerable vapor pressure which causes them to evaporate from solution without leaving an undesirable residue. Some other complexing agents for silver ion such as thiocyanate ion are undesirable because the silver thiocyanate precipitate is approximately as insoluble in water as AgBr.

Microcrystals of silver halides dispersed in gelatin and coated on film base form a system which is capable of demonstrating a great variety of photochemical reactions.

An advantage of such a system is that remarkably small changes caused by light absorption may be observed. For example, a 1 μ cube of AgBr, which contains about 10^{10} silver ions, may be converted completely to silver in a suitable photographic developer when as few as about 10 atoms of silver have been formed by light absorption. Such large amplification factors as this value of 10^9 are responsible for the commercial importance of silver halides in photography.

Some details of precipitation of AgBr microcrystals in gelatin have been given (Berry et al., 1961). Crystals having a variety of shapes and of perfection can be produced. The basic precipitation scheme consisted in adding simultaneously 10 grams of $AgNO_3$ in 30 ml of water solution, and 7.00 grams of KBr in 30 ml of water solution, to a gelatin solution consisting of 3.0 grams of dry gelatin in 100 ml of water kept at 50°C. The two solutions were added dropwise from fine, Pyrex-glass nozzles for 15 min, with vigorous stirring of the mixture. This produces cubes of about 0.1 μ edge length. Larger cubes having an edge length of 0.5 μ are produced when a small amount (0.3 ml) of the solvent ethylenediamine is added to the solution in the precipitation vessel prior to the addition of the $AgNO_3$ and KBr.

Regular octahedra may be produced by precipitation into a solution of gelatin having a concentration of 0.1 N NH_3 and 0.1 N KBr. Generally, the presence of excess bromide ions during precipitation encourages the growth of {111} faces, whereas the {100} faces are formed in the absence of bromide ions. In the presence of excess bromide ions, but in the absence of NH_3, tabular growth occurs. X-ray diffraction measurements have shown that the tabular crystals contain twin planes which act as the growth accelerator (Berriman and Herz, 1957). The twinning has been attributed to the occasional deposition of the complex $AgBr_3^{=}$ in the twinned configuration (Berry et al., 1961).

Recent studies have been reported (Berry and Skillman, 1962)* in which the relative rates of growth were measured from electron micrographs for a variety of AgBr crystals of different shapes and structures. Most of the crystals were perfect cubes, but crystals were also identified which were singly twinned on a (111) plane, doubly twinned on parallel (111) planes, and twinned on a (311) plane. It was found that perfect cubes of different size and the crystals singly twinned on the (111) plane grew at rates such that the volume increase was proportional to the surface area. This showed that a single (111) twin plane is not a growth accelerator and does not have dislocations in it. On the other hand, the doubly twinned crystals and those crystals twinned on a (311) plane grew at accelerated rates. The accelerated growth was attributed to persistent reentrant angles in the case of the double twin and to dislocations resulting from the strain which accompanies twinning on a plane of so high an index as the (311).

Precipitation of AgCl microcrystals in gelatin may be accomplished by the same procedure as for AgBr. Microcrystals of AgCl are ordinarily cubes even in the presence of excess chloride ions. In the absence of added solvent, the AgCl crystals have an edge length of about 0.15 μ.

Crystals of AgBr ranging from 5 to 20 μ have been produced by steam dilution of an aqueous solution saturated at room temperature with KBr and AgBr (Sutherns, 1956). Crystals of this size may be produced in 30 to 40 min and are convenient for optical microscopic observations of photochemical reactions.

Growth of tabular crystals of AgBr, a few millimeters in size, may be accomplished in the following way: dissolve 1.1 gram of AgBr in 25 ml of 48% HBr. At room temperature this gives a nearly saturated solution. If this solution is allowed to stand in a 400-ml beaker that is covered with a cover glass, there will be many triangular crystals on the bottom of the beaker after about two weeks. If a larger volume of solution is used,

* C. R. Berry, and D. C. Skillman (1962), *J. Appl. Phys.* **33**, 1900; *Phot. Sci. and Eng.* (1962), **6**, 159.

larger crystals are formed. With a volume of about 75 ml of solution, long branching dendrites are obtained.

Growth of AgCl crystals from solution in concentrated HCl is not as satisfactory as the growth of AgBr from HBr. This is because AgCl is about twenty times less soluble in concentrated HCl than is AgBr in HBr. On the other hand, ammonia water is a good solvent for AgCl. Crystals of AgCl may be obtained from an NH_3 solution by dissolving 2.5 grams of AgCl in 25 ml of concentrated NH_3 water (15.1 moles/liter) contained in a 400-ml beaker. The rate of vaporization of the ammonia may be reduced to a value that results in crystals 1 to 3 mm in size in two or three days by inverting a 1-liter beaker over the 400-ml beaker.

2. THIN MACROSCOPIC CRYSTALS FROM THE MELT

Utility

Sheet crystals of silver halides, from 20 to 100 μ thick, are particularly suitable in studies of photographic sensitivity since after appropriate sensitization and exposure they can be developed by normal photographic procedures. Furthermore, combining suitable etching techniques with development, the behavior of the latent image can be studied throughout the thickness of the crystal as well as at the surface, adding, in effect, a new dimension to photography. They lend themselves to microscopic observation through the whole thickness of the crystal, so that, for example, the small particles of "print-out silver," formed without development by direct exposure of the crystal to light, can readily be studied throughout the volume of the crystal. One of the most important characteristics of thin crystals of AgCl and AgBr, discovered by Hedges and Mitchell (1953), is that particles of print-out silver separate preferentially at dislocations and dislocation complexes, an observation which initiated the investigation of dislocations by "decoration," with important consequences both with respect to the photochemistry of solids and to the theory of dislocations and plastic deformation.

General Principles of Preparation

The main general principles to be followed in preparing sheet crystals of silver halides suitable for studying photochemical, photoelectric, optical or mechanical properties is to obtain material of the highest possible purity and to avoid subsequent contamination. Almost no conceivable precaution for attaining these ends will prove superfluous. Care should be taken to avoid contamination with dust at all stages of preparation, and to avoid exposure of the final sheets to traces of hydrogen sulfide or other substances likely to form adsorbed layers on the surface. The effect of foreign addenda can be studied by controlling the addition of the impurity, usually at the time of making the sheet.

Other general precautions are (1) during preparation of the material, exposure of the hot silver halide to oxygen is to be avoided, except in specific studies of the effect of this gas. Otherwise, sensitizing and fogging material, probably silver oxide or reduced silver, is introduced. As far as possible, the hot halide is kept in contact with an atmosphere containing the corresponding halogen. When halogen is inadmissible (for example, because of its oxidizing effect on an addendum which must be retained in the reduced form) pure nitrogen or helium may be substituted. Operations such as annealing at high temperature, if carried out in a vacuum, may cause thermal etching of the surface. Hence, they are best carried out in an atmosphere containing, when possible, the halogen, or in an atmosphere of nitrogen or helium.

(2) Liquid silver halides should not be kept in contact with glass at temperatures much above 500°C. Discoloration of the glass observed on cooling shows that reactions between the silver halide and the glass may occur.

Starting Material

The stock material for preparing thin crystals has the form of small pellets made from the purified precipitates that were described in an earlier section. During the preparation of the pellets, further purification is effected by filtration of the molten halide through capillaries of Pyrex glass and by melting for several hours in a vacuum.

The procedure described in the following sections (Clark and Mitchell, 1956) has been used to prepare sheet crystals of AgBr and AgCl by the authors (West and Saunders, 1958, 1959).

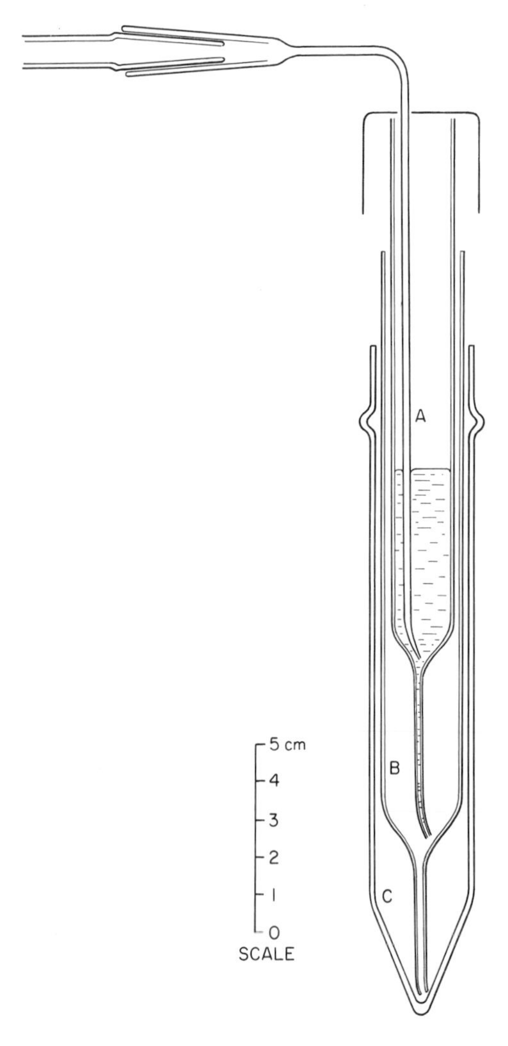

Fig. 1. Apparatus for purification of silver halides by filtration.

Capillary Filtration

Silver halide precipitate prepared from aqueous solutions in glass vessels contains small amounts of hydrated silica. This forms a scum on the surface of the molten halide and must be removed to secure acceptable sheet crystals. The scum can be removed mechanically by running the molten halide through a succession of fine glass capillaries until the last capillary remains clean.

The apparatus, made of Pyrex glass, is shown in Fig. 1. All the glassware is first cleaned in hot concentrated nitric acid, rinsed in distilled water, and dried. The nest of tubes in Fig. 1 is supported at the projection of the outer tube.

The following procedure describes making sheet crystals of AgBr; modifications for AgCl are readily made. A charge of about 50 grams of dry AgBr precipitate is placed in the first Pyrex vessel *A* and a stream of nitrogen (Linde Nitrogen-HP dry) is bubbled through distilled reagent-grade bromine, passing through a tube that reaches to the bottom of the precipitate. Clark and Mitchell (1956) recommend passing hydrogen bromide containing bromine.*

The first object in this part of the preparation is to melt the charge in vessel *A*, and to allow the gas to bubble through the liquid bromide. Heating is conveniently accomplished by means of an oxygen-gas

* It has been found that it is best to generate the gas in an all-glass apparatus, since metallic contamination tends to be greater if the gas is obtained directly from commercial cylinders. Possibly gas from cylinders could be used if it were first liquefied at low temperature and then distilled before passing it into the vessel containing the silver halide. Hydrogen bromide can be conveniently generated from a flask containing 500 grams recrystallized KBr, and 280 cc of 85% orthophosphoric acid, warmed in an electrically heated mantle. The rate of evolution of the gas is controlled by the temperature of the mixture. The gas is passed through a 30-in. column of fresh Drierite and through lengths of glass wool before being bubbled through the bromine.

Hydrogen chloride can be similarly generated by dropping concentrated H_2SO_4 into concentrated reagent-grade HCl and recrystallized NaCl. Chlorine is generated by dropping concentrated HCl on MnO_2. Movable connections in these gas trains are made by ungreased conical joints.

hand torch, which permits freedom in applying the heat that is difficult to obtain otherwise. The temperature is kept below the value at which the flame becomes tinged with the sodium yellow color from the hot glass. Electrical heating has also been employed (see Fig. 6).

Heat is first applied near the bottom of tube *A*, where it narrows to the capillary, so that a little of the solid bromide melts, runs down to the upper part of the relatively cool capillary and freezes, to form a plug that is maintained while the rest of the charge is being melted. The flame is then directed so as to melt the charge, and bromine-saturated nitrogen, or bromine-hydrogen bromide, is bubbled for 30 min through the melt. The plug is then melted, and the liquid bromide allowed to pass through the capillary into the second vessel *B*, leaving minute particles of solid impurity on the lower part of tube *A* and in the capillary. A solid plug forms in the capillary of *B*, but the temperature of the rest of the bromide is maintained above the melting point. The gas inlet tube is withdrawn by disconnecting its conical joint in the generating train, tube *A* is removed, and the inlet tube is replaced so that the gas bubbles through the liquid in *B*. After a few minutes of bubbling, the plug is melted and the liquid passes through the second capillary into the outer tube *C* where it is kept molten. Normally, tube *B* and its capillary are free from scum after this operation, but if they are not, the filtration is repeated after pellets have been made from the molten material in the outer tube.

The pellets are made by pipetting small quantities from the melt onto a cold, clean, freshly flamed Vycor plate. Tube *B* is removed and the gas inlet tube is adjusted so that bromine-saturated gas is passed over the surface of the melt. A clean, Pyrex-glass tube, 3 mm in internal diameter and about 25 cm long, the lower end of which has been heated in the flame, is introduced into the melt. The liquid occupies the lower few millimeters, the top is closed by the finger, and the bromide in the tube is transferred to the plate, on which it solidifies to form a pellet suitable for subsequent handling. The contents of tube *C* are removed in this way.

If the second tube and capillary are free from scum, the pellets are now melted in a vacuum.

Vacuum Melting

The apparatus of Pyrex glass is shown in Fig. 2. Distilled reagent-grade bromine is introduced through the vertical part of the side tube into vessel *A*, which is isolated from the melting tube by a breakable seal. Vessel *A* is connected through traps to a vacuum pump. After cooling *A* in liquid nitrogen, the vessel is evacuated and air is removed from the bromine by carefully raising the temperature until the bromide is partly liquefied, distilling part of the halogen into the traps, freezing again, and repeating several times; *A* is then sealed off from the pump.

The pellets of filtered AgBr are introduced into the melting tube *B*, which is connected through two traps to a mercury diffusion pump. The apparatus is evacuated initially with the traps uncooled and is baked out by the flame; the traps are now cooled in liquid nitrogen, baking is continued for some time, and the temperature of the furnace is brought to about 500°C. Evacuation is continued for 8 hr. A deposit collects in the upper part of *B*, which spectrographic analysis shows to be enriched with respect to the residue in Fe, Cu, Mn, Ni, Pb. (With the exception of Fe, which may be present in the sublimate to the extent of 2 ppm, the other elements mentioned occur in the sublimate to the extent of less than 1 ppm.)

The melting tube is now sealed off from the pumping system and the isolation seal between *A* and *B* is broken by manipulating the glass-enclosed iron bar by a magnet. Bromine vapor at its room-temperature vapor pressure then comes in contact with the melt. After 1 hr, the bromine reservior is sealed off; vessel *B* is opened by breaking off the neck; bromine-saturated nitrogen is

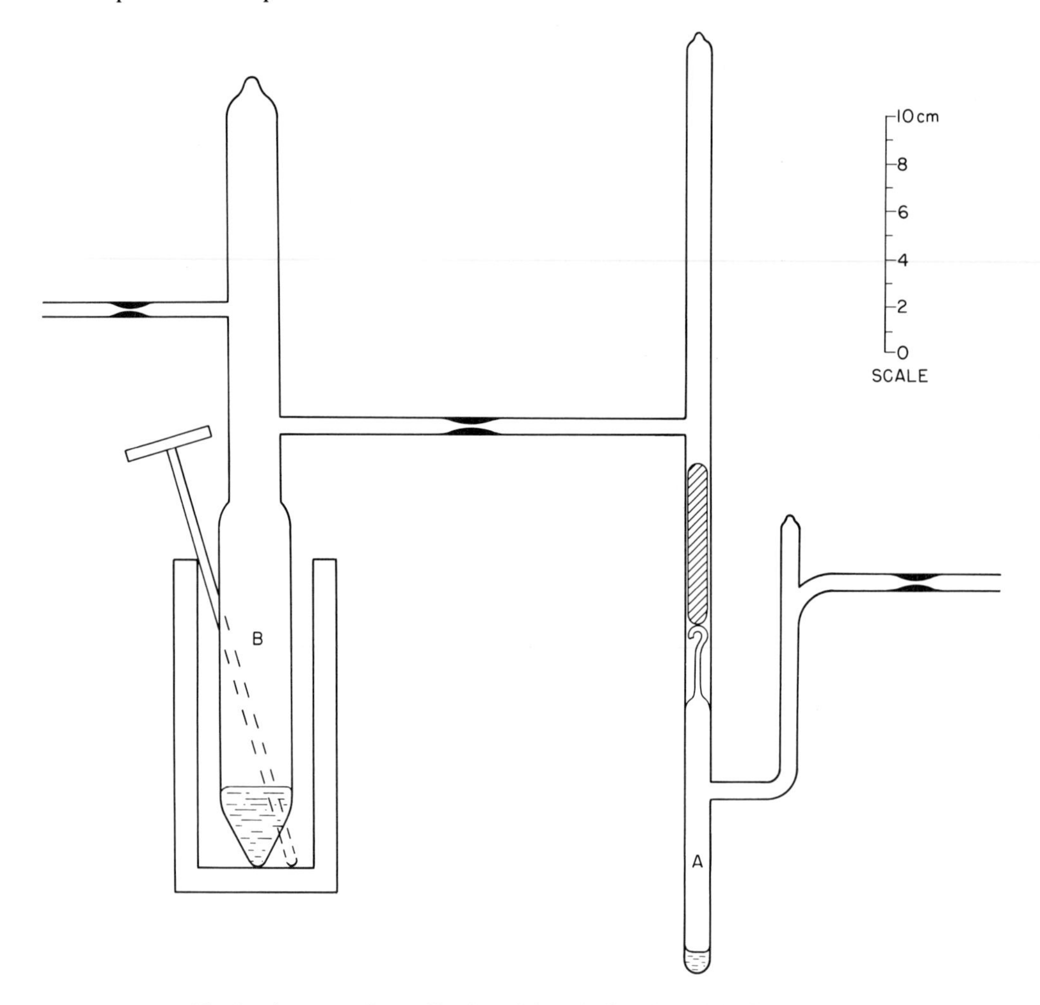

Fig. 2. Apparatus for purification of silver halides by melting in vacuum.

passed over the surface, and the melt is pipetted out to form pellets as before.

Making of Sheet Crystals

Sheets are made by the apparatus shown in Figs. 3 and 4, where *A* and *B* are electrically heated hot plates with a stainless-steel surface, 6 in. square, $\frac{3}{8}$ in. thick, heated by three Chromalox strip heaters (Edwin L. Wiegand Co., Pittsburgh, Pennsylvania) of 400 watts and 120 volts, connected in parallel and clamped to the lower side of the steel plates.

Adjacent to the hot plates and on the same level is a plate of stainless steel *C*, 26 in. long, 6 in. wide, and $\frac{3}{8}$ in. thick, provided at one end with three strip heaters similar to those on *A* and *B*. The three steel plates are mounted on a heat-insulating platform *D*.

The two hot plates are covered by molded Vycor plates, 6 in. square. The crystals are grown between polished Vycor plates of optical quality (Corning 7910, 96% silica ultraviolet-transmitting glass). Plates 8 cm square are convenient for sheets 4 to 5 cm in diameter; with plates 15 cm square, sheets 8 to 10 cm in diameter can be made. The plates are cleaned in 28% NH_3 at room temperature. After rinsing in hot concentrated nitric acid, they are rinsed and dried in dust-free air. The hot plate *A* is carefully leveled, and shims are placed under the feet. The plates to be used in the sandwich are placed on *A*

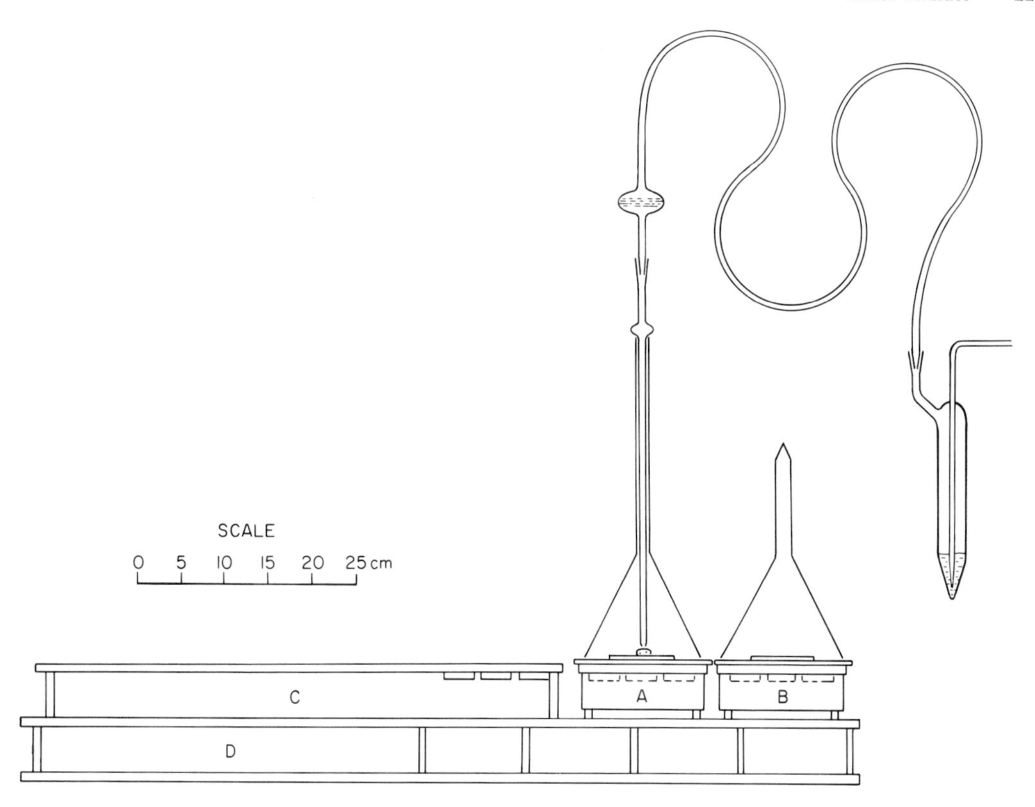

Fig. 3. Apparatus for making sheet crystals of silver halides.

and *B,* cleaned by flaming and protected from dust by inverted glass funnels. The temperature of the hot plates is brought above the melting point of AgBr, the current being controlled by Variacs capable of carrying 10 amperes at 110 volts. Spacers are placed on the Vycor plate on *A* to control the thickness of the sheet. For sheet thicknesses of 50 μ and above, L-shaped spacers made from drawn-down Pyrex-glass rod are used; for spacers 5 to 30 μ thick, platinum foil is used. The spacers having been placed on the plate, the long-stem funnel is brought into the position shown in the diagram, and a vigorous stream of bromine-saturated nitrogen is passed into the funnel through a porous disk in a flexible glass connection. After displace-

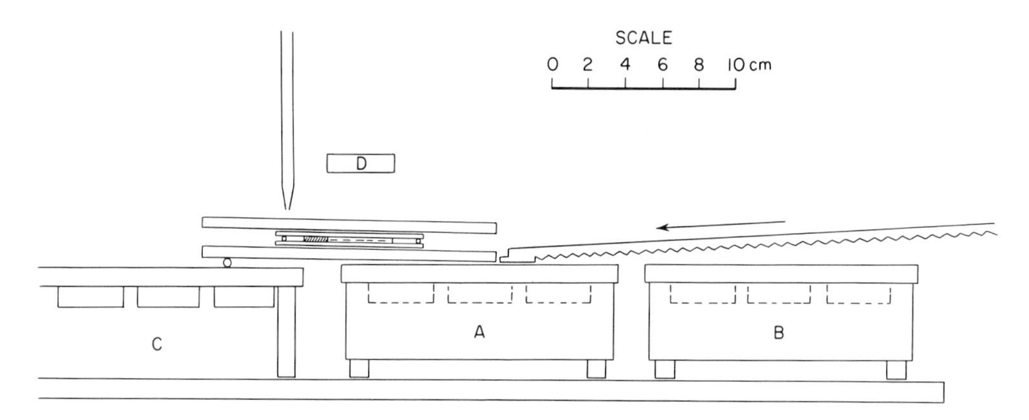

Fig. 4. Formation of sheet crystals of silver halides: detail.

ment of the air from the funnel, the latter, with the glass inlet tube, is raised about 2 in. and the necessary number of pellets is introduced on the plate by ivory-tipped forceps. The funnel is lowered and as vigorous a stream of bromine in nitrogen is introduced as is possible without lowering the temperature of the AgBr below the melting point or displacing the mobile, molten globule. After the pellet is melted, the protective funnel over the plate on *B* is removed. The funnel over *A* is raised, and the Vycor plate on *B,* at a temperature above the melting point of AgBr, is gripped with stainless-steel forceps. The upper face of the Vycor plate as it lies on *B* is placed in contact with the globule, compressing it into a sheet.

If the exclusion of oxygen from the neighborhood of the globule has been adequate, the molten bromide possesses a high surface tension, is very mobile, and is compressed with some difficulty. When the sheet is less than about 100 μ thick, the weight of the upper Vycor plate is inadequate to compress the molten layer to the thickness of the spacer. Additional weight is applied by placing the molded Vycor plate from hot plate *B* on top of the sandwich, as in Fig. 4, whereupon the layer of molten bromide assumes a thickness close to that of the spacer, down to thicknesses of about 20 μ. To form sheets thinner than 20 μ, still greater weights must be applied, to 5000 grams to form sheets 5 μ thick.

In solidification, the aim is to produce not more than a few large, bubble-free single crystals in the sheet by inducing crystallization at as few nuclei as possible and by proper adjustment of the temperature gradient perpendicular to the surfaces of the sheet. The end of plate *C* near the melting plate *A* is raised to a temperature below the melting point of AgBr, so that a temperature gradient exists in the air gap between the two plates. An air jet is directed on the upper plate of the sandwich as shown in Fig. 4, and the sandwich is slowly pushed over the gap by means of a bar screwed to a rack and pinion. The rack pushes the sandwich through the temperature gradient at the rate of 2.5 mm/min.

Crystallization begins in a line perpendicular to the direction of advance when the molten layer reaches the isothermal in the temperature gradient corresponding to the freezing point. The heater *D* is placed above the plates to control the form of the temperature gradient through the thickness of the crystal. If, near the site at which crystallization is occurring, the upper plate becomes cooler than the corresponding part of the lower plate, bubbles appear at the liquid-solid interface in contact with the lower plate, apparently because of gas evolution from the melt; if the upper plate is the warmer of the two, bubbles appear at the upper surface of the sheet. When the isothermals are perpendicular to both upper and lower plates, crystals up to 100 μ thick are usually free from bubbles, but crystals 200 μ thick may then have a layer of bubbles near the center, probably indicating a cusp in the isothermals near the center of the layer. The tendency toward formation of bubbles is greater when hydrogen bromide is used to carry bromine in the process of sheet-making than when nitrogen is used.

When crystallization is complete and the sandwich is wholly on plate *C,* it is, in the course of 10 min or so, moved along the plate until the temperature has dropped to about 100°C. Owing to the difference between the expansion of crystalline silver bromide and the glass plates, the sandwiched crystal is under strain, and the consequent dichroism causes individual crystals to be visible as areas of different brightness when viewed between crossed Polaroid filters against a red safelight. If the material is of high purity, the sheet will contain only a few crystals, crystals 1 cm square or more. When the hot sandwich is plunged into cold distilled water, if silver oxide has not formed, the glass plates soon separate from the silver halide.

Spectrographic analyses of these sheets show about 0.08 ppm of Fe, and Mn, Cu, Pb, Au, B, Cd, Co, Cr, Ni, Si, Sn, and Zn are below detectable limits. Small quantities of foreign material, such as Cd^{++}, Pb^{++}, Tl^{+}, Cu^{+}, or the sulfide, selenide, or oxide of silver, can be incorporated when the sheets are made. The atmosphere in contact with material containing such addenda must be consistent with the chemical properties of the impurities; for example, an atmosphere containing halogen cannot be used in the preparation of sheets intended to contain Tl^{+} or $S^{=}$.

Parasnis and Mitchell (1959) have described special methods for introducing copper salts, and Bartlett and Mitchell (1958) for gold salts, to reveal dislocation structures in silver halides by decoration. To produce silver halide from which thin crystals containing few dislocations can be made, Mitchell (1956) passed a stream of dry hydrogen halide in nitrogen for about an hour through the melt before capillary filtration and fractionally distilled the filtered material during vacuum melting. The final sheet crystals were annealed in an atmosphere of halogen or in a vacuum at temperatures between 200 and 400°C for periods up to 12 hr. The halogen was frozen out into a side tube cooled in liquid nitrogen before the crystals were finally cooled to room temperature. Crystals annealed at a temperature below 350°C were polygonized in the strained state in which they were separated from the glass plates. They were then suitable for observations of decorated dislocation complexes.

Cutting, Mounting, and Orienting Sheet Crystals

Specimens for experiments can be cut from the sheets, preferably by a knife of chemically inert material, such as a platinum-rhodium alloy. If a steel knife is used, it is necessary to treat the cut surfaces with halogen, conveniently in aqueous solution, and rinse them in running distilled water to remove traces of iron halide and reduced silver formed in the cutting operation.

Unannealed sheet crystals prepared in this way have been heavily strained as the sandwich cooled. For some purposes, annealing will be required (see Section 3). On immersion of the unannealed crystals in photographic developer, little fog is formed on the surface, and their photographic sensitivity with respect to developed image formed at the surface as a result of previous exposure to light is very small. Developable latent image is produced, however, in the interior of the sheet by exposure to blue and ultraviolet light, and prolonged exposure forms a visible image of "print-out silver" consisting of minute particles of colloidal silver deposited at dislocations and substructural boundaries in the interior (Hedges and Mitchell, 1953a and b; West and Saunders, 1959). Any photographic sensitivity observed at the surface of the sheets indicates the presence of impurity, usually silver oxide or silver derived from the oxide, resulting from exposure of the hot material to oxygen during the preparation. Material showing some surface fog and sensitivity to light may nevertheless be used for many studies of photographic sensitivity after a minute's immersion in saturated chlorine or bromine water, followed by rinsing in running distilled water. This operation is most conveniently performed on the specimen mounted on a microscope slide for the specific experiment intended.

Mounting is conveniently carried out by Parafilm (Marathon Corportion, Menaska, Wisconsin), a paraffin, wax-crepe rubber composition melting at about 60°C.

The surfaces of thin crystals of AgCl and AgBr prepared as discussed are usually within 15° of a (100) surface, and exact (100) sheets are not uncommon (Hedges and Mitchell, 1953). Occasionally, for no apparent reason, other surfaces, including (111) surfaces, appear. Hedges and Mitchell (1953a and b) described the preparation of sheets in (111) orientation by seeding.

The orientations of thin crystals can be determined by observing the shapes of silver bodies formed by development. The developed bodies (Boissonas, 1950, 1951) consist of minute particles of colloidal silver deposited in pits that are etched into the crystal by the developer. The sides of the pits are {111} planes, and the shape of each silver body, as seen at normal incidence, is that of the section that the pit makes with the surface, square if the surface is a (100) plane, with sides parallel to ⟨110⟩ directions. Triangular or hexagonal shapes indicate (111) surfaces (Boissonas, 1950, 1951; Vuille, 1954; Berry, 1958). The developer *p*-methylamino-phenol-ascorbic acid of James, Vanselow, and Quirk (1953) is convenient for this purpose. To obtain such developed bodies in pure crystals, it is necessary to etch into the interior by 1% potassium cyanide or 4.2 *M* potassium bromide solution (the latter for silver bromide only). West and Saunders (1956, 1959) describe etching procedures for the exploration of latent image throughout the interior of sheet crystals.

Orientation of the surface can also be determined (Mitchell, 1957) by noting the cracks along ⟨100⟩ directions in a (100) surface is produced at −196°C by pressing a fine needle into the surface.

3. GROWTH OF LARGE CRYSTALS FROM THE MELT

Purification by Zone Melting

The starting material for crystal growth from the melt is pure precipitate prepared by the methods described in Section 1. Additional purification is obtained by the technique of capillary filtering as described in Section 2. Recently, zone refining has also been successfully applied to the silver halides. By using such zone-refined material for growth by the Bridgman-Stockbarger method, crystals of extremely high chemical purity have been obtained.

A detailed study of zone melting of the silver halides has been reported by Moser, Burnham, and Tippins (1961). In these experiments, the material was confined in a horizontal quartz boat, and molten zones about 1 cm wide were passed through the ingot at rates as low as 0.6 cm per hour, and as high as 7.5 cm/hr. The faster rate seems to be just as effective as the slower rate in terms of purification per pass. Molten zones were established at every 5 cm along a 50-cm ingot, and the furnace assembly was connected to a drive mechanism that repetitively advanced it a distance of 5 cm and then returned it the same distance in a few seconds. The boat was provided with fittings such that any desired atmosphere could be maintained during the zoning operation. To minimize matter transport, the furnace and the tube were tilted about 4°, the high end being where the zone entered. The distribution of specific impurities was studied by adding several ppm of an element to the starting material and spectrochemically analyzing sections of the ingot after the passage of about 70 zones.

The results of one zone-refining experiment are shown in Fig. 5. Here, about 200 grams of AgCl were refined by the passage of 70 zones, each 1.7 cm wide. During the entire operation, the charge was

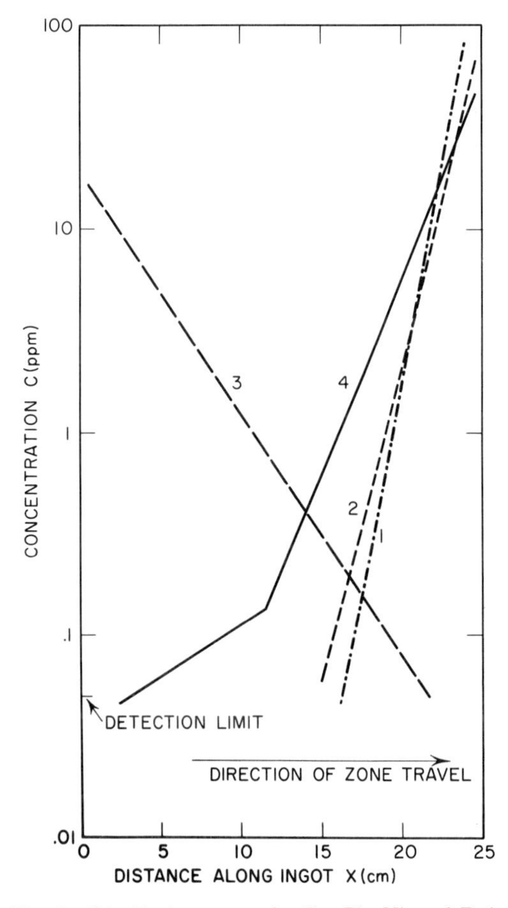

Fig. 5. Distribution curves for Cu, Pb, Ni, and Fe in AgCl zone-refined under a chlorine atmosphere. Curves and initial concentrations are as follows: 1. Cu (5 ppm); 2. Pb (7 ppm.); 3. Ni (3 ppm); 4. Fe (6 ppm).

TABLE 1

DISTRIBUTION COEFFICIENTS OF VARIOUS IMPURITIES IN AgCl AND AgBr

Halide	Impurity	Atmosphere	Distribution Coefficient
AgCl	Cu	Cl_2 or vacuum	0.4–0.6
AgCl	Pb	Cl_2 or vacuum	0.4
AgCl	Ni	Cl_2 or vacuum	1.4
AgCl	Fe	Cl_2	~0.7
AgCl	Fe, Mn, Cd	Vacuum	>1
AgCl	Sn, Al, Sr	Vacuum	<1
AgBr	Cu	Vacuum	<1
AgBr	Fe, Ni, Mn	Vacuum	>1

kept under one atmosphere of chlorine. The results indicate the very effective purification that can be achieved and show that a near-ultimate distribution is achieved with this number of passes. The distribution coefficient defined as the ratio of the concentration of the impurity in the solid to the concentration of the impurity in the liquid at a solid-melt interface, has been determined for a number of impurities in AgCl and AgBr by such zone-refining experiments. The results are summarized in Table 1.

Application of these techniques to pure AgCl results in material of extremely high purity, as shown by the spectrochemical analyses tabulated in Table 2. A variety of direct and indirect tests on material from the central portion of this ingot indicate that the copper and nickel concentration is in each case below 1 part in 10^9 (weight of impurity to weight of AgCl) and that of lead is below 1 part in 10^8. Iron is probably present in a concentration of about 2 parts in 10^8.

The capillary-filtering and the zone-melting operations can be carried out consecutively in a completely closed system. An arrangement recently used in the authors' laboratory is shown in Fig. 6. After the precipitate is loaded in the quartz tube *A*, the system is sealed and the temperature distribution in the vertical furnace *F* is adjusted to melt the precipitate while

TABLE 2

Spectrochemical analysis of nominally pure AgCl ingot, zone-refined under chlorine atmosphere. Seventy zones, at a rate of 3 in./hr, were passed through the 25-cm-long ingot. Values in ppm by weight.

Distance from End of Zone Entry (cm)	Measured Concentration					
	Cu	Fe	Mn	Ni	Pb	Cd, Cr, Co,[d] Sn, Zn, Au
0.2	—	0.15	0.02[c]	0.07	*a*	*a*
2.5	*a*	0.07	*a*	*a*	*a*	*a*
4.3	*a*	0.25	*a*	*a*	*a*	*a*
7.6	*a*	*b*	*a*	*a*	*a*	*a*
11.0	*a*	*b*	*a*	*a*	*a*	*a*
13.4	*a*	*b*	*a*	*a*	*a*	*a*
18.3	*a*	*b*	*a*	*a*	*a*	*a*
21.4	0.04	*b*	*a*	*a*	*a*	*a*
23.2	0.1	0.2	*a*	*a*	0.4	*a*
24.3	0.3	0.15	*a*	*a*	1.0	*a*
Averaged analysis of starting material	*a*	0.3	*a*	*a*	*a*	*a*

[a] The element was not detected.

[b] Fe is not detected in these samples by normal spectrochemical procedures. Special tests indicate that the Fe concentration is about 0.02 ppm. This residual iron may be introduced during preparation of the spectrochemical samples; if so, the zoned crystals are purer than the analysis indicates.

[c] This value for Mn is approximate.

[d] The detection limits for these elements are as follows: Cr, Cu, Pb, Fe, Mn, Ni, Sn: about 0.05; Co: 0.1; Cd, Au: 1.0; Zn: 4.0.

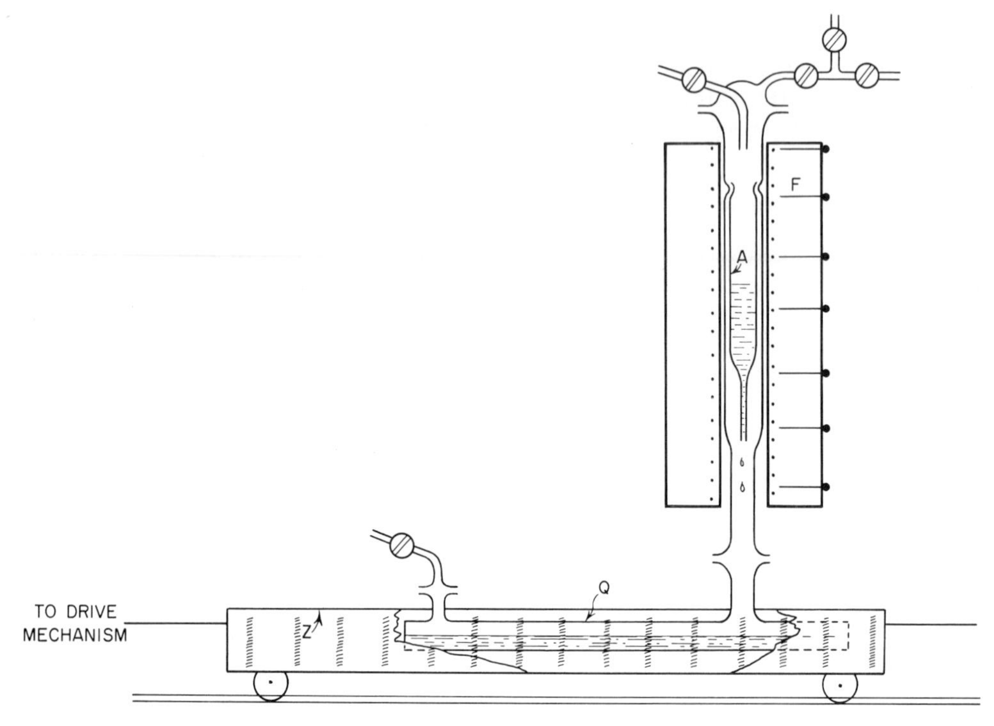

Fig. 6. Arrangement for melting, capillary-filtering, and zone-refining of silver halide precipitate. After initial loading of the precipitate in the quartz tube *A*, the entire operation is carried out in a completely closed system under any desired atmosphere.

the material in the capillary section is kept frozen. At the desired time, the heat input is adjusted to melt the solid plug and permit the molten salt to filter through the capillary and into the horizontal quartz boat *Q*. By appropriate adjustment of the heaters in the horizontal zoning furnace *Z*, the charge is first kept molten to obtain good mixing; then narrow molten zones are established and the zoning operation is begun. The entire operation involves no handling of the material after initial loading and provides complete control of ambient atmosphere.

Bridgman-Stockbarger Growth

The Bridgman-Stockbarger technique of crystal growth has been used by most investigators desiring to produce large, well-defined silver halide crystals (Kremers, 1947; Hofstader, 1949; Yamakawa, 1949; Allemand and Rossel, 1954; Clark and Mitchell, 1956; Nail et al., 1957; Brown, 1958; Süptitz, 1958). The Kyropoulos method has also been used by a few people, but no details have been reported (Hecht, 1932; Van Heerden, 1950). Large AgCl and AgBr crystals are available from several commercial sources (Harshaw Chemical Co., Cleveland, Ohio; Semi-Elements, Inc., Saxonburg, Pennsylvania; Karl Korth, Kiel, Germany). Harshaw and Semi-Elements, Inc. grow ingots up to 5 in. in diameter and 7 in. long by the Bridgman-Stockbarger technique. Since these crystals are prepared primarily for use as optical elements, they are apparently grown from only moderately pure precipitates and under conditions where considerable exposure to light occurs.

By using the Bridgman-Stockbarger technique, useful crystals have been obtained using a wide variety of growing conditions. Most workers have used a vertical furnace consisting of two windings which can be independently controlled. A baffle of some sort is used to minimize interaction between the two windings. Heavy metal liners and windings with many taps for shunting parts of the current have been used to improve uniformity of temperature. The tube containing the molten material may be moved relative to the furnace in several ways. An arrangement used by Brown (1958) is shown in Fig. 7. In practice, the upper furnace is held at a temperature anywhere from 10 to 50° C above the melting point of the silver halide (AgBr, 434°C; AgCl, 455° C), and the lower furnace somewhere below 400°C. Conditions are adjusted so that a flat freezing isotherm is produced at the baffle position. If the rate of growth is kept sufficiently slow, a flat isotherm can be maintained during growth. The temperature gradient normal to the freezing isotherm is of the order of 10° C/cm, although steeper gradients have been used with good results.

Sample tubes constructed of Pyrex glass, quartz, or platinum metal have been used. Nail (1957) has reported high concentrations of sodium in crystals grown in Pyrex, whereas he has found negligible sodium concentrations in crystals grown in quartz vessels. There are also indications of reaction with platinum. Quartz vessels have been most generally used. The vessels are generally of circular cross section 1 to 2 cm in diameter and 10 to 20 cm long. Hofstader (1949) has reported attempts to grow AgCl crystals in square cross-sectional tubing but found that he obtained a polycrystal with many small crystals growing out of the corners. There is evidence that this problem is eliminated by using tubes with slightly rounded corners.

In some cases, prior to crystal growth, the silver halide melt has been treated with halogen gas or hydrohalogen gas in order to convert to halide traces of silver or silver oxide present in the melt. The gas may actually be bubbled through the melt, although it suffices to flow the gas over the melt. Melts kept under chlorine gas show a coloration which disappears when the chlorine is removed by evacuation or displacement with some inert gas. A similar reversible coloration is observed in pure AgCl crystals at temperatures above 300° C and a chlorine pressure of one atmosphere or more (Moser, 1962). When the chlorine-treated crystal or melt is cooled below about 300° C, however, no coloration is observed in pure material even if the chlorine pressure is maintained. The reversible coloration is associated with the introduction of positive holes (defect electrons) into the crystal volume; apparently only negligible concentrations of holes can be introduced or maintained at lower temperatures. Melts or crystals containing certain impurities, however, do show a residual coloration after cooling from high temperature to room temperature in a chlorine atmosphere. Such effects have been extensively studied for copper in AgCl (Moser, Nail, and Urbach, 1959); here, the

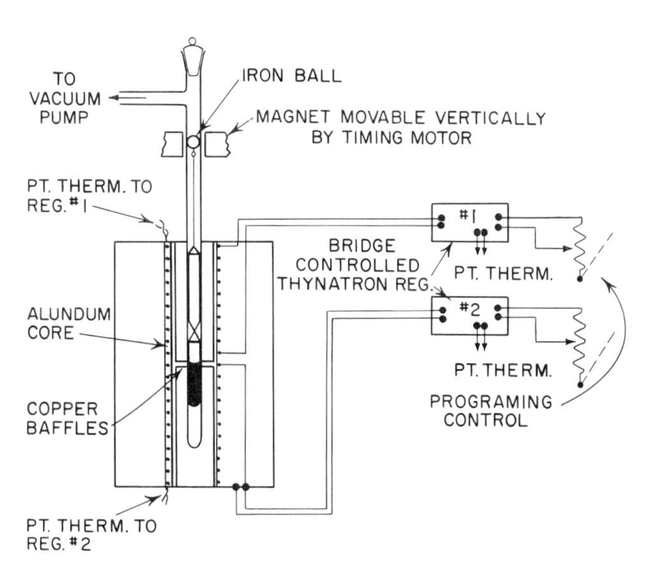

Fig. 7. Temperature-regulated furnace for growing silver halide crystals in a vacuum.

coloration represents the absorption of divalent copper ions.

Growth rates in the range of 1 to 5 mm/hr have been successful. Use of considerably faster growth rates has resulted in polycrystalline samples. Crystals have been grown in inert atmospheres such as helium and nitrogen and in halogen gas atmospheres, but there is little difference in crystal appearance or properties as a result of the particular atmosphere selected. Crystals grown in air or oxygen, however, adhere strongly to the vessel in which they are grown and have abnormal electronic properties (Brown, 1958; Süptitz, 1957; Mitchell, 1957a). Usually, the contraction of the crystal during cooling causes it to pull away from the glass vessel, breaking the vessel. This occurs somewhere between 100 and 200°C.

If reagent-grade silver halide is used as the starting material, the resultant ingot is generally strongly discolored near the top, with a dark scum at the top. By cutting away this top section, and using the lower portion to grow a new crystal, considerable purification is achieved. This procedure has been repeated several times by Hofstader (1949) to obtain pure samples for counting nuclear particles.

Pure crystal ingots are optically clear. The surfaces in contact with the vessel generally show some irregular pitting and small bubbles, especially near the bottom of an ingot. These surface imperfections are nearly absent on crystals grown in vacuum, so they are apparently caused by entrapped gas bubbles. Etching of the ingot cross section with an etchant such as sodium thiosulfate generally delineates a few, large-angle grain boundaries. Occasionally, the entire ingot appears to consist of one crystal. X-ray diffraction studies and density determinations show that these crystals have a mosaic structure arising from a fairly high density of small-angle grain boundaries (Berry, 1955). Crystals grown by the Kyropoulos technique show a lower density of such small-angle boundaries.

Cutting and Polishing

For most physical studies it is necessary to cut specimens from the large ingots and polish the surfaces. Silver halide crystals are soft and react with most metals, but are fairly readily cut by the standard machining techniques used for soft metals. Moeller et al. (1951) carried out an extensive metallurgical investigation of AgCl, and stated that the machinability was similar to that of aluminum. Several workers have used thin tungsten carbide-tipped blades rotating at high speeds. Kerosene may be used as a lubricant, although dry cutting seems to be quite acceptable. Freshly cut surfaces yield very broad x-ray diffraction spots. Sharp patterns are obtained after chemical removal of about 0.05 mm from the surface (Brown, 1958). Acceptable surfaces have been prepared in a variety of ways. Leidheiser (1957) reports polishing samples on metallographic emery paper, followed by polishing on a cotton cloth with levigated alumina, and finally with tin oxide. The sample was then etched in dilute photographic fixer to remove residual strains. Nail et al. (1957) prepared AgCl surfaces by first grinding on ground glass with a water slurry of No. 1000 aluminum oxide grit. This was followed by a short polish etch on soft flannel cloth wetted with a 30% sodium thiosulfate solution. To remove all traces of sodium thiosulfate and its reaction products, the samples are rinsed in alkaline chlorine water, etched for a few minutes in hydrochloric acid, and finally rinsed in doubly distilled water. The samples may be air-dried or blotted with filter paper. Etchants other than sodium thiosulfate can be used and are preferable for certain purposes. Studies by Cook and Leidheiser (1955) and some unpublished work done at the Kodak Research Laboratories show that cyanide solutions used as etchants produce the smoothest surfaces, whereas ammonia solutions tend to produce a mottled roughened surface. Leidheiser has shown that the etching rate and surface appearance depend on crystallographic orientation.

Strains present after the growing, cutting, and polishing operations may be removed by appropriate annealing. Heating near 400°C for several hours, followed by cooling at a rate not exceeding 20°C/hr, produces sections free of optical birefringence. Dur-

Fig. 8. Photograph of a large single crystal of AgCl. This 1-cm thick disk is a cross-sectional cut from a 20-cm long ingot. It has been polished by the techniques described in the text.

ing annealing, the samples may rest on a ground-quartz plate, although magnesium oxide powder, quartz powder, and silver halide sheets have also been used for support.

A photograph of a carefully polished and annealed AgCl crystal is shown in Fig. 8.

Other Halides and Impurity Additions

The procedures described here are applicable to growing and preparing large crystals of AgBr as well as of AgCl. Crystals of AgI have not been grown successfully, apparently because of the phase transition from the high-temperature cubic form to the low-temperature hexagonal form at 146°C. Mixed crystals of AgCl and AgBr and crystals containing a few mole per cent of AgI have been grown. A number of cationic impurities have been deliberately incorporated during crystal growth. The additions are made by adding the element, either as a dry solid or as an aqueous solution of the appropriate salt, to the dry starting precipitate at room temperature. The moisture is removed during the early stages of heating the charge. The distribution of low concentrations of copper and nickel added in this manner is shown in Fig. 9.

The purest crystals have been grown by starting with zone-refined material that had

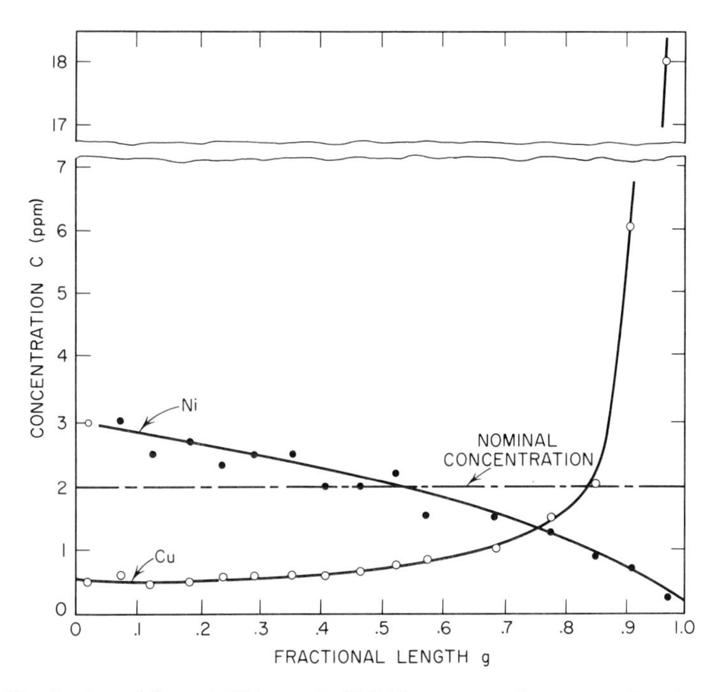

Fig. 9. Distribution of Cu and Ni in an AgCl Bridgman crystal, grown under nitrogen. The measured concentration C is plotted as a function of the fractional length $g = x/L$. The nominal concentration added initially to the melt was 2 ppm.

been subjected to the capillary-filtering operation. The purity of sections from such an ingot is comparable with the analysis reported in Table 2 for the middle sections.

References

Allemand, Ch., and J. Rossel (1954), *Helv. Phys. Acta.* **27,** 519.

Bartlett, J. T., and J. W. Mitchell (1958), *Phil. Mag.,* [8] **3,** 334.

Berriman, R. W., and R. H. Herz (1957), *Nature,* **180,** 293.

Berry, C. R. (1955), *Phys. Rev.,* **97,** 676; (1958), *Sci. et ind. phot.,* [2] **29,** 364.

Berry, C. R., S. J. Marino, and C. F. Oster, Jr. (1961), *Phot. Sci. and Eng.* **5,** 332.

Boissonas, C. G. (1949), *Sci. et ind. phot.* [2], **20,** 361; (1950), *C. R. acad. Sci.,* Paris, **230,** 1270; (1951), *Fundamentals of Photographic Sensitivity,* Butterworth, London, p. 36.

Boyer, S., J. Cappelaere, and J. Pouradier (1959), *J. chim. physique* **56,** 495.

Bridgman, P. (1925), *Amer. Acad.* **60,** 305.

Brown, F. C. (1958), *J. Phys. Chem. Solids* **4,** 206.

Brown, F. C., and F. Seitz (1958), *Photographic Sensitivity,* Vol. 2, Tokyo Symposium, Maruzen Co., Ltd, Tokyo, pp. 11–29.

Clark, P. V. McD., and J. W. Mitchell (1956), *J. Phot. Sci.* **4,** 1.

Cook, F. H. and H. Leidheiser, Jr. (1955), *WADC Technical Report* 54-486, Virginia Institute for Scientific Research, Richmond, Virginia.

Czochralski, J. (1918), *Z. phys. chem.* **92,** 219.

Hecht, K. (1932), *Z. Physik* **77,** 235.

Hedges, J. M., and J. W. Mitchell (1953a) *Phil. Mag.* [7] **44,** 223; (1953b), 357.

Herz, A. H. and J. O. Helling (1961) *J. Colloid Sci.* **16,** 199.

Hofstader, R. (1949), *Nucleonics* **4** (4), 2.

James, T. H., W. Vanselow, and R. F. Quirk (1953), *PSA Journal (Phot. Sci. and Tech.)* **19B,** 170.

Kremers, H. C. (1947), *J. Opt. Soc. Am.* **37,** 337.

Kyropoulos, S. (1926), *Z. anorg. chem.* **154,** 308.

Leidheiser, H. Jr. (1957), *WADC Technical Report* 57-264, Virginia Institute for Scientific Research, Richmond, Virginia.

Meiklyar, P. V. (1957), *NIPFIK* **2,** 389.

Mitchell, J. W. (1957a), *Dislocations and Mechanical Properties of Crystals,* John Wiley and Sons, New York, p. 69; (1957b), *Phil. Mag.* [8] **2,** 1276; (1959), *Progress in Semiconductors,* Vol. 3, John Wiley and Sons, New York, p. 55.

Moeller, R. D., F. W. Schonfeld, C. R. Tipton, Jr., and J. T. Waber (1951), *Trans. Am. Soc. Metals* **43,** 39.

Moser, F. (1962), *J. Appl. Phys. Supp.* **33,** 343.

Moser, F., N. R. Nail, and F. Urbach (1959), *J. Phys. Chem. Solids* **9,** 217.

Moser, F., D. C. Burnham, and H. H. Tippins (1961), *J. Appl. Phys.* **32,** 48.

Nail, N. R., F. Moser, P. E. Goddard, and F. Urbach (1957) *Rev. Sci. Instr.* **28,** 275.

Parasnis, A. S., and J. W. Mitchell (1959), *Phil. Mag.* [8] **4,** 171.

Reinders, W. (1911), *Kolloid Z.* **9,** 10; *Z. phys. chem.* **77,** 677.

Stockbarger, D. C. (1936), *Rev. Sci. Instr.* **7,** 133.

Strock, L. W. (1934), *Z. phys. Chem.* (*B*) **25,** 441.

Süptitz, P. (1957), *Naturwiss.* **24,** 629; (1958), *Z. Physik* **153,** 174.

Sutherns, E. A. (1956), *J. Phot. Sci.* **4,** 83.

VanHeerden, P. J. (1950), *Physica* **16,** 505.

Vuille, R. (1954), *Helv. Chim. Acta* **37,** 2264.

Wakabayashi, Y., and Y. Kobayashi (1957), *J. Soc. Sci. Phot. Japan* **20,** 102.

West, W., and V. I. Saunders (1958), *Wissenschaftliche Photographie* Darmstadt, Verlag Dr Othmar Helwich, p. 48; (1959), *J. Phys. Chem.* **63,** 45.

Yamakawa, K. A. (1949), *Ph.D. Thesis,* Princeton University.

13

Hydrothermal Growth

A. A. Ballman and R. A. Laudise

1. DEFINITION OF HYDROTHERMAL SYNTHESIS

Hydrothermal synthesis is the use of aqueous solvents under high temperature and high pressure to dissolve and recrystallize materials that are relatively insoluble under ordinary conditions. We are concerned here, however, with specific attempts to make large crystals of a material under controlled conditions. "In situ" conversions and chemical reactions do not fall within the scope of this discussion. Such reactions usually produce only microscopic crystals and are primarily designed for the study of phase equilibria. For crystal growth, however, one should be thoroughly familiar with hydrothermal phase equilibria, and such works as *Phase Diagrams for Ceramists* and other recent reviews on hydrothermal synthesis are particularly helpful. (Levin, McMurdie, and Hall, 1956; Ellis and Fyfe, 1957; Laudise and Nielsen (1961) Laudise (1961).

History

Perhaps the earliest hydrothermal experiments in which growth of a material took place on a seed plate were made by DeSenaramount (1851) and by Spezia (1905) over fifty years ago. They were able to increase the thickness of a seed plate by depositing α-quartz from an aqueous medium.

Since these early experiments, hydrothermal systems involving α-quartz have been studied quite extensively, and a large part of our present knowledge comes from these studies.

Advantages and Disadvantages

Although one usually refers to "high temperatures of crystallization" during hydrothermal synthesis, in reality these temperatures are often rather low when compared to the melting point of the material. It is this "relatively" low crystallization temperature that makes the technique an attractive means of crystal growth.

Under hydrothermal conditions a crystal presumably grows under less thermal strain and therefore may contain a lower dislocation density than when grown from a melt where large thermal gradients are apt to be present. The low temperature also permits the growth of low temperature polymorphs which are often unattainable by other means. The growth of α-quartz (the low temperature polymorph of SiO_2) is an especially good example.

Another advantage lies in the use of a closed system where the atmosphere may be controlled to produce oxidizing or reducing conditions. This allows the synthesis of phases difficulty formed or not attainable by other methods.

One of the great attractions of hydrothermal growth is the comparatively rapid rate of growth. Quartz grows as fast as 0.250 in./day on the (0001) plane. This is extremely rapid for growth from a solvent-solute system. Hydrothermal solutions have comparatively low viscosities, together with a large variation of density with temperature at constant average density. This results in rapid convection and very efficient solute transport, permitting large growth rates.

Until recently certain problems have

made the adoption of hydrothermal growth less widespread than other methods. The foremost one is the need for a well-designed, high-pressure vessel with a reliable closure that is capable of withstanding the high pressures generated at operating temperatures. In addition to being structurally strong the vessel should be chemically inert (corrosion resistant) since high purity of the grown crystal is usually of paramount importance, and even slight autoclave attack is intolerable.

Although hydrothermal growth is relatively rapid, experiments are still lengthy. This, plus the inability to observe progress during a run, make unsuccessful experiments costly and discouraging.

2. APPARATUS

Autoclave Design

A variety of autoclave designs and closures is commercially available. The ideal vessel should possess a reliable and easily maintained closure to insure the mechanical success of an experiment. The vessel should also be easily assembled and disassembled. The material of construction should possess high strength characteristics and corrosion resistance to alkaline and acidic solutions.

No one autoclave design is "ideal," and compromises must be made regarding temperature, pressure, and choice of aqueous media. The available systems may best be grouped according to low, medium, or high pressure.

GLASS VESSELS—LOW PRESSURE. For temperatures to about 300°C and pressures approaching 10 atmospheres, heavy-walled tubing of Pyrex or quartz is quite satisfactory. It is particularly advantageous because it permits direct observation of the growth proccss. To datc hydrothcrmal growth has been used for materials requiring temperatures and pressures where glass vessels are not suitable. However, there are many compounds for which glass vessels might be useful in the future.

STEEL VESSELS—MEDIUM PRESSURE. For pressures up to 500 atmospheres, and temperatures near 400°C, steel vessels with a flat plate closure as shown in Fig. 1 are useful. The seal is made by the pressure of a plunger against a metal disk or gasket that butts against the shoulder of the vessel body. Such a vessel is quite easily assembled and the plate gasket requires a minimum of mechanical precision.

Ordinary low-carbon steel may be used for vessels when corrosion is not expected. If corrosion is a problem, silver or platinum liners can be placed in the vessel and used with sealing rings of the same material. These vessels have been used extensively by Morey and his co-workers (1937). For operation at higher temperatures, special alloys such as Inconel X and Udimet permit operation near 700°C.

WELDED CLOSURE—HIGH PRESSURE. A welded closure as described by Walker and Buehler (1950) has been used quite successfully for temperatures up to 450°C and a

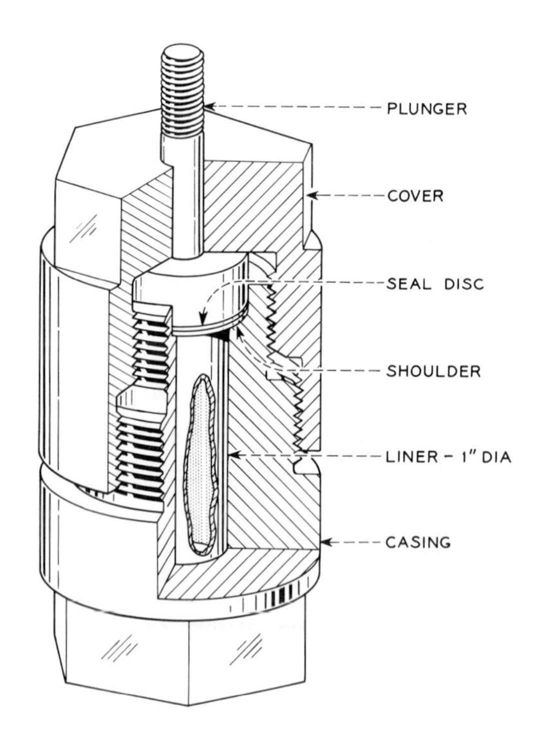

Fig. 1. Autoclave with flat plate closure. (After Morey, 1937.)

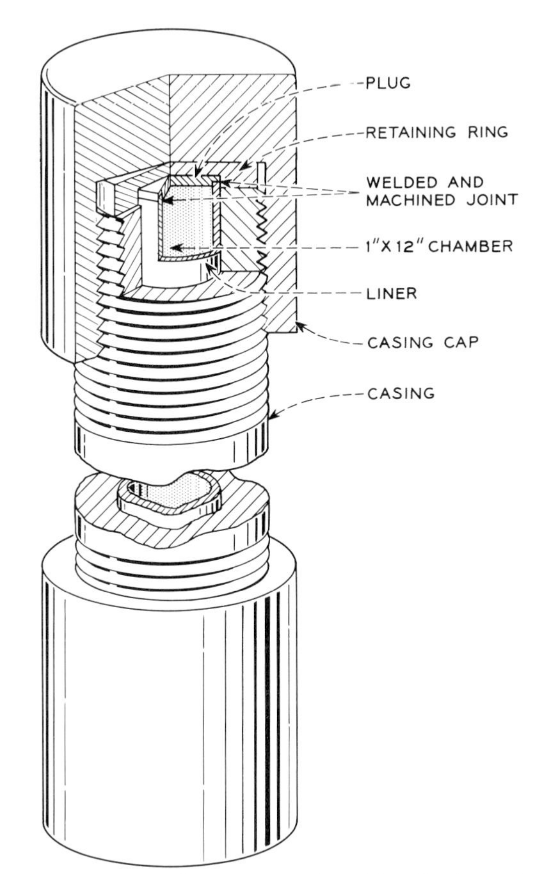

Fig. 2. Welded closure vessel. (After Walker and Buehler, 1950.)

pressure of 3000 atmospheres. The vessel ordinarily consists of an inner liner welded at both ends and an outer tube of heavier construction. Figure 2 shows the construction in some detail.

At the conclusion of an experiment one end of the welded liner is sawed off to remove the charge. The liner is then driven out of the outer casing and discarded, and a new liner inserted. Thus for a sequence of experiments, rather extensive machine shop services are needed. However, welded liners are still relatively inexpensive compared with an entire vessel. The disposable liner permits a good deal of exploratory work to be done in systems where corrosion might ruin a more complex vessel.

MODIFIED BRIDGMAN—HIGH PRESSURE. The closure developed by Bridgman (1914) (as modified in several commercially available vessels) is perhaps the most generally useful design for hydrothermal growth systems (Fig. 3). The initial seal that closes the vessel is made by mechanically tightening the plunger against the deformable gasket by the set screws in the head. The unsupported area closure makes its seal when the force of the generated pressure acts upon the piston, moving it upward against the deformable gasket. Figure 4 illustrates one of several commercially available versions.

Such vessels can be used quite successfully in the 500°C, 3700 atmosphere range, the construction material being chosen to meet the particular experimental conditions.

CONE CLOSURE—HIGH PRESSURE. In principle the cone closure (Fig. 5) is quite similar to the flat plate closure that is used at medium pressures. It is usually reserved for vessels of small internal diameter ($\frac{1}{8}$ in. to $\frac{1}{4}$ in.), and the sealing surface is at a much

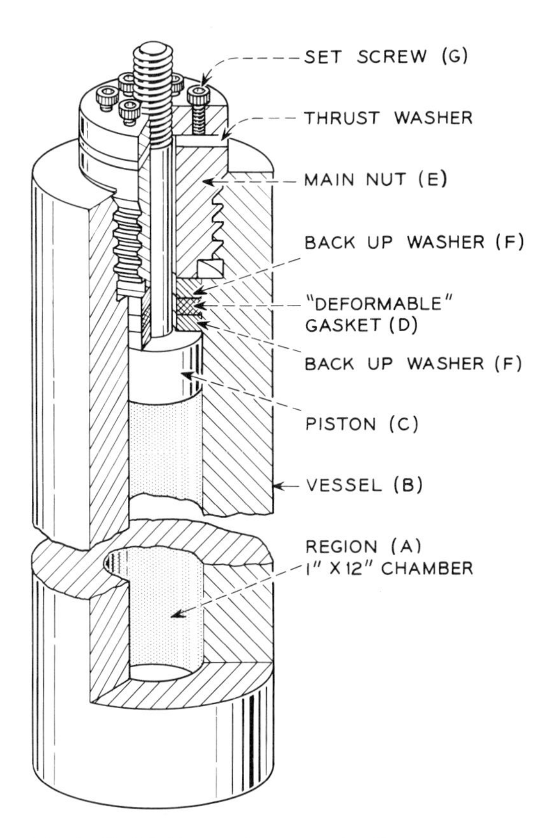

Fig. 3. Full Bridgman closure. (After Bridgman, 1914.)

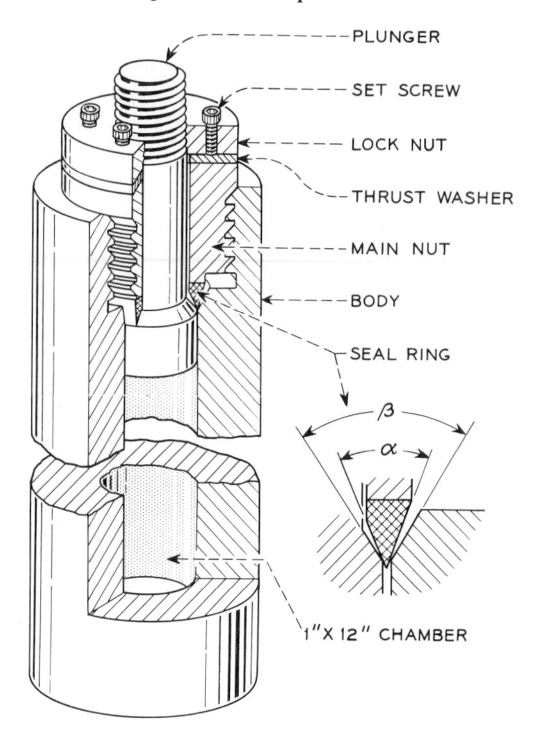

Fig. 4. Modified Bridgman vessel, Autoclave Engineers, Erie, Pennsylvania.

lower temperature than the crystallization zone. The seal is made against a conical seat by a cone-shaped insert that is forced down by a screw cap. The force on the cone seat depends on the torque applied to the screw cap. This limits the closure (at high pressures) to vessels of approximately $\frac{1}{4}$ in. internal diameter or less.

Because the cone seat can be in a relatively cool zone, it permits crystallization at high temperatures and pressures with little creep. Pressures up to 2400 atmospheres can be held quite easily at 750°C and higher temperatures at somewhat lower pressures.

Although vessels of this type are most often used for the study of phase equilibria, crystal growth experiments have been made successfully in them.

INERT SYSTEMS. In each of the pressure systems discussed, except the low pressure glass system, noble metal liners or plating

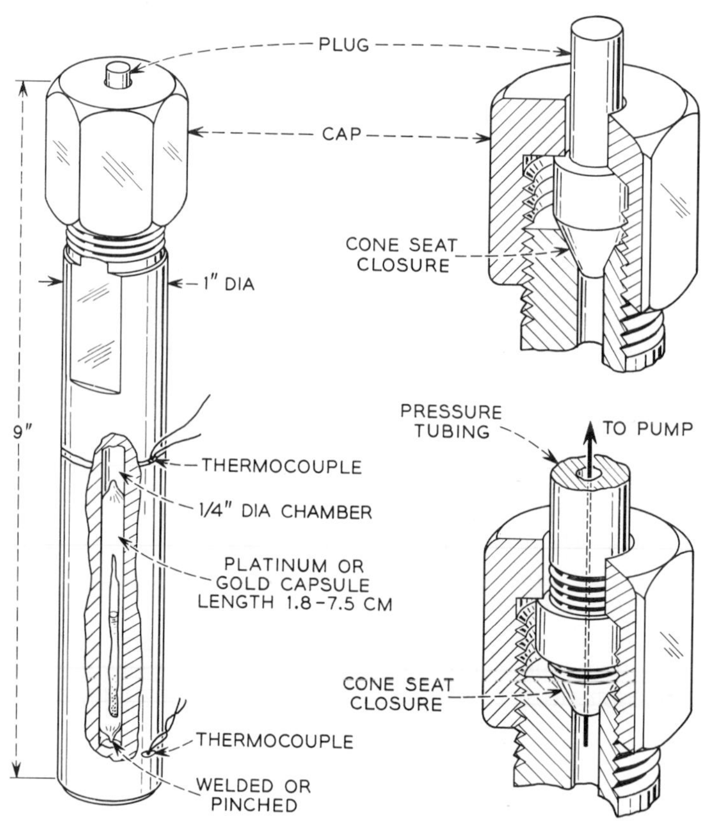

Fig. 5. Reaction vessel with a cold-cone seat closure, Tempress Inc., State College, Pennsylvania. (After Roy and Tuttle, 1956.)

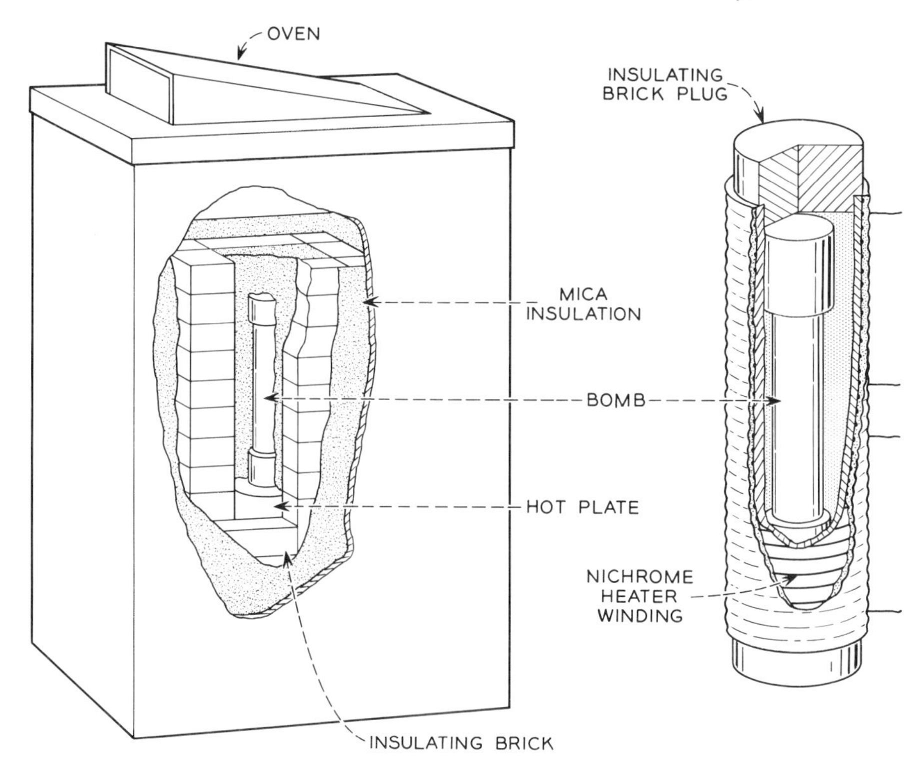

Fig. 6. Furnace for autoclave used in hydrothermal crystal growth.

may be used to prevent or inhibit corrosion. Small platinum, gold, or silver tubes, welded at both ends containing the charge, have been used in both the medium and high pressure vessels. If both the inner noble metal tube and the outer vessel are filled to the same percentage of their free volume (per cent fill, or degree of fill) with the aqueous solvent, the noble metal tube will be in principle under no pressure. This "pressure balancing" technique permits the use of thin-walled tubing.

Although flat plate closures with inert liners have been used quite successfully in the medium pressure range, attempts to design a completely inert system in the high pressure range have not been completely successful. Further improvements are needed before the high pressure range can be fully exploited.

Temperature Control and Measurement

Furnaces for hydrothermal vessels depend on the degree of temperature control required and the temperature gradients to be used for crystallization. Hot plates heated with nichrome strip heaters are often adequate up to about 400°C. For higher temperatures or controlled temperature differentials, ceramic tubes wound with nichrome or Kanthal wire are good. The winding spacing may be varied to produce a variety of temperature gradients.

For rough temperature control a variable transformer, such as "Variac," may be used to regulate the furnace. More sensitive control can be obtained with an on-off type controller, and a further improvement can be obtained by the use of saturable core reactors.

Temperature measurements are usually made by strapping chromel-alumel thermocouples to the outside of the vessel. For internal measurements sheathed thermocouples of small heat capacity have been used. Gradients through the walls of the vessel are surprisingly small (of the order of a few degrees). Figure 6 shows a typical furnace.

PRESSURE MEASUREMENT. The pressure generated in the vessel can be measured by a Bourdon gauge or by a strain-gauge cell. In most laboratory experiments, however, the pressure is estimated from the degree of filling of the free volume in the vessel and the temperature of operation.

Kennedy's (1950) *P-V-T* data for pure water are shown in Fig. 7. Measurements made in the $Na_2O—SiO_2—H_2O$ system (used to grow quartz) indicate a depression from the pure water system of about 15%. The degree of pressure lowering at any given temperature as compared to a pure water system will depend on the mineralizer* concentration and the solubility of the solute. Greater pressure depression occurs for the higher solute and mineralizer concentrations. Figure 8 compares a *P-V-T* curve for the $Na_2O—SiO_2—H_2O$ system with Kennedy's data for pure water.

Safety

The obvious safety hazards associated with hydrothermal crystallization and high pressure equipment can be minimized by reasonable measures.

The entire furnace assembly may be placed in a pit below ground, or a thick-walled barricade can be constructed to deflect possible autoclave fragmentation. Rupture disks should be incorporated in the vessel. They are available in a wide range of bursting pressures. Even for low-pressure, low-temperature experiments, minimum precautions should at least include shielding of personnel from live steam or hot solution in case a vessel fails.

* The term *mineralizer* refers to any component added to the water solution without which crystallization would either not occur or would be extremely slow. The mineralizer most often acts to increase the solubility of the solute by the formation of species not ordinarily present in the water. Very often these new species are formed by complexing with the mineralizer. The increased solubility permits increased supersaturation without spontaneous nucleation and consequently allows more rapid growth rates. In some cases the mineralizer may act as a complexing agent to reduce the molecular size of the species in solution, thereby speeding up the crystallization process.

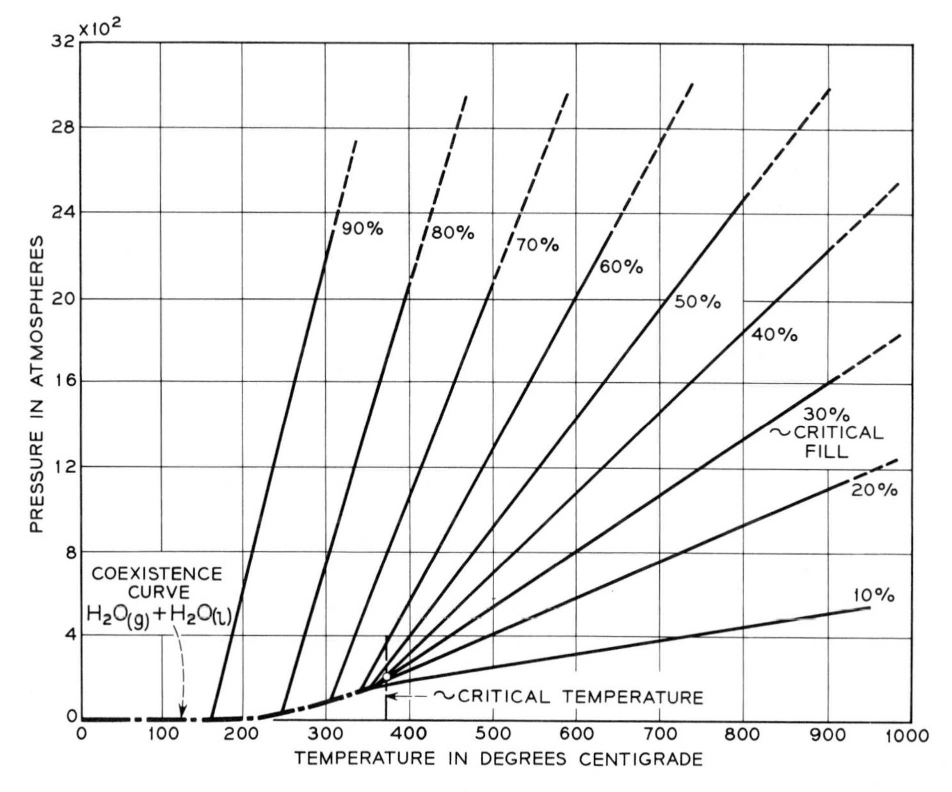

Fig. 7. Pressure-temperature curves for water at constant volume. (After Kennedy, 1950.)

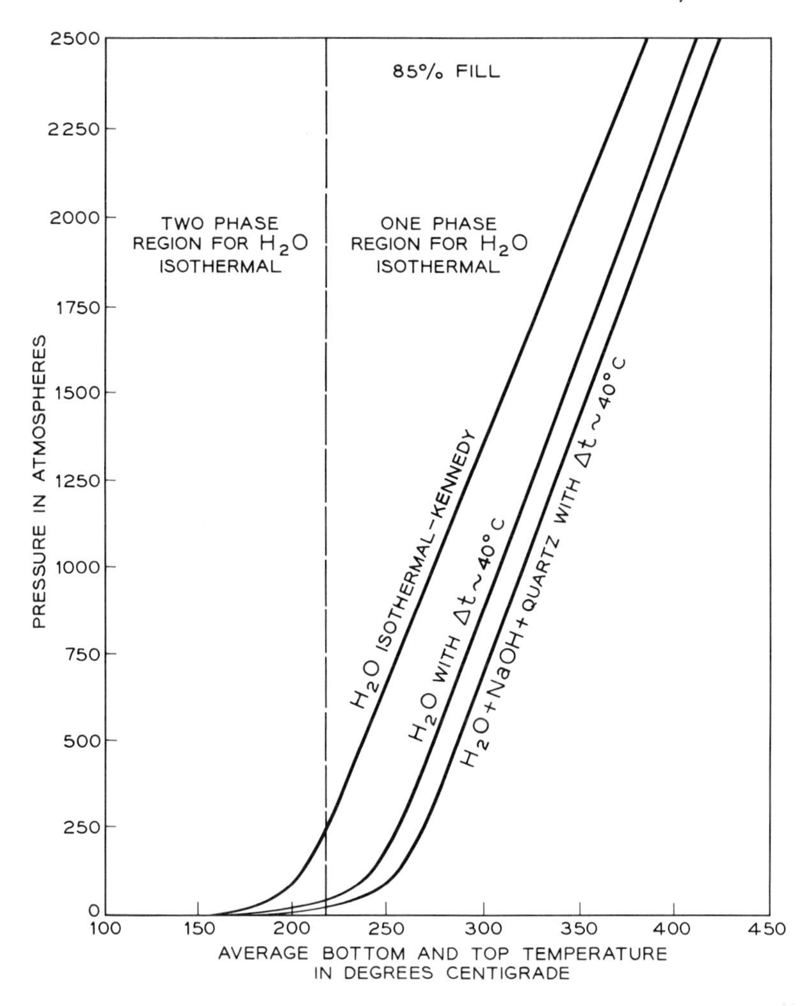

Fig. 8. Pressure-temperature data for three aqueous systems.

3. GROWTH OF QUARTZ

General Considerations

Hydrothermal growth of quartz is similar in many respects to crystal growth from aqueous solutions at room temperature. Nutrient quartz is placed in the hotter bottom or dissolving zone of the vessel, and a frame holding the seed crystals is placed in the upper, or crystallizing, zone. A perforated metal disk, or baffle, separates the two zones into two nearly isothermal regions and promotes the growth of crystals of uniform dimensions throughout the upper zone. After the vessel is charged with solvent to the per cent of fill desired, it is closed and placed in a furnace that gives the desired temperature and temperature differential.

As the temperature of the vessel approaches operating conditions, the nutrient quartz begins to dissolve and saturate the solution. The autoclave (Fig. 9) is cooler at the top, causing the solution there to become supersaturated. The seed plates begin to grow as the supersaturated solution deposits the solid phase. The replenishment required for continuous growth is accomplished by convection currents caused by the temperature differential (Δt). It is influenced by the baffle arrangement, as it transports newly saturated solution to the growing zone and depleted solution away. This continuous cycle of solution and deposition permits the growth of large crystals. Typical operating conditions are:

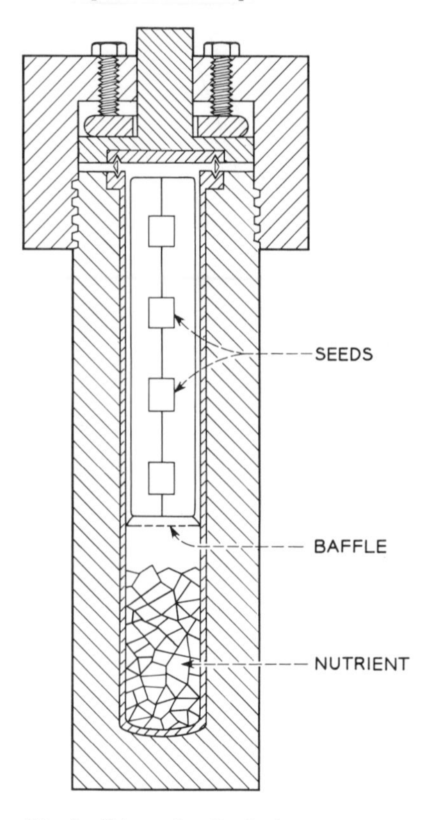

Fig. 9. Schematic of hydrothermal system.

Dissolving temperature	400°C
Crystallizing temperature	360°C
Δt	40°C
Degree of fill	80%
Solution	1.0 M NaOH
Baffle opening	5%
Pressure	21,000 psi

With these conditions, single crystals of quartz have been grown in production vessels weighing up to about 800 grams (Western Electric Company). Figure 10 shows examples.

Impurity Additions

Most procedures for growing quartz use a sodium hydroxide or a sodium carbonate solution as the solvent. Since these strong bases are corrosive under hydrothermal conditions, it is surprising that clear, relatively pure quartz can be grown in steel autoclaves. This, plus the usual occurrence of high purity quartz in nature, attests to the fact that the quartz structure is not very receptive to the inclusion of impurities.

Table 1 gives a list of the usual impurities (Babusci, unpublished work) found in natural and synthetic quartz. The most noticeable difference is in the sodium content that is slightly higher in the synthetic samples.

The effects that impurities may have on acoustic, electrical, and optical properties have led to deliberate attempts to include foreign ions in synthetic quartz. It appears (Stanley and Theokritoff, 1956; Hammond, 1955; Ballman and Flaschen, unpublished work) that impurities which enter the quartz lattice most easily by direct lattice substitution are those ions that possess ionic radii and valence like Si^{4+}. Because of the relatively small ionic radius of silicon, few ions meet the requirements. Substitutions can occur, however, when the ionic radius of the foreign ion is similar to silicon, but the charge is different. In such cases a

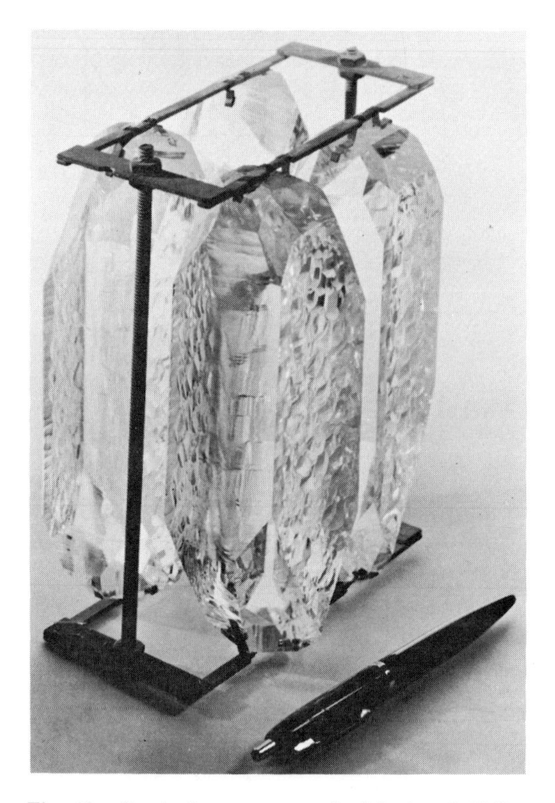

Fig. 10. Synthetic quartz crystals, Merrimack Valley Works, Western Electric Co., North Andover, Massachusetts.

TABLE 1

IMPURITIES IN NATURAL AND SYNTHETIC QUARTZ

Impurity	Brush Synthetic,* %	BTL Synthetic, %	Volcanic Quartz, %	Rose Quartz	Smoky Quartz
Ag	~0.0001	0.0001	0.0001	0.0001	—
Al	~0.003	~0.01	>0.01	>0.01	~0.01
Ca	~0.0003	~0.001	~0.01	0.003	~0.003
Cr	~0.0003	~0.0003	~0.003	—	~0.0003
Cu	~0.001	~0.003	~0.0001	~0.0001	~0.01
Fe	~0.003	~0.001	~0.01	>0.01	~0.001
Mg	<0.001	<0.001	~0.01	~0.001	~0.001
Ti	—	—	~0.003	~0.003	—
Pb	—	~0.003	~0.003	—	—
Na	—	0.003	0.0005	0.02	—

* Clevite Corporation, Cleveland, Ohio. (—)—means not determined.

carrier ion that electrically balances the charge has usually been added along with the ion to be substituted.

Impurities can also be added interstitially. The rather large open channels in the quartz structure are generally considered interstitial sites and appear to permit large ions to be incorporated.

It has been shown (Stanley and Theokritoff, 1956; Hammond, 1955; Ballman and Flaschen, unpublished work) that Ge^{4+}, Sn^{4+}, B^{3+}, Al^{3+}, Ti^{4+}, Pb^{++}, As^{5+}, Li^{+}, Ag^{+}, and Zr^{4+} can be added to synthetic quartz in varying amounts. Of these Al^{3+} and Ge^{4+} appear to enter the lattice substitutionally and the others interstitially.

The crystallographic orientation of the seed on which quartz is grown in the

Fig. 11. X-ray darkened quartz crystal. (After D. L. Wood, in press.)

Fig. 12. Green quartz showing color zoning. (After Ballman.)

presence of impurity ions greatly influences the concentrations of impurity ions that become included. Several investigators (Hammond, 1955; Wood, in press) have shown this directional dependence by x-irradiation, where the intensity of x-ray darkening depends on the impurity concentration. Figure 11 shows the effect of x-irradiation on samples of synthetic quartz where the impurities that cause darkening are Li^+ and Al^{3+}.

A green and a yellow color can be produced by the addition of iron to quartz (Tsinober et al., 1959; Ballman, 1961). The color is arranged in light and dark bands as shown in Fig. 12. Green is thought to indicate included ferrous ions, and yellow to indicate ferric ions.

The effect of impurities on growth rates has not been determined. Preliminary investigations indicate that both the growth rate and the crystallographic perfection decrease, as the impurity level increases.

4. CRITERIA FOR HYDROTHERMAL CRYSTAL GROWTH

Thermodynamic

To begin to grow any material under hydrothermal conditions, one needs to determine whether it is the thermodynamically stable form at the contemplated experimental conditions. Phase equilibria are most often studied in vessels with *cone closures* as described by Roy (Roy and Tuttle, 1956) and his co-workers. Many of the phase diagrams have been compiled in *Phase Diagrams for Ceramists* (Levin et al., 1956). It should be remembered, however, that the addition of an acid or an alkali as a mineralizer, even in relatively low concentrations, has been known to shift the equilibrium temperature for a phase transi-

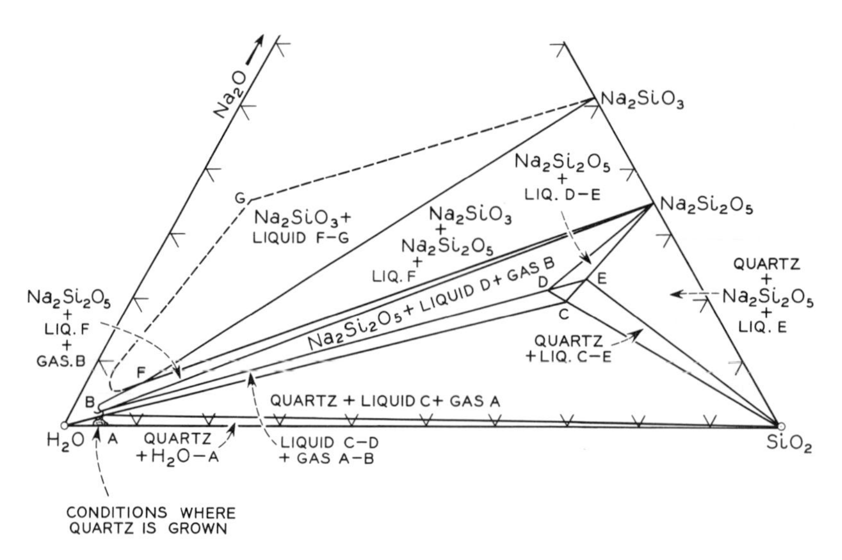

Fig. 13. Part of the system H_2O-SiO_2-Na_2O. (After Morey, 1937.)

tion by as much as 50°C (Laubengayer and Weitz, 1943).

Alpha quartz is the useful piezoelectric modification of silica, but is stable only below 573°C. Only under hydrothermal conditions in the presence of a base does silica have enough solubility to permit rapid crystallization below 573°C. Figure 13 shows the $Na_2O—SiO_2—H_2O$ phase diagram in detail. It appears from most studies that a reasonable solubility for hydrothermal crystallization is ~ 2 to 5 weight per cent. Solubility data for hydrothermal systems are rather sparse, but fortunately they are available for the $Na_2O—SiO_2—H_2O$ (Morey and Hesselgesser, 1952; Laudise and Ballman, 1961) and $SiO_2—H_2O$ (Kennedy, 1950) systems.

A weight loss method for obtaining solubility data lends itself quite well to hydrothermal systems. Weighed samples of the material to be studied are placed in the hydrothermal vessel and held isothermally. A series of weight loss measurements can be made at various temperatures, pressures, and degrees of filling of the vessel.

When two fluid phases are present, solubility determinations can be obtained by a sampling technique (Morey and Hesselgesser, 1952). A sample is withdrawn from the hydrothermal vessel through a suitable valve. If the sample is small enough, the resulting pressure or density changes are not significant.

Solubility data are most helpful in showing particular regions where crystal growth is easiest. Regions where the temperature coefficient of solubility shows a steep slope are ones in which large supersaturations occur for a given temperature gradient. They are usually the best suited for crystallization. The temperature coefficient may either have a positive or a negative value, with growth being carried out for both conditions. The solubilities for silica in water showing regions of both a positive and a negative temperature coefficient are shown in Fig. 14.

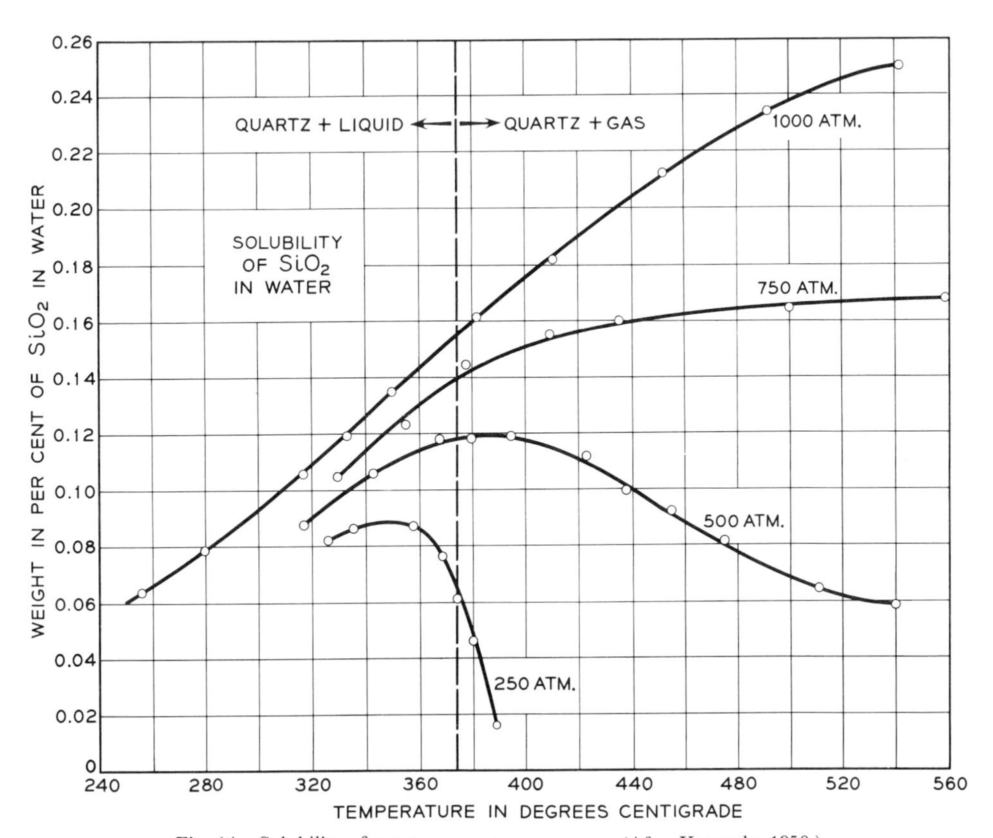

Fig. 14. Solubility of quartz versus temperature. (After Kennedy, 1950.)

Kinetics

A large body of information has been gathered concerning the factors that influence the rate of crystallization of α-quartz. This is one of the few hydrothermal systems in which rate studies have been made, so we shall rely heavily on this information not only to describe the quartz case but also to infer the rate behavior of other systems.

Parameters that influence the rate are the following,

1. Crystallization temperature and temperature difference between the dissolving and the growing zone.
2. Per cent of fill or density.
3. Mineralizer concentration.
4. Seed orientation and the ratio of the seed surface area to the nutrient surface area.

CRYSTALLIZATION TEMPERATURE. Provided that Δt remains constant, the logarithm of the growth rate is nearly proportional to the reciprocal of the crystallization temperature (°K), as shown in Fig. 15.

Figure 16 shows that the rate depends on the temperature difference nearly linearly.* The effective Δt in a given vessel, however, depends not only on the radial and longitudinal heat losses but also on the per cent opening of the baffle. Even though the external Δt is identical in two runs, the internal Δt may be different if the per cent of open baffle is not identical. Figure 17 presents a plot of the rate versus the per cent of open baffle, and shows an increase

* The term *nearly linear* as used here describes conditions under which quartz is usually grown. Rather sharp departures from linearity do occur in particular regions of high temperature and pressure.

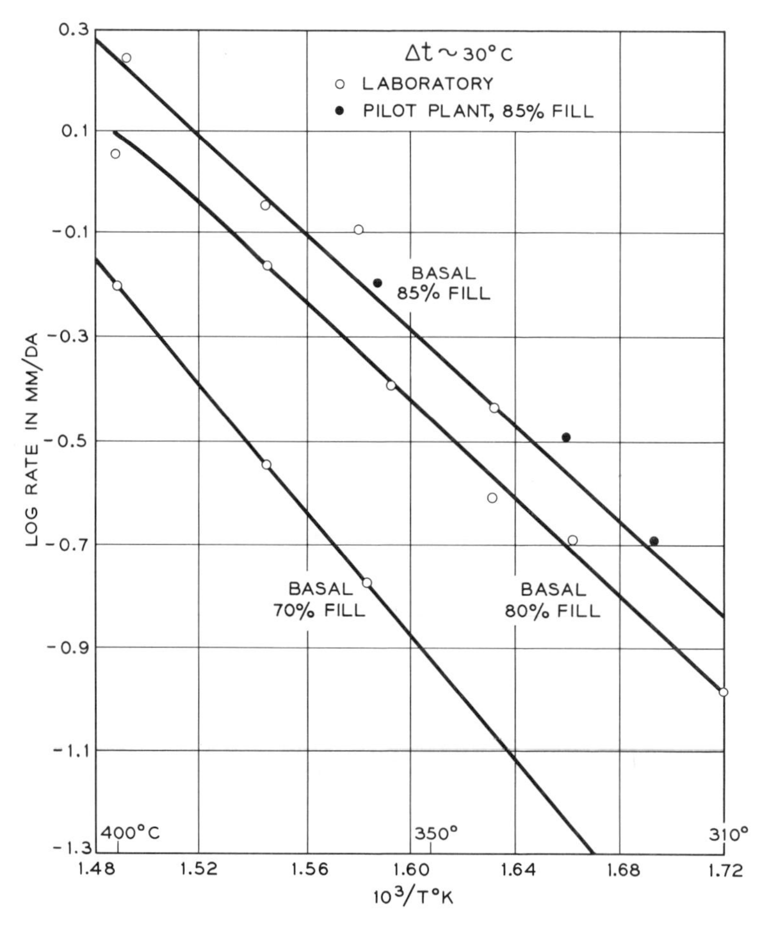

Fig. 15. Log rate versus $1/T$.

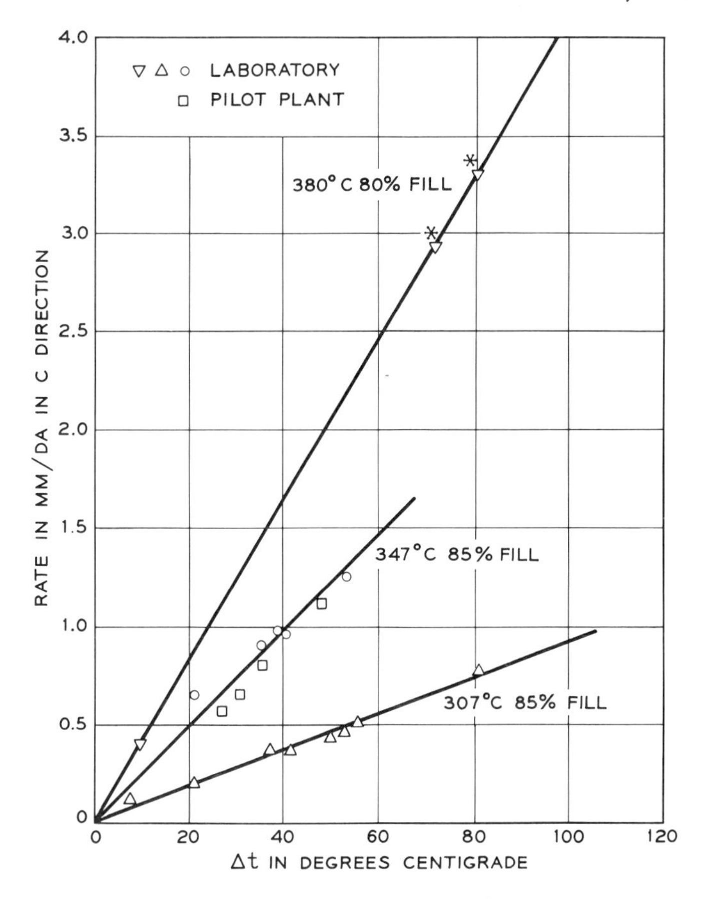

Fig. 16. Rate as a function of Δt.

in rate with decreasing baffle opening. If the rates shown are corrected for the perturbation of Δt caused by the baffle, there is no change in rate as the baffle opening changes. The effect of the baffle is to separate the vessel into two nearly isothermal regions, one for dissolving and the other for growing. The nearly isothermal character of the growing zone is particularly noticeable in production vessels where the growing zones are ~10 ft long, and yet crystals grow at the two ends with equal rates.

PER CENT FILL OR DENSITY. The dependence of the growth rate on the degree of fill (not exceeding ~80%) is nearly linear for NaOH and Na_2CO_3 systems (Fig. 18). The growth rate departs markedly from linearity when fills above 80% are used. For fills from 82 to 87% in NaOH solution at 400°C, small changes in per cent fill cause rather large increases in growth rate. Of particular interest here is the observation that as the fill is increased in this region, resulting in large increases in growth rate, the visual perfection of the grown

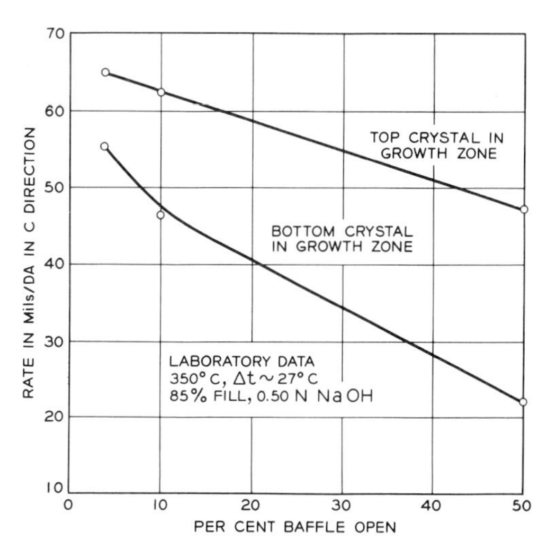

Fig. 17. Rate as a function of baffle opening.

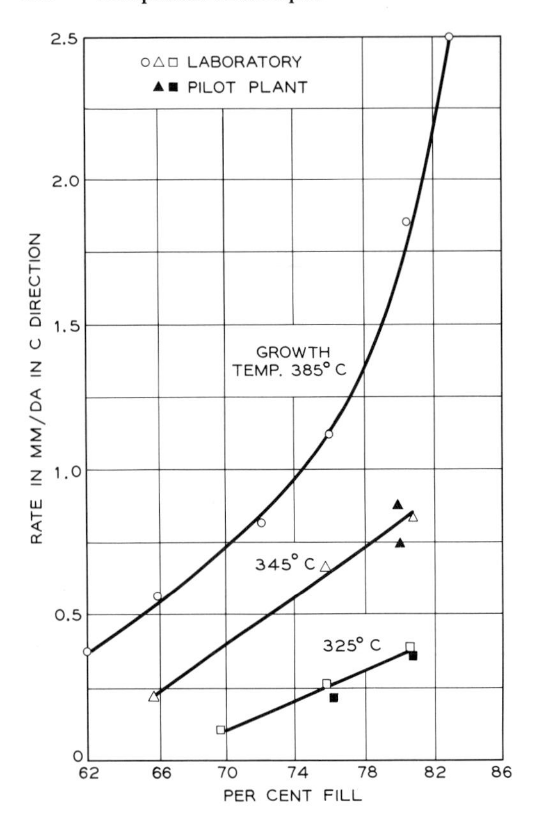

Fig. 18. Rate as a function of density.

crystals improves. In most crystallization systems an increase in rate results in a decrease in perfection.

SOLVENTS AND SOLVENT CONCENTRATION. Perhaps the least known factor is the way solvents influence the growth rate. In pure aqueous solutions (even at 400°C and 25,000 psi.) the solubility of quartz is too low to allow growth to take place in any reasonable time. Alkaline additions, such as NaOH, Na_2CO_3, KOH, and K_2CO_3, are all effective as mineralizers in this pressure and temperature range. Small increases in molarity result in only slight increases in the growth rate, whereas large increases begin to produce additional phases along with quartz. The minimum molarity for good growth is about 0.25 *M* for NaOH. Concentrations about 4.0 *M* for NaOH and 2.0 *M* for KOH form sodium or potassium silicates along with α-quartz.

SEED ORIENTATION. An outstanding example of how the anisotropic character of a material influences a rate process is seen in the dependence of the growth rate on the crystallographic direction. Figure 19 shows this effect in Na_2CO_3 solutions and Fig. 20 shows it in NaOH solution. The ranking of rates for some of the more common faces are

Basal plane* (0001) > minor rhombohedral $(01\bar{1}1)$ > major rhombohedral $(1\bar{1}01)$ > prism $(10\bar{1}0)$

It should be noticeable from the rate versus orientation plots that if the initial experiments on quartz synthesis had been carried out using the prism direction $\langle 10\bar{1}0\rangle$ for seed plates, it would have appeared that quartz grew very little in the temperature and pressure regions studied. Hence, it is necessary to try various orientations when growing a new material.

There appears to be little change of the rate as the surface area of the nutrient is varied, provided the nutrient area exceeds some minimum value. For quartz no effect is observed as long as the ratio of the surface area available for dissolving to the surface area available for growth is greater than five.

From solubility data (Laudise and Ballman, 1961) for quartz, it is possible to calculate ΔS, the supersaturation for a given growth rate. These data show (Laudise, 1954) that the growth rate obeys the empirical equation

$$R = k\alpha\,\Delta S \qquad (1)$$

where R is the growth rate in a particular crystallographic direction, k is a velocity constant for that direction, ΔS is the supersaturation, and α a dimensional conversion constant.

The dependence of the rate on temperature can be explained as the variation of the velocity constant with temperature.

* Although not a naturally occurring face in quartz, the basal plane is most often used in quartz synthesis because of its fast growth rate.

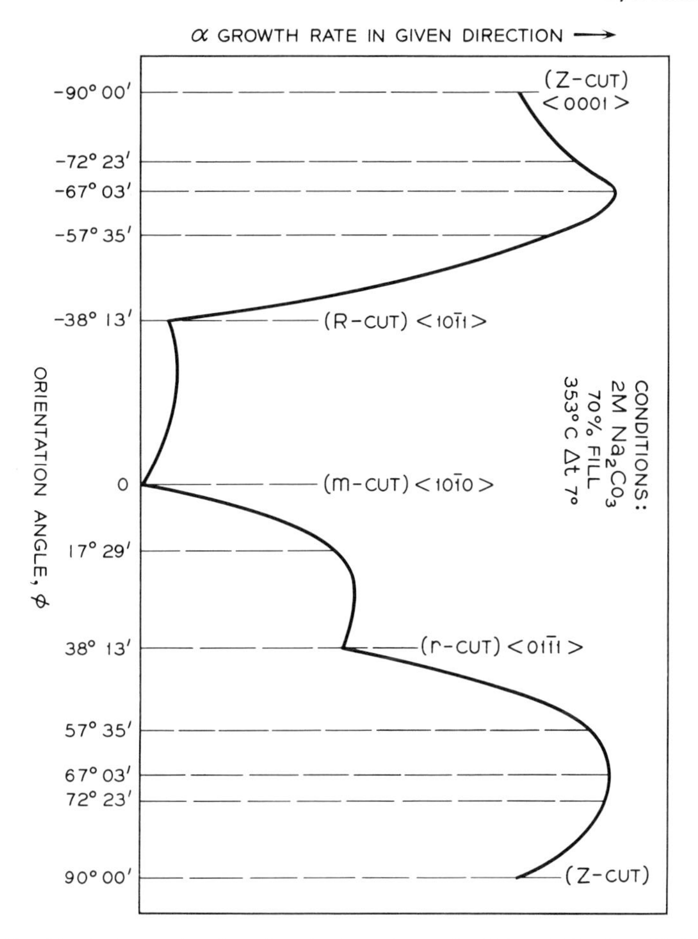

Fig. 19. Rate as a function of seed orientation in Na_2CO_3. (After Jost, 1955.)

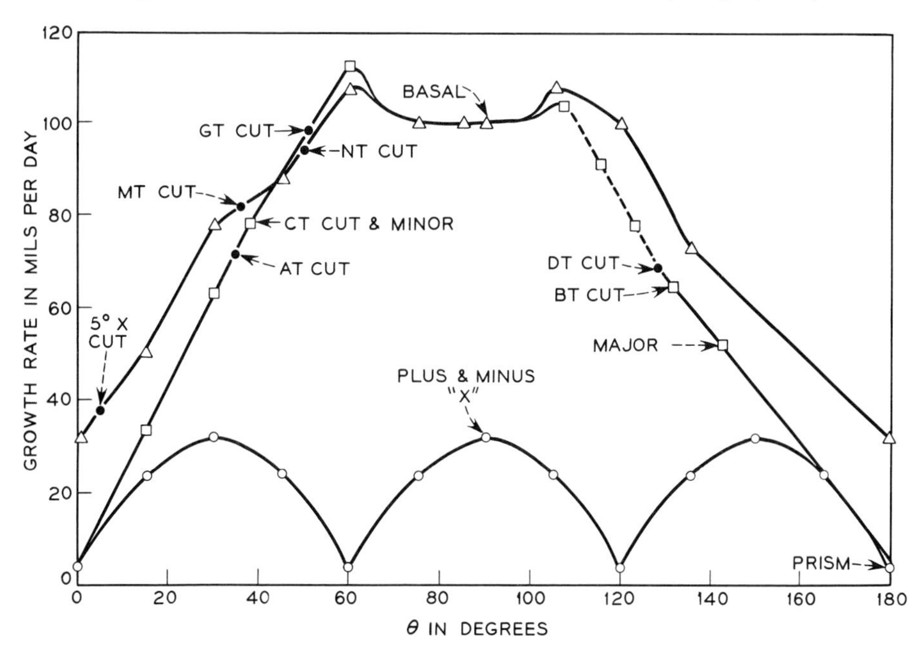

Fig. 20. Rate as a function of seed orientation in NaOH. (After Ballman, unpublished work.)

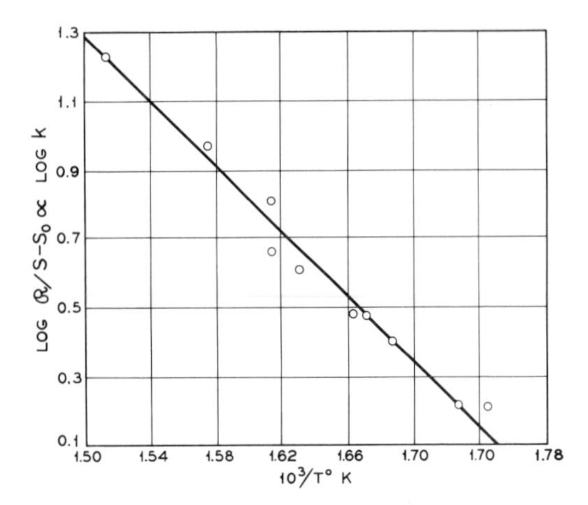

Fig. 21. Log of rate constant versus $1/T$.

Similarly, the effect of Δt and fill on rate can be explained in terms of variations of ΔS. The temperature dependence of the velocity constant obeys the Arrhenius equation (Fig. 21). The energy of activation is directionally dependent, with a typical ΔE being 20 ± 1 kcal/mole on the $\langle 0001 \rangle$ direction.

Since ΔS is to a first approximation linear with Δt, it is not surprising that $\ln R$ is linear with $1/T$ as shown by the data of Fig. 15 (Laudise, 1954), where Δt was held constant. It should be realized, of course, that a simple empirical rate-equation such as (1) does not require a simple reaction mechanism.

From the dependence of growth rate on the above parameters one can make some inferences concerning the growth process.

Dissolution and Transport. Dissolution of the nutrient quartz is not rate limiting when its area is five or more times larger than the growth area since further increases in nutrient surface do not increase the rate. Often the surface area ratio is many times larger than five.

Transport of the dissolved species to the growth zone is done by convection currents whose velocity depends largely on the Δt in the vessel.

Changing the baffle changes the rate of transport, but transport is not rate limiting because no effect on rate is observed after compensation for temperature changes has been made.

Diffusion. In the immediate vicinity of the crystal-liquid interface the solution is depleted as solute is deposited on the seed plate. The depleted region is replenished via diffusion from the surrounding solution. The rate of diffusion could limit the rate. However, diffusion alone could not account for the observed rates since they are anisotropic. Furthermore, the activation energies are rather high for diffusion.

Adsorption and Chemisorption. It is quite possible that the adsorption or chemisorption of the dissolved species onto the seed plate is a rate-limiting step.

Two-Dimensional Diffusion. The chemisorped species may need to migrate to reach its place of permanent bonding. Thus surface diffusion to kinks in steps could limit the rate.

Since both dissolution and transport have already been shown not to be rate limiting, it would appear that the important step occurs in the immediate vicinity of the crystal-liquid interface. This means that bulk diffusion, chemisorption, and/or surface migration are responsible for observed growth rates. Thus hydrothermal crystallization may be limited by consecutive steps of comparable velocity.

5. GROWTH OF OTHER MATERIALS

Hydrothermal techniques for other materials have benefited greatly from the abundance of information about the quartz system. Thermodynamic and kinetic considerations that apply to the quartz system apply also in many other systems, and the rate determining steps are similar as well. On the other hand, solubility and its temperature coefficient will vary, depending on the material. Similarly, the solvent, fill, temperature, and Δt required for reasonable perfection and rates will vary.

A few systems that have been studied in some detail are discussed, and a list is given of systems in which crystallization is known to have occurred either on a seed plate or spontaneously.

α-Al_2O_3

The Al_2O_3—H_2O—NaOH, and Al_2O_3—H_2O—Na_2CO_3 system have been investigated, and the region of phase stability for α-Al_2O_3 has been reported (Laudise and Ballman, 1958). Growth experiments confirm phase equilibria studies in the Al_2O_3—H_2O system previously reported in the literature (Laubengayer and Weitz, 1943), that α-Al_2O_3 (sapphire) is the thermodynamically stable phase in alkaline solutions above ~400°C. Below this temperature diaspore is the stable phase. Figure 22 compares the Al_2O_3—H_2O system with the Al_2O_3—H_2O—NaOH, and Al_2O_3—H_2O—Na_2CO_3 systems.

The solubility of Al_2O_3 in 0.5 *M* NaOH and 1.0 *M* Na_2CO_3 at approximately 405°C and 20,000 psi is 0.2 *M* and 0.25 *M* respectively (Ballman, unpublished work). In both cases increases in temperature and pressure increase the solubility. The increase in supersaturation is much more pronounced, however, in the carbonate solution, and thus with a given temperature differential one expects faster growth in it. A set of conditions for the succesful growth of Al_2O_3 onto seed plates is

Solvent	1.0 *M* Na_2O_3
Crystallization temperature	405°C
Dissolving temperature	435°C
Per cent fill	80

For these conditions crystals have been grown on (0001) seed plates at rates of 0.25 mm per day. A complete study of

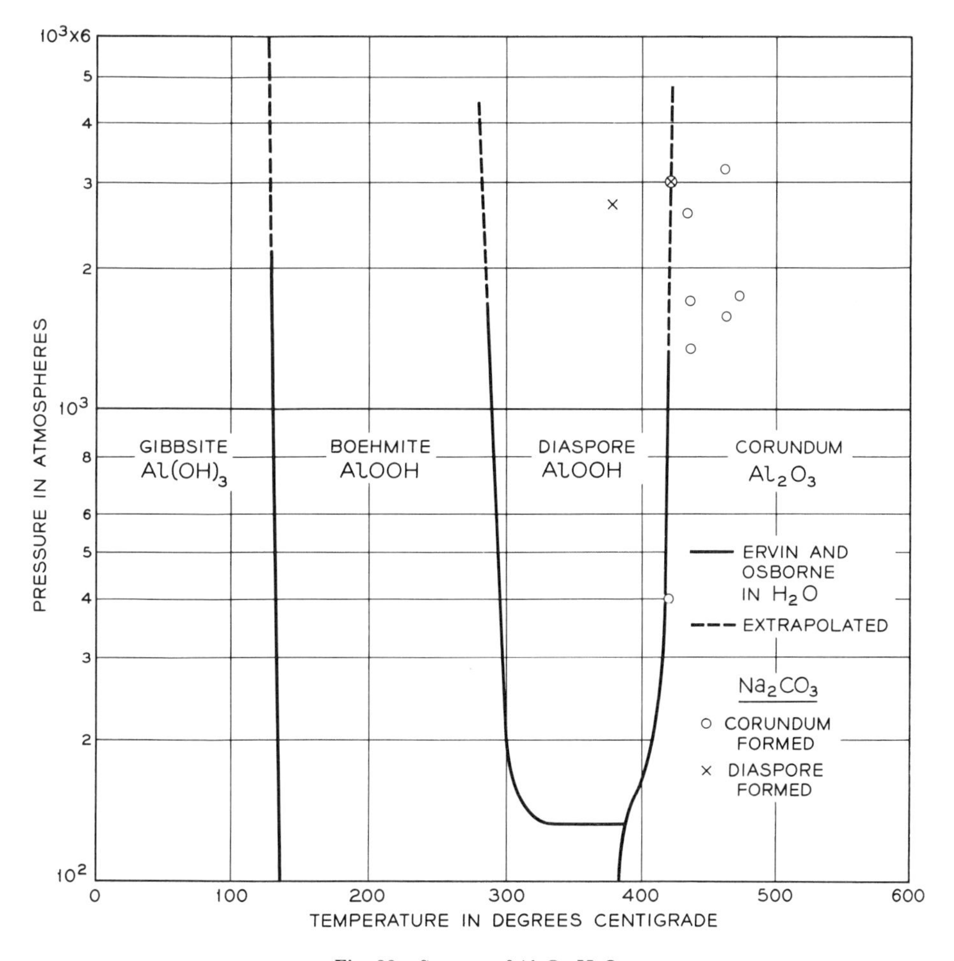

Fig. 22. Systems of Al_2O_3-H_2O.

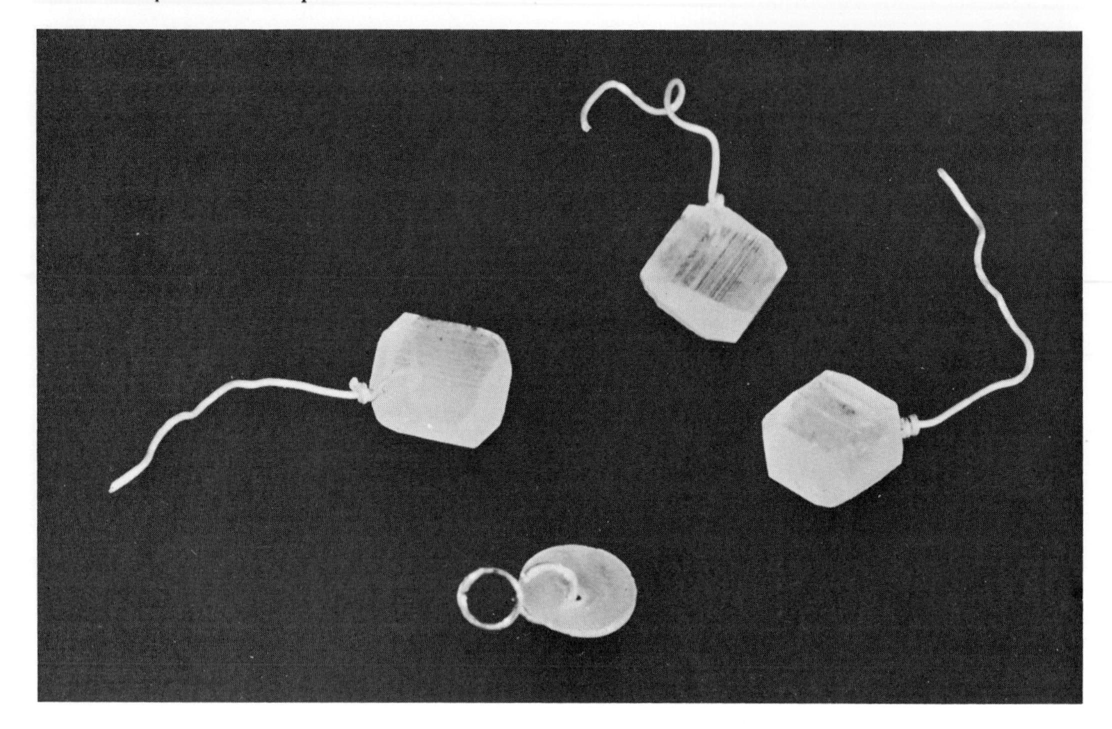

Fig. 23. Hydrothermally grown sapphire.

growth-rate anisotropy has not been made, but it appears that growth is not particularly fast on the (0001) plane.

Small amounts of sodium dichromate added to the solvent color the grown crystals a deep red or ruby. Examples of hydrothermally grown sapphire and ruby are shown in Fig. 23.

A high pressure vessel inert to corrosion by the solvent would be particularly helpful in making the hydrothermal growth of sapphire as economical and practical as the growth of quartz.

Magnetite (Fe_3O_4)

The growth of magnetite on seed crystals has been reported in the literature (Koenig). The best solvent is ammonium chloride, and it is suggested that transport occurs via the halide acting as a mineralizer. Typical conditions are

Solvent	Ammonium chloride 0.5 *M*
Crystallization temperature	515°C
Dissolution temperature	530°C
Per cent fill	50

In almost all experiments excessive nucleation occurred on the autoclave walls, and although growth was rapid initially, it slowed down rapidly after a few days. The rate appears to be affected by the Fe^{3+}/Fe^{2+} ratio. As the ferrous ion increases the rate decreases.

Corrosion was particularly bad, and stainless steel vessels were the only type in which growth on seeds took place. Attack of the vessel walls usually resulted in a layer of magnetite being formed at the site of attack. Thus large areas for crystal growth became available at the expense of the seed plate.

$NiFe_2O_4$

Nickel ferrite has been synthesized, using a diffusion process instead of the usual temperature gradient process (Koenig). The autoclave was mounted horizontally and heated to produce isothermal conditions. NiO was placed at one end of a silver tube within the vessel and Fe_2O_3 at the other. As dissolution took place, the solution (saturated with the components) diffused to the area containing the seed crystals and caused them to grow. Conditions were

Solvent	0.5 $\underline{N}$ NH_4Cl
Nutrient	NiO and Fe_2O_3
Crystallization temperature	475°C
Per cent fill	70–75

Growth was initially rapid and then soon declined. As in the magnetite case, it is probable that reduction of the Fe^{3+} was responsible for the decline in rate. This is supported by the fact that hydrogen was present in the vessel at the end of the experiment.

$AlPO_4$ and $AlAsO_4$

Hydrothermal growth of aluminum phosphate and aluminum arsenate has been reported (Stanley, 1954). These two materials are of particular interest because their temperature coefficients of solubility are negative. Growth is caused in a region of "retrograde solubility," either by slowly raising the temperature of the solution or by a reverse temperature gradient. In either case, the seed plates are suspended in the hot zone of the vessel where the solution is supersaturated.

The piezoelectric modification of $AlPO_4$ is obtainable only in the temperature range 132 to 315°C, so the hydrothermal technique may be the only one available for this material. Experiments with $AlPO_4$ were done in sealed borosilicate-glass vessels placed in low pressure autoclaves. Pressure balancing was used to avoid rupture of the glass vessels at the operating conditions.

Crystals of $AlPO_4$ weighing up to 80 grams were grown by slowly raising the temperature of a solution of phosphoric acid saturated with sodium aluminate. Growth ceased in a few days, so the solution had to be replenished and the cycle repeated.

Growth of $AlAsO_4$ was obtained by forming a temperature gradient along the length of the vessel and placing the seeds in the hot zone. The temperature was not varied with time so that continuous crystallization occurred. Crystals weighing ~14 grams were grown (Fig. 24).

ZnO and ZnS

Crystals of ZnO and ZnS have been grown, and the results of experiments in the ZnO—H_2O—NaOH and ZnO—H_2O—NaOH systems have been reported (Laudise and Ballman, 1960).

Fig. 24. $AlPO_4$ crystals grown hydrothermally. (After Stanley, 1954.)

Both ZnO and ZnS have been grown by a method similar to that used for quartz and Al_2O_3. Both materials tend to nucleate crystals spontaneously at quite low degrees of supersaturation, making controlled deposition onto a seed plate somewhat difficult. Crystals of ZnO have been grown, however, that are quite free of visual imperfections. The color of ZnO is particularly affected by iron. Even slight traces cause a rather deep green coloration.

Although ZnO is the thermodynamically stable phase in the ZnO—H_2O—NaOH system (Huttig and Möldner, 1933) at temperatures above 35 °C, appreciable growth does not occur at temperatures much below 350°C. A successful temperature and pressure region is:

Solvent	1.0 *M* NaOH
Crystallization temperature	400°C
Dissolution temperature	410°C
Per cent fill	80

Growth rates are about 0.25 mm/day in the fastest directions studied. No rate studies have been made for ZnS although the rates are quite similar to those for ZnO.

Of particular interest is the fact that even in a 10 *M* NaOH solution ZnS does not hydrolyze. One might have expected ZnO to be the stable phase under these conditions.

$Y_3Fe_5O_{12}$ and $Y_3Ga_5O_{12}$

Yttrium iron garnet and yttrium gallium garnet have both been crystallized on seed plates in alkaline solutions (Laudise et al., 1961; Laudise and Kolb, 1962). In both cases, a temperature gradient method was used with growth taking place in the cooler zone.

Yttrium gallium garnet saturates congruently and is the stable phase over the temperature range 400 to 500°C at 1000 to 3000 atmospheres.

Yttrium iron garnet saturates congruently in the temperature range from 400 to 750°C, but decomposes to yttrium orthoferrite ($YFeO_3$) at the higher temperatures as reaction times become lengthy (15 days or more).

CdS and PbS

Cadmium sulfide and lead sulfide have been crystallized from sodium sulfide and ammonium polysulfide solutions, resulting in small crystals (Nielsen and Kolb, unpublished work). The use of alkaline solutions causes partial hydrolysis with the production of the respective oxides.

CdO and PbO

Rather large spontaneously nucleated crystals of these substances have been obtained from NaOH solutions (Kolb, unpublished work). No rate information or oriented growth has been reported.

Tourmaline

Synthesis and crystal growth of tourmaline on seed plates have been reported (Smith, 1949). Tourmaline, $R_9Al_3(BOH)_2Si_4O_{19}$, is the stable phase in the temperature region 400 to 500°C at a pressure of ~1000 atmospheres. No rate studies were reported.

ACKNOWLEDGMENTS

The authors are pleased to thank G. T. Kohman of the Bell Telephone Laboratories for his many ideas, and for stimulating their interest in hydrothermal crystal growth.

References

Babusci, D., Unpublished Work.

Ballman, A. A., Unpublished Work; (1961), *Amer. Miner.* **46,** 439.

Ballman, A. A., and S. S. Flaschen, Unpublished Work.

Bridgman, P. W. (1914), *Proc. Am. Acad. Arts and Sci.* **49,** 625.

DeSenarmount, H. (1851), *Ann. Chem. et Phys.* **32,** 129.

Ellis, A. J., and W. S. Fyfe (1957), *Review of Pure and Appl. Chem.* **7,** 261.

Erwin, G., and E. Osborn (1951), *J. Geol.* **59,** (4), 387.

Hammond, D. L. (1955), *Ninth Annual Frequency Control Symposium,* U. S. Army Signal Corps.

Hüttig, G. F., and H. Möldner (1933), *Z. Anorg. Chem.* **211,** 368.

Jost, J. M. (August 1955), *1st Quarterly Progress Report,* U. S. Army Signal Corps, Contract DA-36-039-sc-64689.

Kennedy, G. C. (1950), *Am. J. Sci.* **248,** 540.

Kennedy, G. C. (1950), *Econ. Geol.* **45,** 7, 639.

Koenig, J., *Final Report,* NUAFRC-TR-S7-190 Contract No. AF-19-1419, Air Force Cambridge Research Center.

Kolb, E. D., Unpublished Work.

Laubengayer, A. A., and R. S. Weitz (1943), *J. Am. Chem. Soc.* **65,** 250.

Laudise, R. A. (1961), *Progress in Inorganic Chemistry,* Vol. 3, edited by F. A. Cotton, Interscience, New York; (1954), *J. Am. Chem. Soc.* **81,** 562.

Laudise, R. A., and A. A. Ballman (1958), *J. Am. Chem. Soc.* **80,** 2655; (1960), *J. Phys. Chem.* **64,** 688; (1961), *J. Phys. Chem.* **65,** 1396.

Laudise, R. A., and E. D. Kolb (1962), *J. Am. Ceram. Soc.,* **45,** 51.

Laudise, R. A., and J. W. Nielsen (1961), *Solid State Physics,* Vol. XII, edited by F. Seitz and D. Turnbull, Academic Press, New York.

Laudise, R. A., J. H. Crocket, and A. A. Ballman (1961), *J. Phys. Chem.* **65,** 359.

Levin, E. M., H. F. McMurdie, and F. P. Hall (1956), *Phase Diagrams for Ceramists,* edited and published by the American Ceramic Society.

Morey, G. W., and J. M. Hesselgesser (1952), *Am. J. Sci.,* Bowen Volume, **367.**

Morey, G. W., and E. Ingerson (1951), *Amer. Miner.* **22,** 1121.

Nielsen, J. W., and E. D. Kolb, Unpublished Work.

Roy, R., and O. F. Tuttle (1956), *Physics and Chemistry of the Earth,* **1,** 138.

Spezia, G. (1905), *Acad. Sci. Torino Atti.,* **40,** 254.

Stanley, J. M. (1954), *Ind. Eng. Chem.* **46,** 1684.

Stanley, J. M., and S. Theokritoff (1956), *Am. Miner.* **41,** 527.

Tsinober, L. I., L. G. Chentsova, and A. A. Shternberg (1959), "The Green and Brown Colors of Synthetic Quartz Crystals," *Growth of Crystals,* edited by A. V. Shubnikov and N. N. Sheftal, Vol. 2, Translated by Consultants Bureau, Inc., New York, p. 45 of translation.

Walker, A. C., and E. Buehler (1950), *Ind. Eng. Chem.* **42,** 1369.

Western Electric Company, Merrimack Valley Works and Sawyer Research Products, Eastlake, Ohio.

Wood, D. L., *J. Phys. and Chem. of Solids,* in press.

14

Molten Salt Solvents

R. A. Laudise

1. GENERAL

Molten or fused inorganic salts and oxides are often powerful solvents for refractory materials. Consequently, they can often serve as media for crystal growth. Such solvents are often called *fluxes* by analogy with soldering, brazing, and other metal-joining processes where the flux dissolves oxide films.

In the European literature crystal growth from a flux is often called *growth from the fluxed melt.* The term *growth* from the melt should perhaps be reserved for growth from a one-component system where no solvent is present. However, some ambiguity exists because when it is desirable to effect doping an appreciable dopant concentration is often in the liquid. If this dopant concentration is high enough, we would ordinarily consider that growth was from a solvent-solute system, and if the solvent is appropriate the growth could be called from the flux. Furthermore, it is somewhat arbitrary to consider growth from molten salts separately from growth from other nonaqueous media, particularly since the inorganic solvents used are not all saltlike. Indeed, in some cases, well-formed crystals have been grown from liquid metal solvents such as gallium and mercury and from more "conventional" nonaqueous solvents such as liquid ammonia and alcohol. Nevertheless, a well-developed field exists with its own techniques and region of applicability which can best be described as crystal growth from molten salt solutions or "flux growth." This is discussed here.

Flux growth is defined as the use of a liquid inorganic compound at elevated temperature as a solvent for crystallization. The technique was probably first exploited in the late nineteenth century by French and German preparative chemists such as Fremy and Ebelman. They prepared crystals of sapphire, magnesium oxide, and other refractory oxides from molten salts.

The technique has been used spasmodically since then, but has only recently begun to be fully investigated. Renewed interest has been stimulated by the widespread need for a variety of high-melting inorganic crystals for solid state research and devices.

For conceptual purposes we briefly outline the flux growth of a typical crystal, such as cobalt ferrite, $CoFe_2O_4$. A saturated solution of $CoFe_2O_4$ in PbO contains 0.525 gram $CoFe_2O_4$/gram PbO at 1270°C. The solubility as a function of temperature is known and also that $CoFe_2O_4$ is the stable, solid phase in the system between 1300 and 900°C. Therefore 200 grams of $CoFe_2O_4$ and 668 grams of PbO are weighed into a platinum crucible that is placed in a furnace and held above 1270°C long enough to insure complete solution. Then the temperature is lowered at the rate of 0.5°/hr to a temperature of 900°C. Next, the crucible is removed from the furnace, allowed to cool, and the grown ferrite crystals are separated by leaching away the solidified flux with hot dilute HNO_3. This is the simplest and most generally used method. It requires a knowledge of (1) the solvent power of various molten salts to allow a rational choice of solvent, (2) the phase relationships in the chosen flux so as to insure that the crystal desired is the only (or at least the principal phase) precipitating, and (3) a knowledge of solubility as a function of temperature so that appropriate cooling rates can be chosen. It is, of course, not necessary that (2) and (3) be known explicitly, since a series of experiments can be performed until good crystals are obtained without knowledge of phase equilibria or of solubility. However, analysis of crystal growth mechanisms and indeed even good control of crystal quality are better achieved with this knowledge.

Appropriate furnaces and temperature controllers for temperatures from 1000 to 1500°C should be available. Inert crucibles capable of sustaining these temperatures should be used. The process just described is extremely crude. Adequate control is better achieved by deliberate seeding of the melt. With this discussion as a background, we now consider molten salt crystallization with a view toward pointing out those areas where more experimental work is appropriate as well as those where substantial progress has already been made.

2. PHASE EQUILIBRIA IN MOLTEN SALTS

The chemistry of fused salts has in recent years been a subject of extensive study. Viscosity, vapor pressure, conductivity, and electromotive force measurements have been made. Thermodynamic treatments, which in the nineteenth century were applied to aqueous solutions, have been used to calculate activities, activity coefficients, transport properties, heats of solution, and other properties of the systems studied. However, for experimental reasons, such measurements have usually been confined to rather low temperatures and to molten salt solvents of weak solvent power for refractory materials. Consequently, many such results are not directly applicable to systems of interest in crystal growth, although they should become increasingly useful in the future. For further information, see recent reviews and bibliographies (Blomgren and Van Artsdalen, 1960; Janz, 1961), and the excellent series of papers presented at the 1961 New York Academy of Science Fused Salt Symposium (*Annals of New York Acad. Sci.*, 1960). The recent Russian literature has been summarized admirably in a series of translations (*Soviet Research in Fused Salts, 1949–1955*, 1956). The best compendium of molten salt phase diagrams is in *Phase Diagrams for Ceramists* (Levin et al., 1956).

Let us consider the advantages and disadvantages of molten salt crystal growth. If possible it would be preferable to grow crystals of a substance A by a simple solid-liquid equilibrium. The advantages would be

1. The growth could be rapid since the rate would be limited only by dissipation of the heat of fusion.
2. The purity of the crystal would be high since no components other than A would, in principal, be present.
3. No detailed knowledge of the system would be required except the melting point of A.

If, however, A melts incongruently, decomposes before it melts, has a high vapor pressure at the melting point, or exists as an undesired polymorph at the melting point, crystallization at a temperature lower than the melting point is required. If A were water soluble, aqueous crystallization at room temperature would suggest itself. However, often crystals with high melting points, chemical stability, and low aqueous solubility are desired for applications. For such crystals hydrothermal crystallization is sometimes advantageous, although the high pressures and complexity of the required equipment often make it unattractive. Molten salt crystallization is often attractive because

1. The required equipment is relatively simple.
2. There are so many inorganic solvents it is almost always possible to find an appropriate solvent.
3. Rates of crystallization are quite rapid for a solvent-solute system.

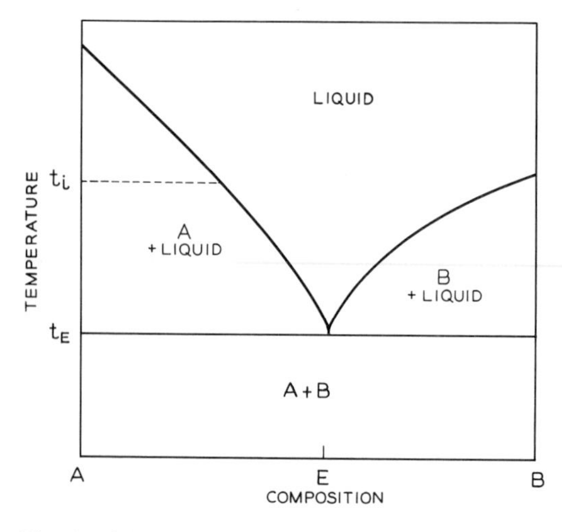

Fig. 1. Schematic phase diagram, congruent melting.

The principal disadvantages are that impurity control is quite difficult and, in contrast to hydrothermal crystallization, the temperatures required are high.

There are several ways of choosing a solvent, but the following has proved useful most often.

Consider the generalized phase diagram of Fig. 1, where A is the material to be crystallized and B is a potential solvent. The determination of phase diagrams and of solubilities in molten salts has been described in several standard works (Levin et al., 1956; Glasstone, 1946). If pure A is desired, A and B must not form a continuous series of solid solutions. Ideally, as shown in Fig. 1, there should be no solid solubility of B in A.

To repress solid solubility, A and B should not be too similar chemically. The solid solubility of B in A is usually decreased if A and B have a common ion. We assume that A is congruently melting, that is, it does not decompose into any other solids before it melts. Then the role of the solvent B can be thought of as lowering the melting point of A in the composition region from A to E, the eutectic. Consequently, since we would usually like to crystallize A at as low a temperature as possible, the eutectic temperature t_E should be low. This suggests, of course, that the melting point of B be as low as possible. To minimize importance of diffusion processes (which will tend to necessitate extremely slow growth rates) it would be advantageous to crystallize A from a melt as rich in A as possible. Assume that t_i is the upper temperature at which crystallization can be carried on; where t_i is set by decomposition, the appearance of an undesired polymorph of A, equipment limitations, etc. It would then be desirable that the melt be as rich in A as possible at t_i. Thus the eutectic composition should lie near A. This is advantageous provided t_e is not increased seriously and that it does not result in the slope of the solubility curve being so steep that nucleation cannot be controlled with the temperature control equipment available.

If A and B form a compound AB that melts congruently and is not appreciably soluble in A, then the phase diagram still resembles Fig. 1 if B is replaced by AB, and the criteria for our choice of B apply instead to the compound AB.

If A melts incongruently and there is still little solid solubility of B in A, we would expect a phase diagram like that of Fig. 2. A can only be crystallized as the primary equilibrium phase between the points α and β from melts containing more B than is contained in A. Very good crystals can be grown from such a system, but since the melt composition differs from the composition of A itself, the possibility of

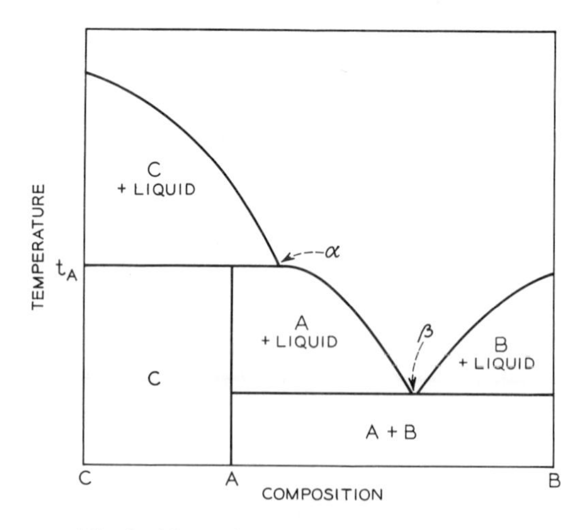

Fig. 2. Phase diagram, incongruent melting.

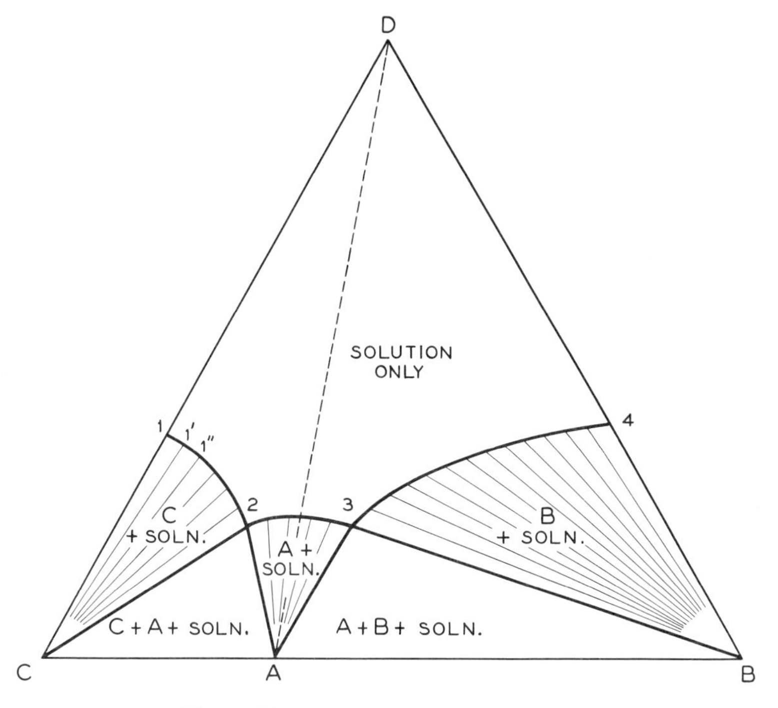

Fig. 3. Phase diagram, congruent saturation.

inclusions of other phases in *A* is increased. Rates of crystallization will probably be more severely diffusion-limited and will need to be slower than for a substance that melts congruently. Consequently, when a substance melts incongruently it is often advantageous to add an additional component to lower the liquidus temperature.

The possible results of adding such an additional component are shown in Fig. 3, where the additional component is *D,* and *A, B,* and *C* are as before. Both Figs. 3 and 4 are cuts at constant temperature *t* through the respective triangular prisms. The criteria involved in the choice of *D* are the same as those applied in the choice of *B* previously, that is, *D* should preferably be low melting and insoluble in *A*. In Fig. 3 the lines *C*-1, *C*-1′, *C*-1″, *C*-2 are tie lines, that is, they connect compositions of coexisting phases. Here, solid *C* with the various compositions of the liquid phase coexists with *C* along the solubility curve for *C* (the line 1-2). In the triangular prism the line 1-2 would, of course, be on the solubility surface at *t*. Similarly, 2-3 is the solubility line for *A* and 3-4 for *B*. Thus in the region 1-2-3-4-*D* there exists only solution and in 1-2-*C* solution plus solid *C,* the composition of the solution being given by the intersection of the appropriate tie line with 1-2. In the region *C*-*A*-2 there exists solution of composition 2 plus solid *C* plus solid *A*. The phases present in the other regions of the phase diagram are as labeled and the regions of Fig. 4 are analogous.

In the systems *CBD* shown in both Fig. 3 and 4 there is no compound formation between *D* and *C* or *D* and *B*. If there were, for our purposes the component *D* could be replaced by the *D*-*C* or *D*-*B* compound and our criteria for the choice of *D* would instead be applicable to the *D* containing compound. For both Figs. 3 and 4 the temperature lies above melting temperature for *D*. This is usual when it is desired to use *D* as a molten salt solvent. In many instances the temperature will also be above the melting point of *B*. Then the region 3-4-*B* will not appear, and the solubility curve for *A* (line 2-3) will intersect the line *A*-*B*. Below the melting temperature of *D* the shape of the region *A*-2-3 which is of

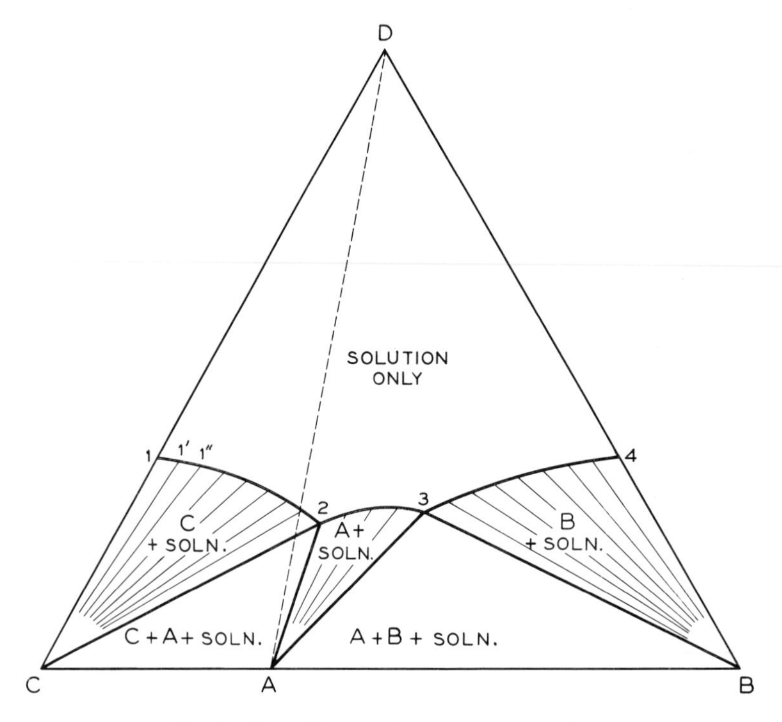

Fig. 4. Phase diagram, incongruent saturation.

principal interest for crystal growth would usually remain like that of Fig. 3 or Fig. 4, but other parts of the diagram would be altered. Ternary phase diagrams of this sort are adequately discussed elsewhere (Ricci, 1951; Findlay et al., 1951; Massig, 1944) and are not considered any further here.

Everywhere along the line *A-D* in both Figs. 3 and 4 the ratio *C/B* is the same as in solid *A*. If the line *A-D* cuts 2-3 (the solubility curve for *A*), *A* can be crystallized from a solution whose composition with regard to the constituents composing *A* is the same as *A*, and *A* is congruently saturating in the solvent *D* as in Fig. 3. If as in Fig. 4 the line *A-D* does not cut 2-3, *A* is incongruently saturating in *D*. If we take the view that the function of *D* is to lower the melting point of *A*, then if *A* is incongruently melting it is quite probable that *A* will be incongruently saturating, at least at temperatures near its melting point.

It is usually advantageous to grow a crystal from a solution in which it is congruently saturating so that difficulties associated with adjusting the ratio *C/D* in the liquid are obviated. Since the liquid and solid compositions are more nearly identical, faster growth rates can be sustained and the likelihood of foreign inclusions in the grown crystal is reduced. Consequently, further solvent modifications are sometimes used to cause recalcitrant crystals to become congruently saturating. Solvents that have high solvent power at as low a temperature as possible are advantageous for incongruent melters because the more one depresses the melting point the greater the likelihood of obtaining congruent saturation. In addition, mixed molten salt solvents can be used. Suppose that instead of using *D* as the solvent for *A* in Fig. 4 we had a mixture of *E* and *F*. We would then be involved with the quaternary system *C-B-E-F*. This could be treated as pseudoternary, provided there was no compound formation between *E* and *F* by considering a given ratio of *E/F* to be a component. If *C* were more soluble than *B* in *E* and if *B* were more soluble than *C* in *F*, some *E/F* ratio should exist at which *A* would be congruently saturating.

Other rationales have been used to

choose suitable molten salt solvents. Chemical reasoning would suggest for instance that (1) acidic oxides would be good solvents for basic oxide crystals provided compound formation does not intervene, (2) common ion effects could be used in insuring that the molten salt does not contaminate the grown crystal, (3) complex formers ought to be good solvents, and (4) chain-breaking cations should be useful in lowering melt viscosity. Although such generalizations are useful, there is no substitute for experience, a careful perusal of the literature in search of analogous systems, and a willingness to make a number of exploratory phase equilibria experiments.

3. EQUIPMENT

Furnaces

The usual temperatures involved range from about 900 to 1300°C. Phase equilibria studies can be made in a strip furnace or even on a hot stage microscope. Actual crystal growth requires that substantial volumes be heated for protracted periods of time. Fortunately, several vendors* manufacture nearly ideal furnaces for growth by slow cooling. A diagram of a typical furnace is shown in Fig. 5.

The furnace is resistively heated by silicon carbide heater elements separated from the growth region by a silicon carbide or other ceramic muffle with high thermal conductivity. A typical furnace whose usable volume is 5 x 12 x 5 in. requires about 6 kw of power. This is usually supplied through a 230-volt auto-transformer whose secondary is tapped in coarse and fine voltage increments. For maximum life of silicon carbide heating elements, 1400°C should not be exceeded for protracted periods. The resistance of silicon carbide changes with use, so the transformer voltage must be adjusted with time. When isothermal conditions or a particular temperature profile is required, it may be necessary to compensate for the fact that all heating elements do not age at the same rate. Temperature is generally measured by Pt versus Pt-10% Rh thermocouples. Temperature is controlled by a millivoltmeter, or potentiometer, and a contactor, or saturable core transformer (Paschkis and Persson, 1960). The

* Hevi Duty Electric Co., Milwaukee, Wisconsin; Burrell Corp., Pittsburgh, Pennsylvania; and Harper Electric Co., Buffalo, New York.

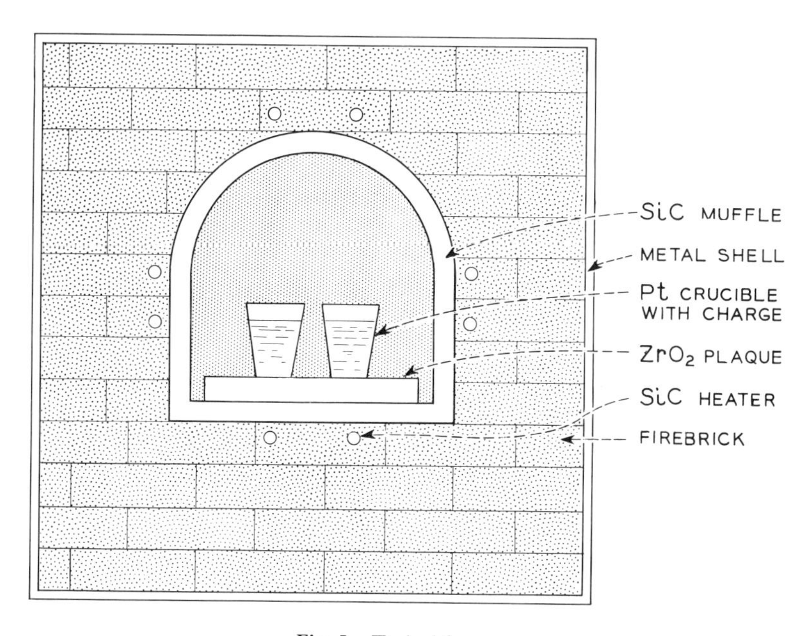

Fig. 5. Typical furnace.

sophistication of the control equipment depends on the desired precision. Programming the temperature with time is usually done by a motor-driven cam attached to the controller. A common error is to choose cooling cycles not commensurate with the precision of temperature control. A cooling rate $d\theta/dt$ of 0.1°/hr for a furnace whose temperature control is estimated at ±5° is obviously impractical.

For growth by a thermal gradient, cylindrical-grooved ceramic-cores* are generally wound with Pt or Pt-Rh wire, which is cemented to the core and surrounded by insulation and/or firebrick and placed within a metal can. Taps are provided along the core for the addition of shunt resistances to control the gradient. For a core of 3 in. ID and 9 in. length, 55 ft of 0.035 in. diameter Pt-20% Rh wire (3 ohm resistance at room temperature and 9 ohm at 1200°C) can dissipate 1 kw of power and safely maintain a temperature of 1200°C nearly indefinitely. The properties of Pt and its alloys that are of interest in furnace design are available in the literature (*Handbook of Chemistry and Physics*, 1960; Motzfeldt, 1959; Vines and Wise, 1941).

Noble Metal Crucibles

For molten salt crystallization an inert container for temperatures to at least 1300°C is required. The most readily available material is platinum. In a few instances, dense ceramic crucibles or even refractory monocrystals have been used. However, then, the crucible material is dictated by the system to be contained (Mackenzie, 1959). The gain in flexibility usually more than outweighs the expense of platinum for laboratory operations.

Crucible geometry is dictated by the furnace configuration and the desired temperature profile. For simple slow-cooling experiments, conventional 50 to 200 cc crucibles are adequate. For obtaining large crystals, cylindrical crucibles as large as a liter have been used. When the solvent is volatile, lids are often crimped, or even welded, onto crucibles. Welding may be accomplished with an oxyhydrogen torch, and "soldering" may be performed with Pd wire as solder since Pt melts at 1773°C and Pd at 1554°C. For thermal gradient experiments, intermediate sizes and shapes are usually used. Platinum baffles, Pt-sheathed stirrers, and seed holders are also required. For most experiments "crucible-grade" (99.5% pure) Pt is satisfactory. When extreme purity is desired, chemically pure (99.99%) Pt might be used. However, since the principal impurities in crucible-grade Pt are 0.3 to 0.5% of deliberately added Ir and Rh, which improve the strength and thermal stability of Pt, this grade is usually preferred (Vines and Wise, 1941).

The care, handling, and cleaning of platinum are adequately described in the literature (Wilson, 1959) and is not discussed here. Platinum equipment is fabricated on a custom basis by several manufacturers.*

4. CRYSTALS GROWN FROM MOLTEN SALTS

We shall not attempt a comprehensive review, but instead will select several representative systems that have been extensively investigated and point out the factors of general applicability and importance.

Growth by Slow Cooling

The principle involved is quite simple. A saturated solution is prepared by holding, or "soaking," a crucible containing flux and the crystal constituents at a temperature slightly above the saturation temperature long enough to effect complete solution. Then the crucible is cooled at a rate $d\theta/dt$ (generally expressed in degrees/hr) through a temperature range

*Alundum, which is obtainable from Norton Co., Worcester, Mass., is typical.

*Englehard Industries, Baker Division, Newark, New Jersey; J. Bishop and Co., Malvern, Pennsylvania; and Texas Instruments Co., Metals and Controls Division, Attleboro, Mass.

where the desired crystal is known to precipitate. The supersaturation will depend upon the values of the temperature coefficient of solubility $ds/d\theta$ in the chosen temperature interval. Usually, $d\theta/dt$ is adjusted so that ds/dt is in the range 0.02 to 0.20 weight per cent/hr. For most values of $ds/d\theta$, this results in $d\theta/dt$ varying between 1 and 5°/hr. Crystals usually nucleate first in the coolest region of the crucible. Even when the most careful efforts are made to keep a crucible isothermal, some region remains slightly cooler and nucleation begins there. It is preferrable that nucleation be confined to as few initial sites as possible. Once initial nucleation has taken place, ds/dt will be low enough so that no new nucleation occurs. Hence, the initial sites will be the only sinks for subsequent growth, and the crystals obtained will be of maximum size. Thus the smaller $d\theta/dt$ the larger the crystals obtained, but a lower limit on $d\theta/dt$ is usually set either by errors in the temperature control or by the time available. Once the temperature range of growth has been traversed, the crucible may be cooled more rapidly (in some cases quenched) to room temperature, and the grown crystals separated by hand sorting and/or chemical leaching out of the flux.

Barium Titanate

The unusual dielectric properties of barium metatitanate were first noticed by Wanier and Solomon (1942). It has five polymorphic modifications as follows (Merz, 1949):

Liquid (congruent melting) $\xrightarrow{1618°}$
Hexagonal (nonferroelectric) $\xrightarrow{1460°}$
Cubic perovskite (nonferroelectric) $\xrightarrow{120°C}$
Tetragonal (ferroelectric) $\xrightarrow{5°}$
orthorhombic (ferroelectric)
$\xrightarrow{-80°}$ rhombohedral (ferroelectric)

The BaO-TiO_2 phase diagram has been determined by Rase and Roy (1955) and $BaTiO_3$ monocrystals have been grown by a number of workers (Blattner et al., 1947a, b; Remeika, 1954; De Vries, 1959; Nielsen et al., 1962; White, 1955; Von Hippel, 1950). The cubic-hexagonal transition is sluggish, but can be influenced by impurities.

The crystal morphology most usually preferred for ferroelectric studies is the "butterfly twin" habit of the cubic modification. It was first grown from molten KF by Remeika (1954). This habit has been described by White (1955), and its genesis is discussed by De Vries (1959) and by Nielsen et al., (1962).

The cubic-hexagonal transition is influenced by the composition of the melt as shown in Fig. 6 (Rase and Roy, 1955). On the TiO_2-rich side of $BaTiO_3$, there is some solid solution of TiO_2 in $BaTiO_3$. Apparently cubic $BaTiO_3$ can more easily accommodate TiO_2 so that it crystallizes in the region *A-B* from TiO_2-rich melts at temperatures above the transition tempera-

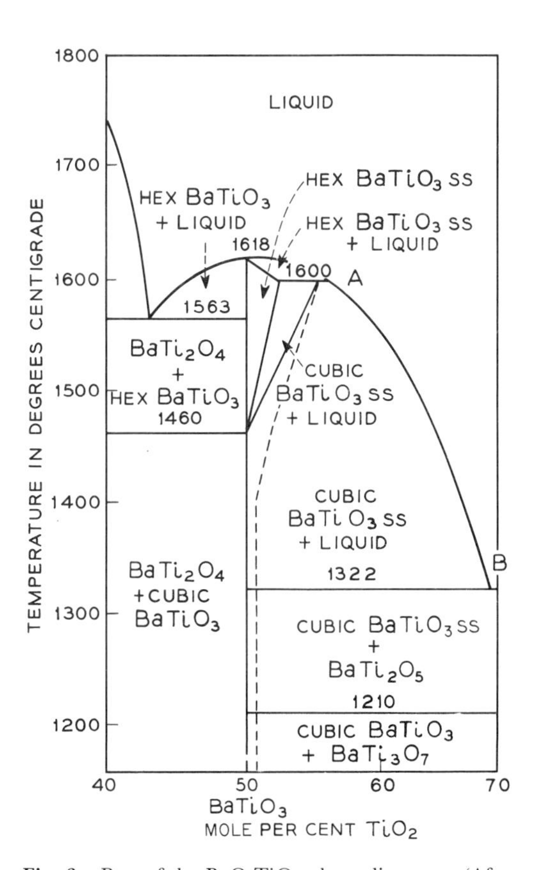

Fig. 6. Part of the BaO-TiO_2 phase diagram. (After Rase and Roy, 1955.)

ture. Platinum that dissolves in the melt under oxidizing conditions may also effect the cubic-hexagonal transition (Linares, private communication). For the growth of crystals of useful size, molten salt growth is used because the desired phase is not the equilibrium form at the melting temperature. Several solvents (including, $BaCl_2$, BaF_2, and KF have been used, but growth from KF has proven most satisfactory because (1) KF does not attack Pt excessively (2) $BaTiO_3$ sinks in KF and consequently nucleates at the bottom of the crucible as well defined butterfly twins, and (3) KF is readily leached with water. The procedure as devised by Remeika (1954) has been used in all subsequent preparations. It consists of soaking a KF melt containing excess $BaTiO_3$ (30 weight per cent $BaTiO_3$—70 weight per cent KF) at a temperature between 1150° and 1200° for eight hours. The crucible is closed with a tightly crimped cover to suppress KF evaporation. At the end of the soaking period, the furnace is cooled at a slow constant rate (20 to 50°/hr) to a temperature between 900 and 1000°. The liquid flux is quickly decanted off at this point and the crystals are annealed by cooling slowly (10 to 50°/hr) to room temperature. Crystals are broken free from the residual flux mechanically and then leached with hot water. Typical butterfly twins that result are shown in Fig. 7. In Fig. 8 the important crystallographic features of the twin are labeled (Nielsen et al., 1962). The twin plane is (111) and the principal surfaces bounding the crystal are (100). Nielsen et al., (1962) found that the twin could form under a variety of conditions and from a variety of solvents, provided an excess of solid $BaTiO_3$ was present in the crucible when nucleation began. It was also necessary that an appreciable number of particles in this excess were smaller than 10 μ. The receipe of Remeika provides excess solid $BaTiO_3$ at the beginning of the cooling cycle only as a result of extensive evaporation during the soak period (see phase diagram of Fig. 9). Unless this excess is present, small

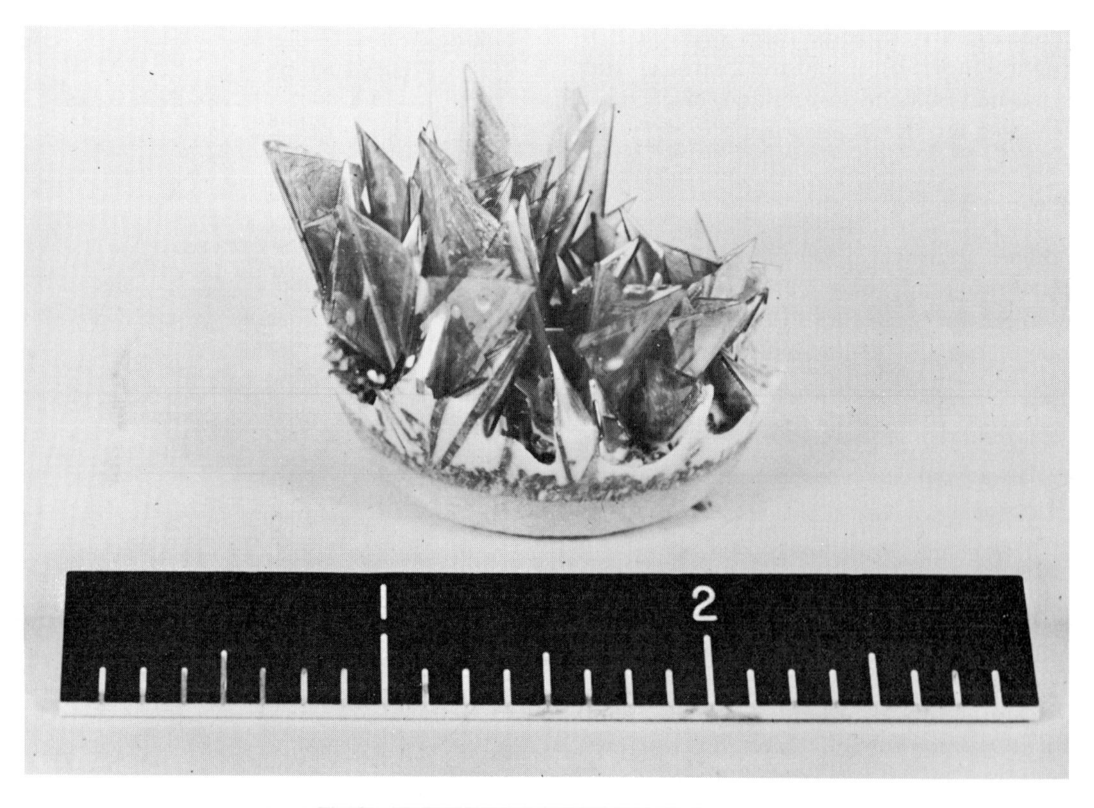

Fig. 7. Typical KF grown $BaTiO_3$ butterfly twins.

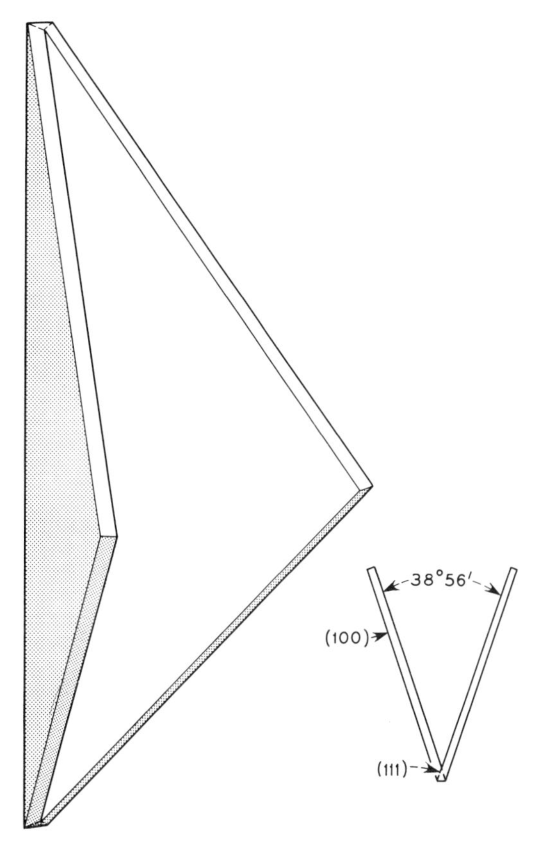

Fig. 8. Crystallographic features of $BaTiO_3$ butterfly twin. (After Nielsen et al., 1960.)

cubic crystals are the inevitable product. The dashed regions of Fig. 9 are speculative and there probably is some solid solution of other phases in the crystallized $BaTiO_3$, particularly at higher temperatures. De Vries (1959, 1961) and De Vries and Sears (1961) have discussed the nucleation and growth of the twin in some detail and point out that the butterfly twin is probably formed when (111) twinning of the (100) habit takes place. Such a twin will have re-entrant angles that grow out of existence through the formation of the butterfly habit. It is probable that the original nucleus is a cube that has been cleaved to reveal a (111) plane (Tanenbaum, unpublished work), and unless small enough $BaTiO_3$ particles are present to make the probability of exposure of this (111) plane high, the steps necessary for twin nucleation cannot begin.

Yttrium Iron Garnet

The discovery of ferromagnetism (Bertaut and Forrat, 1956; Geller and Gilleo, 1957) in the compound yttrium iron garnet,

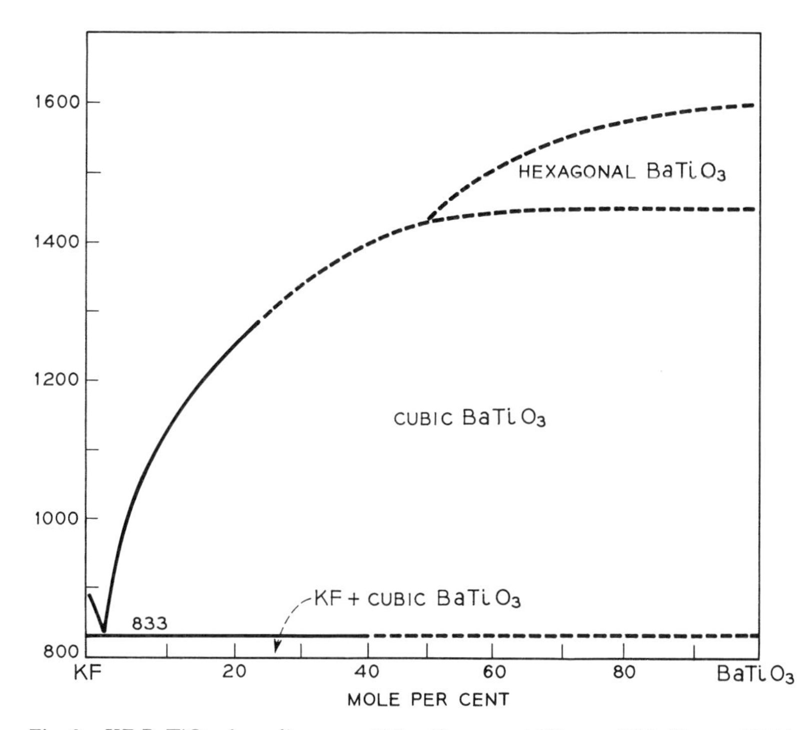

Fig. 9. KF-$BaTiO_3$ phase diagram. (After Karan and Skinner, 1953; Karan, 1954.)

$Y_3Fe_5O_{12}$, (YIG) created a demand for crystals of various rare-earth substituted compounds with the garnet structure. We shall concentrate here on the growth of YIG since it has been studied most intensively. Other rare-earth garnets can be prepared by similar techniques.

The Fe_2O_3—Y_2O_3 phase diagram was first investigated by Nielsen and Dearborn (1958) in air and later by Van Hook (1961a, b) who investigated the effect of oxygen pressure. In the strict sense, because of the reductive decomposition of Fe_2O_3, the system can only be described by the use of three components such as Fe_2O_3—Fe_3O_4—$YFeO_3$. The diagram, as suggested by Van Hook, is shown in Fig. 10. The phase relations in the Fe_2O_3—Fe_3O_4 system are known mainly from the work of Darken and Gurry (1945, 1946) and have been discussed by several authors (Eugster, 1959; Muan, 1958). The ternary model shown in Fig. 10 was shown by Van Hook to be qualitatively correct. The line of intersection of the 0.21 atmosphere oxygen-isobar and the liquidus surfaces determines the compositions of melts in equilibrium with various crystalline phases in air. It should be pointed out that although the traces of each of the line segments $A'A$, AB, and BB' in the Fe_3O_4—Fe_2O_3—$YFeO_3$ plane are a straight line, the line $A'ABB'$ is not straight. The diagram lying along this isobar as determined by Van Hook is shown in Fig. 11. It gives the appearance of a binary system with a "eutectic" at A (points are labeled similarly on Figs. 10 and 11) at 1469° and a "peritectic" at B at 1555°. The "eutectic" and "peritectic" points are actually the intersections of the air isobar with invariant boundary curves of the ternary system. Thus points A and B are only invariant when the oxygen pressure is fixed. Also the compositions given for points A and B are in terms of the starting materials Y_2O_3 and Fe_2O_3 as originally analyzed and are not true compositions of these melts since some oxygen has been lost at high temperature.

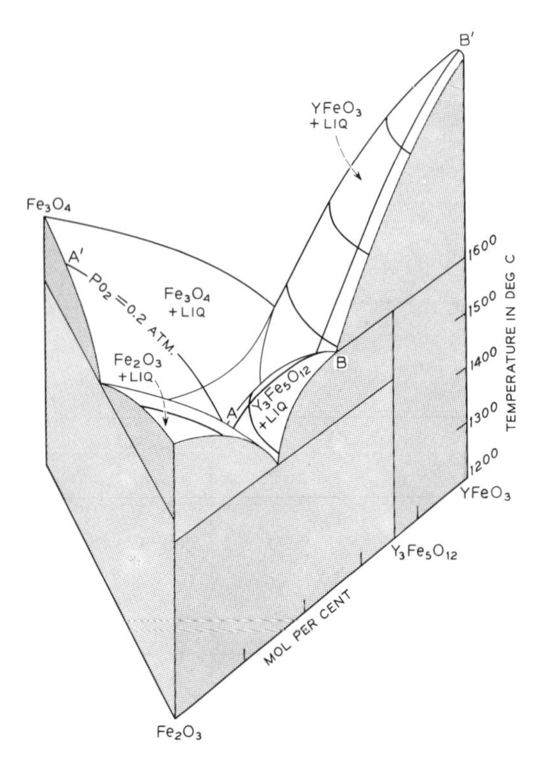

Fig. 10. Ternary model of the system Fe_2O_3-Fe_3O_4-$YFeO_3$, showing assumed liquidus relations. (After Van Hook, 1961b)

It should be noted that garnet melts incongruently in this system. Garnet crystallized in the region A-B is oxygen deficient; thus the concentration of Fe^{++} in it is a function of the crystallization temperature and the oxygen pressure. However, a quantitative description of the garnet solid solubility curve is lacking. Isobars similar to $A'ABB'$ of Fig. 10 will exist for other oxygen pressures, and indeed the isobar for one atmosphere oxygen pressure has been determined, as has the "isobar" for the oxygen pressure in equilibrium with CO_2 (Van Hook, 1961b). The most striking effect is that increasing the oxygen partial pressure widens the garnet stability region. At one atmosphere of oxygen, point A is at 1455° and point B at 1582°. Presumably the concentration of Fe^{++} in garnet is also reduced.

Because of the incongruent melting and reduction problems at high temperatures, it was recognized early that molten salt crystallization provided an attractive method for YIG growth. The first solvent used by Nielsen and Dearborn (1958) was molten PbO. It was found that YIG was incon-

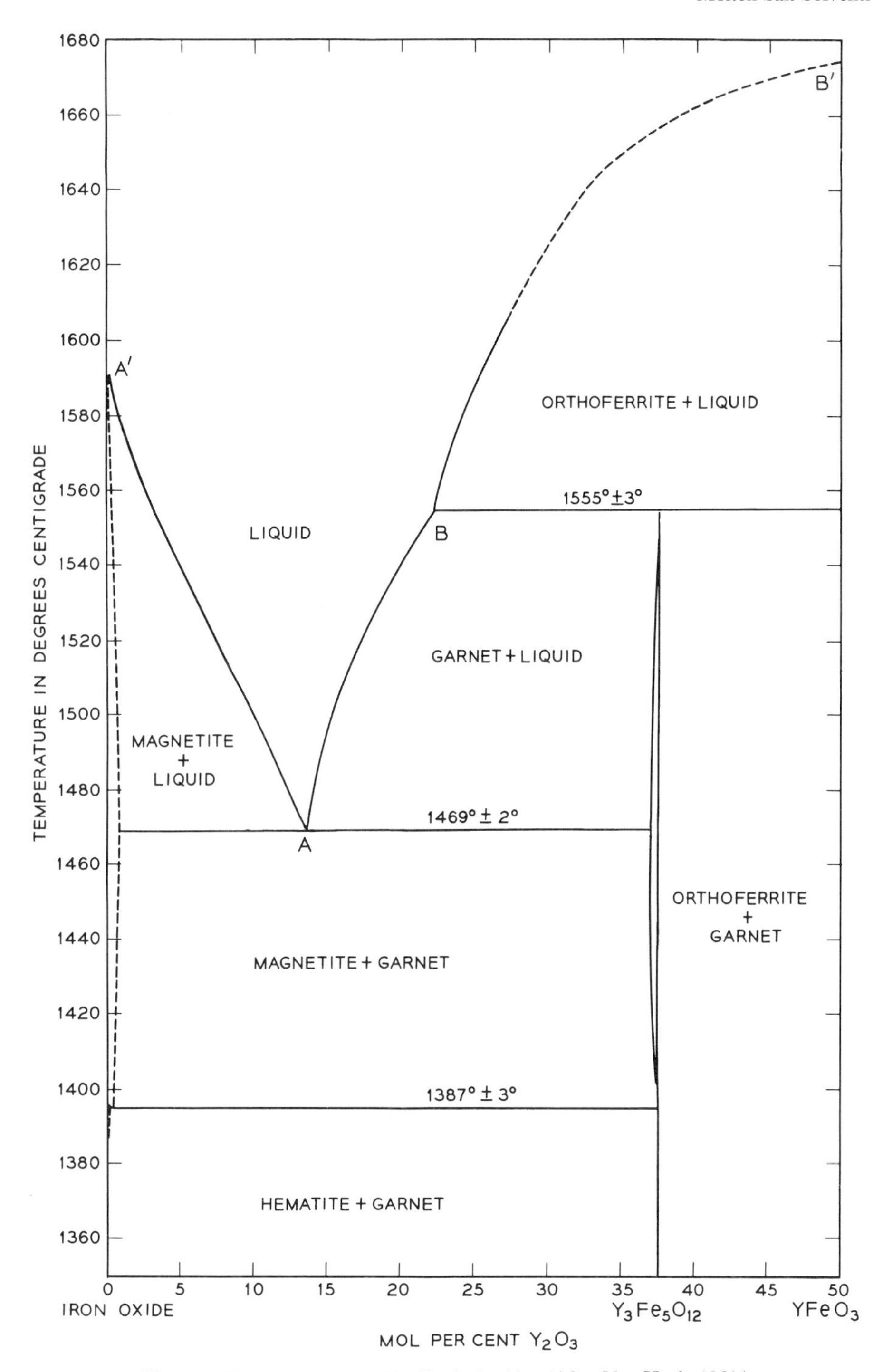

Fig. 11. The system iron oxide, Fe_2O_3 in Air. (After Van Hook, 1961.)

gruently saturating in PbO between about 1350 and 950°C, and part of the pseudo-phase diagram $PbO—Fe_2O_3—YFeO_3$ was determined (Nielsen and Dearborn, 1958). If the Fe_2O_3/Y_2O_3 ratio in the melt was substantially greater than 5/3 (the ratio of the oxides in YIG) YIG could be crystallized as the primary phase. However, the best YIG crystals were grown from extremely Fe_2O_3 rich melts where Fe_2O_3/Y_2O_3 was

12.6. Under these conditions the primary phase precipitating was probably Fe_2O_3 or $PbFe_{12}O_{19}$ (magneto-plumbite). It was necessary to use iron-rich melts so the composition would have a liquidus below about 1350° C, since it was found that platinum crucible life was severely shortened by molten PbO above that temperature. Furthermore, the largest and most perfect YIG crystals were grown when the first phase to precipitate was not garnet. When hematite or magneto-plumbite precipitated first they tended to nucleate on the melt surface and inhibit subsequent YIG growth there. This reduced the number of YIG nuclei and increased the average size of the grown crystals. Because of the volatility of PbO, high supersaturation and preferential nucleation will occur at the melt surface. To inhibit volatilization is was necessary to use crimped crucible lids.

It was found that the melts could usually be quenched to room temperature once they were below about 900° without adversely effecting the grown crystals. Crystals were removed from the melt by leaching in hot HNO_3. The phases were separated magnetically at temperatures above and below the appropriate Curie temperatures. Similar methods were developed for Sm, Er, and Gd, iron-garnets.

Since lead oxide is not an ideal solvent, several other solvents have been used for garnet growth. Nielsen (1960) used $PbO—PbF_2$. He found that YIG could be crystallized at lower temperatures, crystal yields were larger, the melt viscosity was lower, and magneto-plumbite could be avoided as a coprecipitating phase when the solvent was 43 mole % PbO and 57 mole % PbF_2. However, garnet was still not congruently saturating and the high vapor pressure of PbF_2 made evaporation a problem. Nevertheless, crystals of larger size and better microscopic perfection than those obtained in PbO were prepared. Slow cooling of large crucibles (volume > 1 liter) resulted in crystals weighing more than 50 grams.

An interesting solvent system, $BaO-B_2O_3$,

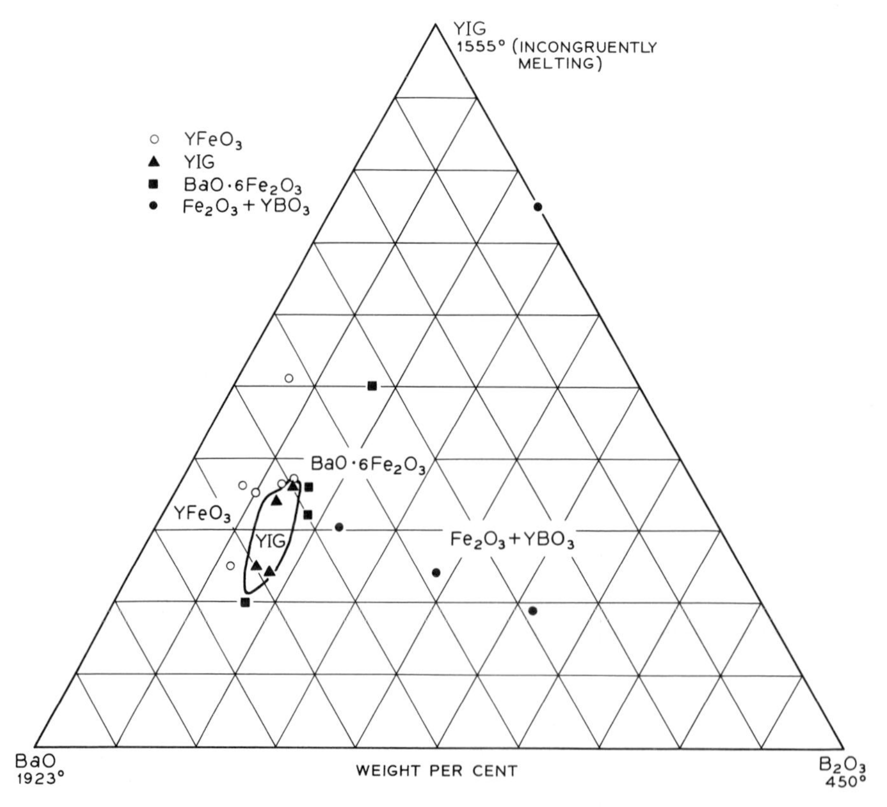

Fig. 12. The $BaO-B_2O_3$-YIG psuedo-ternary system. (After Linares, 1962.)

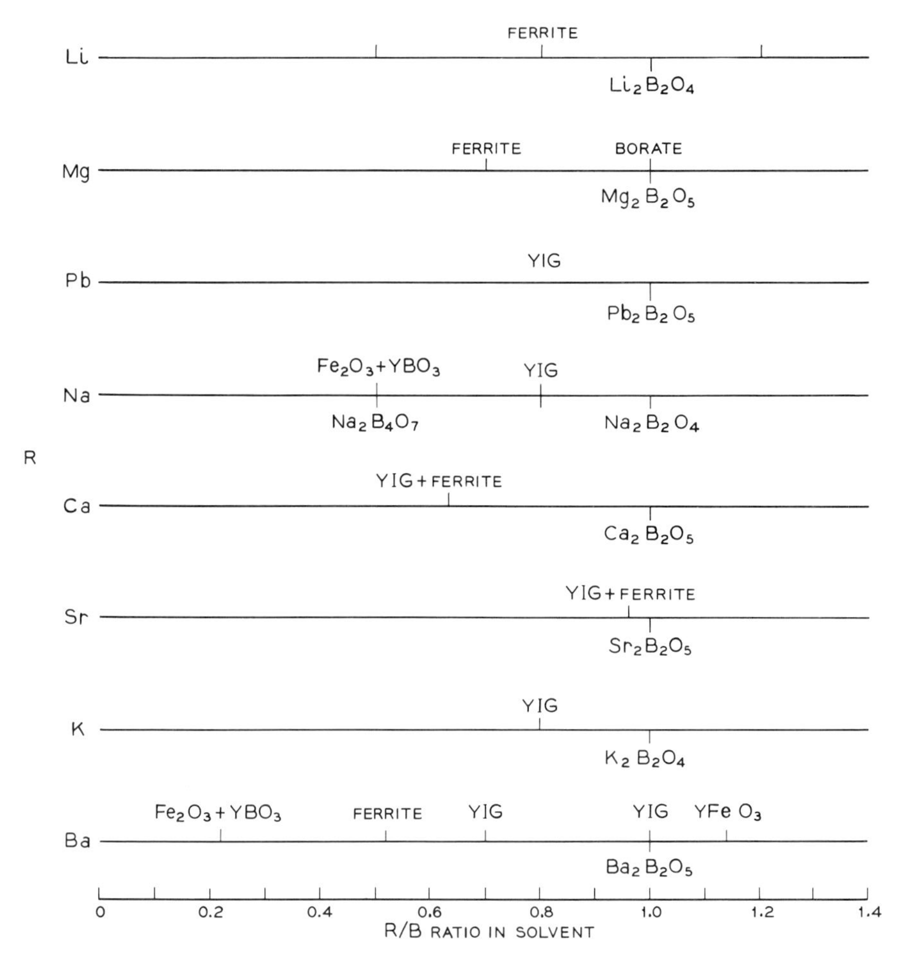

Fig. 13. Schematic diagram showing the effect of substitutions in the BaO-B_2O_3 solvent. (After Linares, 1962.)

was discovered by Linares (1962b). This system is noteworthy in several respects:

1. When the solvent is BaO—0.6 B_2O_3, YIG is congruently saturating.
2. The vapor pressure of the system is low, so that flux evaporation is not a problem.
3. The molten solvent is less dense than crystalline YIG, so that YIG sinks in it, thus permitting growth in a temperature gradient (see the following).
4. Since the solvent does not contain Pb the hazards of noxious Pb vapors are obviated, and reduction of the flux to metallic Pb which corrodes Pt crucibles cannot take place.

Phase equilibria results for this solvent are particularly interesting. The system (Fig. 12) was considered as pseudo-ternary with the components $Y_3Fe_5O_{12}$, BaO, and B_2O_3. Figure 12 shows the identity of the first phase to crystallize as melts of the indicated composition are cooled. The iron content of the stable phase increases with increasing B_2O_3 content in the melt. Thus it would appear that B_2O_3 is a poorer solvent for Fe_2O_3 than BaO and that the proper BaO-B_2O_3 ratio would exist when the ratio of the solubilities of Fe_2O_3 and and Y_2O_3 and Y_2O_3 are the same as their ratio in YIG. Under such conditions (the YIG region of Fig. 12) it is not surprising that YIG is the primary phase to crystallize. That this effect is not unique might be expected from the discussions under Phase Equilibria. As Fig. 13 shows, when BaO is replaced with certain RO and R_2O type oxides, YIG will crystallize when the R/B ratio is between ~0.7 to 1.0. In many of these solvents, a R/B ratio where YIG saturates congruently can be found. This might be expected if the solvent power of RO and R_2O is not strongly dependent on the nature of R. Again the general trend is for more iron to be in the crystallized phase as the flux becomes more B_2O_3 rich. YIG does not crystallize with Li_2O and MgO because of the high stabilities of lithium ferrite and magnesium borate respectively. V_2O_5 and MoO_3 can be substituted for B_2O_3 with similar results. Figure 14 shows the solubility of YIG in BaO—0.6 B_2O_3. Curve *B* shows the effect of Fe_2O_3 in the melt in excess of the ratio $Fe_2O_3/Y_2O_3 = 5/3$. Although treatment of this sort is admittedly crude, it is probably a valid first approach toward the rationalization of molten salt solvent power. Figure 15 shows some YIG crystals grown by slow cooling of a molten salt melt.

Other Crystals

Barium titanate and YIG have been treated in some detail as representatives of the complexities to be expected in molten salt growth. A wide variety of crystals in

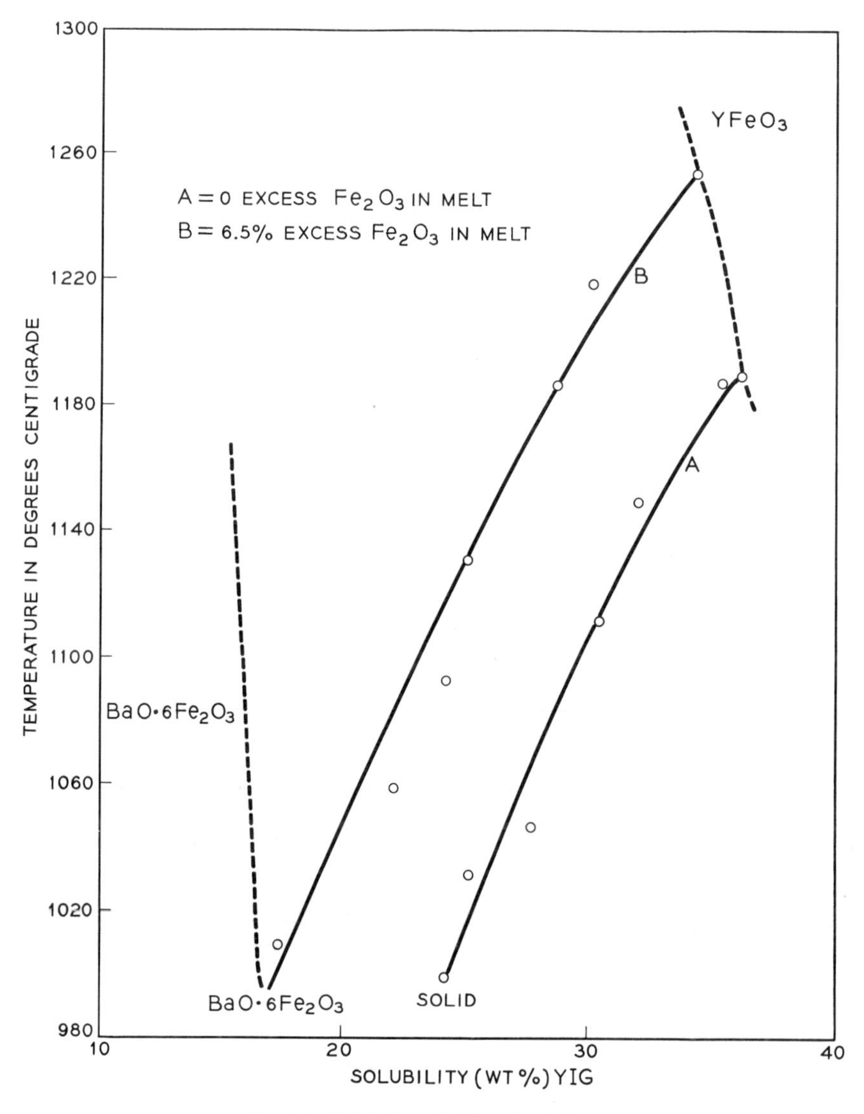

Fig. 14. Solubility of YIG in BaO-B_2O_3.

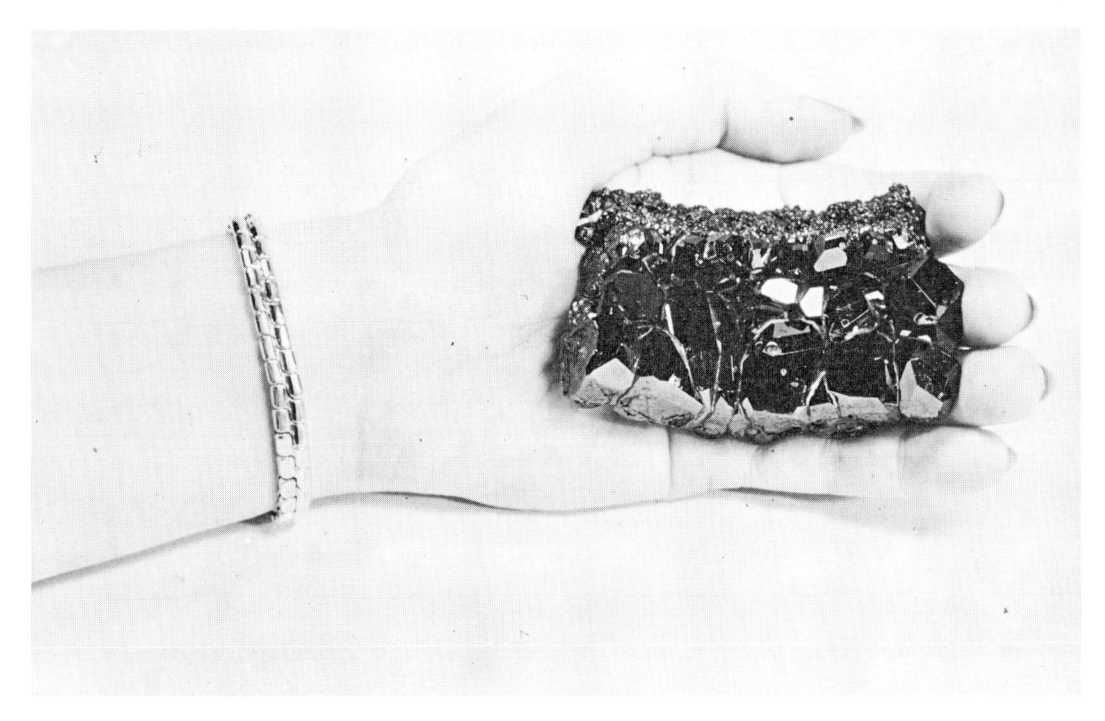

Fig. 15. YIG Crystals grown by seed cooling of PbO melts.

addition to the garnets and barium titanate has been grown from molten salts. The results have been summarized in Table 1. The listing is by no means complete, but should be useful in choosing a solvent and growth conditions by analogy with previous results. The data listed for each crystal are the minimum necessary for specifying the growth conditions. It is hoped that future crystal growers will be careful to specify these conditions more definitely than they have in the past.

Growth on a Seed Crystal

The advantages of control over orientation, growth rate, perfection, and doping make growth on a seed an attractive improvement over growth by random nucleation. Two classes of seeded growth experiments have been performed.

1. Growth by *slow cooling* in the presence of a seed.
2. Growth on a seed in the cool part of a system while excess solute (often called nutrient) is in contact with the solvent in a hotter region of the system. This method is often called, "growth in a thermal gradient," and is in some respects analogous to hydrothermal gradient growth.

Growth in the presence of a seed has been used by Miller (1958) to grow potassium niobate and by Reynolds and Guggenheim (1961) to grow a ferrite. Perhaps the most complete study of this technique is that of Laudise, Linares, and Dearborn (1962) who grew YIG from the $BaO—B_2O_3$ flux previously described. Figure 16 shows a typical furnace arrangement for growth both by slow cooling and in a thermal gradient. In the slow cooling experiments, the stirrer and baffle were not used, and the crucible was maintained as nearly as possible isothermal. The seed was mounted at the end of a platinum-sheathed ceramic rod that projected just below the liquid surface. This rod and the attached seed were rotated by a gear-train motor arrangement. The rotation was 30 sec clockwise, followed by 30 sec counterclockwise. Figure 17 shows the effect of rotation rate on the growth rate of YIG in the $\langle 110 \rangle$ direction at two different cooling rates. For a value of $d\theta/dt$ of 10°/hr, the flat region A-B began at 200 rpm. The

TABLE 1

Crystal	Formula	Flux and Temperature Range of Cooling	Cooling Rate $d\theta/dt$	Typical Charge (Mole %)	Crystals Leached from Flux with	Comments	Ref.
Titanates—Zirconates—Niobates							
Barium orthotitanate	$BaTiO_3$	KF 1200–850°C	1–5°/hr	30% $BaTiO_3$, 70% KF	Hot Water	Excess $BaTiO_3$ present at beginning of cooling cycle. Anneal below 850°. Evaporation of KF aids in causing excess $BaTiO_3$	(1)
Bismuth titanate	$Bi_4Ti_3O_{12}$	Bi_2O_3 1200–~800	2–5°/hr	22% TiO_2, 88% Bi_2O_3	Strong mineral acids		(2)
Potassium ortho-niobate	$KNbO_3$	KF or KCl ~1100°–~800°	~1–10°/hr	50% K_2CO_3, 50% Nb_2O_3 Amount of flux not given.	Hot water	Shirane, Jona, and Pepinsky have used K_2CO_3 as a flux	(3)
Lead ortho-zirconate	$PbZrO_3$	PbF_2 1250–~1000°	50°/hr	50% $PbZrO_3$, 50% PbF_2	Not chemically separable	Growth probably due mainly to evaporation of PbF_2	(4)
Garnets							
Yttrium iron garnet (YIG)	$Y_3Fe_5O_{12}$ (YIG)	1350–930° PbO	1–5°/hr	44% Fe_2O_3, 3.5% Y_2O_3, 52% PbO	Hot 25% HNO_3	Can replace Y_2O_3 with Gd_2O_3, Er_2O_3, or Sm_2O_3 and crystallize analogous rare earth iron garnets	(5)
Yttrium iron garnet (YIG)	$Y_3Fe_5O_{12}$	1260–950° PBO–PbF_2	0.5–5°/hr	22% Fe_2O_3, 8% Y_2O_3, 30% PbO, 40% PbF_2	Hot 25% HNO_3–50% HAC	Can replace Y_2O_3 with Sm_2O_3, Er_2O_3, Gd_2O_3, Tb_2O_3, Dy_2O_3, Ho_2O_3, Eu_2O_3, or Yb_2O_3 and crystallize analogous rare earth iron garnet under generally similar conditions. $Y_3Ga_5O_{12}$ (yttrium gallium garnet—YGG) made by starting with 5.8% Y_2O_3, 14% Ga_2O_3, 39.4% PbO, and 40.8% PbF_2. It was possible to substitute other rare earths in YGG and to prepare crystals which were solid solutions of YIG and YGG	(6)
Yttrium iron garnet (YIG)	$Y_3Fe_5O_{12}$	~1260–900° BaO–B_2O_3	1°/hr	Calculated for saturation at starting temperature according to Fig. 14.	Hot 25% HNO_3		(7)

TABLE 1 (CONTINUED)

Crystal	Formula	Flux and Temperature Range of Cooling	Cooling Rate $d\theta/dt$	Typical Charge (Mole %)	Crystals Leaded From Flux With	Comments	Ref.
Yttrium aluminum garnet (YAG)	$Y_3Al_5O_{12}$	1250–950° $PbO-B_2O_3$	1°/hr	32% Y_2O_3, 5.3% Al_2O_3, 18.5% B_2O_3, and 73% PbO	Hot HNO_3		(8)
Ferrites							
Nickel (etc.) ferrite	RFe_2O_4 R = Ni, Co, Zn, Mg, Cd	PbO ~1300–~800°	~10°/hr	20.5% NiO, 20.5% Fe_2O_3, 59% PbO	25% Hot HNO_3		(9)
Nickel ferrite	$NiFe_2O_4$	$Na_2B_4O_7 \cdot 10H_2O$ 1330–1250°	2°/hr	32.1% NiO, 32.1% Fe_2O_3, 35.8% $Na_2B_4O_7 \cdot 10H_2O$	Hot HNO_3	$Na_2B_4O_7 \cdot 10H_2O$ (borax) dehydrates easily to $Na_2B_4O_7$	(10)
Gallium orthoferrite (nonstoichiometric)	$Ga_{(2-x)}Fe_xO_3$	$Bi_2O_3-B_2O_3$ 1125–500°	4–7°/hr	16.7% Ga_2O_3, 19.6% Fe_2O_3, 27.8% B_2O_3, 35.9% Bi_2O_3	HNO_3-H_2O	The value of x depends on ratio of Fe_2O_3/Ga_2O_3 and on temperature of crystallization	(11)
Yttrium (etc.) ortho-ferrite	$RFeO_3$ R = Y, La, Pr, Sm, Er, Gd	PbO 1300–800°	~30°/hr	50% R_2O_3, 50% Fe_2O_3 Assumed weight solubility of ferrites ~15%.	Hot HNO_3	R = Y, La, Pr, Nd, Sm, Er, Gd; Fe was replaced with Al^{+3}, Sc^{+3}, Ga^{+3}, Co^{+3}, and Cr^{+3}	(12)
Others							
Magnesium (etc.) tungstate	RWO_4 R = Mg, Zn, Cd, Ca, Sr, Ba	$Na_2W_2O_7$ 1250–700°	2–3°/hr	25–40% RWO_4, 75–60% $Na_2W_2O_7$	Conc NaOH	R = Mg, Zn, Cd, Ca, Sr, and Ba. Doping with transition metal and rare earth ions possible	(13)
Zinc sulfide (β)	ZnS	ZnF_2 1000–800°	1–5°/hr	21% ZnS, 79% ZnF_2	Hot Conc NH_4OH	βZnS was found to crystallize above the reported $\alpha-\beta$ transition	(14)
Zinc oxide	ZnO	PbF_2 1150–800°	1–10°/hr	30% ZnO, 70% PbF_2	Not Chemically separable	PbF_2 volatility high. Habit was (001) plates	(15)

(1) J. P. Remeika, 1954. (2) L. G. Van Uitert and L. Egerton, 1961. (3) B. T. Matthias and J. P. Remeika, 1951, Shirane et al., 1955. (4) F. Jona et al., 1955. (5) J. W. Nielson and E. F. Dearborn, 1958. (6) J. W. Nielsen, 1960. (7) and (8) R. C. Linares, 1962. (9) J. P. Remeika, 1958. (10) J. K. Galt et al., 1950. (11) J. P. Remeika, 1960. (12) J. P. Remeika, 1961. (13) L. G. Van Uitert and R. R. Soden, 1960. (14) R. C. Linares, unpublished work. (15) J. W. Nielsen and E. F. Dearborn, 1961.

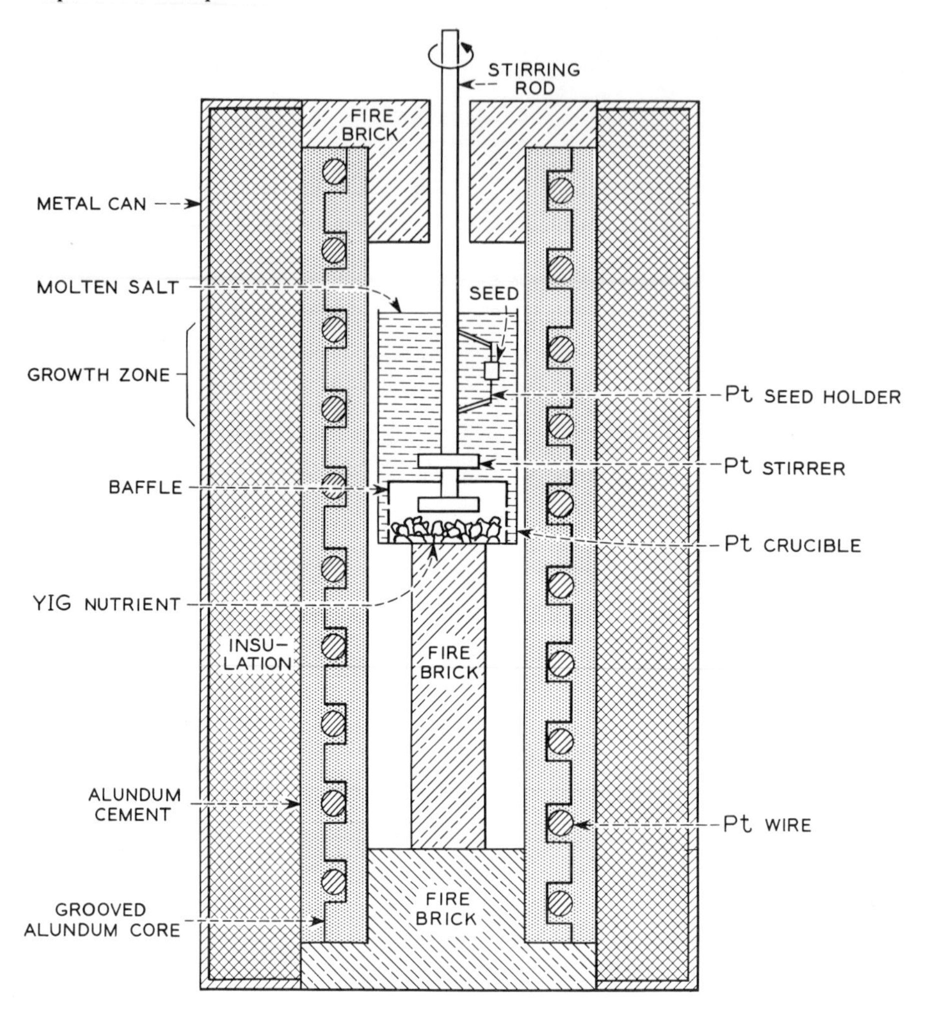

Fig. 16. Furnace arrangement for growth of YIG on seeds. (After Laudise, et al., 1962.)

value of A has been called the critical stirring speed. It increases with $d\theta/dt$, and the mass of solution (M). Once the critical stirring speed was exceeded, the rate was found to be nearly linear with $d\theta/dt$ and with M.

Spontaneous nucleation was negligible for most conditions, so the seed was the principal sink for YIG. Most experiments were begun at 1200°, but the rate was nearly independent of the temperature at which cooling began (between 1200 and 1100°). The rate was also independent of the elapsed time of the experiment between several hours and several days.

One would expect that diffusion would limit the rate of growth from a molten salt solvent. Therefore the rate depends on the stirring speed since stirring shortens the diffusion path close to the growing interface. Furthermore, it would be expected that the effect of increasing the stirring rate would reach a limit as in Fig. 17. The critical stirring speed increases with saturation, apparently because the diffusion step tends to have increasing importance. A greater critical stirring velocity is required the greater the cooling rate, since supersaturation increases with cooling rate. Surprisingly, it was also observed that greater critical stirring velocities were required the larger the mass of the melt. The seed is the principal sink on which excess solute deposits, so apparently the larger the melt mass the longer the diffusion path.

Since, for high stirring rates the rate of

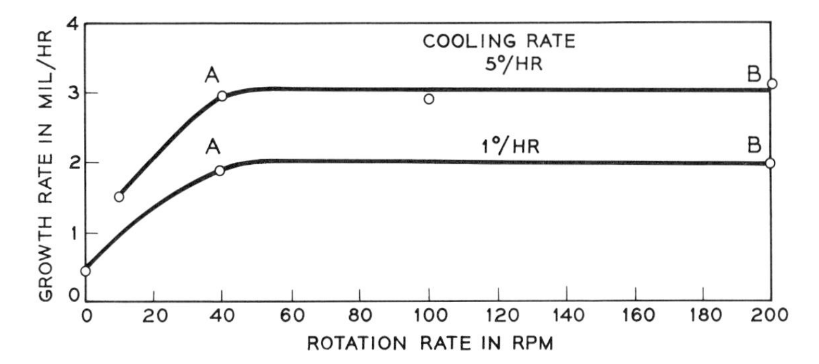

Fig. 17. Growth rate of YIG *vs* stirring rate. (After Laudise et al., (1962).)

growth is linear with cooling rate and with melt mass, some discussion is required. It can be assumed that once the critical stirring speed is exceeded diffusion is not rate limiting. Furthermore, since little spontaneous nucleation is observed elsewhere in the system, the seed is the principal sink for excess YIG. Then for steady state growth:

$$\frac{dx}{dt} = RA\rho \qquad (1)$$

where dx/dt is grams YIG deposited/hr or grams YIG in the entire melt mass caused to be supersaturated per hour, R is the rate of growth of the YIG seed, that is, the rate in the $\langle 110 \rangle$ direction when the (110) face is growing, A is the area of the growing face, and ρ the density. We can, however, also write

$$\frac{dx}{dt} = \left(\frac{d\theta}{dt}\right)\frac{ds}{d\theta}M \qquad (2)$$

where $d\theta/dt$ is the cooling rate, $ds/d\theta = \beta$, the temperature coefficient of solubility, and M the melt mass. The solubility data of Fig. 14 show β to be nearly independent of temperature. By substitution,

$$R = \frac{M\beta}{A\rho}\left(\frac{d\theta}{dt}\right)$$

Thus R increases linearly with melt mass and cooling rate.

Equation (1) can be validated directly by the use of growth rate and supersaturation data (Fig. 18), the value of dx/dt being calculated in each case from β values obtained in solubility studies (Linares, 1962b). The rate curve of Fig. 18 does not extrapolate to zero at $dx/dt = 0$ probably because the growth rate runs were not begun at equilibrium. Opening the furnace to insert a seed significantly lowered the melt temperature, and even after the melt had regained thermal equilibrium when the seed was introduced and stirring began the temperature dropped slightly again. Although such effects have been minimized, they have not as yet been entirely eliminated.

The furnace and stirrer arrangement in the temperature gradient experiments is shown in Fig. 16. Two hundred revolutions per minute, which exceeds the critical speeds for the expected supersaturations, was adopted as the standard stirring rate. In early experiments, with small diameter crucibles and small stirrer blade area, the rate fell off markedly during a run. However, with a large area double-impeller and a 3 in.-diameter crucible, the decrease in rate with time was reduced by more than a factor of two.

Dissolution of YIG became rate limiting in small crucibles as the nutrient sintered together in the bottom of the crucible so as to decrease its surface area. The rate recovered when powdered YIG with large surface area was added to a crucible. Of course, stirring below the baffle promoted dissolution. Figure 19 shows some typical YIG crystals grown on a seed.

With adequate stirring below the baffle, control over nucleation, and a reasonable size of crucible, dissolution and transport should cease to be rate controlling, so the interfacial reaction could be studied. This

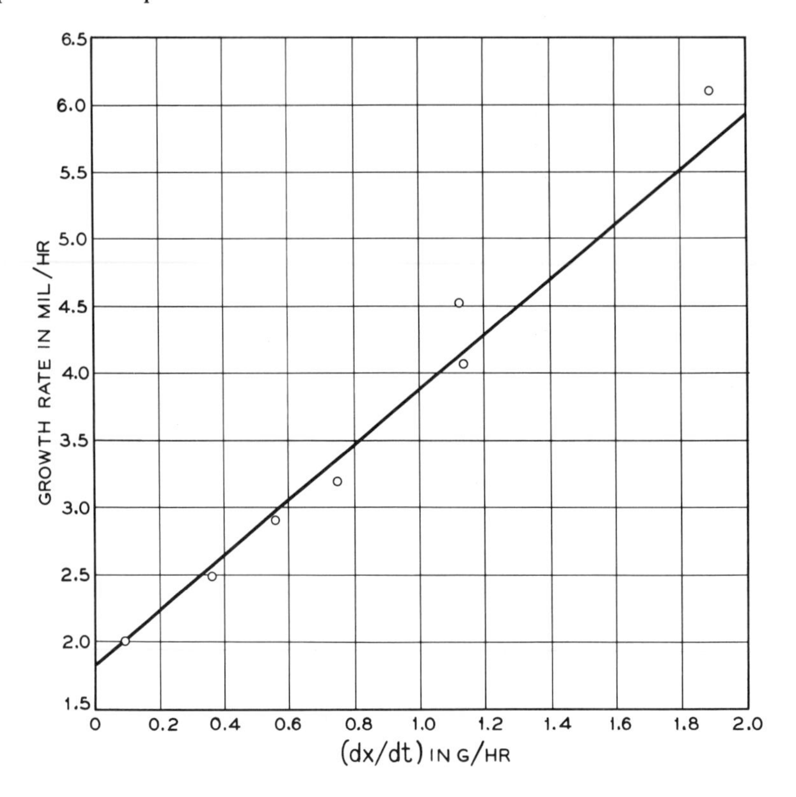

Fig. 18. Growth rate of YIG versus dx/dt. (After Laudise et al., (1962).)

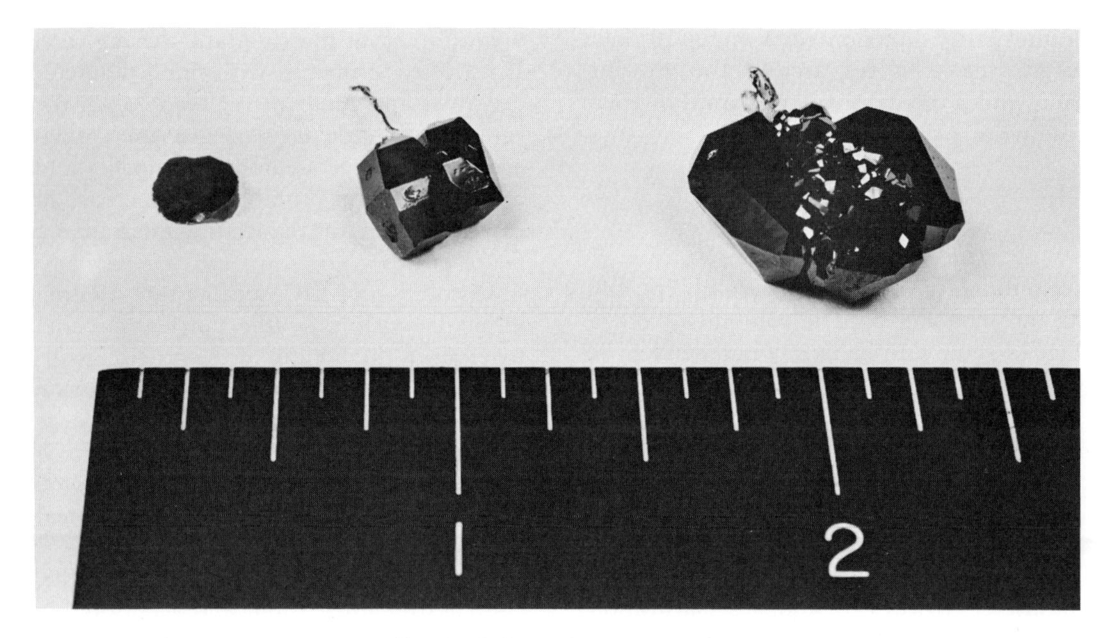

Fig. 19. YIG crystals grown on seeds.

would be valuable in understanding the growth mechanism.

ACKNOWLEDGMENTS

The author would like to thank J. P. Remeika and J. W. Nielsen who first stimulated his interest in molten salt crystal growth. R. C. Linares is to be particularly thanked for his assistance in writing this discussion, and for providing access to some of his unpublished work.

References

Annals of the New York Academy of Science (1960), **79** 761–1098 (Art. 11).

Bertaut, F., and F. Forrat (1956), *Comptes Rendu Acad. Sci. (Paris)* **242,** 382.

Blattner, H., B. Matthias, W. Merz (1947a), *Helv. Phys. Act.* **20,** 225.

Blattner, H., B. Matthias, W. Merz, and P. Scherrer (1947b), *Experiments* **3,** 148.

Blomgren, G. E., and E. R. Van Artsdalen (1960), "Fused Salts" in *Annual Review of Physical Chemistry,* Vol. 11, edited by H. Eyring, C. J. Christensen, and H. S. Johnston, Annual Reviews, Palo Alto, California, p. 273ff.

Darken, L. S., and R. W. Gurry (1945), *J. Am. Chem. Soc.* **67,** 1398; (1946), **68,** 798.

Eugster, H. P. (1959), *Researches in Geochemistry,* P. H. Abelson, John Wiley and Sons, New York, p. 397.

Findlay, A., A. Cambell, and N. Smith (1951), *The Phase Rlue,* Dover Publications, New York, p. 277ff.

Galt, J. K., B. T. Matthias, J. P. Remeika (1950), *Phys. Rev.* **79,** 391.

Geller, S., and M. Gilleo (1957), *Acta Cryst.* **10,** 239.

Glasstone, S. (1946), *Textbook of Physical Chemistry,* Second Edition, Van Nostrand, New York, pp. 693–814.

Handbook of Chemistry and Physics (1960), 42nd Edition, Chemical Rubber Publishing Co., Cleveland, Ohio, pp. 3344, 3360.

Janz, George (1961), *Bibliography on Molten Salts,* Second Edition, Rensselaer Polytechnic Institute, Troy, New York.

Jona, F., G. Shirane, and R. Pepinsky (1955), *Phys. Rev.* **97,** 1584.

Karan, C., and B. J. Skinner (1953), *J. Chem. Phys.* **21,** 2225.

Karan, C. (1954), *J. Chem. Phys.* **22,** 957.

Levin, E., H. McMurdie, and F. Hall (1956), Editors, *Phase Diagrams for Ceramists,* American Ceramic Society, Columbus, Ohio.

Laudise, R. A., R. C. Linares, and E. F. Dearborn, (1962), *J. Appl. Phys.* **33S,** 1362.

Linares, R. C., Private Communication.

Linares, R. C., (1962a) *J. Am. Ceram. Soc.* **45,** 119. (1962b), *J. Am. Ceram. Soc.* **45,** 307.

Linares, R. C., Unpublished Work.

Mackenzie, J. D. (1959), in *Physico-Chemical Measurements at High Temperatures,* Edited by J. O'M Bockris, J. L. White, and J. D. Mackenzie, Butterworths, London, p. 334.

Massig, G. (1944), *Ternary Systems,* Dover Publications, New York.

Matthias, B. T., and J. P. Remeika (1951), *Phys. Rev.* **82,** 727.

Mertz, W. J. (1949), *Phys. Rev.* **76,** 1221.

Miller, C. E. (1958), *J. Appl. Phys.* **29,** 233.

Motzfeldt, K. (1959), *Physico-Chemical Measurements at High Temperatures,* Edited by J. O'M Bockris, J. L. White, and J. D. Mackenzie, Butterworths, London, p. 55.

Muan, Arnulf (1958), *Am. J. Sci.* **256,** 171.

Nielsen, J. W. (1960), *J. Appl. Phys. Suppl.* **31,** 515.

Nielsen, J. W., and E. F. Dearborn (1960), *J. Phys. Chem.* **64,** 1762; (1958), *J. Phys. Chem. Solids* **5,** 202.

Nielsen, J. W., R. C. Linares, and S. E. Koonce (1962), *J. Am. Ceram. Soc.* **45,** 12.

Paschkis, V., and J. Perrson (1960), *Industrial Electric Furnaces and Appliances,* Second Edition, Interscience, New York, Vol. I, p. 295ff.

Rase, D. E., and R. Roy (1955), *J. Am. Ceram. Soc.* **38,** 110.

Remeika, J. P. (1954), *J. Am. Chem. Soc.* **76,** 940; (August 19, 1958), U. S. Patent 2,848,310; (1960), *J. Appl. Phys.* **31,** *Suppl.* 263S; (1961), *Phys. Rev.* **122,** 757.

Reynolds, G. F., and H. Guggenheim (1961), *J. Phys. Chem.* **65,** 1655.

Ricci, J. E. (1951), *The Phase Rule and Heterogeneous Equilibrium,* Van Nostrand, New York.

Shirane, G., F. Jona, and R. Pepinsky (1955), *Proc. I.R.E.* **43,** 1758.

Soviet Research in Fused Salts—1956, Parts 1, 2 (1958); *Soviet Research in Fused Salts*—1949–1955, Sections 1–5 (1959), Consultants Bureau, New York.

Tanenbaum, M., Unpublished Work.

Van Hook, H. J. (1961a), *J. Am. Ceram. Soc.* **44,** 208; (January 1961b), *Single Crystal Growth of Garnet Type Oxides,* Parts I and II, Research Division, Raytheon Co., Waltham, Mass.; Reports 1 and 2, AF 19(604)5511, prepared for Electronics Research Directorate, U. S. Air Force, Bedford, Mass., January, 1961.

Van Uitert, L. G., and L. Egerton (1961), *J. Appl. Phys.* **32,** 959.

Van Uitert, L. G., and R. R. Soden (1960), *J. Appl. Phys.* **31,** 328.

Vines, R. F., and E. M. Wise (1941), *The Platinum Metals and Their Alloys,* The International Nickel Co., New York.

Von Hippel, A. (1950), *Rev. Mod. Phys.* **22,** 221.

De Vries, R. C. (1959), *J. Am. Ceram. Soc.* **42,** 547; (June 1961), *General Electric Research Lab. Rept.* No. 61-RL-2767M, General Electric Research Information Section, The Knolls, Schenectady, New York.

De Vries, R. C., and G. W. Sears (1961), *J. Chem. Phys.* **34,** 618.

Wanier, E., and A. N. Solomon (1942), *Titanium Alloy Manufacturing Co., Elect. Rept.* **8;** (1943), **9** and **10.**

White, E. A. D. (1955), *Acta Cryst.* **8,** 845.

Wilson, C. L. (1959), *Comprehensive Analytical Chemistry,* Vol. IA, edited by C. L. Wilson and D. W. Wilson, Elseiver, New York, pp. 27–32.

SOLIDIFICATION

General Principles

15

Principles of Solidification

W. A. Tiller

The term *solidification* is commonly understood to imply the formation of a crystalline phase from its molten form. The phase transformation is driven by the extraction of heat from the melt, and the progress of the transformation is properly separated into two parts: (1) the initial nucleation of crystals and (2) the growth of these initial nuclei by the accretion of atoms from the melt. Much has been written concerning the initial nucleation of crystals both in this book and elsewhere (Turnbull, 1956; Pound, 1958), so that only slight mention of it will be made here. The main concern of this chapter is with the growth of a crystal once it has been formed. Attention is given to three important aspects of crystal growth from the melt: (1) solute manipulation during crystal growth, (2) solid-liquid interface morphologies, and (3) defects introduced during crystal growth.

1. CRYSTAL NUCLEATION

During the nucleation process, nuclei grow by adding single atoms to a growing cluster (embryo) of atoms having the configuration of the solid. This may happen homogeneously somewhere in the volume of the melt or heterogeneously on the surface of a foreign solid. Since, as appears later, homogeneous nucleation can be considered as one extreme of heterogeneous nucleation, and since the latter usually occurs in practice, the brief theoretical discussion given here considers only heterogeneous nucleation.

Following Pound (1958), a cap-shaped embryo forming on a perfect substrate is presented schematically in Fig. 1. It exists at a temperature T below the melting temperature T_m. Assuming that an equilibrium distribution exists among embryos containing i atoms and those consisting of one atom, the familiar relationship between the concentration $n(i)$ of embryos of size i and their Gibbs free energies of formation ΔG_i can be derived as

$$n(i) = n(1)\, e^{-\Delta G_i/kT} \tag{1}$$

where $n(1)$ is the number of single atoms of the supersaturated phase in contact with 1 cm^2 of substrate surface.

Both the volume free energy change $\Delta G_v = \Delta H(1 - T/T_m)$ and the specific interfacial free energies γ contribute to ΔG_i as can be appreciated from Fig. 2. Here ΔH is the enthalpy of fusion and θ is the equilibrium contact angle of the crystal c on the substrate s so that

$$\gamma_{ms} = \gamma_{cs} + \gamma_{mc} \cos\theta \tag{2}$$

Using the surface and volume of the cap,

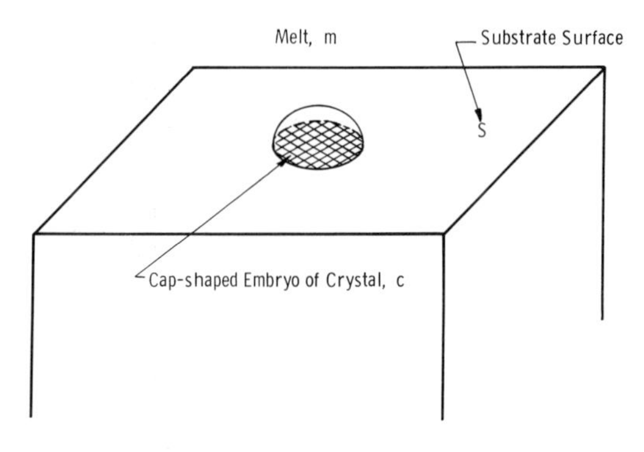

Fig. 1. Schematic cap-shaped embryo of a crystal C forming on a perfect substrate s from a melt m.

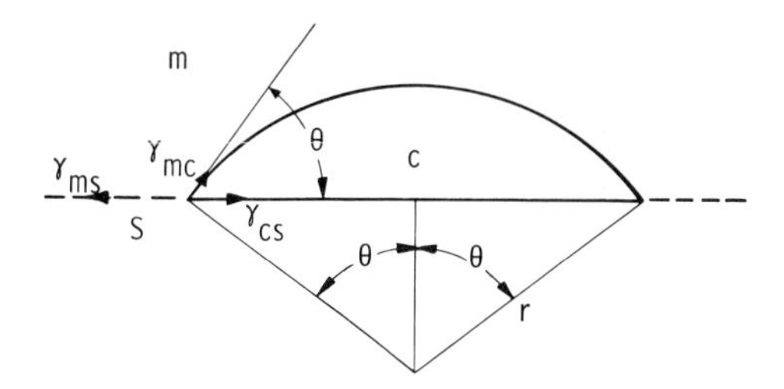

Fig. 2. The surface energy balance at an embryo-substrate interface.

$$\Delta G_i = \gamma_{mc}[2\pi r^2(1 - \cos\theta) - \pi r^2 \cos\theta(1 - \cos^2\theta)] + \left(\frac{\pi r^3}{3}\right)(2 - 3\cos\theta + \cos^3\theta)\Delta G_v \quad (3)$$

Equation (3) exhibits a maximum ΔG^* at $r = r^*$, so that clusters of solid having $r < r^*$ are unstable, whereas for $r > r^*$ the clusters can grow steadily with a decrease in free energy and are therefore stable. It is clear that formation of the critical size nucleus r^* requires a statistical fluctuation to produce the requisite increase in free energy. Maximizing Eq. (3) with respect to r at constant θ and T, we obtain

$$r^* = \frac{2\,\gamma_{mc}}{\Delta G_v} \quad (4)$$

Substituting into Eq. (3),

$$\Delta G^* = \frac{16\pi\gamma^3{}_{mc}}{3\Delta G_v{}^2}(2 + \cos\theta)\frac{(1 - \cos\theta)^2}{4} \quad (5)$$

and

$$n(i^*) = n(1)\,e^{-\Delta G^*/kT} \quad (6)$$

If $\theta = 0$ in Eq. (5) there is complete wetting of the solid by the crystal and $\Delta G^* = 0$ (actually, θ cannot equal zero since the mimimum height of a nucleus is one-lattice spacing). At the other extreme, if $\theta = 180°$ we have the case of "no wetting," which is the case of "homogeneous" nucleation. Then,

$$\Delta G^* = \frac{16\pi\gamma^3{}_{mc}}{3\Delta G_v{}^2} \quad (7)$$

$n(1)$ is the number of atoms per unit volume and $n(i^*)$ is the number of nuclei per unit volume. Usually, the contact angle θ is less than 180° and the rate of nucleation of new crystals I per cm²/sec on the substrate is given by

$$I = \nu n(i^*) \quad (8)$$

where ν is the frequency at which a single atom from the melt joins a critical nucleus to make it a stable crystal, so that

$$\nu = n_s{}^*\epsilon\nu_L e^{-\Delta G_D{}^*/kT} \quad (9)$$

In Eq. (9) $n_s{}^*$ is the number of atoms of liquid in contact with the surface of the critical nucleus ($n_s{}^* \sim 100$), ϵ is the probability of a given atomic jump direction ($\epsilon \sim \frac{1}{6}$), ν_L is the lattice vibrational frequency ($\nu_L \sim 10^{13}$ sec^{-1}), and $\Delta G_D{}^*$ is the free energy of activation for diffusion in the melt ($\Delta G_D{}^* \sim kT$). Combining Eqs. (1), (2), and (9) we have

$$I = n_s{}^*\epsilon\nu_L n(1)e^{-(\Delta G_D{}^* + \Delta G^*)/kT} \quad (10)$$

In Eq. (10) I increases very rapidly with decrease in temperature from a negligible to a very large value with a small decrease in temperature as illustrated in Fig. 3. Thus one may speak of a critical supercooling for an appreciable nucleation rate on a particular catalytic surface. This means that there is only a small probability that a given melt can be cooled below some critical temperature T_c without crystal formation.

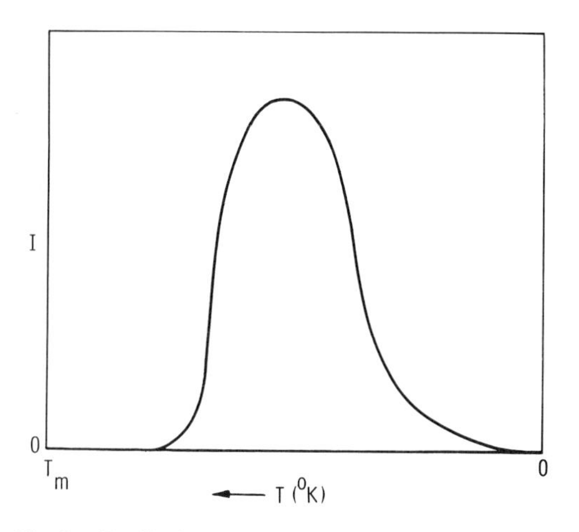

Fig. 3. Qualitative variation of the rate of nucleation of new crystals I as a function of substrate temperature T.

2. SOLUTE MANIPULATION

The type of solute distribution that one wishes to have in a crystal determines the freezing conditions to be used. The conditions will also depend on whether a conservative or a nonconservative system is being considered, that is, whether the solute or solvent has high vapor pressure. Finally, the conditions depend on whether one is interested in short range or long range solute distributions; that is, one may wish to purify an ingot or to make a crystal of constant solute concentration or a crystal with a particular solute distribution.

The behavior that allows solute manipulation is the fraction of solute in the solid different from that in the liquid phase at equilibrium. A measure of this difference is the partition coefficient k defined as the ratio of the solute concentration in the freezing solid C_S to that in the bulk liquid, C_∞. It may have three values: (1) the equilibrium value k_0, (2) unity, or (3) the effective value k which falls somewhere between k_0 and unity. Solute having $k_0 > 1$ raise and solutes with $k_0 < 1$ lower the melting point of the solvent as illustrated in Fig. 4. In both cases, freezing begins when the temperature of the liquid is reduced to T, where solid of concentration k_0C_0 forms.

At constant pressure, a pure liquid may be in equilibrium with a pure solid at only one temperature, the melting temperature. Here, atoms leave the liquid and become part of the solid at some characteristic rate R^f. At the same time, atoms leave the solid and become part of the liquid at a characteristic rate R^m, which equals R^f at

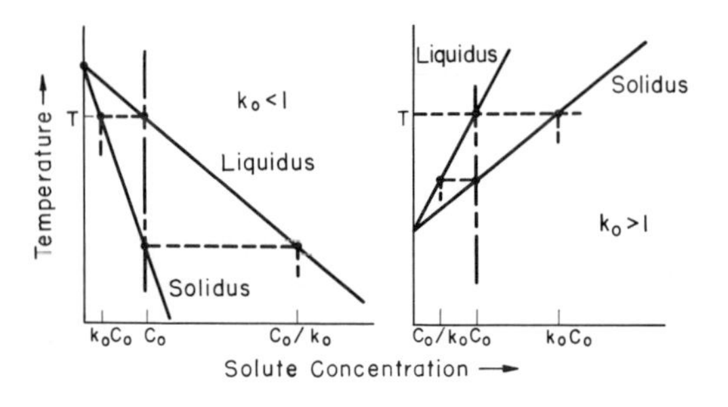

Fig. 4. Portions of phase diagrams for the cases where the solute either lowers ($k_0 < 1$) or raises ($k_0 > 1$) the melting point of the alloy relative to the pure solvent.

the melting temperature. These two rates describe the kinetics of exchange between the liquid and the solid phase. Therefore, the equilibrium partition coefficient $k_0{}^A$ of a solute A is given by

$$k_0{}^A = \frac{X_A{}^s}{X_A{}^L} = \frac{R_A{}^f}{R_A{}^m} \tag{11}$$

where $X_A{}^i$ is the mole fraction of solute A in phase i.

It has been shown by Thurmond (1959) that the interface partition coefficient $k_i{}^A$ should decrease with increase of the freezing rate V in the following manner:

$$k_i{}^A = \frac{k_0{}^A}{1 + \dfrac{V}{R_A{}^m}} \tag{12}$$

Furthermore, enrichment or depletion of the solute concentration in the interface layer often occurs because of adsorption effects. Under equilibrium conditions, the interface partition coefficient is still $k_0{}^A$ since the bulk solid can communicate with the bulk liquid through the interface layer. However, at normal freezing rates $k_i{}^A$ may be changed from $k_0{}^A$ if the kinetic coefficients $R_A{}^m$ and $R_A{}^f$ are altered by the presence of the adsorbed layer. In this discussion the equilibrium value k_0 will be used to represent k_i since Eq. (12) and adsorption effects cannot be adequately evaluated at present.

The equilibrium partition coefficient $k = k_0$ is rarely achieved in a practical freezing process, although it may be approached closely. For brevity, we discuss only the case of $k_0 < 1$. Then, unless the freezing rate is very small or mixing of the liquid is great, the solute rejected by the freezing solid accumulates at the solid-liquid interface (Fig. 5). The concentration in the freezing solid is k times that in the bulk of the liquid ahead of the diffusion boundary layer of thickness δ_c. A general expression relating k to k_0 and the freezing conditions (Burton et al., 1953; Pfann, 1957) is

$$\frac{k}{k_0} = \frac{1}{k_0 + (1 - k_0)e^{-(\delta_c V/D_L)(\rho_S/\rho_L)}} \tag{13}$$

where V is the growth rate, δ_c is the thick-

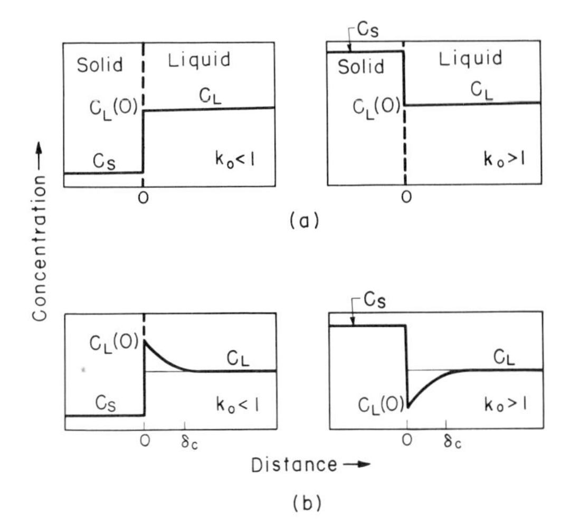

Fig. 5. Solute concentration in the liquid C_L ahead of the solid-liquid interface for $k_0 < 1$ and $k_0 > 1$: (*a*) complete mixing, $k = k_0$, (*b*) partial mixing, k/k_0.

ness of the diffusion boundary layer, D_L is the diffusivity of the solute in the liquid, and ρ_S/ρ_L is the ratio of the densities of the solid and liquid. Since $\rho_S/\rho_L \approx 1$, if $(V\delta_c/D)$ is large, $k = 1$, whereas if $(V\delta_c/D)$ is small, $k = k_0$. Thus, control of the gross parameter $(V\delta_c/D)$ allows one to control k between the limits $k_0 \lesseqgtr k \lesseqgtr 1$. We shall assume throughout that $\rho_S/\rho_L = 1$.

One procedure for controlled freezing is to cool a long molten bath progressively from one end; another is to melt a small portion of a long solid rod and slowly move this molten portion through the remainder of the rod. The first procedure is called *normal freezing;* the second is called *zone-melting* (Pfann, 1957). Each leads to characteristic solute distributions in the resulting solid which may be altered by changing the freezing procedure in one of two ways: (1) changing the solute concentration of the bulk liquid during freezing (by changing the liquid volume, by adding or removing a component from the liquid, or by using a charge of variable concentration), (2) changing V or δ_c to influence the interface diffusion layer.

Normal Freezing

Freezing that proceeds along a cylinder from one end is the method generally used for preparing monocrystals. The main factors are the fraction of the total volume that is solid, and the rejected solute that accumulates in the liquid fraction. This increased solute concentration in the liquid produces an increasing solute concentration in the solid, leading to segregation along the solid cylinder (called *normal segregation*).

For constant freezing conditions, four distinct segregation curves can result from normal freezing as illustrated in Fig. 6. (1) If the freezing is slow enough, diffusion in the liquid and solid eliminates all concentration gradients. This case, called *equilibrium freezing,* leads to no segregation, but is never realized in practice because diffusion rates in the solid are too slow. (2) The freezing is slow enough for mixing to erase all concentration gradients in the liquid but fast enough to make diffusion in the solid negligible. This case of *complete mixing,* leads to the maximum possible segregation and may be closely approached in practice. (3) The freezing is rapid enough so that only liquid diffusion affects the solute distribution at the interface. This case, called *no mixing,* leads to only slight segregation and is often realized in practice. (4) The solute distribution is affected both by diffusion and convection. This case, called *partial mixing,* leads to intermediate segregation and is the usual case in practice.

PARTIAL MIXING. The case of partial mixing is important in practice and will be dis-

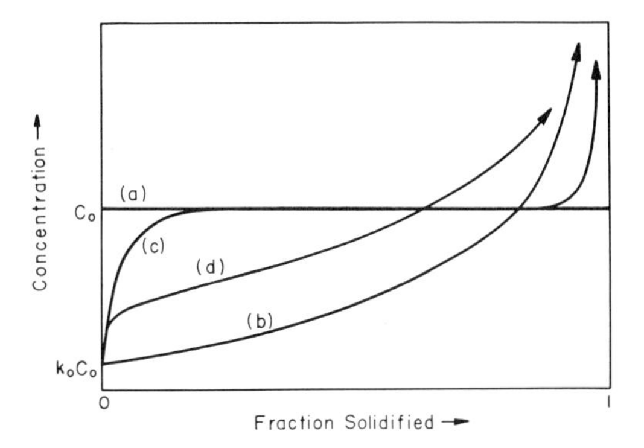

Fig. 6. Solute distributions in a solid bar frozen from liquid of initial concentration C_0, for (*a*) equilibrium freezing, (*b*) complete mixing, (*c*) no mixing, and (*d*) partial mixing.

cussed in some detail. Three situations are worthy of note: (1) k = constant, conservative system, (2) k = variable, conservative system, and (3) k = constant, nonconservative system.

k Equals Constant, Conservative System. For the case of $k_0 < 1$, the solid that freezes from the liquid will be purer than the liquid, so that rejection of solute into the liquid will occur at the solid-liquid interface. Because time is required for diffusion of solute into the liquid, an enriched layer will form just ahead of the advancing interface. Stirring of the liquid limits this enrichment to a layer of thickness δ_c. If δ_c and the freezing velocity V are held constant, the appropriate value of k is soon developed in the layer, the liquid interface concentration becoming $C_i = (k/k_0)C_\infty$. If k remains constant throughout the freezing process, the distribution of the solute in the solid C_s is given as a function of the fraction of liquid solidified g by

$$C_s = kC_0(1 - g)^{k-1} \qquad (14)$$

where C_0 represents the initial concentration of solute in the liquid. Curves of C_s/C_0 as a function of g for various k are shown plotted in Fig. 7 ($x_1 = g$).

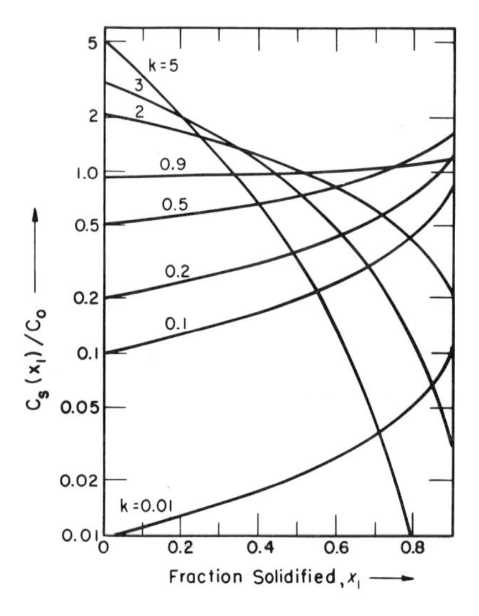

Fig. 7. Relative solute concentration in the solid $C_s(x_1)/C_0$ as a function of distance along a bar x_1 for normal freezing of a bar of unit length and for solutes with different k-values (C_0 = initial liquid concentration).

k Equals Variable, Conservative System. If k is allowed to vary with g and the total solute content is held constant, various segregation curves result that differ from those of Fig. 7. For example, it is possible to produce linear or sinusoidal curves with the proper $k(g)$ variation (Johnston and Tiller, 1962). One can vary $k(g)$ through either $\delta_c(g)$ or $V(g)$. Assuming that the solute distribution in the boundary layer adjusts immediately to accord with a new k, the value of $k(g)$ needed to produce any $C_s(g)$ may be determined. If solute is conserved,

$$\left(\frac{C_s(g)}{k} - C_0\right)(1 - g) = \int_0^g [C_0 - C_s(g)]dg \qquad (15)$$

Therefore

$$k = C_s(g)\Big/\left[C_0 + \left\{\int_0^g [C_0 - C_s(g)]dg\right\}\Big/\{1 - g\}\right] \qquad (16)$$

All constants entering the latter expression must be so chosen that k never lies outside the limits $k_0 \leq k \leq 1$. To demonstrate the validity of the method, the results of experiments on Pb-Sn alloys designed to give a linear distribution and a portion of a cosine distribution are shown in Fig. 8 ($x_1 = g$).

k Equals Constant, Nonconservative System. Suppose the solute is volatile, so that there is an exchange of solute between the liquid and gaseous phases. Since we are dealing with a nonconservative system, to calculate $C_s(g)$ for constant k, the change of liquid concentration with g must be considered (van den Boomgaard, 1955). Let C_{eq} be the concentration of the liquid is equilibrium with the gas phase. The reaction rate dC_L/dt of the liquid and gas phases is directly proportional to the number of atoms entering the melt in time dt through the surface and is inversely proportional to the melt volume. The rate can be described by

$$\frac{dC_L(g)}{dt} = \frac{K_R}{h}[C_{eq} - C_L(g)] \qquad (17)$$

where K_R is the reaction constant and h is the ratio of volume to surface area of the melt. By calculating the change in con-

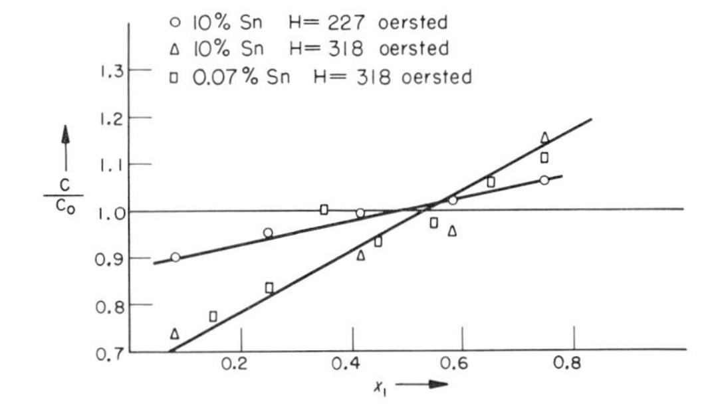

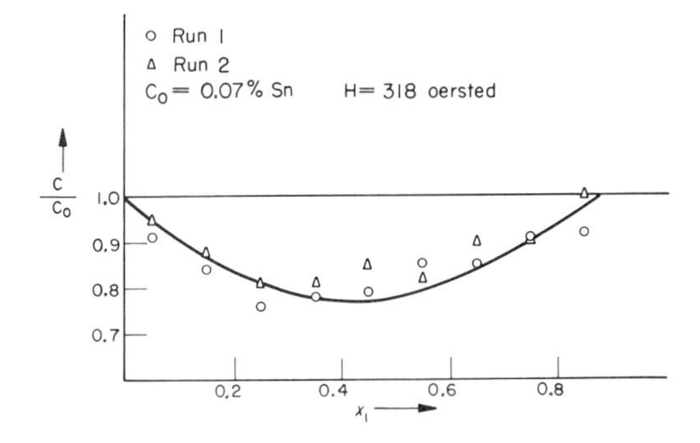

Fig. 8. Experimental test of programmed linear and cosine solute distributions in the solid C/C_0 as a function of distance x_1 for a bar of unit length (Pb-Sn alloys; H = magnetic field of stirring device).

centration of the melt (when the interface is at g) due to partitioning of solute with the gas phase, the form $C_s(g)$ may be found as a solution of the equation

$$\frac{dC_s}{dg} + \left(\frac{k-1}{1-g} + \frac{K_R}{hV}\right) C_s = \frac{K_R}{hV} C_{\text{eq}} \quad (18)$$

The solution is an integral equation which can be evaluated if the constants h, K_R, and V are known. If $K_R = 0$, the solution is given by Eq. (14).

NO MIXING. A single phase alloy forming from a melt with no-mixing is also of interest. The following assumptions are generally made:

1. Diffusion in the solid is negligible.
2. Convective mixing in the liquid is negligible.
3. k_0 is constant.
4. The interface separating solid and liquid is planar, and perpendicular to the axis of the specimen.
5. V is constant.

Since the solute concentration in the solid is related to that in the liquid by k_0, the solute distribution in the solid may be obtained from the concentration in the liquid as a function of interface position. If the original melt is of uniform concentration C_0, the high solute content in the boundary layer must increase until the concentration freezing into the solid is also C_0. Then the concentration in the liquid at the interface will be C_0/k_0. No further variation with time will occur if the growth conditions remain fixed. This solute distribution in the liquid may be calculated from the diffusion equation with a coordinate system that moves with the interface. Two conditions are: (1) the infinity conditions $C_L = C_0$ and (2) the interface flux condition $V(1 - k_0)C_L = -DdC_L/dx$ (Tiller et al., 1953). The result is

$$\frac{C_L}{C_0} = 1 + \frac{q}{k_0} e^{-(V/D)x} \quad (19)$$

where $q = 1 - k_0$ and x is distance from the interface.

Equation (19) is plotted in Fig. 9 for several values of k_0. Note that the concentration decreases with distance from the interface rapidly for high rates of growth and that the amount of impurity carried

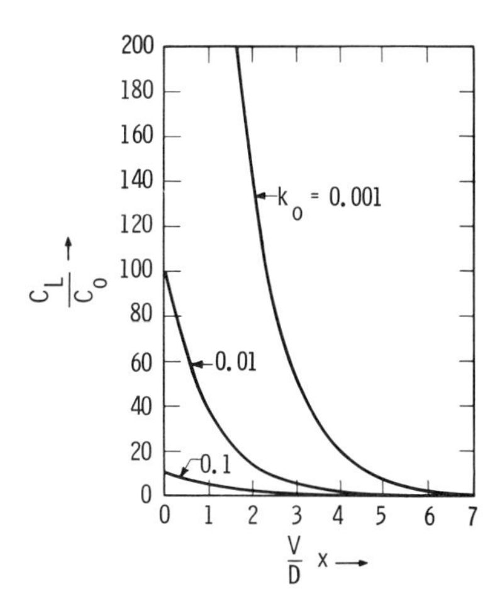

Fig. 9. The relative solute distribution in the liquid, C_L/C_0, for "no mixing," as a function of the dimensionless parameter Vx/D, for various k_0 (V = the constant freezing velocity, D = solute diffusion coefficient in the liquid, C_0 = initial solute concentration in the liquid, and x is distance in the liquid ahead of the freezing interface).

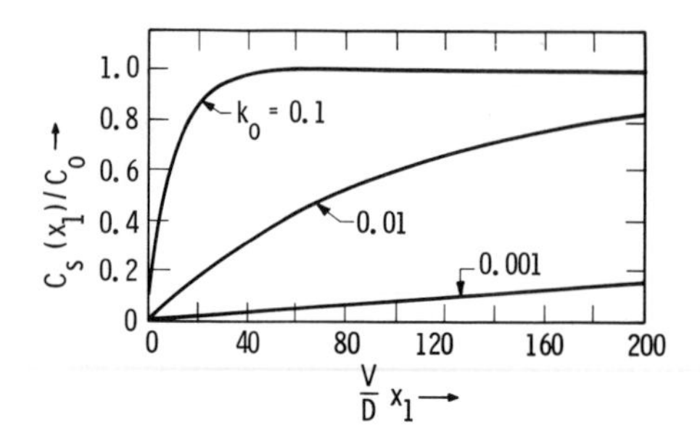

Fig. 10. Initial transient solute-distribution in the solid $C_s(x_1)/C_0$ for "no mixing" as a function of the dimensionless parameter $V\,x_1/D$, for various k_0. (x_1 = distance from the beginning of the crystal, V = constant.)

ahead of the interface (integral of Eq. 19) increases with decreasing rate of growth. The effective thickness of the solute-rich layer is D/V. Since $D \sim 10^{-5}$ cm^2/sec, this "characteristic distance" is between 10^{-1} and 10^{-4} cm for practical freezing rates.

During the formation of the steady state solute distribution in the liquid, an "initial transient" distribution is developed in the solid having the form shown in Fig. 10 (Smith et al., 1955). The "characteristic distance" of $C_s(x_1)$, given by D/k_0V, may be several centimeters long. x_1 is the distance from the beginning of the solid.

Increasing or decreasing the freezing rate during solidification of a crystal produces *intermediate transient* solute distributions which consist of bands of high or low solute concentration respectively (Tiller et al., 1953; Smith et al., 1955). As may be seen from Fig. 9, if V changes abruptly to $V_1(V_1 > V)$, the solute distribution in the liquid must become steeper. When C_L attains its new steady state distribution $C_L(0)$ must once again be C_0/k_0. The amount of solute in the enriched layer is now less than it was before, the difference having been deposited in the solid during the transient. Conversely, if $V_1 < V$, a band of solute depleted solid is produced. If the velocity fluctuates between V and V_1, the solute concentration in the solid fluctuates with the same periodicity. In Fig. 11, $C_s(x_2)/C_0$ is plotted against $(V_1/D)x_2$, where x_2 is distance from the point of abrupt change.

As the solid-liquid interface, carrying ahead of it the steady state solute distribution C_L, approaches the end of the specimen at constant speed V, the solute represented by the area under the C_L curve must appear in the final portion of solid. This makes a *terminal transient* (Smith et al., 1955) which is plotted in Fig. 12 as $C_s(x_3)/C_0$ versus $(V/D)x_3$ for various k. x_3 is distance from the end of the solid.

The foregoing discussion has dealt with constant freezing velocity and planar interfaces. Important differences arise when these conditions are not met as illustrated by two examples: (1) planar interface, $V = \beta t^{-1/2}$, where β = constant and t =

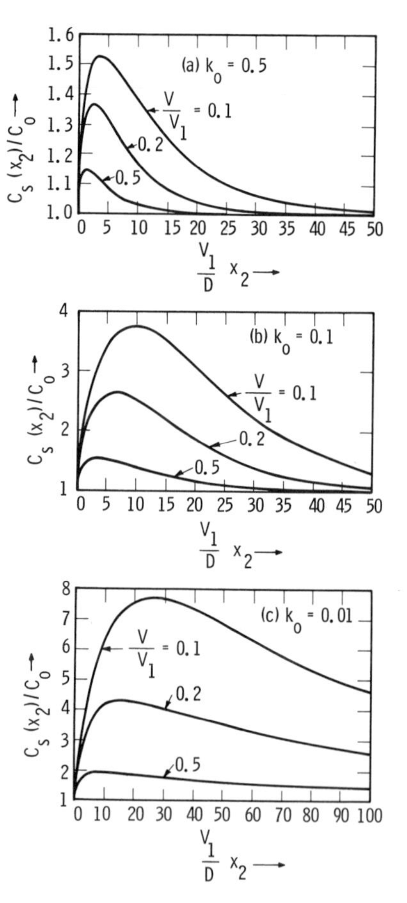

Fig. 11. Intermediate transient solute-distribution in the solid $C_s(x_2)/C_0$ for "no mixing" as a function of the dimensionless parameter V_1x_2/D, for various k_0, and V_1/V = 0.1, 0.2, and 0.5. (V = constant freezing velocity before the change to the new constant freezing velocity V_1, and x_2 is distance in the solid from the point of change in freezing rate.)

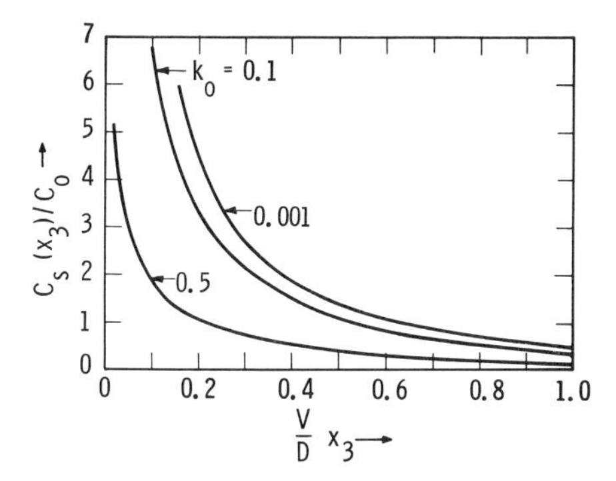

Fig. 12. Terminal transient solute-distribution in the solid $C_s(x_3)/C_0$ for "no mixing" as a function of the dimensionless parameter Vx_3/D for various k_0. (x_3 = distance in the solid from the end of the crystal.)

time and (2) paraboloidal interface, V = constant. For example (1), the diffusion equation predicts a constant interface concentration C_i throughout the freezing of a long bar (except for a terminal transient) (Tiller, 1962). The magnitude of C_i is given by

$$\frac{C_i}{C_0} = \frac{k}{k_0} = \frac{1}{1 - \sqrt{\pi}(1 - k_0)(\beta/\sqrt{D})\, e^{\beta^2/D} \operatorname{erfc}(\beta/\sqrt{D})} \tag{20}$$

which is plotted in Fig. 13. We can see $C^* = C_i/(C_0/k_0) < 1$ and that a crystal can be grown with constant concentration over most of its length by maintaining

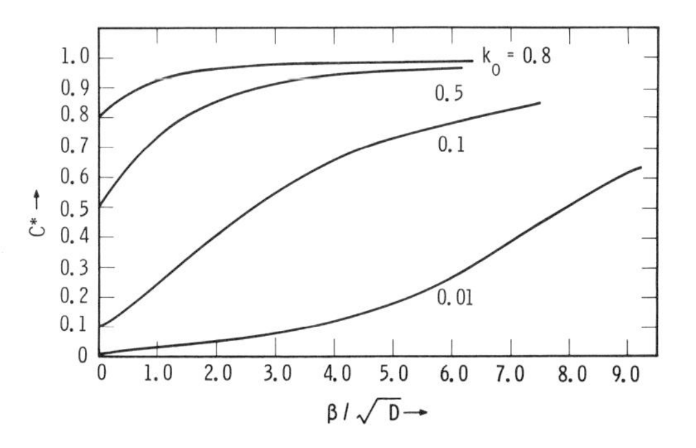

Fig. 13. Relative solute distribution in a one-dimensionless liquid at a planar freezing interface $C^* = C_i/(C_0/k_0)$ as a function of the dimensionless parameter $\beta/\sqrt{D}$ for various k_0, and for $V = \beta t^{-1/2}$ (t = time and β = constant).

$V = \beta t^{-1/2}$. The value of C_i will depend upon the parameters β and D. This type of growth law occurs when a superheated metal is poured into a long mold having insulated sidewalls and one end maintained at a constant low temperature.

The diffusion equation for Example (2) (Bolling and Tiller, 1961) again predicts an isoconcentrate surface for a solid with the shape of a paraboloid growing at constant velocity in its axial direction (this approximates a growing dendrite). Here

$$\frac{C_i}{C_0} = \frac{k}{k_0} = \frac{1}{1 + (1 - k_0)\left(\dfrac{V\rho}{2D}\right) e^{V\rho/2D} E_i\left(\dfrac{V\rho}{2D}\right)} \tag{21}$$

where E_i is the integral exponential function and ρ is the radius of curvature of the tip of the paraboloid. In Fig. 14, k/k_0 is plotted versus the natural variable $V\rho/2D$ for several k_0, showing that for $k_0 < 1$, $C_s/C_0 < 1$, and for $k_0 > 1$ $C_s/C_0 > 1$. Thus this type of growing body is either depleted or enriched in solute with respect to the bulk liquid.

For these "no mixing" cases, when the

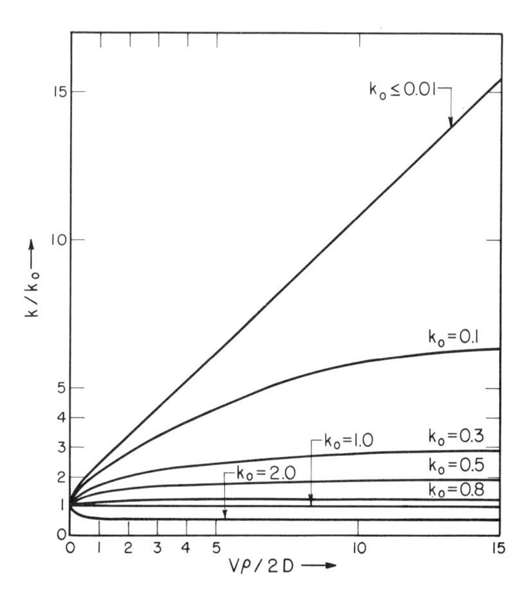

Fig. 14. Relative solute partition coefficient $k/k_0 = C_i/C_0$ in a paraboloidal body freezing at a constant axial velocity V as a function of the dimensionless parameter $V\rho/2D$ for several values of k_0. (ρ = radius of curvature of the tip of the paraboloid.)

freezing velocity is constant and the solid-liquid interface is planar, the steady state average value of k is unity. However, when the freezing velocity is not constant as in Example (1) or when the interface is not planar as in Example (2) the effective partition coefficient is not unity and $C_s/C_0 \neq 1$.

Zone Melting

In this procedure (Fig. 15) only part of the charge is melted at any one time, and both melting and freezing are carried out progressively. This provides important practical advantages in obtaining solute control. The process variables are (1) zone length, (2) charge length, (3) initial distribution of solute in the charge, (4) vapor pressure, (5) mixing conditions, and (6) zone velocity. If (1) to (6) remain constant during a zone passage, a variety of solute distributions will result in the solid, depending on the starting conditions. Furthermore, if these variables are changed during the zone passage, an almost limitless array of solute distributions may be produced.

NO MIXING. Provided the zone length l is greater than the total thickness of the solute distribution in the liquid C_L, the solute distribution in the solid $C_s(x_1)$ will be identical with that found for normal freezing. Perturbations of the freezing conditions also give the same results for the normal freezing provided the initial charge is of constant concentration. In practice, the zone length l is usually several centimeters long, whereas the thickness of the solute boundary layer is only a fraction of a centimeter.

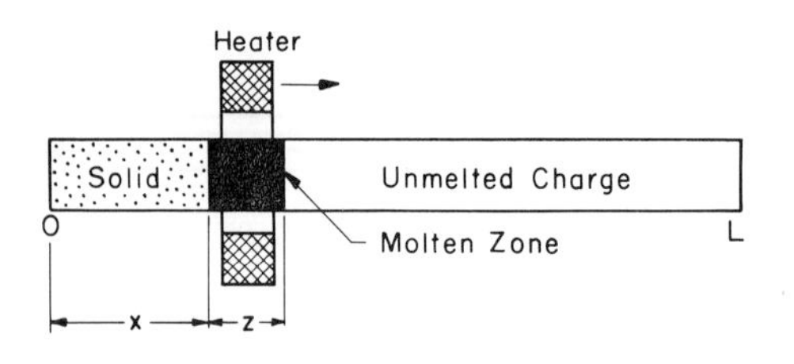

Fig. 15. Schematic arrangement for zone melting.

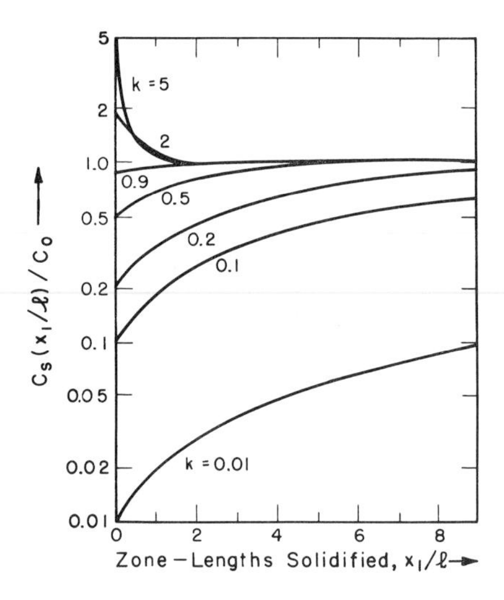

Fig. 16. Relative concentrations in the solid after single-pass zone melting; $C_s(x_1/l)/C_0$, as a function of the number of zone lengths x_1/l, for several values of k(l = constant zone length, k = constant, C_0 = constant charge concentration).

PARTIAL MIXING. There are six cases of interest under this heading. They are (1) single-pass constant conditions, (2) single-pass variable k, (3) single-pass variable charge concentration, (4) single-pass exchange with the vapor, (5) multiple-pass constant conditions, and (6) temperature gradient zone melting.

Single-Pass Constant Conditions. If a molten zone of length l is passed through a charge of constant cross section and constant concentration, C_0, at constant k, the resulting solute distribution in the solid, $C_s(x_1)$, is given by (Pfann, 1957)

$$\frac{C_s(x_1)}{C_0} = \left\{1 + (1 - k)\exp\left(-\frac{k\rho_s}{\rho_L}\frac{x_1}{l}\right)\right\} \tag{22}$$

where ρ_s/ρ_L is the ratio of the density of the solid to the density of the liquid. In Fig. 16, Eq. (22) has been plotted for $C_s(x_1)/C_0$ as a function of the number of zone lengths x_1/l, for several k with $\rho_s/\rho_L = 1$ and for a bar ten zone lengths long. Note that (1) as k increases, C_s/C_0 approaches unity at lower values of x_1/l. For a given k, the yield of uniform material can be increased by increasing the rod length or by reducing l and (2) the difference between the mean and initial concentrations is proportional to $(1 - k)$ as in normal freezing.

Zone melting of a uniform rod results in an initial and a final section where $C_s(x_1)/C_0$ is considerably different from unity. An improvement can be made by passing the zone in the reverse direction. Even if k is very small, a substantially uniform concentration exists in all but the last zone after the reverse pass.

Single-pass variable k. By relaxing the assumption that k be held constant, many new solute distributions can be obtained (Johnston and Tiller, 1962). If one wishes to produce a certain $C_s(x_1)$ it is necessary to know the requisite $k(x_1)$. From conservation of solute requirements

$$k(x_1) = \frac{C_s(x_1)}{C_\infty + \dfrac{1}{l}\displaystyle\int_0^{x_1} [C_\infty - C_s(x_1)]\, dx_1} \tag{23}$$

Single-Pass Variable Charge Concentration. If the initial solute distribution in the charge is given by $C_0(x_1)$ for constant k and l, the conservation of solute condition gives for $C_s(x_1)$

$$\left[\frac{C_s(x_1)}{k} - \int_{x_1}^{x_1\ l} C_0(x)dx\right] = \int_0^{x_1} [C_0(x) - C_s(x)]\, dx \tag{24}$$

Two simple cases are of interest (Pfann, 1957): (1) $C_0(x_1) = C_0/k$ for $0 \leq x_1 \leq l$ and $C_0(x_1) = C_0$ for $x_1 > l$. This case is called *zone leveling;* it eliminates the initial transient region in single pass zone melting so that $C_s = C_0$ throughout the bar except for the final zone to freeze. (2) $C_0(x_1) = C_0$ for $0 \lesseqgtr x_1 \lesseqgtr l$ and $C_0(x_1) = 0$ for $x_1 > l$. This is called *starting charge only.* The concentration of the solid is given in all except the last zone by

$$\frac{C_s(x_1)}{C_0} = ke^{-kx_1/l} \tag{25}$$

In Fig. 17, $C_s(x_1)/C_0$ is plotted versus x_1/l for several values of k. It should be noted from Eq. (25) that for $k = 0.01$ $C_s(x_1)/C_0$ decreases only about 10% in ten zone lengths.

Single-Pass Exchange with Vapor. If a molten zone of constant volume is moved at

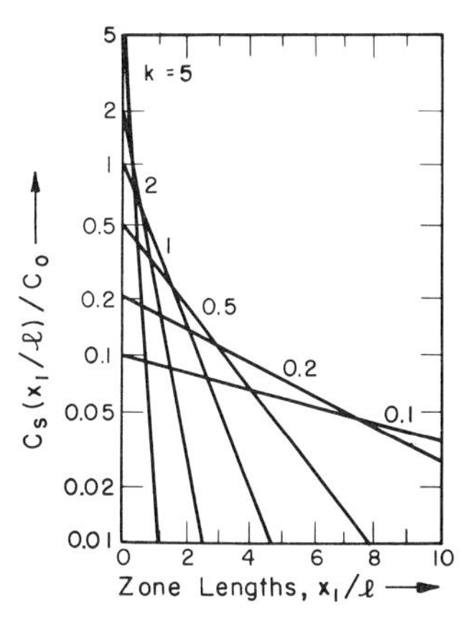

Fig. 17. Relative concentration in the solid after the "starting charge only" method of zone melting: $C_s(x_1/l)/C_0$ as a function of the number of zone lengths x_1/l for several values of k.

constant velocity through a charge of initial solute concentration C_0 under a constant solute vapor pressure B, three effects produce a change in $C_s(x_1)$ (van den Boomgaard, 1955). These are (1) reaction between liquid and vapor, (2) segregation at the freezing end of the molten zone, and (3) dissolution of solute at the leading end of the zone.

The concentration of the liquid will change as long as the number of solute atoms entering the zone differs from the number leaving it. After a certain time, however, the concentration stabilizes. The stable concentration will depend upon l, its volume, its velocity, and the solute partial pressure.

For an infinite rod $C_s(x_1)$ reaches the constant value C_{st}, given by

$$\frac{C_{st}}{C_0} = \frac{\Psi[C_s(\text{eq})/C_0] + 1}{\Psi + 1} \tag{26}$$

where $\Psi = K_R l/hVk$ and $C_s(\text{eq}) = kC(\text{eq})$ (see Eq. 7). A plot of C_{st}/C_0 versus Ψ is given in Fig. 18.

Multiple-Pass Constant Conditions. Recalling the solute distribution after the first pass (Fig. 16) for $k < 1$, consider the passage of a second zone through it. The zone accumulates solute as it moves through the initial region and leaves behind an impoverished and longer initial region. When the front of the zone reaches the

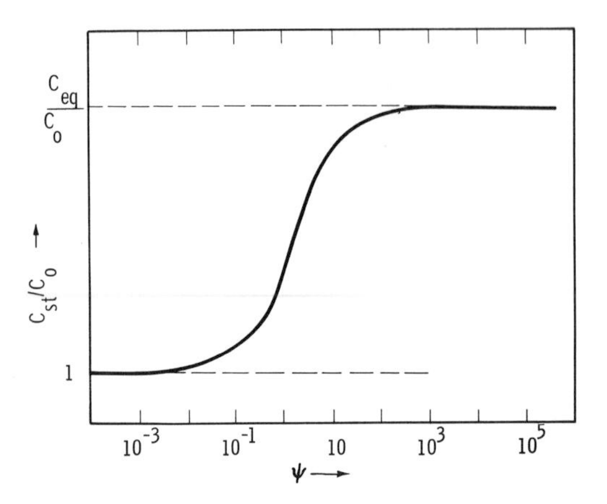

Fig. 18. Relative stationary-state concentration in the solid C_{st}/C_0 during the zone melting of an infinite rod containing a volatile solute; as a function of the dimensionless parameter, $\Psi = K_R l/hVk$ (C_{eq} = equilibrium concentration of the liquid, K_R = reaction constant between the liquid and gas phases, and h = ratio of volume to surface area of the melt).

edge of the last zone length, the slope of the solute distribution will begin to rise sharply. Thus the enrichment of the final part of the ingot progresses backwards one zone length during the second pass and one additional zone length for each succeeding pass. Multiple passes therefore lower the initial concentration, raise the final concentration, and decrease the length of the intermediate region (Pfann, 1957).

The basic equation relating the change in solute concentration in the moving zone to the difference between the fluxes of solute entering and leaving the zone has been derived independently by Lord (1953) and by Reiss (1954). In Fig. 19, $C_n(a)/C_0$ is plotted from the analytical treatment (Lord, 1953) of a semiinfinite bar versus the number of zone lengths l for the first eight passes with $k = 0.10$ and 0.25. For a finite bar more than $a + 8$ zones long the results are the same as in Fig. 19.

After many passes, the solute distribution in an ingot stabilizes and no further separation occurs. At this stage, the forward convective flux of solute is, at every value of x_1, opposed by an equal backward flux. For a charge of length L initial concentration of the solute C_0 and zone length l the ultimate distribution C_s after an infinite number of zone passes is given by

$$C_s = Ae^{Bx} \tag{27}$$

where $k = B/(e^{Bl} - 1)$ and $A = C_0BL/(e^{BL} - 1)$. Ultimate distributions calculated from Eq. (27) have been plotted in Fig. 20. These are very steep for small k; for example, if $k = 0.1$, $L = 10$, and $l = 1$, the ultimate concentration C_s at $x = 0$ is approximately 10^{-14} C_0. Since the concentration at $x = 0$ is decreased by a factor $\sim k$ in each pass, at least fourteen passes are required to reach the ultimate

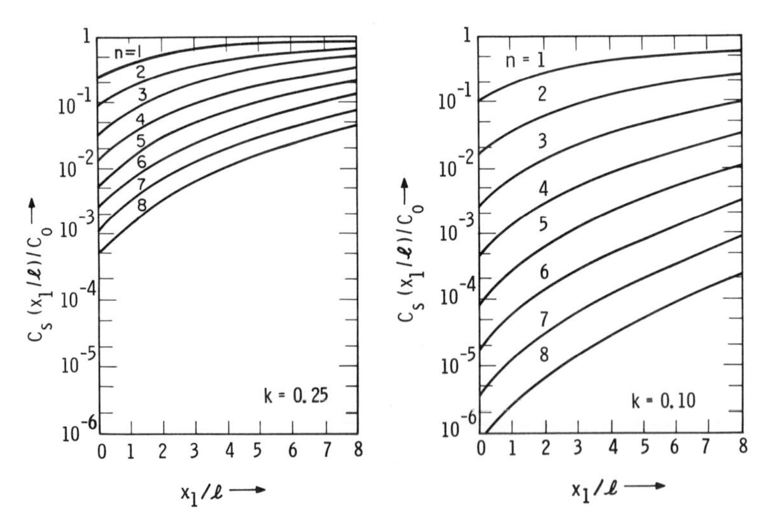

Fig. 19. Relative solute distribution in a semiinfinite solid $C_s(x_1/l)/C_0$ as a function of the number of zone lengths x_1/l for $k = 0.1$ and 0.25 after n zone passes.

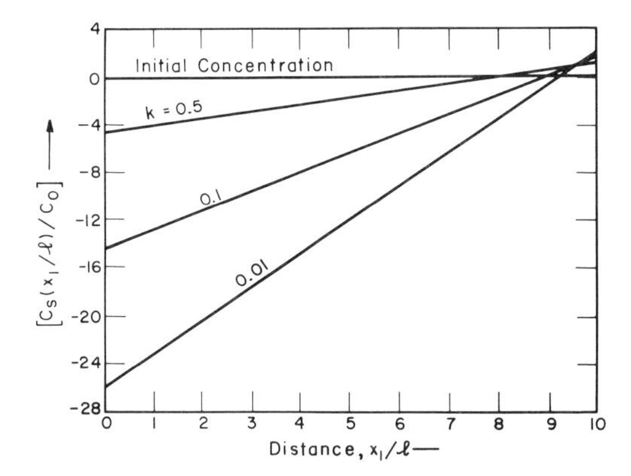

Fig. 20. Ultimate relative solute distribution in a semi-infinite solid; $C_s(x_1/l)/C_0$ as a function of the number of zone lengths, x_1/l.

distribution. Multiple-pass zone melting, called *zone-refining,* produces considerable purification of a portion of the charge and has become a powerful tool for purifying materials.

It is of interest to minimize the time and expense required to obtain a desired yield of material of specified purity. This requires optimization of the parameters: number of passes n, zone length l, interzone spacing i, travel rate v, and diffusion boundary layer thickness δ_c. Small l is desired because it produces a better separation, especially for a large number of passes. Small l and i reduce the time per pass. Large V also reduces the time per pass but it gives $k = 1$. One can circumvent this problem by stirring to reduce δ_c and thus k. If $V\delta_c/D \approx 1$, the optimum zone refining efficiency is achieved (Harrison and Tiller, 1961).

Temperature Gradient Zone Melting. This method differs from the others in these three ways: (1) the zone size, (2) the manner of moving a zone, and (3) usually the amount of solute in the zones. The zones are usually very small; they may be sheet zones, wire zones, or dot zones whose smallest dimensions may be of the order of mils. The zones have a high solute content, making them liquid at temperatures well below the temperature of the solid through which they move. They are caused to move by a stationary temperature gradient impressed across the entire charge (Pfann, 1957). The arrangement is illustrated in Fig. 21.

Consider the partial phase diagram of Fig. 21. A thin layer of solute is sandwiched between blocks of the solvent. The ensemble is placed in a temperature gradient with the solute layer hotter than the lowest melting temperature of the system and the highest temperature T_h lower than the melting point of the solvent. Enclosed in solvent, the solute layer melts and dissolves some solvent, thus expanding its dimensions. Dissolution of solvent at both ends of the layer continues until at temperature T_1 the cooler interface reaches the liquidus at concentration C_1. It continues at the hotter interface, at temperature T_2, until the liquidus is approached at concentration C_2. A solute concentration gradient then exists in the zone, and dissolution at T_2, followed by diffusion in the zone and freezing at T_1, causes the molten layer to move through the block leaving a solute distribution in the solid $C_s(x_1)$ given by

$$C_s(x_1) \approx -m_s[T_M - T_s(x_1)] \qquad (28)$$

where m_s is the solidus slope of the phase diagram (assumed constant), T_M is the melting temperature of the pure solvent, and $T_s(x_1)$ is the temperature distribution in the charge.

3. SOLID-LIQUID INTERFACE MORPHOLOGY

The morphology of an interface under a certain set of growth conditions will depend

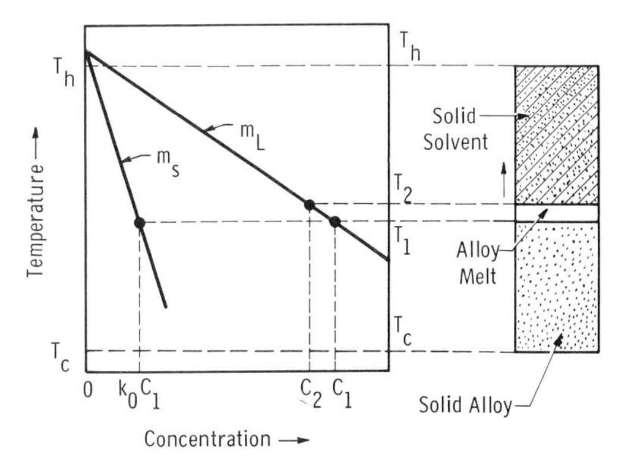

Fig. 21. Partial phase diagram plus the solid and liquid charge used in temperature gradient zone melting.

on: (1) the variables that influence the free energies of the phases, that is, temperature distribution T, solute distribution C, and interface curvature K, (2) mechanical equilibrium of surfaces, that is, grain boundaries, external surfaces, and internal phase boundaries, and (3) the atomic kinetics of the freezing process and any anisotropy of the atomic kinetics.

Several interface morphologies may be able to satisfy a set of these factors; however, one will respond most rapidly to the thermodynamic driving force and will therefore dominate the others. This will be the observed morphology in practice since the others will be overgrown. Thus to predict the expected interface morphology, one needs to solve for the time-dependent shapes and to choose the one that enables the transformation to proceed at the highest rate (consistent with the growth conditions).

It has been shown (Bolling and Tiller, 1961) that some criterion is needed to specify completely the problem of predicting the steady state interface shape. In the time dependent problem this criterion is not needed since the surviving morphology is the one producing the fastest growth rate. More will be said about the criterion. Usually one can expect to predict only the time-average interface morphology, neglecting small-scale time dependent features. This time-averaging process is assumed in what follows.

The steady state growth rate at any point of an interface whose surface-normal makes an angle ϕ with respect to a primary crystallographic direction is

$$V(\phi) = \mu(\phi)\,\delta T(xyz) = \{T_E[C(xyz)] - T(xyz) - \frac{\gamma(xyz)}{\Delta S}\,\mathrm{K}(xyz)\}\,\mu(\phi) \qquad (29)$$

where V is the freezing velocity, μ is the atomic kinetic coefficient, δT is the supercooling at the interface, $T_E[C(xyz)]$ is the equilibrium freezing temperature of a planar surface for the liquid of concentration C at the point (xyz) on the surface, γ the solid-liquid interfacial energy, and ΔS the entropy of fusion per unit volume. Equation (29) relates the freezing velocity to the departure from equilibrium δT at (xyz) on the surface. A series of supplementary boundary conditions must also be satisfied, both in the two phases and on the surface S at (xyz). These conditions are

1. For solute i in the solid and liquid

$$\nabla^2 C^i + \left(\frac{V}{D^i}\right)\frac{\partial C^i}{\partial z} = 0 \qquad (30a)$$

where D^i is the diffusion coefficient in the appropriate phase.

Equation (30*a*) is subject to the following boundary conditions on S:

$$C_S{}^i = k_0{}^i C_L{}^i \quad \text{and} \quad V_n(1 - k_0{}^i)C_L{}^i = - D_L\frac{\partial C_L{}^i}{\partial n} + D_S\frac{\partial C_S{}^i}{\partial n} \qquad (30b)$$

where V_n is the surface-normal component of the freezing velocity.

2. For heat flow in the solid and liquid

$$\nabla^2 T + \left(\frac{V}{\alpha}\right)\frac{\partial T}{\partial z} = 0 \qquad (30c)$$

where α is the thermal diffusivity in the appropriate phase. Equation (30*c*) is subject to the following boundary conditions on S:

$$T_S = T_L \quad \text{and} \quad -K_S\frac{\partial T_S}{\partial n} = -K_L\frac{\partial T_L}{\partial n} + V_n\,\Delta H \qquad (30d)$$

where ΔH is the enthalpy of fusion per unit volume and K is the thermal conductivity of the appropriate phase.

3. The equilibrium temperature of a planar surface having a concentration $C^1, \ldots C^i, \ldots C^n$ of solute $1, \ldots i, \ldots n$ can be found from the phase diagram so that

$$T_E = f\left(\sum_i C_L{}^i\right) \qquad (30e)$$

where f represents some functional relationship as determined from the phase diagram.

The stable interface morphology is the one which satisfies both Eqs. (29) and (30) simultaneously everywhere on S as well as the criterion (mentioned earlier) that the surface S will adopt that morphology, S_{opt}, which allows its extremities to have maximum velocity. This criterion may be written

$$\left(\frac{dV_{\text{ext}}}{dS}\right)_{S_{\text{opt}}} < 0 \tag{31}$$

That is, the freezing velocity of the crystal extremities decreases as the interface shape changes from the optimum. There are two important cases to consider: (1) for an isothermal and isoconcentrate melt the extremities of a growing crystal should be directed and shaped so as to allow them to propagate with maximum velocity and (2) for a nonisothermal system, in which the temperature changes steadily to cause freezing at a constant rate V; a change in interface morphology will occur only if the new morphology freezes more rapidly than the old during the short transient period before freezing again occurs at a rate V. The net result of such a morphology change is to allow the crystal extremities to project further into the melt. Two ways such a change can arise are (1) the interface temperature of the extremities is increased by the shape change or (2) the latent heat content of the extremities is decreased by the shape change.* In general, these may occur simultaneously. Before considering observations of important interface morphologies, a brief discussion of atomic kinetics is given.

Atomic Kinetics

A recent theory by Cahn and Hilliard (1958) suggests that the interface between two similar fluids is diffuse; this is expected to occur also at the interface between a crystalline solid and its melt. Thus the change in density does not occur abruptly at a plane but is spread over several atomic spacings. The thickness of the transition layer is determined by balancing the energy associated with the density gradient and the energy required to form material of intermediate composition.

More recently, Cahn (1960) has considered the motion of diffuse interfaces. At small driving forces, the free energy perturbation required to propagate a crystal face homogeneously is greater than the driving force. Then the surface may advance only by the lateral motion of steps originating either at a screw dislocation surface or at a site of two-dimensional nucleation. At high driving forces, the surface can advance normal to itself without needing steps, that is, every element of surface can continually advance.

Several investigators (Turnbull, 1950; Hillig and Turnbull, 1956; Jackson and Chalmers, 1956) have shown that for uniform interface motion, the interface velocity at small undercoolings should be linearly related to the undercooling δT at the interface

$$V = \mu_1 \, \delta T \tag{32}$$

The rate constant μ_1 in these treatments is in the range 1 to 100 cm/sec/°C. In Turnbull's treatment,

$$\mu_1 = \lambda \frac{kT}{h} \cdot \frac{\Delta S}{RT} \cdot e^{-\Delta F_A/RT} \tag{33}$$

where λ is the interatomic spacing normal to the interface, ΔF_A is the free energy of activation for the freezing process, and the other constants have their usual meanings.

When the rate of formation of surface steps limits the interface velocity, two prime cases arise: (1) two-dimensional nucleation and (2) steps at screw dislocations. For two-dimensional nucleation, Eq. (10) gives the nucleation rate with

$$\Delta G^* = \frac{\pi \gamma^2{}_{mc}}{\Delta G_v} \tag{34}$$

Thus the velocity is

$$V = \mu_2 e^{-\mu_3/\delta T} \tag{35}$$

where μ_2 and μ_3 are approximately constant. For a screw dislocation source, the velocity is (Hillig and Turnbull, 1956)

$$V = \mu_4 \delta T^2 \tag{36}$$

where $\mu_4 = 3(\Delta S)^2 D_L / 4\pi V_m R T \gamma$ and V_m is the molecular volume.

Following this brief description of atomic kinetics, we consider five important morphologies: (1) layer structures, (2) surface

* A surface morphology is stable if shape perturbations of smaller size cannot grow on it. Such perturbations cannot grow on the maximum velocity body or the maximum extent body by definition.

intersections, (3) cells, (4) dendrites, and (5) eutectics.

Layer Structures

If we consider the growth of a microscopically smooth interface at a constant velocity V (Fig. 22*a*) some undercooling δT_a of the interface is necessary. The relationship between V and δT depends on the crystallographic orientation of the interface. We expect that growth in the higher index directions occurs at a much smaller δT than in the lower index directions for the same V. In Fig. 22*b* is illustrated the formation of a layered interface consisting of segments of a low index plane and high index plane. Here, growth occurs primarily in the high index direction. The interface morphology *b* will be stable compared to *a* if $\delta T_b < T_a$, that is, *a* will become unstable if $\mu(\phi)\delta T_a > V/\cos\phi$, where ϕ is the angle between the high index direction and the net growth direction. The morphology *b* will also be stable with respect to *a* if a vacancy supersaturation develops for *b* relative to *a* that lowers the latent heat of fusion, ΔH of *b* relative to *a*. The net effect of the change of ΔH and T_i is to allow *b* to penetrate further into the melt than *a* and for *b* to be the stable morphology.

The important point here is that stepped

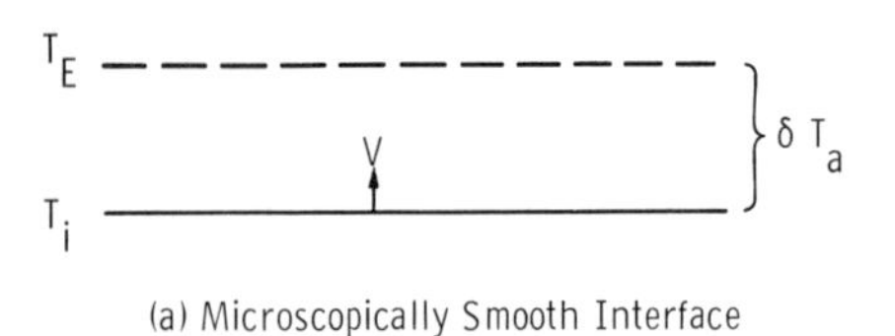

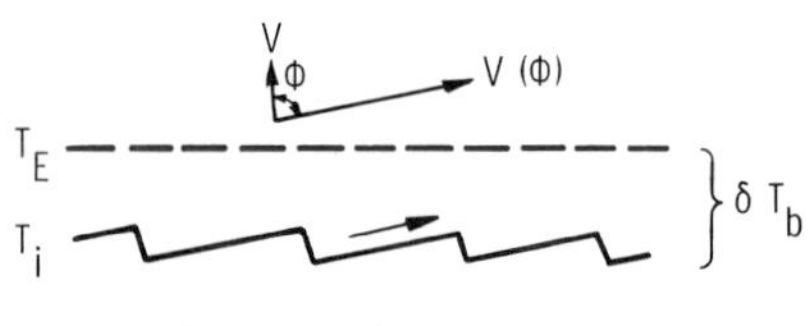

Fig. 22. Schematic illustration of the undercooling δT needed for freezing at the rate V. (*a*) Microscopically smooth interface. (*b*) Microscopically layered interface. (T_E = the equilibrium temperature and T_i = the average interface temperature.)

interfaces may arise as a consequence of an anistropy of atomic kinetics in pure materials. Such interfaces are very common in growth from solution and by electrodeposition, and have been observed in a variety of metals and compounds during growth from the melt (Elbaum and Chalmers, 1955; Rosenberg, 1956). An example of layers growing on a pure Pb crystal is shown in Fig. 23 (it has not been conclusively proved that such layers existed during growth, but this author thinks they did).

The consequences of any interface morphology arise from the distribution of chemical and physical imperfections that the particular morphology introduces into the crystal. As can be appreciated from the section on solute manipulation, layer growth may lead to periodic regions of chemical segregation when an impure melt is frozen. An important nonequilibrium effect arises when solute or solvent atoms adsorb at the surface of the solid. If the surface excess cannot be accommodated easily by the crystal, δT will become large during growth with adsorption. Four steps describe the situation: (1) adsorption of solute or solvent atoms on the interface either uniformly or on preferred planes, (2) inhibition of the sources of layers until a large departure from equilibrium develops followed by sudden source operation and layers rapidly traversing the interface, (3) trapping of excess solute in the solid followed by diffusion back into the interface, and (4) solute enrichment in the liquid at the traveling layer edge. Our present state of knowledge (Chernov, 1960; Trainor and Bartlett, 1961) allows analytic treatment of steps (3) and (4), and a qualitative interpretation of the rate dependence of k_0' as in Fig. 24. However, steps (1) and (2) cannot be described analytically. Therefore a complete assessment of adsorption and layer trapping effects is not yet possible.

Facet development at certain crystallographic orientations will be a macroscopic consequence of selective interface adsorption. Adsorption causes a decrease of the surface free energy γ. If there is

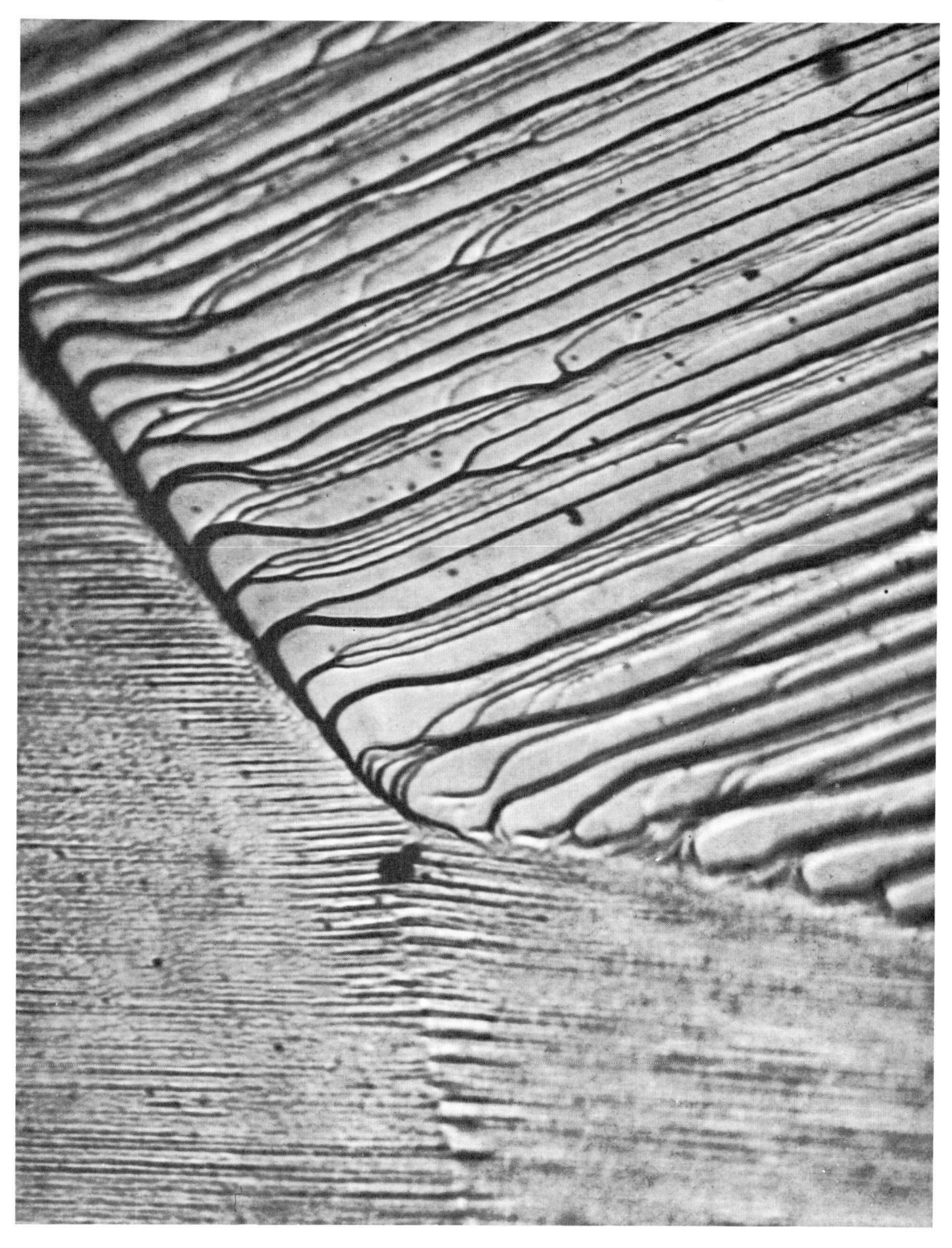

Fig. 23. Layer structure on the decanted interface of a high purity lead bicrystal (375 X).

some surface *a* of original surface energy γ upon which selective adsorption occurs, then after adsorption $\gamma_a < \gamma$. This will cause the *a* surface to become a singular orientation (Frank, 1958), and a facet will form on the interface. This is illustrated in Fig. 25 where (*a*) represents γ as a function of an orientation including the $\langle a \rangle$ direction. Figure 25*b* shows the effect of adsorption at the $\langle a \rangle$ pole. It produces a cusp

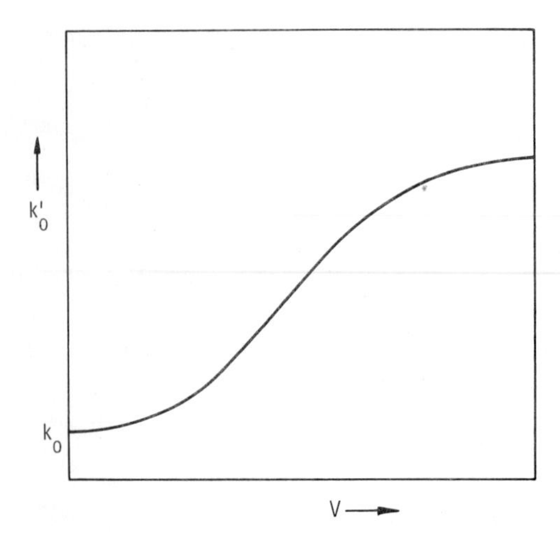

Fig. 24. Schematic variation of the apparent parition coefficient k'_0, with freezing velocity V as a result of interface adsorption.

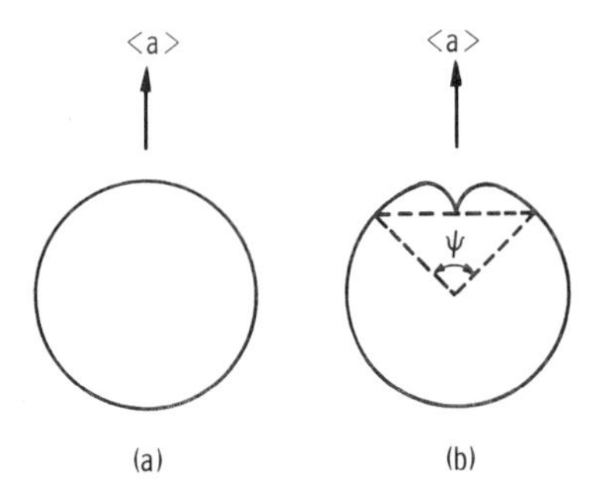

Fig. 25. The formation of a cusp at the $\langle a \rangle$ direction in a polar surface energy plot because of selective adsorption. (*a*) Before adsorption. (*b*) After adsorption.

such that all surfaces within the angle Ψ will disappear to expose facets having the $\langle a \rangle$ direction.

If the equilibrium isotherm T_E is such that the interface is curved, exposing a range of orientations to the liquid (including $\langle a \rangle$), then facets may form at the $\langle a \rangle$ pole. In Fig. 26*a* we see that if the T_E isotherm is hemispherical and convex to the liquid, a flat can form at the $\langle a \rangle$ pole no matter what the angle this makes with the specimen axis. The size of the flat will be such that $\gamma \cos \theta = \gamma_a$ (provided the supercooling at the flat is not so large that layer sources on the flat operate rapidly). If the T_E isotherm is paraboloidal and convex to the liquid, we can see from Fig. 26*b* that the size of a flat (holding θ constant) will decrease rapidly as the angle Ψ between $\langle a \rangle$ and the specimen axis increases (the relative magnitude of the two principal radii of curvature changes greatly as Ψ increases). Finally, if the T_E isotherm is concave to the liquid, flats cannot form since they would be superheated.

The stability conditions indicate that facets can sometimes develop on a surface, but not always. Thus, to eliminate facet formation, one should either make the T_E isotherm concave to the liquid or paraboloidal and convex with large Ψ.

Surface Intersections

Again a solution of Eq. (29) is sought, this time in the vicinity of a grain boundary or an external surface (that is, a crucible surface or free surface). The additional

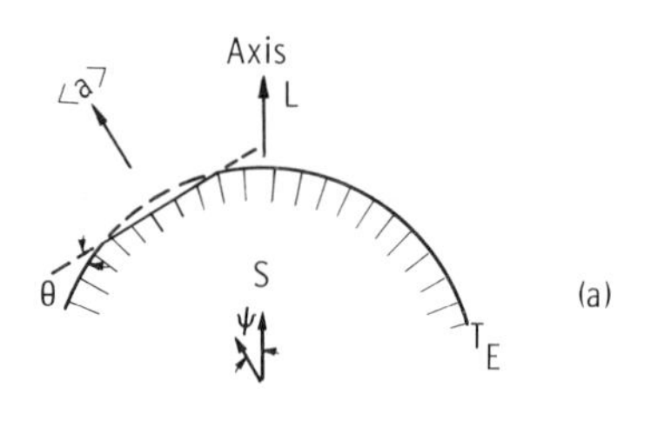

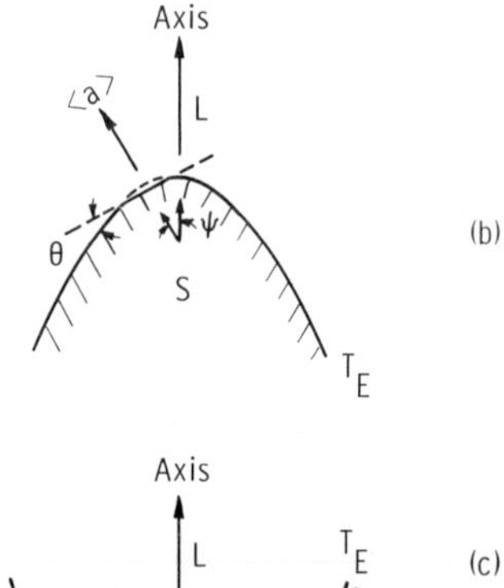

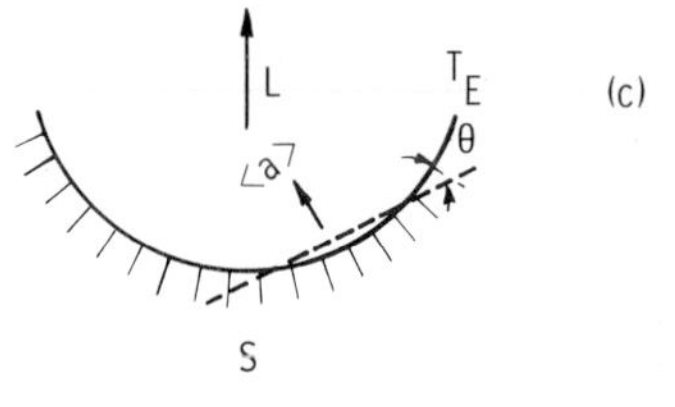

Fig. 26. Facet formation on the solid-liquid interface of a crystal as a function of interface shape (L = liquid, S = solid, θ = equilibrium contact angle between the facet and the rest of the interface, and ψ = the angle between the crystal axis and the $\langle a \rangle$ direction.

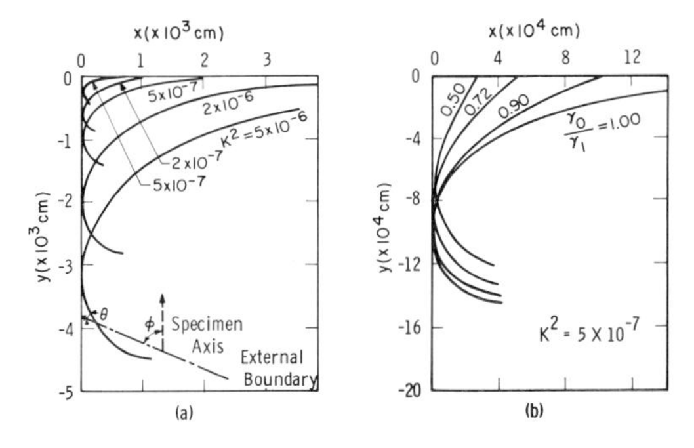

Fig. 27. Calculated solid-liquid interface shapes (two-dimensional) at an intersection with a grain boundary or an external boundary at $x = 0$; $y = 0$ represents the main interface. Two cases: (*a*) isotropic surface energy, γ_1 and (*b*) a singular surface at $y = 0$ with energy, γ_0 the remaining surface isotropic with energy γ_1. (ϕ = angle between the specimen axis and the external boundary, θ = equilibrium contact angle between the external boundary and the solid-liquid interface, and $K^2 = \gamma_1/G\Delta S$, where ΔS is the entropy of fusion per cubic centimeter, and G is the constant temperature gradient in the liquid in the y direction for $y > 0$).

condition that must be applied at the point of intersection of the interface and the surface is that of "surface tension balance." The equation has been solved for a pure material at equilibrium (Bolling and Tiller, 1960), and the interface shape depends on the average solid-liquid interfacial energy γ, the temperature gradient in the liquid at the interface G, the entropy of fusion per unit volume ΔS, and whether there are any surfaces of lower free energy than γ. Plots of a two-dimensional interface shape $x = f(y)$ are shown in Fig. 27*a* as a function of $K^2 = \gamma/G\ \Delta S$ for isotropic γ, and in Fig. 27*b* for anisotropic γ.

The angle between the external boundary and the specimen axis ϕ is normally zero, but not during seeding operations or changes in thickness. Both ϕ and the contact angle θ are important because both the degree of supercooling at surface intersection points and the volume of supercooled liquid increase as ϕ increases and θ decreases, thereby increasing the probability of stray crystal nucleation.

Contact angles and interface shapes must also be considered in analyzing the kinetics of crystal growth on substrates. For example, in the growth of ice (Lindemeyer et al., 1959), it was found that the growth rate is markedly dependent on the substrate material.

The most interesting case of an internal surface intersection occurs for a bicrystal consisting of one grain that has isotropic surface energy γ_1 (Fig. 27*a*), and the other having a singular surface (γ_0) exposed at $y = 0$ and an isotropic surface energy γ_1 for $y < 0$ (Fig. 27*b*). Thus, although the energies of the interfaces near the root of the groove are the same, the groove shape will be anisotropic and the grain boundary will make some angle to the specimen axis rather than being parallel to it. This leads to the encroachment of the isotropic grain by the grain with the singular surface. Such an effect will lead to the development of a preferred orientation in pure materials (Bolling and Tiller, 1960; Bolling et al., 1962), and provides a reasonable interpretation of the $\langle 111 \rangle$ preferred orientation in castings of zone-refined lead.

Cell Formation

An example of an important interface morphology that sometimes arises when a crystal is being grown from an alloy or impure melt is illustrated in Fig. 28 for a lead crystal. It was first postulated by Rutter and Chalmers (1953) that this cellular substructure (observable by rapid decanting of the solid-liquid interface) results from the formation of a zone of "constitutionally supercooled" liquid ahead of the interface.* Solute enrichment occurs at the freezing interface as illustrated in Fig. 29. From this solute distribution C_L and the phase diagram, one can plot the equilibrium temperature distribution T_E of this liquid, as also illustrated in Fig. 29. Superimposed on this latter plot are several actual temperature distributions T_A. If $T_E > T_A$, the liquid ahead of the interface is at a temperature below its equilibrium temperature and is said to be constitutionally supercooled. The condition for this is

* Their condition was predicted from mathematical considerations by G. P. Ivantsov, *Doklady Akademic Nauk U.S.S.R.* **81**, 179, 1951.

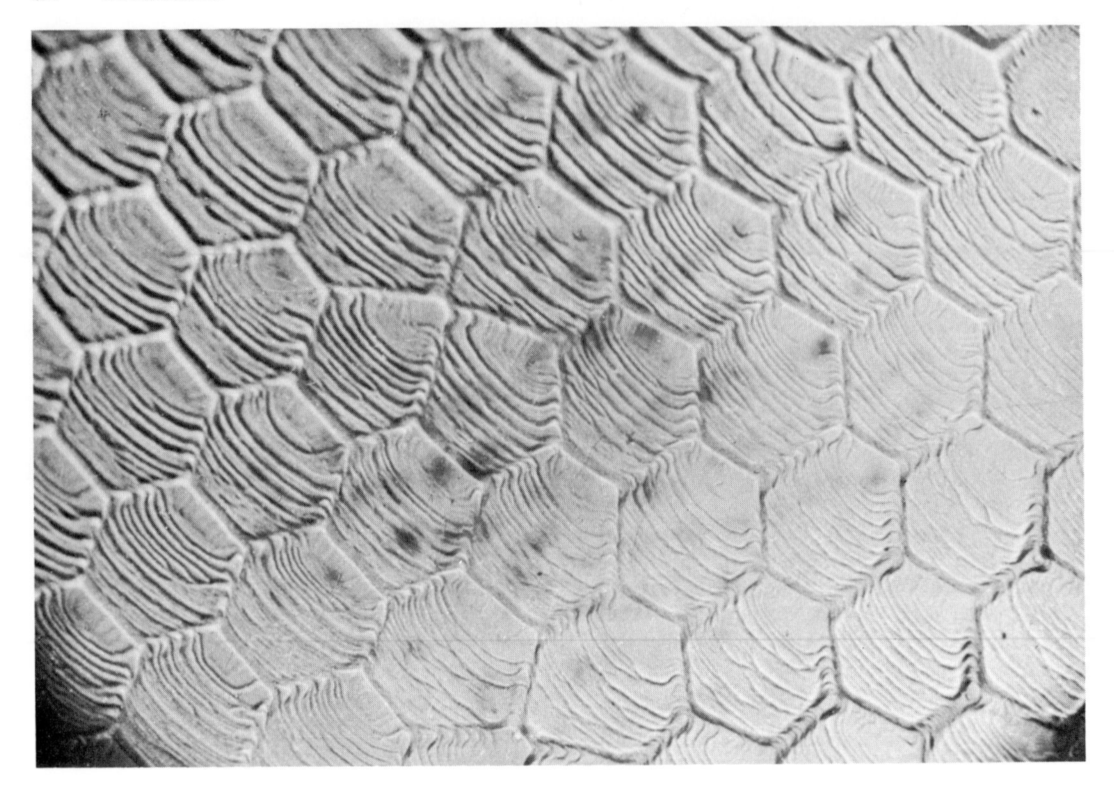

Fig. 28. Photomicrograph of a decanted interface of lead + 5×10^{-4} weight per cent silver alloy crystal illustrating the cellular interface morphology (285 X).

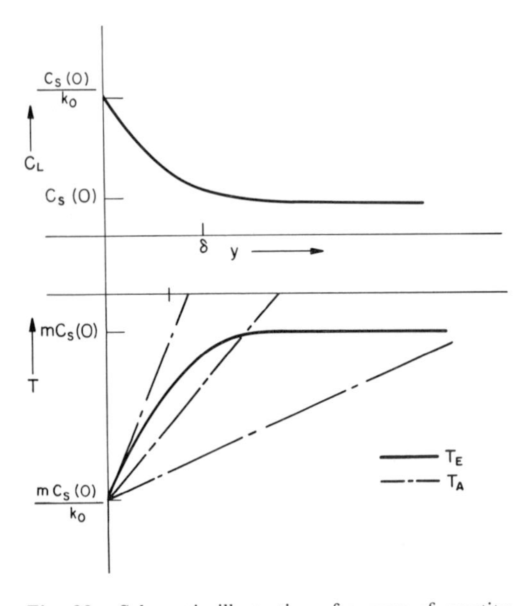

Fig. 29. Schematic illustration of a zone of constitutional supercooling in the liquid ($T_A < T_E$) due to solute enrichment within the layer of thickness δ ahead of interface (m = liquidus slope, T = temperature, and y = distance in the liquid).

$$\frac{G}{V} \lesssim \frac{mC_s(0)(1 - k_0)}{k_0 D_L} \tag{37}$$

where G = the temperature gradient in the liquid at the interface, m = the liquidus slope, and $C_s(0)$ is the solute concentration in the solid at the interface. For typical values, $G = 10°\text{C/cm}$, $m = -3°\text{C/atomic}$ per cent, $k_0 = 0.5$, $D_L = 10^{-5}$ cm²/sec, and $C_s(0) = 10^{-2}$, 10^0, 10^1 atomic per cent; constitutional supercooling will exist if V is greater than 3×10^{-3}, 3×10^{-5}, and 3×10^{-6} cm/sec, respectively. Since normal crystal growth rates are about $10^{-4} - 10^{-2}$ cm/sec, we see that for crystals containing greater than 0.5 atomic per cent solute with $k_0 \gtrsim 0.5$ constitutional supercooling will usually exist.

The consequence of a zone of supercooled liquid ahead of the interface is that a planar interface becomes unstable with respect to an interface having a cellular morphology as shown in Fig. 28. For a pure material, if a curved projection forms on the interface, its equilibrium temperature is decreased according to the Gibbs-Thomson relationship, and the surface must revert to the planar state. However, with an alloy liquid, the curved projection also produces lateral diffusion of solute, depleting the local interface concentration, and increasing the equilibrium temperature of the projection.

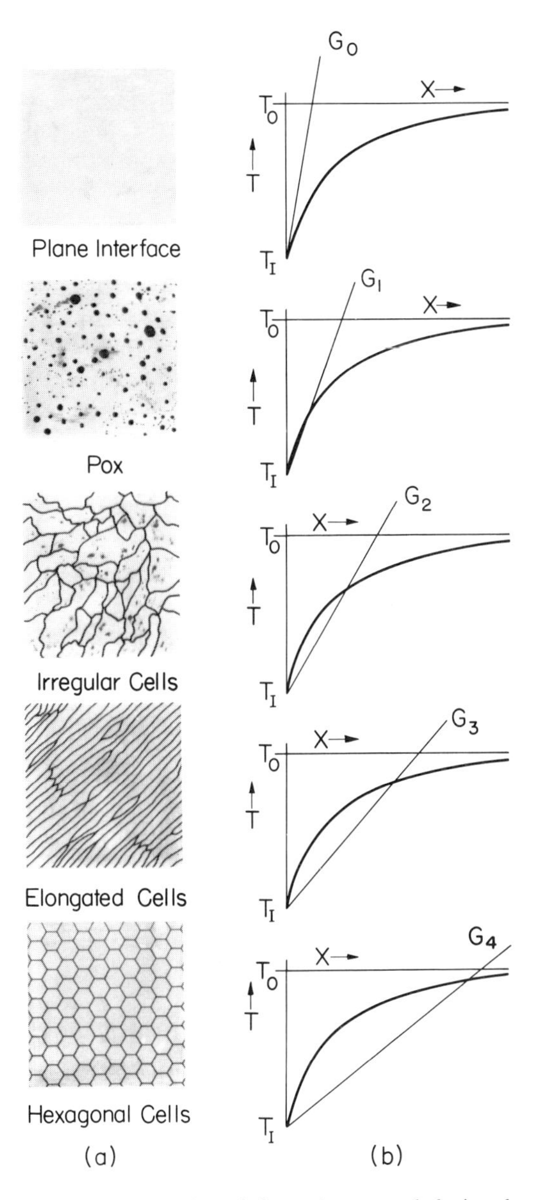

Fig. 30. Illustration of the various morphologies observed on decanted interfaces of tin and lead as a function of the degree of constitutional supercooling that would have existed ahead of a planar interface.

This segregation also decreases ΔH. If the net effect of the changes in T_i and ΔH is to allow the tip of the cells to move ahead of the planar interface position, Eq. (31) will be satisfied for a cellular interface and this will become the stable morphology.

Equation (37) has been experimentally verified with lead and tin as solvent materials (Walton et al., 1955; Tiller and Rutter, 1956; Holmes, Rutter, and Winegard, 1957). As might be expected from the foregoing, a finite degree of constitutional supercooling was necessary before the cellular morphology became the stable form. The various transition morphologies as a function of constitutional supercooling are presented in Fig. 30.

If a sinusoidal shape perturbation of variable wavelength and small amplitude is applied to a smooth interface, the amplitude is found to grow with time when the following condition is obeyed (for an unstirred melt when capillarity is neglected):

$$\frac{\overline{G}}{V} < -\frac{mC_\infty(1 - k_0)}{Dk_0} \tag{38}$$

$\overline{G} = (K_L G_L + K_S G_S)/(K_L + K_S)$, where K represents thermal conductivity. Equation (38) is very similar to Eq. (37). It states that one must consider a "lumped" temperature gradient in the "constitutional supercooling" relationship in order for a perturbation to grow in the absence of capillarity effects. If one includes the effects of capillarity, a particular wavelength grows in amplitude at a greater rate than all the rest and a smaller value of $\overline{G}$ than that given by Eq. (38) is required to stabilize the perturbation.

A schematic drawing of the stable cell shape is given in Fig. 31, consisting of a cap region of width $(a - \delta)$ over most of the projection and groove regions of width δ over the remainder. The cap region has been analyzed for elongated cells (not radially symmetric) using Eq. (29) and assuming $V = 0$. The shapes predicted (Bolling and Tiller, 1961) are as shown in Fig. 32. These calculations assume that $G_0' = dT_L/dy - m dC_L/dy$ is constant on the cap surface, and K_c = the curvature at the

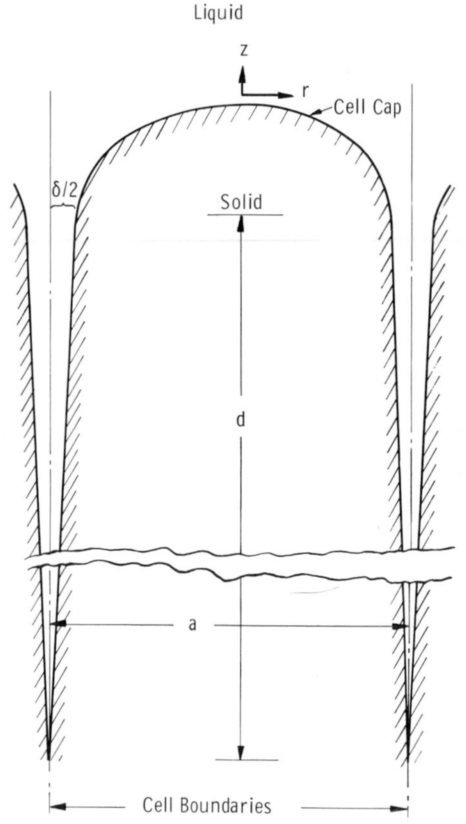

Fig. 31. Schematic cut through a two-dimensional (or a hexagonal) cell illustrating the dimensions of the cell grooves d relative to the cell width a.

tip of the cell. These are consistent with the shapes observed in practice. An approximate equation for the cell size in a quiescent liquid is

$$a \approx \frac{D}{2\pi V}\left\{1+\left[1+\frac{16\pi^2 V}{D}\left(\frac{0.6\gamma}{\Delta S G_0'}\right)^{1/2}\right]^{1/2}\right\} \tag{39a}$$

and is fitted to the data of Boček et al. (1958) in Fig. 33. In addition to the variation of a with V and G, we can see that a will increase as γ increases (Boček et al., 1958) and that a will increase as C_∞ increases, since dC_L/dy increases and G_0' decreases as C_∞ increases. This also is in agreement with practice (Tiller, 1955; Kratochvíl et al., 1960).

The depth of the cell boundary grooves d is usually very large compared to a. Measurements of the groove roots in ice have shown that $d \approx 1$ to 2 cm, whereas $a \approx 2 \times 10^{-3}$ to 5×10^{-3} cm. The solute distribution at a cellular interface has not yet been calculated, but an approximate determination of d can be made. Let us denote the average concentration in the liquid over the cell cap as C_δ. Then from (1) conservation of solute (that is, the average solute concentration freezing out in the solid is C_∞), and (2) the description of a steady state solute distribution in the grooves (that is, inflow minus outflow equals incorporation) we find that

$$d = \frac{2D}{G_s(1-k_0)}\left\{\frac{\dfrac{-mC_\infty(1-k_0)}{k_0 D} - \dfrac{G_s}{V}}{\phi\dfrac{\delta}{a} + \dfrac{k_0}{1-k_0}}\right\} \tag{39b}$$

where $\phi(\delta/a)$ is the ratio of cell boundary area per unit interfacial area, $\phi = 1$ or 2 for two-dimensional and hexagonal cells respectively and $\phi \approx 2$ to 4 as the cells become dendritic. Denoting C_δ as $\alpha C_\infty/k_0$ and ΔC as the average concentration difference

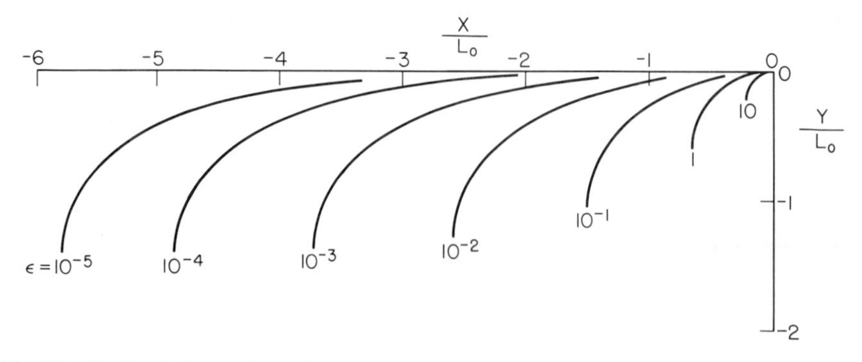

Fig. 32. Predicted shape of two-dimensional cells, where $Y = 0$, $X = 0$ represents the tip of the cell cap, and $Y > 0$ is completely liquid ($L_0^2 = \gamma/\Delta S G_0'$ and $\epsilon = L_0^2 K_c^2/2$, where G_0' is an effective temperature gradient and K_c is the curvature of the tip of the cell).

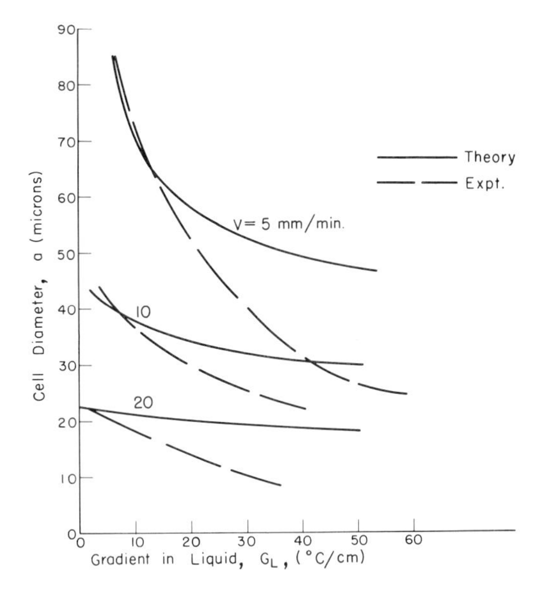

Fig. 33. Comparison between theory and experiment for two-dimensional cells in zinc.

between the solid in the cap region and at the groove root, we can obtain

$$\Delta C = -\frac{k_0 G_s d}{m} \tag{40}$$

and

$$\alpha = 1 - \phi \frac{\delta}{a} \frac{\Delta C}{2C_\infty} \tag{41}$$

From Eq. (39), we see that if G_L is decreased to produce constitutional supercooling at a certain value of V and then decreased by a factor of ~ 2, the numerator in the bracket is about G_s/V in magnitude. One generally finds that $\delta/a \approx 0.05$ to 0.1; thus if $\phi = 1$ and $k_0 \approx 0.1$, $d \approx 10D/V \approx 10a$. If C_∞ is increased by an order of magnitude, $d \approx 100a$. From Eq. (40), if $d \sim 0.1$ to 0.5 cm, $\Delta C \sim 0.1$ to 1.0 weight per cent for $k_0 \approx 0.5$. Thus since cells can be formed with C_∞ as small as 10^{-2} weight per cent, ΔC can be ten times the average concentration for $k_0 \approx 0.5$. It is observed that certain solutes in water, having $k_0 < 10^{-3}$, produce cellular segregation with $\Delta C \approx$ eutectic concentration.

As the solute concentration increases in the liquid, the cells project so far into the liquid that they become unstable with respect to lateral perturbations. This produces the onset of side-branching, and is described as the "dendritic breakdown" condition. It is found experimentally that it depends on the value of $G_L/V^{1/2}$ (Tiller and Rutter, 1956). By qualitative reasoning (Tiller, 1957), the following condition was suggested to describe the breakdown condition:

$$\frac{G_L}{V^{1/2}} = A' a_\theta \frac{C_\infty}{k_0} \tag{42}$$

where A' is a constant and a_θ is the cell width for the orientation θ.

Dendritic Growth

The treelike forms that are often observed growing from the melt, solutions, the vapor, and during electrodeposition are called *dendrites*. There is a tendency for dendritic growth whenever the difference in free energy between the nutrient and the solid ΔG is positive and increases with distance ahead of the interface. In a pure melt, the dendritic morphology can become predominant only if the temperature gradient in the liquid is negative at the interface. For alloy liquids, the dendritic morphology can develop with a positive temperature gradient, provided the solute distribution is such that $d(\Delta G)/dz > 0$ at the interface (see p. 294).

The characteristic temperatures involved in dendrite growth are shown in Fig. 34. Here T_i is the interface temperature and T_∞ is the bath temperature. The tempera-

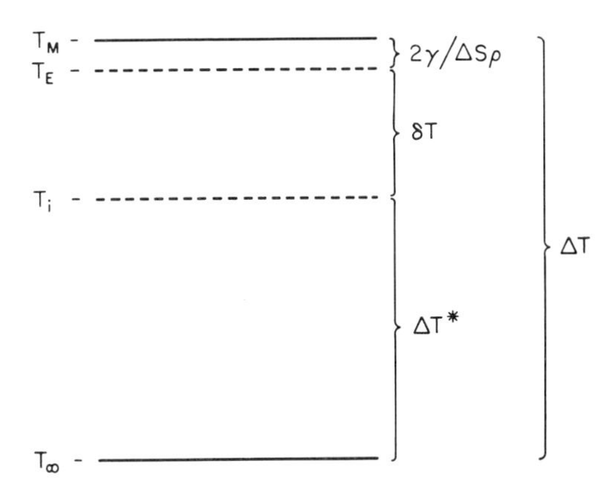

Fig. 34. The important temperatures involved in dendritic growth for a pure material (T_M = melting temperature of the pure material and T_∞ = the bath temperature).

ture difference δT controls the rate of deposition onto the solid. The temperature difference ΔT^* controls the rate of heat flow away from the tip of the dendrite and therefore controls the rate of dissipation of latent heat into the liquid. Consequently it also controls the axial freezing rate V. Therefore T_i must have a value such that the deposition rate equals the macroscopic freezing rate as determined by the heat transfer rate. By noting that $\Delta H \approx \Delta S T_i$, where ΔH is the latent heat of fusion and ΔS is the entropy of fusion ($\Delta S = \Delta H/T_M$), we find that thermal transport and solute transport are described by the same kind of differential equation, and the same kind of boundary conditions (see Eq. 30). Thus both the temperature distribution and the solute distribution have the same shape about any dendrite. Therefore knowledge concerning dendritic growth from solution is applicable to growth from the melt.

It is found that a paraboloid of revolution growing at constant velocity is an isotherm and an isoconcentrate (Ivantsov, 1956). However, an isothermal shape is not stable since $(T_E - T_i)$ increases with distance along the surface away from the tip, whereas V in the surface normal direction must decrease for shape stability. Hence either the growth rate must be very anisotropic or the surface is not isothermal. The latter is usually true.

The solution for the nonisothermal dendrite is much more complicated (Temkin, 1960). However, the radius of curvature ρ of the tip of the dendrite which satisfies Eq. (29) can be described in terms of the axial freezing velocity, the material parameters, and the atomic kinetics (Bolling and Tiller, 1961b, 1962).

$$\rho = P\left(\frac{[e^x E_i(-x)]^2}{x[e^x E_i(-x)]^2 - \dfrac{C}{\Delta H}\Delta T}\right) \tag{43}$$

In Eq. (43), x equals $V_{\max}(\rho/2\alpha_L)$; α_L equals thermal diffusivity of the liquid, E_i equals exponential integral function, C equals specific heat of the solid, ΔT equals the bath supercooling, and P is a constant. The value of ρ from Eq. (43) that also satisfies Eq. (31) is most stable, and thus describes a dendrite having the maximum velocity of propagation. In Table 1, calculated values of ρ and μ that give a good fit with the experimental results for dendrites of nickel (Walker, 1960), tin (Rosenberg and Winegard, 1954), and ice (Jackson, 1961) are given. It appears that nickel, tin and perhaps ice all exhibit a diffuse interface

TABLE 1

Dendrite	a'	C_L/L (°C^{-1})	α_L (cm²/sec)	$\Delta K/K_L$	$\gamma/\Delta S$ (cm–°C)	ΔT_{exp} (°C)	V_{exp} (cm/sec)	ρ_{calc} (cm)	μ_{calc} (cm/sec/°C)	δT_{calc} (°C)
Nickel	0	2×10^{-3}	0.1	+1	10^{-5}	123	2×10^3	6×10^{-6}	175	11.4
						87	1×10^3	8×10^{-6}	175	5.7
						45	$.25 \times 10^3$	1×10^{-5}	200	1.25
					1.7×10^{-5}	87	1×10^3	8×10^{-6}	450	2.22
Tin	0	4×10^{-3}	0.2	+1	6.5×10^{-6}	13.3	30	5.4×10^{-5}	25	1.2
						7.8	10	7.6×10^{-5}	20	0.5
						3.7	2.5	2.7×10^{-4}	14	0.179
					10^{-5}	13.3	30	5.4×10^{-5}	32	0.94
					6.5×10^{-6}	13.3	10	1.7×10^{-4}	7.5	1.33
						7.8	3.33	2×10^{-4}	4	0.833
						3.7	0.83	4.6×10^{-4}	3	0.276
Ice	0	1.25×10^{-2}	1.44×10^{-3}	+2	1.5×10^{-6}	4.77	1.0	2.2×10^{-5}	7.5	0.133
						3.67	0.5	2.6×10^{-5}	4	0.125
						1.57	0.056	10^{-4}	1.25	0.045
					2.4×10^{-6}	4.77	1.0	2.2×10^{-5}	100	0.01
						3.67	0.5	N.S.*	N.S.	
						1.57	0.056	N.S.	N.S.	

* N.S. = no solution

mechanism of atomic deposition even though μ is slightly temperature dependent. From Table 1, it can also be seen that $\delta T < 0.1\Delta T$; this is in sharp contrast to the prediction of $\delta T \sim 0.5\Delta T$ for an isothermal dendrite tip.

Although the calculations of Table 1 have been made assuming no anistropy of atomic kinetics this is fictitious. If the solid exhibited no anisotropic characteristics, a dendrite of given shape could grow equally fast in any crystallographic direction. The existence of crystallographic dendritic growth demonstrates that the solid is anisotropic. From Eq. (31), we would expect the dendrite to grow in the direction with greatest axial velocity. This is the dendrite with minimum ρ for constant δT.

The two most important crystal faces during dendritic growth are (1) the face growing in the axial direction (0-direction) and (2) the slowest growing crystal faces in the system (usually the most close-packed faces, ϕ-direction). These faces make symmetrical angles ϕ with the specimen axis. For a diffuse interface,

$$\rho = \frac{2\gamma(0)}{\Delta S \delta T(0)} \left[\frac{1 - \dfrac{\gamma(\phi)}{2\gamma(0)} \cos\phi(1 + \cos^2\phi)}{\dfrac{\mu(0)}{\mu(\phi)} \cos\phi - 1} \right] \tag{44}$$

The growth direction that minimizes ρ in Eq. (44) for constant $\delta T(0)$ is the preferred growth direction. This direction is one that (1) maximizes $\mu(0)/\mu(\phi)$, (2) minimizes ϕ, (3) maximizes $\gamma(\phi)/\gamma(0)$, and (4) minimizes $\gamma(0)$. Since μ may be expected to decrease as the reticular density of the faces increases, one expects that ρ is minimized by growth in the direction that is symmetrically bounded by the faces of greatest packing and making the smallest angle ϕ with the 0-face. This is consistent with experiments for most metallic systems (Weinberg and Chalmers, 1952). For different atomic kinetics, ρ is minimized by the same parameters as in Eq. (44) (Bolling and Tiller, 1961).

Fast growth, induced by the presence of a self-perpetuating step, may result in a ribbon or lathelike dendrite. For example, if a dendrite material with the diamond structure contains at least 2-{111} closely spaced twins (Fig. 35), it will readily propagate in the ⟨211⟩ direction as a flat strip with {111} flat surfaces rather than as a rod in the ⟨100⟩ direction (Wagner, 1960; Hamilton and Seidensticker, 1960). The re-entrant corners formed by the twins at the dendrite cap allow faster growth to proceed for a given δT than do the faces of the ⟨100⟩ rod-form. The lathe-form grows in the ⟨211⟩ direction because this is the direction of best balance between minimum ρ and minimum δT.

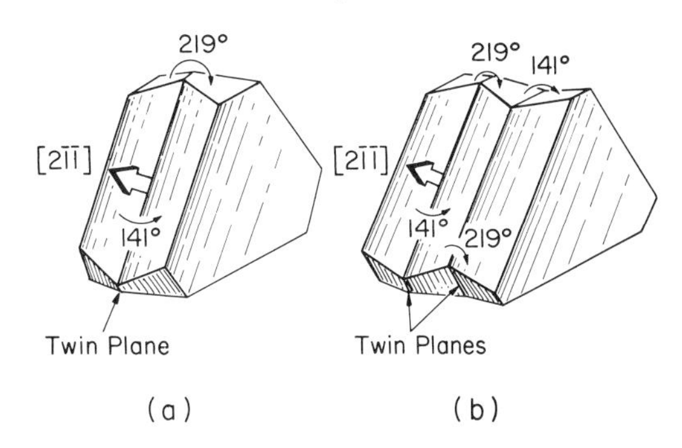

Fig. 35. Schematic dendrite tip-morphology and the angles between the important crystal faces for lathlike dendrites of germanium and silicon containing twin planes. (*a*) Single twin-plane. (*b*) Two twin-planes.

Some distance back from the tip of a rod-shaped dendrite, the rod should approach the shape of a cylinder and conditions should lead to the development of an unstable shape. These conditions correspond to those that developed the original dendrite trunk. Branches form in those directions that give maximum growth velocity, these being the same as the trunk orientation or some equally preferred direction.

Eutectics

The two phase mixtures that result from eutectic solidification assume a wealth of possible macro- and micromorphologies. *Micromorphology* refers to the shape and arrangement of the individual particles of the two phase mixture. One of the phases is continuous and the other has the form of (1) plates in parallel array, (2) rods in parallel array, (3) globules of regular shape, and (4) particles of irregular shape. *Macromorphology* refers to the shape and arrangement of groups of particles of both phases. These groups of particles are called *colonies* and are, in fact, cells or dendrites. These cells or colonies in eutectic alloys form via the same mechanisms that cells form in single phase alloys (Weart and Mack, 1958).

We should like to know why some eutectics form parallel arrays of lamellae, rods, or globules of the discontinuous phase and why some do not. We should also like to know if there is a transition from one morphology to another under certain freezing conditions. To provide some answers to these questions, the constraints that are imposed upon the steady state growth of the lamellae illustrated in Fig. 36 have been considered (Tiller, 1958). There are five primary constraints: (1) steady state heat flow, (2) steady state diffusion, (3) interface contact angles, (4) kinetic equilibrium characteristics, and (5) nucleation characteristics of the two phases.

CONSTRAINTS. The interface is approximately isothermal for small lamellar spacings (Tiller, 1958), and the most probable interface shape that satisfies mechanical equilibrium at the three phase junctions is shown in Fig. 37. In order to appreciate constraint (2), imagine for the moment lateral diffusion in the liquid to be pro-

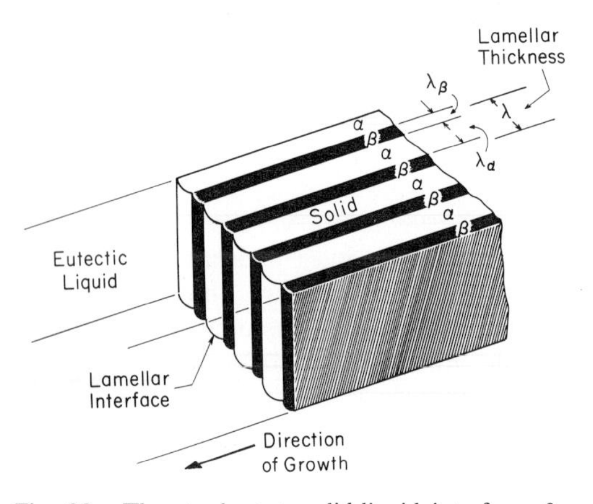

Fig. 36. The steady state solid-liquid interface of a schematic lamellar eutectic.

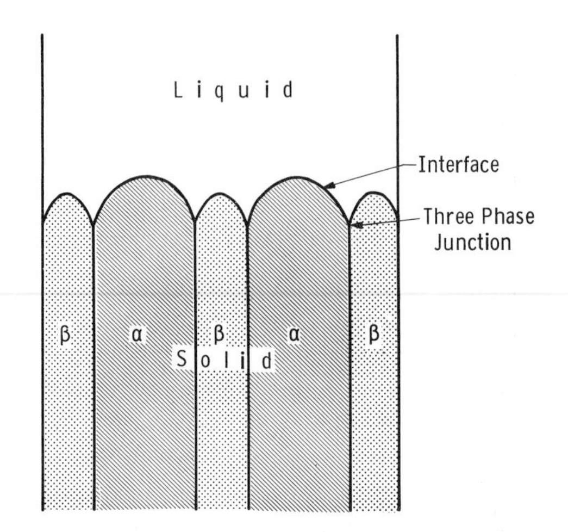

Fig. 37. Drawing of the steady state solid-liquid interface-morphology that satisfies mechanical equilibrium at the three phase junctions.

hibited (for example, by placing impermeable membranes coplanar with the α β-phase boundaries). Then ahead of the lamallae the steady state solute distributions in the liquid will be similar to those given by Eq. (19). Here enrichment of the B-constituent occurs ahead of the α-phase and depletion ahead of the β-phase. The inverse occurs for the A constituent. Because of the different slopes of the liquidus lines and the different concentrations at the interfaces of the α and β lamellae, one phase would usually lead the other by an appreciable distance if nucleation were inhibited.

If the imaginary membranes are held in place, the lamellae will be unstable since the interface of one phase will be greatly supercooled with respect to the other. This will create a tendency for nucleation to occur, either on the solid or in the liquid. Removing the lateral diffusion restriction by removing the membranes causes the interface concentrations to approach each other, and the lead distance between the tips of the lamellae will decrease. The nucleation tendency keeps the interface concentrations below some critical values, so a stable lamellar width λ is formed.

LAMELLAR SPACING. The interlamellar spacing is determined by applying Eqs.

(29), (30), and (31). Semiquantitative treatment (Tiller, 1958) has shown that only two of the constraints are seriously effected as λ varies: (1) the concentration difference between the tips of the lamellae increases as λ increases, causing both T_i and ΔH to decrease and (2) as λ decreases, the surface energy contribution $\gamma_{\alpha\beta}/\lambda$ to the free energy of the system increases causing T_E to decrease, and thus T_i to decrease. By neglecting the ΔH effect and maximizing T_i with respect to variation of λ, one finds that

$$\lambda = \epsilon(\gamma_{\alpha\beta})^{1/2} V^{-1/2} \qquad (45)$$

where ϵ is a constant. Experiments performed on the Pb-Sn eutectic system (Chilton and Winegard, 1961) and the Al—$CuAl_2$ eutectic system (Chadwick, 1961) have substantiated the $V^{-1/2}$ dependence of λ in Eq. (45). From the slopes of their λ versus $V^{-1/2}$ plots, one finds reasonable values for $\gamma_{\alpha\beta}$.

ROD AND GLOBULAR FORMS. For eutectics exhibiting the same ratio of volumes of the two phases $V^\alpha/V^\beta = \theta_\beta{}^\alpha$, the rod morphology allows shorter diffusion paths between phases than the lamellar form. This lowers the concentration differences between the tips of the phases, favoring the rod form. When $\gamma_{\alpha\beta}$ is isotropic, and $\theta_\beta{}^\alpha$ is large, the rod form will be observed. However, the departure of the α-β-phase boundaries from a particular crystallographic plane generally causes an increase in $\gamma_{\alpha\beta}$. This favors the lamellar form when $\gamma_{\alpha\beta}$ is strongly anisotropic. Some eutectic systems form a globular morphology because stable lamellae or rods are not consistent with mechanical equilibrium at the three phase junctions. Then discontinuous particles may nucleate on the continuous phase, but their continued growth will not lead to a stable morphology (Tiller, 1958). Instead, repeated nucleation of the discontinuous phase occurs at all rates of growth. The globular morphology is favored by small values of solid-liquid interfacial energy γ_{mc} and large values of $\gamma_{\alpha\beta}$.

A very different class of eutectics results when the primary phase is unable to serve as the nucleating agent for the secondary phase. Then, the secondary phase is nucleated heterogeneously in the liquid, and its randomly oriented particles grow rapidly to absorb the supercooling in an irregular fashion. Such systems have been termed *anomalous eutectics.*

MORPHOLOGY TRANSITIONS. A morphology that is stable at one freezing velocity V may transform to another one as V changes. This effect is greatly enhanced by the presence of a ternary solute. If the solute has very different values of partition coefficient and liquidus slope in the two phases, a transition from lamellar to rod to globular morphology may be observed (Tiller, 1958).

4. DEFECTS

This section is concerned with departures of the lattice from physical homogeneity, and thus will deal mainly with physical defects introduced during crystal growth. The defects to be considered are: (1) chemical inhomogeneities, (2) dislocations, (3) growth twins, (4) voids and (5) stray crystals.

Chemical Inhomogeneities

Many long-range solute distributions may be produced in a crystal, depending on the conditions of freezing (Section 2). A second kind of chemical inhomogeneity is produced by fluctuating growth conditions. This produces a fluctuating solute distribution in the solid with the isoconcentrates parallel to the solid-liquid interface (Goss et al., 1956). These fluctuations can be large in magnitude ($\Delta C \sim 10^{-1}C_0 - 10C_0$) and they may occur over small distances ($\Delta X \sim 10^{-2} - 1$ cm).

Small fluctuations of solute concentration ($\Delta C \sim 10^{-5} - 10^{-3}$ atomic per cent) may ocur over very small distances ($\Delta X \sim 10^{-5} - 10^{-4}$ cm) when the growing interface consists of layers. These fluctuations lie parallel to the layer planes (Tiller, 1958). Fluctuations of the same type, but sometimes several orders of magnitude larger, may be produced if the layer motion is not uniform because of chemical adsorption onto the layer plane.

Crystal growth of the cellular type leads to large solute fluctuations ($\Delta C \sim C_0 - 10C_0$) occurring over small distances ($\Delta X \sim 10^{-4} - 10^{-3}$ cm) (Tiller, 1958). This type of segregation may occur in a two-dimensional or three-dimensional network, depending upon the particular interface morphology, the network size a being approximately $10^{-3} - 10^{-2}$ cm.

Because of diffusion in the solid the values of ΔC observed after growth will be less than the values just quoted. However, homogenization will occur at a slower rate than might be expected because of the creation of dislocations at the region of segregation during homogenization (to be discussed later).

A final type of chemical inhomogeneity arises when the solid is supersaturated with respect to a certain solute which precipitates, causing large-scale solute fluctuations ($\Delta C \sim C_0 - 10^4 C_0$) occurring over small distances ($\Delta X \sim 10^{-4}$ cm).

Dislocations

Several mechanisms have been suggested to account for dislocation formed in crystals grown from the melt: (1) thermal stresses, (2) constitutional stresses, (3) mechanical stresses, (4) vacancy supersaturation, and (5) probability accidents. The final number of dislocations in the crystal may also be augmented by (6) certain multiplication and annihilation mechanisms. Their interaction also leads to (7) the formation of dislocation networks in the solid.

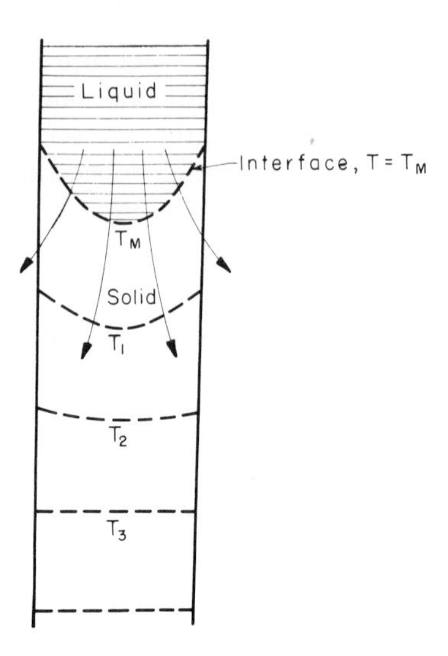

Fig. 39. Possible temperature isotherms in a crystal during freezing and cooling.

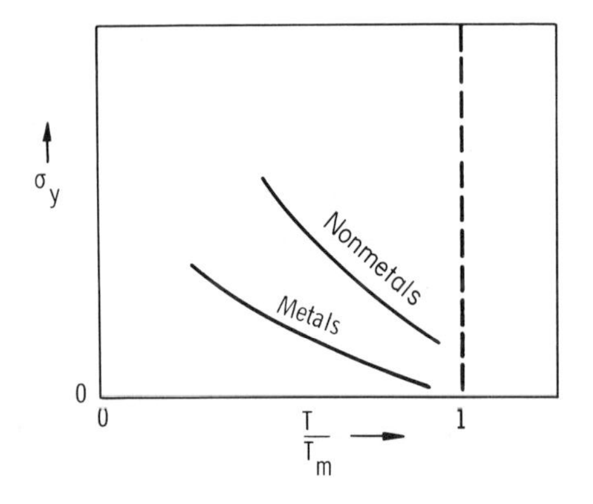

Fig. 38. Schematic dependence of the yield stress of metals and nonmetals on reduced temperature T/T_m (T_m = the melting temperature of the material).

No matter how a stress σ is put into a crystal, it produces a proportional amount of elastic strain ϵ ($\epsilon = \sigma/E$), where E is Young's modulus. Depending on the yield stress σ_y some plastic strain may also be produced, and this will introduce dislocations. The yield stress near the melting temperature T_m is nearly zero for metals, but may be larger for covalent or ionic solids (Fig. 38). Therefore a given stress will tend to put more dislocations into metal than into nonmetal crystals [$n \sim (\sigma - \sigma_y)/Eb$, where b is the Burgers vector].

THERMAL STRESSES. During crystal growth and cooling to room temperature, radial temperature differences are usually produced in the crystal (Fig. 39). Thermal contraction then causes the outer surface to compress the core of the crystal and put itself into tension. Plastic flow may occur to relax the stresses. For example, if ΔT is the temperature difference between the outer surface and the center, and α is the thermal expansion coefficient, then $\sigma \simeq \alpha E\ \Delta T$. Thus if $\Delta T \approx 50°C$, $\alpha \sim 10^{-5}$, and $E \sim 10^4\ kg/mm^2$, $\sigma \sim 5\ kg/mm^2$, a value well above the yield stresses of most hot materials.

Penning (1958) has shown that if a crystal is grown with no radial temperature distribution and a constant axial temperature gradient, no thermal stresses are produced. However, if a nonzero radial temperature gradient exists (with constant axial temperature) a symmetrical stress about the crystal axis is created. If it is large enough, this will cause predicable dislocation patterns, for example, if the crystal has a ⟨111⟩ axis and has the diamond or fcc structure, a hexagonal pattern as in Fig. 40*a* is produced on the cross section. Since no shear stress is produced on the cross-sectional plane, no slip occurs on planes parallel to it. If both a radial temperature gradient and a nonlinear axial gradient exist in the crystal during growth, a shear stress is produced on the axial and cross-sectional planes. A triangular slip pattern as in Fig. 40*b* may be produced by slip on planes perpendicular to the crystal axis.

After crystal growth, cooling plays a large part in the introduction of dislocations. In fact, more dislocations may be put into a crystal by cooling stresses than by the thermal stresses existing during growth. These stresses can be large enough

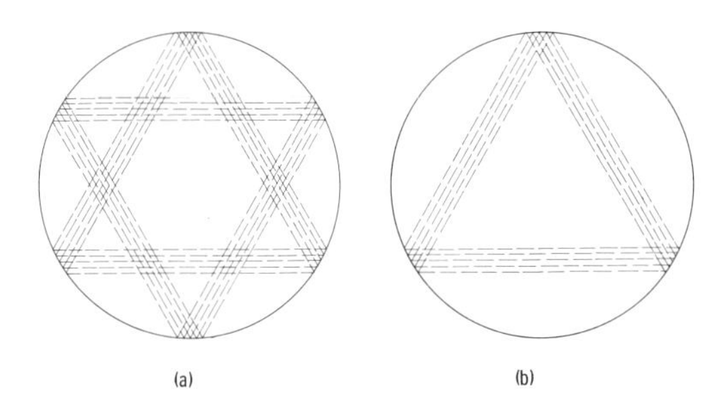

Fig. 40. Predominant slip traces on a section perpendicular to the specimen axis for a crystal of either the diamond or the face-centered cubic structure grown with a ⟨111⟩ specimen axis. (*a*) Nonzero radial temperature gradient, constant axial temperature gradient. (*b*) Nonzero radial temperature gradient, nonlinear axial temperature gradient.

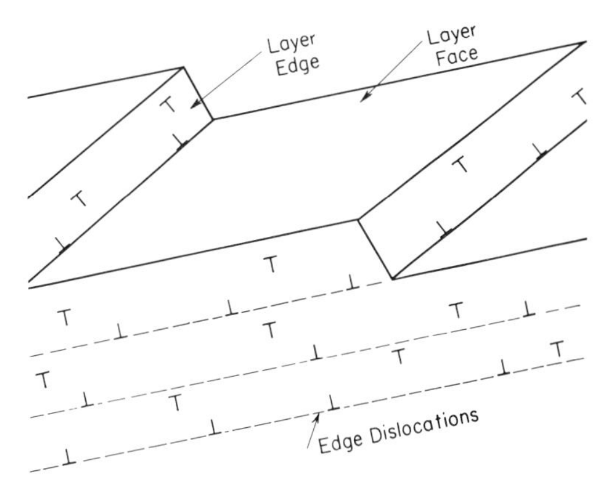

Fig. 41. Possible dislocation configuration in a crystal produced by segregation of solute during layer growth.

to crack some materials. Thermal stresses can be expected to increase as the thermal conductivity of the material decreases. For poor thermal conductors the yield stress is usually large, however, so the plastic strain for a given surface cooling rate may be relatively independent of the material. The density of dislocations produced by thermal stresses may be $n \sim 10^4 - 10^5/cm^2$. At lower temperatures, residual stresses opposite in sign to those that caused plastic flow at high temperatures may exist when the crystal has reached a uniform temperature.

CONSTITUTIONAL STRESSES. When microscopic regions of solute segregation are produced during growth, the different lattice parameters in the segregated regions may cause dislocations to form at the boundaries of the segregate. The stress is given by

$$\sigma = \Delta C \left(\frac{\Delta\lambda}{\lambda}\right) E \tag{46}$$

where ΔC is the atom fraction difference in concentration across the segregated region, λ is the lattice parameter of the solvent, and $\Delta\lambda$ is the difference in lattice parameter between solute and solvent. The dislocations introduced during layer growth have the configuration shown in Fig. 41 and the density

$$n \approx \frac{4}{bd}\left\{\frac{\Delta C \Delta\lambda}{\lambda} - \epsilon_e\right\} \tag{47}$$

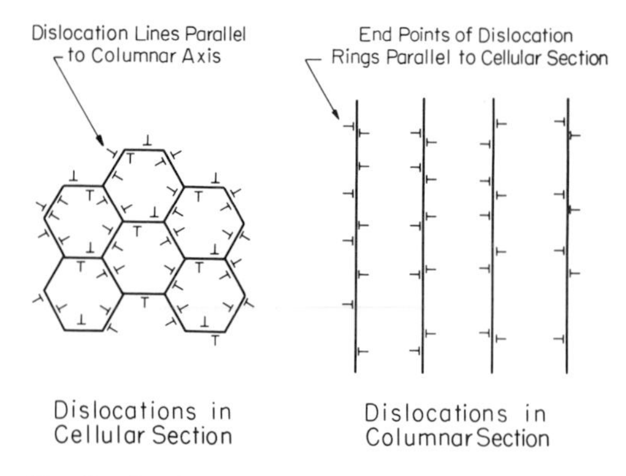

Fig. 42. Possible dislocation configurations in a crystal produced by segregation of solute during growth with the cellular interface morphology.

where d is the spacing between the layers and ϵ_e is the elastic strain. For steady state growth $n \sim 10^3 - 10^5$ lines/cm^2, whereas for periodic layer motion in the presence of adsorption, n may be of the order of 10^6 to 10^7 lines/cm^2. The dislocations introduced by cellular segregation have the configuration shown in Fig. 42 and the density given by Eq. 47 with $2/a$ substituted for $1/d$ (a = cell width). Here $n \sim 10^6 - 10^{10}$ lines/cm^2, depending primarily on the magnitude of ΔC. For long-range segregation ΔC is one-half the concentration difference across 1 cm of crystal length and $d = 1$. For growth-fluctuation-induced segregation ΔC is one-half the average amplitude of the solute fluctuation and d is the wavelength of the fluctuations. In this case $n \sim 10^0 - 10^6$ lines/cm^2.

MECHANICAL STRESSES. The presence of insoluble particles in the melt may lead to complex dislocations with a large screw component as shown in Fig. 43. Koslovskii (1958) postulated that as the edge of a growing crystal layer captures a particle in its path, a shearing of the layer edge may occur. This might be due to epitaxial or elastic stresses as a result of the particle capture. Once the length of the resulting step exceeds the dimensions of a critical nucleus, the step grows and winds around the particle to form a spiral. It was observed experimentally that the probability of forming a spiral growth by particle capture increases to a constant value as the growth rate of the crystal increases. Thus the number of dislocations produced by these stresses should depend on the concentration and type of insoluble particles as well as the freezing rate. The number can be significantly reduced by zone-refining since this tends to dissolve the insoluble particles.

A second type of mechanical stress that may put dislocations into a crystal arises only in materials that expand on freezing and contain deep cell-boundary grooves. As the liquid in these grooves expands on freezing, it is partially constrained by the cells and so becomes stressed. A constraint due to crucible walls causes similar stresses.

VACANCY SUPERSATURATION. If the concentration of vacancies in a solid exceeds the thermal equilibrium number, there is a tendency for the vacancies to anneal out by condensing into aggregates. Dislocation loops will form from the vacancy aggregates if the excess free energy of a dislocation loop is less than that for an equivalent vacancy aggregate. It is estimated that dislocation loops will nucleate at a loop radius of $3b$ to $5b$ for metals and several times this for semiconductors or nonmetals

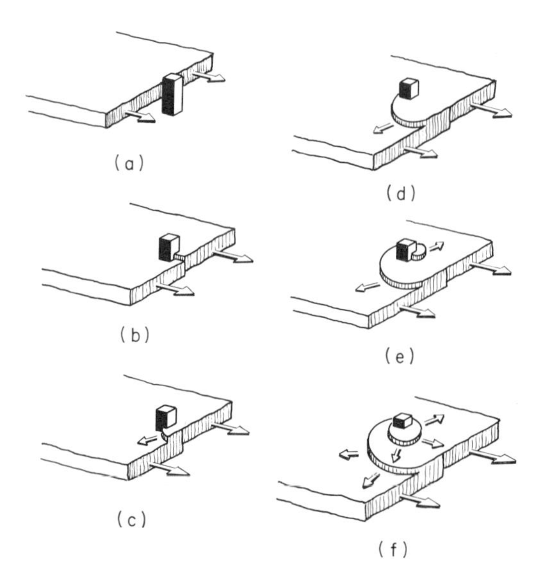

Fig. 43. Capture of a foreign particle by a growing crystal resulting in the formation of a complex screw dislocation.

having larger moduli of rigidity. Once a loop has formed, the vacancy supersaturation necessary to keep the loop growing to a larger size by climb can be calculated. Since all loops formed by vacancy condensation pass through this early stage of growth, this yields a lower limit of the vacancy supersaturation necessary for loop nucleation.

Once a dislocation-loop is formed, a vacancy condensing on it will increase its radius, and hence its strain energy E_r. At the same time, the free energy associated with vacancy supersaturation F_v will decrease. A dislocation loop of radius r tends neither to grow nor shrink when (Schoeck and Tiller, 1960; Bolling, 1961)

$$\frac{\Delta T}{T_m} = \frac{b}{r}\left[\frac{Gb^3}{6(1-\nu)U_f}\left(\ln\frac{r}{b} + Z + 1\right) - \frac{\phi kT}{U_f}\right] + \frac{\gamma_{st}b^2}{U_f} - (q-1) \tag{48}$$

where $\Delta T = T_m - T$, G is the rigidity modulus, ν is Poisson's ratio, U_f is the energy of formation of a vacancy, $Z \approx 1.8$ and accounts for the core energy of the dislocation, $\phi = 3$ and accounts for the self-entropy of the dislocation, k is Boltzmann's constant, γ_{st} is the stacking fault energy, and q accounts for a nonequilibrium vacancy concentration C_x frozen in at the solid-liquid interface ($C_x = \beta e^{-U_f/qkT_m}$) and is thus equal to unity for equilibrium. Equation (48) assumes that there are no vacancy sinks in the crystal, so that at the temperature T the vacancy concentration is C_x. For the case where (1) $q = 1$, (2) the self-entropy is neglected, and (3) no stacking fault is formed in the dislocation loop, a plot of $\Delta T/T_m$ versus r/b is given in Fig. 44 for several values of the parameter $\alpha = Gb^3/6(1-\nu)U_f$. A typical value of α for metals is 0.6, so that dislocation rings of $r = 15b$ are growing only at temperatures below $\Delta T/T_m \approx 0.15$. If other vacancy sinks exist in the solid, for example, grown-in dislocation lines, $\Delta T/T_m$ must increase and may become as large as 0.5 for certain growth conditions (Schoeck and Tiller, 1960). Elbaum (1960) has extended this treatment to germanium and silicon and finds, mainly because G is larger for these materials than for metals, that $\Delta T/T_m$ is increased from 0.15 to about 0.5. Because of the self-entropy term in Eq. (48), an equilibrium concentration of small loops should be present in metal crystals in the vicinity of T_m (a much smaller number should be present in covalent crystals). If sufficient vacancy supersaturation exists at T_m, these loops will grow as crystal growth proceeds (this requires that, with increase in r, the change in the increase in F_v due to

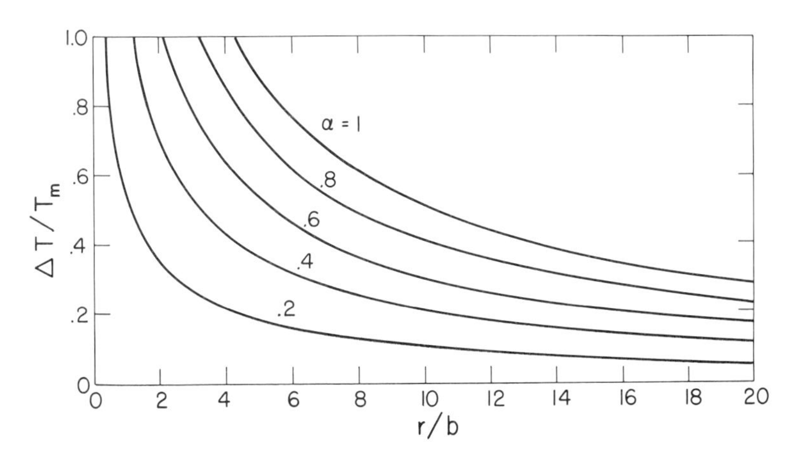

Fig. 44. Relative undercooling $\Delta T/T_m$ of a vacancy supersaturated crystal required to keep a dislocation ring of radius r from shrinking for several values of α (b = Burgers vector and α is a material parameter).

the decrease in self-entropy is less than the change in F_v due to the decrease in the vacancy supersaturation).

Since 10^{17} to 10^{20} vacancies/cc are frozen into the lattice during crystal growth, if they formed vacancy disks 1 atom thick and 1 μ in diameter, the resulting total length of dislocation line/cc in the crystal would be about 10^6 to 10^9 cm. In order to prevent dislocations from being introduced by this mechanism, it is advisable to keep ΔT small during growth. Cooling to room temperature should then occur at a rate slow enough so that vacancy diffusion to the external surface or to internal sinks will prevent the development of a critical vacancy supersaturation.

PROBABILITY ACCIDENTS. Dislocation introduction by a probability accident is indicated in Fig. 45 (Bolling, 1961). At any temperature below T_m, a distribution of two-dimensional embryos will exist on growth layers. They may be stacked incorrectly on the layer plane, or lie on top of an adsorbed layer. As the next growth layer advances, it may capture some of these embryos in their disoriented configurations, producing dislocations in the lattice. It is not easy to calculate the number of dislocations introduced in this way, but one might expect it to be small.

ANNIHILATION AND MULTIPLICATION. Annihilation of dislocations of opposite sign is not expected to reduce the dislocation density by more than a factor of two. On the other hand, dislocation multiplication during crystal growth may cause a large increase in the dislocation density. If a dislocation terminates in the solid-liquid interface, it may propagate down the entire crystal, especially if the axial direction is energetically favorable for a dislocation line.

To account for a given dislocation density one must consider both generation within a volume element and propagation into this element from the adjacent element lying closer to the seed. Propagation in the axial direction plus generation will cause a continual increase in the dislocation density. For metals, one expects the dislocation line energy to be quite isotropic; thus the propagation mechanism and solid-liquid interface shape will be important. For most nonmetals, more anisotropic line energy is expected, so growth with certain axial orientations will be preferred to minimize dislocation propagation. For a crystal of given diameter, the larger the minimum angle between the growth axis and the slip planes the greater will be the probability that dislocations generated in a unit volume emerge at the outer surface of the crystal. Thus for diamond cubic or fcc materials, growth in the $\langle 100 \rangle$ and $\langle 111 \rangle$ directions are more favorable than $\langle 110 \rangle$ or $\langle 211 \rangle$ growth (Dash, 1959).

Dislocations can be moved towards the interface, by thermal stresses, by climb due to vacancy supersaturation, or by climb and glide caused by the interface image stresses. The climb velocity V_0 in a vacancy field is given by (Schoeck and Tiller, 1960)

$$V_0 \approx 0.5 \frac{D}{b}\left(\frac{C_x}{C_0} - 1\right) \tag{49}$$

where the self-diffusion coefficient $D =$

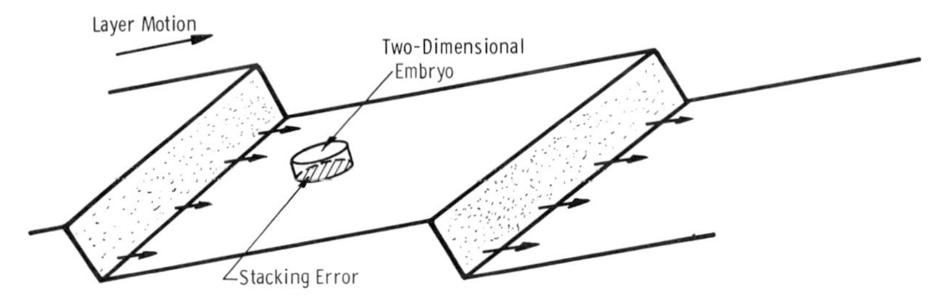

Fig. 45. A possible means of forming a stacking fault and a dislocation loop by layer trapping at the solid-liquid interface.

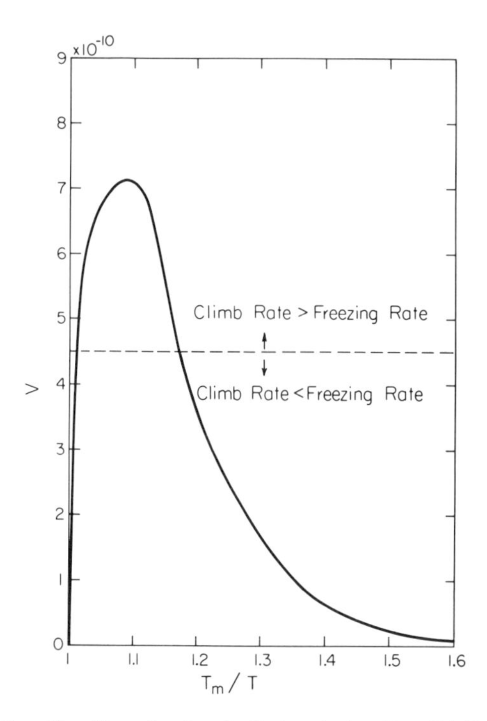

Fig. 46. Plot of reduced climb velocity $V = V_0 b/D_0$ versus the dimensionless temperature T_m/T for a vacancy-supersaturated crystal (V_0 = edge dislocation climb velocity, b = Burgers vector of the dislocation, and D_0 = the diffusion constant).

$C_0 D_v$, D_v is the vacancy diffusion coefficient, and C_0 is the equilibrium vacancy concentration at temperature T. With $C_x = C_0(T_m)$, a reduced velocity, $V = V_0 b/D_0$, $(D = D_0 e^{-U_d/RT})$ is plotted versus T_m/T in Fig. 46. We see that there is a maximum in V at $\Delta T/T_m \approx 0.1$. The climb velocity vanishes for small T because the vacancies becomes immobile and it also vanishes at the melting point T_m because the supersaturation disappears there. The dislocations move to within about 0.5 cm of the interface and thereafter move at the same rate as the interface. In order for the dislocations to climb into the interface under this driving force, a vacancy supersaturation $C_x/C_0(T_m)$ must exist. From Eq. (49), $V_0(T_m) \approx 20[C_x - C_0(T_m)]$; thus if $C_x/C_0(T_m) = 2$ and $C_0(T_m) = 10^{-5}$, $V_0(T_m) = 2 \times 10^{-4}$ or 2×10^{-3} cm/sec respectively. If the freezing velocity is less than this, these dislocations will terminate in the interface. Thus slow freezing velocities for a fixed vacancy supersaturation favor dislocation penetration of the interface. However, for V_0 small, C_x/C_0 is expected to be unity and to increase as V_0 increases.

The image force on a dislocation parallel to and at a distance z from the solid-liquid interface is $Gb^2/4(1 - \nu)z$ or about $10^{-5}/z$ dynes/cm. Thus if z is much greater than 1 μ, the force attracting a dislocation into the interface will be very small. Within such a distance the climb velocity V_0 should be less than $20C_x$, but the image force may be sufficient to cause segments of dislocation lines to penetrate the interface by gliding if the material is a metal.

ARRAY FORMATION. Dislocations put into a crystal by various mechanisms will interact with each other behind the interface to form arrays. This tends to happen because the energy of an array is less than that of the single dislocations that constitute it. If the arrays divide the crystal into rodlike elements (Fig. 47), they may also lower the free energy of the crystal by decreasing the excess vacancy concentration. The dislocations in the arrays will interact with each other and climb, absorbing vacancies in the polygonization-type process.

Frank (1956) has developed an alternative mechanism using vacancy platelets to

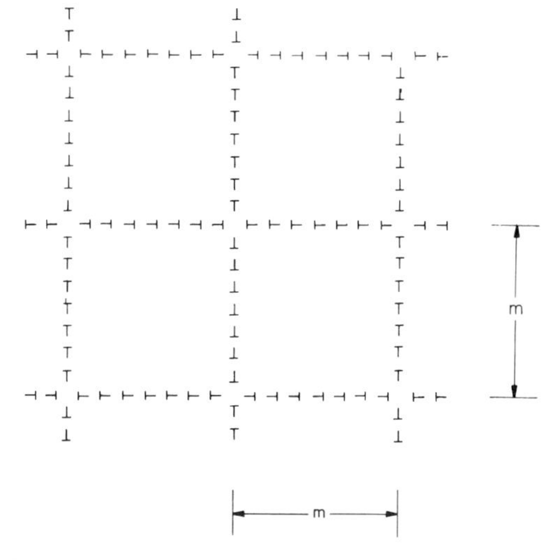

Fig. 47. Dislocation configuration in the "striation-type" array.

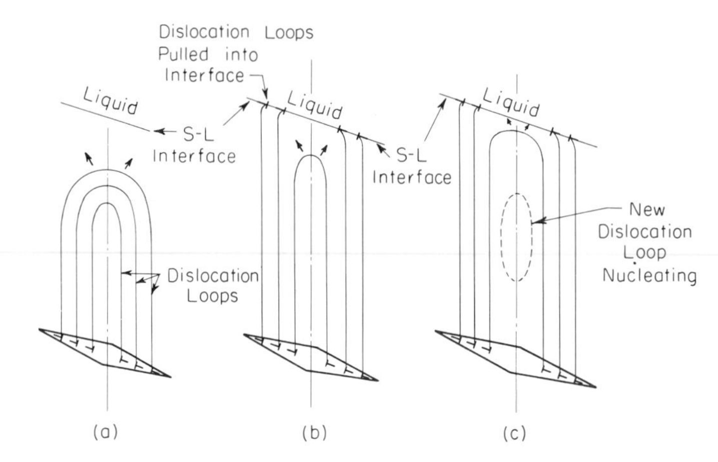

Fig. 48. Dislocation configuration during formation and after motion into the interface from Frank's mechanism of striation formation.

account for the "striation" substructure of Fig. 47. The dislocations in the arrays could not have been formed by the collapse of vacancy disks without some multiplication mechanism because the total vacancy concentration in the crystal is two to three orders of magnitude too small. Therefore Frank (1956) postulated the following: Edge dislocation loops, formed by vacancy condensation at some distance behind the interface, grow in size to reduce the vacancy supersaturation. Those loops that lie in planes parallel to the growth axis grow towards the interface and catch up to it. Once a dislocation has entered the interface it will contribute to the misorientation of the growing solid without further vacancy condensation. This is illustrated in Fig. 48. The elastic interaction between edge dislocations should cause them to assemble into parallel arrays as shown.

The ends of the loops are in an unstable position and are attracted to the interface. The dislocations can no longer remove vacancies by lengthening, and the vacancy excess tends to move them apart. A new dislocation loop then forms preferentially in the central region. Thus rotation in the lineage boundaries continually increases until the rate of production of new loops is balanced by the rate of pair annihilation. This mechanism can only operate if there is sufficient vacancy supersaturation at the freezing interface to cause climb.

However it might be formed, an array of side m cm (Fig. 47) will draw vacancies and dislocations from an area m^2 cm^2, and the larger is m the closer together are the dislocations in the array wall. Since the rate of dislocation climb is a strong function of the distance of separation, the rate of vacancy annihilation will increase as m increases. However, when m becomes too large, the vacancies do not have time to diffuse to a boundary from within the area m^2. Thus we would expect the most stable array to be the largest to which vacancies can diffuse in the time allowed. This array size m is given by (Frank, 1956)

$$m^2 = D_v k T^2 / V_m V G_s \qquad (50)$$

where D_v is the diffusion coefficient of vacancies, V is the freezing velocity, V_m is the specific volume, and G_s the temperature gradient in the solid. Equation (50) gives $m = 0.1$ cm when $V = 10^{-3}$ cm/sec.

These conclusions fit the data on "striation"-type arrays which consist of low angle grain boundaries ($\Delta\theta \sim 1° - 5°$) and which divide a crystal into rodlike elements 0.5 mm to 5 mm on the side (Teghtsoonian and Chalmers, 1951, 1952).

Twins

Two mechanisms have been postulated to account for the formation of growth twins. One attributes them to thermal and other

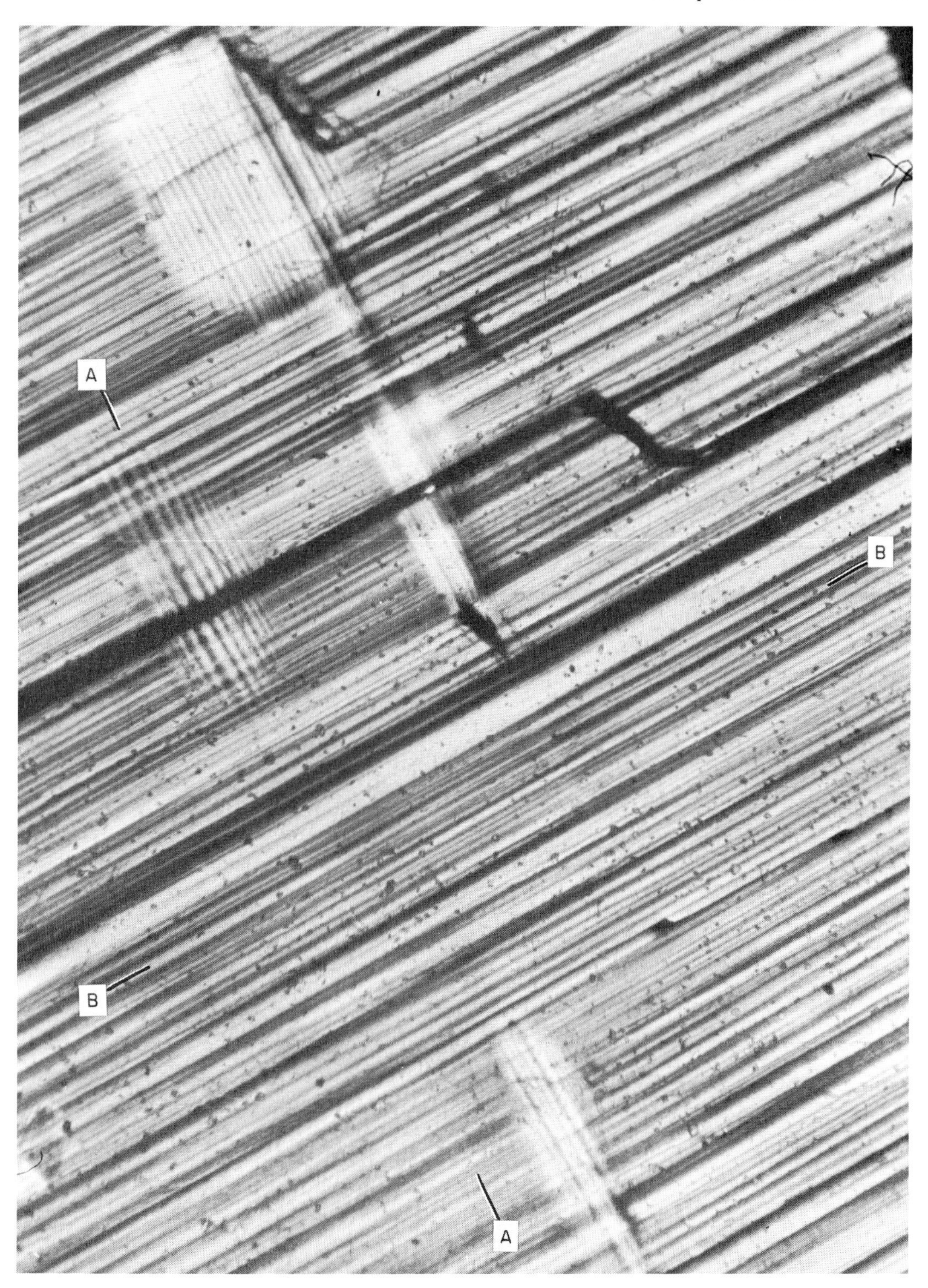

Fig. 49. Photomicrograph of a polished section from a $2BaO:TiO_2:P_2O_5$ crystal illustrating parallel growth twins, *B-B*, and groups of mechanical twins, *A-A* (1000 X).

stresses that arise during the freezing process. The other maintains that twins arise from the chance occurrence of stacking faults at the solid-liquid interface. It is necessary to have some driving force for twin formation, but little energy is associated with coherent twin boundaries in most materials of simple structure.

During the growth of crystals of $2\,BaO:TiO_2:P_2O_5$, it has been found (Harrison and Tiller, 1962), that two types of twins form (Fig. 49). The deformation twins *A-A* result from thermal stresses, and their number increases as the thermal stresses increase. The growth twins *B-B* form at the solid-liquid interface, providing an array of re-entrant corners that allow rapid freezing.

Solidification of germanium and silicon occurs preferentially in a direction lying in a $\{111\}$ plane (Billig, 1954). If the growing crystal is oriented so that by twinning a $\{111\}$ plane is brought more nearly parallel to the heat flow direction, twinning often occurs. The reason appears to be that germanium solidifies by the extension $\{111\}$ planes, and this is facilitated by aligning a set of $\{111\}$ planes parallel to the direction of heat flow, that is, perpendicular to the interface. An additional investigation (Bolling et al., 1956) indicates that the driving force for twin formation is the development of supercooling in the liquid adjacent to the interface. When small amounts of constitutional supercooling exist in the liquid, cells do not form as might be expected; instead twin lamellae appear on the interface.

Void Formation

Since the interface concentration during growth from an unmixed melt is C_0/k_0 and from a mixed melt is C_s/k_0, the concentration may be large enough to cause the nucleation of a second phase. The partition coefficient k_0 may be very small ($10^{-3} - 10^{-2}$); thus, it is possible for a small initial concentration in the liquid ($C_0 \sim 10^{-3}$ atomic per cent) of a gaseous constituent to cause the nucleation of gas bubbles on the interface. This occurs during the freezing of ice crystals due to the buildup of dissolved oxygen and due to carbon monoxide formation in rimmed steel. The gas bubbles inhibit the growth of the solid beneath the cap of the bubble, and voids are left in the solid as the interface advances. At rapid freezing rates, an array of bubbles is formed, whereas at slow freezing rates long pores are formed as the individual bubbles are fed by diffusion in the interface layer (Fig. 50).

When a growth system contains a volatile solute and interface enrichment is allowed to occur, the vapor pressure of the solute may become large enough to cause nucleation of vapor bubbles on the interface or an external surface, producing characteristic voids and a "Swiss cheeselike" structure. This can only be eliminated by effective mixing of the liquid.

Vapor bubbles can also be nucleated at the liquid-container interface if the liquid is significantly hotter than other parts of the container or if the container is a good nucleation catalyst. Such voids can be seen on the exterior of a crystal.

Voids can also be produced near cell boundaries, especially in crystals solidified at high rates. This happens because macroscopic liquid flow through long narrow channels is required to feed the groove roots between cells. At low freezing rates this fluid flow occurs readily, but at high freezing rates it is inhibited by the viscosity of liquid. The effect occurs only when the specific volume of the solid is less than that of the liquid ($\Delta V_{sL} > 0$).

Stray Crystal Formation

Two additional factors that determine success in crystal growth are (1) the macroscopic interface shape and (2) the degree of constitutional supercooling existing in the liquid ahead of the interface. Consideration of the interface morphology in the vicinity of an external boundary, for example, at the crucible surface, shows that a small volume of supercooled liquid exists in this region. Both the degree of supercooling and the volume of supercooled liquid are greater if the interface is concave rather than convex to the liquid; so the probability of stray crystal formation is greater for con-

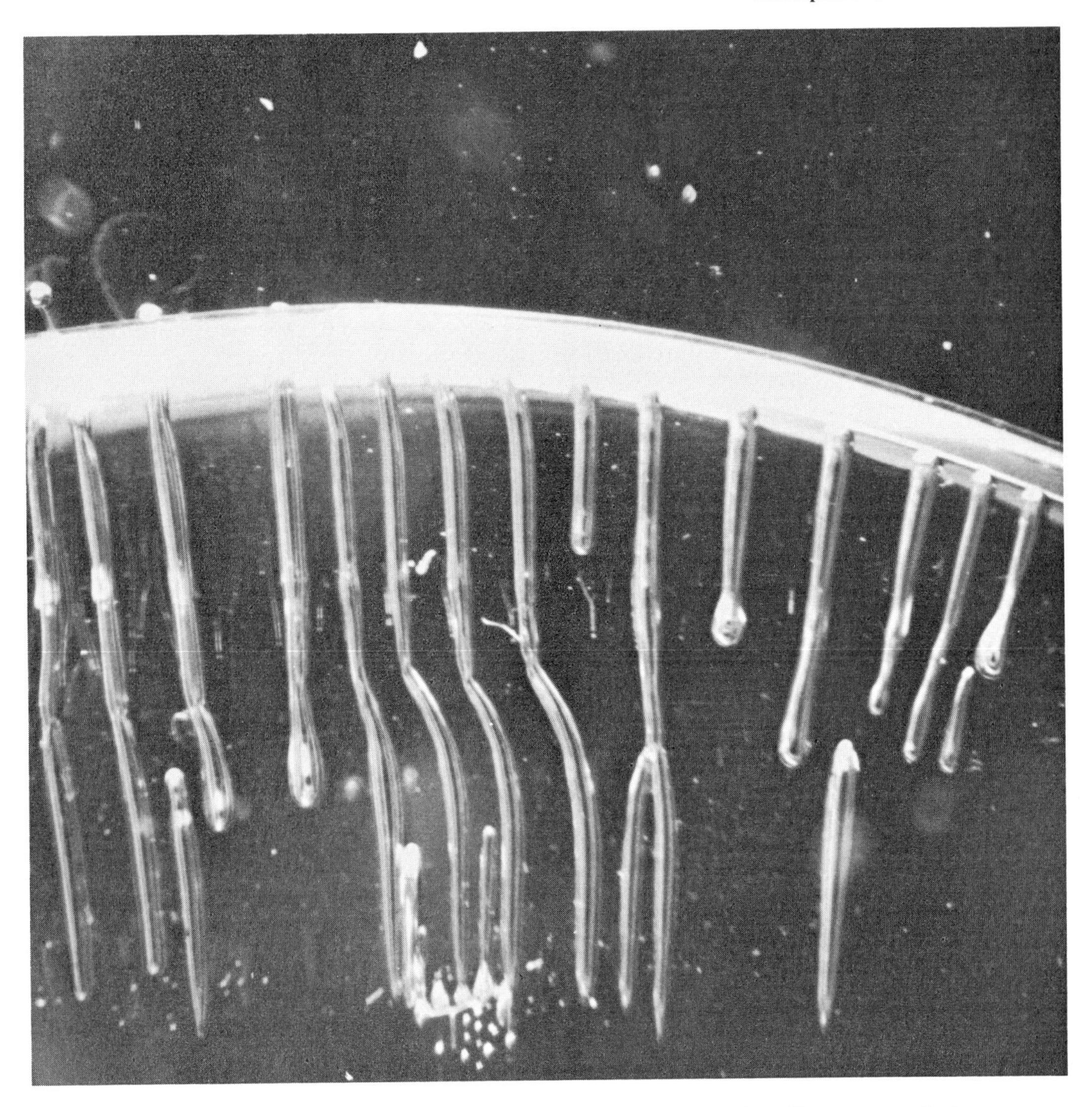

Fig. 50. Photomicrograph of growing ice crystal illustrating the formation of long pores at the ice-water interface because of dissolved oxygen buildup at this interface (15 X).

cave than for convex interfaces. Also for concave interfaces, stray crystals tend to grow into the existing crystals because the grain boundary tends to align itself perpendicular to the interface. When a stray crystal forms on a convex interface, it cannot grow to any appreciable size because the existing crystal encroaches on it.

If heterogeneous nuclei are present in the liquid that can operate at a supercooling of δT_c many stray crystals will form wherever the degree of constitutional supercooling exceeds δT_c. The condition for stray crystal formation is similar to that for cell formation

$$\frac{G}{V} < -\frac{mC_s(0)(1-k_0)}{Dk_0} - \frac{\delta T_c}{V\delta_c} \quad (51)$$

If this condition is satisfied, stray crystals will form ahead of the interface and inhibit individual crystal growth.

5. SUMMING UP

In this chapter, no attempt has been made to discuss the considerable amount of experimental work in this area; this will be treated in the subsequent chapters by other authors. However, there has been discussions of those principles of solidification that are of primary interest and usefulness to workers involved in the growth of crystals from the melt. To this end, the importance of the four primary controllable variables in any crystal growth experiment, that is, thermal, constitutional, hydrodynamic, and crystallographic, has been illustrated. Furthermore, it has been illustrated how the constraints imposed by these four variables on the growth process leads to the incorporation into the crystal of many different macroscopic and microscopic solute distributions and physical defects.

ACKNOWLEDGMENT

This work was partially supported by the U. S. Air Force Office of Scientific Research, AF49(638)-1029.

References

Boček, M., P. Kratochvil, and M. Valouch (1958) *Czech. J. Phys.* **8,** 557.

Billig, E. (1954), *J. Inst. Metals* **83,** 53.

Bolling, G. F., 1961, Private Communication.

Bolling, G. F. and W. A. Tiller (1960a), *J. Appl. Phys.* **31,** 1345; (1960b), *J. Appl. Phys.* **31,** 2040; (1961a), *Metallurgy of Elemental and Compound Semiconductors,* A.I.M.E., Vol. 12, Interscience, New York, p. 97; 1961b), *J. Appl. Phys.* **32,** 2587; (1962), *J. Appl. Phys.*

Bolling, G. F., W. A. Tiller, and J. W. Rutter (1956), *Can. J. Phys.* **34,** 234.

Bolling, G. F., J. Kramer, and W. A. Tiller (1962), *Trans. A.I.M.E.,* (submitted).

van den Boomgaard, J. (1955) *Phillips Res. Repts.* **10,** 319.

Burton, J. A., R. C. Prim, and W. P. Slichter (1953), *J. Chem. Phys.* **21,** 1987.

Cahn, J. W. (1960), *Acta. Met.* **8,** 554.

Cahn, J. W., and J. E. Hilliard (1958), *J. Chem. Phys.* **28,** 258.

Chadwick, J., (1961), Private Communication.

Chernov, A. A. (1960), *Doklady Acad. Nauk SSSR* **132,** 818.

Chilton, J. P., and W. C. Winegard (1961), *I. Inst. Met.* **89,** 162.

Dash, W. C. (1959), *J. Appl. Phys.* **30,** 459.

Elbaum, C. (1960), *Phil. Mag.* **5,** 669.

Elbaum, C. and B. Chalmers (1955), *Can. J. Phys.* **33,** 196.

Frank, F. C., 1956, *Deformation and Flow of Solids,* Madrid Conference Springer, Berlin; (1958), *Growth and Perfection of Crystals,* John Wiley and Sons, New York, p. 304.

Goss, A. J., K. E. Benson, and W. G. Pfann (1956), *Acta. Met.* **4,** 332.

Hamilton, D. R., and R. G. Seidensticker (1960), *J. App. Phys.* **31,** 1165.

Harrison, D. E., and W. A. Tiller (1962), *J. Appl. Phys.* **33,** 2451.

Harrison, J. D., and W. A. Tiller (1961), *Trans. A.I.M.E.* **221,** 649.

Hillig, W., and D. Turnbull (1956), *J. Chem. Phys.* **24,** 219.

Holmes, E. L., J. W. Rutter, and W. C. Winegard (1957), *Can. J. Phys.* **35,** 1223.

Ivantsov, G. P. (1958), *Growth of Crystals,* Consultants Bureau Inc., New York.

Jackson, K. A., and B. Chalmers (1956), *Can. J. Phys.* **34,** 473.

Jackson, K. A. (1961), *Elemental and Compound Semiconductors,* A.I.M.E., Vol. 12; Interscience, New York, p. 120.

Johnston, W. C., and W. A. Tiller (1962), *Trans. A.I.M.E.* **224,** 214.

Koslovskii, M. I. (1958), *Acad. Sci. U.S.S.R.* **3,** 209.

Kratochvíl, P., P. Lukac, and M. Valouch (1960), *Czech. J. Phys.* **B10,** 48.

Lindemeyer, C. S., G. T. Orrok, K. A. Jackson, and B. Chalmers (1959), *J. Chem. Phys.* **27,** 822.

Lord, N. W. (1953), *Trans. A.I.M.E.* **197,** 1531.

Penning, P. (1958), *Phillips Res. Repts.* **13,** 79.

Pfann, W. G. (1957), *Zone Melting,* John Wiley and Sons, New York.

Pound, G. M. (1958), *Liquid Metals and Solidification,* A.S.M., Cleveland.

Reiss, H. (1954), *Trans. A.I.M.E.* **200,** 1053.

Rosenberg, A. (1956), *Ph.D. Thesis,* University of Toronto.

Rosenberg, A., and W. C. Winegard (1954), *Acta. Met.* **2,** 242.

Rutter, J. W., and B. Chalmers (1953), *Can. J. Phys.* **31,** 15.

Schoeck, G., and W. A. Tiller (1960), *Phil. Mag.* **5,** 43.

Smith, V. G., W. A. Tiller, and J. W. Rutter (1955), *Can. J. Phys.* **33,** 723.

Teghtsoonian, E., and B. Chalmers (1951), *Can. J. Phys.* **29,** 370; (1952), *Can. J. Phys.* **30,** 338.

Temkin, D. E. (1960), *Sov. Phys. Doklady* **5,** 609.

Thurmond, C. D. (1959), *Semiconductors,* Rheinhold, New York, p. 145.

Tiller, W. A. (1955), *Ph.D. Thesis,* University of Toronto; (1956), *Can J. Phys.* **34,** 729; (1958a), *J. Appl. Phys.* **29,** 611; (1958b), *Liquid Metals and Solidification,* A.S.M., Cleveland; (1962), *Trans. A.I.M.E.* **224,** 448.

Tiller, W. A., K. A. Jackson, J. W. Rutter, and B. Chalmers (1953), *Acta. Met.* **1,** 428.

Tiller, W. A., and J. W. Rutter (1956), *Can. J. Phys.* **34,** 96.

Trainor, A., and B. E. Bartlett (1961), *S. S. Electronics* **2,** 106.

Turnbull, D. (1950), *Thermodynamics in Physical Metallurgy,* A.S.M., Cleveland; 1956, *Solid State Physics,* III, Academic Press, New York.

Wagner, R. S. (1960), *Acta. Met.* **8,** 57.

Walton, D., W. A. Tiller, J. W. Rutter, and W. C. Winegard (1955), *Trans. A.I.M.E.* **203,** 1023.

Walker, J. (1960), Private Communication.

Weart, H. W., and D. J. Mack (1958), *Trans. A.I.M.E.* **212,** 664.

Weinberg, F., and B. Chalmers (1952), *Can. J. Phys.* **30,** 488.

Specific Substances

16

Low Melting Point Elements

A. J. Goss

This chapter considers the experimental problems of crystal growth, particularly of elements with melting points less than 1100°C. The previous chapter has considered nucleation and the general principles of solidification, so that maximum emphasis is now placed on practical information. Previous publications that have been found most useful include reviews of the growth of metal crystals by Holden (1950), Chalmers (1953), and by Honeycombe (1959). A comprehensive survey has also been made by Hurle (1962). Other discussions have also been helpful: of the growth interface by Chalmers (1954), the solidification of metals by Martius (1954), and a more recent review by Winegard (1961).

The experimental arrangement for crystal growth must be such that the change of phase takes place smoothly, preferably without low angle boundaries or strain in the crystal. The thermal conditions that control solidification and the impurity and crystallographic factors that modify the growth are briefly considered in Section 1. The choice of experimental methods is so large that the possible mechanical arrangements are indicated only schematically in Section 2. Long experience has shown that high purity starting materials are required (imperfections and their relations to impurities have been described by Elbaum (1959)) and methods for cleaning and purifying elements are outlined in Section 3. Mold material is also of great importance; thus suitable materials with their properties and main impurities are listed in Section 3. (Methods of growth from the melt without a mold, that is, crystal pulling and float-zone techniques are dealt with in other chapters.) The shapes of crystals that may be grown from the melt are essentially simple, for example, square or circular in cross section with a vertical mold and half-round with a horizontal mold where surface tension controls the shape of top surface. A brief description of some early growth experiments is given in Section 4.

The main summaries of practical detail are given in Section 5. Each element is considered in turn, and methods for purification, crystal growth, and etching are given. A number of special methods for alloys, dendritic crystals, bicrystals, and other forms are discussed in Section 6. Finally, but not least important, the techniques for optical orientation and cutting of crystals are reviewed in Section 7.

1. THERMAL CONDITIONS, IMPURITY EFFECTS, AND CRYSTALLOGRAPHIC FACTORS

All these factors play a part in the growth process and, depending on conditions, one or another may dominate.

Thermal Conditions

The change of phase is primarily controlled by thermal conditions. Kuznecov and Saratovkin (1934) considered the pertinent heat flow equations, but did not consider the limiting conditions for monocrystalline growth. Roscoe (1934) took the condition to be G_l greater than zero, that is, no supercooling and obtained*

* Where G is temperature gradient, R is growth rate, ρ is density, L is latent heat of melting, and K is thermal conductivity.

$$\frac{G_{\text{mold}}}{R} > \frac{\rho L}{K_s(1 + \sqrt{K_l/K_s})} \qquad (1)$$

For cadmium, he found the least values from experiment to be ten times greater than the theoretical value. An alternative treatment was made by Pomeroy (1952) who assumed that a planar solid-liquid interface is necessary. His condition for monocrystalline growth is

$$\frac{G_{l,s}}{R} > \frac{\rho_s \rho_l L}{\rho_l K_s - \rho_s K_l}$$

Since $\rho_s \simeq \rho_l$, this gives

$$\frac{G_{l,s}}{R} > \frac{\rho L}{K_s - K_l}$$

which does not consider the change of gradient between solid and liquid. Goss (1953b) considered the simple heat flow from the liquid plus the latent heat of solidification and equated it to the heat flow in the solid.

$$G_l K_l + RL\rho = G_s K_s \qquad (2)$$

Temperature gradients were measured in six metals, and it was shown that this equation was satisfied for slow rates of growth. For no supercooling to occur, that is, $G_l > 0$, Eq. (2) is

$$\frac{G_s}{R} > \frac{\rho L}{K_s}$$

which is similar to Eq. (1). The corresponding heat flow equations for a crystal pulled from a melt by the Czochralski method have been given by Pohl (1954).

If an extended volume of crystal is considered, radiation losses need to be included. Billig (1955) gives

$$R_{\max} = \left(\frac{2\sigma K_m T_m{}^5}{3r\rho^2 L^2}\right)^{1/2} \qquad (3)$$

assuming that the melt temperature is much greater than its surroundings. Hurle (1960) has pointed out that this is not true for low melting point metals, so for them $R_{\max}$ is less than given by Eq. (3). A similar relation has been given by Maslennikov (1955) and compared with experimental data for Pb, Sn, and Zn.

These heat flow theories are obviously idealized, but they indicate the advantage of using small rates of growth. When the thermal conductivity is low, or latent heat and density are both high, slow growth rates are even more important.

Impurity Effects

The presence of a second element as impurity or addition enhances the problems of growth. Rutter and Chalmers (1953) have pointed out that the melting point ahead of a freezing interface is modified by the impurity distribution in the melt, and hence "constitutional" supercooling can occur. This is discussed in more detail in Chapter 15. With an impurity with segregation coefficient k^* less than one, the impurity distribution ahead of the freezing interface is as shown in Fig. 1. Tiller et al. (1953) give the condition for no supercooling as

$$\frac{G_l}{R} > \frac{mC_l(1 - K^*)}{dk^*} \qquad (4)$$

The stages in development of imperfections caused by constitutional supercooling have been shown by Tiller and Rutter (1956). First, a "pox" on the interface is detected with corrugations on the surface. The next stage is a well-defined cell structure (Fig. 2). This has been made visible on many interfaces (Blaha, 1953; Pond and Kessler, 1951; Smialowski, 1937) and traditionally the appearance of the surface of a cast metal has been used as a test of purity in the foundry. When $k^* < 1$ the impurity concentrates in the boundaries of the cell structure. As the degree of super-

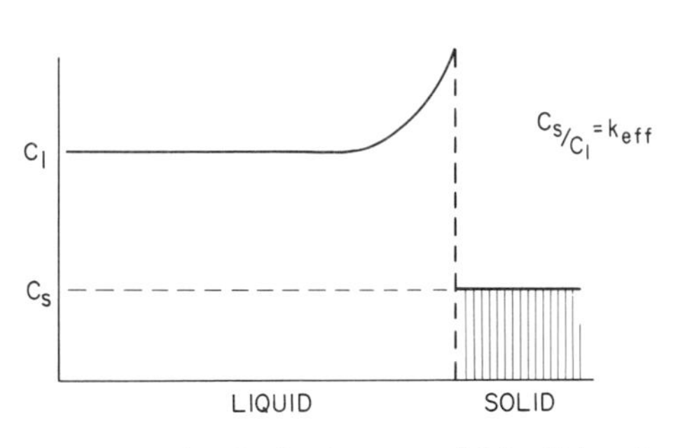

Fig. 1. Impurity distribution near solid-liquid interface. $k < 1$

Fig. 2. Cell structure on growth interface. (From R. B. Pond, 1951, *Trans. AIME,* p. 1158.)

cooling increases (with increase of R or C_1 or decrease of G_1) the cells become dendrites (Tiller and Rutter, 1956). In some materials new crystals will nucleate before dendritic growth occurs. Sparingly soluble impurities have the most marked effect on crystal growth (Goss, 1953a), but zone-refining is particularly effective in removing such impurities. A small rate of growth is effective in avoiding constitutional supercooling just as it is helpful for heat flow reasons.

The various structural features that appear as a result of impurities have been dealt with by Rutter (1958) as imperfections, by Elbaum (1960) as substructures, and by Hirsch (1956) as mosaic structures.

The growth of alloy crystals depends on the segregation coefficient and the factors given in Eq. (4). For lamellar eutectics, Tiller (1958) has analyzed the heat flow, but there is little published information on systematic experimentation (Section 6). Since the growth of alloy crystals is currently of interest to the semiconductor industry, new information may soon become available.

Crystallographic Effects

In addition to thermal and impurity effects on growth, crystal structure also plays a part. The evidence for crystallographic effects is not always direct, but it is none the less extensive.

Chalmers and Martius (1952) concluded that growth takes place in the presence of a normal temperature gradient by steps on the interface. Examination of interfaces on lead (cubic structure) has shown steps on the growth front of 0.001 mm height when the interface is within 15° of (100) or (111) (Elbaum and Chalmers, 1955). When the surface is nearly parallel to (111), facets are found (Atwater et al., 1955) even when the melt is vibrated.

Preferred growth orientations for cubic metals may also be taken as evidence of crystallographic effects. For fast rates of growth in pure lead, the [111] orientation is preferred, whereas the addition of 5×10^{-4} weight per cent silver favors the [100] orientation (Rosenberg and Tiller, 1957). Aluminum crystals grown by the Czochralski method (Lainer et al., 1960) showed the effect in reverse, [100] growth going to [111] upon addition of zinc.

The diamond-type structure (represented by germanium among low melting point elements) has isotropic thermal conductivity, but shows evidence of very strong crystallographic forces, in particular, the stability of the close-packed (111) plane. Growth flats on the sides of pulled germanium crystals are well known (Shashkov, 1961), and (111) facets have often been observed, for example, Robinson (1952). In Fig. 3 a fine example is shown of a crystal grown in three months by Trumbore et al. (1959). Observations analogous to those for lead have been made of facets on Ge interfaces (Wilkes, 1959 and in this laboratory). Recent work by Dikhoff (1960) has shown very clearly the presence of (111) facets during growth and orientation-dependent impurity segregation.

For anisotropic elements such as bismuth, gallium, and selenium, strongly preferred growth directions have frequently been observed (Section 5). However, these elements are also thermally anisotropic, so they show combined effects.

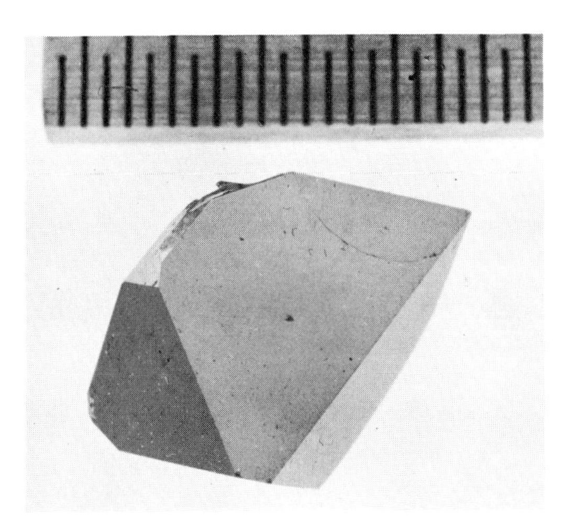

Fig. 3. Germanium crystal with (111) growth faces. [From F. A. Trumbore (1959), *J. Phys. Chem. Solids* **11**, opposite p. 240.]

2. EXPERIMENTAL METHODS—SCHEMATIC

Growth of homogeneous crystals is controlled by the movement of the melting point isotherm through the specimen. Three basic methods may be used: (1) movement of the furnace, (2) movement of the sample, and (3) movement of the isotherm by temperature control.

1. Movement of the furnace is often the simplest to carry out and is well recommended for laboratory use. One advantage is the absence of vibration in the melt, although the writer is somewhat sceptical of the significance of small vibrations if the movement is essentially smooth. Constant vibration has been purposely applied by Atwater et al. (1955), Eisenhower (1961), and others without obvious crystal degradation.
2. Movement of the specimen is convenient when large furnaces or high-frequency heating systems are used. When no suitable mold material for the melt is known, the crystal pulling technique (Czochralski) is essential.
3. Movement of the isotherm has advantages when very slow growth rates are required, even as low as mm/10^4 min. In some laboratories temperature programmers may be more readily available than furnace-moving equipment.

The physical arrangement of the furnace and the melt is a matter of choice, depending on the available resources and the type of crystal to be prepared. Usually, simple horizontal arrangements (Fig. 4) or the corresponding vertical arrangements are used. The horizontal methods are convenient with boat-shaped molds and seeding is simply arranged, but the cross sections of the crystals so obtained are asymmetric. It is also difficult to avoid a difference in thermal conditions above and below the melt.

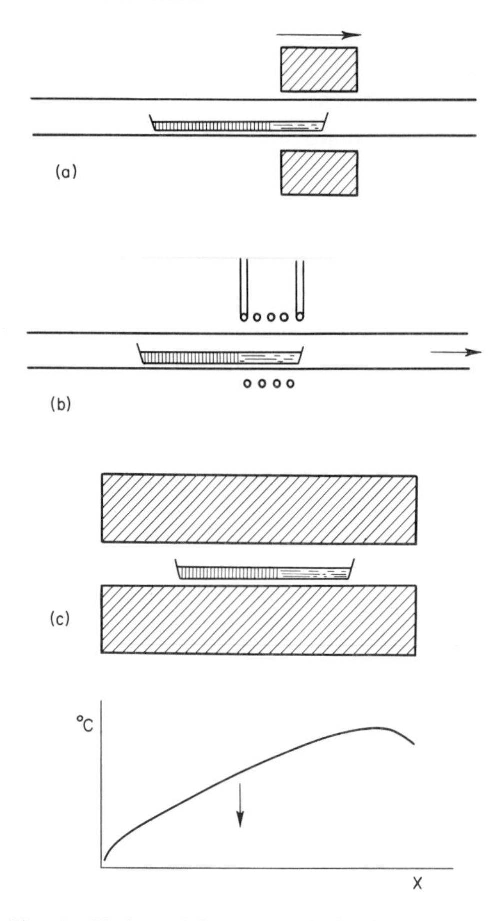

Fig. 4. Horizontal furnace methods of growth. (*a*) Furnace movement. (*b*) Sample movement. (*c*) Temperature control.

The vertical methods give symmetrical crystals and uniform conditions, but seeding is more difficult and any expansion on solidifying can strain the crystal. The soft mold method (Noggle, 1953) overcomes the expansion problem.

A growth rate of ~1 mm/min is seen in Section 5 to be useful for elements with normal thermal characteristics. When growth rates of 0.1 mm/min or less are necessary (for example, with alloys), the stability of the furnace temperature becomes important. The product of the growth rate and the temperature gradient gives the rate of temperature change

$$G \times R = \frac{dT}{dt}$$

Therefore if the gradient at the melting point is 20°/cm and the temperature stability is 1°/min, growth-rate fluctuations of 0.5 mm/min can occur. Faster rates or better control can reduce this. In Ge-Si alloys the results of thermal fluctuations are clearly seen at a growth rate of 1 mm/hr (Goss et al., 1956). See Fig. 5. Thus temperature control must be kept in mind when a growth method is being developed.

A review of the various growth methods has been given by Buckley (1951). For more practical details on furnaces, the book by Lawson and Nielsen (1958) is helpful. Chalmers (1953) has summarized the techniques used in his researches.

Seeds

Seeds are usually used when the orientation of a crystal is to be predetermined. It is particularly simple with a horizontal boat to fit a seed to the cross section of the mold by etching or grinding. This prevents the melt from running beyond the end of the seed. The procedure is to melt some of the seed (1 mm or more) in contact with the main bath to insure good wetting and hence continuity. Seeds of smaller cross section than the main bath are normal for the Czochralski process. Use of highly per-

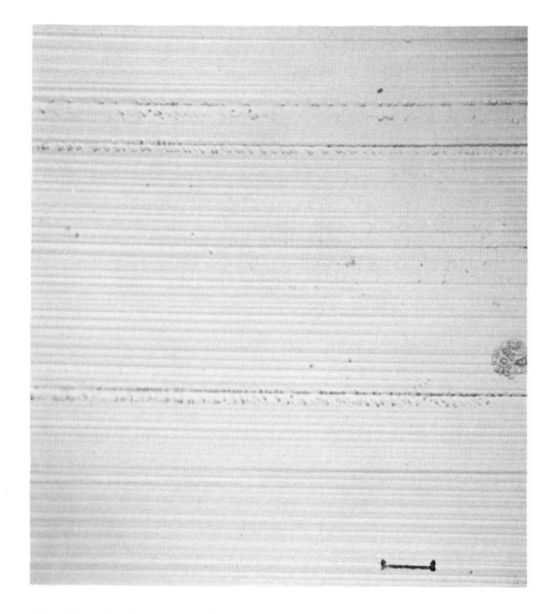

Fig. 5. Striae in a Ge-Si alloy crystal grown with fluctuating temperature. (marker denotes 50 μ). (From Goss et al. 1956, *Acta Met* 4, p. 332.)

fect seeds insures that no defects in the crystal originate in the seed. By reducing the cross section of the growing crystal after seeding, perfection may be improved by growing the imperfections that grow from the seed out of the sides of the crystal. Dash (1959) used this technique with good success for silicon, and it has since been applied to other materials.

Selective cutting of polycrystalline ingots may be used to provide initial seeds. These can be increased in size by careful growth procedures. If no seed is used, the number of initial nuclei may be minimized by using a pointed ingot, and self-selection may be produced by reducing the size of the growth cross section (Bridgman, 1925).

3. PURE MATERIALS: ELEMENTS AND MOLDS

Two essentials for monocrystalline growth are pure elements and pure mold materials. Also, the latter must not react with the liquid. In the last ten years very considerable improvements have been made in the purity of both elements and molds, due in no small measure to the influence of semiconductor research.

Elements

Pure elements are available from many manufacturers and agents. A list of some of these is given by Seybolt and Burke (1953). The chemical processing of the elements is considered to be outside the present scope. However, there are important physical methods of purification that are often convenient for the crystal grower and may even use the growth apparatus. These are discussed briefly.

1. As-received elements although "pure" are often contaminated with surface oxides, etc. Many experimenters have removed these by the simple expedient of melting the element and running it through a small hole, or better through a length of capillary, into a clean receiver containing a vacuum or inert gas. The dross, oxide, and other solids are left behind. After this treatment many metals look like very clean mercury. An apparatus used for this cleaning is shown by Lawson and Nielsen (1958).

2. If the purity of an element is not high enough, simple melting and freezing from one end [that is, "normal freezing" (Pfann, 1958)] will effectively segregate any impurity with a high or low value of k^*, the segregation coefficient. Further purification may be obtained by removing the end of the ingot and refreezing the remainder. Those elements most likely to be troublesome in the growth process are the most readily removed. Soluble impurities will remain. This process is often used for arsenic and gallium (see Section 5).

3. Impurities with less favorable values of k^* can be removed best by Pfann's zone-refining methods (1958). Pfann has given useful segregation curves in his book. In carrying out zone-refining, the mold material is sometimes a problem, as for example with copper and gold, for which the purification is commonly limited by mold contamination. However, with many elements (for example, Al, Bi, Sb, Ge, Pb, Te, and Sn) simple forms of zone-refining gives considerable purification. Zone-refined material is usually ideal for crystal growth. In fact, the measure of refinement is the presence of large crystals in the purified specimen.

4. Vacuum pumping can be used to remove impurities with high vapor pressures (for example, Cd, Zn, and volatile oxides). Stirring is necessary to expose fresh surfaces and to avoid trapping gas under liquid pressure. A suitable apparatus design with a cold trap in direct line with the melt is important.

5. Vapor distillation is used for the purification of volatile elements. Commercial processes for the preparation of high purity Cd, Hg, and Zn are well known.

Mold Materials

A mold that will not react and hence become attached to the solid is important in order to avoid spurious nucleation. It should be remembered that an element may have different wetting characteristics when clean than with a thin oxide coating.

Pyrex

Pyrex is a good mold material for very low-melting-point elements. Its softening point, about 600°C, is high enough for Cd, Pb, Sn, Zn, and others. It may be removed by cracking, but this may damage the crystal. Removal by dissolution of the Pyrex is a slow process so a coating of Aquadag (colloidal graphite) is often used to avoid sticking and allow easy removal. Alternatively, precision-bore Pyrex tubing may be used for easy removal or a split mold. Pyrex has prime advantages in being cheap, easily fabricated, and conveniently connected with ground joints for vacuum pumping.

Composition: approximately $80\frac{1}{2}$% silicon oxide, 13% boron oxide, 2% aluminum oxide, 4% sodium oxide, and $\frac{1}{2}$% potassium oxide.

Thermal conductivity: 0.002 cgs units.

Cleaning: 5% HF, 33% HNO_3, 2% Teepol, and 60% water (Crawley, 1953).

Vycor

A higher range of temperatures can be covered with Vycor which has a softening point of ~1000°C. Vycor is much more expensive than Pyrex, not so easy to fabricate, and ground joints for it are less readily available.

Composition: $96\frac{1}{2}$% silicon dioxide, 3% boron oxide, and $\frac{1}{2}$% aluminum oxide.

Thermal conductivity: ~0.003 cgs units.

Silica

Silica is more than adequate for melting any element considered in this chapter. It softens at about 1200°C, is chemically inert (only HF attacks it of the common acids), and available in very high purity forms, for example, Spectrosil, synthetic silica.

Composition: Typical analyses of Thermal Syndicate materials give the following in ppm: (*a*) Spectrosil grade. (*b*) Transparent Vitreosil.

	Al	Sb	As	B	Ca	Cu	Ga	Fe	Mn	P	K	Na
(*a*)	<0.02	0.0001	0.0002	0.01	0.1	0.0002	0.004	0.1	0.001	0.001	0.004	0.04
(*b*)	50–60	0.23	—	<1	0.4	0.01	—	0.74	0.026	0.01	—	4

Thermal conductivity: 0.003 cgs units (transparent)

Cleaning: 1 HF with 10 HNO_3; soak for 1 hr; soak and rinse in deionized water for the same period.

Carbon coating can be used on silica as on Pyrex, and sandblasting of the surface improves adhesion.

Alumina

Alumina can be obtained in prefabricated forms, in which case a binder has been used during firing or as very fine powder; this is available in high purity. Alumina is stable to very high temperatures. An important application of the powder is the soft mold technique (Noggle, 1953). The powder is tamped around a shaped ingot and contains the shape when the ingot is molten, but allows expansion during solidification.

High purity material, for example, from Aluminium Industrie A.G., Switzerland, is quoted as having the following impurities:

Silicon dioxide <20, titanium dioxide <10, iron oxide <10, phosphorus pentoxide <10, and sodium oxide <10 ppm.

Alumina has specific use for aluminum, but has also been used with other elements, for example, Bi.

Graphite

Graphite is obtainable in very high purity (99.999%) and can be used to high temperatures. It is easily machined, although this is not popular in the workshop. A hard surface can be obtained on graphite of proper grain size, and a final high-temperature firing provides cleaning. Graphite has high thermal conductivity and has found general application for metals with melting points at about 1000°C that also have high thermal conductivity: Cu, Ag, and Au. Graphite is sometimes also preferred for lower melting-point elements.

Thermal conductivity: 0.4 (0°C) to 0.2 (800°C) cgs units.

Impurities: PT grade graphite from Le Carbone (GB) Ltd., is quoted as having a maximum ash content of 10 ppm with boron <1, molybdenum <1, calcium <1, sodium <1 ppm, and traces of magnesium and silicon.

Other Mold Materials

Pyrophyllite is an aluminum silicate that can be machined and then fired at 1050°C before use. A good finish together with low thermal conductivity is obtained (thermal $k = 0.003$ cgs). Stainless steel has been used for magnesium melts; silicon carbide, aluminum nitride, and boron nitride have also been suggested. Recently, the possibility of coating molds with refractory materials by plasma spraying has been discussed, but no applications to low melting-point elements are known.

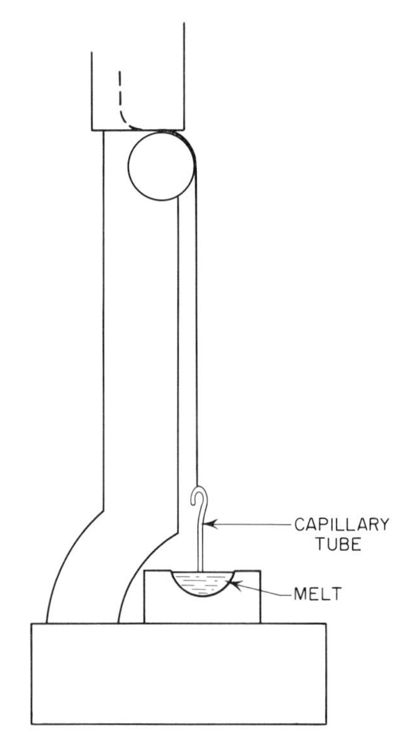

Fig. 6. The Czochralski method of growth by pulling the solid from a melt.

4. EARLY EXPERIMENTS

The first experiments with monocrystals of the elements were made with those that showed external crystallographic features, principally Bi. Perrot (1898) used slabs of Bi cut from a block as did others (Lowndes, 1901). Experiments with Na and K (Baker, 1913) and with Hg (Andrade, 1914) showed the mechanism of deformation of monocrystals and were among the first to use artificial crystals. Tammann (1921) has described many of the early experiments on crystal growth, particularly organic melts. The study of growth by pulling dates from Czochralski (1917), who determined maximum rates of withdrawal of seeds of Pb, Sn, and Zn from the melt. He lowered a capillary tube into the surface of a melt, and upon lifting it out obtained seed crystals. The melt temperature and pulling rate are the main factors used to control the crystal dimensions. Modern adaptations use many refinements (Blaha, 1953; Marshall and Wickham, 1958). The *Czochralski method* is used in this text to describe vertical pulling methods (Fig. 6).

In the following decade the variety of growth methods was increased by several workers, of whom Obreimov and Schubnikov (1924) and Bridgman (1925) deserve special mention. They used a vertical furnace with the melt held in a mold. Obreimov and Schubnikov used an air blast to cool the bottom of the mold followed by reduction of the furnace temperature. Bridgman lowered the mold so as to solidify the melt as it emerged from the furnace. The term *Bridgman method* is used to describe methods in which the melt is lowered through a furnace.

Horizontal methods were also developed. Kapitza (1928), trying to avoid mechanical strains inherent in vertical arrangements, used a flat horizontal mold and isotherm control for freezing. He grew crystals of Bi in this way, as did Goetz (1930) who moved the metal horizontally through a stationary furnace. It is often easier to move the furnace as Hasler (1933) and others did.

Historical aspects of crystal growth are described in considerable detail by Buckley (1951). They provide a noteworthy reminder of the extensive work that has been done.

5. CRYSTAL GROWTH OF ELEMENTS

This section deals with purification, crystal growth, and etching of Al, Sb, As, Bi, Cd, Cu, Ga, Ge, Au, In, Pb, Li, Mg, Hg, K, Se, Ag, Na, Te, Tl, Sn, and Zn. *Preferred orientation* refers to the crystallographic axis bearing a special relationship to the cylindrical growth axis. The following groupings of elements should be noted because a method is often applicable to all members of a group:

(1) Cu, Au, and Ag
(fcc structure)
(2) Na, K, and Li
(bcc structure)
(3) Cd, Hg, and Zn
(hexagonal structure)
(4) In and β-Sn
(tetagonal structure)
(5) Bi and Sb
(rhombohedral structure)
(6) Se and Te
(hexagonal chain structure)

L = latent heat (cal gram^{-1})

K_s, K_l = thermal conductivity, solid, liquid (cal sec^{-1} cm^{-1} °C^{-1})

Aluminum

M.P. = 660.2°C, $L = 96$, $K_s = 0.52$ R.T., $K_l = 0.25$.

PURIFICATION. Aluminum oxide forms very readily and is difficult to remove; filtration in a vacuum would appear advantageous. Zone-refining is very effective in producing super-purity aluminum (Wernick et al., 1959; Albert, and other authors in the same reference, 1960). Production of high purity aluminum has been reviewed by Pearson and Phillips (1958).

GROWTH. Both solidification and strain-anneal methods have been used. Special attention has been paid to crystal perfection. The soft mold technique due to Noggle (1953; Noggle and Koehler, 1955) is preferred. High purity Al is held in vacuum (10^{-6} mm) in a mold consisting of alumina tamped firmly around the metal. The crystal is grown by moving the furnace (700°C) at a rate of 1.2 mm/min. A seeding technique has been described by Alden (1960) for this process. The resulting crystals show about the same degree of perfection as strain-anneal crystals (Noggle and Koehler, 1955). Similar experiments by Kelly and Wei (1955) suggest that the melt grown crystals are slightly imperfect containing low-angle boundaries of 1 to 20 min of arc. In all cases a low conductivity mold is used so the heat flow tends to be axial. Highly perfect crystals (1 mm diameter with 10^4 dislocations/cm^2) have been grown by Elbaum (1960) using the Czochralski method; and 0.15 mm diameter dislocation-free crystals (Howe and Elbaum, 1961).

Aluminum crystals often show mosaic structure (Ovsienko and Sosnina, 1956), with misorientations of 20 to 50 min increasing as the growth rate increases from 0.1 to 6.0 mm/min. Aust and Chalmers (1958) have shown that a tilted interface suppresses the mosaics. Changes in orientation of pulled crystals have been reported by Marinelli and Blaha (1956). These may be associated with the effect of 0.2% zinc addition, which changes the preferred orientation from [100] to [111] (Lainer et al., 1960).

Cahn (1950), using a horizontal method, showed by temperature measurements made in the melt that monocrystals were not obtained when supercooling occurred. High conductivity molds reduced the supercooling.

ETCHING. A number of workers have described etches for aluminum. Weerts (1928) used them (Section 7) for optical orientation determination. An authoritative account of etch pits and their use for finding orientations is given by Tucker and Murphy (1953). Excellent photomicrographs are shown by Kostron (1953), and similar ones by Basinka et al. (1954). Etch pits in aluminum are difficult to produce at pure dislocation sites according to Wyon (1955).

Antimony

M.P. = 630.5°C, $L = 38.3$, $K_s = 0.04$, $K_l = 0.05$.

PURIFICATION. Zone-refining is useful for antimony (Schell, 1955; Tanenbaum, 1954; Wernick et al., 1958). As, Sn, Bi, and Pb are difficult to remove (Vigdorovich, 1960).

GROWTH. There is little published work on the growth of Sb crystals. Bridgman (1929) used a quartz mold, lowered the tube through the furnace, and then ground off the quartz. The horizontal method was used by Rausch (1947). Zone-refined Sb obtained by Wernick et al. (1958) was grown in monocrystalline form in carbon-coated quartz. Epstein (1962,* private communication), using pyrophyllite molds, has grown seeded crystals at rates of ~1 mm/min.

ETCHING. The etching of Sb to give dislocation etch pits has been described by Wernick et al. (1958).

Arsenic

M.P. = 814°C at 36 atmospheres pressure.

PURIFICATION AND GROWTH. Weisberg and Rosi (1960) have described vapor zone-refining for As, but this appears to be less effective than distillation from a lead-arsenic melt (Whelan et al., 1960). Purification by the growth of crystals of As from the melt has been described in an RCA Report (1959). Crystals 10 cm long were grown under a pressure of 60 atmospheres in thick-walled quartz, using a growth rate of 0.16 mm/min and the Bridgman method. Very effective purification was obtained giving spectrographically detectable impurities of <1 ppm.

Bismuth

M.P. = 271°C, $L = 12.5$, $K_s = 0.02$ R.T., $K_l = 0.04$.

Vacuum filtration is used to clean the surface of oxides in addition to chemical polishing (Lovell and Wernick, 1959). Zone-refining is effective (Wernick, 1957; Hurle, 1960) for segregating Ag, Ca, Cu, Fe, Mg, Ni, Pb, and Sn. Forty-five zone passes at 0.76 mm/min reduced detectable impurities to <10 ppm (Wernick et al., 1957).

GROWTH. There are many early papers, in the 1920's especially, but there is little information relevant to the growth of highly pure and perfect bismuth crystals from this period.

Bismuth expands on solidification (3.3%), so the soft mold technique is essential as demonstrated by Vickers and Greenough (1956) and by Hurle (1960). Tamped alumina, or bismuth oxide, is used plus vacuum and a growth rate of 1 to 2.8 mm/min. The [111] axis strongly prefers to lie normal to the growth axis. This can be ascribed in part to the marked thermal anisotropy (K_s perpendicular to [111] = 0.022 cgs and K_s parallel to [111] = 0.013 cgs) (Goetz, 1930).

Bismuth crystals have been grown by pulling from the melt (Sajin and Dulkina, 1955; Packman, 1960; Porbansky, 1959). A melt in vacuum is usually used with a growth rate of 0.5 mm/min. A strong preference for the same orientation is found. Dislocation densities of 10^5 dislocation pits/cm^2 are typical.

ETCHING. The polish-etch (as described) is used, followed by a solution of 1% iodine in methyl alcohol (15 sec) to show dislocations (Lovell and Wernick, 1959). For orientation determination Yamamoto and Watanabe (1954) have described some etches (nonideal). Hurle and Weintroub (1959) have found that a combination of etching (50% HNO_3 for 30 sec) for optical checking together with x-ray examination is necessary.

Cadmium

M.P. = 321°C, $L = 13.2$, $K_s = 0.25$ (300°C), $K_l = 0.1$.

* Epstein, S. (1962), *J. Electrochem. Soc.*, **109**, 738.

PURIFICATION. Cadmium of 99.999% purity is commercially available. The writer found that vacuum filtering cleaned oxide from the metal. For purification, Silvey et al. (1961) found distillation to be very effective and Aleksandrov and Verkin (1960) used fifteen zone passes at 15 mm/hr in argon.

GROWTH. Although experiments on Cd crystals have been done for the past forty years, there are relatively few papers dealing with growth in detail. Roscoe (1934) and Andrade and Roscoe (1937) described the growth of $\frac{1}{2}$ mm diameter wires, both of pure Cd, and with 0.1% Pb added. The wires were sealed in vacuo in quills and a furnace gradient of 20°C/cm at a rate of 0.7 mm/min were best (see Section 2, Eq. 1). Difficulty in producing crystals with the basal plane perpendicular to the rod axis was observed by Boas and Schmid (1929), Gruneisen and Goens (1924), and Tsuboi (1929). The writer found similar problems when growing rods 4 mm diameter in tubes of 5 to 6 mm bore at rates of 0.5 to 1 mm/min. Aquadag (carbon) prevents wetting of glass molds, and a pressure of 10 mm reduces vaporization. Bridgman (1929) and also Gwathmey et al. (1948) found Cd more difficult to grow than other metals. De Grinberg (1961), using the same method, obtained a high yield of randomly oriented crystals (rates ≤ 0.4 mm/min). A detailed study of Zn and Cd by Blaha (1953) gives special information on cell structures and other imperfections.

ETCHING. Evaporation of the solid in vacuo gives good etch pits. Andrade (1949, 1950) has found an angular range of 20 min for the basal planes so produced. The writer found that HCl gives good optical reflections. The surfaces developed by HCl and vacuum etching have been studied by German (1954). Dislocation etch pits (density, $5 \times 10^3/cm^2$) were shown by Wernick and Thomas (1960). Predvoditelev (1960) has described electrolytic etching and polishing.

Copper

M.P. = 1083°C, $L = 49$, $K = 0.94$ R.T.

PURIFICATION. Zone-refining of copper does not appear to be very effective, either in carbon molds (Kunzler and Wernick, 1958) because of impurity pick-up or with the floating zone-refining method (Le Hericy, 1960). Impurity determination is difficult.

GROWTH. From over a score of references on crystal growth, a number of common features emerge. High purity copper (for example, copper from the American Smelting and Refining Company of 99.999% purity) and high purity graphite are suitable materials. A split mold avoids damage to the soft crystal after growth. A nitrogen or vacuum atmosphere is used with the Bridgman technique. Rates of solidification range from 6 mm/hr (Read, 1940), 16 mm/hr (Gwathmey, 1940), to 50 mm/hr (Wernick and Davis, 1956). Faster rates of 110 mm/hr gave lineage (Greninger, 1935) and dendrites. The orientations of crystals so grown appear to be random (Glocker and Graf, 1930; Petrauskas and Gaudry, 1949) although Davey (1925) reported the cube axis parallel to the growth axis.

Apparatus has been described by Gold (1949) consisting of a molybdenum furnace and a mercury float for lowering the melt. A much simpler method has been described by Wernick and Davis (1956). Lazarus and Chipman (1951) used isotherm control. Rowland (1951) used powdered carbon for the soft mold method. Gow and Chalmers (1951) used the horizontal technique with a carbon boat to grow mono- and bicrystals (8 in. × 1 in. × $\frac{3}{8}$ in.) at rates not greater than 120 mm/hr.

A unique shape (a cylinder 5.5 mm diameter with $\frac{3}{4}$ mm thick wall) was grown by Rebstok (1956). The effect of alloy additions on solidification of polycrystalline copper was described by Northcott (1938, 1939).

ETCHING. Lovell and Wernick (1959a) have used a chemical polish at 80°C, followed by etching to reveal dislocations in copper. More recently, Young (1961) has studied etch pitting in copper at both pure and impure dislocations.

Gallium

M.P. = 29.8°C, $L = 19$, $K_s = 0.08$ R.T., $K_l = 0.07$.

PURIFICATION. High purity Ga (99.999% pure) was prepared by Hoffman (1934) using chemical methods, electrolysis, and recrystallization. Studies reported by RCA (1959), and Zimmerman (1954) have shown that repeated directional freezing and vacuum heat treatment in quartz at 650°C are particularly effective in producing 99.999% purity. Richards (1956) used the ingenious device of zone melting gallium chloride.

GROWTH. Few published papers are available, although Hoffman (1934), Powell (1949), and Zimmerman (1954) have described unusual ease of crystal growth. Powell (1951) has described the growth of 99.999% Ga in glass molds at 27 to 28°C using a suitable seeding technique. A similar method was used by Horner (1961) and a variation that used a Teflon mold to avoid sticking was used by the Radio Corporation of America (1959).

ETCHING. Powell (1951) used dilute HCl as an etchant.

Germanium

M.P. = 937°C, $L = 112$, K_s M.P. = 0.07, $K_l \simeq K_s$.

Because of the large number of papers on Ge, considerable selection has had to be done. The most useful papers from the viewpoint of crystal growth are considered to be those by Rosi (1958), Billig (1956), Cressell and Powell (1957), and a complete but noncritical review by Tanenbaum (1959). A complete survey of the metallurgy of Ge up to 1958 is given in Shashkov's book (1961).

PURIFICATION. Purification by zone-refining for Ge is the classic application of this technique, and was essential for much semiconductor work. Pfann (1958) quotes impurities as 1 in 10^{10} of electrically detectable impurities, 1 in 10^7 of many other impurities, plus perhaps 100 ppm of oxygen and hydrogen (Thurmond et al., 1957). Zone-refining is carried out in carbon, or carbon-coated quartz, boats in dry nitrogen. Induction heating of the zones gives good stirring, and hence better segregation of impurities than d-c heating; although for Ge the latter is also effective. Refined material is available in quantity from commercial suppliers. Production on a large scale, with reference to sources of Ge and to the chemical purification processes, is discussed by Wilson (1959). The liquid solubilities of impurities have been given by Thurmond (1960), solid solubilities by Trumbore (1960), and theoretical and experimental segregation coefficients by Weiser (1958).

GROWTH. Growth of Ge has been extensively reported, but details are often lacking. Because of the very high purity of semiconductor grade Ge and considerable research efforts, highly perfect crystals have been made (giving nearly perfect X-ray diffraction, and containing few or no detectable dislocations).

CZOCHRALSKI METHOD. First reports of Ge crystals pulled from the melt were made by Teal and Little (1950). Later papers have described the apparatus in more detail. Lehovec et al. (1953) used a vacuum apparatus with a split graphite heater (10 kva power), and a graphite crucible for large crystals. Details of the apparatus as well as techniques for growing Ge pure and "doped" single crystals have been given by Bradley (1958). Marshall and Wickham (1958) have described a good machine suitable for Ge or Si having a chamber of water-cooled Mg-Al alloy pumped with a 6 in. vacuum pump. The graphite element is heated by direct current; maximum 15 kw. Mechanical arrangements allow pulling, seed rotation, crucible rotation, and vertical adjustment of the crucible for fine control of thermal conditions. A 12 in. × 6 in. window gives excellent visibility. Wilkes (1959) has described a somewhat similar machine for 5 kg crystals Wilkes has emphasized cleanliness problems.

A modification of normal pulling methods is the floating crucible technique; Leverton (1958), and Goorissen et al. (1960). A graphite crucible floats on a large Ge melt and through a capillary allows a constant liquid level to be maintained (Fig. 7). A similar result is obtained by using a large graphite crucible into which Ge is fed at the same time as the crystal is being pulled (Priest, 1956). Both techniques give crystals of high perfection, uniform cross section, uniform impurity, and therefore good electrical characteristics.

The effect of pulling conditions on Ge crystals was studied in detail by Rosi (1958). Growth rates of 0.1 to 6 mm/min were used. One noteworthy result is that a curved solid-liquid interface gives a high dislocation density. Billig (1956) has also discussed this effect and its relation to the radial temperature distribution. By using a very thin seed, a "necked" section and a flat interface, Okkerse (1960) has obtained dislocation free Ge, following Dash's success for Si.

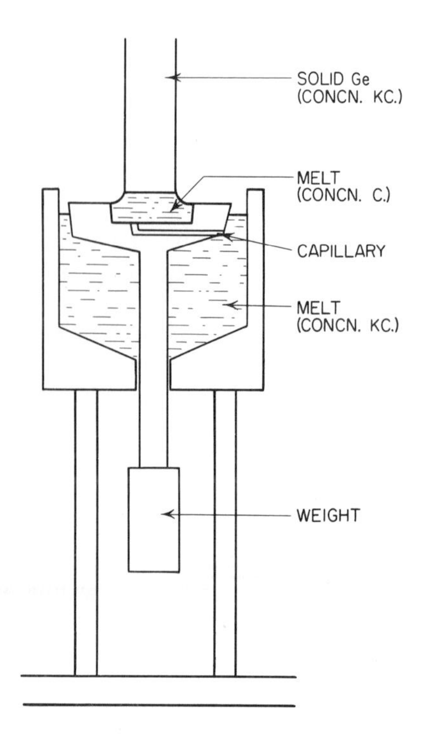

Fig. 7. The floating crucible technique for Ge.

Fluctuations during growth are difficult to avoid. In semiconductors, changes of electrical properties provide sensitive indications of impurity fluctuations (Camp, 1954; Burton et al., 1953). Pfann (1960) has suggested that the addition of two suitable impurities can be used to compensate this effect. Differences between impurity segregation on true (111) faces and on the rest of an interface have been clearly demonstrated by Dikhoff (1961).

The dendritic mode of growth for Ge is considered in Section 6.

HORIZONTAL GROWTH METHODS. The growth of Ge crystals of high perfection in a horizontal boat has been dealt with especially by Bennett and Sawyer (1956) and Cressell and Powell (1957). In both cases, a boat of quartz coated with carbon was used in a gas ambient. Bennett and Sawyer moved the boat through an induction coil. A long after-heater kept thermal gradients low in the plastic region behind the interface. Slow growth rates of 0.8 to 8 cm/hr were used. Cressell and Powell moved the furnace at 6 to 15 cm/hr and used d-c heating with a water-cooled muffle to control the isotherms. The latter process gave more perfect crystals, although variations in purity across the cross section were greater because of inadequate stirring. A comparison of methods by Eisenhower (1961) points to the success of the horizontal method for producing large quantities of highly perfect Ge. Similar techniques are used by Societé Generale Metallurgique de Hoboken, Belgium, the largest producers of Ge in the world.

ETCHING. Markings in Ge surfaces noticed by Corey were identified by Vogel et al. (1953) as etch pits that form at the surface terminations of dislocations in Ge. Recent papers of special interest include those of Holmes (1959) listing etchants and Riessler (1960) containing excellent photographs of etched germanium and showing the orientation dependence of etch rates. For determining crystal orientations, Coughlin (1959) has used the cleavage planes of Ge or surfaces etched with molten indium (see Section 7). Electroetching has been used by Scofield (1959) to give precise specimen thickness. Dislocation pits form in 1 HF:4 HNO_3 acid, or HF plus H_2O mixtures (Batterman, 1957; Irving, 1960).

SPECIAL TECHNIQUES FOR GERMANIUM. Horn (1958) has described "melted layer growth," which is complicated by the fact that both pulling rate and melting rate must be controlled. Trumbore and Porbansky (1960) have used solvent evaporation from Ge-As melts (18 hr) and from a Ge-In melt (Trumbore et al., 1959) to give high concentration alloy-crystals as in Fig. 3. Bolling (1954) states that alloying in Ge may lead to twinning. Similar observations for Ge-Si alloys are discussed in Section 6. Growth for specific devices is discussed in semiconductor reference books.

Gold

M.P. = 1063°C, L = 15, K_s = 0.70 (R.T.).

PURIFICATION. Gold of 99.99% purity is available, and methods of purification are given in the *Liquid Metals Handbook* (1952). Experiments on zone-refining in graphite (Wernick et al., 1959) showed some purification but also contamination from the graphite.

GROWTH. Most experimenters have used graphite molds, sometimes split ones (Sawkill and Honeycombe, 1954). Andrade and Henderson (1951) found horizontal methods of growth to be unsuccessful, so they used the vertical Bridgman method, lowering the mold at a rate of 1.6 mm/min.

Hemispherical crystals of gold and other metals with clean surfaces have been made by Menzel et al. (1955) using a strip heater in high vacuo.

ETCHING. Aqua regia was used by Sawkill and Honeycombe (1954) and also Gwathmey (1948). An alternative is $KCN(NH_4)_2S_2O_8$ used by Okkerse (1956).

Indium

M.P. = 156.4°C, L = 6.8, K_l = 0.10 (M.P.), K_s = 0.06.

PURIFICATION. Electrolytic refining (Baxter and Alter, 1933) is successful for purifying In for crystal growth (see also the RCA report, 1959).

GROWTH. Gwathmey et al. (1948) grew crystals of 99.9% In in glass under vacuo. They were 25 mm × 8 mm diameter, with a 15 mm sphere on the end. The Bridgman technique was used with the furnace at 200°C and a lowering rate of 0.2 mm/min. Some damage resulted when the glass molds were cracked off. Pyrex lined with graphite was used by Dickey (1959). The horizontal method with a traveling furnace was used by Goss and Vernon (1952). Indium of 99.9% purity as 3 mm rods was sealed into 4 mm glass tubes with 10 mm pressure of air. Oxide films retained the rod shapes. Crystals were grown at 0.5 mm/min, the maximum temperature being 175°C.

ETCHING. Gwathmey et al. (1948) used HCl, or HCl plus dilute HNO_3, but Goss and Vernon (1952) preferred HCl plus $KClO_3$ to give etch pits suitable for optical orientation. Smithells' (1955) etch has also been used.

Lead

M.P. = 327.4°C, L = 5.9, K_s = 0.07 (300°C), K_l = 0.039.

PURIFICATION. Filtering of liquid Pb *in vacuo* is useful for removing the oxide coating. A Pyrex apparatus may be used with a coating, such as Aquadag, to prevent adhesion of Pb. Heritage and Lawson (1961) have used repeated evaporation in hydrogen and also zone-refining. The latter has also been used by Tiller and Rutter (1956). A review of the purification of Pb appears in *Metallurgical Reviews* (Greenwood, 1961).

GROWTH. Czochralski (1917) pulled 1 mm crystals at the rate of 140 mm/min. Using the same method, Tsuboi (1929) found some preference for a [100] growth orientation, but this may be an impurity effect. The Bridgman technique has been used extensively with a variety of mold materials, such as glass, graphite, and alundum plus asbestos. Rates of 5 to 25 mm/hr are used, somewhat slower than usual for metals, but Pb has relatively low thermal conductivity (Swift and Tyndall, 1942; Neurath and Koehler, 1951; Prasad and Wooster, 1956).

Chalmers and co-workers have carried out numerous experiments with Pb using a horizontal mold (Chalmers, 1953). The growth of bicrystals was described by Chalmers (1949). Nucleation on the surface of Pb gives (111) planes in the surface (Atwater and Chalmers, 1955), with facets on the crystal surface (Atwater et al., 1955), and steps of 0.001 mm height at the growth interface (Elbaum and Chalmers, 1955). For rapid growth in a chilled mold, Rosenberg and Tiller (1957) have shown that a (111) orientation is preferred with pure Pb, and (100) is preferred in the presence of 5.10^{-4} weight per cent silver. Relations between growth conditions and mosaic structure has been discussed by Sekerka et al. (1960). Effects of tin and mercury additions have been described (Tiller and Rutter, 1956; Holmes et al., 1957).

The orientations of crystals grown in Pyrex molds with Apiezon oil coatings were found to be random for rates of 0.2 to 30 mm/min (Goss and Weintroub, 1952).

ETCHING. Swift and Tyndall (1942) got good results with 2 HNO_3, 3 acetic acid, and 20 water for $2\frac{1}{2}$ hr. Prasad and Wooster used 3 H_2O_2, 2 acetic acid, and 1 water for a few minutes. Other etchants are given by Yamamoto and Watanabe (1957) and by German (1954); the latter also used a vacuum etch.

Lithium

M.P. = 179°C, $L = 158$, $K_l = 0.09$.

GROWTH. Handling and purification of liquid Li are discussed in the *Liquid Metals Handbook* (1952). See also Fedorov and Shamray (1960) as well as Hoffman (1960). Monocrystals were reported by Bowles (1951) using the Bridgman technique with a thin wall (0.015 in.) steel mold. Slow cooling (2°C/hr) was used by Champier (1956) with a nickelplated iron mold in argon. Stainless steel molds with a high temperature gradient and cooling at 30°C/hr has been preferred by Nash and Smith (1959).

ETCHING. Butyl and amyl alcohols give macroetching (Bowles, 1951).

Magnesium

M.P. = 651°C, $L = 82.2$, $K_s = 0.37$ (R.T.).

PURIFICATION. Vacuum distillation gives high purity Mg (Nichols, 1955), and may also be used for crystal growth.

GROWTH. The methods used for Cd and Zn would appear to be useful when applied to Mg. A horizontal mold in hydrogen at four atmospheres pressure was used by Mochalov (1936). Zone melting was adapted by Labzin and Bazhenov (1958) for growth of Mg.

A vertical graphite mold and cooling *in situ* has also been used (Burke and Hibbard, 1952; Reed-Hill and Robertson, 1957). The cooling rate was 5°C/hr, and a crystal 1 in. diameter and 4 in. long took 2 days to grow (Long and Smith, 1957). Random orientations were observed for 25 such crystals (Nichols, 1955).

ETCHING. Cahn (1949) has given an etch for magnesium.

Mercury

M.P. = $-38.9°C$, $L = 2.8$, $K_l = 0.02$ (at 0°C).

PURIFICATION. Distillation in a quartz-still produces very high purity Hg (Greenland, 1937); see also the *Liquid Metals Handbook,* p. 125 (1952).

GROWTH. Andrade's (1914) technique has hardly been modified since this time. A mold of glass is used (Sckell, 1930; Reddeman, 1932) and slowly lowered into a freezing mixture. Andrade and Hutchings (1935) used a rate of 2 mm/min. Greenland (1937) used 3 to 10 mm/min with high purity Hg in a tapered mold, with alcohol as lubricant, to grow 1 to 2 mm diameter crystals. Reddeman (1932) observed that the hexagonal axis grew parallel to the maximum temperature gradient. He used bent tubes to provide seeded growth.

Potassium

M.P. = 63.7°C, $L = 14.6$, $K_s = K_l = 0.1$.

PURIFICATION. The handling and the purification of K are dealt with in the *Liquid Metals Handbook* (1952). Cleaning is done by filtering in glass. This is followed by distillation (Chow, 1938).

GROWTH. Baker (1913) observed slip bands in the monocrystals he obtained after cooling K in capillary glass tubes. Crystals in the form of wires 1.5 to 2 mm diameter and 12 cm long were made by Andrade and Tsien (1937). Glass tubes, coated with Apiezon oil to prevent sticking, were filled with K and lowered from a furnace (2° above the melting point) at 2 mm/min. A similar method was used by Chow (1938).

Selenium

M.P. = 217 − 221°C, $L = 16.4$, $K =$ 0.001 (R.T.).

PURIFICATION. Distillation of Se in quartz apparatus is effective in removing many impurities (Nijland, 1954; Eckart, 1954). Oxidation and reduction have also been used in combination with distillation (Henkels and Maczuk, 1953). A German patent claims that adding Mg to the liquid reduces the Te content of distilled Se. Heritage and Lawson (Nielsen and Heritage, 1959) found zone-refining ineffective. They used the formation and decomposition of hydrogen selenide to obtain spectographically pure Se.

GROWTH. Several allotropic forms occur (see Nijland, 1954; Henisch, 1957 for reviews), although the stable hexagonal form consisting of a regular parallel array of screw-chain molecules is the only recognized one. Growth of crystals is very difficult.

The melt probably consists of randomly oriented atomic chains (Ioffe and Regel, 1960) and normally supercools to a glass. Growth of crystalline Se from this supercooled solid occurs as a function of temperature and impurities (Borelius, 1949; reviewed by Henisch, 1957). Borelius found that dendritic growth could show a growth ratio of 70:1, depending on crystallographic direction. This results from the highly anisotropic structure.

It is possible to avoid supercooling, and Henkels (1949) has used a double container with methyl salicylate (B.P. = 223°C) outside and naphthalene (B.P. = 218°C) on the inside. After lying a few weeks between these two, a mass of pure Se contains small hexagonal crystals. X-ray examination (Henkels, 1951) showed some mosaic structure, but not the twins found in vapor-grown Se crystals. The transition of Se from liquid to crystal at pressures up to 4000 atmospheres has been studied by Nasledov and Kozyrev (1954). High pressure can increase the rate of crystallization at some temperatures. Crystals of 5 mm dimensions have been grown from the melt by Kozyrev (1958) using a cooling rate of 0.1 mm/hr. See also Moenke-Blankenburg (1959).

Silver

M.P. = 960.5°C, $L = 26$, $K_s = 1.0$ (R.T.).

PURIFICATION. This is carried out by electrorefining (*Liquid Metals Handbook,* 1952). Zone-refining under argon is also useful to remove oxygen (Kunzler and Wernick, 1958).

GROWTH. Graphite crucibles are usually used, the high thermal conductivity of silver being advantageous. Silica molds have been reported as unsatisfactory. Vertical and horizontal methods have been used with helium atmospheres and growth rates of 4 mm/min with no apparent problems.

ETCHING. An etchant for Ag that is suitable for optical orientation has been given as concentrated H_2SO_4 at 50 to 60°C, or 50% H_2SO_4 saturated with $K_2Cr_2O_7$ (Yamamoto and Watanabe, 1953). Thermal etching of Ag in argon with 10% oxygen at 600°C (5 to 10 min) has been described by Machlin (1956). Thermal etch pits do not correlate with dislocations (Hirth, 1958).

Sodium

M.P. = 97.8°C, $L = 27$, $K_s = 0.28$ (at M.P.), $K_l = 0.20$ (at M.P.).

PURIFICATION. Distillation may be used for purification (Horsley, 1953) although the removal of sodium oxide may be a problem. Glass or steel is suitable as a low-temperature mold with a dry nitrogen atmosphere. The *Liquid Metals Handbook* (1952) gives much useful information.

GROWTH. Not much published information is available. Andrade and Tsien (1937) described growth by lowering a tube containing Na out of a jacket 2°C above the M.P. (100°C) at 2 mm/min. Dawton (1938) made crystals simply by sucking molten Na into a glass tube. Quimby and Siegel (1938) used Pyrex tubes sealing in the Na under helium. About one-third of their specimens (10 cm × 0.47 cm) were monocrystals after the tubes were lowered through a heater at 0.8 mm/min.

Tellurium

M.P. = 450°C, $L = 7.3$, $K_s = 0.014$ (R.T.), $K_l = 10 \times K_s$ (M.P.).

PURIFICATION. Some controversy exists regarding the purification of Te. Weidel (1954) found that distillation and zone-refining were unsatisfactory. He used the thermal decomposition of H_2Te. Shvartsenau (1960), on the other hand, has found that zone melting in quartz under hydrogen at 6 cm/hr gave greater than 99.999% purity, although Mg and Si were not removed. For extreme purity several hundred zone passes have been used (Blakemore, 1960). Multiple distillation has been used by several workers, with and without zone-refining (Bottom, 1952; Davies, 1957; Semenkovich, 1958; Shvartsenau, 1960).

GROWTH. Te, like Se, has a hexagonal structure containing chains of atoms. However, there is only one form of solid Te, and supercooling does not present the problems that arise with Se. Ioffe and Regel (1960) have discussed the liquid forms of Se and Te. Crystals have been grown using the horizontal method with carbon-coated quartz (Lovell et al., 1958). The Czochralski method was used by Weidel (1954) to make 1 cm × 8 cm crystals at rates of 0.1 to 0.4 mm/min in hydrogen. The same method was used by Davies (1957) with rates of 0.3 to 0.8 mm/min using a quartz-lined graphite susceptor. By necking down the seed from 2 to 3 mm to $\frac{1}{2}$ mm, Davies was able to grow highly perfect crystals (Blakemore, 1960). Tellurium is optically active; both laevo and dextro forms can be prepared (Blakemore and Nomura, 1961).

ETCHING. Blakemore and Nomura (1961) have described a chemical polish for Te crystals. A slow etch (~1 μ/min) in concentrated H_2SO_4 at 150°C produces better dislocation pits than the fast etch used by

Lovell et al. (1958). An alternative dislocation etch has been given by Blum (1961).

Thallium

M.P. = 303°C, $L = 5.0$, $K_s = 0.09$, $K_l = 0.06$.

PURIFICATION. Filtering through a capillary in vacuum is useful (Pomeroy, 1952), but no real study of purification is known.

GROWTH. Thallium exists in a close-packed hexagonal form up to 230°C, and in body centered cubic form from 230 to 303°C, with little volume change at the transition.

Early papers by Rao and co-workers (1936, 1938) reported crystal growth in glass tubes lowered through a furnace (330°C) at 3 cm/hr. Difficulties encountered with a horizontal method by Erfling (1939) were attributed to the structure change. Pomeroy (1952) used 99% Tl, and found that crystals only 2 cm long could be obtained in a horizontal furnace at a rate of 1 mm/min. Apiezon oil prevented adhesion to the mold.

Recent experiments have been more successful. Crystals of 99.9% Tl, $\frac{3}{8}$ in. diameter and 3 to 4 in. long, were grown by the Bridgman technique in carbon-coated Pyrex (Shirn, 1955). Spherical crystals were prepared in split graphite molds by Alexopoulos (1955) with $G > 20$°C/cm, and cooling at 2°C/min.

ETCHING. Prompt examination is necessary because of oxidation. Etchants of NaOH (Pomeroy, 1952), or dilute H_2SO_4 (Rao, 1936) have been used.

Tin

Tin occurs in two allotropic forms: *alpha,* stable below 13.2°C (gray tin); cubic structure and *beta,* stable from 13.2°C to the M.P. = 231.9°C (white tin); body centered tetragonal structure.

WHITE TIN. $L = 14.5$, $K_s = 0.14$ (200°C), $K_l = 0.08$.

Purification. Zone melting is effective (Pfann, 1958) and filtering *in vacuo* removes the oxide. Aleksandrov (1960) used a combination of zone-refining and vacuum heating to prepare very pure Sn.

Growth. Tin has been popular for crystal growth experiments since the beginning (Czochralski, 1917). The most comprehensive work is that of Chalmers and co-workers. Chalmers (1935) grew Sn crystals in glass molds by lifting them slowly out of the melt. He observed preferred growth perpendicular to the *c*-axis. Later (1937), he used a horizontal method with glass molds and also studied bicrystals using a plaster-of-Paris mold (1940). Growth interfaces in tin were revealed by Chalmers (1949) by agitating the melt so as to freeze ripples along them. Crystals were grown at rates up to 20 mm/min. Puttick and King (1949) used glass and graphite molds; less perfect crystals were obtained with the latter. Cahn (1949) made temperature measurements within Sn, and found that monocrystals were always produced when no negative gradient was detected at solidification. Goss and Weintroub (1952) studied growth at rates of 0.5 to 20 mm/min with a range of temperature gradients. They found a preferred orientation for slow growth rates as did Yamamoto and Watanabe (1951). Sn also shows a marked tendency for mosaic growth (Fig. 8) with misorientation angles up to 5°. This has been studied in detail by Teghtsoonian and Chalmers (1951, 1952). Chalmers' method of tilting the interface combined with a slow growth rate (0.25 mm/min) has been used to prevent mosaic growth in Sn (House and Vernon, 1960).

Pond and Kessler (1951) clearly showed the presence of fine lines on Sn surfaces and characteristic hexagonal networks on the cross sections. Rutter and Chalmers (1953) examined this substructure in great detail and showed its relation to impurities. Impurity effects were also discussed by Goss (1953a), and the increased effects of less soluble impurities were shown. Impurities in 99.95% Sn cause preferred growth along [110], whereas purer Sn (99.998%) has no preferred growth direction (Esin and

Fig. 8. Mosaic structure in tin.

Kralina, 1960). This may explain the results of Yamamoto and Watanabe (1959a) who found [110] dominant for fast growth rates and random orientation for slow growth rates in contradication to earlier work.

Spherical single crystals of tin have been reported by Bolognesi (1953).

Etching. McKeehan and Hoge's (1935) etchant (potassium chlorate and HCl) gives sharp optical reflections.

GRAY TIN. Many attempts have been made to transform white tin into gray tin crystals, but the increase in volume (25%) makes this very difficult. Becker (1958) found that small crystals (millimeter dimensions) could be made by holding white tin near the transformation temperature (13°C). Ewald and Tufte (1958) appear to have solved the problem by using Hg as solvent and precipitating crystals at -20°C onto seeds formed in the apparatus at -30°C. Residual resistivity measurements indicated a very low solubility of Hg in the Sn ($\sim$0.001%). Crystals $\sim$1 cm in size with well-developed faces were grown at $\sim$1 in./month.

Zinc

M.P. = 419.5°C, $L = 24.4$, $K_s = 0.26$ (R.T.), $K_l = 0.14$.

PURIFICATION. High purity zinc is prepared by repeated distillation. An analysis of the zone-refining in it has been made by Kyotani (1957).

GROWTH. Zinc has often been studied, and over fifty papers deal with its crystal growth. Selected practical results will be described.

Czochralski (1917) found that a rate of 100 mm/min could be used with Zn. Horizontal methods have also been used. Pyrex molds coated with carbon or graphite molds are often used. Slow growth rates in the range 0.1 to 0.8 mm/min are most successful (Gruneisen and Goens, 1924; Jillson, 1950; Goss and Weintroub, 1952; Rozhansky et al., 1957) provided the temperature gradient is not too low (Jillson, 1950; Cinnamon and Martin, 1940). Under these conditions, crystals of all orientations are obtained (Jillson, 1950; Goss and Weintroub, 1952; Kuznecov and Saratovkin, 1933; Rozhansky et al., 1957; Yamamoto and Watanabe, 1959).

Faster rates of growth favor an orientation, in common with Cd, where the basal planes lie parallel to the growth direction. The orienting effect of a capillary at the seed end of a crystal showed that the basal planes grew normal to the cold surface (Palibin and Froiman, 1933). The preferred orientation of the basal planes has also been described by Chigvinadze (1955) as parallel to lines of thermal currents. The effect of furnace design was shown by Kreuchen (1931).

The presence of Cd impurity tends to improve the crystal quality according to Poppy (1934) and Schilling (1935). Cinnamon (1934) and later Cinnamon and Martin (1940) have given values of G/R for good growth, with 0.001% iron impurity. The presence of Cd may be advantageous by increasing the solubility of a third element. The effects of added impurities, especially on the cell structure in Zn, has been described by many workers. Straumanis (1929) was among the first. Cell structures in Zn have been described in detail by Blaha (1953), and recently, dislocations associated with the cell structure have been shown (Damiano and Tint, 1961).

Large crystals, greater than 1 kg in weight have been grown by Bushmanov (1957) and Smialowski (1937). A soft mold, made of ZnO, has been used by Rybalko et al. (1960).

ETCHING. Thermal etching gives very sharp pits that may be used for orientation. Etching with 50% HCl is not particularly good. Yamamoto and Watanabe (1950, 1955) have given a number etches for use in optical orientation.

Gilman (1956) has described a technique for making dislocation etch pits in zinc in the presence of 0.1% Cd. Stepanova and Urusovskaya (1960) have confirmed Gilman's results, and also shown that a trace of iron will decorate dislocations.

6. SPECIAL METHODS

ALLOY CRYSTALS. Monocrystals of dilute alloys have often been prepared. Nearly all Ge crystals are "doped" with very dilute additions. However, for higher concentration alloys (for example, 1 to 10% in the melt) no reports of systematic studies have been found.

DILUTE ALLOYS (nominally less than 1% alloy). Crystals of Ge (and other semiconductors) are required with precisely controlled "doping" concentrations in order to control the electrical conductivity. For elements with $k^* \ll 1$ (In and Sb in Ge), zone melting gives uniform concentration because losses from a zone are very small. The floating crucible technique can be used for crystal pulling from a doped bath. This yields very uniform crystals with 10^{-6} to 10^{-3} atomic per cent of the added element. At concentrations of 10^{-1} atomic per cent, twins may form (Bolling et al., 1956). The problems of these "high concentration" alloy crystals have been discussed by Trumbore (1959), Hurle (1961), and see Grubel (1961). When $k^* \simeq 1$, crystal pulling or normal freezing gives more uniform alloys, or alternatively, the reverse passes of a molten zone can be used (Pfann, 1958).

Growth has been studied in dilute alloys of lead with additions of Sn, Ag, and Au (Tiller and Rutter, 1956), and of tin (Walton et al., 1955); also in dilute alloys of Sn with Bi, Sb, and Pb (Plaskett and Winegard, 1959). The theory of constitutional supercooling has been completely confirmed and reviewed by Elbaum (1959). Observations of the effects of 0.01 to 1% of Cd, Zn, Pb, Ag, In, and Sb on Sn were made by Goss (1953a). Sn-Pb eutectics with trace additions were studied by Tiller (1957).

Studies of other dilute alloys were relatively few.

CONCENTRATED ALLOYS (nominally greater than 1% alloy). For general use, making monocrystals by solidification is not recommended because of segregation problems.

Usually, fine structure, cells, or dendrites are present. Growth by strain annealing of the homogenized alloy is preferred.

However, certain alloys (principally those of Cu, Ag, and Au) solidify into very satisfactory crystals. Brass crystals, α and β, have often been prepared. Graphite molds with the Bridgman method have been frequently used, although horizontal growth has also been successful. Growth at rates of 0.5 to 5 mm/min is followed by annealing at high temperatures. The growth of brass crystals by Burghoff (1930) is described in unusual, but helpful, detail. Crystals 20 cm × 2 cm diameter of Cu 70%, Zn 30% were grown at 0.8 mm/min with the furnace moving up around a graphite mold. Wires 1.5 mm diameter and 20 cm long were grown by Ardley and Cottrell (1953) with Cu containing 1, 5, 10, 15, and 30% Zn. Sealed silica quills were used and a horizontal traveling furnace.

Graf (1931) describes the use of induction heating (8 kc/*s*) for preparing Au-Ag, Au-Cu, and Cu-Pd alloy crystals, 2 to 12 mm diameter and 5 to 6 cm long. For elements completely soluble in one another, Graf considered the preparation to be without difficulty once the cooling conditions were established. A number of others who have grown Cu and Au alloy crystals tend to confirm this view, but experiments with Ge-Si and Cu-Ni disagree. Other alloy crystals that have been prepared are In—20% Tl (Burkart and Read, 1953); Sb—3½ weight per cent Sn (Hart, 1937); and Te-Se alloys (Loferski, 1954; Nussbaum, 1954).

Although Ge crystals can be grown at 10 cm/hr (Cressell and Powell, 1957), the addition of a few per cent of Si reduces the maximum rate to 0.1 cm/hr if nucleation is to be avoided, even if high purity elements are used. Experiments have also been made with alloys of Sb-Bi and Cu-Ni, (Goss, unpublished work). Both systems have complete solid solubility as in the Ge-Si system. Crystals could be grown at 2 cm/hr with Bi—5% Sb, but only 0.15 cm/hr was tolerable to get satisfactory results from Cu-Ni 5%. The purity does not approach that of the Ge—Si system, however. Very slow growth rates were used by Trumbore et al. (1959) for Ge containing several % (Al + Ga). Problems of nucleation and growth in most alloy systems remain to be explored.

Eutectic crystallization, which may be considered as a special alloy, has been reviewed by Scheil (1954).

Dendritic Crystals

Although dendritic surface markings have been commented on for many years (Buckley, 1951), the details for low melting point metals have only recently received attention. Columnar crystals of copper alloys were examined by Northcott and Thomas (1939). The [100] axis was found to lie parallel to the direction of growth. Alexander (1948) reviewed dendritic growth and pointed out that the dendritic spacing increased as the rate of freezing increased. The thermal flow is of great importance as, for example, in dendrite spacing, which may be compared with the thermal conductivity of the solid:

Metal	Cu	Al	Mg	Sb
Dendrite spacing	0.15 mm	0.2 mm	0.25 mm	0.3 mm
Thermal conductivity, cgs units	0.9	0.5	0.3	0.1

With lead, the effect of increasing the solute concentration of Sn from ½ to 7% has been related to the onset of dendrites (Morris et al., 1955). Critical values of G and R are found for each solute concentration. The transition from cell structure to dendrites was shown by Tiller and Rutter (1956), and the effect of Mg in Pb by E. L. Holmes et al. (1957). In these cases the importance of G/R can be related to the supercooling caused by impurities at the growth interface.

For pure metals, dendritic growth has been studied in great detail by Weinberg and Chalmers (1951, 1952). Pure Pb and also Sn and Zn were studied. A thermocouple in the lead indicated that supercooling of the metal is necessary for dendritic growth. Dendrite directions were shown to be ⟨100⟩ for Pb, ⟨110⟩ and ⟨1$\bar{1}$0⟩ for tin, and ⟨10$\bar{1}$0⟩ for zinc. These orientations have been tabulated by Schlipf and Seeger (1954) and shown not to be in the close-packed direction but normal to the next densest plane, that is, (100) in cubic, (110) and (1$\bar{1}$0) in tetragonal, and (10$\bar{1}$0) in hexagonal metals. The rate of growth, as suggested earlier, appears to be limited only by thermal dissipation. Rosenberg and Winegard (1954) observed rates of 3 cm/sec with 3°C undercooling and 25 cm/sec with 10.5°C undercooling.

Dendritic growth also occurs in pure Ge, the crystals containing twin lamellae, with ⟨211⟩ axes and [111] outer faces (Billig and Holmes, 1955, 1957). Very long thin crystals about 0.25 mm thick and 1.5 mm wide can be pulled (Bennett and Longini, 1959). These are in the form of long ribbons grown at ~10 cm/min. The outer faces of these ribbons are optically flat, with detectable steps up to 5000 A high (Longini, 1960). The role of twins in providing re-entrant corners on the growth interface which can be propagated indefinitely, without the need for surface nucleation, has been adequately described (Wagner, 1960; Hamilton and Seidensticker, 1960). The detailed structure of Ge dendrites has been shown by Faust and John (1960, 1961). Lever et al. (1960) have described a simple puller for making Ge ribbon dendrites.

Other metals can also be prepared by fast growth rates in dendritic form, for example, Ag (Faust and John, 1961a), and Al (Herenquel, 1949). Lamellar growth of crystals, including rapid growth, has been surveyed by Graf (1951).

Other Special Methods

Polycrystals have been grown from multiple seeds in order to produce well-defined boundaries between the crystals. Bicrystals of Ag, Al, Cu, Ge, Sn, Pb, and Zn have been made using normal growth conditions. The method of Gow and Chalmers (1951) is frequently used. In an earlier paper, Chalmers (1949) discussed the growth of bicrystals of Sn and Pb using surface ripples as time markers (Fig. 9). Precise control of boundaries in making Ge bicrystals has been described by Mataré and Wegener (1957). The effect of the growth rate on Al bicrystals has been reported by Rybalko (1956). Bicrystals of Zn have also been grown (Gilman, 1953).

Spherical crystals are useful in studies of

Fig. 9. Tin bicrystal with ripples formed at some positions of the solid-liquid interface.

orientation effects. Gwathmey et al. (1945, 1948) has specialized in techniques for spherical crystals. He gives details of machining, etching, and other methods for making spheres. Spherical crystals of Sn, Zn, and Cd have been grown by Bolognesi (1953, 1955). See also Yamamoto and Watanabe (1951a). Hemispheres of Cu, Ag, and Au have been grown on strip filaments in high vacuum by Menzel et al. (1955).

Finally, there may be some interest in very large crystals. Smialowski (1937) grew large spherical crystals, weighing 6.4 kg in copper and 1.48 kg in zinc. Semiconductor crystals weighing ~1 kg and 5 kg crystals (Wilkes, 1959) are impressive.

A method of growth that has been of interest for semiconductors is temperature-gradient zone melting, described in detail by Pfann (1958). The molten alloy zone moves through the solid because of the temperature gradient and the resulting solubility gradient or, as Pfann says, the melt climbs up the liquidus. This type of zone can be modified by Peltier effects.

7. PROCESSING

Orientation (Optical Methods)

Barrett (1952) has described various X-ray methods for orientation determination in detail. Some references to optical methods are also given. For a review of detailed X-ray examination of orientation, and therefore of fine structure, see Hirsch (1956).

Optical techniques were widely used in the 1920's. Bridgman (1925) examined many metal crystals by the simplest of methods. Czochralski (1925) also used optical reflection patterns for metallographic examination. A globe and scale arrangement was described by Yoshida (1927), and special methods for several metals, including Al, were given by Potter and Sucksmith (1927). These early methods have been considerably developed.

The arrangement used by Chalmers (1935) for crystals of Sn is shown in Fig. 10. An adaptation of this method was used by Goss (1953) to scan along the length of a crystal while observing the reflection pattern on a curved screen. This arrangement is shown schematically in Fig. 11.

Optical methods are not without difficulties (Barrett and Levenson, 1940). A good etching method must be available so that well-defined reflections occur, and spurious reflections from the crystal surface from rounded etch pit surfaces or from double reflections are minimized. Sb and Bi are especially difficult, for example. The usual accuracy is 0.5 to 1° although semiconductor crystals give better accuracy. Many etching solutions have been described by Yamamoto and Watanabe. These are given in Section 5.

Rowland (1951a) has described a simple Pyrex sphere, 15 cm diameter for rough orientation observations. Using a liquid of high refractive index, Frank and Oliver (1951) have obtained wide angle reflections. For semiconductor crystals (especially Ge and Si) optical methods give an accuracy of

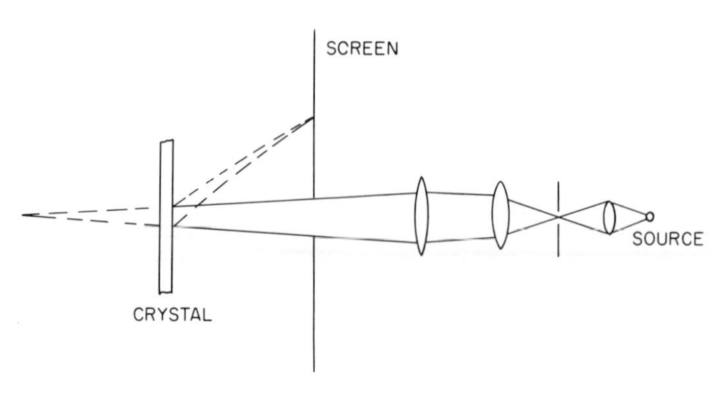

Fig. 10. Optical orientation method.

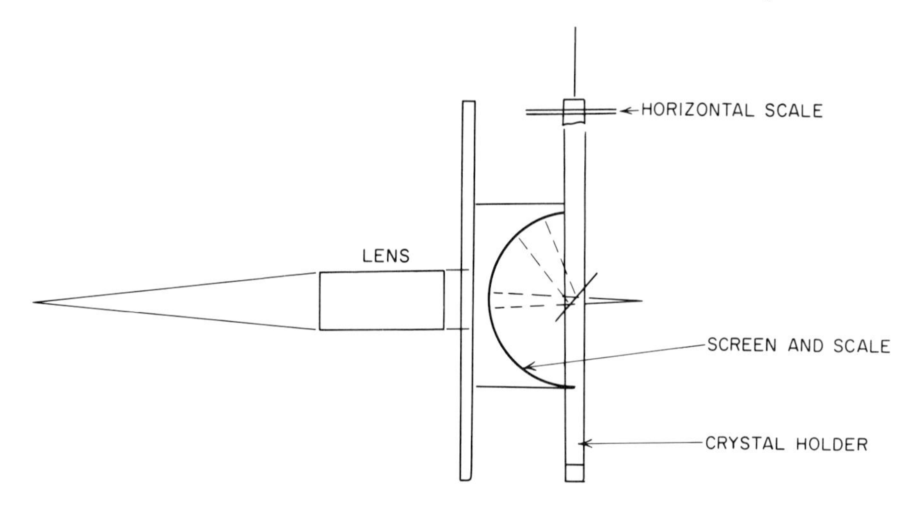

Fig. 11. Optical method for examining long crystals.

12 min (Coughlin et al., 1959; Schwuttke, 1959). Even higher accuracy is claimed by Wolff et al. (1957) and Marshall (1959).

A study of particular interest is that of Tucker and Murphy on etch pits in Al crystals (1953). Slip-bands can be used, in deformed crystals, to determine orientation (Roscoe and Hutchings, 1933). Etching methods for revealing dislocations have been recently reviewed by Regel et al. (1960) with comprehensive references.

Cutting

Pure crystals are often very ductile, so delicate methods are necessary to avoid cutting damage. Ge and Si are notable exceptions, since diamond cutting saws are needed, but the crystals are damaged to depths of 0.005 cm or more. Etching is used to remove damaged layers. Considerable effort has been spent developing methods for strain-free cutting and some of the results are described.

A jeweler's saw can be used, followed by careful etching. Childs (1950) describes the procedure for tin. When cutting at particular angles is required, the methods described by Sato et al. (1957) and by Vaughan (1958) may be useful.

Mechanical damage can be avoided by using a thread dipped in acid as the cutter. McGuire and Webber (1949) described a simple apparatus using a reciprocating fiberglass thread. However, it took 5 hr to cut 1 cm of tin, and the thread had to be replaced frequently. A thread of Saran as used by Maddin and Asher (1950) was

TABLE 1

TABLES OF CRYSTALLOGRAPHIC ANGLES USEFUL FOR ORIENTATION ANALYSIS

Cubic system	Bozorth (1925)
Hexagonal metals: Cd, Mg, and Zn	Salkovitz (1951)
	Taylor and Leber (1954)
Antimony and Bismuth	Salkovitz (1956)
	Vickers (1957)
Tin	Chalmers (1935)
	Nicholas (1951)
Indium	Goss and Vernon (1952)
	Chandrasekhar and Veal (1961)

better, but cutting rates were still low (for example, 12 hr to cut 0.75 in. diameter of Mg and 7 hr to cut 0.25 in. diameter of Ag). Wayson (1961) describes a servo-controlled cutter. Rates less than 1 mm/hr are usual (Yamamoto and Watanabe, 1956). Multiple acid cutting (for example, 12 slices at a time) has been described by Kinman and Hayward (1955). Cutting is speeded up by applying voltage with the metal as the anode. In this manner, Piontelli et al. (1955) cut tin and Metzger (1958) used piano-wire (current of 1 amp/cm of cut) to cut Al and Cu crystals at the rate of 1 cm/hr, but the cuts were not sharp. Electrical spark-erosion cutting (condenser discharge) has also been used for Al by James and Milner (1953). More recently, spark cutting has been applied to In, Sn, Bi, Al, and Cu by Chandrasekhar (1961), but the rates are slow, 1 mm/hr. Electromachining by spark-erosion is also possible. Cole et al. (1961) claim a 10 μ-in. finish at a cutting rate of 1 mm/min. Specialized techniques for electrolytic machining of semiconductors have been discussed by Uhlir (1955, 1956). Holes 2 mil diameter and 40 mil deep were drilled in Ge in 10 min.

Specimens of Fe and Cu with a smooth cylindrical shape made by electroturning have been described by Farmer and Glaysher (1953). Surface layers may also be removed by sandblasting (Pepinsky, 1953) or an acid wheel (Avery et al., 1958). The removal of damaged layers on Cu by oxidation and oxide removal has been used by Petrauskas et al. (1949) to prepare Cu crystals 5 in. × 0.13 in. × 0.003 in.

A general review of electrolytic polishing methods has been given by McTegart (1956). Electroerosion machining has been dealt with by Livshits (1960). Young and Wilson (1961) have shown that acid cutting and polishing do not put dislocations into Cu.

References

Albert, P. (1960), *Nouvelles Proprietes Physiques et Chimiques des Metaux de tres haut Pureté,* Paris (C.N.R.S.).

Alden, T. H. (1960), *Rev. Sci. Inst.* **31,** 897.

Aleksandrov, B. N. (1960), *F. M. i M.* **9,** 53.*

Aleksandrov, B. N., and B. I. Verkin (1960), *F. M. i M.* **9,** 362.

Alexander, B. (1948), *Carnegie Institute of Technology, D.Sc. Thesis.*

Alexopoulos, K. D. (1955), *Acta Cryst.* **8,** 235.

Andrade, E. N. da C. (1914), *Phil. Mag.* **27,** p. 869; (1949), *Nature* **164,** 536; (1950), *Proc. Phys. Soc.* **63B,** 198.

Andrade, E. N., and P. J. Hutchings (1935), *Proc. Roy Soc.* **A148,** 120.

Andrade, E. N., and Tsien (1937), *Proc. Roy Soc.* **A163,** 1.

Andrade, E. N., and R. Roscoe (1937), *Proc. Phys. Soc.* **49,** 152.

Andrade, E. N., and C. Henderson (1951), *Phil. Trans. Roy. Soc.* **244A,** 177.

Ardley, G. W., and A. H. Cottrell (1953), *Proc. Roy. Soc.* **A219,** 328.

Atwater, H. A., and B. Chalmers (1955), *J. Appl. Phys.* **26,** 918.

Atwater, H. A., A. R. Lang, and B. Chalmers (1955), *Can. J. Phys.* **34,** 234.

Aust, K. T., and B. Chalmers (1958), *Can. J. Phys.* **36,** 977.

Avery, D. H., M. L. Ebner, and W. A. Backofen (1955), *Trans. A.I.M.E.* **212,** 256.

Baker, B. B. (1913), *Proc. Phys. Soc.* **25,** 235.

Barrett, C. S., and L. H. Levenson (1940), *Trans. A.I.M.E.* **137,** 76.

Barrett, C. S. (1952), *Structure of Metals,* Second Edition, McGraw-Hill, New York.

Basinska, S. J., J. J. Polling, and A. Charlesby (1954), *Acta Met.* **2,** 313.

Batterman, B. (1957), *J. Appl. Phys.* **28,** 1236.

Baxter, G. P., and C. M. Alter (1933), *J. Am. Chem. Soc.* **55,** 1943.

Becker, J. H. (1958), *J. Appl. Phys.* **29,** 1110.

Bennett, D. C., and B. Sawyer (1956), *Bell System Tech. J.* **35,** 637.

Bennett, A. I., and R. L. Longini (1959), *Phys. Rev.* **116,** 53.

Billig, E. (1955), *Proc. Roy. Soc.* **A229,** 346; (1956), *Proc. Roy. Soc.* **A235,** 37.

Billig, E., and P. J. Holmes (1955), *Acta Cryst.* **8,** 353; (1957), *Acta Met.* **5,** 53.

Blaha, F. (1953), *Acta Physica Austriaca* **8,** 141.

Blakemore, J. S., K. C. Nomura, and J. W. Schultz (1960), *J. Appl. Phys.* **31,** 2226.

Blakemore, J. S., and K. C. Nomura (1961), *J. Appl. Phys.* **32,** 745.

Blum, A. I. (1961), *Soviet Physics—Solid State,* **2,** 1509.

Boas, W., and E. Schmid (1929), *Z. Phys.* **54,** 16.

Bolling, G. F., W. A. Tiller, and J. W. Rutter (1956), *Can. J. Phys.* **34,** 234.

Bolognesi, G. (1953), *Compt. Rend.* **236,** 2414; (1955), *Rev. Met.* **52,** 909.

Borelius, G. (1949), *Arkiv. Fysik* **1,** 305.

Bottom, V. E. (1952), *Science* **115,** 570.

Bowles, J. S. (1951), *Trans. A.I.M.E.* (*J. Met.*) **189,** 44.

Bozorth, R. M. (1925), *Phys. Rev.* **26,** 390.

Bradley, W. W. (1958), *Transistor Technology.* I, van Nostrand, New York, p. 130.

* *Note: F.M.iM.* is the abbreviation for *Fizika Metallov i Metallovedenie.*

Bridgman, P. W. (1925), *Proc. Am. Acad. Arts Sci.* **60,** 305; (1929) *Proc. Am. Acad. Arts Sci.* **63,** 351.
Buckley, H. E. (1951), *Crystal Growth,* Chapman Hall, London.
Burghoff H. L. (1930), *Yale University, M.S. Thesis.*
Burkart, M. W., and T. A. Read (1953), *J. Metals* **5,** 1516.
Burke, E. C., and W. R. Hibbard (1952), *J. Metals* **4,** 295.
Burton, J. A. et al. (1953), *J. Chem. Phys.* **21,** 1992.
Bushmanov, B. N. (1957), *F.M.iM.* **4,** 310.
Cahn, R. W. (1949), *Ph.D. Thesis,* Cambridge University, Cambridge; (1949), *J. Instr. Met.* **76,** 121.
Camp, P. R. (1954), *J. Appl. Phys.* **25,** 459.
Chalmers, B. (1935), *Proc. Phys. Soc.* **47,** 733; (1937), *Proc. Roy. Soc.* **A162,** 120; (1940), *Proc. Roy. Soc.* **A175,** 100; (1949), *Proc. Roy. Soc.* **A196,** 64; (1953), *Can. J. Phys.* **31,** 132; (1954), *Trans. A.I.M.E.* (*J. Met.*) **6,** 519.
Chalmers, B., and U. M. Martius (1952), *Phil. Mag.* **43,** 686.
Champier, G. (1956), *Compt. rend.* **243,** 657.
Chandrasekhar, B. S. (1961), *Rev. Sci. Inst.* **32,** 368.
Chandrasekhar, B. S., and B. V. Veal (1961), *Trans. A.I.M.E.* **221,** 202.
Chigvinadze, D. M. (1955), *Zhur. Tehkn. Fiziki* **25,** 805.
Childs, B. G. (1950), *J. Sci. Instr.* **27,** 102.
Chow, Y. S. (1938), *Ph.D. Thesis,* University of London.
Cinnamon, C. A. (1934), *Rev. Sci. Instr.* **5,** 187.
Cinnamon, C. A., and A. B. Martin (1940), *J. Appl. Phys.* **11,** 487.
Cole, M., I. A. Bucklow, and C. W. B. Grigson (1961), *Brit. J. Appl. Phys.* **12,** 296.
Coughlin, B. J. et al. (1959), *J. Sci. Inst.* **36,** 144.
Crawley, R. H. A. (1953), *Chem. and Ind.* (*Rev*), No. 45, p. 1205.
Cressell, I. G., and J. A. Powell (1957), *Progress in Semiconductors.* II, Heywood, London, p. 137.
Czochralski, J. (1917) *Z. Phys. Chem.* **92,** 219; (1925), *Z. anorg. Chem.* **144,** 131.
Damiano, V. V., and G. S. Tint (1961), *Acta Met.* **9,** 177.
Dash, W. C. (1959), *J. Appl. Phys.* **30,** 459.
Davey, W. P. (1925), *Phys. Rev.* **25,** 248.
Davies, T. J. (1957), *J. Appl. Phys.* **28,** 1217.
Dawton, R. H. V. M. (1938), *Proc. Phys. Soc.* **50,** 483.
de Grinberg, D. (1961), *Can. J. Phys.* **39,** 1919.
Dickey, J. E. (1959), *Acta Met.* **7,** 350.
Dikhoff, J. A. M. (1960), *Solid State Electronics* **1,** 202.
Eckart, F. (1954), *Arbeitstag Festkorperphysik* II, p. 126.
Eisenhower, W. D. (1961), *Western Elec. Eng.* **5,** 18.
Elbaum, C. (1959), *Progress Met. Phys.* **8,** 203; (1960), *J. Appl. Phys.* **31,** 1413.
Elbaum, C., and B. Chalmers (1955), *Can. J. Phys.* **33,** 196.
Erfling, H. D. (1939), *Ann. Phys.* **34,** 139.
Esin, V. O., and A. A. Kralina (1960), *F.M.iM.* **7,** 305.
Ewald, A. W., and O. N. Tufte (1958), *J. Appl. Phys.* **29,** 1007.
Farmer, M. H., and G. H. Glaysher (1953), *J. Sci. Instr.* **30,** 9.
Faust, J. W., and H. F. John (1960), *J. Electrochem. Soc.* **107,** 562; (1961a), *J. Electrochem. Soc.* **108,** 855–868; (1961b), *J. Electrochem. Soc.* **108,** 109.
Fedorov, T. F., and F. I. Shamray (1960), *Izvest. Akad. Nauk. sssR Met. i Topl.* (6), p. 56.
Frank, F. C. and D. S. Oliver (1951), *Research* **4,** 341.
German, S. (1954), *Z. Metkde.* **45,** 484.
Gilman, J. J. (1956), *J. Metals* **8,** 998.
Glocker, R., and L. Graf (1930), *Z. anorg. allgem. Chem.* **188,** 232.
Goetz, A. (1930), *Phys. Rev.* **35,** 193.
Gold, L. (1949), *Rev. Sci. Instr.* **20,** 115.
Goorissen, J., and F. Kartsensen (1959), *Z. Metkde.* **50,** 47.
Goss, A. J. (1953a), *Proc. Phys. Soc.* **66B,** 65; (1953b), *J. Sci. Inst.* **30,** 283; (1953c), *Proc. Phys. Soc.* **66B,** 525.
Goss, A. J., and S. Weintroub (1952), *Proc. Phys. Soc.* **65B,** 561.
Goss, A. J., and E. V. Vernon (1952), *Proc. Phys. Soc.* **65B,** 905.
Goss, A. J., K. E. Benson, and W. G. Pfann (1956), *Acta Met.* **4,** 332.
Gow, K. V., and B. Chalmers (1951), *Brit. J. Appl. Phys.* **2,** 300.
Graf, L. (1931), *Z. Phys.* **67,** 388.
Graf, L. (1951), *Z. Metkde.* **42,** 336 and 401.
Greenland, K. M. (1937), *Proc. Roy. Soc.* **A163,** 28; *Ph.D. Thesis,* University of London.
Greenwood, J. M. (1961), *Met. Rev.* **6,** 279.
Greninger, A. B. (1935), *Trans. A.I.M.E.* **117,** 75.
Grubel, R. O. (1961), *Metallurgy of Compound and Elemental Semiconductors,* Interscience, New York.
Gruneisen, E. and E. Goens (1924), *Z. Phys.* **26,** 235.
Gwathmey, A. T., and A. F. Benton (1940), *J. Phys. Chem.* **44,** 35.
Gwathmey, A. T., H. Leidheiser, and G. P. Smith (1945), *Nat. Adv. Cttee. Aeronautics, Tech. Note* 982; (1948), *Nat. Adv. Cttee. Aeronautics, Tech. Note* 1460.
Hamilton, D. R., and R. G. Seidensticker (1960), *J. Appl. Phys.* **31,** 1165.
Hart, H. M. (1937), *Phys. Rev.* **52,** 130.
Hasler, M. F. (1933), *Rev. Sci. Inst.* **4,** 656.
Henisch, H. K. (1957), *Rectifying Semiconductor Contacts,* Clarendon, Oxford.
Henkels, H. W. (1949), *Phys. Rev.* **76,** 1737; (1951), *J. Appl. Phys.* **22,** 916.
Henkels, H. W., and J. Maczuk (1953), *J. Appl. Phys.* **24,** 1056.
Herenquel, J. (1949), *Rev. de Met.* **46,** 309.
Heritage, R. J., and W. D. Lawson (1961), Private Communication.
Hirsch, P. B. (1956), *Prog. in Metal Phys.* **6,** 236.
Hirth, J. P., and L. Vassamillet (1958), *J. Appl. Phys.* **29,** 595.
Hoffman, E. E. (1960), *ASTM Symposium on Newer Metals,* 1959, 195.
Hoffman, J. I. (1934), *J. Research Natl. Bur. Standards* **13,** 665.
Holden, A. N. (1950), *Trans. Amer. Soc. Metals* **42,** 319.
Holmes, E. L. et al. (1957), *Can. J. Phys.* **35,** 1223.
Holmes, P. J. (1959), *Proc. I.E.E.* **B106,** 861; (1959), *Acta Met.* **7,** 283.
Honeycombe, R. W. K. (1959), *Met. Rev.* **4,** 1.
Horn, F. H. (1958), *J. Electrochem. Soc.* **105,** 393.

Horner, P. (1961), *Nature* **191,** 58.
Horsley, G. W. (1953), *Report No. AERE M/R* 1152, AERE Harwell, England.
House, D. G., and E. V. Vernon (1960), *Brit. J. Appl. Phys.* **11,** 254.
Howe, S., and C. Elbaum (1961), *J. Appl. Phys.* **32,** 742.
Hurle, D. T. J., and S. Weintroub (1959), *Proc. Phys. Soc.* **76,** 163.
Hurle, D. T. J. (1960), Ph.D. Thesis, University of Southampton, (Part I and II).
Hurle, D. T. J. (1961), *Solid State Electronics* **3,** 37; (1962), *Prog. in Materials Sci.* **10,** (1962), Pergamon Press, New York.
Ioffe, A. F., and A. R. Regel (1960), *Progress in Semiconductors,* Vol. 4, Heywood, London.
Irving, B. A. (1960), *J. Appl. Phys.* **31,** 109.
James, J. A., and C. J. Milner (1953), *J. Sci. Instr.* **30,** 386.
Jillson, D. C. (1950), *J. Metals* **2,** 1006.
John, H. (1958), *J. Electrochem. Soc.* **105,** 741.
Kapitza, P. (1928), *Proc. Roy. Soc.* **A119,** 358.
Kelly, A., and C. T. Wei (1955), *J. Metals* **7,** 1041.
Kinman, T. H., and C. Hayward (1955), *B.T.H. Activities* **26,** 137.
Kostron, H. et al. (1953), *Z. Metkde.* **44,** 17.
Kozyrev, P. T. (1958), *Zhur. Tekhn. Fiziki* **28,** 500.
Kreuchen, K. H. (1931), *Z. physik. Chem.* **A155,** 161.
Kunzler, J. E., and J. H. Wernick (1958), *Trans. AIME* **212,** 856.
Kuznecov, V., and D. Saratovkin (1934), *C. R. Leiningrad (N.S.)* **1,** 251.
Kyotani, M. (1957), *Nippon Kinzoku Gakkai*-Si **21,** 241.
Labzin, V. A., and V. V. Bazhenov (1958), *F.M.iM.* **6,** 941.
Lainer, D. I., R. L. Petrusevich, and E. S. Sollertinskaya (1960), *F.M.iM.* **9,** 535.
Lawson, W. D., and S. Nielsen (1958), *Preparation of Single Crystals,* Butterworths, London.
Lazarus, D., and D. R. Chipman (1951), *Rev. Sci. Instr.* **22,** 211.
Le Hericy, J. (1960), *Nouvelle Prop. Phys. et Chimiques des Metaux de très haut pureté,* C.N.R.S., Paris, 221.
Lehovec, K. et al. (1953), *Rev. Sci. Instr.* **24,** 652.
Lever, R. F. et al. (1960), *Rev. Sci. Instr.* **31,** 1334.
Leverton, W. F. (1958), *J. Appl. Phys.* **29,** 1241.
Liquid Metals Handbook (1952), edited by R. Lyon (Second Edition) U. S. Govt. Printing Office.
Livshits, A. L. (1960), *Electro-Erosion Machining of Metals,* Butterworths, London.
Loferski, J. J. (1954), *Phys. Rev.* **93,** 707.
Long, T. R., and C. S. Smith (1957), *Acta Met.* **5,** 200.
Longini, R. L. (1960), *J. Appl. Phys.* **31,** 1204.
Lovell, L. C., J. H. Wernick, and K. E. Benson (1958), *Acta Met.* **6,** 716.
Lovell, L. C., and J. H. Wernick (1959a), *J. Appl. Phys.* **30,** 234; (1959b), *J. Appl. Phys.* **30,** 590.
Lowndes, L. (1901), *Ann. Physik.* **6,** 146.
Machlin, E. S. (1956), *Dislocations and Mechanical Properties of Crystals,* John Wiley and Sons, New York, p. 165.
Maddin, R., and W. R. Asher (1950), *Rev. Sci. Instr.* **21,** 881.
Marinelli, S., and F. Blaha (1956), *Acta Met.* **4,** 443.
Marshall, K. H. J. C., and R. Wickham (1958), *J. Sci. Instr.* **35,** 121.
Marshall, S. (1959), *B.T.H. Activities* **30,** 170.
Martius, U. M. (1954), *Progress Met. Phys.* **5,** 279.
Maslennikov, B. M. (1955), *Zhur. Tekhn. Fiziki* **25,** 933.
Mataré, H., and H. A. R. Wegener (1957), *Z. Phys.* **148,** 631.
Mc. Tegart, W. J. (1956), *Electrolytic and chemical polishing of metals,* Pergamon Press, New York.
McGuire, T. R., and R. T. Webber (1949), *Rev. Sci. Instr.* **20,** 962.
Menzel, E., W. Stössel, and M. Otter (1955), *Z. Phys.* **142,** 241.
Metzger, M. (1958), *Rev. Sci. Instr.* **29,** 620.
Mochalov, M. D. (1936), *J. Tech. Phys. USSR* **6,** 605.
Moenke-Blankenburg, L. (1959), *Jenaer Jahrbuch* **I,** 211.
Morris, W. et al. (1955), *A.S.M. Trans.* **47,** 463.
Nash, H. C., and C. S. Smith (1959), *J. Phys. Chem. Solids* **9,** 113.
Nasledov, D. N., and P. T. Kozyrev (1954), *Zhur. Tekhn. Fiziki* **24,** 2124.
Neurath, P. W., and J. S. Koehler (1951), *J. Appl. Phys.* **22,** 621.
Nicholas, J. F. (1951), J. Metals, **3,** 1142.
Nichols, J. L. (1955), *J. Appl. Phys.* **26,** 470.
Nielsen, S., and R. J. Heritage (1959), *J. Electrochem. Soc.* **106,** 39.
Nijland, I. M. (1954), *Philips Reas. Rept.* **9,** 259.
Noggle, T. S. (1953), *Rev. Sci. Instr.* **24,** 184.
Noggle, T. S., and J. S. Koehler (1955), *Acta Met* **3,** 260.
Northcott, L. (1938), *J. Instr. Met.* **62,** 101; (1939), *J. Inst. Met.* **65,** 173.
Northcott, L., and D. E. Thomas (1939), *J. Inst. Met.* **65,** 205.
Nussbaum, A. (1954), *Phys. Rev.* **94,** 337.
Obreimov, I., and L. Schubnikov (1924), *Z. Phys.* **25,** 31.
Okkerse, B. (1956), *Phys. Rev.* **103,** 1246; (1960), *De Ingenieur,* **72,** 21.
Ovsienko, D. F., and E. I. Sosnina (1956), *F.M.iM.* **3,** 374.
Packman, J. E. (1960), *J. Instr. Met.* **89,** 112.
Palibin, P. A., and A. I. Froiman (1933), *Z. Krist.* **85,** 322.
Pearson, T. G., and H. W. L. Phillips (1957), *Met. Rev.* **2,** 305.
Pepinsky, R. (1953), *Rev. Sci. Inst.* **24,** 403.
Perrot, L. (1898), *C. R. Acad. Sci. Paris* **126,** 1194.
Petrauskas, A. A., L. A. Cromer, and J. E. Macdonald (1949), *Rev. Sci. Inst.* **20,** 961.
Petrauskas, A. A., and F. Gaudry (1949), *J. Appl. Phys.* **20,** 1257.
Pfann, W. G. (1958), *Zone melting,* John Wiley and Sons, New York; (1960), *Solid State Phys. in Electronics and Telecom.* I, Academic Press, London, p. 17.
Piontelli, R., B. Rivolta, and G. Sternheim (1955), *Rev. Sci. Inst.* **26,** 1206.
Plaskett, T. S., and W. C. Winegard (1959), *Can. J. Phys.* **37,** 1555.
Pohl, R. G. (1954), *J. Appl. Phys.* **25,** 668.

Pomeroy, C. D. (1952), *M.Sc. Thesis,* University of London.

Pond, R. B., and S. W. Kessler (1951), *Trans. AIME, J. Metals* **3,** 1156.

Poppy, W. J. (1934), *Phys. Rev.* **46,** 815.

Porbansky, E. M. (1959), *J. Appl. Phys.* **30,** 1455.

Potter, H. H., and W. Sucksmith (1927), *Nature* **119,** 924.

Powell, R. W. (1949), *Nature,* **164,** 153; (1951), *Proc. Roy. Soc.* **A209,** 525.

Prasad, S. C., and W. A. Wooster (1956), *Acta Cryst.* **9,** 38.

Predvoditelev, A. A., and N. A. Tyapunina (1959), *Physics Metals Metal.* **7,** 55.

Priest, H. F. (1956), *Handbook of Semiconductor Electronics,* edited by L. P. Hunter, McGraw-Hill, New York, pp. 6–27.

Puttick, K. E., and R. King (1949), *Royal Aircraft Est. Rept., No. Met.* 45.

Quimby, S. L., and S. Siegel (1938), *Phys. Rev.* **54,** 293.

Rao, S. R., and K. C. Subramanian (1936), *Phil. Mag.* **21,** 609.

Rao, S. R., and A. S. Narayanaswami (1938), *Phil. Mag.* **26,** 1018.

Rausch, K. (1947), *Ann. Phys.* (Liepzig) **1,** 190.

RCA Report (1959), PB 161541, Office of Technical Services, Dept. of Commerce, Washington, D.C.

Read, T. A. (1940), *Phys. Rev.* **58,** 371.

Rebstock, H. (1956), Thesis, Technische Hochschule, Stuttgart.

Reddeman, H. (1932), *Ann. d. Phys.* (Liepzig) **14,** 139.

Reed-Hill, R. E., and W. D. Robertson (1957), *Acta Met* **5,** 728.

Regel, V. R. et al. (1960), *Soviet Phys. Crystal.* **4,** 895.

Richards, J. L. (1956), *Nature,* **177,** 182.

Riessler, W. (1960), *Z. angew. physik.* **10,** 433.

Robinson, S. A. et al. (1952), *Science* **116,** 362.

Roscoe, R. (1934), *Ph.D. Thesis,* University of London.

Roscoe, R., and P. J. Hutchings (1933) *Phil. Mag.* **16,** 703.

Rosenberg, A., and W. C. Winegard (1954), *Acta Met.* **2,** 342.

Rosenberg, A., and W. A. Tiller (1957), *Acta Met.* **5,** 565.

Rosi, F. D. (1958), *R.C.A. Review* **19,** 349.

Rowland, P. R. (1951), *Trans. Farad. Soc.* **47,** 193; (1951), *J. Sci. Inst.* **28,** 61.

Rozhansky, V. N., N. V. Dekartova, and I. A. Bakeeva (1957), *F.M.iM.* **4,** 527.

Rutter, J. W. (1958), *Liquid Metals and Solidification,* A.S.M., New York, p. 243.

Rutter, J. W., and B. Chalmers (1953), *Canad. J. Phys.* **31,** 15.

Rybalko, F. P. (1956), *F.M.iM.* **3,** 184.

Rybalko, F. P., M. A. Bainov, and L. M. Katanov (1960), *F.M.iM.* **9,** 796.

Sajin, H. P., and P. Y. Dulkina (1955), *Internat. Conf. Peaceful Uses of Atomic Energy, Paper* 8/P/637.

Salkovitz, E. I. (1951), *J. Metals* **3,** 64; (1956), **8,** 176.

Sato, H., J. Getsko, and W. E. Hickman (1957), *Rev. Sci. Inst.* **28,** 58.

Sawkill, J., and R. W. K. Honeycombe (1954), *Acta Met.* **2,** 854.

Scheil, E. (1954), *Z. Metkde.* **45,** 298.

Schell, H. A. (1955), *Z. Metkde.* **46,** 58.

Schilling, H. K. (1935), *Physics,* **6,** 111.

Schlipf, J., and A. Seeger (1954), *Acta Met.* **2,** 546.

Schwuttke, G. H. (1959), *J. Electrochem. Soc.* **106,** 315.

Sckell, O. (1930), *Ann. d Phys.* (Leipzig) **6,** 932.

Scofield, B. T. (1959), *J. Sci. Inst.* **36,** 371.

Seybolt, A. U., and J. E. Burke, *Experimental Metallurgy,* John Wiley & Sons, 1953.

Sekerka, R. F., G. F. Bolling, and W. A. Tiller (1960), *Canad. J. Phys.* **38,** 883.

Semenkovich, S. A., and N. N. Astashev (1958), *Zh. Tekhn. Fiziki* **28,** 725.

Shashkov, Y. M. (1961), *The Metallurgy of Semiconductors;* (1960), Trans. by Consulants Bureau, New York.

Shirn, G. A. (1955), *Acta Met.* **3,** 87.

Shvartsenau, N. F. (1960), *Soviet Phys. Solid State,* **2,** 797.

Silvey, G. A., V. J. Lyons, and V. J. Silvestri (1961), *J. Electrochem. Soc.* **108,** 653.

Smialowski, M. (1937), *Z. Metkde.* **29,** 133.

Smithells, C. J. (1955), *Metal Ref. Bk.,* Vol. 1. Butterworths, London.

Stepanova, V. M., and A. A. Urusovskaya (1960), *Soviet Phys. Crystallography* **4,** 867.

Straumanis, M. (1929), *Z. anorg. allgem. Chem.* **180,** 1.

Swift, I. H., and E. P. T. Tyndall (1942), *Phys. Rev.* **61,** 359.

Tammann, G. (1921), *Lehrbuch d. Metallographie, Leipzig, Voss* translation (1925).

Tanenbaum, M. (1959), *Semiconductors,* edited by H. B. Hannay, Reinhold, New York, p. 87.

Tanenbaum, M., A. J. Goss, and W. G. Pfann (1954), *J. Metals* **6,** 762.

Taylor, A., and S. Leber (1954), *J. Metals* **6,** 190.

Teal, G. K., and L. B. Little (1950), *Phys. Rev.* **78,** 647.

Teghtsoonian, E., and B. Chalmers (1951), *Can. J. Phys.* **29,** 370; (1952), **30,** 388.

Thurmond, C. D., W. G. Guldner, and A. L. Beach (1957), *J. Electrochem. Soc.* **103,** 603.

Thurmond, C. D., and M. Kowalchik (1960), *Bell System Tech. J.* **39,** 169.

Tiller, W. A. (1957), *Acta Met.* **5,** 56; (1958), *Liquid Metals and Solidification,* A.S.M., New York.

Tiller, W. A. et al. (1953), *Acta Met.* **1,** 428.

Tiller, W. A., and J. W. Rutter (1956), *Can. J. Phys.* **34,** 96.

Trumbore, F. A., E. M. Porbansky, and A. A. Tartaglia (1959), *J. Phys. Chem. Solids,* **11,** 239.

Trumbore, F. A., and E. M. Porbansky (1960), *J. Appl. Phys.* **31,** 2068.

Trumbore, F. A. (1960), *Bell System Tech. J.* **39,** 205.

Tsuboi, S. (1929), *Mem. Coll. Sci. Kyoto. Imp. Univ.* **12,** 223.

Tucker, G. E. G., and P. C. Murphy (1953), *J. Inst. Met.* **81,** 235.

Uhlir, A. (1955), *Rev. Sci. Instr.* **26,** 965; (1956), *Bell System Tech. J.* **35,** 333.

Vaughan, T. B. (1958), *J. Sci. Inst.* **35,** 147.

Vickers, W., and G. B. Greenough (1956), *Nature* **178,** 536.
Vickers, W. (1957), *Trans. AIME, J. Metals* **9,** 827.
Vigdorovich, V. N., V. S. Ivleva, and L. Krol' (1960), *Izvest. Akad. Nauk. SSSR Met i Topl.* (1), p. 44.
Vogel, F. L. et al. (1953), *Phys. Rev.* **90,** 489.
Wagner, R. S. (1960), *Acta Met.* **8,** 57.
Walton, D. et al. (1955), *J. Metals* **7,** 1023.
Wayson, A. R. (1961), *Rev. Sci. Instr.* **32,** 967.
Weerts, J. (1928), *Z. tech. Phys.* **9,** 127.
Weidel, J. (1954), *Z. fur Naturforschung* **9a,** 697.
Weinberg, F., and B. Chalmers (1951), *Can. J. Phys.* **29,** 382; (1952), **30,** 488.
Weisberg, L., and F. D. Rosi (1960), *Rev. Sci. Instr.* **31,** 206.
Weiser, K. (1958), *J. Phys. Chem. Solids* **7,** 118.
Wernick, J. H., and H. M. Davis (1956), *J. Appl. Phys.* **27,** 149.
Wernick, J. H., K. E. Benson, and D. Dorsi (1957), *J. Metals* **9,** 996.
Wernick, J. H. et al. (1958), *J. Appl. Phys.* **29,** 1013.
Wernick, J. H., D. Dorsi, and J. J. Byrnes (1959), *J. Electrochem. Soc.* **106,** 245.
Wernick, J. H., and E. E. Thomas (1960), *Trans. AIME* **218,** 763.
Whelan, J. M., J. D. Struthers, and J. A. Ditzenberger, (1960), *J. Electrochem. Soc.* **107,** 982.
Wilkes, J. G., (1959), *Proc. I.E.E.* **B106,** 866.
Wilson, J. M., (1959), *Research* **12,** 47.
Winegard, W. C., (1961), *Met. Reviews* **6,** 57.
Wolff, G. A., J. M. Wilbur, and J. C. Clark, (1957), *Z. Electrochem.* **61,** 101.
Wyon, G., (1955), *Phil. Mag.* **46,** 1119.
Yamamoto, M., and J. Watanabe, (1950), *Res. Rep. Res. Inst. Tohoku Univ.* **A2,** 81 and 270; (1951a), **A3,** 165; (1951b), **A3,** 655; (1954), **A6,** 233; (1955), **A7,** 329; (1956), **A8,** 230.
Yamamoto, M., and J. Watanabe, (1953), Nippon Kinzoku Gakkai-Si, **17,** 424; (1957), **21,** 732; (1959a), **23,** 675; (1959b), **23,** 679.
Yoshida, (1927), *Jap. J. Phys.* **4,** 133.
Young, F. W., (1961), *J. Appl. Phys.* **32,** 192.
Young, F. W., and T. R. Wilson, (1961), *Rev. Sci. Inst.* **32,** 559.
Zimmerman, W., (1954), *Science,* **119,** 411.

17

High Melting Point Elements

H. W. Schadler

The Bridgman and Czochralski techniques have been most useful for growing crystals of metals with melting points below about 1500°C, whereas the floating zone melting technique has been better for metals with melting points above 1500° and up to 3400°C. The general features of the Bridgman and Czochralski techniques have been considered by Goss and others. Therefore only the specific details applicable to the growth of crystals of Ni and Co and alloys of Fe, Ni, and Co are considered here.* The major emphasis is on the floating zone melting technique with specific consideration of equipment design, experimental procedure, and growth conditions.

The advantages that melt-grown crystals have in comparison with crystals grown from the solid state are primarily ones of degree rather than of kind. Control of the orientation of the crystal is easier, and the purification and crystal growth processes can be combined. With the exceptions of titanium (Darnell, 1958) and cobalt (Cioffi et al., 1937), metals and alloys which undergo a solid state transformation cannot be obtained in monocrystalline form from the melt and must be grown by solid state techniques.

1. THE BRIDGMAN AND CZOCHRALSKI TECHNIQUES FOR Ni, Co; AND ALLOYS OF Ni, Co, AND Fe

General descriptions of the Bridgman and Czochralski techniques are given by other authors in this volume and by Tanenbaum (1959). These techniques are limited to metals and alloys with melting temperatures below about 1540°C because crucible materials that will not react with the melt are not available for higher temperatures. The Bridgman technique has been used for Ni and Ni alloys (Pearson, 1953; Takaki et al., 1954; Sestak, 1957; Butuzov and Dobrovensky, 1958; Menzel and Otter, 1959; Gow and Chalmers, 1951, Walker et al., 1949; Meissner, 1959); Fe alloys (Walker, 1949; Cioffi et al., 1937; Hall, 1957; Takaki et al., 1954); and Co (Walker et al., 1949; Cioffi et al., 1937 and 1939). Rutter and Sawyer (1961) used a magnesium oxide crucible and the Czochralski technique for Co. The aforementioned authors used a variety of crucible materials including alundum, alumina, silica, and porcelain. Since the details of the crucible and furnace design are somewhat unique for these materials, two experimental systems are described in detail, those of Pearson (1953) and Hall (1957).

Pearson (1953), using a molybdenum wound hydrogen furnace and an independently evacuated Morgan recrystallized alumina tube, was able to grow Ni crystals. He states that the equipment and technique could be used for Co, Fe-Si alloys, and Ni-Fe alloys. The details of the furnace design can be obtained from his paper. Taking advantage of the natural temperature gradient of the furnace, Pearson placed the recrystallized alumina tube in the furnace so that the top of the charge was at the hottest point. The furnace was heated slowly to 1100°C to avoid cracking

*See Table 1 for brief description of specialized or proprietary techniques.

TABLE 1

Element or Alloy (melting point °C)	References	Comments
Ni (1455)	Pearson (1953)	Bridgman technique, vertical recrystallized Al_2O_3 boat, vacuum, growth rate of 0.1 to 0.2 mm/min. No seeding
	Gow and Chalmers (1951)	Bridgeman technique, horizontal alundum boat, argon, growth rate of 2 mm/min. Crystals seeded. Bicrystals were also produced
	Takaki et al. (1954)	Bridgman technique, vertical procelain crucible, vacuum, growth rate 0.4 to 10 mm/min. Crystals seeded
	Walker, Williams, and Bozorth (1949)	Bridgman technique, vertical Norton ATM409 or RA1139 alundum crucible, hydrogen, growth rate of about 0.05 mm/min. No seeding
	Butuzov and Dobrovensky (1958)	Zone fusion, horizontal fused-silica crucible, vacuum, growth rate of 0.5 mm/min
	Sestak (1957)	Bridgman technique, vertical sintered Al_2O_3 boat, growth rate 0.15 to 0.4 mm/min
	Calverly (1959)	Electron bombardment floating zone melting. Single crystals do not always result (for 0.235-in. diameter rods, 3 mm zone length, zoned down at 3 mm/min. Surface tension of Ni is low, $I = 90$ ma, $V =$ 1300 volts)*
Ni–Co alloys	Meissner (1959)	Bridgman technique in a vertical, Al_2O_3 boat, vacuum. Long time 1000°C anneal. Co content varied no more than 0.3% from end to end
Ni–Fe alloys	Pearson (1953) Walker et al. (1949) De Carlo (1961)	Electron bombardment floating zone refining; $\frac{3}{16}$-in. diameter rod zoned at 3 mm/min. $I = 50$ ma, $V = 2200$ volts
Ni–Cu alloys	Walker et al. (1949)	
Co (1495)	Rutter and Sawyer (1961)	Czochralski technique, grown from a graphite crucible. Crystal pulled at 0.2 mm/min
	Pearson (1953) Cioffi and Boothby (1939)	See Walker et al. (1949)
Gd (1312)	DeCarlo (1961)	Electron bombardment floating zone, $\frac{1}{4}$-in. crystal, $I = 30$ ma, $V = 1000$ volts, grown at 3 mm/min
	Rutter and Sawyer (1961)	Czochralski technique using a Ta crucible and gettered argon
Fe–30 Ni	Cioffi, Williams, and Bozorth (1937)	
Fe–Si	Pearson (1953) Takaki et al. (1954), Hall (1957)	Bridgman technique—see text
Fe–Al	Hall (1957), Walker et al. (1949)	
Fe–Mo	Walker et al. (1949)	

TABLE 1 (continued)

<table>
<tr><th>Element or Alloy (melting point °C)</th><th>References</th><th>Comments</th></tr>
<tr><td>Fe (not single crystal)</td><td>DeCarlo (1961)</td><td>Electron bombardment floating zone $\frac{1}{4}$-in. diameter rod. $I = 50$ ma, $V = 1900$ volts. Franklin Institute has successfully purified Fe by induction floating zone melting 3-in. diameter rods. This work is as yet unpublished and the coil design is proprietary at the present time</td></tr>
<tr><td>Ti (1668)</td><td>Darnell (1958)</td><td>Induction floating zone melting of 1 cm bar. Zone length 1 cm. Zone speed 1 mm/min. Argon Atmosphere. Apparently the slow growth rate permits the α-β phase transformation to proceed uniformly behind the liquid-solid interface. See the supporting evidence of E. C. Broch, *Phys. Rev.* **100,** 1297 (1955)</td></tr>
<tr><td>B (2300)</td><td>Franklin Inst. (unpublished)</td><td>Induction floating zone refining in argon atmosphere</td></tr>
<tr><td>Pt Group Metals
Pd (1554)
Pt (1773)
Rh (1996)
Ir (2454)
Ru (2500)
Os (2700)</td><td>Rhys (1959); Allred, Hines, and Goering (1959); Allen (1961a,b)</td><td>Electron bombardment floating zone melting of rods from $\frac{1}{8}$ to 2-in. diameter. Zone speed 0.2 mm/min Pt, Rh, and Ru, and 1 mm/min for Pd, Os, and Ir. Rh was a single crystal. The rest gave large-grained ($\frac{1}{8}$ to $\frac{1}{2}$-in. long) polycrystals. Gas evolution is problem with Rh; and Pd and Os have a high vapor pressure. $\frac{1}{4}$-in. Pt required $I = 90$ ma, $V = 2200$ volts</td></tr>
<tr><td>Refractory Metals
V (1775)
Cb (2415)
Mo (2625)
Ta (3000)
Re (3170)
W (3410)</td><td>Calverly, Davis and Lever (1957), Calverly (1959), Carlson (1959), Belk (1960), Buehler (1958), Geach and Jones (1959), Lawley (1959), Orehotsky and Opinsky (1959), Schadler (1959), Witzke (1959), Sell (1960), Allen (1961)</td><td>Electron bombardment and induction floating zone melting of rods from $\frac{1}{8}$ to $\frac{3}{8}$-in. diameter in vacuum. Zone speeds of from 1 to 4 mm/min both up and down. Zone lengths comparable to the rod diameter. Induction units were usually 450 kc Ecco 10 kw. Electron beam power requirements after Lawley (1959):
<table>
<tr><th>Metal</th><th>Diameter (inches)</th><th>Power (watts)</th></tr>
<tr><td>W</td><td>0.180</td><td>400</td></tr>
<tr><td>Ta</td><td>0.150</td><td>300</td></tr>
<tr><td>Mo</td><td>0.150</td><td>180</td></tr>
<tr><td>Fe</td><td>0.250</td><td>130</td></tr>
<tr><td>V</td><td>0.240</td><td>125</td></tr>
</table>
Single crystals always result (except Fe) and seeding is possible

Power requirements after DeCarlo (1961):
<table>
<tr><th>Metal</th><th>Diameter (inches)</th><th>I (ma)</th><th>V (volts)</th></tr>
<tr><td>Mo–Re</td><td>$\frac{3}{16}$</td><td>100</td><td>2500</td></tr>
<tr><td>W</td><td>$\frac{1}{8}$</td><td>180</td><td>3100</td></tr>
<tr><td>Ta</td><td>$\frac{1}{8}$</td><td>110</td><td>2800</td></tr>
<tr><td>Nb</td><td>$\frac{1}{8}$, $\frac{3}{16}$</td><td>90</td><td>2500</td></tr>
</table>
</td></tr>
</table>

TABLE 1 (continued)

Element or Alloy (melting point °C)	References	Comments
Refractory Metal Alloys		
Mo–Re	Jones (1959), Lawley (1959),	Segration was not a problem
Ta–Cb	Calverly (1959), DeCarlo (1961),	
Mo–W	Belk (1959),	
Ta–V	Calverly (1959)	*V* volatilizes, producing a Ta-rich alloy

* Parenthetical data due to V. J. DeClarlo at General Electric Research Laboratory.

SUPPLEMENTARY TABLE

Specialized Techniques	References
Spherical crystals of 70% Fe-30% Ni	Cech and Turnbull (1956)
Drop melting of refractory metal powders. Powder is melted with an electron beam and then falls onto a pedestal	Petrov and Shashkov (1957)
Spherical crystals of Ni formed by melting Ni on a W or Mo ribbon. The ribbon is coated with Al_2O_3 or BeO	Menzel and Otter (1959)
B, V, Fe, Zr, NiAl spheres formed by sparking metal chips in a water-cooled copper crucible using a W electrode	Ray and Smith (1958)
Single crystals of refractory metals have been grown using an arc flame fusion process. The process is analogous to the Verneuil combustion flame fusion technique except an inert electric arc is substituted for the oxy-hydrogen flame	Linde Division of Union Carbide and Carbon, Crystal Products Department, Chicago, Illinois

the crucible, and then rapidly to about 1500°C. After holding just above the melting point of nickel for 2 hr the temperature was lowered at a rate of about 5°C/hr until the entire melt had solidified. (Since the temperature gradient of the furnace was about 12°C/in. at 1455°C the crystal was grown at a rate of about 0.17 mm/min.) After solidification was complete, the cooling rate was increased to 15 to 20°C/hr down to 1200°C; then the furnace was rapidly cooled to room temperature. If an alloy crystal were being produced by this technique, an homogenizing anneal at a temperature just below the melting point could be used to reduce the composition gradient resulting from solidification. Although Pearson had no trouble obtaining a monocrystal, the possibility of nucleating more than one crystal at the start of solidification could be reduced by using a crucible with a pointed bottom (see Chalmers, 1952).

Hall (1957) used induction heating and alundum crucibles to produce alloy crystals of Fe-Al and Fe-Si from the melt. His technique is different from that of Pearson (1953), Walker, Williams, and Bozorth (1949), and Cioffi and Boothby (1939) in that he used a pointed crucible and a heat sink at the tip of the crucible to promote the nucleation of a single grain. A schematic diagram of his equipment and the details of the heat sink are shown in Fig. 1,

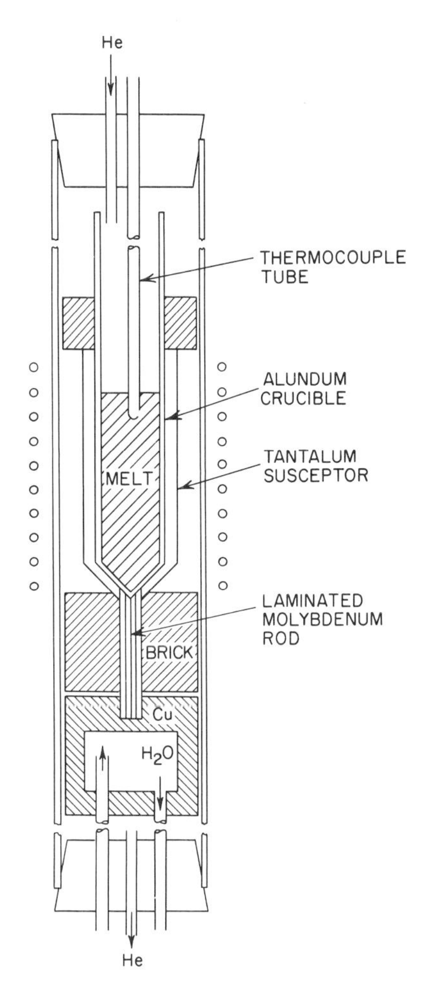

Fig. 1. Furnace and crucible design used by Hall (1957) for growth of single crystals of Fe-Al and Fe-Si alloys. The drawing was taken from Hall's paper.

which was taken from his paper. The crucible with its surrounding Ta susceptor is contained in a Vycor tube that is surrounded by a water-cooled induction coil and contains a He atmosphere. The crucible is held in contact with a 3-in.-long laminated cylinder of Mo that is cooled at the opposite end by contact with the water-cooled copper block. A laminated cylinder is used because the eddy current heating is less than for a solid cylinder. The power input was held constant after the entire melt reached 20 to 40°C above the melting temperature. Crystals were grown by lowering the Vycor tube assembly through the induction coil at a rate of 0.1 mm/minute.

Neither Pearson nor Hall attempted to grow oriented crystals.

Crystals of controlled orientation have been grown by Gow and Chalmers (1951) and Takaki et al. (1954). The former grew oriented Ni crystals in a horizontal boat of alundum using the technique of seeding which Goss has previously described in this book. An argon atmosphere in a graphite resistance furnace was used with a growth rate of 2 mm/min. Takaki et al. (1954) have grown oriented crystals of Ni and silicon-iron using a vertical tube furnace and seeds. In their case, the seed was placed in the bottom of the crucible and held in a porcelain sleeve to give the desired orientation. It was partially melted by careful control of the placement of the crucible in a temperature gradient, and the crystal was grown by lowering the crucible through the furnace. For the silicon-iron crystals, Takaki et al. found that the composition varied from 3.2 to 4.0% silicon within the first 2 cm of the crystal but was relatively constant over the remaining 6 cm of its length.

Table 1 contains a summary of the metal and alloy crystals grown by these techniques along with pertinent references.

2. FLOATING ZONE MELTING TECHNIQUES

Melt grown monocrystals of high-melting temperature elements and alloys were made possible by the development of the floating zone melting technique by Keck and Golay (1953). Prior to that time crystals of elements and alloys with melting points above about 1500°C were produced by solid-state techniques (see Chapter 23). Since its introduction, the floating zone melting technique has been used to produce high-purity crystals of semiconductors and many high-temperature metals and alloys. The term *floating zone* as used here applies to a vertical zone; the basic concept is depicted in Fig. 2*a*. A molten zone is established (the means is described later) between two vertical cylindrical rods and is supported by the surface tension of the liquid. The mechanical stability of such a configuration

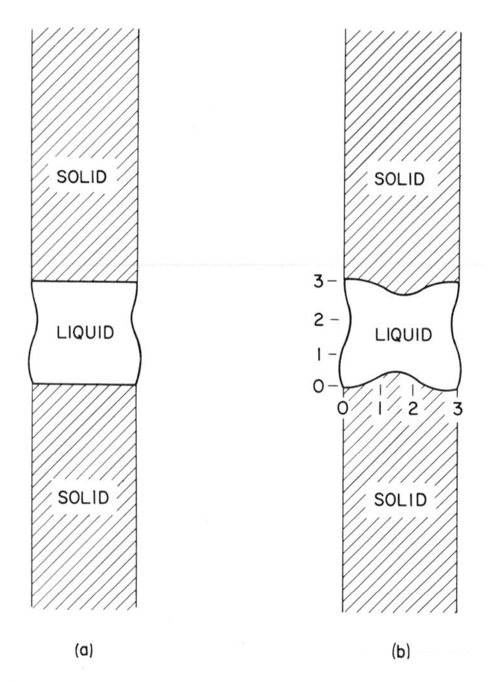

Fig. 2. The concept of a floating zone. (*a*) Idealized model of a floating zone. (After Keck, Green, and Polk, 1953.) (*b*) Shape of the molten zone in a ⅛ in. diameter tungsten rod. Numbers are distance measured in millimeters. (See text.)

is a function of the radius of the rods, the length of the molten zone, and surface tension and density of the liquid, gravity, and the shape of the liquid-solid interface. Two treatments of the interrelation of these parameters have been made and are described next.

Stability of Floating Zones

Heywang (1954, 1956) and Keck, Green, and Polk (1953) have treated the stability of vertical floating zones using the idealized zone shown in Fig. 2*a*. Keck, Green, and Polk describe the shape of a stable molten zone for various lengths of zones and radii of rods, and Hegwang derives quantitatively the relationship among zone length, rod radius, surface tension, and density of the liquid. The conclusions of both treatments are qualitatively the same. (The stability of a horizontal zone that is supported electromagnetically has been treated by Pfann and Hagelbarger (1956) and is not considered here.)

Both treatments assume (1) cylindrical rods held vertically, (2) a flat, horizontal solid-liquid interface, (3) complete wetting of the horizontal solid surfaces by the liquid, (4) no change in density on melting, and (5) no motion of the zone. For a zone traveling under steady state conditions (the quantity of metal solidifying being exactly equal to the quantity melting per unit time) conditions (4) and (5) are compensating. Heywang derives a differential equation in terms of the radius of curvature of the liquid, zone length, rod radius, surface tension, and density by expressing the total energy of the system in terms of these parameters and then minimizing the total energy. Exact solutions are possible for zone lengths which are small with respect to rod radius (that is, large radius rods) and for very small rods, but an approximate solution is obtained for intermediate cases. For simplicity, Heywang presents the results as normalized zone length L versus normalized zone radius r_0 in the graphical form reproduced in Fig. 3. The results are applicable to any metal if the parameters in the normalizing factor $\sqrt{dg/\gamma}$ are known. d is the density of the liquid, γ is the surface tension, and g is the acceleration of gravity. The condition for stability is that zone length and rod radius lie below the curve. The solution shown in Fig. 3 is for the condition that the radii of the rods above and below the molten zone are equal and that the zone is traveling up. Heywang also treats cases where the rod radii are not equal, but they are not applicable to the floating zone-refining technique since it is usually used for growing single crystals.

The conclusions of the Keck, Green, and Polk (1953) treatment are pertinent here. For a moving zone in which the growing crystal is of constant radius, they point out that the vertical tangent to the liquid surface at the solidifying interface must be perpendicular to the radius vector. For a given zone length they show in a series of line drawings how this steady state situation is reached. If the initial volume of liquid is too small, the solidifying rod will first decrease in radius and then increase until the steady state is reached. Experiment verifies this prediction. Generally, if there is too much liquid in the zone, the liquid will bulge badly and even spill out; and if there is too little liquid, surface tension will pinch off the zone. Also for a given rod radius the stability of the molten zone decreases with increasing zone length.

Operating experience with floating zone techniques has indicated that although the Heywang treatment can be used as a guide in designing apparatus, other experimental parameters influence the final conditions, namely, variations in power input caused by impedance changes in the circuit, the direction of zone travel, mechanical vibrations, and the shape of the solid-liquid interface. Surface tension data are not usually available for the high melting temperature metals, but can be measured by the drop weight method (see Calverly, 1959). Surface tension is not independent of purity and can vary considerably from one zone pass to the next. Since heat flow considerations determine the shape of the solid-liquid interface, it is never the idealized flat surface assumed in the theory.

Figure 2*b* shows schematically the shape of the solid-liquid interface in a $\frac{1}{8}$ in.-diameter tungsten rod (Schadler, 1959). The molten zone was established using the same power as is required in growing tungsten crystals, and the zone was held in a fixed position for several minutes. The liquid solidified as one crystal and the shape of the interface was convex with respect to the liquid. In general, however, the shape will be determined in a complex way by the emissivity and thermal conductivity of the material. The shape of the liquid-vapor surface was approximately as shown, the figure being drawn to emphasize the tangent requirement of Keck et al. (1953). Note that the zone length in this case is about equal to the diameter of the rod, a rule that usually follows for small-diameter rods.

The direction of zone travel affects the stability of molten zones in a way which is not fully understood. Keck, Green, and Polk (1953) comment that for silicon the zone is more stable if the direction of travel is upward. For metals both directions of zone travel have been used, but some investigators report greater stability and better crystal surfaces for upward motion of the zone, and other investigators the reverse. For the molten zone shape shown in Fig. 2*b* and the electrode configuration in Fig. 5*b*, Schadler finds that upward motion of the zone is impossible. As the zone is moved upwards, the bulge at the bottom of the zone increases in size until the liquid spills out of the zone. M. Adams (1961) reports that he encounters no such problem in melting $\frac{1}{8}$-in. rods of W, Mo, and Nb, and in fact prefers upward motion of the zone. The reasons for this discrepancy are not

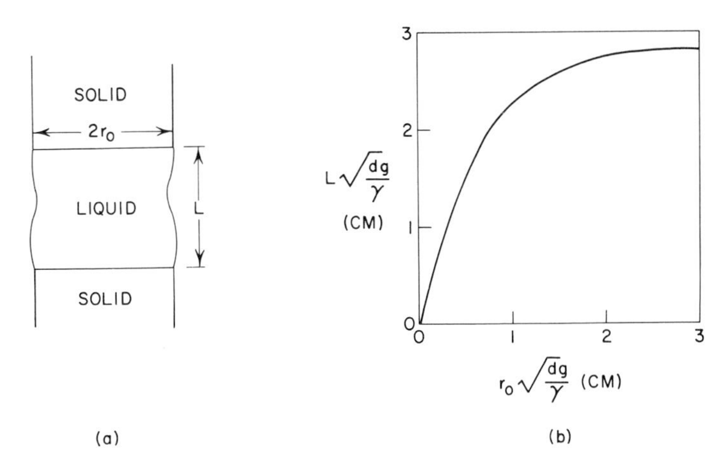

Fig. 3. The stability of a floating zone for rods of equal radii taken from Heywang (1956). L = zone length, r_0 = rod radius, d = density of the liquid, γ = the surface tension, and g = the acceleration of gravity.

known, but are probably related to undetermined differences in zone length and heat flow conditions present in different designs of the emitter configuration. This author would suggest that, in the absence of previous experience, downward direction of zone travel be tried first.

In an effort to increase the stability of a floating molten zone in Fe, Oliver and Shafer (1960) passed a d-c current in excess of 380 amperes/cm^2 along the length of the rod. As described by Pfann and Hagelbarger (1956), the current creates an internal compressive pressure which is opposite in sign to the hydrostatic pressure of the liquid zone and hence acts to increase the stability of the zone. Current densities as high as 700 amperes/cm^2 were used by Oliver and Shafer with no evidence of instability. This technique deserves more attention.

Heating Techniques

Induction, electron bombardment, and radiation heating have been used to produce a floating zone. The technique adopted by particular investigators seems to be a result of local circumstances. If a new piece of equipment is to be built, careful consideration should be given to designs which permit the use of both electron bombardment and induction heating. Electron bombardment probably affords more direct control of the length of the molten zone, and because of the vacuum required, 10^{-4} mm Hg, it may yield purer crystals. For materials with high vapor pressures at their melting points, induction heating in an inert atmosphere is to be preferred. Also induction heating offers the possibility of levitation, and thus the melting of larger diameter samples. Heating by radiation has not been used for metals. Power equipment for electron bombardment is cheaper to build or buy than induction equipment, but this difference is often offset by the availability of induction generators in most laboratories. The following sections review the experimental equipment and techniques used for electron bombardment and induction floating zone melting.

ELECTRON BOMBARDMENT. Heating with electron beams is not a new technique. It has been used for many years in vacuum tubes and has been a problem in x-ray tubes. It was used by Hultgren and Pakkala (1940) to melt alloys in a crucible and was first used for floating zone melting by Calverly, Davis, and Lever (1957). A recent review of electron bombardment techniques has been published by Lawley (1961). The concept of electron bombardment floating zone melting is shown in Fig. 4. Electrons, thermionically emitted from a hot filament, are accelerated by a d-c potential of 1 to 20 kv toward a specimen where they give up their excess kinetic energy as heat. The excess kinetic energy is dissipated within a distance equal to the mean free path of the electron in the specimen, and subsequent heating of the interior of the specimen is by conduction. Electron bombardment units have been operated with either the specimen or the cathode at ground potential. The latter reduces the amount of shielding required to protect personnel from shocks. The basic features of electron bombardment floating zone melting equipment are the vacuum system, the power supply and control, and the emitter with the attendant focusing shields. The electron bombardment floating zone melting technique has been used for a wide variety of metals and alloys, and the only

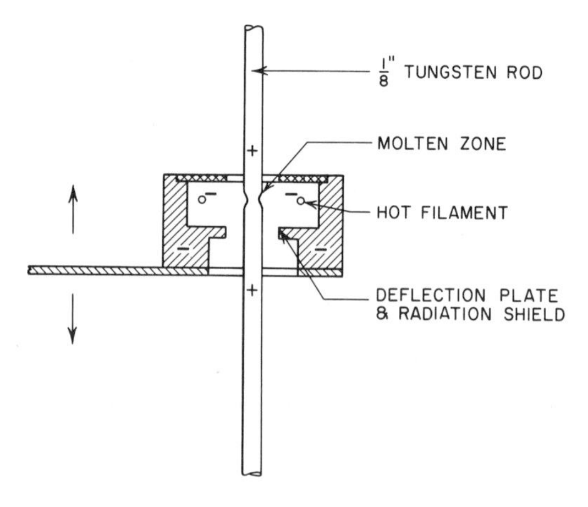

Fig. 4. Schematic drawing of the geometric arrangement among filament, specimen, and focusing shields for electron bombardment floating zone melting.

severe limitation is the vapor pressure of the metal at the melting point. If the equilibrium vapor pressure at the melting point is less than 10^{-1} mm Hg or so, the method is satisfactory.

The design of vacuum systems is a subject which has been adequately treated elsewhere, but several requirements of vacuum systems to be used for floating zone melting are worth emphasizing. For electron bombardment melting a vacuum of at least 10^{-4} mm Hg is required. Hence the pumping capacity of the system must be sufficient to maintain this vacuum in the face of the leak rate of the system and outgassing of the specimen and apparatus. For melting small diameter rods ($<\frac{3}{8}$ in.) various speed diffusion pumps with various pumping speeds have been used. Although the pumping speed at the system is always much less than the capacity of the pump (at least a factor of two where proper baffling and a nitrogen trap are used), most people quote the speed at the pump. Buehler (1958) used a 600 liters/min pump, Carlson (1959) a 100 liters/min pump, and Calverly (1959) estimates his pumping speed at 1200 liters/min. In general, the higher the pumping speed the less susceptible will be the operation of the system to sudden releases of gas from the sample. Provision must be made for power leads into the system (see Calverly, 1957, for E. B. requirements and Buehler, 1958 or Wernick et al., 1959, for induction heating) and some means for changing the relative position of specimen and power source. For ultimate vacuums up to 10^{-7} mm Hg, an ordinary Wilson seal is adequate for introducing rotary or sliding motion. A rotating bellows seal is often an improvement if rotary motion is needed. A variable speed drive is advisable. The vacuum system can be all metal with "O-ring" seals (for example, the Calverly or MRC unit)* or a combination of metal and glass as in standard evaporation units. Always a liquid nitrogen trap and adequate baffles to prevent back streaming from the diffusion pump are required. A sighting port is essential. Components of the system that are likely to become hot should not be built from metals or alloys that have a high vapor pressure. For example, brass is poor because of the volatility of its zinc. Last, it is convenient to have a valve between the system and the liquid nitrogen trap (or water-cooling on the oil reservoir of the diffusion pump) and a separate forepump line to the system to reduce the waiting time when the system is opened.

Success with electron bombardment melting depends on the design of the emitter and the focusing plates. One arrangement of emitter (filament or cathode), specimen (anode), and focusing plates is shown schematically in Fig. 4. Some variations of this design are illustrated in Fig. 5. Figure 5*a* shows the original Calverly design and one that has been widely used. Figure 5*b* is the design used by Carlson (1959) and Schadler (1959). The enclosed molten zone shown in Fig. 5*b* reduces the deposit of metal over the vacuum system, but may have the disadvantage that the effective pressure is higher around the molten zone.

Several cathode designs have been used. The simplest is the wire loop shown in Fig. 6*a*. Calverly also suggested the ribbon design shown in Fig. 6*b*, and Belk (1960) used a circular wire loop rolled to flatten the wire. This last design is advantageous in that it reduces the projected area of the filament that the specimen sees and hence increases the filament life. Metal volatilizing from the sample, and droplets thrown from the sample by the bursting of gas bubbles, deposit on the filament and lead to failure. They cause nonuniform changes in the cross-sectional area of the filament, alloying of the filament, and the weight of the deposit sometimes displaces the filament from its symmetrical position between the focusing plates. Sell (1960) has calculated that a molten zone of W 4.6 mm in diameter and 4 mm long moving at 3 mm/min will volatilize about 1% of the sample.

In most cases, commercial tungsten wire or sheet is used for the cathode, although

* Electron bombardment floating zone refining equipment is currently sold by Materials Research Corp., Yonkers, New York and by Miles-HiVolt Ltd., Shoreham-by-Sea, Sussex, England.

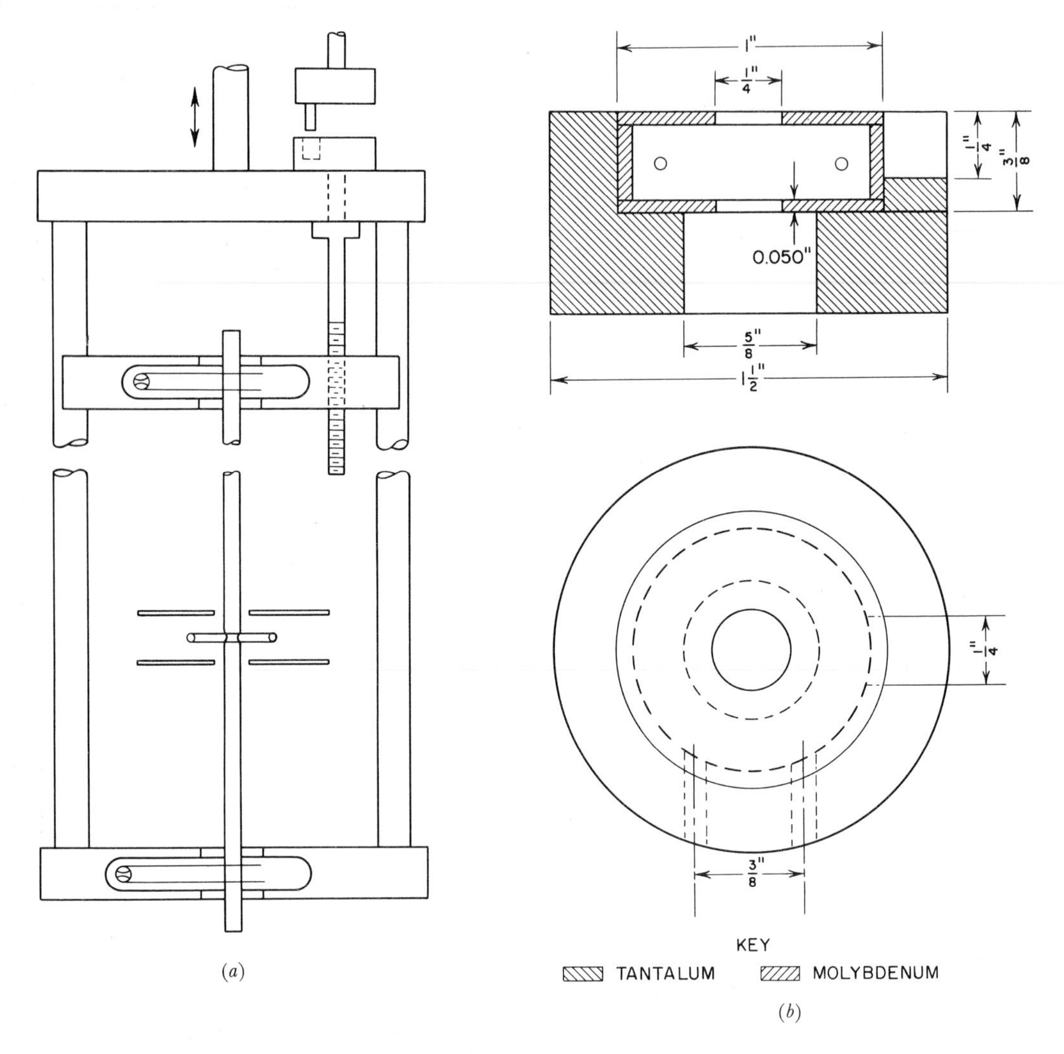

Fig. 5. Details of focusing shields and cathode arrangement. This system was designed for ⅛ in. diameter W rods but has since been used for all refractory metals and Ni with modification of the hole in the center of the Ta shields. This hole is usually made to be twice the diameter of the rod to be melted. [(*a*) After Calverly, 1957. (*b*) After Carson, 1959 and Schadler, 1959.]

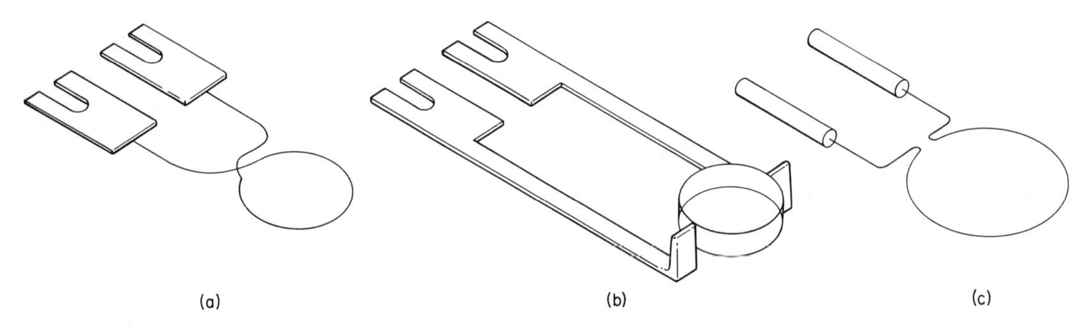

Fig. 6. Cathode designs. (*a*) 0.010 in. diameter W wire spot welded to Ni lugs. (*b*) 2 min × 0.001 – 0.002 in. W-tape. (*c*) 0.015 in. diameter W wire slip fit into an 0.018 in. diameter hole in a 0.100 in. diameter Mo cylinder. [(*a*) and (*b*) after Calverly, 1957.]

tantalum could be used. The dimensions of the cathode depend on the power source. Usually tungsten wire of 0.010 to 0.020 in. diameter has been used. The loop diameter depends on the size of the sample and the geometry of the focusing arrangement. For $\frac{1}{8}$-in. diameter rods and the focusing configuration showing in Fig. 5*b*, a filament about 15 mm in diameter proved to be satisfactory.

The filament can be held in a variety of ways. Calverly (1957) spot-welded the tungsten to nickel lugs which were bolted to a steel block. The scheme illustrated in Fig. 6*c* is satisfactory and permits easy filament replacement. The 0.015-in. tungsten wire slips into a 0.018-in. hole in the 0.100-in. molybdenum cylinder which is held in a copper grip.

Another type of emitter configuration that avoids the problem of short filament life and possible contamination of the specimen by the cathode is the design illustrated in Fig. 7. It was brought to the author's attention by Calverly and was originally built by E. E. Martin, Linfield Research, MacMinville, Oregon. Because of the longer electron path indicated, this design requires higher voltages than the previous designs ($>$5000 volts).

M. Adams (1961) has described a modification of the design illustrated in Fig. 7. The filament is placed closer to the lower focusing plate, and the upper plate is made in the form of a truncated cone with the apex up. The molten zone will form here above the filament, and the amount of metal collected by the filament will be less than where the zone is in the plane of the filament, since metal droplets ejected from the zone by bursting gas bubbles have a horizontal trajectory initially.

Eaton (1960) and Gruber (1961) have described the use of electron guns for floating zone melting, but these designs have not been widely used to date.

One of the advantages of electron-bombardment heating over induction heating is the relative ease with which the length of the molten zone can be changed. Since the electrons travel along the steepest voltage gradients in paths approximately orthogonal to the equipotential lines (neglecting inertial effects), the shape of the beam can be controlled by regulating the shapes of the equipotential lines. For a fixed distance be-

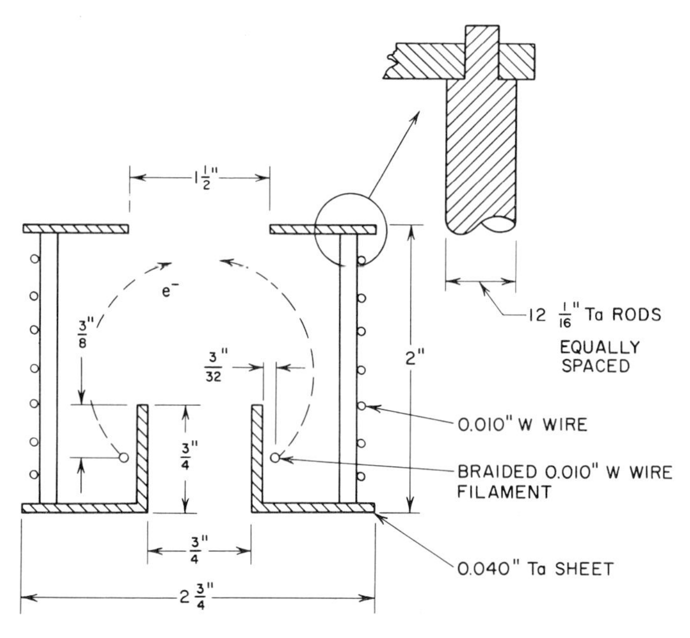

Fig. 7. Suggest design for emitter configuration which permits longer filament life and less chance for the filament to contaminate the sample. (Originally used by E. E. Martin.)

tween cathode and anode, this is usually accomplished by decreasing the vertical separation of the focusing plates (see Fig. 6). It could also be done by putting the focusing plates at a negative potential with respect to the cathode, but this technique has not been widely used.

Figure 8 is a block diagram of the basic electric circuit required for electron-bombardment melting, taken from Calverly. The essential features are the emitter and focusing arrangement already discussed, the power source, the filament power source, and some means of controlling the power which are (1) emission control, where the bombardment current is controlled by controlling the temperature and thus the thermionic emission from the cathode, and (2) voltage control, where the bombardment current is held constant and the voltage varied. The second approach is more difficult and, as discussed earlier, would vary the area bombarded. Consequently the first approach has been used almost exclusively.

For emission control, the d-c power supply should have a variable output of 0 to 5000 volts and 0 to 500 ma, but it need not be well regulated. Regulated units have been built that deliver 10,000 volts and 0.5 ampere (Birback and Calverly, 1959). The greater stability that results is not necessary for most purposes. The d-c supply should have a current limiting switch to protect it from short circuits. The cathode is supplied from a controlled a-c supply that should be capable of delivering 2 to 20 amperes, depending on the filament length and diameter. (Thermionic emission begins from ordinary W at about 2100°K where the resistivity is about 60 $\mu\Omega$-cm.) The filament will require about 50% more power at the end of its life than initially.

The basic scheme used to control the emission current is to compare it with a standard that can be varied by the operator. The amplified difference signal controls the filament power through a thyratron circuit such as Calverly's, shown in Fig. 9*a,* through a saturable reactor like the circuit shown in Fig. 9*b,* or through a motor-driven Variac like the circuit of Rocco and Sears (1956) shown in Fig. 9*c.* Since rapid changes in the impedance of the emission circuit occur because of gas evolution from the sample, the fast thyratron or the saturable reactor circuits are the best.

Specific power requirements are deter-

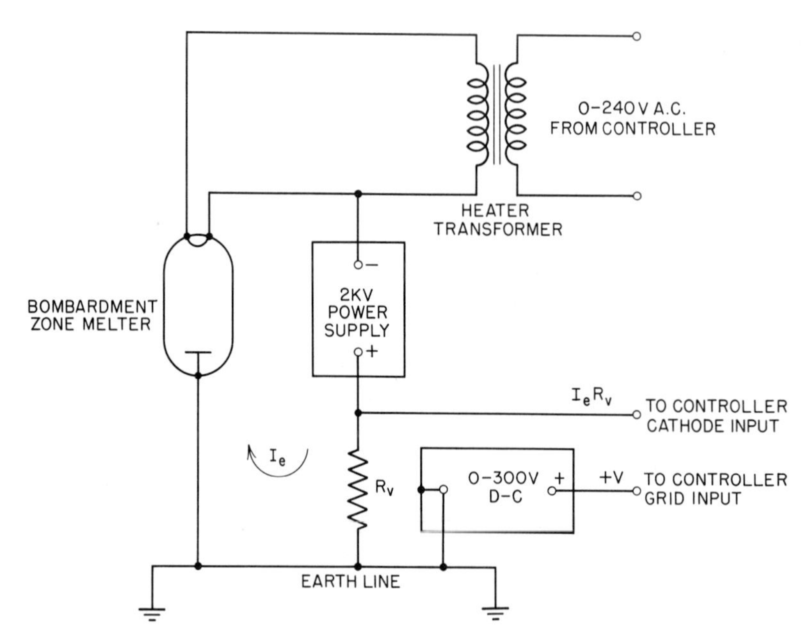

Fig. 8. Block schematic diagram of electric circuit used for electron bombardment zone melting. (After Calverly, 1957.)

mined by the design of the filament and focusing configuration and the melting point, emissivity, thermal conductivity and size of the bar being melted. Since control is accomplished through thermionic emission, the applied voltage should be great enough so that the bombarding current is emission limited. In practice, this ideal condition is not necessary (see Appendix A). Experience regarding power requirements is summarized in Table 1. Belk (1959) found empirically that the power required is

$$P = Ad + Bd^2$$

where d is the rod diameter, A is a constant proportional to the fourth power of melting temperature, and B is proportional to the thermal conductivity. According to Belk, a $\frac{1}{8}$-in. molybdenum rod requires 260 watts, and a $\frac{1}{2}$-in. molybdenum rod requires 2140 watts for a zone speed of 3 mm/min. He also states that it takes 10% more power to melt through than it does to melt the surface of a $\frac{1}{8}$-in. rod.

INDUCTION HEATING. Since induction heating is a common laboratory practice, the reader is referred to Seybolt and Burke (1953) for a review of the subject. As in electron beam melting, a good vacuum system is necessary to protect the specimen from oxidation and contamination. A controlled atmosphere of an inert gas may be used, but a leak-tight system is essential. Since the samples to be melted are relatively small, radio frequency is needed and the induction coil must be closely coupled to the sample. This usually means that there should be nothing between the coil and the specimen, but this means that the coils can be shorted by condensation from the sample (Buehler, 1958). Coil designs used by Wernick, Dorsi, and Byrnes (1959) are illustrated in Figs. 10*a, b,* and *c*. Figure 10*d* shows another design that is frequently used. The problem of controlling the length of the molten zone is somewhat difficult for induction heating, and, in the absence of previous experience, the designs shown should be tried. Another design that has been very successful for reducing the zone length in silicon (Silverman, 1961) could probably be applied to metals, particularly to large diameter rods.

RADIATION HEATING. Radiation heating has been mentioned in the literature as a means of producing a floating molten zone (Emeis, 1954). Here the sample is surrounded by a cylinder of a higher melting point metal about equal in length to the zone desired. This cylinder is then heated by induction or by electron bombardment, and it heats the sample by radiation. The technique has not been applied to metals because it is difficult to control. It has been used successfully for solid state crystal growth.

Operating Methods and Problems

SAMPLE HOLDERS. The simplest way to hold a specimen is in a spring clip device like the one shown in Fig. 5*a*. The mild steel V-block was designed to minimize heat loss at this point. This design is satisfactory if the molten zone does not come too close to the holder. A special chuck described by Wernick, Dorsi, and Byrnes (1959) permits vertical motion of the sample and thus allows for thermal expansion.

It is often desirable to move the specimen ends vertically with respect to each other, and to be able to rotate one end of the specimen. See Buehler (1958) for a design. The vertical motion permits restarting a broken zone and compensation for volatilization losses. The rotation is desirable to reduce the effective distribution coefficient in zone-refining (see Pfann, 1958). Green (1961) claims that rotation of both ends in opposite directions produces straighter crystals.

MOTION OF THE ZONE. Relative motion of specimen and heater has been accomplished both by moving the heater (see Carlson, 1959) and by moving the sample (see Fig. 5*a*). The former leads to a simpler design if provision for vertical motion and rotation of the specimen is to be made. Also it reduces the tendency for mechanical vibration of the zone, and it permits a smaller vacuum system for the same length of product. Motion of the heater has been

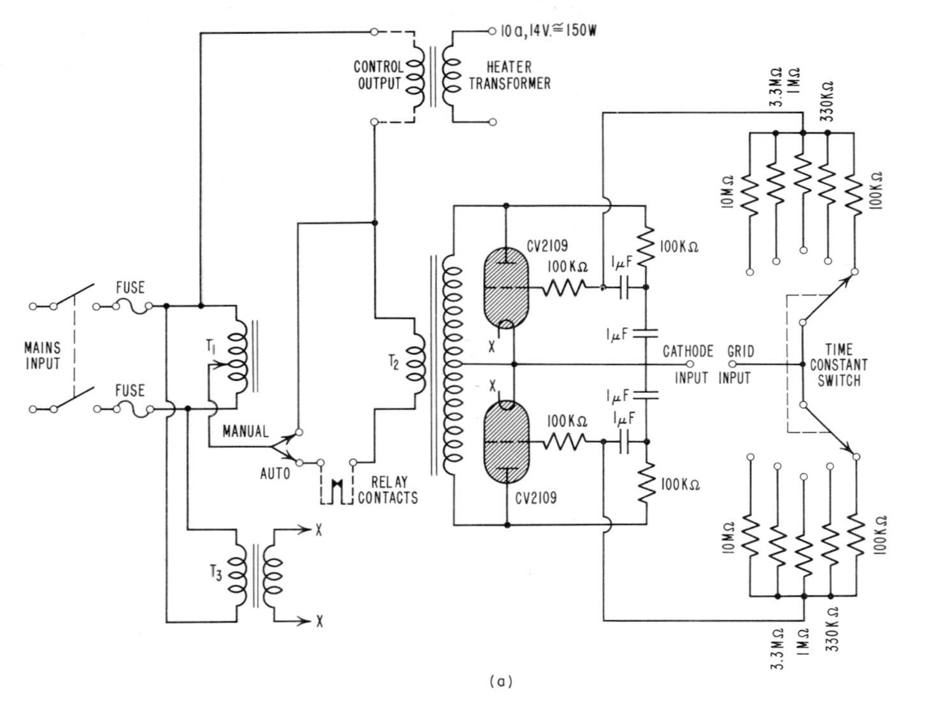

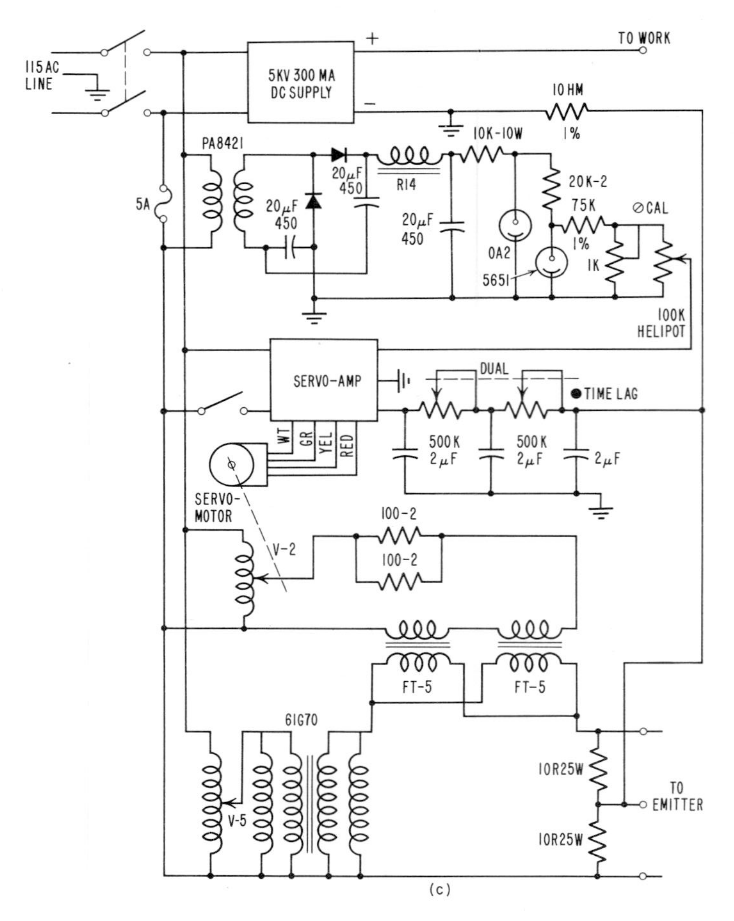

Fig. 9. Details of control circuits used for emission limited control. (*a*) Thyration control (Calverly, 1957). (*b*) Saturable reactor control (Dickey at G. E. Research Lab). (*c*) Mechanically driven variac control (Sears and Rocco, 1956).

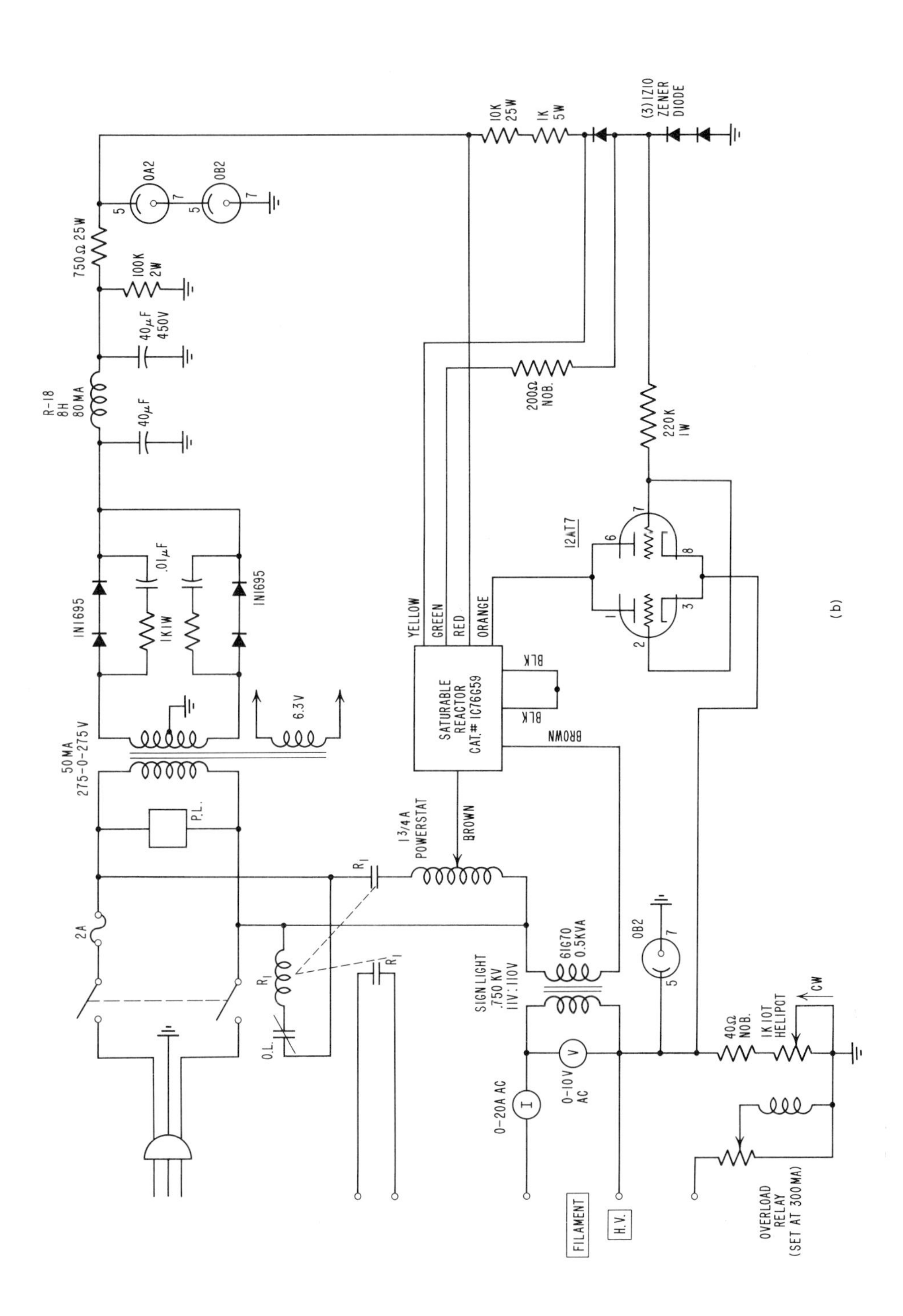
50 MA
275-0-275 V
IN1695
1K1W
.01μF
R-18
8H
80 MA
40μF
40μF
450V
100K
2W
750Ω 25W
OA2
OB2
10K
25W
1K
5W
(3) 1Z10
ZENER
DIODE
200Ω
NOB.
220K
1W
12AT7
YELLOW
GREEN
RED
ORANGE
SATURABLE
REACTOR
CAT.# 1C76G59
BLK
BLK
BROWN
BROWN
6.3V
2A
P.L.
13/4A
POWERSTAT
O.L.
SIGN LIGHT
.750 KV
11V:110V
61G70
0.5KVA
0-20A AC
0-10V
AC
40Ω
NOB.
1K10T
HELIPOT
CW
OVERLOAD
RELAY
(SET AT 300MA)
FILAMENT
H.V.

(b)

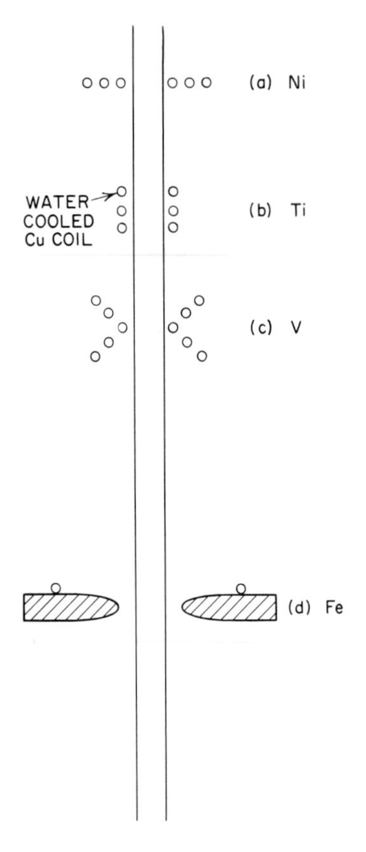

Fig. 10. Schematic illustrations of successful induction coils used in floating zone melting of various metals. (*a*) Ni. (*b*) Ti. (*c*) V. (After Wernick, Dorsi, and Byrnes, 1959.) (*d*) A common design consisting of a slotted copper plate with the central hole surrounded by a single turn of water-cooled copper tubing.

produced by sliding or ball-bearing lead-screw drives, and by chain drives. Metal vapor deposition on screws and bearings must be avoided for reliable operation.

STARTING A ZONE MELTING PASS. Starting a floating zone can be troublesome unless extreme care is used in aligning the specimen. Two separate rods cause less trouble. The ends of the rods should be aligned and almost touching each other initially. First, the ends of the rods are welded together; then the molten zone can be moved with little fear that the rods will change position.

SEEDING. A crystal of a specific orientation can be grown by the technique just outlined if a seed crystal is used in place of the top rod (for zone motion down). If a seed crystal of the desired orientation is not available, a simple goniometer (Fig. 11) can be used. Since the angle α in Fig. 11 is usually limited, a series of seeding operations may be required. The surface of the seed should be parallel to the top of the lower rod. To insure propagation of the seed crystal rather than some stray crystal in the lower rod, it is best to weld the seed to the lower rod and then start the zone within the seed crystal.

If no seed crystal is used, a crystal produced by floating zone melting is of random orientation (Carlson, 1959; Belk, 1959; Buehler, 1958; Geach and Jones, 1959). For very high melting metals (W, Ta, Mo, Mo—Re, Nb) monocrystals are usually produced by single zone melting passes. Nickel does not always grow as a mono-

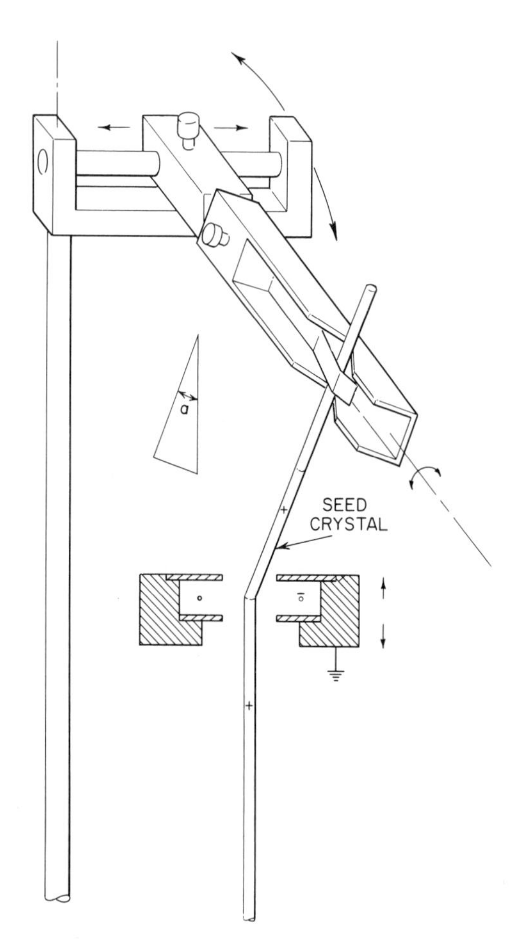

Fig. 11. Simple gonimeter arrangement for holding seed crystals.

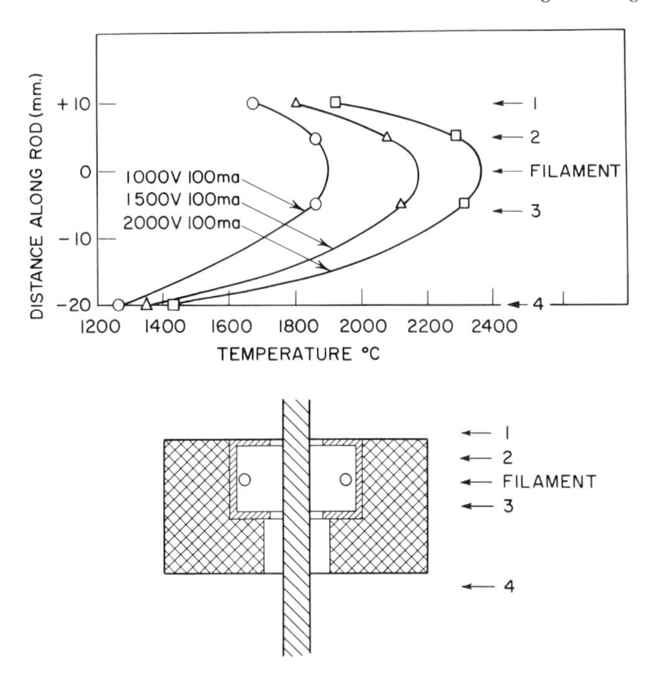

Fig. 12. Temperature gradient along a $\frac{1}{8}$-in. diameter W rod heated by electron bombardment. Temperature was measured with an optical pyrometer and the readings corrected for emissivity and the glass sighting point.

crystal (Calverly, 1959), and some difficulty has been experienced with rhodium and iridium (Allen, 1961). Apparently the combination of a very high melting point and a steep temperature gradient is sufficient to cause the growth of monocrystals.

It is important for the complete cross section of the rod to be molten. When melting is complete, Buehler (1958) and others have noted that a vibration of the molten zone will vibrate out of phase with the solid rods. Also, one end of the specimen can be rotated.

THERMAL STRESSES. Solidification of the final zone contracts the rod and can subject the crystal to high stresses. This can be avoided by increasing the power to the final zone so as to cut through the rod by volatilization.

TEMPERATURE GRADIENTS. The steep temperature gradients encountered in floating zone melting are important for the easy growth of monocrystals. An example of the temperature variation along the length of a $\frac{1}{8}$-in. tungsten rod is shown in Fig. 12. The temperature was measured with an optical pyrometer (corrected for emissivity and the thickness of the glass). Note that the gradient is in excess of 1000°C/cm.

Calverly has analyzed an idealized heat flow model of the floating zone technique. His approximation ignores radial differences in temperature. He concludes, "most of the heat incident on the molten surface is dissipated by radiation from the neighborhood of the molten zone, and the heat flow to the ends of the rod is slight in comparison. The very steep temperature gradient which undoubtedly exists at the solid-liquid interface may help to explain the comparative ease with which single crystals are produced. The exact form of the isotherms near the molten zone is of considerable practical interest, but is scarcely amenable to simple theoretical treatment."

ZONE SPEEDS. The speeds at which monocrystals of the high melting point metals can be grown are usually in excess of the speeds most desirable for zone refining (see Pfann, 1958). The speeds used represent a compromise between what zone refining practice requires and what is practical. Loss of the sample by volatilization reduces the recovery rate for zone speeds of 0.01 to 0.1 mm/min. In practice, zone speeds of from 1 to 4 mm/min are used. Table 1 gives the speeds used for specific metals.

STARTING MATERIAL. Almost any rodlike shape can be melted by the floating zone technique if the material is relatively dense and gas free. The stability of the molten zone is greatly influenced by mechanical disturbances, and hence by sudden changes in the volume of the liquid. The sudden escape of physically entrapped gas is a frequent source of trouble. The sudden release of gas can initiate glow discharges or act to momentarily short the heater to the specimen if molten metal is thrown from the zone by the gas. Both events lead to power loss to the sample and often to complete failure of the operation. As Sell (1960) has shown, excess gas in the specimen often becomes concentrated in the molten zone, but also can be left as gas pockets in the solid.

Belk (1960) and Geach and Jones (1959) have expressed a preference for the use of rods made by powder metallurgy because the specimen can be outgassed in the solid state. Powder metallurgy has been used to produce material for both metal and alloy crystals. Swaged rods made from cast or sintered bars have been a good source of pure metals. Control of gas content is important.

It is more difficult to get rods of a desired alloy than of a pure metal. In addition to powder metallurgy, two other methods have been used to produce alloy crystals. Lawley (1959) and Jones (1959) produced crystals of molybdenum-rhenium alloys by wrapping a molybdenum rod with a suitable quantity of rhenium wire, and carefully melting the two together. Spot welding of the wire to the rod at frequent intervals prevents unraveling. DeCarlo (1961) used 0.030-in. wires of pure niobium and tantalum packed into a bundle and twisted to produce a monocrystal.

Melting an alloy to produce crystals by floating zone melting (or by the Bridgman or Czochralski techniques) produces crystals whose compositions vary along their lengths. Most workers have tolerated such composition variations or minimized them by zone melting the crystals in two directions. Zone leveling (Pfann, 1958) has not been used extensively, although Oliver and Shafer (1960) have used it to introduce B into iron.

Simultaneous Crystal Growth and Purification

Floating zone melting supplies high purity crystals both because of the clean conditions under which the crystal is grown and because of zone refining. Since there is no crucible to act as a source of contamination and vacuum surrounds the zone, the liquid metal can divest itself of impurities with little chance for recontamination. Metallic impurities are removed by preferential volatilization, and interstitial impurities are removed by sacrificial volatilization or simply as a gas. Preferential volatilization is a possible means of purification if the following ratio is greater than one:

$$\frac{p_A^\circ \alpha_A}{p_B^\circ \alpha_B} > 1$$

where p_A° is the equilibrium vapor pressure of the solute element over the pure metal A, α_A is the thermodynamic activity of A in B, p_B° is the equilibrium vapor pressure of the solvent B over pure B, and α_B is the thermodynamic activity of B, all at the melting temperature and composition of the material. If this condition is satisfied, element A will leave the molten zone faster than element B and purification will result. In practice, only the equilibrium vapor pressures are usually known. The most reasonable assumption to make, to predict whether or not preferential volatilization will occur, is to assume Raoult's law (unless there is an intermetallic compound in the binary system). Smith (1958) gives a more detailed discussion of this means of purification during melting.

Interstitial impurities can sometimes be removed by sacrificial volatilization. Some metals tend to give up oxygen as the metal suboxide, and Smith (1958) lists Mo, Cb, B, W, Zr, and Hf as being among those likely to be purified in this way. Ti, V, Be, Cr, Fe, and Ni would not be so purified. Nitrogen may simply diffuse out when the partial pressure of N_2 in the system is low enough to reduce the solubility of N_2 in the liquid. Smith postulates that carbon is removed by the reaction:

$$\mathrm{MeO} + \mathrm{MeC} \rightleftarrows \mathrm{Z\,Me} + \mathrm{CO}\uparrow$$

and subsequent diffusion of CO out of the metal. At present there is not enough experimental data to assess the importance of these purification processes, but as proof of interest in this subject note the recent compilation of data by Gleiser (1961) on the "Free Energies of Formation of Gaseous Metal Sub-Oxides."

Purification by the zone refining mechanism is also important. Smith and Rutherford (1957) have determined the distribution of P in Fe using the P_{32} tracer. Oliver and Shafer (1960) determined the distribution of B in Fe. Wernick, Dorsi, and Byrnes (1959) used induction floating zone refining in a vacuum to determine the mechanism of purification of Ti, Ni, and V. They concluded on the basis of residual resistivity ratios (also used by Seraphim et al., 1960 and reviewed by Kunzler and Wernick, 1958) and chemical spectrographic analyses that Ni and Ti were purified by preferential volatilization, and that V was purified by both preferential volatilization and zone refining. Buehler (1958) found the same to be true for the purification of molybdenum, and Carlson (1959) for tungsten. Smith and Rutherford also showed that zone refining was important in the removal of Au_{98} and Zn_{65} from Ti. Using the eddy current technique of Bean, DeBlois, and Rodbell (1959) for measuring the resistivity ratios, Schadler (1959) showed that some purification of W occurred by zone refining. It is concluded that both volatilization and zone refining contribute to the purity of materials melted by the floating zone technique.

Determination of Crystal Orientation and Perfection

Techniques for determining whether a specimen is a single crystal and what its orientation is are well documented. This short section cites a few of the existing references. Etching is a simple way to determine whether a specimen is a single crystal. Kehl (1949) gives a list of etchant for the common metals. In the absence of a known etchant, a mixture of strong acids is often very useful. For the refractory metals a 1:1 mixture of HNO_3:HF is quite satisfactory. The Laue back-reflection X-ray technique (see Barrett, 1952) can also be used.

Determining the degree of perfection of a crystal is more difficult, unless a dislocation etch-pitting technique exists for the metal or alloy under consideration. For a list of etch-pitting techniques see Lovell and Wernick (1958) and Newkirk and Wernick (1961). The recent symposium on *Direct Observations of Imperfections in Crystals* (Newkirk and Wernick, 1961) also discusses X-ray techniques for observing dislocations.

The details of perfection and substructure which have been documented for monocrystals are too numerous to report here. The reader is referred to the aforementioned symposium and to articles like the one by Geach and Jones (1958 or 1959).

SUMMARY TABLE

Table 1 page 344, contains a list of high melting point metals and alloys from which monocrystals have been produced as well as the pertinent references, techniques used, and summaries of the growth conditions.

ACKNOWLEDGMENTS

The author wishes to thank his associates at the General Electric Research Laboratory and especially J. W. Rutter and V. J. DeCarlo for their discussion of this manuscript and for providing much of the operating data contained in Table 1.

Appendix A

EMISSION LIMITED OPERATION IN ELECTRON BOMBARDMENT

Since control of the power input in electron bombardment floating zone refining is usually accomplished by controlling the temperature of the cathode, optimum control will be experienced when the unit is operated in the emission limited rather than the space charge limited range. For any given arrangement of emitter, focusing plates, and sample, the voltage required to prevent space charge limitation can be determined experimentally by measuring the emission current as a function of applied voltage for various cathode temperatures. Figure A.1 illustrates data obtained for the emitter configuration shown in Fig. 5*b* and a 3 mm diameter anode. Since the space charge limited current is proportional to the three-half power of the voltage (see Spangenberg, 1948), the data taken at low voltages can be used to predict the theoretical space charge limited current at higher voltages. In Fig. A.1 this is the curve designated $I = 3.79 \times 10^{-4}\ V^{3/2}$.

The curves designated 11.0^{A} to 12.5^{A} are experimental data taken with the filament current indicated held fixed. If the filament current was the only source of power to the filament, these curves should be almost horizontal at high voltages because the emission current is now limited by the filament

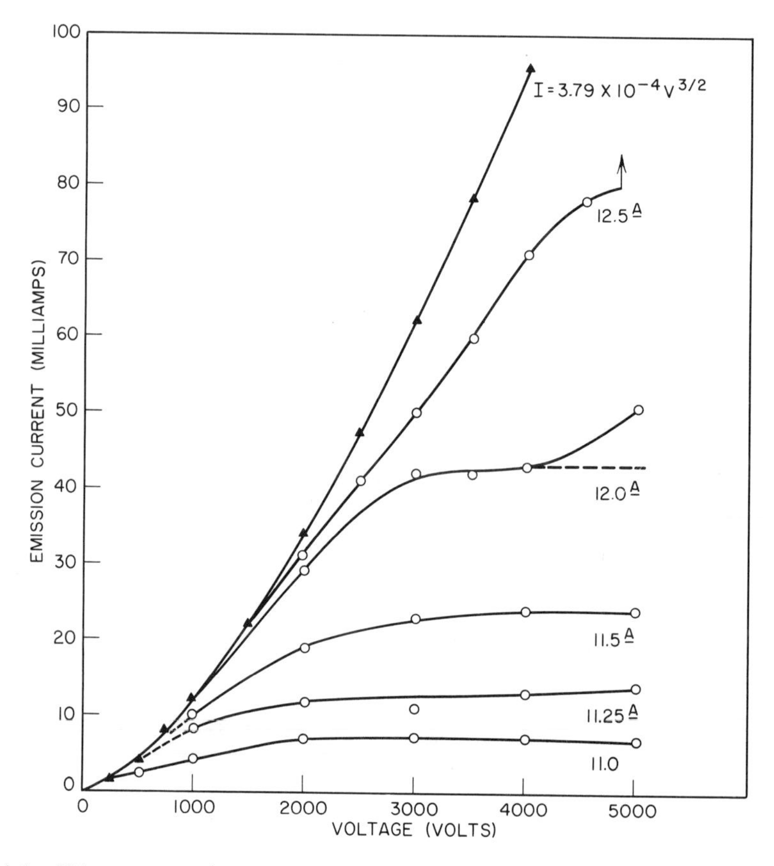

Fig. A.1. Voltage-current characteristics of the emitter configuration shown in Fig. 5*b*. Bombardment specimen was a $\frac{1}{8}$-in. diameter wood.

temperature. In fact, if the anode becomes hotter than the filament, the filament will be back heated by radiation. This is the reason that the curves designated 12.0^{A} and 12.5^{A} show the upward turn at high voltages.

The relation between emission current and applied voltage is useful to insure operation in the emission-limited range and hence maximum sensitivity of the control system. Once the power required to melt a sample is established by noting at what current and voltage a zone is established, the minimum voltage for emission-limited operation can be determined from such data. It should be added, however, that many good single crystals have been grown using emission control without attention to whether or not emission was really limiting the current.

References

Adams, M. (1961), Personal Communication.

Allen, B. C. (1961a), Personal Communication, crystals grown for R. W. Douglas, Thesis, *A Study of the Deformation and Fracture of Rhodium and Iridium,* Ohio State University, May 1961.

Allen, B. C. (1961b), to be published in *J. Inst. Met.* 1961.

Allred, W. P., R. C. Hines, and H. L. Goering (1959), *First Symposium on Electron Beam Melting,* March 20, 1959, Boston, Mass., or *Symposium on Electron Bombardment Techniques,* S.E.R.L. Labs., Baldock, Herts., England (March 1959), A. Calverly, editor.

Barrett, C. S. (1952), *Structure of Metals,* 2nd ed., McGraw-Hill Book Co., New York.

Bean, C. P., R. W. DeBlois, and L. B. Nesbitt (1959), *J. Appl. Phys.* **30,** 1976.

Belk, J. A. (1959), *J. Less Common Metals* **1,** 50.

Birbeck, F. E., and A. Calverly (1959), *J. Sci. Instr.* **36,** 460.

Broch, E. C. (1955), *Phys. Rev.* **100,** 1619.

Buehler, E. (1958), *Trans. Met. Soc. AIME* **212,** 694.

Butuzov, V. P., and V. V. Dobrovensky (1958), *First Conf. on Crystal Growth,* Consultants Bureau, p. 252.

Calverly, A. (1959), *Symposium on Electron Bombardment Techniques,* S.E.R.L. Labs., Baldock, Herts., England.

Calverly, A., M. Davis, and R. Lever (1957), *J. Sci. Instr.* **34,** 142. See also U.S. Patent No. 2,809,905 granted to M. Davis and R. Lever.

Carlson, R. G. (1959), *J. Electrochem. Soc.* **106,** 49.

Cech, R. E., and D. Turnbull (1959), *Trans. Met. Soc. AIME* **206,** 124.

Chalmers, B. (1952), *Modern Research Techniques in Physical Metallurgy,* ASM, Cleveland, Ohio, p. 170.

Cioffi, P. P., and O. L. Boothby (1939), *Phys. Rev.* **55,** 673.

Cioffi, P. P., H. J. Williams, and R. M. Bozorth (1937), *Phys. Rev.* **51,** 1009.

Darnell, F. J. (1958), *Trans. Met. Soc. AIME* **212,** 356.

DeCarlo, V. J. (1961), Personal Communication.

Eaton, N. F. (1960), *J. Less Common Metals* **2,** 104.

Emeis, R. (1954), *Z. Naturforsch.* **9A,** 67.

Geach, G. A., and F. O. Jones (1958), *Plansee Proceedings,* p. 77.

Geach, G. A., and F. O. Jones (1959), *J. Less Common Metals* **1,** 56.

Gleiser, Molly (1961), *Trans. Met. Soc. AIME* **221,** 300.

Gow, K. V., and B. Chalmers (1951), *Brit. J. Appl. Phys.* **2,** 300.

Green, G. W. (1961), *J. Sci. Instr.* **38,** 167.

Gruber, H. (1961), *Z. Metallk.* **52,** 291.

Hall, R. C. (1957), *Trans. Met. Soc. AIME* **209,** 1267.

Heywang, W. (1956), *Z. Naturforsch.* **11,** 238.

Heywang, W., and G. Ziegler (1954), *Z. Naturforsch.* **9A,** 561.

Hultgren, R., and M. H. Pakkala (1940), *J. Appl. Phys.* **11,** 643.

Jones, F. O. (1959), *Symposium on Electron Bombardment Techniques,* S.E.R.L. Labs., Baldock, Herts., England (March 1959), edited by A. Calverly.

Keck, P. H., and M. J. E. Golay (1953), *Phys. Rev.* **89,** 1297.

Keck, P. H., M. Green, and M. L. Polk (1953), *J. Appl. Phys.* **24,** 1479.

Kehl, G. L. (1949), *Metallographic Laboratory Practice,* McGraw-Hill Book Co., New York.

Kunzler, J. E., and J. H. Wernick (1958), *Trans. Met. Soc. AIME* **212,** 856.

Lawley, A. (1959), *Electronics* **32,** September 4, p. 39.

Lawley, A. (1961), *Introduction to Electron Beam Technology,* R. Bunshah, editor, John Wiley and Sons, New York.

Lovell, L. E., F. L. Vogel, and J. H. Wernick (1958), *Met. Progr.* **75,** 96 (May 1958).

Meissner, J. (1959), *Z. Metallk.* **50,** 207.

Menzel, E., and M. Otter (1959), *Naturwissen.* **46,** 66.

Oliver, B. F., and A. J. Shafer (1960), *Trans. Met. Soc. AIME* **218,** 194.

Orehotsky, J. L., and A. J. Opinsky (1959), *First Symposium on Electron Beam Melting,* March 20, 1959, Boston, Mass.

Pearson, R. F. (1953), *Brit. J. Appl. Phys.* **4,** 342.

Petrov, P. A., and Yu. M. Shashkov (1957), *Izvestiya Akademii Nauk SSR, Otdelenie Tekhnicheskikh Nauk SSR,* No. 5, 102 (May 1957). Henry Brutcher Translation No. 4074.

Pfann, W. G. (1958), *Zone Melting,* John Wiley and Sons, New York.

Pfann, W. G., and D. W. Hagelbarger (1956), *J. Appl. Phys.* **27,** 12.

Ray, A. E., and J. F. Smith (1958), *Acta Cryst.* **11,** 310.

Rhys, D. W. (1959), *Symposium on Electron Bombardment Techniques,* S.E.R.L. Labs., Baldock, Herts., England (March 1959), A. Calverly, ed.

Rocco, W. A., and G. W. Sears (1956), *J. Sci. Instr.* **27,** 1.

Rutter, J. W., and T. F. Sawyer (1961), Personal Communication.

Schadler, H. W. (1959), *First Symposium on Electron Beam Melting,* March 20, 1959, Boston, Mass.

Sell, H. G. (1960), *Physical Metallurgy of Tungsten and*

Tungsten Base Alloys, Westinghouse Electric Co., WADD Rept. No. 60-37, p. 28ff.

Seraphim, D. P., J. I. Budnick, and W. B. Ittner (1960), *Trans. Met. Soc. AIME* **218,** 527.

Sestak, B. (1957), *Ceskoslovensky Casopis pro Fysiku* **7,** No. 2, 202.

Seybolt, A. U., and J. E. Burke (1953), *Procedures in Experimental Metallurgy,* John Wiley and Sons, New York.

Silverman, S. J. (1961), *J. Electrochem. Soc.* **108,** 585.

Smith, H. R. (1958), *Vacuum Metallurgy,* R. Bunshah, editor, Reinhold Publishing Co., New York, p. 221.

Smith, R. L., and R. L. Rutherford (1957), *Trans. Met. Soc. AIME* **209,** 478.

Spangenberg, K. R. (1948), *Vacuum Tubes,* McGraw-Hill Book Co., New York, p. 176.

Takaki, H., S. Nakamura, Y. Nakamura, J. Hayashi, K. Fusakawa, and M. Aso (1954), *J. Phys. Soc. Japan* **9,** 204.

Tanenbaum, M. (1959), *Methods in Experimental Physics,* Vol. 6, Part A, K. Lark-Horowitz and V. A. Johnson, Academic Press, New York (1959), p. 86ff.

Walker, J. G., H. J. Williams, and R. M. Bozorth (1949), *Rev. Sci. Instr.* **20,** 947.

Wernick, J. H., D. Dorsi, and J. J. Byrnes (1959), *J. Electrochem. Soc.* **106,** 245.

Wernick, J. H., and J. B. Newkirk (1961), *Symposium on Direct Observations of Imperfections in Crystals,* Winter AIME Meeting 1961, to be published.

Witzke, W. R. (1959), *Trans. Vacuum Metallurgy Conference.* R. Bunshah, editor, New York University Press, New York, p. 140.

18

Semiconducting Compounds

W. D. Lawson and S. Nielsen

In this chapter methods of making crystals of semiconducting compounds (other than the III–V compounds) are discussed. Particular emphasis is on the chalcogenides, but mention is also made of other compounds. The chalcogenides are the compounds between metals and the chalcogens: O, S, Se, and Te. The discussion is restricted to compounds and alloys that can be crystallized from the melt.

Crystals can be grown from the melt by the Bridgman-Stockbarger technique, the Kyropoulos technique, zone melting, and the Czochralski (or pulling) technique. For compounds with a low vapor pressure at the melting point, these techniques are quite adequate. However, for compounds with relatively high vapor pressures some modification of apparatus is required. A further problem is that many compounds can show a marked deviation from stoichiometry. This affects the electronic and optical properties in the same way as impurities and defects do in an element. Growth of perfect crystals therefore requires attention to impurity defects, to structural or lattice defects, and to the stoichiometric ratio of the components.

We have therefore subdivided the chapter into sections describing the various stages in compound preparation and crystal growth, illustrating this with some typical examples. The first section deals with crucible materials, the second with purification of elements, the third with compound preparation, the fourth deals with crystal growth, and a final one with the control of stoichiometry.

1. CRUCIBLE MATERIALS

Many refractory materials (for example, oxides, sulfides, borides, carbides, nitrides, and fluorides) are potentially useful crucible materials. These include for example Al_2O_3, MgO, BeO, ZrO_2, ThO_2, CaO, CeS_2, SiC, TiC, BN, AlN, and CaF_2. However, they are not easily available in the high state of purity required. In practice only silica or carbon are used since they are available in the required purity and can be easily shaped.

Silica is now commercially available in a semiconductor grade of high purity. This is a synthetic material prepared by oxidizing a pure volatile halide or organic compound of silicon in a flame. The metallic impurities in the final silica are of the order of ppm, which is a major improvement over the silica made from selected natural quartz or sand that was previously used. The principal impurity is chemically combined hydrogen which is difficult to remove but is not serious for most applications. Silica is inert to many elements and compounds, notable exceptions being Mg, Ca, Ba, Sr, Al, Si, the rare earths, the many transition elements, many oxides, and fluorides. Reaction with such materials is progressively more rapid at higher temperatures. Silica is readily shaped and sealed in the same way as glass, using suitable oxypropane burners (Conaboy, 1960). It can be sealed to glass through commercial graded seals. Tubes can be made sufficiently accurate for the use of "O" ring joints (Conaboy, 1959). Silica can be used at temperatures up to

1200°C or slightly higher, although in air some devitrification occurs. Conventional tubing is vacuum tight and withstands pressures up to 25 atmospheres. Thicker walled tubing is available for higher pressures (for example, tubes $\frac{1}{2}$ in. I.D., $\frac{1}{8}$ in. wall thickness withstand about 100 atmospheres).

Silica as-received is usually contaminated by surface dust, grease, and inorganic chemicals. These must be removed by appropriate cleaning procedures. Dust and grease are probably best removed by brushing with detergent and hot water. Absorbed inorganic chemicals and adsorbed detergent can be removed by washing with acids containing HF, mixtures of HF with HNO_3, and HF with HN_4F, being quite effective. Traces of acid must then be removed by repeated washing with deionized water. Since water does not drain easily from the tube at this stage, it is usual to dry the tube by washing with a pure volatile organic solvent, followed by a vacuum or high temperature bake to remove all traces of organic solvent. A final recommended treatment is to heat the tube to the softening point in air until the silica is very transparent and free from striations. This leaves the surface very smooth, removing etch pits etc., that might otherwise act as nucleation sites during crystal growth. The advantage of a thorough cleaning process is illustrated in Fig. 1 showing photographs of two PbTe crystals. One was grown in a well-cleaned crucible and the other in an inadequately cleaned crucible.

Vycor and other borosilicate glasses with high silica content have lower softening temperatures than pure silica, thus making them easier to work. However, they cannot be recommended because they may introduce impurities, particularly boron.

Elements that cannot be reacted in silica can often be contained in graphite, of which very pure grades are available. Graphite is easily machined and can be conveniently heated by induction. It is inert to most molten metals, except carbide formers, notably Si, B, Al, and Fe. In oxygen-free atmospheres graphite can be used to temperatures of at least 2500°C. Since it is porous to gases*, it is usually used inside a silica or metal outer tube. After machining or shaping, it is advisable to clean the graphite by heating at 1500 to 2000°C in vacuo for several hours. This removes much of the surface contamination. More rigorous cleaning methods have been suggested, such as boiling in acids followed by repeated boiling in deionized water until no increase in the conductivity of the water can be detected or by heating in chlorine at high temperature. In both methods, a final high temperature vacuum bake is advisable. The latter alone is usually adequate.

Silica has also been successfully protected with a very thin carbon layer. Two techniques are normally used. Carbon can be deposited pyrolytically onto silica to form a

* A strong, light, pure, nonporous form of "vitreous carbon" has recently been developed by the Plessey Co., Caswell, England.

Fig. 1. PbTe crystals. *Left*: grown in thoroughly cleaned crucible. *Right*: grown in inadequately cleaned crucible.

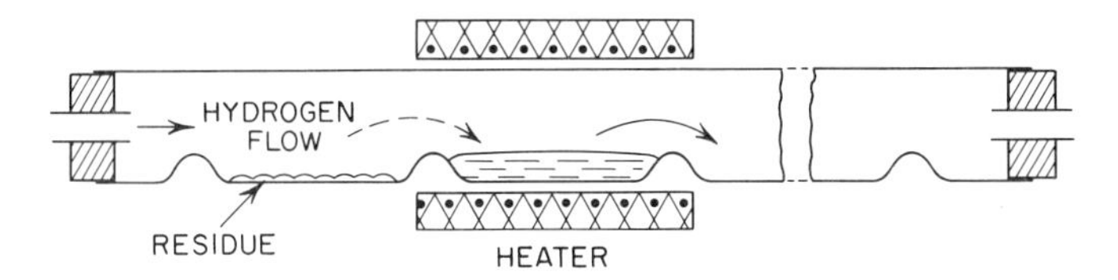

Fig. 2. Metal distillation apparatus.

dense mirrorlike and firmly bonded coating. This is conveniently done by washing a clean tube with an organic solvent such as acetone or methanol and heating rapidly to over 700°C. Inside the tube there is insufficient oxygen to burn the solvent on the walls, and it is pyrolyzed to carbon. This can be repeated to give a thicker coating. Carbon can also be deposited as a loose film by burning benzene with a smoky flame near the object to be coated. Use of very pure solvent is recommended.

2. PURIFICATION OF THE ELEMENTS

Most elements are now available commercially in purities greater than 99.99%. If necessary, further purification can be attempted by zone melting, crystal growth, distillation or vaporization, intermediate compound formation, or direct chemical treatment.

The elements Cu, Ag, Au, Mg, Zn, Cd, Al, In, Sn, Pb, Ge, Si, Bi, Te, for example, can be purified to varying extents by zone melting. This has not been effective for S and Se or, in general, for any material whose liquid structure differs little from that of the solid. The technique is only moderately effective for the low melting elements Cd, Pb, and Sn and especially for Ga which shows such pronounced supercooling. For these, slow crystal growth may be more satisfactory. Not all impurities can be removed by zone melting. This follows if the distribution coefficient k is near unity (that is, if the impurity is about equally soluble in the solid and liquid phases), or if vapor transfer of impurities from the molten zone back to the solid occurs (Hulme and Mullin, 1957), or if the presence of a second phase such as surface dross or scum causes back transfer.

Many elements can be purified by distillation, depending on relative vapor pressures. Volatile impurities (such as Zn, Cd, etc.) can be removed from Ge, Bi, Sn by high temperature vacuum-baking. Baking for long periods at about 10^{-5} Torr may lead to some oxidation of the element and this should be removed either by heating in hydrogen at an appropriate temperature or as described in the following. Elements having reasonable vapor pressures at temperatures up to 1200°C can be distilled in quartz tubes of the type shown in Fig. 2. The distillation may be carried out in vacuo or often with advantage in a low pressure of hydrogen introduced through a Pd or Ni leak. Hg, S, Se, Te can be distilled in conventional apparatus.

There are many possibilities for purification by intermediate compound formation. In this method the element is converted to a compound that is more easily purified than the element itself, and then the element is reformed. Typically the element is converted to the halide or hydride which is decomposed after purification. Halide and hydride intermediates are usually more satisfactory than more complex compounds (for example, carbonyls, etc.) because the latter may introduce impurities in the decomposition stage. Also, traces of halide or hydride remaining in the purified element can usually be removed by a high temperature vacuum-bake. As and Se (Nielsen and Heritage, 1959) are among the elements that have been purified via the hydride intermediate.

Purification by direct chemical action can be a convenient first step for elements with low melting points. The element is melted in an acid or molten flux. Purification of Hg is best achieved by shaking with dilute nitric acid, washing with water, and finally slowly distilling (to avoid splash-over) in quartz apparatus. Purification to a few parts in 10^8 has been estimated for this procedure (Lawson et al., 1959). Impurities have been removed from sulfur (Murphy et al., 1960) by oxidation in a H_2SO_4—HNO_3 mixture, followed by distillation.

3. PREPARATION OF COMPOUNDS

The most convenient and satisfactory method for making intermetallic compounds is by simultaneously fusing together appropriate amounts of the constituent elements. It is not usually necessary to weigh the elements accurately, for any excess of either element can be removed at a later stage or during crystal growth. After putting the element in a silica tube, the tube is evacuated and sealed (Fig. 3). The tube is then heated sufficiently to allow complete interaction. Homogenization is helped by agitation or stirring (the authors have rocked the furnace at about 10 cycles/min with good results). When the reacting elements have a high vapor pressure, such as in HgS, HgSe, etc., it is advisable to increase the furnace temperature very slowly, and to allow considerable time for reaction, taking at least a day to bring the temperature up to the melting point of the compound. This avoids an excessively high pressure rise in the silica tube, and the long time insures that pockets of uncombined elements are eliminated. Zone levelling (Pfann, 1958) is also a convenient method of insuring complete reaction and a homogeneous ingot. Alternatively, a hot zone can be passed along the crucible repeatedly, starting with a low temperature and increasing at each pass to the melting point of the compound. If one of the elements is volatile, the temperature increase should be slow, and if the compound has a high vapor pressure at its melting point an ambient heater should be used as in Fig. 4. The compound can be zone melted in the same apparatus.

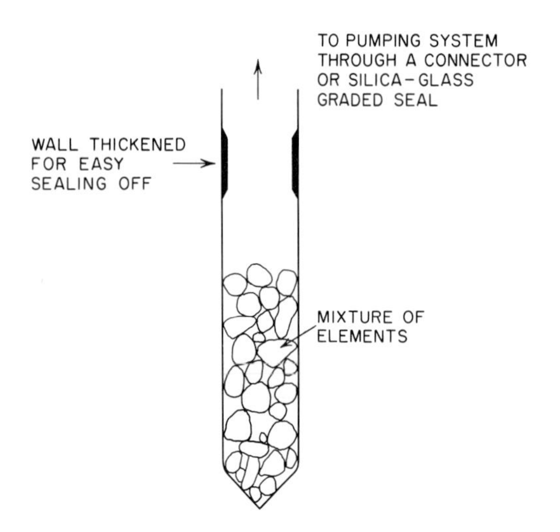

Fig. 3. Silica crucible for reacting elements.

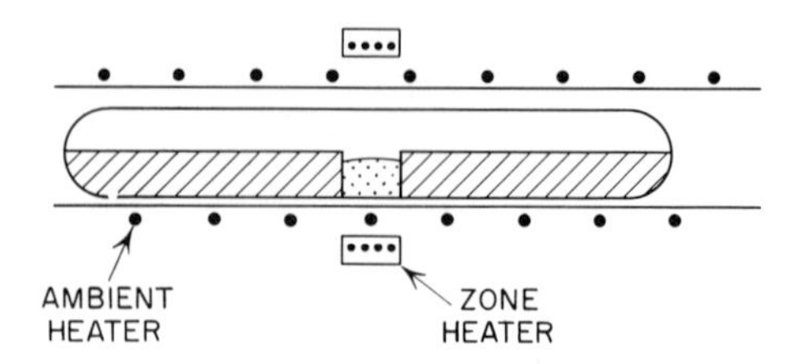

Fig. 4. Zone melting apparatus with ambient heater.

Where reaction with silica is likely to occur, the elements can be reacted in carbon-coated silica or in carbon tubes or boats. The boats, together with an inert gas or hydrogen, can be sealed into silica tubes. Subsequent operations can be carried out under pressures up to a few atmospheres, depending on the initial pressure and the temperatures in the system. Greater pressures can be generated if the far end of the reaction tube is cooled (for example, with liquid nitrogen) while the gas is being sealed in.

It is important to introduce the elements into the reaction tube with minimum contamination. Obviously, clean conditions and a dust free atmosphere are essential. Soft metals such as In, Pb, Sn, and Tl, may be cut with a sharp knife or secateurs and given a bright polish etch. Suitable etches for metals are to be found in various metallurgical handbooks (Smithells, 1955; Lyman, 1948). Pieces can be weighed and transferred to the reaction tube with silica, stainless steel, or polythene forceps in the cleanest conditions available. Brittle materials such as Te, Ge, and Sb can be broken into smaller pieces with an agate pestle and mortar and the largest convenient pieces weighed and transferred to the reaction tube. Metals such as Ag, Cu, Fe, etc., must be sawed and etched before weighing.

There are two obvious sources of impurity contamination using this technique, apart from accidental pick-up of impurities. The first is that etched materials invariably have traces of the etchant on the surface and the second is that many metals quickly form a surface layer of oxide. These problems can be overcome by alternative methods of introducing the elements into the reaction tube. One useful technique is the Y-tube method (Lawson, 1952) (Fig. 5). The procedure is as follows. The Y-tube is connected, through side tubes on the arms of the Y, to a supply of pure, dry hydrogen at one side and to a vacuum system or hydrogen burner at the other side. The arms of the Y are open-ended at first and appropriate amounts of the elements are put into the two arms which are then sealed off. The tube is next pumped out and the reaction crucible heated for thorough outgassing. The hydrogen flow is then started and the elements melted in. Any unreduced oxides tend to remain behind, and some impurities are flushed out of the system. Finally, the hydrogen stream is stopped and the reaction tube is evacuated and sealed. This method of preparation has been very successful in making selenides and tellurides of the low melting metals Pb, Sn, Tl, Bi, Sb, In.

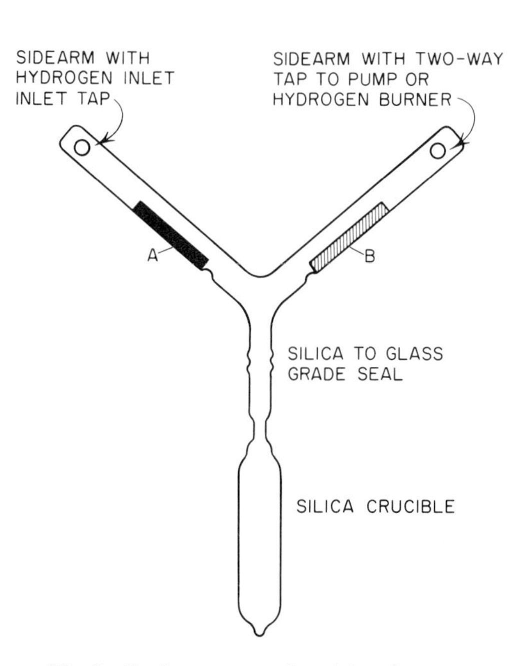

Fig. 5. Y-tube apparatus for mixing elements.

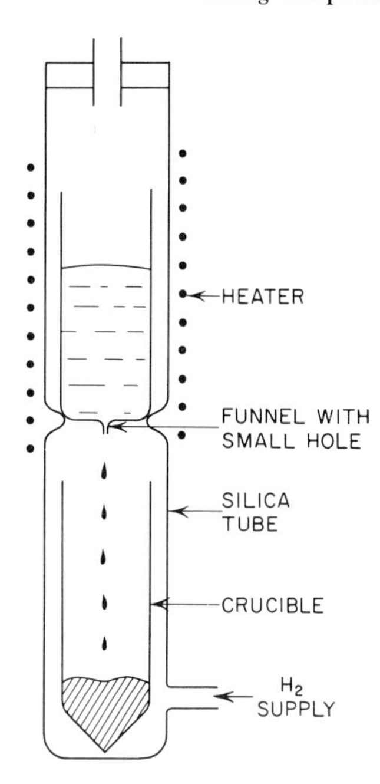

Fig. 6. Filtering funnel for removing scum.

Another useful apparatus is a funnel with a very small aperture (Fig. 6). The metal is melted in hydrogen in the silica funnel until it drops through the small hole leaving dross and scum behind.

Other methods involve distillation or some other melting procedure. Figure 7 illustrates one in which rods of zone-refined elements *A* and *B* are cleaned by melting in hydrogen and then mixed by rotating the crucible. Finally, the compound can be zone melted in hydrogen or in vacuo. Stoichiometric excess introduced in any of

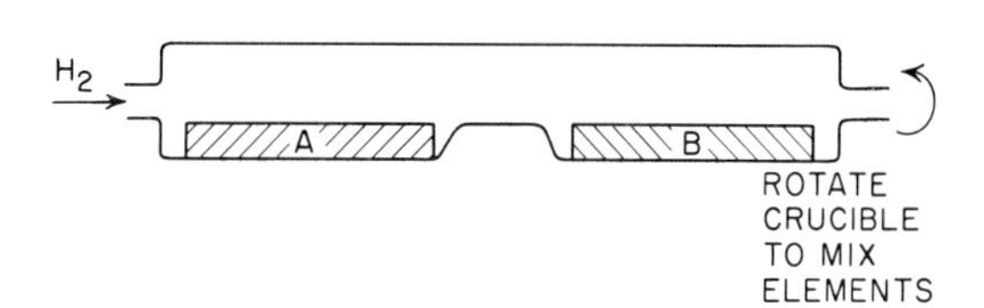

Fig. 7. Reaction tube.

these melting-in methods would be removed during zone melting of the compound or subsequent crystal growth.

When one element is volatile (for example, in Mg_2Sn), the method illustrated in Fig. 8 can be used. The elements are placed in the crucibles as shown, in this case in graphite boats inside a silica tube. The Sn is cleaned in the hydrogen stream by heating to about 1000°C, and the Mg is then distilled over by heating it to about 900°C. If the Sn is held at about 400°C, the formation of Mg_2Sn takes place. The tube can then be sealed at X and Y for subsequent refining or crystal growth.

These methods are most suitable if it is intended to purify further the compound in the reaction tube or to grow a crystal in the reaction tube by the Stockbarger technique. When crystal pulling is to be the succeeding step, it may be more convenient to melt the elements together in an open crucible that can then be used for crystal growth. Then the elements should be introduced as large pieces to minimize surface contamination. However, the elements can also be put into the crucible in two stages: in a first stage the metal is introduced and baked at a high temperature in vacuo or in hydrogen to remove volatile impurities and oxide, and in the second stage the more volatile element is melted in. As surface scum is most objectionable when pulling from the melt this two-stage method is often advisable.

Ternary compounds and ternary alloys, which have come into prominence recently because of their potential thermoelectric applications, can be prepared much the same as the binary compounds. It may, however, be advisable to make them from the binary compounds rather than from the elements. The binary compounds can be purified by zone refining or crystal growth, putting in

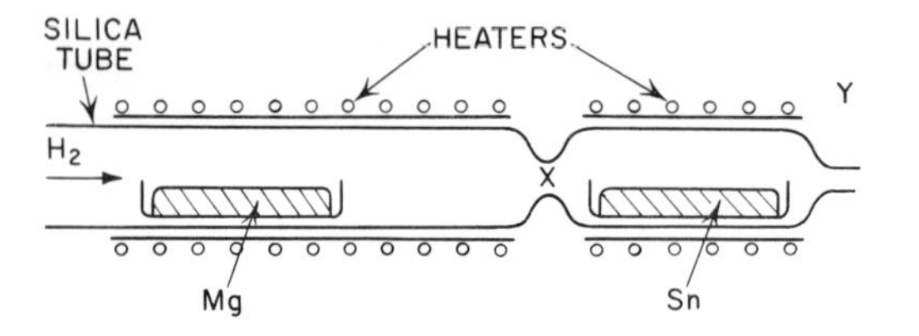

Fig. 8. Formation of Mg_2Sn.

another purification step. We have also found that ternary alloys prepared from compounds are often more homogeneous than those prepared from the elements, and it is possible that mixing is easier in this way. In some cases the possibility of explosions is reduced. For example, when Hg., Cd, and Te are reacted, the exothermic reaction of Cd with Te can heat the uncombined Hg, causing high pressure (Lawson et al., 1959).

Methods other than by direct fusion are also available. For instance, selenides and sulfides can be prepared by passing H_2S or H_2Se over hot or molten metal. A disadvantage arises if the compound has a high melting point because a solid crust of compound then forms on the surface. This can be overcome to some extent by agitating the metal. This method can be useful with very high-melting metals (for example, Cr and Mn), and if the metal can be obtained in the form of powder or flakes reaction can be virtually completed.

Chemical reaction between solutions or solutions and gases can be used. PbS can be easily prepared by bubbling H_2S through a solution of lead nitrate (or acetate), filtering, and drying. Care is required in the drying process to prevent oxidation. Selenides of Zn, Cd, Cu, Pb, and Hg have been made by the reduction of selenites in solution with hydrazine (Benzing et al., 1958). This type of process produces the compound in a finely divided state which may be advantageous for some purposes, such as luminescence screens, but is generally unsuitable for crystal growth because of oxidation or adsorption on the large surface area. Also traces of the reagents are likely to be included. An advantage is that the products are likely to be stoichiometric. Chemical methods for the preparation of SrSe and CaSe have also been described, but the purity achieved was not high.

4. CRYSTAL GROWTH

Review of Methods

For compounds and alloys that can be prepared in sealed silica tubes, crystal growth by the Stockbarger method (1936)

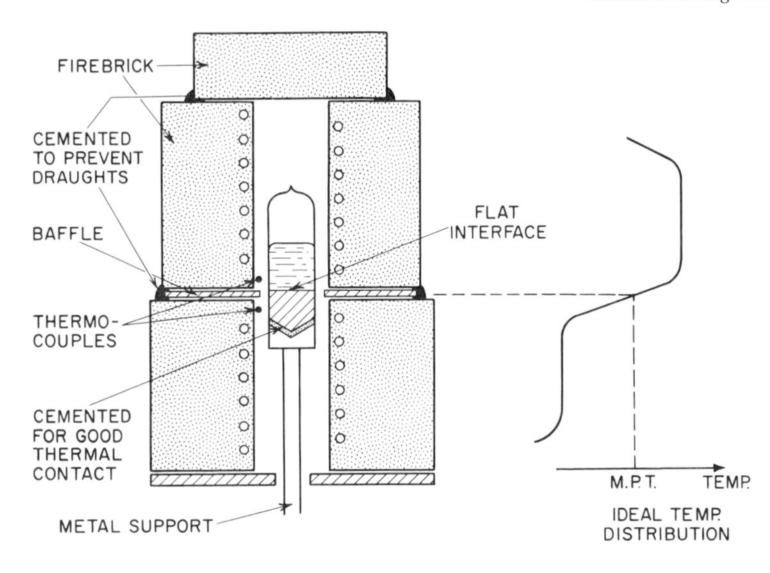

Fig. 9. Crystal growth by Stockbarger's method.

is very convenient. The silica tube is dropped vertically through a freezing gradient (Fig. 9). The upper furnace is maintained at about 50°C above the melting point of the compound and the lower furnace about 50°C below the melting point. The tube usually has a pointed bottom that enters the freezing zone first. This encourages crystal growth to start at a single site. Pear-shaped or conical ends seem to be the most suitable. Materials that do not react with silica, but melt above its softening point, 1200°C, and have a vapor pressure greater than 1 atmosphere, can be grown in silica crucibles if the latter are enclosed by close-fitting ceramic or carbon containers as in Fig. 10. The silica crucible takes up the shape of the container and forms a gas-tight seal.

Tipped crucibles of many refractories are available for materials that are impossible to prepare in silica or carbon.

The temperature conditions of Fig. 9 can best be obtained by using two furnaces separated by a baffle. Two heaters wound on the same tube usually give less steep temperature gradients. An auxiliary heater, or increased turns of the winding, at the lower end of the upper furnace can be used to increase the gradient. A temperature gradient of about 25°C per cm is usually adequate, but in special cases a sharper gradient may be advisable, for example, when constitutional supercooling (Rutter and Chalmers, 1953) is likely to occur.

Various methods have been proposed for supporting the crucible and lowering it through the temperature gradient. The writers prefer to support the crucible on a metal rod driven by a lead screw mechanism because this is mechanically stable and gives a regular, easily controlled growth rate. Dropping rates varying from 0.1 to 10 cm/hr have been used, the lower rates usually being regarded as advantageous for better crystals, although not always essential. Rates around 1 cm/hr are most commonly used.

An additional advantage of support on a metal rather than a ceramic rod or suspen-

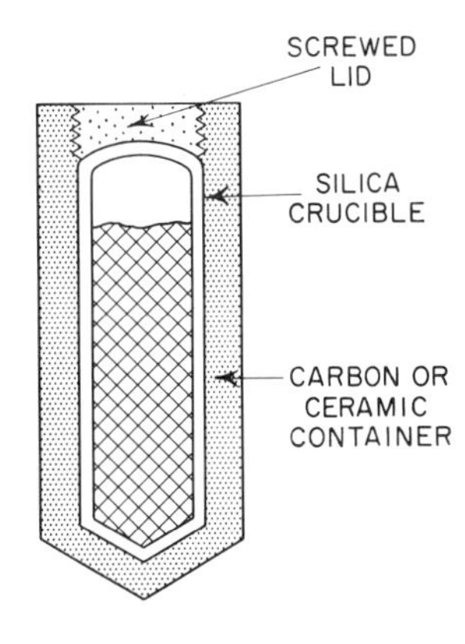

Fig. 10. Sealed silica crucible in close-fitting ceramic or carbon sheath.

sion of the crucible from above is that heat flow is encouraged vertically downwards through the body of the crystal. The isotherms within the furnace, including the charge, are then more likely to remain horizontal, thus maintaining a flat, horizontal solid-liquid interface. If the solid-liquid interface becomes concave (for example, when the dropping rate is too rapid), the resultant sideways cooling can cause spurious nucleation. A polycrystalline ingot as in Fig. 11 usually results.

If the furnace conditions are uniform across a horizontal section, there is no advantage in rotating the crucible and this should be avoided. Vibration and other mechanical disturbances should be minimized during crystal growth.

The power input to the furnaces should be stabilized. For many purposes this gives adequate temperature control. Input power variations will raise or lower the temperatures of both furnaces giving effective variations in growth rate. These can lead to crystallographic imperfections and inhomogeneities of composition, particularly in the growth of mixed or doped crystals. Thus for more perfect crystals close control of temperature is essential.

Instead of using two separate furnaces or windings, the crucible can be dropped through the natural temperature gradient of a single furnace as in Fig. 12. This is the basis of the Bridgman method. Its main disadvantage is that the crystal cools in a nonuniform temperature region. Nevertheless, the method suffices for many preparations where crystal perfection is not of prime importance. Alternatively, the freezing gradient can be made to move up the charge simply by allowing the furnace to cool. This method requires no mechanical movement, but the rate of growth and particularly the temperature gradient are more difficult to control. Nevertheless, the technique is

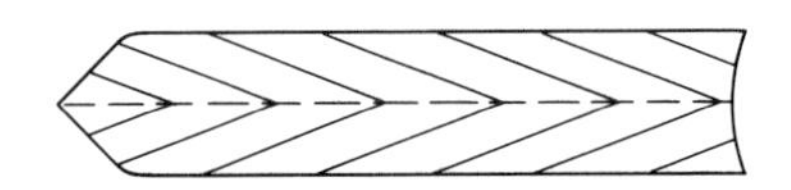

Fig. 11. Polycrystalline ingot, effect of sideways cooling.

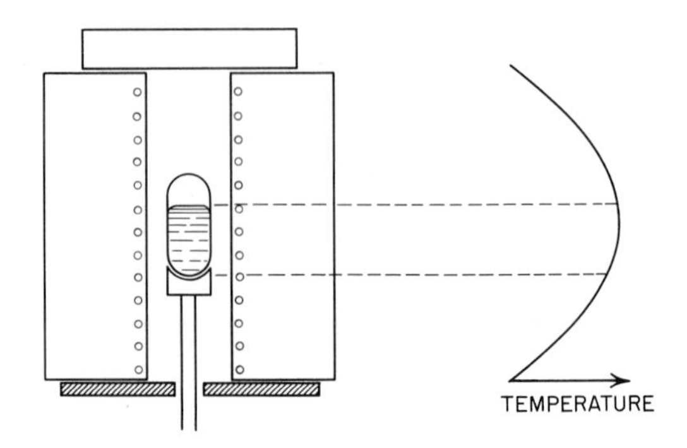

Fig. 12. Bridgman method of crystal growth.

useful, especially for growth within a pressure chamber where mechanical movement may be difficult.

Crystals of many compound semiconductors have been grown by these methods, including PbS, PbSe, PbTe, SnS, ZnSe, ZnTe, CdSe, CdTe, Bi_2Te_3, Mg_2Sn etc.

In the Stockbarger and other methods described, crystal growth starts from the bottom of the melt. It is therefore most successful with materials whose solid-liquid density ratios are greater than one. When this ratio is less than one, there is a tendency for crystal nuclei to float to the top of the melt. One objection to the method is that considerable constraint may be put on the crystal by the crucible walls, particularly if adhesion or keying occurs. The low coefficient of expansion enhances the value of silica crucibles in this respect since the crystal shrinks away from the wall. However, less constraint is placed on the crystal if a horizontal method is used. For example, the two furnaces of Fig. 9 can be arranged horizontally and a tube or boat drawn horizontally through the freezing gradient.

The zone melting technique has been widely used and is capable of producing excellent crystals (Bennett and Sawyer, 1956). A molten zone traverses the ingot from the seed crystal (Fig. 13) which can be oriented. This method also permits some control of stoichiometry as discussed in the next section.

Perhaps the most perfect semiconductor crystals have been grown by pulling from the melt. The apparatus has been improved, refined, automated to an extreme degree, and is available in many forms commercially. Crystals of many of the compounds discussed in this chapter can be made by pulling. For compounds with a relatively low vapor pressure (for example, Bi_2Te_3), the crystals can be pulled directly from the melt in a protective gas at atmospheric pressure. Apparatus for pulling crystals has been described in other chapters. A typical problem is unwanted nucleation from scum of the melt. This usually consists of traces of oxide; an apparatus has been designed to wipe the scum from the melt prior to crystal pulling (Allred, Mefferd, and Willardson, 1960).

Crystals of compounds with a high vapor pressure may also be prepared by pulling if it is done in a completely sealed system. Apparatus has been designed to operate at pressures up to about 20 atmospheres either by enclosing the crucible and pull rod in a silica envelope and arranging to use a magnetic coupling to drive the pull rod (Gremmelmaier, 1956) or by using a piston system where the pressure is maintained between two close fitting silica or graphite pistons moving inside a silica tube (Moody and Kolm, 1958). This latter type of system was devised for crystals of GaAs, but could be applied to many chalcogenides.

Little has been reported so far about the effect of growth conditions on the properties or crystallographic perfection of the compound crystals covered in this chapter. Recently however, Cosgrave et al. (1958) have reported the effects of freezing conditions on the thermoelectric properties of $BiSbTe_3$ crystals doped with 0.56% Se. Crystals were grown at rates varying from about 0.01 to 5 cm/hr in temperature gradients of 25°C/cm, 50°C/cm, and 250°C/cm. The steepest gradient was achieved by mounting a vertical tube furnace directly over a circulating water tank and maintaining a high temperature close to the surface of the water with an auxiliary heater.

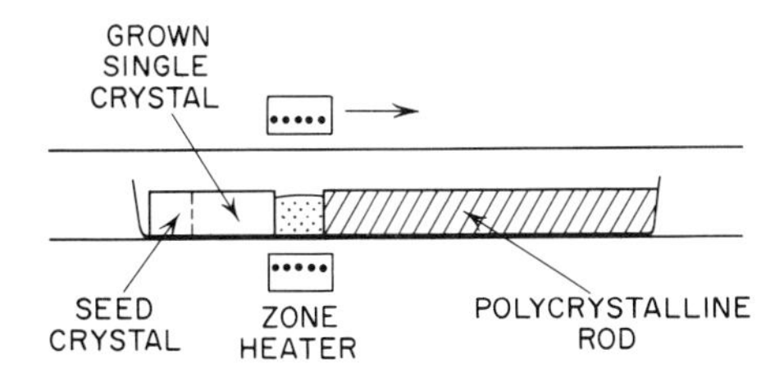

Fig. 13. Crystal growth by zone melting.

As might be expected crystals grown at slower rates in high temperature gradients were more perfect and exhibited lower resistivities; they also had higher thermoelectric powers. Specimens grown at higher rates and lower gradients were often not monocrystalline and contained inhomogeneities.

Scanlon (1957) compared the etch pit patterns on natural PbS crystals with those on synthetic crystals grown by the Stockbarger method. He found that the pattern on the natural crystals was much more regular, etch pits being arranged mainly in parallel and perpendicular lines. The pattern on synthetic crystals was more or less random and any alignment of etch pits that occurred was along irregular curves. This seems to indicate a greater degree of crystalline perfection in the natural crystals. Measurements of carrier lifetime appeared to substantiate this conclusion.

PbSe crystals grown from the melt also show random etch patterns, whereas the etch pits on crystals grown from the vapor (Coates, Lawson, and Prior, 1960) are arranged in regular arrays as for natural PbS crystals. Subsequent annealing of the melt-grown crystals can polygonize the random arrangement to some extent into ordered lines.

Growth of Specific Compounds

PbSe and PbTe (Lawson, 1951) have been prepared in silica crucibles in the Y-tube apparatus of Fig. 5 followed by crystal growth in a Stockbarger-type furnace. Yamade (1960) has prepared CdTe by first purifying the elements, by distillation, and then growing the crystals in a sealed silica tube crucible drawn to a capillary at the bottom to select only one seed. The crucible was then suspended in the upper furnace of a Stockbarger arrangement, heated to 1150°C for 3 hr, then lowered through the freezing plane at 2 cm/hr. Crystals

10 mm in diameter and 50 mm long resulted. Kroger and de Nobel (1955) grew crystals of the same compound by a horizontal technique, pulling a carbon-coated silica boat containing CdTe through a sharp temperature gradient at 1 cm/hr. A seed crystal was used. The stoichiometry of the crystal was controlled by maintaining a vapor pressure of Cd over the boat from an accurately controlled reservoir of Cd. CdTe has also been grown by Lawson et al. (1959) by zone melting in a high ambient temperature. The elements were first purified by repeated distillation along a silica tube in a stream of hydrogen, as in Fig. 2, and Te was further purified by zone-refining. Hall and resistivity measurements on the Te indicated high purity. The elements were then reacted in a sealed evacuated silica crucible, and the compound zone-refined in a high ambient temperature. Crystals were grown in pointed silica crucibles in the zone refining apparatus using a zone rate of 1 cm/hr. Crystals of HgTe and of the mixed compounds HgTe-CdTe were also grown. Mercury was first purified by shaking to an emulsion for several minutes with dilute nitric acid to remove base impurities and filtering, followed by "quiet" distillation in a silica apparatus. Radiotracer checks showed this purification process to be very effective, reducing the concentration of most impurities to below 0.1 ppm. HgTe was prepared by reacting the elements in thick-walled sealed silica crucibles of 10 mm bore, wall thickness 3 mm, capable of withstanding internal pressures of about 150 atmospheres. The compound was zone-refined in the same crucible using an ambient heater to maintain the solid within 50°C of the melting point to prevent undue loss of material from the molten zone. The mixed crystals CdTe-HgTe were prepared by melting together in thick-walled silica crucibles the appropriate weights of the compounds rather than directly from the elements. The latter method tended to result in explosions which were probably due to high pressures from uncombined mercury.

Bismuth telluride and related compounds have been grown by a variety of methods. Ainsworth (1956) used Stockbarger-type furnaces in both vertical and horizontal arrangements, and also tried pulling crystals from the melt in a stream of hydrogen at a pressure just above atmospheric to prevent evaporation of Te. He zone-refined the elements initially and prepared the compound in silica crucibles. In the first two methods, segregation of Te to the end of the crystal occurred. In the pulling method, the melt composition was varied from 5% excess Te to 2% excess Bi, but all the crystals turned out *p*-type with high thermoelectric power.

Satterthwaite and Ure (1957) grew the same compound in a Stockbarger apparatus using Vycor crucibles drawn down to a narrow tube at the bottom for seed selection. The first gram of an 80-gram ingot to solidify was used for measurements, and thus a uniform composition was achieved. From results of electrical measurements the phase diagram was constructed around the composition Bi_2Te_3, making the assumption that each excess Bi atom gave a singly ionized acceptor and each excess Te atom a singly ionized donor.

Harman et al. (1958) have also prepared Bi_2Te_3, Sb_2Te_3, and As_2Te_3 by a variety of methods. Te was first purified by distillation in a hydrogen stream onto a cold finger, Bi and Sb by zone melting, and As by sublimation in a hydrogen stream or into molten lead to remove sulfur that is not removed by vacuum sublimation. Bi_2Te_3 crystals grown by the Stockbarger method again showed segregation of Te. Crystals were also grown by zone melting with the ingot maintained at a high ambient temperature. As_2Te_3 was grown by a method of vertical zone melting in a closed capsule. In this way vapor phase transfer of impurities or stoichiometric excesses were avoided.

The compounds ZnS, ZnSe, ZnTe and CdS, CdSe form an interesting group of materials with considerable technological importance because of their useful luminescent and photosensitive properties. Several methods for growing crystals of these compounds have been reported, including growth from the vapor described in another chapter. They cannot be melted at ordinary pressures because they dissociate, but techniques for growing the crystals from the melt under high pressure have been devised. Fischer (1959) describes a very simple method where he melted ZnSe in sealed silica crucibles within close-fitting thick-walled graphite containers as in Fig. 10. The melting point of ZnSe was previously determined to be 1515° ± 20°C and the vapor pressure at the melting point about 1.8 atmospheres. Silica is softened at 1500°C but is not attacked by ZnSe and the dissociation pressure simply molds the silica tube to the graphite container which then takes the stress. Crystals of ZnSe were then grown by cooling the assembly in a temperature gradient. The method was also suitable for CdS which has a dissociation pressure of about 2.5 atmospheres at the melting point, 1475° ± 15°C. Fischer also grew crystals of ZnS and mixtures containing ZnS in an autoclave, the material being contained in zirconia crucibles sealed into silica tubes. The pressure within the autoclave was adjusted to slightly higher than the decomposition pressure of the material in the crucible so that the silica tube collapsed around the zirconia crucible. Silica, although soft at these temperatures, still formed a relatively gas-tight lining, enabling the compounds to be melted in well-defined atmospheres.

Heinz and Banks (1956) describe a pressure vessel in which they grew crystals of CdSe from the melt under a pressure of argon. The compound was melted in a graphite crucible, heated with a molybdenum coil heater, the whole assembly being contained within a thick steel pressure vessel fed with argon from a cylinder. Crystals were grown by allowing the crucible to cool slowly from the bottom, and a narrow seed selection hole was drilled in the crucible to prevent random seeding. An argon pressure of 100 atmospheres was maintained within the chamber. Although the high pressure apparatus is not essential for the growth of CdSe from the melt, it is suitable for melting many of the compounds that have high dissociation pressures at the melting point.

Medcalf and Fahrig (1958) also describe a high pressure furnace in which they grew CdS crystals from the melt under argon pressures around 100 atmospheres. In their case the heater was a multiple split graphite cylinder and they melted charges of about 150 grams of CdS in a high purity graphite crucible fitted with a graphite lid. The melt was solidified from the bottom in the natural gradient of the heater by turning down the power. Crystals were also grown under argon pressures of about 17 atmospheres, but these contained more voids and noncrystalline regions than crystals grown under higher pressures. Also loss of material from the crucible was greater, 25% as compared with 5%. Concentration profiles of impurities in slowly grown crystals indicated that Si, Ca, Cu, and Pb has segregation coefficients less than 1.0; Zn, Mn, and In were not segregated, and Mg was found to have a segregation coefficient greater than 1.0. In later experiments crystals were grown from Cd and S which had been purified both by distillation and zone refining. Cd was distilled three times in a cast iron retort. This reduced the concentrations of Zn and Pb to about 20 ppm, which was subsequently reduced below 1 ppm by zone refining. After zone melting a 1 kgm ingot six times in a graphite boat, the impurity content along half the ingot could not be detected spectrographically. Sulfur was distilled four times in a clear silica apparatus. This reduced Fe, Zn, Al, and Cu to near the limits of spectrographic detection, but Si, Ca and Mg were still present. Removal of Si and Ca by a modified form of zone melting was claimed. CdS was then formed from these purified elements by reaction of their vapors in helium carrier gas inside a silica reaction chamber at 900°C.

Many other compound semiconductors and solid solutions have been grown by these methods (see references). Also, a very wide range of compounds composed of three or more elements have been prepared, but not always as monocrystals. Some idea of the variety of compositions possible is given by Wernick and Wolfe (1960), who list 58 three-element compounds with general formulas (1) ABC_2, where A = Cu, Ag, Zn, Cd, Hg; B = Fe, Al, Ga, In, Si, Ge, Sn, Sb, Bi, Te; C = S, Se, Te, P, As, (2) A_3BC_4, where

A = Cu; B = As, Sb; C = S, Se, and (3) AB_2C_4, where A = Pb; B = As, Sb; C = S. Other possibilities are given in the last six references. The interest in these more complex compounds arises from their possible value as thermoelectric materials; such compounds may have good electrical conductivity and at the same time have poor thermal conductivity.

5. CONTROL OF STOICHIOMETRY

A compound is stoichiometric when the constituent atoms are present in a simple ratio 1:1, 1:2, 2:3, etc. Some semiconducting compounds appear to crystallize from the melt with an exact stoichiometric ratio of atoms as far as can be determined by present measurements. InSb is a typical example. Other compounds (for example, PbSe) can exist as a single phase over a range of composition. This deviation from stoichiometry affects the electrical properties of the compound in a similar way to impurities. In PbSe, for example, excess Pb acts as a donor impurity giving n-type material, and excess Se gives p-type material. It will be realized that the problem of stoichiometry is only a part of the wider problem of the incorporation of defects (vacancies, etc.) into the growing crystal (Lawson and Nielsen, 1958). However, we restrict ourselves here to stoichiometry and to methods designed to offer some degree of control.

Figure 14 shows a phase diagram for

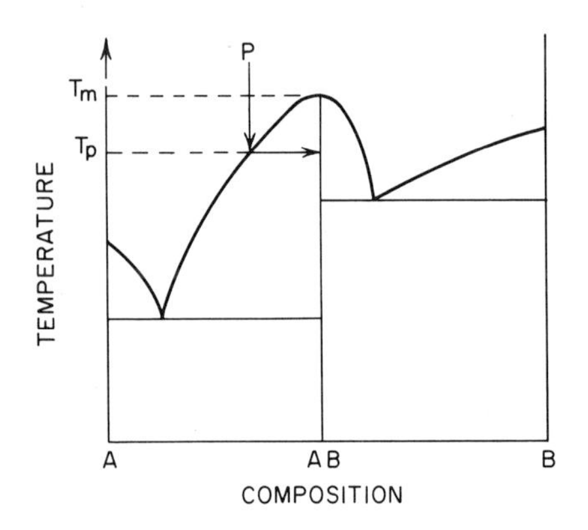

Fig. 14. Phase diagram for system A-B where compound AB solidifies from the melt in exact stoichiometric ratio only.

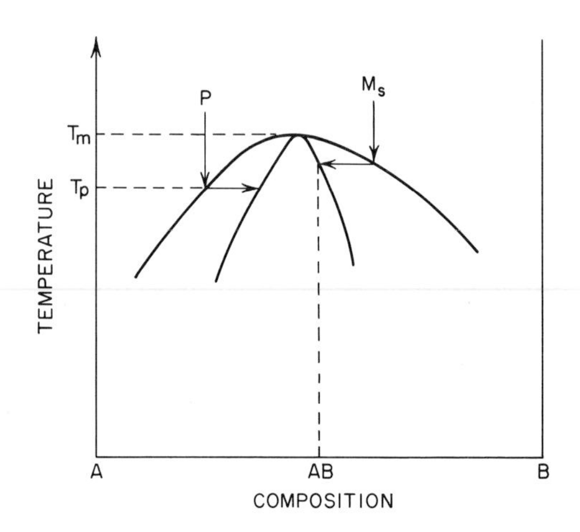

Fig. 15. Phase diagram for system A-B where compound near AB can solidify in other than exact stoichiometric ratio.

compound AB which exists as a single phase only over a very narrow range of composition. Melt of composition AB solidifies to solid of composition AB at the melting point T_m. Solid of this composition is also formed even if the melt composition is appreciably different from AB. For example, AB crystallizes from melt of composition P at a temperature T_p which is lower than T_m. If the compound has an inconveniently high vapor pressure at the melting point T_m, it can be grown at a lower temperature T_p, at which the vapor pressure may be more manageable. This is a special case of growth from solution rather than from the melt, so slow growth rates are necessary to allow solute to diffuse to the growing crystal face. Several III–V compounds have been crystallized from nonstoichiometric melts, for example, InSb, GaAs, and GaP. It is possible that Mg_2Sn could also be grown in this way. Hurle (1961) shows that the solvent or stoichiometric excess in the melt can cause constitutional supercooling and emphasizes that slow growth rates and steep temperature gradients are essential for good crystals.

The phase diagrams of the chalcogenides [for example, PbS (Bloem and Kroger, 1956), PbSe (Goldberg and Mitchell, 1954), PbTe (Miller et al., 1959), CdS, CdTe (de Nobel, 1959)] do not appear to be of the type discussed. Perhaps this is to be expected because the free energies of cation and anion vacancies differ more for the chalcogenides than for the III–V compounds. The phase diagram takes the form of Fig. 15 where the composition with maximum melting point is not stoichiometric and growth from a non-stoichiometric melt *P* does not result in a stoichiometric compound. On the other hand, if the melt composition can be maintained at M_s, then the resulting compound will be more nearly stoichiometric.

The melt composition can be controlled for those compounds having a reasonable vapor pressure by controlling the composition of the vapor phase. Figure 16 illustrates the type of apparatus that has been used. The composition of the vapor phase is controlled by the vapor pressure of the more volatile component *B*. This is controlled by the temperature T_1 of a reservoir of *B*, where T_1 is the lowest temperature in the system. If equilibrium between the melt and vapor is established rapidly, the composition of the melt is determined solely by the vapor pressure of *B*, hence by temperature T_1. The composition of the solid is determined by the composition of the melt at the solid-melt interface, and this can be appreciably different from the average melt composition. Slow growth rates, high temperature gradients, and stirring in the melt are therefore recommended. Slow growth rates and stirring also encourage the rapid attainment of equilibrium between melt and vapor.

The theory of this type of reaction has been discussed by van den Boomgaard (1957) and experimental verification obtained with CdTe (Kroger and de Nobel, 1955) by controlling the temperature of a Cd reservoir. The technique can, of course, be applied to the other methods of growth. The appropriate vapor pressure may be determined by trial and error, stoichiometry being conveniently deduced from electrical measurements. The degree of control required depends on the system and may be very severe. This is discussed briefly for the case of PbS following the work of Bloem and Kroger (1956).

The vapor-solid-melt diagram (the *R-T-X* diagram) of the Pb—S system, in the region of the compound PbS, was determined by Bloem and Kroger. The main points of interest can be represented by projections onto the *P-T* face (Fig. 17), and the *T-X* face (Fig. 18). The three-phase line in Fig. 17 indicates the range of solid compositions that can be prepared from the melt when the solid, melt, and vapor coexist in equilibrium. The compound with the maximum melting point 1400°K contains 6×10^{18} excess Pb atoms/cc, whereas the stoichiometric compound *O* melts at 1350°K. The contour lines in the region bounded by the three-phase line represent the composition of the solid PbS (in terms of excess Pb or S) in equilibrium with a particular pressure of sulfur vapor at a particular temperature. These contour lines approach the stoichiometric line *O* asymptotically, indicating that the higher the temperature of formation the more accurately must the sulfur

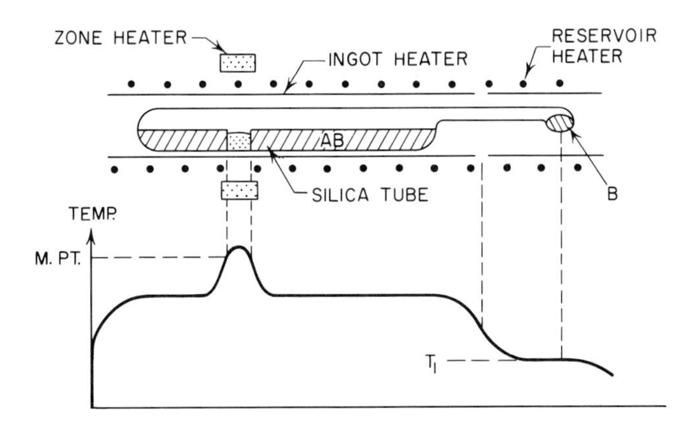

Fig. 16. Control of stoichiometry.

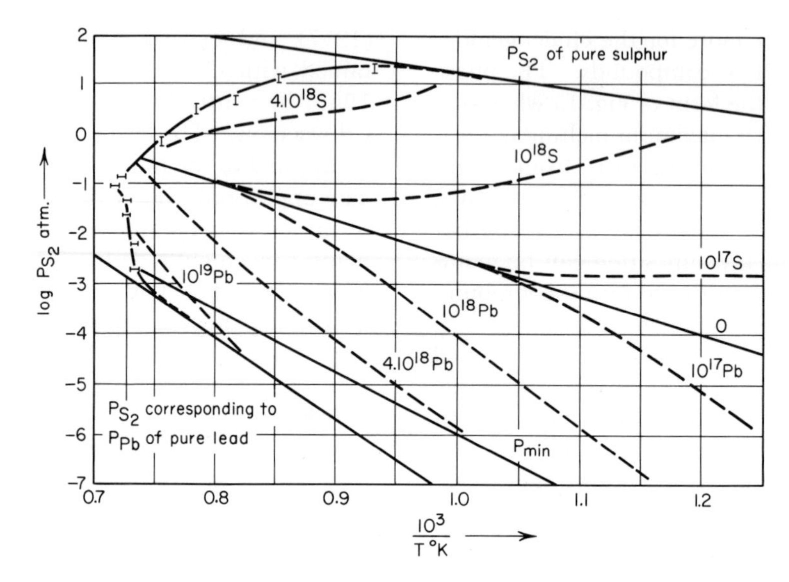

Fig. 17. *P-T* diagram for PbS. (After Bloem and Kröger.)

vapor pressure be controlled. From this diagram it can be seen that there is little hope of preparing crystals of PbS from the melt with less than 10^{18} carriers/cc. In Fig. 18 the composition is plotted in terms of the log of the absolute values of the deviation from stoichiometry in order to expand the region of interest. Again the composition of the compound with the maximum melting point shows that there is little hope of a closer approach to stoichiometry by, for example, growing crystals from a melt containing excess of Pb or S. Figure 17 shows clearly that a closer approach to stoichiometry is possible only by either growing the crystal at a lower temperature (that is, from the vapor or from a flux) or by a solid state diffusion process.

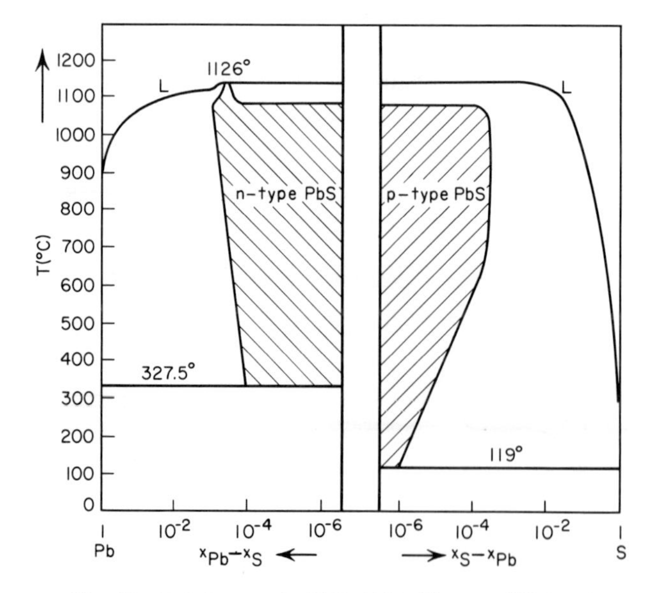

Fig. 18. *T-X* diagram for PbS. (After Bloem and Kröger.)

References

Addamiano, A., and M. Aven (1960), "Some properties of ZnS crystals grown from the melt," *J.A.P.* **31,** 36.

Ainsworth, L. (1956), "Single crystal bismuth telluride," *Proc. Phys. Soc.* **B69,** 606.

Albers, W., C. Haas, and F. van den Maeson (1960), "Preparation and electrical and optical properties of SnS crystals," *J. Phys. Chem. Soc.* **15,** 306.

Allred, W. P., W. L. Mefferd, and R. 'K. Willardson (1960), "Preparation and properties of aluminum antimonide AlSb," *J. Electrochem. Soc.* **107,** 117.

Asanabe, S., and A. Okazaki (1960), "Electrical properties of GeSe," *J. Phys. Soc. Jap.* **15,** 989.

Bennett, D. C., and B. Sawyer (1956), "Preparation of single crystals of germanium of exceptional uniformity and perfection by zone melting," *Bell Syst. Tech. J.* **35,** 637.

Benzing et al. (1958), "Synthesis of selenides and tellurides, I Reduction of selenites by hydrazine," *J. Amer. Chem. Soc.* **80,** 2657.

Black, J., E. M. Conwell, L. Seigle, and C. W. Spencer (1957), "Electrical and optical properties of some $M_2^{VB}N_3^{VIB}$ semiconductors (As_2Te_3, Sb_2S_3, Sb_2Se_3, Sb_2Te_3, Bi_2S_3, Bi_2Se_3, Bi_2Te_3)," *J. Phys. Chem. Solids* **2,** 224.

Bloem, J. (1956), "Controlled conductivity in PbS crystals," *Phillips Res. Rept.* **11,** 273.

Bloem, J., and F. A. Kroger (1956), "The *P-T-X* phase diagram of the lead-sulphur system," *Z. Phys. Chem.* **7,** 1.

van den Boomgaard, J. (1955), *Philips Res. Rept.* **10,** 319; (1956), **11,** 27; (1956), **11,** 91.

van den Boomgaard, J., F. A. Kroger, and H. J. Vink (1957), "Zone melting of decomposing solids," *J. Electronics* **1,** 212.

Bridgman, P. W. (1925), "Certain physical properties of single crystals of W, Sb, Bi, Te, Cd, Zn, and Sn," *Proc. Amer. Acad. Arts and Sciences* **60,** 305.

Brixner, L. H. (1960), "Structure and electrical properties of some new rare earth arsenides, antimonides, and tellurides," *J. Inorg. and Nuclear Chem.* **15,** 199.

Bube, R. H., and E. L. Lind (1959), "Photoconductivity of GaSe crystals," *Phys. Rev.* **115,** 1159.

Bube, R. H., and E. L. Lind (1960), "Photoconductivity in gallium sulfo-selenide solutions," *Phys. Rev.* **119,** 1535.

Busch, G., et al. (1961), "Structure, electric and thermoelectric properties of $SnSe_2$," *Helv. Phys. Act.* **34,** 359.

Busch, C., and V. Winkler (1953), "Electrical Conductivity and Hall effect in intermetallic compounds (Mg_2Sn, Mg_2Ge, Mg_2Si, Mg_3Sb, CdSb, ZnSb)," *Helv. Phys. Act.* **26,** 395.

Busch, G., and P. Junod (1957), "Electrical properties of Ag_2Se," *Helv. Phys. Act.* **30,** 470.

Coates, D. G., W. D. Lawson, and A. C. Prior (1961), "Single crystal photoconductive detectors in PbSe," *J. Electrochem. Soc.* **108,** 1038.

Conaboy, J. (1959), "Precision ends for fused silica tubes," *J. Sci. Instr.* **36,** 148.

Conaboy, J. (1960), "A high temperature burner for working silica," *J. Sci. Instr.* **37,** 35.

Cosgrove, C. J., J. P. McHugh, and W. A. Tiller (1958), "Effect of freezing conditions on the thermoelectric properties of $BiSbTe_3$ crystals," *J.A.P.* **32,** 621.

Fischer, A. G. (1959), "Preparation and properties of ZnS-type crystals grown from the melt," *J. Electrochem. Soc.* **106,** 838.

Goldberg, A. E., and G. R. Mitchell (1954), "Occurrence of natural *p-n* junctions in PbSe," *J. Chem. Phys.* **22,** 220.

Goodman, C. H. L. (1957), "A new group of compounds with diamond-type (chalcopyrite) structure," *Nature* **179,** 828.

Gorgunova, N. A., and N. N. Federova (1959), "Solid solutions in the ZnSe—GaAs system," *Soviet Phys. Solid State* **1,** 307.

Gremmelmaier, R. (1956), *Z. Naturforsch.* **11a,** 511.

Harman, T. C., M. J. Logan, and H. L. Goering (1958), "Preparation and electrical properties of HgTe," *J. Phys. Chem. Solids* **7,** 228.

Harman, T. C., B. Paris, S. E. Miller, and H. L. Goering (1957), "Preparation and some physical properties of Bi_2Te_3, Sb_2Te_3 and As_2Te_3," *J. Phys. Chem. Solids* **2,** 181.

Heinz, D. M., and E. Banks (1956), "Growth and properties of a large single crystal of CdSe," *J. Chem. Phys.* **24,** 391.

Hulme, K. F., and J. B. Mullin (1957), "Role of evaporation in zone refining InSb," *J. Electronics and Control* **3,** 160.

Hurle, D. J. (1961), "Constitutional supercooling during crystal growth from unstirred melts," I. Theoretical. *Solid State Electronics* **3,** 37; II. Experimental. **3,** 142.

Kroger, F. A., and D. de Nobel (1955), "Preparation and electrical properties of CdTe single crystals," *J. Electronics* **1,** 190.

Lawson, W. D. (1951), "A method of growing single crystals of PbTe and PbSe," *J.A.P.* **22,** 1444.

Lawson, W. D. (1952), "Oxygen-free single crystals of lead telluride, selenide and sulphide," *J.A.P.* **23,** 495.

Lawson, W. D., and S. Nielsen (1958), *Preparation of Single Crystals,* Butterworth, London, Chapters 8 and 9; (1959), *Preparation of Single Crystals,* Butterworth, London and Academic Press, p. 65.

Lawson, W. D., S. Nielsen, E. H. Putley, and V. Roberts (1955), "The preparation, electrical and optical properties of Mg_2Sn," *J. Electronics* **1,** 203.

Lawson, W. D., S. Nielsen, E. H. Putley, and A. S. Young (1959), "Preparation and properties of HgTe and mixed crystals of HgTe-CdTe," *J. Phys. Chem. Solids* **9,** 325.

Medcalf, W. E., and R. H. Fahrig (1958), "High pressure high temperature growth of CdS crystals," *J. Electrochem. Soc.* **105,** 719.

Miller, E., K. Komarck, and I. Cadoff (1959), "Stoichiometry of lead telluride," *Trans. Met. Soc. AIME* **215,** 882.

Moody, P. L., and C. Kolm (1958), "Syringe-type single crystal furnace for materials containing a volatile constituent," *Rev. Sci. Instr.* **29,** 1144.

Murphy, T. J., W. S. Clahaugh, and R. Gilchrist (1960), "Preparation of sulphur of high purity," *J. Res. N.B.S.* **64A,** 355.

Nielsen, S., and R. J. Heritage (1959), "A method for the purification of selenium," *J. Electrochem. Soc.* **106,** 39.

de Nobel, D. (1959), "Phase equilibria and semiconducting properties of CdTe," *Philips Res. Rept.* 14, 357; (1959), **14,** 431.

Pfann, W. G. (1958), *Zone Melting,* John Wiley and Sons, New York, Chapter 7.

Rodot, M., and H. Rodot (1960), "Band structure and dispersion mechanism in monocrystalline tellurides and selenides of mercury," *Comptes Rendus* **205,** 1447.

Rutter, J. W., and B. Chalmers (1953), "A prismatic substructure formed during solidification of metals," *Can. J. Phys.* **31,** 15.

Satterthwaite, C. B., and R. W. Ure (1957), "Electrical and thermal properties of Bi_2Te_3," *Phys. Rev.* **108,** 1164.

Silvey, G. A. (1958), "Zn_3As_2 a semiconducting compound," *J.A.P.* **29,** 220.

Smith, A. L., R. D. Rosenstein, and R. Ward (1947), "Preparation of SrSe and its properties as a base material for phosphors," *J. Amer. Chem. Soc.* **69,** 1725.

Smithells, C. J. (1955), *Metals Reference Book,* Butterworth, London. Lyman, T. (1948), *Metals Handbook,* Amer. Soc. for Metals.

Stockbarger, D. C. (1936), "Large single crystals of lithium fluoride," *Rev. Sci. Instr.* **7,** 133.

Strauss, A. J., and A. J. Rosenberg (1961), "Preparation and properties of $CdSnAs_2$," *J. Phys. Chem. Solids* **17,** 278.

Scanlon, W. W. (1957), "Lifetime of carriers in lead sulphide crystals," *Phys. Rev.* **106,** 718.

Wernick, J. H., and R. Wolfe (1960), "Three-element semiconductor materials," *Electronics* **33,** 103.

Wernick, J. H., S. Geller, and K. E. Benson (1958), "Constitution of the $AgSbSe_3$—$AgBiSe_2$—$AgBiTe_2$ system," *J. Phys. Chem. Solids* **7,** 240.

Woolley, J. C., and B. Ray (1960), "Solid solutions in $A^{II}B^{IV}$ tellurides," *J. Phys. Chem. Sol.* **13,** 151.

Woolley, J. C., and B. Ray (1960), "Effects of solid solutions of Ga_2Te_3 with $A^{II}B^{IV}$ tellurides," *J. Phys. Chem. Solids* **16,** 102.

Yamada, S. (1960), "On the electrical and optical properties of *p*-type CdTe crystals," *J. Phys. Soc. Japan* **15,** 1940.

Zhuze, V. P., V. M. Serguva, and E. L. Shtrum (1958), "Semiconducting compounds with general formula ABX_2," *Soviet Phys.-Tech. Phys.* **3,** 1925.

19

Group III-V Compounds

L. R. Weisberg

Considerable attention has been given in recent years to the binary intermetallic compounds known as the III-V compounds. These compounds are composed of elements of Group III-A (B, Al, Ga, In, Tl) and Group V-A (N, P, As, Sb, Bi) of the Periodic Table. Interest in this small group of materials is a result of their wide range of properties, ranging from near metallic to insulating. General reviews describing them have already appeared (Hilsum and Rose-Innes, 1961; Willardson and Goering, 1962; Hannay, 1959; Cunnell and Saker, 1957; Welker and Weiss, 1956; Pincherle and Radcliffe, 1957; Saker and Cunnell, 1954; Hulme and Mullin, 1962) and some relevant properties are listed in Table 1, together with the properties of Ge and Si. Phase diagrams for some of the compounds are shown in Fig. 1. Actually, the systems include the gas phase although it is not indicated, and the variation of vapor pressure in such a system (typified by GaAs) is shown schematically in Fig. 2. No compound is yet known to exist between B-Sb, B-Bi, Al-Bi, Ga-Bi, Tl-N, Tl-P, or Tl-As.

Some of the III-V compounds have been of only passing interest (such as InBi, In_2Bi, in addition to $TlSb_2$, $Tl_{11}Sb_5$, and various Tl-Bi phases) so that no effort has been made to grow their crystals. Several others have such high melting points and dissociation pressures that they have never been melted in a controlled fashion. As might be inferred from Table 1, these include BN, AlN, GaN, InN, BP, AlP, BAs, and AlAs. These compounds have been prepared mainly by chemical reactions in the liquid or vapor phase at temperatures well below the melting points. Solidification from solution in a component element (that is, off stoichiometry) has also been employed, but is discussed later only briefly since it is properly a precipitation phenomenon. Crystal growth by solidification has been most successfully employed for InSb, GaSb, AlSb, InAs, GaAs, InP, and GaP. Large crystals (>1 cm^3) have been grown for all of these except GaP; however, coarse-grained specimens of GaP have been grown.

It has been known for over a decade that these seven compounds are semiconductors (Welker, 1952). The most widely studied has been InSb not only because of its unusual properties (high mobility and low effective mass) but also because it is the simplest to prepare. The most important commercially is GaAs. It combines in one material a band gap exceeding that of Si, and mobilities equivalent to those of Ge (Jenny, 1958). It therefore has potential for use in high temperature semiconductor and electroluminescent devices. Because of its higher dissociation pressure, GaAs is considerably more difficult to grow than InSb and therefore has required greater attention.

1. GENERAL CRYSTAL GROWTH CONSIDERATIONS

The main difficulty is that many of the III-V compounds decompose at the melting point releasing the volatile component. Thus growth must take place under a pressure of the volatile component to maintain the stoichiometric composition in the melt, and procedures such as the Verneuil process (1904) are unsuccessful. A second diffi-

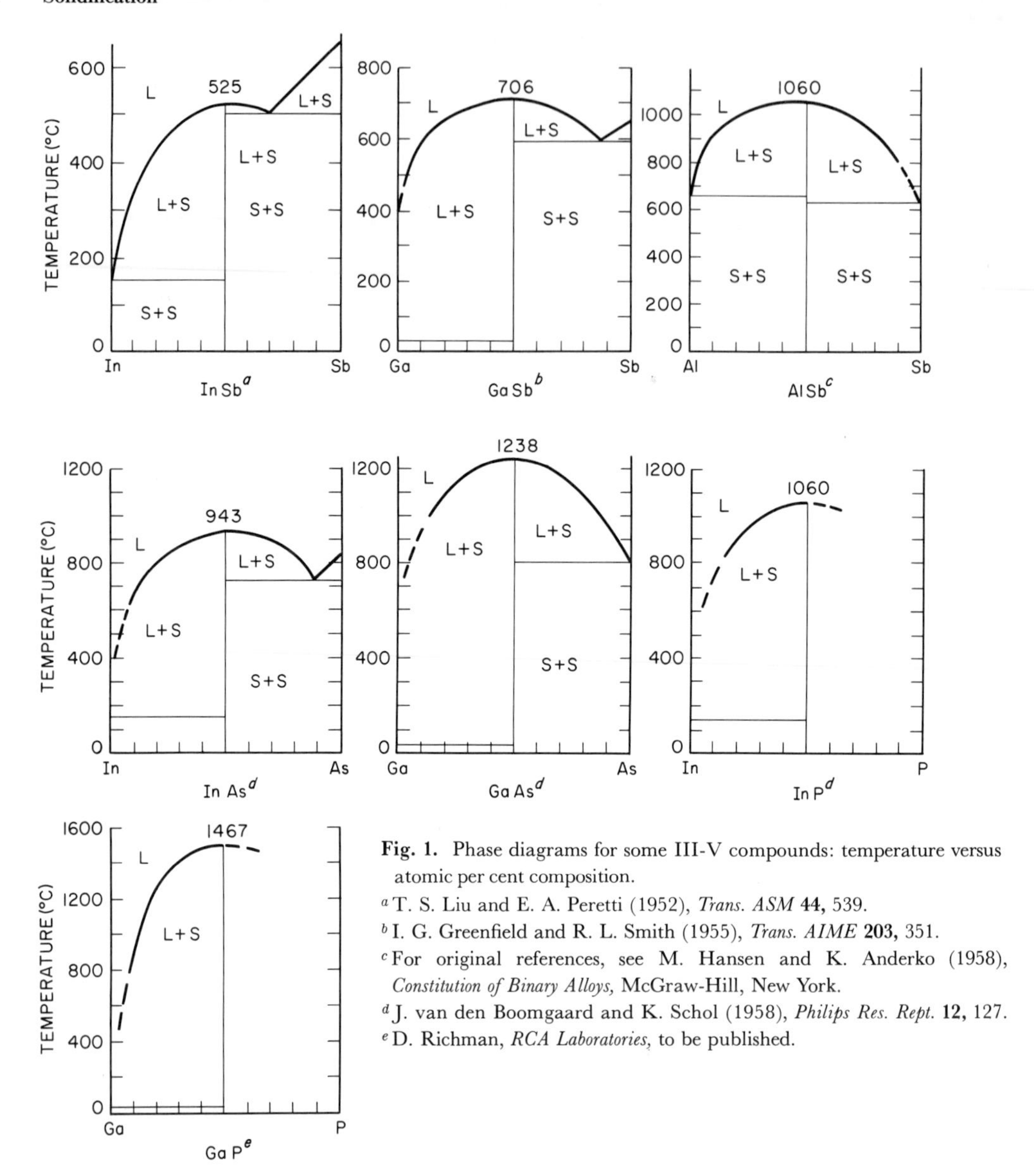

Fig. 1. Phase diagrams for some III-V compounds: temperature versus atomic per cent composition.

[a] T. S. Liu and E. A. Peretti (1952), *Trans. ASM* **44**, 539.

[b] I. G. Greenfield and R. L. Smith (1955), *Trans. AIME* **203**, 351.

[c] For original references, see M. Hansen and K. Anderko (1958), *Constitution of Binary Alloys*, McGraw-Hill, New York.

[d] J. van den Boomgaard and K. Schol (1958), *Philips Res. Rept.* **12**, 127.

[e] D. Richman, *RCA Laboratories*, to be published.

culty is that some of the constituent elements are extremely reactive, such as Al, As, and P. This has necessitated, for example, the use of special crucibles (such as one III-V compound to contain another) and the elimination of all metal parts in systems containing As or P. Third, some of the compounds have high melting points. This by itself is not an unduly severe problem, but in conjunction with the other restrictions can lead to considerable difficulties. Fourth, the compounds expand upon freezing and therefore are badly strained when constrained by crucibles. Therefore, the vertical Bridgman (1925) or Stockbarger (1936) technique is not suitable.

It will be assumed that the reader is already familiar with the salient features of solidification techniques for Ge and Si. Growth from a stoichiometric melt is conveniently divided into two categories depending on whether the compound has a low or high dissociation pressure at its melting point. The first category (low dissociation pressures) has as its prototype InSb, and also includes GaSb and AlSb. The second category has GaAs as its prototype, and also includes InAs, InP, and GaP.

Not only is the crystalline quality of concern but so are the physical properties, which can be greatly influenced by impurities and possibly deviations from stoichiometry. This problem is particularly acute for semiconductors since defects in amounts even below 1 ppm, or about 5×10^{16} atoms/cm^3, can drastically affect the electrical and optical properties. Therefore parts of the following discussion concern improvement of purity rather than crystallinity. For example, although crystals of InSb are easily prepared, considerable effort has been required to reduce electrically active impurities to the 10^{13}/cm^3 range. There is no unambiguous evidence to date that there is appreciable solid solubility of the component elements in their compounds. Therefore no extreme precautions have been needed to maintain precise stoichiometry in the melt. However, large deviations from stoichiometry will affect the crystallinity since any excess of one of the elements is rejected into the melt during solidification. This locally reduces the freezing point, causing nucleation of new crystals. The excess might also be entrapped in the solid ingot as a second phase.

2. COMPOUNDS WITH LOW DISSOCIATION PRESSURES AT THEIR MELTING POINTS

Indium Antimonide

Both zone-refining (Pfann, 1958; Tanenbaum and Maita, 1953; Harman, 1956; Hulme, 1959) and Czochralski crystal growth (Czochralski, 1917; Gremmelmaier and Madelung, 1953) (crystal pulling) for InSb can be easily carried out in an apparatus essentially identical to that used for Ge and Si. The elements are usually placed in a quartz

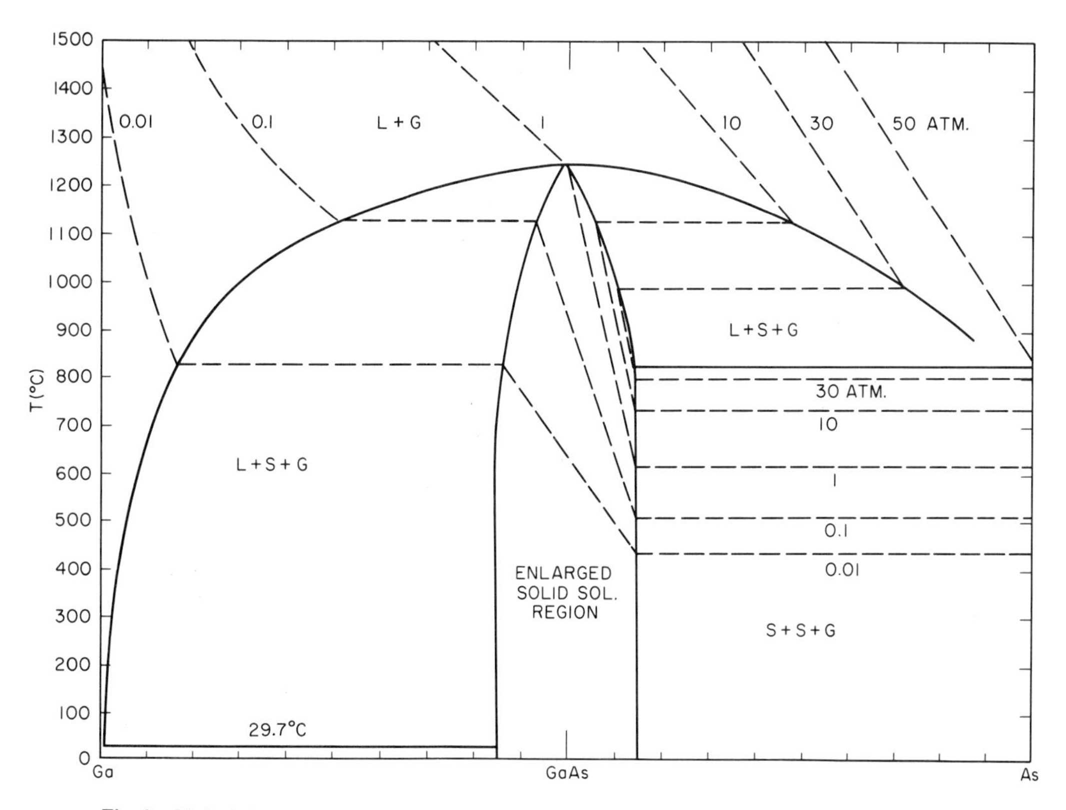

Fig. 2. Variation of vapor pressure with temperature and composition in GaAs, shown schematically.

TABLE 1

SELECTED PROPERTIES OF III–V COMPOUNDS, GERMANIUM, AND SILICON[a]

Material	Band Gap at R.T. (ev)	Melting Point °C	Vapor Pressure at M.P. (atm)	Crystal Structure
Si	1.11	1420	10^{-6}	Cubic (diamond)
Ge	0.67	936	10^{-9}	Cubic (diamond)
BN	10 (Calc)[b]	—	High	Cubic (zinc-blende)
BN	4.6 (Calc)[c]	>2700[i]	High	Hexagonal (wurtzite)
AlN	>5[d]	>2900[j]	>80[j]	Hexagonal (wurtzite)
GaN	3.25[d]	>1700[j]	>200[j]	Hexagonal (wurtzite)
InN	3.5 (Calc)[e]		High	Hexagonal (wurtzite)
BP	5.9[f]	>1300[f]	>24[g]	Cubic (zinc-blende)
AlP	2.42[g]	>2000[g]	High	Cubic (zinc-blende)
GaP	2.23	1467[k]	35[k]	Cubic (zinc-blende)
InP	1.25	1060	25[k]	Cubic (zinc-blende)
BAs	2.6 (Calc)[e]	>1000[f]	High	Cubic (zinc-blende)
AlAs	2.16	>1700[l]	High	Cubic (zinc-blende)
GaAs	1.40	1238	0.9	Cubic (zinc-blende)
InAs	0.33	943	0.33	Cubic (zinc-blende)
AlSb	1.74[h]	1050[h]	<0.02	Cubic (zinc-blende)
GaSb	0.70[h]	706[h]	$<4 \times 10^{-4}$	Cubic (zinc blende)
InSb	0.17	525	$<10^{-5}$	Cubic (zinc-blende)
InBi	Metallic	107	Very low	Tetragonal
TlBi	Metallic	230	Very low	Cubic CsCl

[a] No references are given for common values occurring throughout the literature or available in M. Hansen and K. Anderko (1958), *Constitution of Binary Alloys,* McGraw-Hill, New York.
[b] L. Kleinman and J. C. Phillips (1960), *Phys. Rev.* **117,** 460.
[c] R. Taylor and C. A. Coulson (1952), *Proc. Phys. Soc.* **A65,** 834.
[d] E. Kauer and A. Rabenau (1959), *Z. fur Naturforsch* **12a,** 942.
[e] Computed according to F. N. Hooge (1960), *Z. Phys. Chem.* **24,** 275.
[f] F. V. Williams and R. A. Ruehrwein (1960), *J. Am. Chem. Soc.* **82,** 1330.
[g] H. G. Grimmeiss, W. Kischio, and A. Rabenau (1960), *J. Phys. Chem. Solids* **16,** 302.
[h] J. F. Miller, H. L. Goering, and R. C. Himes (1960), *J. Electrochem. Soc.* **107,** 527.
[i] A. Fischer (1958), *Z. fur Naturforsch.* **13a,** 105.
[j] A. G. Fischer, Unpublished Work.
[k] D. Richman, RCA Laboratories, to be published.
[l] V. N. Vertoprakh and A. G. Grigor'eva (1958), *Izvest. Vysshikh Ucheb. Zavedenii, Fiz.* **5,** 133.

boat or crucible, and synthesis occurs as the material is heated above its melting point. Because of the low melting point (compared to Ge and Si), more care is needed to establish proper gradients, as for example to maintain a given zone width during zone refining. Zone-refining results in very pure, coarse-grained material. To grow large monocrystals, the Czochralski technique is used. Hydrogen is usually the atmosphere, but a vacuum can aid in removing volatile impurities (Hulme and Mullin, 1957).

Four main features distinguish the growth of InSb from Ge. First, several impurities in it have segregation coefficients close to unity (Mullin, 1958; Strauss, 1959) so they are difficult to remove. Second, the oxide does not volatilize at the melting point. It is essential therefore to avoid contamination between purification and crystal pulling. To do this, special apparatus has been constructed to allow zone-refined InSb to be transferred to the pulling crucible without removal from a closed system, or crystals have been pulled directly from the refining boat (Bourke et al., 1959). To reduce oxide contamination, specially purified gases must be used with a minimum of plastic or rubber tubing in the gas lines to keep out water vapor. To eliminate oxide particles on the melt surface, Allred and Bate (1961) used a double crucible. The second crucible had a small hole in its bottom and was held above the main crucible. After the InSb was melted, the second crucible was lowered into the melt so that InSb entered it through the small hole. The oxide adheres to the outer wall of the inner crucible and is trapped.

A third distinguishing feature is that impurities in InSb segregate anisotropically (Allred and Bate, 1961; Mullin and Hulme, 1961, Trainor and Bartlett, 1961; Hulme and Mullin, 1962). This is attributed to the formation of (111) facets on the solid-liquid interface. Solidification occurs much more rapidly parallel to the facets, thereby trapping impurities in the solid. Crystals pulled in a $\langle 111 \rangle$ direction exhibit a high concentration of Se and Te along the central axis as compared with the edges. The effective segregation coefficient of Te can vary from <1 outside the $\{111\}$ facet, to >1 in the facet. Therefore it may be advisable to avoid the $\langle 111 \rangle$ orientation.

The fourth distinguishing feature is that the III-V compounds are polar. Therefore planes such as the (111) and $(\bar{1}11)$ are not equivalent, the two planes being composed of different atomic species. In InSb, the surfaces terminating with In are designated as $(\bar{1}11)$, whereas those with Sb are called (111). These faces differ both chemically (Gatos and Lavine, 1960) and physically (Warekois et al., 1960). Gatos, Moody, and Lavine (1960) have studied the Czochralski growth of InSb in the $\langle 111 \rangle$ direction. They report that when the end of a seed terminated with Sb atoms, monocrystals were invariably obtained. In contrast, when the end of a seed terminated with In, twinning or large-grain formation frequently occurred. In addition, the (111) and (111) faces differ in growth rates (Ellis, 1959), thereby leading to asymmetrically shaped crystals when a $\langle 110 \rangle$ seed is used. The polarity is easily determined by etching techniques (Gatos and Lavine, 1960), so it is simple to choose a desirable orientation.

Gallium Antimonide

Next to InSb, GaSb is the easiest compound to prepare. One additional precaution relates to the greater reactivity of Ga as compared with In. For example, above 800°C, Ga reacts quite rapidly with quartz (Foster and Kramer, 1960), thereby introducing Si and O_2 into the melt. However, gallium oxide can be removed from Ga without danger of Si contamination from quartz by heating for 2 to 3 hr in a vacuum at 650°C (Ekstrom and Weisberg, 1961). Graphite or carbon-coated quartz can also be used for crucibles; however, carbon might be soluble in GaSb.

Some of the precautions mentioned for InSb for maintaining the highest purity have not been taken for GaSb. This is because carrier concentrations below about 1×10^{17} cm^{-3} have never been achieved in GaSb despite extensive zone refining. This same carrier concentration is obtained in as-grown crystals without previous zone refining. Whether this results from a deviation from stoichiometry (Detwiler, 1955; Hall and Racette, 1961; Owens and Strauss, 1961) or an impurity is still to be determined. In any case, zone refining is not an essential (or effective) step in making high purity GaSb, as it is for InSb.

Aluminum Antimonide

Growth of AlSb is quite similar to that of InSb and GaSb. Of the three, AlSb has the highest melting point and highest dissociation pressure. Thus, during growth, there is a slight loss of Sb from the melt unless preventative steps are taken (Herczog et al., 1958). To slow down the escape of Sb, growth is usually carried out in a protective atmosphere of H_2, A, or He. In addition, constrictions can be placed in the escape path. In the Czochralski technique, this can be accomplished (Herczog et al., 1958) by placing an extension tube on top of the crucible that has only a small opening for the pulling rod to pass through. Alternatively, a disk can be attached to the seed holder that nearly touches the inner walls of the extension tube (Herczog et al., 1958).

A second and more serious problem is that Al is quite reactive. Crucible materials such as C, BN, Si_3N_4, Ta, MgO, and Al_2O_3 were tested (Herczog et al., 1958; Allred et al., 1958). All but Al_2O_3 either reacted with or contaminated the AlSb melt. Two other possibilities are AlN and BeO. Although AlSb does not react with Al_2O_3, wetting does occur, so that good crystals have not yet been made in horizontal zone refining (Schell, 1955) or Bridgman crucibles (Allred et al., 1958). The wetting problem is avoided in the Czochralski method, and large AlSb crystals are readily grown (Herczog, et al., 1958; Allred et al, 1958; Gremmelmaier and Madelung, 1953; Becker and Oshinsky, 1954; Allred et al., 1960).

In addition, Al is difficult because of its tenacious oxide film. Subsequent to etching, the oxide film immediately reforms. The oxide particles act as nucleation centers for new grains and prevent good monocrystals from being grown. Therefore, several methods have been developed to remove the oxide. Schell (1955) whirled the crucible containing molten Al, thereby driving the oxide particles to the sides. Herczog, Haberecht, and Middleton (1958) removed the oxide by means of a probe. Allred, Mefferd, and Willardson (1960) lowered a paddle into the melt, froze the top surface onto the paddle, and then removed the paddle. They also reported that the oxide film can be removed by heating at 1000°C for 4 hr in a vacuum.

Soon after the oxide is removed, Sb is put into the crucible. This reduces the reactivity of the melt, and the surface subsequently remains free of oxide provided purified gases are used (Allred et al., 1960). Herczog (1958) described a purification train for He; for H_2 diffusion through Pd is best. Despite such precautions, as-grown crystals are nearly always covered with a glasslike coating of Sb_2O_3, indicating the presence of oxygen during growth (Allred et al., 1960).

3. COMPOUNDS WITH HIGH DISSOCIATION PRESSURES AT THEIR MELTING POINTS

General Considerations

For application to GaAs, InAs, InP, and GaP, the growth apparatus must undergo major revision. Growth must take place in either a completely sealed system or in a system with only a small leak. The coolest section of the vessel must be maintained above the condensation temperatures of the volatile species, which are in the range of 560 to 600°C. Because of the high reactivities of the volatile species, the growth vessel must be constructed mainly of quartz. Metals such as Ta or Mo are not severely attacked, but nevertheless contaminate the crystal. Alumina, sapphire, and nonpermeable ceramics are difficult to work, have low thermal shock resistance, and are too impure.

One of the simplest procedures for synthesis of the compound is to place the components at the opposite ends of a quartz tube (Gans et al., 1953). The less volatile constituent is heated to its melting point, and then the more volatile element is slowly heated, causing it to diffuse along the tube and react with the other element. The coolest section of the vessel acts as a reservoir for excess amounts of the more volatile constituent. Under these conditions, the pressure varies exponentially with the temperature. To avoid this high temperature sensitivity, the excess quantity of the volatile constituent can be made barely sufficient to fill the vessel with the required pressure. Then, since all the molecules will be in the gaseous phase, the pressure will vary only linearly with temperature. At times, it may be desirable to have nearly isothermal conditions. Then the excess of volatile constituent must be reduced further to avoid explosion. In this case, it is also possible to use no excess at all, but instead to allow the compound to decompose slightly to provide the necessary pressure for preventing further dissociation.

Gallium Arsenide

Large crystals of GaAs can be routinely prepared by several techniques, including the Czochralski, floating zone, and horizontal procedures. Each provides good crystals with dislocation densities in the range of 10^3 to 10^5 cm^{-2} (Abrahams and Ekstrom, 1959; Richards, 1960). Similar to GaSb, horizontal zone refining of GaAs has not produced significant improvements in total purity, and therefore it is not required. It should be noted that in the Czochralski and horizontal cases, the compound can be synthesized directly in the apparatus, whereas for the floating zone case, it must first be prepared using one of the other two methods.

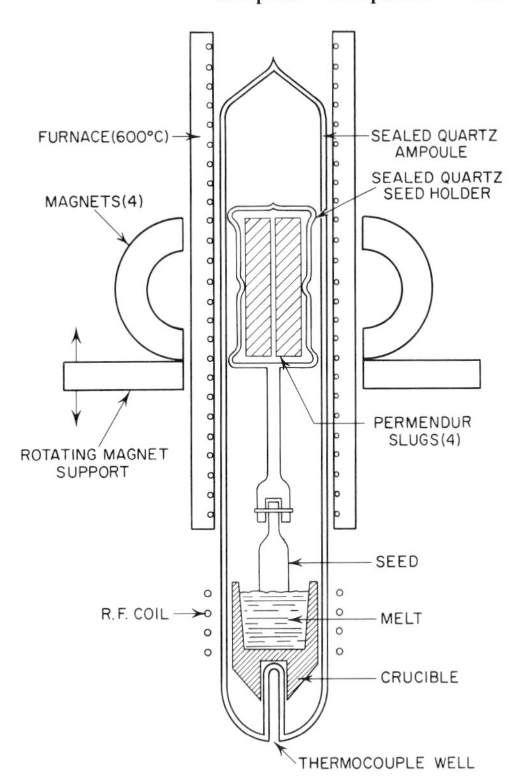

Fig. 3. Apparatus for the growth of GaAs single crystals by the magnetic Czochralski technique.

CZOCHRALSKI TECHNIQUE. Large crystals of GaAs were first made by Gremmelmaier (1956) in an apparatus similar to that shown in Fig. 3. The essential elements (seed holder, seed, crucible, and melt) are all contained in a sealed quartz ampoule. Coupling between the drive mechanism and the seed holder is accomplished magnetically. Hence the method is often referred to as the "magnetic Czochralski technique." Sealed within the seed holder are four magnetic slugs made of an alloy with a high Curie temperature such as Permendur. A thin heater surrounds the ampoule, extending from above the crucible (to allow for visibility) to the top of the ampoule. Four magnets are located outside this furnace. It is convenient to use r-f induction heating. The upper portion of the ampoule serves as a reservoir of condensed As.

The main difficulty is that the rotating seed holder vibrates excessively because of rough sliding contacts and the nonuniform cross section of the ampoule. To avoid this, the outer quartz tube must have an accurate and smooth bore which is made by shrinking it onto a mandrel or by grinding. This is difficult and expensive, so two alternates have evolved. In the first, a small-diameter, precision-ground quartz-tube was included along the central axis of the ampoule (Woolston, 1959) (Fig. 4*a*). The seed holder rides on this center rod and a bearing of graphite or BN. This apparatus has been used successfully for four years (Herkart, private communication). A second and simpler variation (Weiser and Blum, 1960) is shown in Fig. 4*b*. Here the seed holder rotates within two bearing sleeves made of either graphite or BN. These bearings do not themselves rotate, but move only up and down along one side of the quartz ampoule. Therefore it does not matter whether the ampoule is out-of-round, but the inner surface of the ampoule should be smooth along its length, which is usually so. This procedure does not provide as smooth rotation as the previous one.

A second difficulty with the magnetic technique is that the ampoule must be sealed before and cut open after each run. To avoid this, demountable joints can be included in the system, but such joints involve

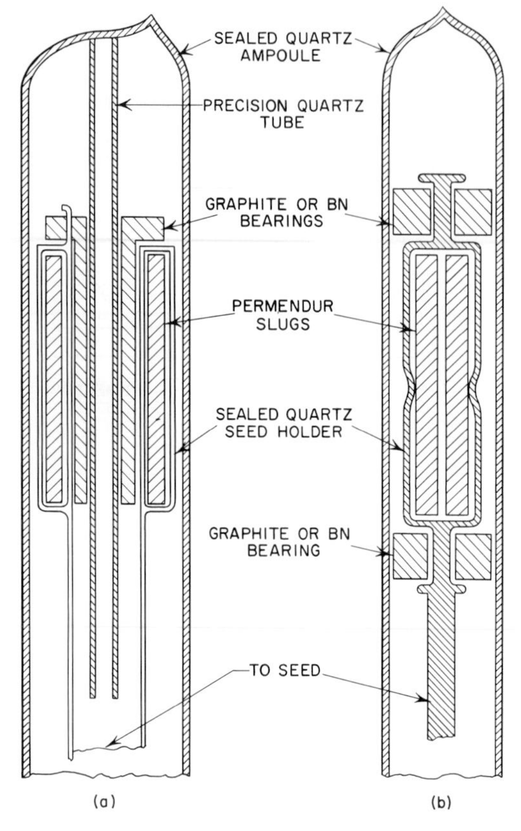

Fig. 4. Methods of construction of the seed holder and ampoule to eliminate vibration in the magnetic Czochralski apparatus.

close-fitting quartz parts and usually have small residual leaks. When such leaks can be tolerated, it is more convenient to replace the magnetic coupling with a direct rod. To prevent excessive loss of As, this rod must slide within the ampoule with the fitting tolerance reduced to at least 0.001 in., necessitating precision machining. One such apparatus (Fig. 5) is called a "syringe-type" puller (Moody and Kolm, 1958). Leakage of the volatile constituent is contained at the top of the ampoule. An important advantage is that this apparatus is demountable. Recently, a similar version but more complex apparatus has been described (Wickham, 1961). For improved smoothness of rotation, it may be desirable to coat the quartz plunger with graphite. The major problem with this type of apparatus is the jamming of the syringe caused by condensation between the plunger and the walls.

A third directly coupled puller has been constructed (Richards, 1957) using a pool of Ga for sealing. However, the pressure on both sides of the Ga must be equalized; otherwise As can still bubble through the Ga. This apparatus is somewhat awkward and the seal eventually becomes jammed because As reacts with the Ga in the seal at 650°C at an appreciable rate (Hilsum and Rose-Innes, 1961, p. 98).

In all of these methods, rapid exchange of As between the vapor phase and the melt can cause turbulence and sometimes spattering. To reduce this problem, it is advisable to use a partial pressure of an inert gas such as argon (Weisberg, et al., 1959). Also, visibility is lost due to deposition of GaAs

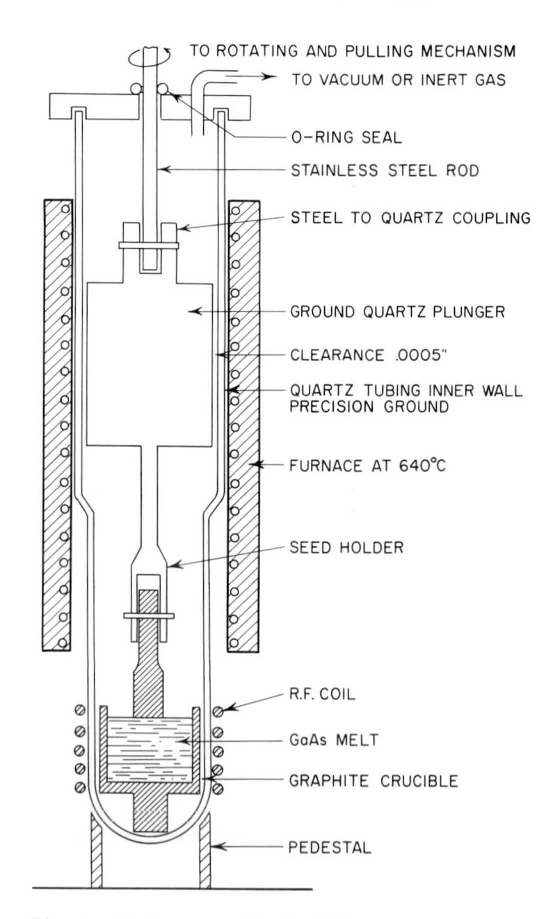

Fig. 5. Syringe-type Czochralski apparatus for the growth of GaAs employing a directly coupled seed holder permitting a slow leak.

on the ampoule walls. This is avoided by heating the walls close to the crucible with a transparent furnace, which is constructed by winding platinum wire on a quartz tube longitudinally so it is not heated by the induction coil (Weisberg et al., 1959). In these systems, crystals can be grown as fast as 4 cm/hr, provided a heat sink such as a tantalum rod sealed in quartz (Herkart, private communication) conducts heat from the seed.

As with InSb, GaAs grown in a ⟨110⟩ direction has unsymmetric flats (Ellis, 1959) because of the noncentrosymmetric zincblende structure. Concerning the difference of perfection between GaAs crystals grown in ⟨111⟩ and ⟨$1\bar{1}1$⟩ directions, Moody, Gatos, and Lavine (1960) find results similar to those for InSb. They claim that crystal growth in ⟨111⟩ directions is difficult, and results in higher dislocation densities than growth in ⟨$\bar{1}\bar{1}\bar{1}$⟩ directions. In contrast, Herkart found, during the growth of over 50 crystals, that crystals grew equally well in either direction. This was also reported by Richards (1960). This suggests that differences caused by polarity occur only when growth conditions are nonoptimum.

An advantage of the Czochralski technique over horizontal procedure is that it makes little difference whether or not the crucible is wet by the melt. Quartz, BN, AlN, Al_2O_3, and BeO crucibles have been used with nearly equal success with respect to crystallinity (Herkart, private communication). Carbon crucibles have also been used successfully, but carbon is soluble in GaAs (Goldstein and Weisberg, 1960) to a level of about 1×10^{17} cm^{-3}. For induction heating of nonconducting crucibles a graphite susceptor can be used or, to avoid the presence of carbon, direct coupling to the melt.

FLOATING ZONE TECHNIQUE. Floating zone-refining of GaAs is considerably more difficult than for Si because of zone instability caused by two effects. First, the surface tension/density ratio for GaAs is much less favorable than that for Si. Second, the composition of the GaAs melt varies along a zone due to the variation of temperature. In addition, the induction coil is external to the quartz ampoule, so that coupling conditions are unfavorable. Furthermore, the frequent attention of an operator is needed. In spite of these difficulties, GaAs has been floating zone-refined for several years (Weisberg, et al., 1959; Whelan, 1958; Whelan and Wheatley, 1958; Cunnell et al., 1960). By minimizing fluctuations in the inductive power and by maintaining uniform temperatures, as many as fourteen zone passes have been made on ingot without interruption (Weisberg et al., 1959).

To produce a floating zone, it is necessary to hold a GaAs ingot rigidly at the top and bottom within a quartz ampoule. This was first accomplished by Whelan (1958) using

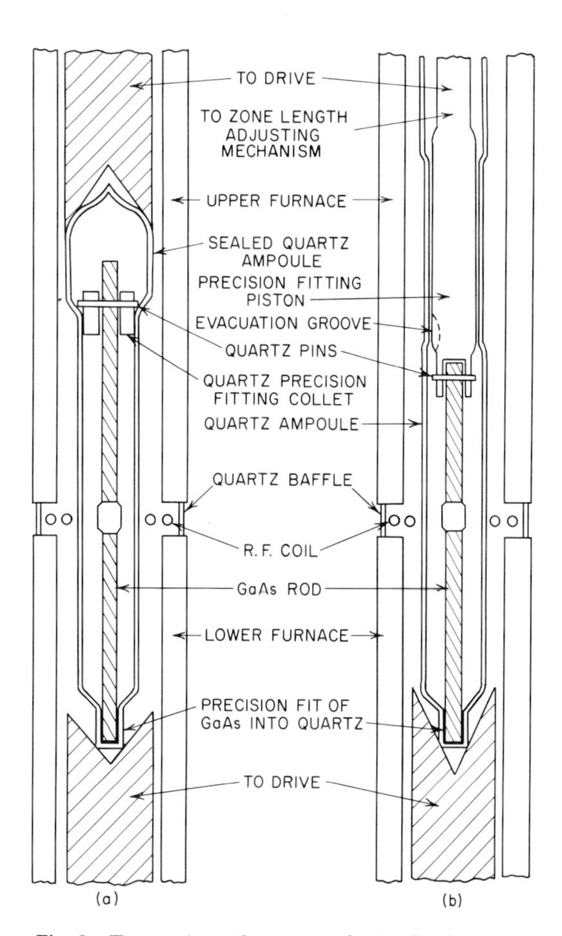

Fig. 6. Two versions of apparatus for the floating zone-refining of GaAs. (*a*) With a sealed ampoule. (*b*) With a "piston" seal.

an apparatus like that shown in Fig. 6*a*. For the best induction coupling, a coil consisting of either one or two concentric turns is operated at 4 mc. The latter provides better coupling but makes the zone size more sensitive to power fluctuations. Silverman (1961) has described a coil that should be advantageous. It has reversed turns above and below a central turn to concentrate the magnetic field. To move the zone through the ingot, it is best to fix the coil and move the ampoule. Motion of the coil can cause power variations as the flexible connecting cables move.

As with Czochralski growth, deposition of a GaAs film on the ampoule walls causes a loss of visibility. This problem has been overcome (Whelan and Wheatley, 1958) by placing a slotted quartz tube within the quartz ampoule, so that the ampoule wall becomes coated only behind the slot. After a thick coating forms, the inner tube is rotated slightly by manual manipulations, thereby exposing clear regions. Condensation can also be inhibited by heating the ampoule walls with a suitably constructed transparent furnace.

A difficulty of this apparatus is that no adjustment of the zone dimensions can be made during a run. Therefore, if a nonuniformity develops, the ingot thickens at that point and eventually causes the molten zone to spill out. A syringe-type seal (Cunnell and Wickham, 1960), as shown in Fig. 6*b*, solves this problem. The tolerance between the piston and outer tube is $<5\ \mu$ over a length of 7 cm, thereby reducing the loss of As to a few milligrams per hour while still allowing motion. In addition, the system has the advantage of being demountable. It can be evacuated in situ by withdrawing the piston to expose an evacuation channel.

An alternative auxiliary heater system has been devised by Vegter (private communication). The region surrounding the ampoule is evacuated, thereby reducing heat losses so that noninsulated resistance wires wound on quartz tubing can be used. This makes the entire quartz ampoule visible.

A deterrent to crystal growth in a floating zone is the presence of a floating oxide film. The formation of this film can be minimized by etching the GaAs in HCl heated to 75°C, with a few drops of HNO_3 added to start the reaction (Vegter, private communication).

HORIZONTAL GROWTH. The main advantage of horizontal growth is that crystals can be grown automatically with little supervision. Growth of these compounds is relatively easy to carry out. A boat containing the compound is sealed into a quartz tube and is moved through at least two different temperature zones (Folberth and Weiss, 1955), similar to the growth of CdTe (Kroger and de Nobel, 1955). Alternatively, a zone melting procedure can be applied (Richards, 1960; Weisberg et al., 1959; Edmond et al., 1956). Both procedures will be discussed. The critical factor in either procedure is the interaction between the boat and the melt since this can seriously affect the final crystallinity. Materials such as BN, AlN, BeO, Al_2O_3 (also sapphire), and zirconia are not suitable (Herkart, private communication; Cunnell et al., 1960) since the melt wets them causing adherence of the ingot. Graphite or carbon-coated quartz (Weisberg et al., 1959) are best since little or no wetting occurs. However, carbon is soluble in GaAs (Goldstein and Weisberg, 1960). Also, carbon can cause Si to enter the GaAs (Weisberg et al., 1959; Knight, 1961) since carbon reacts with the quartz walls through transporting agents.

The purest material is grown in quartz (Weisberg et al., 1959). The boat should be sandblasted and an HF etch should be avoided (Weisberg et al., 1962). It has also been found that it is preferable to grow crystals from an As-rich melt, at $1\frac{1}{2}$ to 2 atmospheres (Weisberg et al., 1962). The use of a Ga-rich melt results in pronounced supercooling and a much greater tendency for polycrystallinity. There is a slight tendency for crystals to grow preferentially in the $\langle 310 \rangle$ direction, but the major axes such as the $\langle 111 \rangle$, $\langle 100 \rangle$, and $\langle 110 \rangle$ are avoided (Weisberg et al., 1962). The maximum rate of growth is determined by the temperature gradients in the system.

Without the inclusion of special heat sinks, the growth rate is limited to ~2 cm/hr.

The purity of the resulting GaAs crystal depends strongly on the maximum temperature to which the melt is heated in contact with quartz (Ekstrom and Weisberg, 1961). Also, the melting point of GaAs is very close to the softening point of quartz (~1300°C). Therefore furnaces must be designed to operate very near the melting point of GaAs. Although the quartz is kept hot, it has been shown (Ekstrom and Weisberg, 1961) that there is no appreciable diffusion of atmospheric gases through it during growth. However, about 10^{18} molecules are released into the ampoule by outgassing of the quartz.

A typical horizontal growth apparatus (Weisberg et al., 1959) is shown in Fig. 7. In other versions (Cunnell et al., 1960; Hilsum and Rose-Innes, 1961, pp. 102, 103) a third section has been included, but this is not essential. The crystal is grown by moving the entire furnace so that the ampoule moves from the hotter section (~1250°C) to the cooler section (~610°C).

The zone melting of unstable solids has been considered by van den Boomgaard (1955) and van den Boomgaard, Kroger, and Vink (1955). The necessary apparatus is quite similar to the one mentioned except that three temperature zones are needed (Folberth and Weiss, 1955). Two outer sections are maintained at 610°C while a small center section melts a zone. One successful design by Richards (1960) is shown in Fig. 8. The ampoule is contained in a uniformly heated tube and the molten zone is formed by additional insulation around part of the tube. The zone moves with the insulation. By varying the amount and position of the insulation, the solid-liquid interface can be made either concave, convex, or planar. A carbon-coated quartz boat is used which enhances the crystallinity but limits the purity to about 10^{17} cm^{-3}. Growth rates in

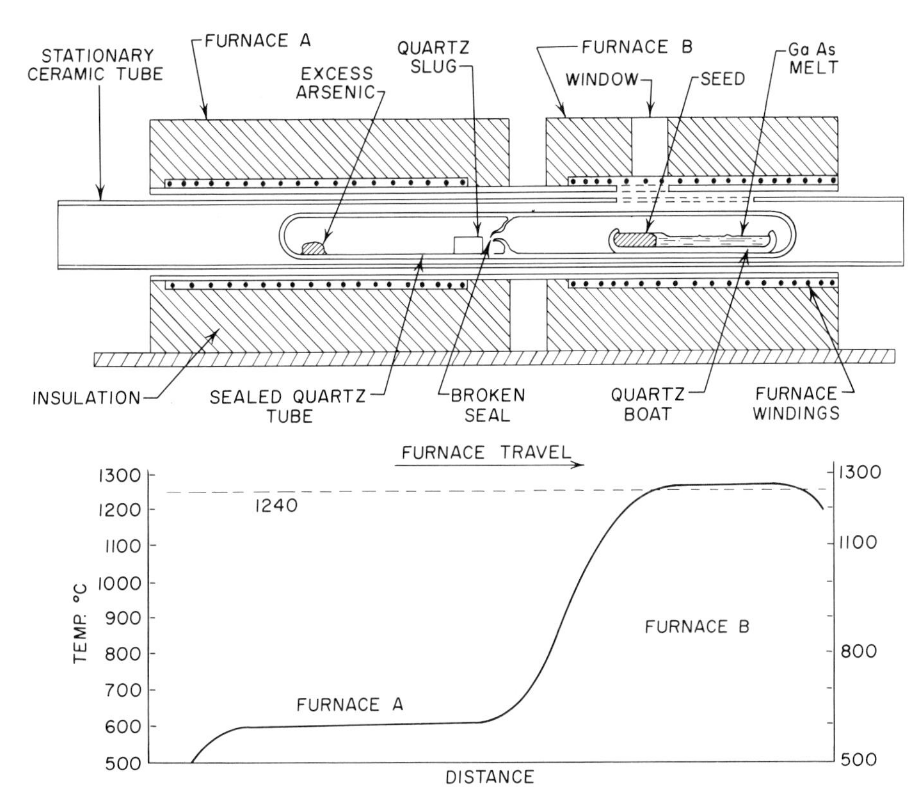

Fig. 7. Horizontal Bridgman growth apparatus for GaAs and furnace temperature profile. (After Weisberg, Rosi, and Herkart, 1959.)

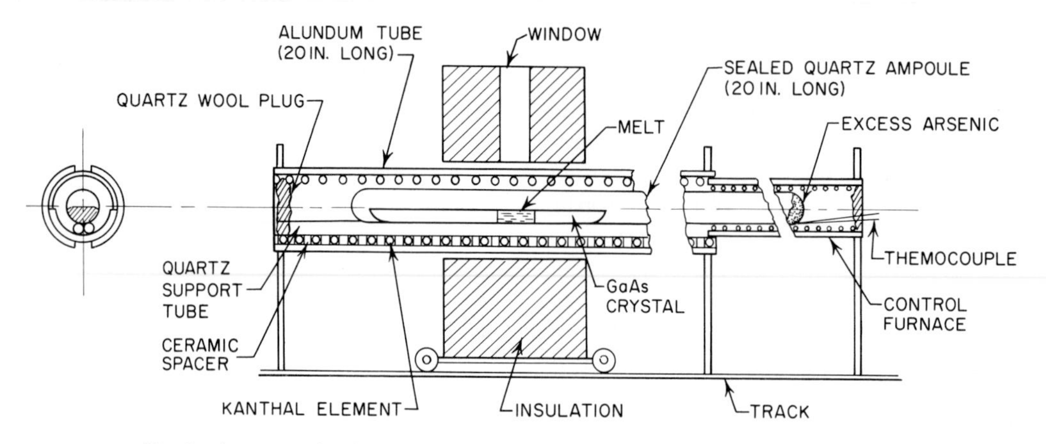

Fig. 8. Apparatus for the growth of GaAs single crystals by horizontal zone melting.

the range of 1 to 5 cm/hr were used. Richards also reported (1960) that unseeded crystals are oriented within about 10° of a [111] direction. It was found that interfaces concave into the solid resulted in lineage, whereas convex interfaces removed the lineage but introduced slip. Richards etched to find dislocation densities, and found that densities as low as 10^3 cm^{-2} could be readily obtained.

Attempts have been made to grow GaAs crystals horizontally using induction heating (Weisberg et al., 1959; Cunnell et al., 1960). The advantage would be that the ampoule walls could be maintained well below 1000°C. However, little success has been achieved. Direct coupling to the melt in quartz boats has resulted in poor crystallinity (Weisberg et al., 1959), and this was not improved upon with an open graphite boat. The best results have been achieved with the material contained in a closed tubular graphite-container (Weisberg et al., 1959). Screening the induction field from the melt resulted in more uniform and more easily controlled temperatures and an improved solid-liquid interface shape. However, the best ingots still had crystallites only ~1 cm by a few mm^2.

Indium Arsenide

Growth of InAs is very similar to that of GaAs, and because of its lower melting temperature and dissociation pressure is somewhat simpler. Therefore its growth does not require detailed discussion. Large monocrystals have been made by the magnetic Czochralski technique (Gremmelmaier and Madelung, 1953). The horizontal growth of InAs is more difficult than that of GaAs, perhaps because of its greater heat of fusion (Richman, to be published). As with GaAs, the quartz boats should be coated with carbon or silicon monoxide, or roughened by abrasive-blasting (Hilsum and Rose-Innes, 1961, pp. 102, 103; Harman et al., 1956).

The purest InAs has been made without zone refining (Effer, 1961). Therefore floating zone refining has not yet been reported, and horizontal zone refining, although used (Harada and Strauss, 1959), can be avoided.

Indium Phosphide

The high dissociation pressure of InP introduces an explosion hazard. However, quartz tubing with a 1 cm inner diameter and with heavy walls (3 mm) can withstand internal pressures up to at least 50 atmospheres (Guire, private communication). In such tubes, large crystals of InP have been made by directional freezing (Weisberg et al., 1959; Guire, private communication; Richman, private communication). Because of explosion difficulties with larger tubes, not Czochralski growth, horizontal zone refining, or floating zone refining have yet been used with stoichiometric melts.

Even with thick tubes, explosions occur in about one in ten growth attempts (Richman, private communication) because of flaws in the quartz. This makes it impractical to use wire wound furnaces and carefully aligned and adjusted parts. The most successful method used to date for InP is the gradient freeze technique (Lawson and Nielsen, 1958 p. 17) shown in Fig. 9 (Weisberg et al., 1959; Guire and Weiser, 1959). This method is simple because the insulation bricks and Globars can be assembled in only half an hour and there are no moving parts in the furnace. A temperature gradient is maintained along the melt in Section B. Solidification occurs when the temperature of Section B is slowly lowered and freezing starts at one end of the melt and progresses along it. InP does not adhere to quartz, so the outer tube can serve as a melt container. The use of graphite and carbon-coated quartz boats has not improved the crystallinity (Richman, private communication).

To eliminate the exponential temperature

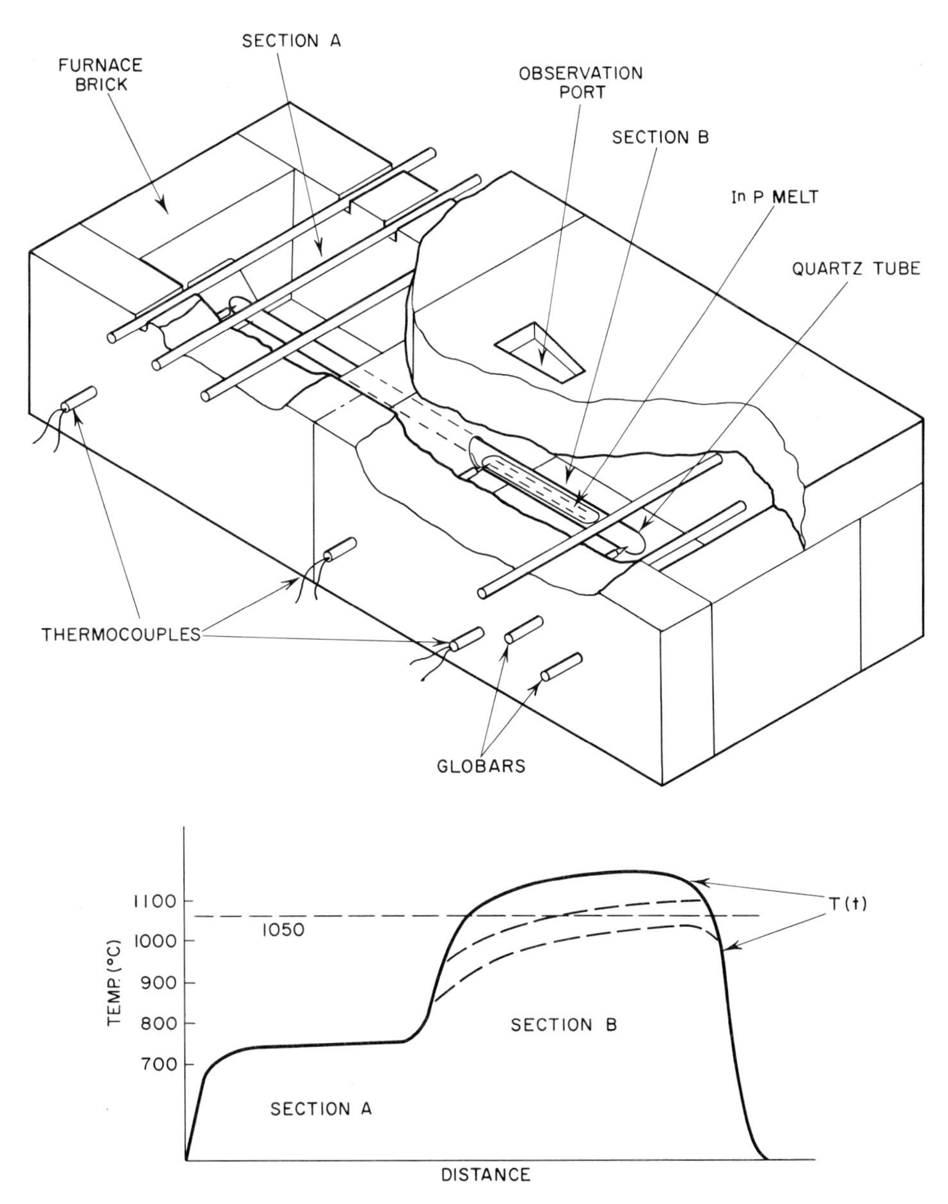

Fig. 9. Gradient freeze growth apparatus for InP, and furnace temperature profile. (After Weisberg, Rosi, and Herkart, 1959.)

dependence of the P vapor pressure, the cooler end of the ampoule is maintained above the critical point of P and a calculated excess of P is used. The crystallinity becomes poor when the P pressure is decreased below about 15 atmospheres. At stoichiometry, large monocrystals have been grown at velocities as high as 1 cm/hr. A difficulty is that InP has a marked tendency to supercool by as much as 50°C and then to freeze suddenly. To avoid this, seeding can be used or, before starting growth, the temperature can be lowered rapidly until about half the ingot is solidified, and then the ingot can be melted back until only the tip remains solid. The compound is usually synthesized directly in the ampoule prior to growth. To avoid explosions during heating, it is advantageous to heat the In above the melting point of InP before raising the P pressure, so the charge always remains molten.

Gallium Phosphide

Because of its high melting point and dissociation pressure, the growth of GaP is so difficult that no large crystals have yet been reproducibly prepared. The gradient freeze method cannot be used for GaP because quartz is quite soft at the melting point of GaP. Frosch and Derick (1961) have had the most success. They prepared a crystal 20 mm long by 5 mm^2 using a floating zone process essentially the same as Whelan's (Whelan and Wheatley, 1958) for GaAs but having several modifications. Because of the high pressures involved, heavy-walled (3 mm) quartz ampoules having an inner diameter of only 12 mm were employed. The induction frequency was increased to 12 Mc, and a specially designed three-turn coil was used, with the lower turn wound in the reverse direction. An improved coil design has since been reported by Silverman (1961). Coarse-grained ingots were regularly produced by Frosch and Derick (1961) using growth rates of 0.15 to 0.5 mm/min, temperatures close to 1500°C, and P pressures from 20 to 35 atmospheres.

Ingots of GaP have also been produced in horizontal boats using off-stoichiometric conditions. Davis (1960) describes the solidification of GaP in a graphite boat under 3 to 20 atmospheres of P, in which grains as large as $\frac{1}{2} \times \frac{1}{2} \times 1\frac{1}{2}$ cm were produced. Frosch and Derick (1961) provided the important modification of enclosing the apparatus in an autoclave. This allowed pressures as high as 35 atmospheres at a temperature of 1500°C to be used. However, large crystals were not produced. Gershenzon, Frosch, and Mikulyak (1961) found high carbon contents in GaP ingots made in graphite boats. They concluded that carbon causes spurious nucleation, making it difficult to grow large crystals. The greater success with floating zones was attributed to removal of the carbon during refining (Gershenzon et al., 1961).

Polycrystalline GaP has also been solidified from a nearly stoichiometric melt (Richman, private communication) using the method of Fischer (1959). The Ga and P were enclosed in a sealed quartz tube within a tightly fitting graphite (or BN) crucible. Above 1200°C, the quartz becomes plastic, and because of the P pressure expands against the inner wall of the crucible. Thus the quartz tube serves mainly as a gas-tight lining, whereas the crucible withstands the high pressures.

It should be realized that although growth procedures for the other III-V compounds are quite well established, for GaP marked improvements are expected in the relatively near future, making the techniques described here ephemeral.

4. MIXTURES OF III-V COMPOUNDS

Many pairs of III-V compounds having a common element are completely miscible. These include InSb-GaSb (Ivanov-Omskii and Kolomiets, 1959; Koester and Ulrich, 1958), InSb-AlSb (Baranov and Goryunova, 1960), GaSb-AlSb (Woolley et al., 1956; Miller et al., 1960), GaAs-InAs (Woolley and Smith, 1957; Abrahams et al., 1959), GaAs-AlAs (Strambaugh et al., 1960) InAs-AlAs (Borshchevskii and Tret'yakov, 1960). InAs-InP (Koester and Ulrich, 1958; Folberth, 1955; Bowers et al., 1959), InAs-InSb (Woolley and Smith, 1957), GaAs-GaSb

(Borshchevskii and Tret'yakov, 1960), and GaAs-GaP (Folberth, 1955; Welker, 1955; Johnson and Towns, 1961). As Weiser has pointed out (1961) this can be understood by realizing that unlike atoms in the pseudobinary systems are never nearest neighbors. For example, in InSb-GaSb, both the In and Ga atoms are completely surrounded by Sb atoms.

Little effort has been made to grow large crystals of these alloys. One reason is that growth is quite difficult because new grains often nucleate when the alloy composition fluctuates. Another is that the commercial importance of these alloys is mainly for thermoelectric applications (Abrahams et al., 1959; Bowers et al., 1959; Johnson and Towns, 1961) where good crystallinity is not necessary. What is mainly desired is uniform composition, which is best obtained by horizontal leveling (Pfann, 1958; Ivanov-Omskii and Kolomiets, 1959). Vibrational mixing has also been used (Borshchevskii and Tret'yakov, 1960). The growth procedures are modifications of those used for single compounds, so further discussion of them is not warranted.

5. GROWTH FROM SOLUTION IN A COMPONENT ELEMENT

Phase diagrams for III-V compounds have been shown in Fig. 1. These systems have three features in common: (1) in the liquid state, the component elements are completely miscible, (2) no solid solubility of the component elements in the compound is shown, and (3) one of the constituents has a relatively low melting point. The phase diagrams for the other III-V compounds such as BP, AlP, and the nitrides are not as simple and contain intermediate compounds. Nevertheless the diagrams are not overly complicated and appear to be of the eutectic variety. Because of these three features, it is quite easy to grow the III-V compounds from a solvent consisting of the low melting point element. After solution is formed, its temperature is slowly reduced, or the concentration of the solute is increased and the compound precipitates out.

The main advantage of this method is that compounds with high melting points or dissociation pressures can be grown at greatly reduced temperatures and pressures. The compounds are not usually contaminated by the solute element since its solubility is low. The melt can be decanted after growth, and the residual solvent removed from the compound by acid dissolution or by mechanical means.

Many of the III-V compounds have been prepared by precipitation from solution. Wolff, Keck, and Broder (1954) prepared platelets of InSb, GaSb, AlSb, GaAs, InAs, InP, and GaP using slow (about 2°C/hr) cooling rates. The platelets were as large as $1 \times 1 \times 0.1$ cm. Solution growth has also been used for BP (Popper and Ingles, 1957; Stone and Hill, 1959; Williams and Ruehrwein, 1960), BAs (Williams and Ruehrwein, 1960), AlP (Grimmeiss et al., 1960) and AlN (using an autoclave) (Adams et al., 1960).

Improved crystallinity and faster growth rates can be obtained by maintaining the solution composition close to stoichiometry and employing growth techniques described in previous sections. Of course, different temperatures and pressures and a nonstoichiometric melt must be used. For example, GaAs has been grown from both a Ga-rich and an As-rich solution using both the vertical Bridgman (Vieland and Skalski, 1961) and horizontal Bridgman (Weisberg et al., 1959) techniques. GaP has been prepared from solution by the vertical Bridgman technique (Miller et al., 1960), the Czochralski technique (Miller et al., 1960; Stambaugh et al., 1961), and by horizontal zone refining (Herkart, private communication; Davis, 1960; Gershenzon et al., 1961). InP has been grown from a nonstoichiometric melt by the gradient freezing technique (Weisberg et al., 1959), horizontal zone refining (Folberth and Weiss, 1955), the Czochralski technique (Harman et al., 1958), and floating zone refining (Herkart, private communication); the last at a P vapor pressure of 15 atmospheres. The method of solute enrichment has also been used for solution growth of GaP (Stambaugh et al., 1961) and InP (Harman et al., 1958). Instead of reducing the temperature, the solution is held in a temperature gradient, and the vapor pressure of the volatile species is slowly increased. In all these cases, the resulting ingots have been polycrystalline. Solution growth has not been more widely applied to the III-V compounds since it has never provided large crystals.

References

Abrahams, M. S., and L. Ekstrom (1959), in *Properties of Elemental and Compound Semiconductors,* Interscience, New York, p. 225.

Abrahams, M. S., R. Braunstein, and F. D. Rosi (1959), *J. Phys. Chem. Solids* **10**, 204.

Adams, I., J. W. Mellichamp, and G. A. Wolff (1960), *J. Electrochem. Soc.* **107**, 60C.

Addamiano, A. (1961), *J. Electrochem. Soc.* **108**, 1072.

Allred, W. P., and R. T. Bate (1961). *J. Electrochem. Soc.* **108**, 258.

Allred, W. P., W. L. Mefferd, and R. K. Willardson (1960), *J. Electrochem. Soc.* **107**, 117.

Allred, W. P., B. Paris, and M. Genser (1958), *J. Electrochem. Soc.* **105**, 93.

Baranov, B. V., and N. A. Goryunova (1960), *Fiz. Tverdogo Tela* **2**, 284.

Becker, J. H., and W. Oshinsky (1954), *Physica* **20**, 1073.

Van den Boomgaard, J. (1955), Philips Res. Rep. **10**, 319.

Van den Boomgaard, J., F. A. Kroger, and H. J. Vink (1955), *J. Electronics* **1**, 212.

Borshchevskii, A. S., and D. N. Tret'yakov (1960), *Soviet Phys.-Solid State* **1**, 1360.

Bourke, R. C., S. E. Miller, and W. P. Allred (1959), *J. Electrochem. Soc.* **106**, 61C.

Bowers, R., J. E. Bauerle, and A. J. Cornish (1959), *J. Appl. Phys.* **30**, 1050.

Bridgman, P. W. (1925), *Proc. Am. Acad. Sci.* **60**, 305, 385, 423.

Cunnell, F. A., J. T. Edmond, and W. R. Harding (1960), *Solid-State Electronics,* **1**, 97.

Cunnell, F. A., and E. W. Saker (1957), "Properties of the III-V Compound Semiconductors," in *Progress in Semiconductors,* John Wiley and Sons, New York, Vol. 2, pp. 37–65.

Cunnell, F. A., and R. Wickham (1960), *J. Sci. Instr.* **37**, 410.

Czochralski, J. (1917), *Z. physik. Chem.* **92**, 219.

Davis, R. E. (1960), in *Properties of Elemental and Compound Semiconductors,* Interscience, New York, p. 295.

Detwiler, D. P. (1955), *Phys. Rev.* **97**, 1575.

Edmond, J. T., R. F. Broom, and F. A. Cunnell (April 1956), *Report on Meeting on Semiconductors,* Rugby, England, p. 109, London, The Physical Society.

Effer, D. (1961), *J. Electrochem. Soc.* **108**, 357.

Ekstrom, L., and L. R. Weisberg (to be published), *Conf. on Ultrapurity of Semiconductor Materials,* Boston, April 11–13, 1961.

Ellis, S. G. (1959), *J. Appl. Phys.* **30**, 947.

Fischer, A. G. (1959), *J. Electrochem. Soc.* **106**, 839.

Folberth, O. G. (1955), *Z. Naturforsch.* **10a**, 502.

Folberth, O. G., and H. Weiss (1955), *Z. fur Naturforsch.* **10a**, 615.

Foster, L. M., and R. A. Kramer (1960), *J. Electrochem. Soc.* **107**, 189C.

Frosch, C. J., and L. Derick (1961), *J. Electrochem. Soc.* **108**, 251.

Gans, F., J. Lagrenaudie, and P. Seguin (1953), *Compt. rend.* **237**, 310.

Gatos, H. C., and M. C. Lavine (1960), *J. Phys. Chem. Solids* **14**, 169; (1960), *J. Appl. Phys.* **31**, 743.

Gatos, H. C., P. L. Moody, and M. C. Lavine (1960), *J. Appl. Phys.* **31**, 212.

Gershenzon, M., C. J. Frosch, and R. M. Mikulyak (1961), *J. Electrochem. Soc.* **108**, 53C.

Goldstein, B., and L. R. Weisberg (1960), *Bull. Am. Phys. Soc.* **5**, 407.

Gremmelmaier, R. (1956), *Z. Naturforsch.* **11a**, 511.

Gremmelmaier, R., and O. Madelung (1953), *Z. Naturforsch.* **8a**, 333.

Grimmeiss, H. G., W. Kischio, and A. Rabenau (1960), *J. Phys. Chem. Solids* **16**, 302.

Guire, R. J., RCA Laboratories, Private Communication.

Guire, R. J., and K. Weiser (January 1959), U. S. Patent 2,871,100.

Hall, R. N., and J. H. Racette (1961), *J. Appl. Phys,* **32**, 856.

Harada, R. H., and A. J. Strauss (1959), *J. Appl. Phys.* **30**, 121.

Harman, T. C. (1956), *J. Electrochem. Soc.* **103**, 128.

Harman, T. C., J. I. Genco, W. P. Allred, and H. L. Goering, (1958), *J. Electrochem. Soc.* **105**, 731.

Harman, T. C., H. L. Goering, and A. C. Beer (1956), *Phys. Rev.* **104**, 1562.

Hannay, N. B., Editor (1959), *Semiconductors,* Reinhold Publishing Corp., New York.

Herczog, A., R. R. Haberecht, and A. E. Middleton (1958), *J. Electrochem. Soc.* **105**, 533.

Herkart, P. G., RCA Laboratories, Private Communication.

Hilsum, C., and A. C. Rose-Innes (1961), *Semiconducting III-V Compounds,* Pergamon Press, New York.

Hulme, K. F. (1959), *J. Electronics and Control* **6**, 397.

Hulme, K. F., and J. B. Mullin (1957), *J. Electronics and Control* **3**, 160; (1962), *Solid State Electronics* **5**, 211.

Ivanov-Omskii, V. I., and B. T. Kolomiets (1959), *Soviet Phys.-Solid State* **1**, 834.

Koester, W., and W. Ulrich (1958), *Z. Metallk.* **49**, 365.

Kroger, F. A., and D. de Nobel (1955). *J. Electronics* **1**, 190.

Jenny, D. A. (1958), *Proc. IRE* **46**, 717.

Johnson, R. E., and G. Towns (1961), *J. Electrochem. Soc.* **107**, 189C.

Knight, J. R. (1961), *Nature* **190**, 1001.

Lawson, W. D., and S. Nielson (1958), *Preparation of Single Crystals,* Butterworth Publications, London, p. 17.

Miller, J. F., R. C. Himes, and H. L. Goering (1960), *J. Electrochem. Soc.* **107**, 189C.

Miller, J. F., H. L. Goering, and R. C. Himes (1960), *J. Electrochem. Soc.* **107**, 527.

Moody, P. L., H. C. Gatos, and M. C. Lavine (1960), *J. Appl. Phys.* **31,** 1696.

Moody, P. L., and C. Kolm (1958), *Rev. Sci. Instr.* **29,** 1144.

Mullin, J. B. (1958), *J. Electronics and Control* **4,** 358.

Mullin, J. B., and K. F. Hulme (1960), *J. Phys. Chem. Solids* **17,** 1; (1961), **19,** 352.

Owens, E. B., and A. J. Strauss (1961), in *Ultrapurification Semiconductor Materials,* Macmillan, New York, p. 340.

Pincherle, L., and J. M. Radcliffe (1957), "Semiconductor Intermetallic Compounds," in *Advances in Physics,* **5,** 271.

Pfann, W. G. (1958), *Zone Refining,* John Wiley and Sons, New York.

Popper, P., and T. A. Ingles (1957), *Nature,* **179,** 1075.

Richards, J. L. (1957), *J. Sci. Instr.* **34,** 289; (1960), *J. Appl. Phys.* **31,** 600.

Richman, D., RCA Laboratories, to be published.

Richman, D., RCA Laboratories, Private Communication.

Saker, E. W., and F. A. Cunnell (1954), "Intermetallic Semiconductors," *Research (London),* **7,** 114.

Schell, H. A. (1955), *Z. Metall.* **46,** 58.

Silverman, S. J. (1961), *J. Electrochem. Soc.* **108,** 585.

Stambaugh, E. P., J. I. Genco, and R. C. Himes (1960), *J. Electrochem. Soc.* **107,** 65C.

Stambaugh, E. P., J. F. Miller, and R. C. Himes (1961), in *Metallurgy of Elemental and Compound Semiconductors,* Interscience, New York, p. 317.

Stockbarger, D. C. (1936), *Rev. Sci. Instr.* **7,** 133.

Stone, B., and D. Hill (1959), *Bull. Am. Phys. Soc.* **4,** 408.

Strauss, A. J. (1959), *J. Appl. Phys.* **30,** 559.

Tanenbaum, M., and J. P. Maita (1953), *Phys. Rev.* **91,** 1009.

Trainor, A., and B. E. Bartlett (1961), *Solid-State Electronics* **2,** 106.

Vegter, Philips Research Laboratories, Private Communication.

Verneuil, M. A. (1904), *Am. chim. (Phys.)* **8,** 320.

Vieland, L. J., and S. Skalski (1961), in *Metallurgy of Elemental and Compound Semiconductors,* Interscience, New York, p. 303.

Warekois, E. P., M. C. Lavine, and H. C. Gatos (1960), *J. Appl. Phys.* **31,** 1302.

Weisberg, L. R., F. D. Rosi, and P. G. Herkart (1959), in *Properties of Elemental and Compound Semiconductors,* Interscience, New York, p. 25.

Weisberg, L. R., J. Blanc, and E. J. Stofko, (1962), *J. Electrochem. Soc.* **109,** 642.

Weiser, K., (1961), in *Metallurgy of Elemental and Compound Semiconductors,* Interscience, New York, p. 371.

Weiser, K., and S. Blum (1960), *J. Electrochem. Soc.* **107,** 189C.

Welker, H. (1952), *Z. fur Naturforsch.* **7a,** 744.

Welker, H. (1955), *J. Electronics* **1,** 181.

Welker, H., and H. Weiss (1956), "Group III-Group V Compounds," in *Solid-State Physics,* F. Seitz and D. Turnbull, Eds., Academic Press, New York, Vol. 3, pp. 1–78.

Wickham, R. (1961), *J. Sci. Instr.* **38,** 396.

Whelan, J. (1958), Bell Labs. Record, **36,** 267.

Whelan, J. M., and G. H. Wheatley (1958), *J. Phys. Chem. Solids* **6,** 169.

Willardson, R. K., and H. L. Goering (1962), *Preparation of III-V Compounds,* Reinhold Publishing Corp., New York.

Williams, F. V., and R. A. Ruehrwein (1960), *J. Am. Chem. Soc.* **82,** 1330.

Wolff, G., P. H. Keck, and J. D. Broder (1954), *Phys. Rev.* **94,** 753(A).

Woolley, J. C., and B. A. Smith (1957), *Proc. Phys. Soc. (London)* **B70,** 153; (1958), 372.

Woolley, J. C., B. A. Smith, and D. G. Lees, (1956), *Proc. Phys. Soc.* **B69,** 1339.

Woolston, J. R. (1959), as referred to by L. R. Weisberg, F. D. Rosi, and P. G. Herkart, in *Properties of Elemental and Compound Semiconductors,* Interscience, New York, p. 27.

20

Verneuil Method

W. H. Bauer and W. G. Field

The Verneuil technique for growing crystals is a special modification of growth from the melt. Material in the form of a fine powder is continuously added to the molten top of a crystal with heat usually being provided by an oxy-hydrogen flame. For convenience, the powder is introduced into the gas at some point above the flame to insure even distribution upon the crystal. Other heat sources are now being utilized to improve the quality of the crystals or to grow materials that are incompatible with a gas flame.

A chief virtue of the Verneuil method is that no crucible is required to hold the melt. This permits the growth of crystals from substances having very high melting points and from those for which no nonreactive crucible material has been found.

The method was first described by Verneuil (1902), and for nearly fifty years no important changes were made in the basic technique. Only larger crystals having various colors represented the improvements during this period. These were used for gems, various mechanical applications including watch and instrument bearings, and as knife edges for balances. Modern solid state electronics demands a variety of high quality crystals for the successful operation of devices, such as the ruby laser, and for research. This has stimulated many advances in crystal growth methods, including the Verneuil method. Most of the improvements in the technique have been made in the last fifteen years.

Before 1947, only two materials, alumina and spinel, were being grown by the Verneuil technique. Today the number of different substances successfully grown as monocrystals has nearly reached one hundred. Rutile was first grown in 1947 (*Science Newsletter*), followed several years later by scheelite (Zerfoss et al., 1949). The first successful growth of an incongruently melting material was mullite by Bauer, Gordon, and Moore (1950). Several ferrites and many of the silicate materials were also grown (Bauer, Gordon, 1951). Yttrium-iron garnet, another incongruently melting material, was added to the list by Rudness and Kebler in 1959.

1. APPARATUS FOR THE METHOD

Although the Verneuil method is an important means of crystal growth, it has some disadvantages. Two of the worst are high thermal gradients and composition of the flame atmosphere. Most improvements have come from two directions: improvements in making the gas-flame and the use of various other sources of heat.

Improvements in the original flame method have mostly involved two aspects: improved nozzles and control of certain operating parameters. Attempts have also been made to anneal crystals while they are growing and to use gases other than oxygen and hydrogen. These improvements are discussed later.

Various heat sources which have been studied include radiant heat sources such as a resistively (or inductively) heated metal ring, radio-frequency and direct-current plasma torches, arc-imaging techniques, electron bombardment, and several others.

Feed Materials and Purity

The key to the growth of high quality crystals is the behavior of the feed material and therefore the process by which the feed is made. Optimum feed is characterized by high purity, low density, and small-grain size. The feed should be anhydrous, micro- (or crypto-) crystalline to amorphous in size, and should have been completely reacted chemically.

These characteristics are important for the following reasons. Impurities lead to imperfections such as bubbles, cracks, and colloids that decrease the transparency. Fine size and low density of the particles insure instantaneous melting and preclude gas entrapment caused by solid feed penetrating the molten surface. Chemical combination of the feed materials is necessary for homogeneous melting and prevention of volatilization of certain components. It prevents chemical reaction from occurring in the viscous molten surface which might cause frothing or boiling. Chemical combination also promotes intimate chemical mixing and homogeneity of the physical structure.

These four general types of processes produce feed material: (1) oxide blending, (2) double salt, (3) coprecipitation, and (4) gel methods. The oxide-blending method is closely related to standard methods for preparing polycrystalline ceramics. Homogeneity is obtained by mechanical mixing rather than by chemical means. The component oxides (or the oxides obtained by reacting their carbonates, nitrates, sulfates, etc., at their decomposition temperatures) are intimately mixed by ball milling or blending, then sintered at sufficient high temperatures to produce the desired reaction, and finally pulverized to the desired grain size. This method is susceptible to contamination, but allows a wide range of compositions to be produced.

The double-salt or alum method is most desirable from the point of view of purity, homogeneity, and density. The alum ($Me_2^{3+}(NH_4)_2(SO_4)_4 \cdot 24\ H_2O$) or pseudo-alum ($Me^{++}(NH_4)_2 \cdot 6\ H_2O$) (where Me is the desired metallic cation) is fired to promote reaction and drive off the volatiles. This method has long been used to make feed for the synthesis of corundum and spinel gems and more recently for light-maser rubies. Various compounds can be added to the alum prior to firing to obtain new substances. During firing, the alum melts in its water of hydration, mixes and reacts with the other oxides, and produces an open-structured mass that only requires sieving.

The coprecipitation or oxalate method has been used for producing feed for the synthesis of ferromagnetic spinels (ferrites, $MeO \cdot Fe_2O_3$). It is susceptible to contamination and is not often used.

The laboratory production of silica aerogels or the use of commercial gels such as Cab-O-Sil is essential in the synthesis of silicates in order to obtain anhydrous silica (SiO_2) of high purity and low density. The gel can then be mixed with other chemical compounds by one of the previous methods to make the desired end product.

Feed powders are usually stored in some type of drying apparatus, often at low temperatures (100°C) to prevent adsorption of water from the atmosphere. Modifications of the foregoing procedures are often required, and time-temperature firing studies are especially useful in optimizing the characteristics. Microscopic and X-ray inspection of the feed is useful for controlling the chemical composition and physical structure. Various references (Bauer, Gordon, and Moore, 1950; Bauer and Gordon, 1951; Merker, 1947; Bauer, Mertz, and Baba, 1958) discuss powder preparation in more detail.

Maintenance of purity is also a problem. Impurities may come from the gas, from the nozzle, or be present in the powder. Recently, all-ceramic nozzles have been developed by the authors. These may be made of dense polycrystalline material of the same composition as the crystal being grown. If they are of high purity, the possibility of contamination from this source is greatly reduced, and one effective material is recrystallized alumina. The importance of adequate powder preparation and purity maintenance cannot be overemphasized. They often determine directly whether good crystals can be grown. Inadequate methods have sometimes caused failures with materials that are in fact suitable for the Verneuil method.

An interesting variation is the addition of a vapor to the flame. The vapor is decomposed by the heat of the flame to provide feed material. Silicon has been successfully grown this way.

Mechanical Arrangements

The basic components of the Verneuil apparatus, shown in the schematic diagram in his paper (Verneuil, 1904), have changed little in many years of application. The feeding hoppers, oxy-hydrogen burner, and lowering mechanism have been modified many times and automated and even replaced by newer devices; but with conventional gas heating, the general principle and apparatus are still retained. Merker's FIAT Report (Merker, 1947) reviewed the process as practiced in Europe at that time. The practice in the Soviet Union has been reviewed more recently by a detailed account of the work of Popov (1959).

Various systems for feeding the powder include vibrating feeders, air pick-up devices, screw feeders, and many variations of tapping a screen-bottomed hopper. Although claims have been made that intermittent tapping feeders are most desirable because they provide a time interval for stabilization of the molten pool, some other feeder designs have been more suitable for certain materials. Figure 1 shows a vibratory feeder designed to drop powder directly into a feeding tube, thereby eliminating sticking along the sloping walls of conventional hopper feeders.

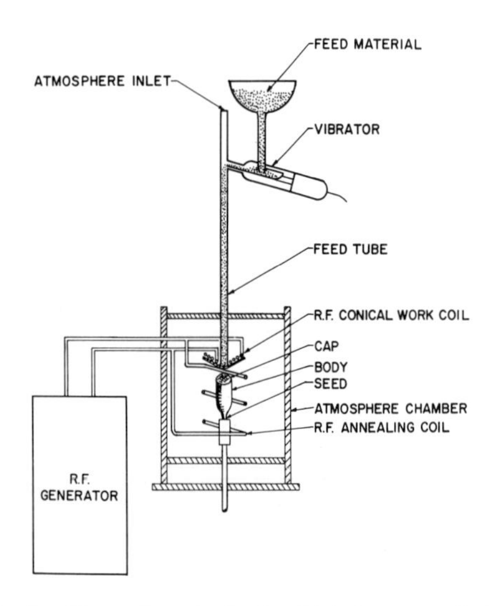

Fig. 1. Verneuil method using high frequency induction heating with vibratory feeder.

Lowering mechanisms have included racks and pinions, hydraulic and pneumatic pistons, mechanical jacks, and a variety of split-nut lead screws. Any mechanism with smooth action that can be closely controlled at rates from 3 mm to 50 mm/hr is satisfactory. In addition, a centering mechanism is desirable and in some cases rotation of the growing crystal is necessary. Detailed descriptions of mechanisms are given in the references.

Heat Sources

Problems associated with the original gas flame method have resulted in many attempts to utilize other heat sources. The more successful of these include radiation heaters, induction heating, plasmas, and arc images. Although others have been or are being tried, the discussions are limited to these.

GAS FLAMES. Oxy-hydrogen is the flame that is usually used, although others are in limited or experimental use. In Europe, small amounts of illuminating gas are added to the flame in production plants that grow crystals for use as gems or instrument bearings.

Three other combinations that have been tried consist of oxygen combined with either illuminating gas, acetylene, or carbon monoxide. These mixtures require mixing prior to burning and usually make dirty flames. They have not been used successfully for growing high quality crystals. Other possible high temperature flames, such as the fluorine flame, have not been used.

Although many different gas mixtures

and burner designs have been tried, the original oxy-hydrogen mixture has the most general success.

Since oxygen and hydrogen should not be mixed prior to burning, the design of a Verneuil nozzle or burner can be quite simple. For many years, most burners have consisted of just two concentric tubes: oxygen being supplied to the inner tube and hydrogen to the outer. There are many minor variations. In some the tubes are cylindrical, in others they may be curved inward, or the gas may be supplied through small ports to the outer tube of the nozzle. A particular design (Bauer) is shown in the lower left of Fig. 2. Another modification that operates well is the multitube burner. In this design a number of small tubes are nested inside a larger tube. Oxygen is supplied through the smaller tubes and hydrogen through the interstices between tubes. This tends to produce a large flame front of fairly uniform temperature. The all-ceramic burner mentioned has this form.

The tricone burner is an important design developed by Merker (see Fig. 2). Three concentric tubes are used with the inner one carrying hydrogen and the other two tubes oxygen. This produces a pointed flame that is useful for starting growth when a seed crystal is not available. It is also useful for growing crystals that tend to become reduced by the flame of the normal two-tube burner, such as nickel oxide. The additional oxygen also tends to cool the flame, making this burner useful for materials with relatively low melting points.

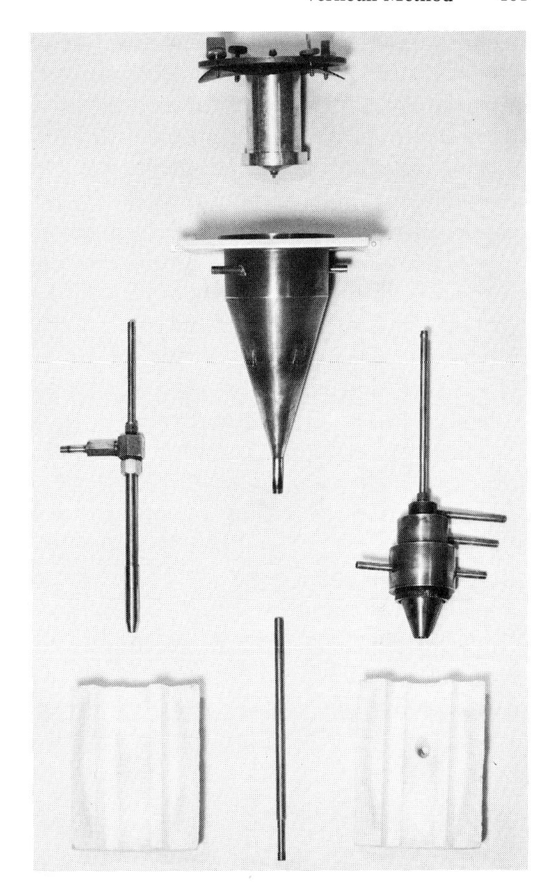

Fig. 2. Feed hoppers (top), burners (middle), and muffles (bottom) for Verneuil crystal growth.

RADIATION HEATERS. The first succcssful use of radiation heating was by Keck, Levine, Broder, and Lieberman (1954). A metal susceptor was placed around the top of the crystal and heated by induction. Because of the difficulty of finding a susceptor material that would operate at high temperatures in an oxidizing atmosphere, inert gases were needed to protect the susceptor.

A modification of Dickinson and Field (1956) is based on the same idea. However, the metal ring is heated by electrical resistance. This eliminates the radio-frequency generator and results in a very simple device whose temperature can be adjusted by a Variac. Several ferrite compositions have been successfully grown as monocrystals by this method. As in Keck's method, an inert atmosphere is needed to obtain reasonable life of the heater.

PLASMA HEATING. The use of plasmas, that is, streams of ionized gas, has opened several new avenues for crystal growth. By substituting a plasma torch for the oxy-hydrogen burner, extremely high temperatures can be attained, and growth without the chemical complications caused by the products of combustion can be accomplished. Also, growth in highly oxidizing, reducing, or other desired atmospheres becomes possible. Therefore in recent years several groups have been working with plasma techniques (Reed, 1961; Bauer and Longo, 1960).

Although patents and articles on heating with electrical discharges go back to 1900, it was only in the early 1950's that the demand for high temperature materials accelerated the development of plasma heating. Direct-current plasma torches became commercially available in about 1955. Several versions of both a-c and d-c torches have been developed specifically for crystal growth so that they can sustain plasmas having a variety of gas compositions.

The d-c plasma torch consists of an electric arc contained within a small tube through which gas is blown. The current is carried by thermonic or high field cathodes and water-cooled anode nozzles. These torches are characterized by extremely high currents (hundreds to thousands of amperes) and low voltages (10 to 100 volts). Power levels up to 5000 and temperatures up to 30,000°F have been reported.

A number of designs have been developed that involve either gas or liquid stabilization. In the older liquid-stabilized torches water is swirled in the arc region, producing an inherently dirty plasma because of the high electrode consumption. This would not seem to be desirable for crystal growth.

Gas stabilization methods include vortex stabilization, gas sheath stabilization, wall stabilization, and magnetic stabilization. Of these, gas sheath stabilization is most commonly used for crystal growth. In this design, the arc path goes from a tungsten cathode to a hollow water-cooled anode. The arc remains within the nozzle and is prevented from striking the wall immediately by a sheath of gas that is considerably larger than the arc in diameter. Further downstream the arc strikes through the

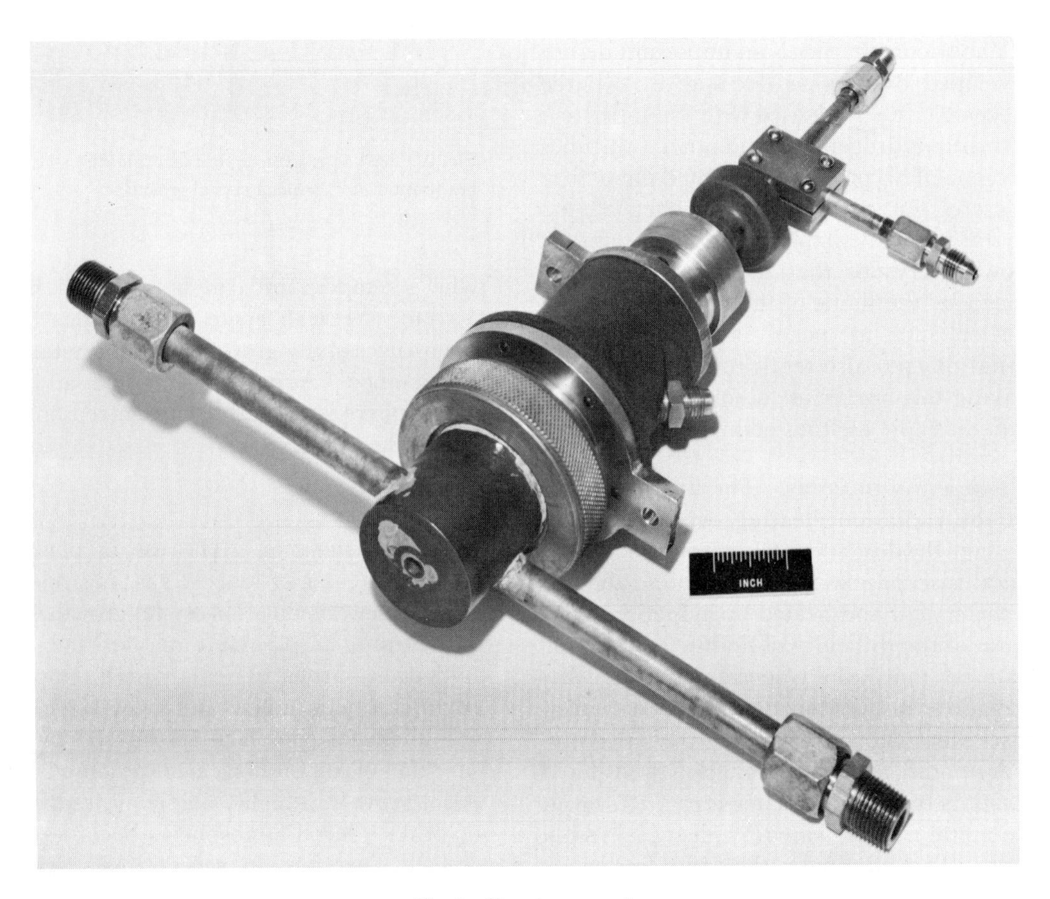

Fig. 3. D-c plasma torch.

sheath, but then it has passed considerable energy into the gas stream. Figure 3 shows a sheath-stabilized torch that was especially designed to operate with a low velocity of flow suitable for crystal growth. The water-cooled tungsten inner-electrode is the cathode the insulated main body, and the water-cooled copper anode can be seen. Located below the $\frac{1}{4}$-in. diameter exit port, but not shown, would be another chamber for powder feed. This design has been operated at power levels up to 60 kw.

These devices operate normally with a molten tungsten tip which is a source of contamination. Plasmas in H_2, N_2, Ar, He, air, and O_2 can be formed, but only for limited duration in the oxidizing gases.

Several a-c plasma torches have been used for crystal growth, including both attached and electrodeless types. The single electrode torch, in which the end of the plasma is in contact with a metallic electrode, was first described in 1928 by Zilitinkovitch. Cobine and Wilbur (1951) described a series of experiments using this. With the frequencies and power levels used by them, heat could not be obtained from monatomic gases, but large amounts of heat could be delivered using polyatomic gases or even traces of them in monatomic gases such as argon. Heating appeared to result from disassociation and subsequent recombination processes. Energetic electrons cause diatomic molecules to break apart, and upon recombination the energy of disassociation is given to the target.

In 1958, an experimental program began to develop a torch for crystal growth (Field, 1958). In the resulting torch, the discharge is initiated by touching the electrode tip with an insulated carbon rod or some other conductor. The torch may be operated vertically upward so that convection currents provide stabilization, or can be operated vertically downward. In the downward position, control can be accomplished by gas flowing between the inner and outer conductors or by jets of gas emanating from the inner conductor tip. This latter arrangement (the "salt shaker" design) surrounds the plasma with six jets of gas, thereby eliminating wandering of the plasma. It is also good for directing the feed powders into the molten pool. It has been operated at frequencies ranging from 30 to 3000 Mc. Plasmas have been sustained with O_2, NO, CO_2, He, Ar, SO_2, and air, but heating was feeble with the monatomic gases.

Fig. 4. Inductively coupled radio-frequency plasma torch.

Electrodeless designs present the best approach for obtaining high purity conditions. In their simplest form, electrodeless torches consist of quartz tubes open at one end with gas injected at the other end. A r-f coil surrounds the tube for inductive coupling; plane-parallel plates or parallel rings enclose it for capacitive coupling.

The inductively coupled torch is started by heating a graphite rod or wire loop within the r-f field until gaseous breakdown has occurred. The hot probe is then withdrawn. For the capacitively coupled torch, a graphite rod is used to draw an arc from the tube wall adjacent to the high potential electrode.

Both torches require a regenerative action for continuous operation; otherwise the initial plasma would be swept away and extinguished. Although convection currents might suffice at very low gas velocities, vortex stabilization produced by a tangential gas inlet is convenient for normal operation. Vortex flow also helps to cool the walls and, in the capacitive version, to prevent arcing.

Figure 4 shows an inductively coupled torch that consists of three concentric quartz tubes with the induction coil just below the middle tube exit. The outer orifice provides an avenue for the swirling gas while the low velocity flow through the middle orifice makes a laminar stream that aids the high velocity center-jet (which carries the feed powder) to reach the molten pool of the growing crystal.

Inductively coupled torches have been operated at 5 Mc and 50 Mc with argon and argon mixtures containing up to 50% of oxygen or 20% air, He, or H_2. Gas flow rates are 5 to 20 liters/min. Power levels up to 3 kw have been produced in the plasmas, and practical temperatures (that is, temperature measured for induced molten pools or probes) of 2800 to 3000°C have been reached.

Two capacitively coupled torches have been reported using plane-parallel plates or parallel rings. Water-cooled quadrant electrodes, 1 in., 3 in., and 6 in. long and rings spaced as close as 4 cm apart, have been used on fused silica tubes 24 to 50 mm in diameter.

Ring electrodes give electric field lines that are longitudinal rather than transverse, making arcing and quartz punctures less likely. The spacing of the rings can be varied for control of field intensities. Capacitive torches have operated at 5 Mc, 30 Mc, and 80 Mc at power levels up to 3 kw. Temperatures have been obtained up to 2100°C. Although a capacitively coupled plasma starts most easily in Ar, starting in air, N_2 and O_2 can be accomplished. Normally, starting is done in Ar and then the polyatomic gas is introduced. The flow rate is 3 to 30 liters/min.

Crystals grown by various plasma techniques include the following:

Cb	TiC	Al_2O_3	Zr_2O_3 (stab.)
Mo	CbC	Al_2O_3 + Cr	$NiFe_2O_4$
Ta	$MoSi_2$	TiO_2	$MnFe_2O_4$
Va	TiB_2	Ti_2O_3	$CoFe_2O_4$
W		TiO	

INDUCTION HEATING. Marino and Bauer (1960) have developed a method that relies on eddy current losses for heating, which has been used for growing ferrite and other crystals (Fig. 5). A chief asset is that it allows a wide latitute of growth atmospheres.

Suitable work coils and generator frequency are the main requirements of this technique. Normal coils are not suitable for providing the thermal conditions needed for crystal growth. The best results have

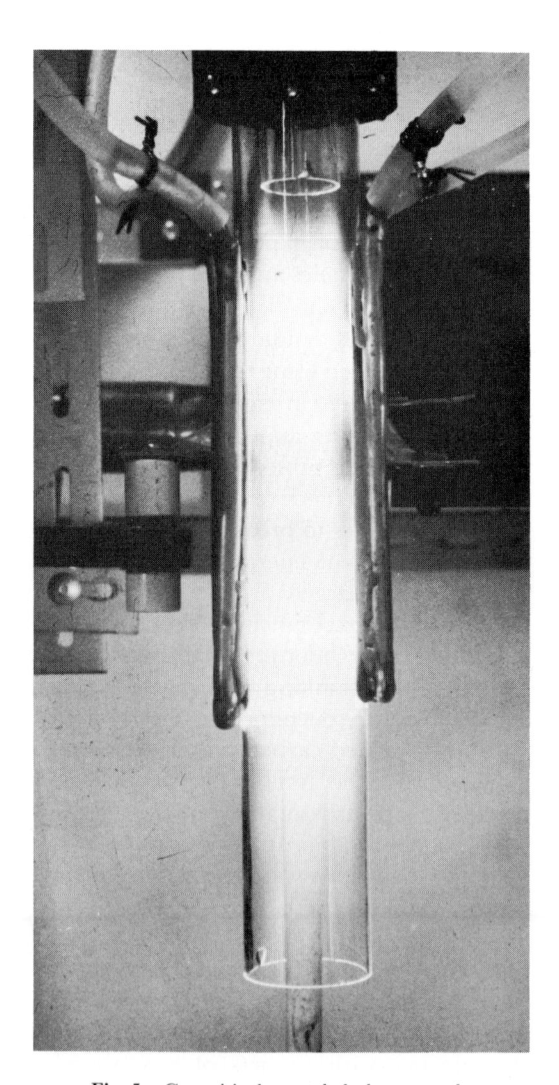

Fig. 5. Capacitively coupled plasma torch.

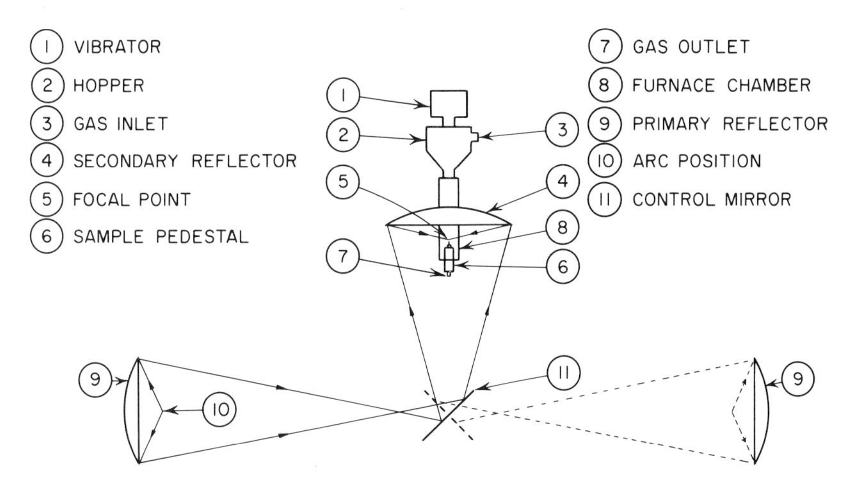

Fig. 6. Optical system of arc-image crystal growth furnace.

been obtained with a five-turn conical concentrating coil having an outer diameter of 3 in., a $\frac{1}{4}$ in. central hole, and a conical plate with an apex angle of 120°. The toroidally shaped field produces a molten pool that can be sufficiently controlled for crystal growth. Rapid stirring activated by the r-f field equalizes the temperatures within the melt and prevents development of a central cold spot that might result from the stream of gas and feed particles. Rotation of the crystal during growth compensates for asymmetries in the field caused by the slot in the coil.

The highly agitated melt requires relatively dense feed particles to overcome the tendency to float on the molten surface and be swept to the side by the stirring action.

Inductive coupling to the crystal depends on its size and the operating frequency. The minimum size seed for adequate coupling at 5 Mc is 2 to 3 mm diameter. Coupling increases markedly with frequency. Thus a seed that could only be slightly heated at 5.2 Mc and 10 kw could be melted at 12.8 Mc and 10 kw, but a large molten pool suitable for crystal growth could be maintained at 25 Mc with only 1 kw.

To provide annealing, a loosely coupled three-turn helical coil ($\frac{9}{16}$ in. ID, $1\frac{7}{16}$ in. in length) is attached in parallel below the conical coil.

With a suitable enclosure, growth in a variety of atmospheres was accomplished, allowing crystals of Ge, Si, NiO, TiO_2, Yttrium-iron garnet, and a variety of ferrites to be grown.

ARC-IMAGES. This heat source consists of a high temperature electric arc and an optical means for focusing its emission in usable form. De la Rue and Halden (1960) used two theatre-type arc sources, or projectors (Fig. 6). Each consisted of an arc and parabolic reflector to produce a parallel beam of light. Since the carbon electrodes used in the arc are short lived, two projectors were used with a mirror that could be flipped to transfer from one to the other during replacement of the electrodes. The reflected light from the mirror was focused by another parabolic reflector. Absorption of this light by an object can easily cause its temperature to rise to ~3000°C. Small crystals of Al_2O_3, stabilized zirconia, and rutile have been grown.

A modification of this technique that should be good uses the clamshell arc-image source described by Ploetz (1959). In Fig. 7 two such clamshell arrangements are shown. Each consists of two parabolic reflectors with a hole cut out of the center of each reflector. A source of light can be placed just outside one reflector and focused at the same distance outside the other reflector. The optical efficiency of this design almost equals the standard arrangement. With two clamshells as shown, the heat fluxes are added

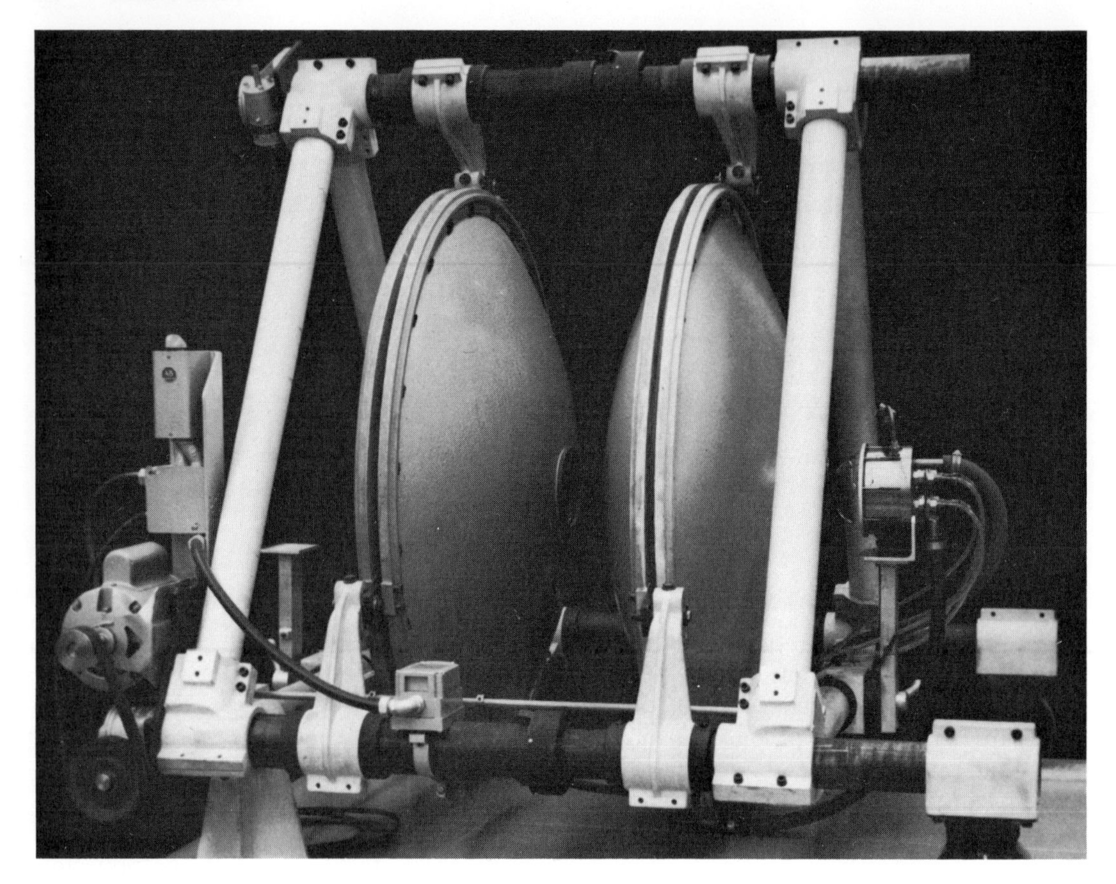

Fig. 7. Two "clamshell" arc-image heaters arranged to supply heat to one object.

and higher temperatures can be obtained. Using high temperature tungsten-lamps temperatures up to 1500°C have been reached, and temperatures up to 2000°C seem feasible. Lamps have the advantages of simplicity, long life, and simple electronic control. By properly contouring the tungsten, it is also possible to shape the heating zone. For controlling the atmosphere, the growth chamber can be surrounded with a transparent shield. Quartz does very well; the light passing through a tube heats it very little because it is not focused on the tube.

2. ANNEALING

Crystals are usually subjected to high thermal gradients during growth by the Verneuil method, so annealing of the crystals is important. This can be done either after the crystal is grown or by surrounding the crystal with an additional heater which tends to keep the entire crystal at a temperature near the annealing point. Thermal strains are thereby greatly reduced during growth, and when the main burner is stopped, annealing can continue with the minimum of thermal shock. Hence, this latter method is by far the best. The additional source of heat also reduces the heating requirements of the main burner. Therefore burners capable of growing larger crystals become possible.

Heating rods that can operate in oxidizing atmospheres up to 1500°C are commercially available. They can be placed around a crystal or separated from it by a ceramic shield. The latter method gives a more uniform temperature distribution and is recommended. Special rod heaters for temperatures up to 1700°C are becoming available. Temperatures up to 2000°C have been produced by gas-fired arrangements; also by heaters operating in inert atmospheres, but such arrangements are difficult to instrument.

Annealing during growth can also be achieved by inductive or capacitive heating.

3. COMPARISON WITH OTHER METHODS

The Verneuil method usually does not produce crystals of as high quality as other methods of growth from the melt because of the presence of the gas flame or other localized heat sources. Its products are usually characterized by large mechanical strains and variations from chemical stoichiometry; control of ionic valence is also a problem.

In spite of its shortcomings, however, the method has a number of favorable features. Chief among these is that it allows crystals to be grown of high temperature materials for which no crucible material exists. The method occupies an important place in solid state research. Nearly all the ferrite crystals that have been studied have been grown by this means. It has also been used for most other oxide crystals including gems such as sapphire, ruby, spinel, garnet, and rutile, as well as zirconium, magnesium, calcium, nickel, and cobalt oxides. Still another important class of crystals for which it is used is the hard metals: carbides, nitrides, borides, silicides, and beryllides. At present, its main commercial importance is for synthetic gems, optical parts, and bearing surfaces.

References

Bauer, W. H., I. Gordon (1951), *J. Am. Ceram. Soc.* **34,** 250–254. Reports N7-ONR-454/2, Jan. 48-Dec. 51.

Bauer, W. H., I. Gordon, C. H. Moore (1950), *J. Am. Ceram. Soc.* **33,** 140–143.

Bauer, W. H. and C. V. Longo (October, 1960), *"R. F. Plasma Torch for Crystal Growth," Air Force Cambridge Research Conference on Verneuil Crystal Growth Reports,* Contract AF 19(604)-5902, June 1959–1961.

Bauer, W. H., K. M. Mertz, P. D. Baba (June, 1958), *Tech. Rept. No. 1* ONR Contract No. Nonr-404(02) NR 032-347.

Cobine, J. D., and Wilbur, D. A. (1951), *J. Appl. Phys.* **22,** 835–841.

De la Rue, F. E. and F. A. Halden (1960), *Rev. Sci. Instr.* **31,** 35.

Dickinson, S. K., W. G. Field, (October, 1956). *Proceedings of Conference on Magnetism and Magnetic Materials (AIEE)* Boston, Mass.

Field, W. G. (1958), *Proceedings of NSIA-ARDC Conference on Molecular Electronics,* Washington, D. C. 13–14.

Keck, P. H., S. B. Levine, J. Broder, R. Lieberman (1954), *Rev. Sci. Instr.,* **25,** 298–299.

Marino, H. A and W. H. Bauer (October, 1960), *Verneuil Crystal Growth Using R.F. Induction Heating,* Air Force Cambridge Research Conference on Verneuil Crystal Growth.

Merker, L. (1947), *FIAT Final Report* #1001.

Ploetz, G. P. (1959), *Status Report Electronic Material Sciences Laboratory,* Air Force Cambridge Research Center, 1 December 1959, AFCRC-TR-59-360.

Popov, S. K. (1959), *Growth of Crystals,* Edited by Shubnikov and Sheftal, Consultants Bureau, New York, p. 103.

Reed, B. (1961); *J. Appl. Phys.* **32,** 821 and 2534. *Science Newsletter* (1947), **52,** No. 16, 243.

Verneuil, A. (1902), *Compt. Rend.* **135,** 791–794.

Verneuil, A. (1904), *Annales de Chimie et de Physique,* Huitieme Serie, III, 20–48.

Zerfoss, S., L. R. Johnson, O. Imber (1949), *Phys. Rev.* **75,** 320.

Zilitinkovich, S. J. (1928), *Telegr. i Telef. bez Provod.* **9,** Nr. 6, 51.

RECRYSTALLIZATION

General Principles

21

Alkali Halides

R. W. Dreyfus

During the last five years, considerable progress has been made in improving the purity and perfection of alkali halide crystals. Previously, the generally accepted standards for "pure" alkali halide crystals have been the products of the Harshaw Chemical Company.* The fact that these crystals were about 10 to 100 times purer than analytical reagent-grade materials and were relatively strain free discouraged most experimenters from preparing their own material. Recently, however, the demand for improved crystals has increased. The result has been major improvements in the field of purification before growth rather than in the growth process itself. Therefore this chapter deals largely with various specific pretreatments before considering the crystal growth process.

Electrical measurements have shown that residual impurities, particularly divalent metallic ions, limit studies of both nominally pure and lightly doped (<50 ppm) crystals. For example, in "pure" NaCl and KCl, it had not been possible to observe intrinsic electrical conductivity below about 400°C because of impurity ions. Also, conductivity (Dreyfus and Nowick, 1962) and dielectric polarization measurements (Haven, 1954; Dreyfus, 1961) on intentionally doped crystals have been limited to relatively high doping levels because it was necessary to override effects due to background impurities. However, precipitation of intentionally added impurities may occur (Dreyfus and Nowick, 1962; Watkins, 1959) when the doping concentration is above about 10 ppm, making lower doping levels desirable.

A demand for purer crystals has also been generated by studies of color centers. Impurity anions and radicals, as well as cations, affect colorability, bleaching, and fundamental absorption edges (Bron, 1960; Etzel and Patterson, 1958; Kerkhoff, 1960; Rolfe et al., 1961; Otterson, 1960; Haupt, 1959). Effects of strain on the coloring rate have also been noted (Bron, 1960; Gordon and Nowick, 1956; Nowick, 1958). Thus studies of intrinsic colorability require better crystals. Other studies (for example, of photoconductivity and mechanical deformation studies, Johnston, 1962) have also been handicapped by relatively poor crystals.

Some of these experiments provide quantitative measures of purity. For NaCl and KCl, the relationship between ionic conductivity and divalent metallic impurities has been relatively well established (Etzel and Maurer, 1950; Kelting and Witt, 1949; Lidiard, 1957). Other types of impurities play a less well-defined role (Gruzensky and Scott, 1961). The electrical conductivity at relatively low temperatures is proportional to the impurity content (in particular, that of divalent metallic ions) under many conditions. The initial rate of *F*-center production by X-rays at room temperature has also been used to estimate relative purities. Broadening of the fundamental optical absorption-edge often indicates the presence

* Harshaw Chemical Co., Cleveland Ohio. This company has been willing to select samples of above average purity when so requested by experimenters. NaCl and KCl crystals supplied by this source have usually had a metallic-impurity level of about 1 ppm, primarily consisting of aluminum, magnesium, and calcium chlorides.

of anion impurities (Etzel and Patterson, 1958; Kerkhoff, 1960; Rolfe et al., 1961; Otterson, 1960; Haupt, 1959; Kobayashi and Tomiki, 1960). Infrared absorption bands caused by radicals at anion sites have been used to identify and measure such radicals (Etzel and Patterson, 1958; Morgan and Staats, 1962). Measurements of electrical conductivity and F-center production (along with conventional spectroscopic analysis) have been the primary methods for estimating divalent metallic-ion concentrations. Since the objective of many purification schemes is to eliminate cation impurities, these methods have been emphasized.

1. PURIFICATION

Zone Refining

A number of methods for purification have been applied to the alkali halides especially zone refining (Pfann, 1958) because of its success with many other materials. This method was not successful at first, the difficulty being that the melt adhered strongly to all containers. Even graphite boats often would not withstand the strains set up by the large thermal contractions. The cause of this adhesion problem was apparently first recognized by Wartenberg (1953, 1957), who found that alkali oxides or hydroxides were to blame. Gründig (1960) was the first to eliminate these compounds and then to successfully zone refine KBr and KCl. (The method used to eliminate the oxides and hydroxides are discussed in detail later, since it constitutes a distinct step in the purification process; reaction with such gases as HCl or Cl_2 is required.)

Following the original work on zone refining, supplementary results have been forthcoming (Kanzaki et al., 1962; Kanzaki and Kido, 1960; Inoue and Mizuno, 1961; Anderson et al., 1960). The actual operating parameters do not differ much from those for semiconductors. Typical ingot dimensions range from 1 to 10 cm^2 in cross section by 10 to 50 cm in length. From one to three hot zones pass over the boat simultaneously at a rate of 0.6 to 10 cm/hour. The width of the molten zone is usually 2 to 6 cm. Graphite or quartz boats have been used, although quartz boats fracture occasionally when cooled below about 500°C. Platinum boats are not used because they react with the deoxidizing atmospheres. Neither quartz nor very high-purity graphite-boats give evidence in electrical measurements of causing contamination.

Most zone refinement has dealt with KCl, where the primary objective has been to eliminate divalent impurities. The predominant impurities in analytical reagent-grade KCl are: a) $CaCl_2$ (3 to 50 ppm); b) the other alkali metals (30 to 100 ppm), and c) transition, trivalent, or heavy metals (20 to 100 ppm). The effect of zone refining is least for $CaCl_2$ for which the rejection ratio is about 0.4 (solid to liquid concentration ratio). Electrical measurements indicate that Ca^{++} and other electrically active ions are decreased to 0.001 to 0.01 ppm after 50 to 70 zone passes. No final limit to this purification has appeared yet, but the process is tedious because of the relatively large rejection ratio of 0.4 for calcium. The anticipated purification for a particular apparatus can be found from Pfann's graphs (1958). Sodium ions are rejected from KCl with a ratio of 0.1, thereby resulting in very rapid elimination (Inoue and Mizuno, 1961). Polyvalent metals other than calcium are also usually rejected rapidly. In addition, gas flowing over the melt can be used to eliminate these latter impurities. If the gas flows from the pure to the impure end of the boat, any material that evaporates from the boat is moved in this same direction (Anderson et al., 1960). Many of these impurities have boiling points lower than 1000°C and hence are evaporated and then carried away. The few elements that remain usually have a smaller rejection ratio than Ca^{++} so that after a few zone passes the latter is the predominant remaining impurity.

Some additional complications of zone refining are (1) for KCl purified in graphite, particles of graphite have been observed. The use of pyrolytic graphite (Beels and De Sutter, 1960; Taylor, private communication) at least partially overcomes this contamination. (2) Gas bubbles tend to form in the solid as a result of the conditioning-gas treatment. (3) Zone-refined material is polycrystalline, with grains about 1 cm^3 in size that appear to be highly strained.

The results the author had for NaCl are different from those for KCl in one important sense. Even after as many as sixty zone passes, no reduction in the Ca^{++} ion content (about 10 to 30 ppm) is observed. This was long after the other metallic impurities had been reduced below spectroscopic limits. Apparently the rejection ratio of NaCl for Ca^{++} ions is essentially unity.

It now appears that potassium bromide (Gründig, 1960), potassium iodide (Haupt, 1959), silver chloride (Moser et al., 1961; Süptitz, 1962), silver bromide (Süptitz, 1962), and cesium iodide (Besson et al., 1962) can be zone refined. There has not as yet been an investigation of zone refining for lithium fluoride. Visual inspection of small LiF ingots indicates that at least some of the impurities remain in the liquid phase until crystallization is almost complete (Welber, 1963; also see *Note,* 1963).

Chemical Methods

Purification by strictly chemical treatment of aqueous solutions has been used occasionally. These methods may be very effective, but they usually require too much time, equipment, and experience to be suitable when only small quantities of pure material are required. Two general categories exist, depending on whether purification begins with the starting materials to be reacted later to form the alkali halide or with purification of the halide itself. All the possibilities are too numerous to describe here; they are discussed in many chemical treatises. We shall discuss only techniques that have been specifically used prior to crystal growth. The purification of $KHCO_3$ or $NaHCO_3$ and their later reaction with purified HCl are described by Kobayashi and Tomiki (1960). A method for a variety of alkali halides has been described in a Harshaw Chemical Company patent.* The use of an ion exchange resin has recently proved successful for eliminating polyvalent metallic ions both in our laboratory and others (Fredericks and Hatchett, 1962).† A dilute aqueous NaCl or KCl solution is passed through a chelating resin. In addition, anion purity is subsequently improved by using an anion exchange resin (Fredericks and Hatchett, 1962).‡ This author estimates that the $CaCl_2$ content can be reduced to about 1 ppm by the preceding treatments. Fractional crystallization of the resulting NaCl solution may further decrease the $CaCl_2$ level. Even without ion exchange purification, this latter treatment is sufficient to reduce the $CaCl_2$ concentration to $\sim$1 to 5 ppm.

Removal of $O^=$ and OH^-

Residual water exists in the purified material resulting from all the preceding chemical processes as well as in the reagent-grade material. If this water is not removed before the halide is heated near the melting point, hydrolysis will take place. The result is that some OH^- and $O^=$ ions remain in the salt, whereas Cl^- ions escape as HCl. These anions not only contaminate the material but also make zone refining impossible because of the adhesion problem mentioned previously. Treatment with gaseous Cl_2 or HCl (Wartenberg, 1953, 1957; Gründig, 1960) eliminates these impurities. The bromides behave analogously (Gründig, 1960), but the author has experienced somewhat more difficulty with the iodides, primarily because of the lower reactivity of iodine. The treatments are usually carried out either slightly above or below the melting point for periods up to 48 hours (Gründig, 1960; Otterson, 1961). The greater surface area of powdered solid apparently compensates for higher reaction rates in the liquid state. The partial pressure of the gas is usually 10 mm to 760 mm of Hg, and commercial mixtures containing 2% Cl_2 and 98% N_2 are quite satisfactory for treating NaCl and KCl. It is often convenient to apply this gas treat-

* U. S. Patent No. 2,640,755 by John O. Hay, dated June 2, 1953.

† For example, Chelex 100, Bio-Rad Laboratories, Richmond, California.

‡ AG2-X10 Bio-Rad Laboratories, *loc. cit.*

ment to powder lying in a zone-refining apparatus.

There is a simpler way to reduce the $O^{=}$ and OH^- impurity level, but it may not result in complete elimination. This consists of drying the halide very carefully in a vacuum of $<10^{-2}$ mm Hg. In this laboratory the powder is heated for one hour at $\sim 100°C$ temperature intervals from ambient up to 500°C. At the end of each one-hour period, dry inert gas (say 500 mm pressure of argon) is put into the apparatus before raising the temperature. When this gas is pumped off after about one minute, it presumably carries water vapor out of the crevices in the powder. It is believed that this water would not normally diffuse out by itself. This treatment has been used successfully to prepare for the addition of reactive doping agents such as vanadium dichloride.

Gas occlusions may exist in halides after treatments with HCl, Cl_2, or similar gases. It is possible to minimize this difficulty however. A few helpful factors are mentioned: treating the halide with a relatively low partial and/or total gas pressure of HCl, etc., is helpful, particularly in the later stages of purification. The use of graphite, instead of quartz, boats eliminates some of the major gaseous contaminants (Dekeyser, private communication), namely the silicon oxy-chlorides. Both these modifications decrease, however, the effectiveness for oxygen removal.

Sublimation and Distillation

Vacuum sublimation or distillation rejects a variety of impurities (Kanzaki et al., 1962; Kanzaki and Kido, 1960). If purification is terminated with a step of this type, the problem of gas or graphite occlusion should be eliminated completely. In addition, both these processes reject $CaCl_2$ and other metallic chlorides. The first condensate to be collected is normally improved by 5:1 or 10:1. If the distillation is carried out in a quartz column, any remaining OH^- ions may react with the quartz and thus be further eliminated (Etzel and Patterson, 1958; Kerkhoff, 1960; Rolfe et al., 1961; Otterson, 1960). Of the two methods, distillation columns have the advantage that purification equivalent to many operations can be obtained within a single apparatus. This advantage is somewhat offset by the fact that the quartz apparatus often fractures during cooling. Therefore the equipment should be simple enough for easy repair. Sublimation, although it normally leaves the equipment intact, is usually only feasible as a one-step-at-a-time arrangement. For both techniques, the rate of material transport is satisfactorily high.

Electrolysis

The alkali halides can be purified by passing a relatively large d-c electric current through them (Kerkhoff, 1951, 1956). For high conductivity this necessitates a high temperature, and purification can be obtained in either the solid or liquid state. Polyvalent ions in these structures have higher mobility than the alkali metals because their motion is facilitated by the high electrostatic force per ion and by the greater probability of a vacancy being nearby (Lidiard, 1957).

2. CRYSTAL GROWTH

Actual crystal growth is usually done by the Kyropoulos method (1926), which has been highly developed by work in the semiconductor field, and only a slight apparatus modification adapts it for the present purpose (Patterson, 1962). Crystals are also produced by the Bridgman technique (Kremers, 1940).

The general apparatus requirements are: only quartz, graphite, or platinum tubes, crucibles, etc., are directly exposed to the hot halide fumes. This tends to avoid metallic chloride formation that could contaminate the melt. Inert gas at nearly atmospheric pressure is normally maintained over the melt to inhibit evaporation as well as to avoid the formation of hydroxides and oxides.

Figure 1 shows schematically how these features are incorporated in a typical crystal grower. Highly precise temperature control

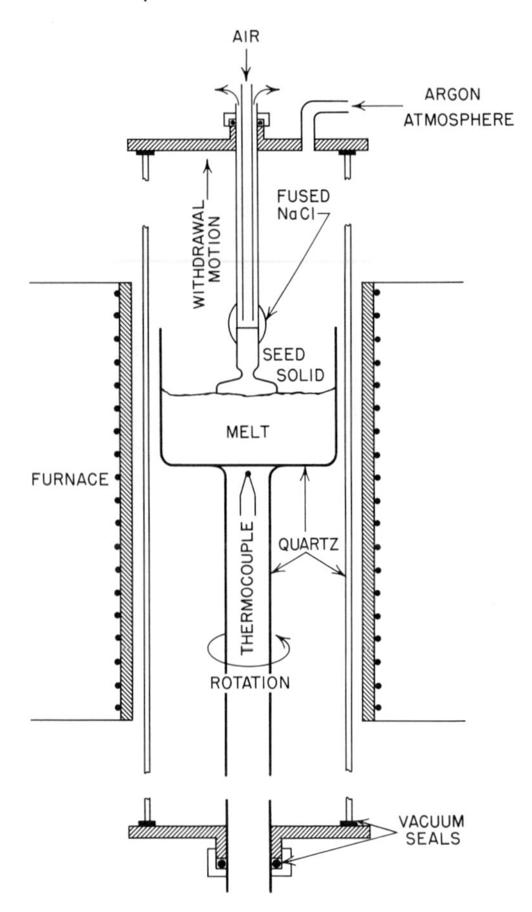

Fig. 1. Growth apparatus for alkali halide crystals.

is not required for crystal growth, presumably because of the low thermal conductivity of the alkali halides as compared with semiconductors. The low thermal conductivity also facilitates the formation of large temperature gradients which, combined with large thermal expansion coefficients, often leads to badly strained crystals. Thus considerable care must be taken to avoid sudden temperature changes.

The seed-crystal holder that is used is somewhat unique (Bean, 1952). In the apparatus of Fig. 1, the seed is wired to the bottom of a sealed-off, $\frac{1}{4}$ in. O.D. platinum tube. Concentrically mounted inside this tube is another tube of $\frac{1}{8}$ in. O.D. This allows cooling air or water to be brought close to the top of the seed crystal. Such drastic cooling is necessitated by the seed's low thermal conductivity. Cooling is further enhanced by briefly dipping the seed and about $\frac{1}{4}$ in. of the platinum tube into the molten alkali halide just prior to crystal growth to form a heat conducting path (Bean, 1952). The resulting polycrystalline layer on the bottom of the seed is removed by melting back part of the seed.

Rotation of the seed (or better, of the melt) at perhaps 2 rpm is useful for smoothing out thermal gradients at the surface of the melt. A typical velocity of seed withdrawal is $\frac{1}{4}$ in. per hour, but the crystal growth rate may be several times this because the liquid level in the crucible lowers as solidification proceeds. The latter helps prevent the crystal from growing laterally and fusing to the crucible. Experience shows that it is possible to grow crystals for days at a time with little attention.

Improvement of the perfection of alkali halide crystals has received only limited attention. Etch pit counts show that LiF crystals from the Harshaw Chemical Co. (Johnston, 1962), and those carefully grown using the Kyropoulos technique (Kyropoulos, 1926; Washburn and Nadeau, 1958; Gilman and Johnson, 1962) have dislocation densities of $\sim 5 \times 10^4/cm^2$. Similar numbers have been given for Harshaw NaCl (Mendelson, 1962), but some crystals have been obtained (Mitchell, private communication) with as few as 500 dislocation/cm^2. To obtain these, not only careful growth was needed, but also it was necessary to etch away the surface (after cleaving) and anneal the remainder. For KCl, workers (Kanzaki et al., 1962) have counted about 2×10^6 dislocations per cm^2 in Harshaw crystals. Crystals grown in a zone refining boat had similar dislocation densities (Kanzaki et al., 1962).

ACKNOWLEDGMENTS

The author was aided in the design, construction, and operation of his equipment for purification and crystal growth by K. W. Asai. Thanks are also due to A. S. Nowick and W. E. Bron for helpful suggestions. This work was supported in part by the U.S. Atomic Energy Commission.

References

Anderson, S., J. S. Wiley, and J. Hendricks (1960), *J. Chem. Phys.* **32**, 949.

Bean, C. (1952), Thesis, University of Illinois, Urbana, Illinois.
Beels, R., and W. DeSutter (1960), *J. Sci. Instr.* **37,** 397.
Besson, H., D. Chauvy, and J. Rossel (1962), *Helv. Phys. Acta.* **35,** 211.
Bron, E. W. (1960), *Phys. Rev.* **119,** 1853.
Dekeyser, W., Private Communication.
Dreyfus, R. W. (1961), *Phys. Rev.* **121,** 1675.
Dreyfus, R. W., and A. S. Nowick (1962), *Phys. Rev.* **126,** 1367.
Etzel, H., and R. Maurer (1950), *J. Chem. Phys.* **18,** 1003.
Etzel, H. W., and D. A. Patterson (1958), *Phys. Rev.* **112,** 1112.
Fredericks, W. J., and J. L. Hatchett, *1962 International Symposium on Color Centers in Alkali Halides,* Stuttgart, Germany.
Gordon, R. B., and A. S. Nowick (1956), *Phys. Rev.* **101,** 977; A. S. Nowick (1958), *Phys. Rev.* **111,** 16.
Gründig, H. (1960), *Z. Physik* **158,** 577.
Gruzensky, P. M., and A. B. Scott (1961), *J. Phys. Chem. Solids* **21,** 128.
Haupt, U. (1960), *Z. Physik* **157,** 232.
Haven, Y., (1954), *Report of the Bristol Conference on Defects in Crystalline Solids,* The Physical Society, London, 1955, p. 261.
Inoue, M., and H. Mizuno (1961), *J. Phys. Soc. Japan* **16,** 128.
Johnston, W. G. (1962), *J. Appl. Phys.* **33,** 2050.
Kanzaki, H., and K. Kido (1960), *J. Phys. Soc. Japan* **15,** 529.
Kanzaki, H., K. Kido, and T. Ninomiya (1962), *J. Appl. Phys.* **33,** 482.
Kelting, H., and H. Witt (1949), *Z. Physik* **126,** 697.
Kerkhoff, F. (1951), *Z. Physik* **130,** 449; and M. Chemla, (1956), *Ann. Phys.* **1,** 959.
Kerkhoff, F. (1960), *Z. Physik* **158,** 595; J. Rolfe, F. R. Lipsett, and W. J. King, (1961), *Phys. Rev.* **123,** 447, and D. A. Otterson (1960), *J. Chem. Phys.* **33,** 227.
Kobayashi, K., and T. Tomiki (1960), *J. Phys. Soc. Japan* **15,** 1982.
Kremers, H. C. (1940), *Ind. Eng. Chem.* **32,** 1478.
Kyropoulos, S. (1926), *Z. Anorg. u. allgem. Chem.* **154,** 308.
Lidiard, A. B. (1957), in *Handbuch der Physik,* edited by S. Flugge, Springer-Verlag, Berlin, Vol. 20, p. 246.
Mendelson, S. (1962), *J. Appl. Phys.* **33,** 2175.
Mitchell, J. W., Private Communication.
Morgan, H. W., and P. A. Staats (1962), *J. Appl. Phys.* **33,** 364.
Moser, F., D. C. Burnham, and H. H. Tippins (1961), *J. Appl. Phys.* **32,** 48.

Note: 1963 (added in proof). Zone refining of commercial grade LiF and RbCl has recently been carried out in this laboratory. Spectroscopic analysis indicates that the total metallic impurity content has been reduced below 10 ppm in both substances.

Otterson, D. A. (1961), *J. Chem. Phys.* **34,** 1849.
Patterson, D. A. (1962), *Rev. Sci. Instr.* **33,** 831.
Pfann, W. G. (1958), *Zone Melting,* John Wiley and Sons, New York.
Suptitz, P. (1962), *Ann. Physik* **9,** 133.
Taylor, A., Private Communication.
Wartenberg, H. v. (1953), *Z. anorg. u. allgem. Chem.* **273,** 257; (1957), *Z. angew. Chem.* **69,** 258.
Washburn, J., and J. Nadeau, *Acta Met.* **6,** 665 (1958), and J. J. Gilman and W. G. Johnston (1962), "Dislocations in Lithium Fluoride Crystals," *Solid State Physics,* edited by F. Seitz and D. Turnball, Vol. 13, p. 147.
Watkins, G. D. (1959), *Phys. Rev.* **113,** 79 and 91.
Welber, B. (1963) Private Communication.

Principles of Recrystallization

W. G. Burgers

To define precisely what is meant by recrystallization is less simple than it might appear. In its most general sense recrystallization involves a change in the crystalline state of a solid material. Such changes may be of widely different kind and extent. They may vary from local changes on an atomic scale, detectable only by a change in definite physical, mechanical, or magnetic properties, to the formation and growth on a "visible" scale of new crystallites in a specimen. The term "recrystallization" is usually reserved for this second category of structural changes, especially when they take place within the range of thermodynamic stability of a given phase. Structural changes due to allotropic transitions or to the transgression of a boundary line in the phase diagram of a multicomponent system (for example, the processes occurring during precipitation of a solute from a supersaturated solid solution) are not considered because of their very complicated characters.

In practice then, recrystallization encompasses those phenomena which bring about a change in the number, size, shape, orientation and, in a certain measure, the state of perfection of the constituent crystallites of a specimen.

In this chapter we discuss recrystallization with the object of the present book in view, viz., the preparation of large, single crystals. Preparing a large crystal by recrystallization amounts in practice to the transformation in the solid state of a fine-grained polycrystalline specimen into one consisting of a single, or at the most a few, large crystals. This can be achieved if one succeeds in making one of the grains consume all others, or, alternatively, if by proper nucleation a new grain is formed which then grows at the cost of all others. From the energetic point of view such a process must be favorable because a polycrystalline structure does not represent the metal in its most stable thermodynamic state. In fact, even a well-annealed specimen consists of individual lattice domains,* which are differently oriented with respect to each other, and therefore are separated by transition zones, the domain boundaries. In these zones the atoms do not fit precisely into either one of the adjoining lattice regions but occupy intermediate positions. Such boundaries represent zones of higher free energy, and energy could therefore be gained if their total extent were reduced. It is this that the metal tends to achieve if upon heating to a sufficiently high temperature, the atomic vibrations increase enough that movements of the atoms become possible within reasonable times. The final state is reached if the whole specimen is transformed into one lattice region of constant orientation, that is, into one crystal.

It is the purpose of what follows to set forth the conditions under which such transformation becomes possible. Thus it is necessary to discuss the general "laws" which apply to the two fundamental parts of the recrystallization process: (1) nucleation and (2) growth of crystals.

1. SOME BASIC RECRYSTALLIZATION PHENOMENA

Several aspects of nucleation and growth have been studied by specific experiments carried out with specimens prepared espe-

* We do not consider here whether these domains are crystallites proper or, for example, mosaic blocks.

cially for investigating the property in question; for example, specimens containing only one large crystal with a definite lattice orientation or fine-grained specimens with a prechosen preferential orientation of the constituent crystallites. Such specimens have usually been prepared by precisely the methods that are discussed in Section 4, and which are a result of the experiments themselves. In order to understand how these specimens have been obtained we first discuss some general rules concerning recrystallization phenomena. A considerable number of these have been deduced from experiments with the metal aluminum. Because of its relatively low recrystallization temperature (melting point about 660° C) and its resistance against oxidation when heated in air, aluminum is a good material to use in recrystallization experiments.

A normal piece of metal, whether it is a plate, rod, or wire, is usually fine-grained and built up of crystallites (grains) often a few hundredths of a millimeter in diameter. Having started from a casting, it is the product of a series of alternating cold-working and annealing treatments. During cold-working, the original crystallites formed from the melt are deformed and distorted by gliding (and possibly twinning or kinking), so that a fiberlike texture is obtained in which the individual crystallites can often not be clearly discerned (Fig. 1). The distortions put the metal into a state of increased free energy. Upon heating at a suitable temperature, this surplus energy may be released by the formation in the solid state of many new "perfect" crystallites.*

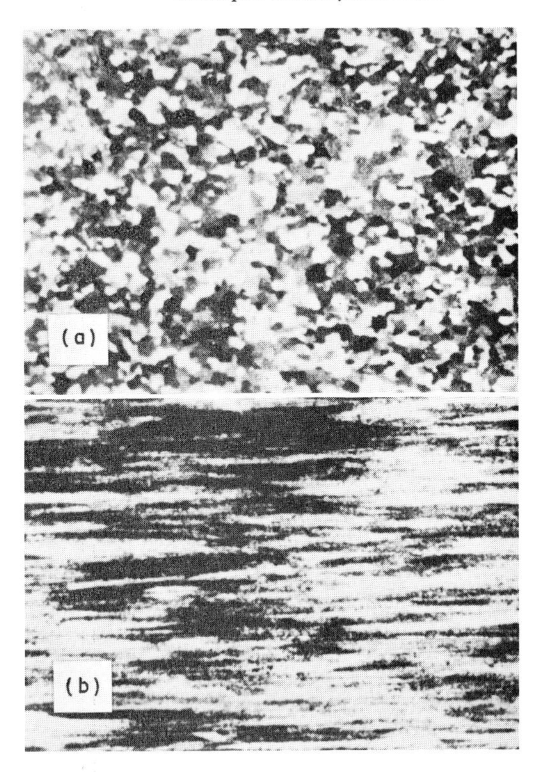

Fig. 1. Distortion of crystallites by rolling. (*a*) Fine-grained recrystallized aluminum testpiece. (*b*) Fibrous structure due to rolling (5 X). (After Sachs, 1934.)

A polycrystalline fine-grained specimen may either be quasi-isotropic or may possess a more or less pronounced preferential orientation of the crystallites, that is, a texture. The latter state is the more common one, as in the course of the deformation process the active glide planes and glide directions tend to become parallel to the principal directions of the deformation (direction of rolling or drawing, for example). This results in a preferential orientation. Because of the special character of the recrystallization nucleation process (see Section 3), the texture is almost never completely obliterated by a subsequent recrystallization. It is therefore rctaincd to some extent in the annealed material, although its crystallographic characteristics may be considerably altered.

It is important to realize that very pronounced ("sharp") textures can often be obtained by cold-working a monocrystal in a definite direction. One starts from a specimen containing only one lattice orientation so that throughout it the same glide planes and directions may be active. Recrystallization then usually yields new crystallites with closely similar orientations (within about 5 to 10°). The resulting texture approaches a (pseudo) monocrystalline state. The orientation of the texture finally

* The crystallites are made visible by etching with a suitable reagent which attacks definite lattice planes preferentially. For example, hydrofluoric acid etches the cube planes {100} in aluminum. Each crystallite acquires a stepped surface formed by {100} planes varying in direction from crystallite to crystallite. Incident light is reflected by the individual crystallites in different directions, thus making them visible either directly or under the microscope.

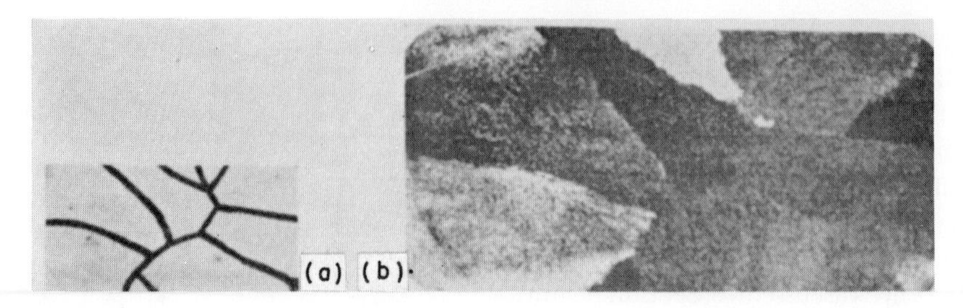

Fig. 2. Formation of preferred orientations due to recrystallization. (*a*) Drawing of the original aluminum specimen, containing a few large crystals, the boundaries of which are indicated. (*b*) The same specimen after rolling to about one-fourth of its original thickness, followed by primary recrystallization; the whole specimen is now fine-grained, yet the domains originally occupied by the individual crystals can still be recognized. This is due to preferential orientation of the new crystallites formed within each original crystal. (After van Arkel, 1932.)

obtained depends on the orientation of the original crystal with respect to the directions of deformation. This dependence is clearly illustrated in Fig. 2. It shows the effect of rolling and recrystallizing an aluminum specimen that consisted originally of a small number of large crystals. After recrystallization, although the whole specimen is now fine-grained, the various regions originally occupied by the large crystals can still be discerned after etching because of preferential orientation of the new crystallites formed within each original crystal. X-ray Laue photographs taken at two spots lying in two different regions confirm this.

A fine-grained polycrystalline specimen as just described can be transformed into a specimen with larger crystallites either by heat treatment or by plastic deformation followed by heat treatment. Very often, depending on such factors as original grain size, impurity content, and texture, only prolonged heating is needed to increase the grain size (Fig. 3). The course of the process has been followed by repeated etchings between the steps of the annealing treatment. For a noncubic metal like zinc or tin, etching is not necessarily required because its optical anisotropy allows growth to be followed directly under the microscope in polarized light (Brinson and Moore, 1951; Barten, 1954). In some cases, moreover, direct observation of growth in heated specimens is possible in an electron emission microscope (Burgers and Ploos van Amstel, 1937, 1938; Rathenau and Baas, 1951).

From such experiments it is found that the increase in grain size takes place by the growth of certain crystallites at the expense of their neighbors, the latter becoming reduced in size or completely consumed. Usually this process proceeds to some extent and then nearly stops, the grain size remaining constant during very prolonged further heating. The growth stops because the driving force decreases with increasing grain size and because a state of pseudoequilibrium is attained in which the tensions acting in the boundaries become mutually balanced. The process is called *normal* or *continuous grain growth,* also *normal coarsening* or *coalescence.*

It sometimes happens, however, that prolonged heat treatment of a fine-grained matrix does not yield general coarsening. Instead only a few crystals grow at the expense of all (or nearly all) of the primary matrix grains. The latter do not themselves increase in size. Then one speaks of abnormal, discontinuous, or exaggerated grain growth, also, particularly in the German literature, of secondary recrystallization.* An example for aluminum is shown in Fig. 4. In such cases, structural conditions must exist in the matrix which are capable of creating some kind of "contrast" between one lattice domain (or at the most a small number) and all the others, thus enabling the special

* The latter expression refers to the fact that the fine-grained matrix in which the abnormal grain growth takes place is usually itself obtained by "primary" recrystallization of a cold-worked specimen.

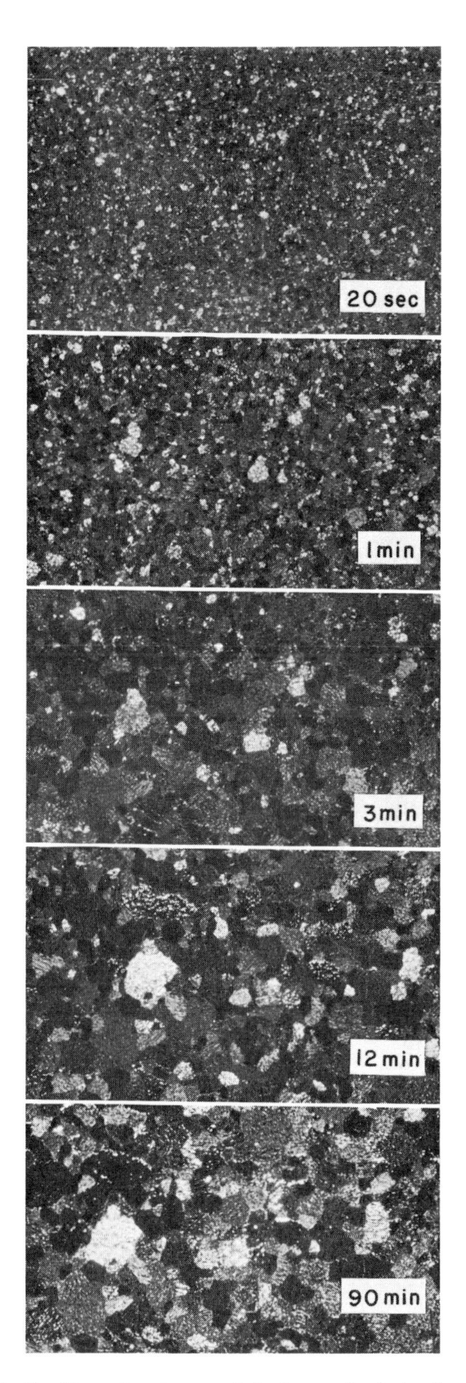

Fig. 3. Normal grain growth in fine-grained aluminum heated at 600°C for various times (10 X).

ones to grow continuously and finally to occupy the whole specimen. We come back to this phenomenon in Section 2.

A more widely applicable method for altering the grain size of a specimen involves the use of a precise amount of plastic deformation prior to annealing. If, for example, a fine-grained aluminum specimen, whose grain size is not affected by direct heating, is extended by a few per cent and then heated at about 600°C, a small number of new grains appear which grow during prolonged heating at the expense of the fine-grained matrix (Fig. 5). The number of new grains, and thus their size, depends on the amount of prior deformation. To illustrate this, Fig. 6 shows a series of aluminum strips which before deformation and heating were identical (grain size of circa 0.01 mm). They were subjected to increasing amounts of extension (2 to 10%), followed by heating at 600°C for a few hours. The new grain size is largest for the smallest degree of extension. Figure 7, taken from Karnop and Sachs (1930), shows intermediate drawings of a specimen after increasing periods of heating. The new grains form at different times, and each grows into the matrix until mutual impingement of neighboring crystallites occurs.

The "birth" of the new crystallites occurs at places in the matrix where the extension has caused strong local deformation. For example, such places are the grain boundaries of the original crystallites which act as barriers to continuous glide. This can be easily demonstrated (Fig. 8) by a slightly extended specimen containing only a few easily visible, original boundaries. It is consistent with this that the formation of new crystals can be locally stimulated by local deformations produced by pin pricks or notches (Fig. 9). Subsequent heating causes the formation of new crystals at the special places.

There is one more point that is important for the considerations discussed in the following sections. A crystal growing in a polycrystalline specimen has a definite lattice orientation with respect to the main directions of its external shape (for example, the axis of a rod, the surface, and direction of rolling of a plate, etc.). This orientation is fixed when the crystal nucleates and is usually unknown beforehand. As will appear later on (Section 2), it may be useful or necessary to produce a crystal with a definite orientation with respect to these

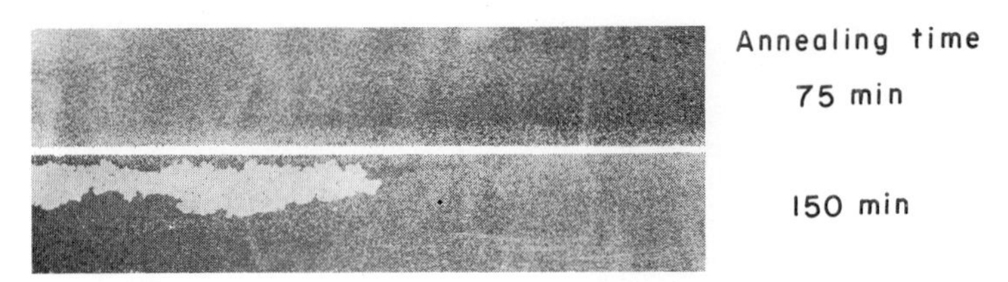

Fig. 4. Abnormal ("discontinuous") grain growth in fine-grained aluminum after heating at 600°C (approximately natural size).

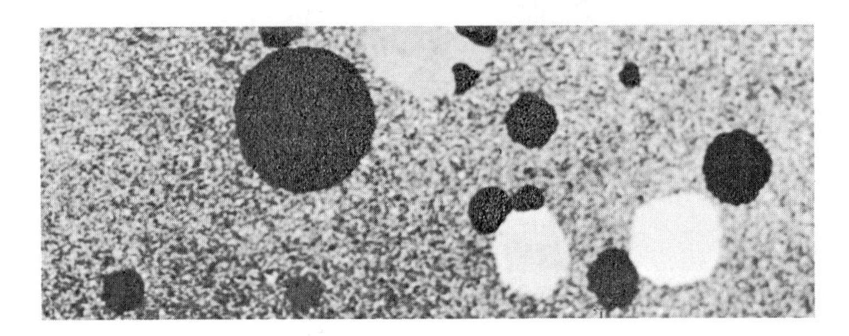

Fig. 5. Circular crystallites formed on recrystallization in a fine-grained aluminum matrix without preferential orientation of the constituent grains (approximately natural size).

directions. For example, one may require a crystal with a cube face {100} parallel to the surface of a strip.

For crystals grown from the melt, specific orientations can be obtained by introducing into the melt a "seed" held at the required orientation. For growth in the solid this procedure has been modified by Rutter and Aust (1961) by placing the "seed crystal" and the matrix in which it must grow end-to-end in a crucible and welding them together at the junction. On further heating, the "seed crystal" may grow into the matrix in the desired orientation.

One can also alter the relative orientation of the matrix and the growing crystal by bending part of the unconsumed matrix and making the crystal grow "around the corner" as shown schematically in Fig. 10 (Tiedema, 1949). In this way the relative orientations can, at least in principle, be altered in any desired way. To apply this method, a crystal needs to be grown partially into the cold-worked deformed matrix (by introducing only part of it into a furnace). Then recrystallization is interrupted, the bending done, and recrystallization continued. In order to obtain a definite orientation relationship between the crystal lattice and the specimen, bending (or possibly twisting) must be done around definite axes through definite angles. To fix these, the orientation of the growing crystal with respect to the directions of the specimen must be determined (for example, from Laue photographs). The actual deformation should be done with the aid of a special jig (see Fig. 11). This method, originally due to Fujiwara (1939, 1941), was applied by Tiedema (1949) and Weik (1953) for aluminum, and by Dunn (1949a) for silicon-iron. Also "curved" monocrystals, to be used as monochromators for X-rays, can be prepared in this way (Despujols, 1952; Hägg and Karlsson, 1952).

A closely related method for growing oriented crystals from the melt was developed and applied for lead crystals by Brugman and Tiedema (1952).

2. CRYSTAL GROWTH (BOUNDARY MIGRATION)

Direction of Displacement

The growth of crystals within solid metals takes place under conditions quite different from those prevailing during growth from solutions or vapors. The growing surface is not exposed to a bombardment of free atoms or molecules. Instead, it must capture these particles from a surrounding matrix where each of them occupies a fixed position. Growth is thus brought about by movement of the boundary layer between two non-parallel lattice regions. In a well-annealed specimen the driving force for such movements is the reduction of the extra free energy stored in these layers. Given a distinct boundary, one way to achieve this is to shorten its length. This leads to the expectation (Fig. 12) that a curved boundary will move in the direction of its center of curvature. Examples of such movements are given in Figs. 13 and 14. The former (due to Beck and Sperry, 1950) refers to grain growth in pure aluminum, the latter (Ferro, Sari, and Venturello, 1959) to growth in silicon-iron. In both cases, successive posi-

Fig. 6. Set of fine-grained aluminum testpieces, subjected to increasing amounts of extension, followed by recrystallization; the number of the new grains increases with the deformation (approximately natural size).

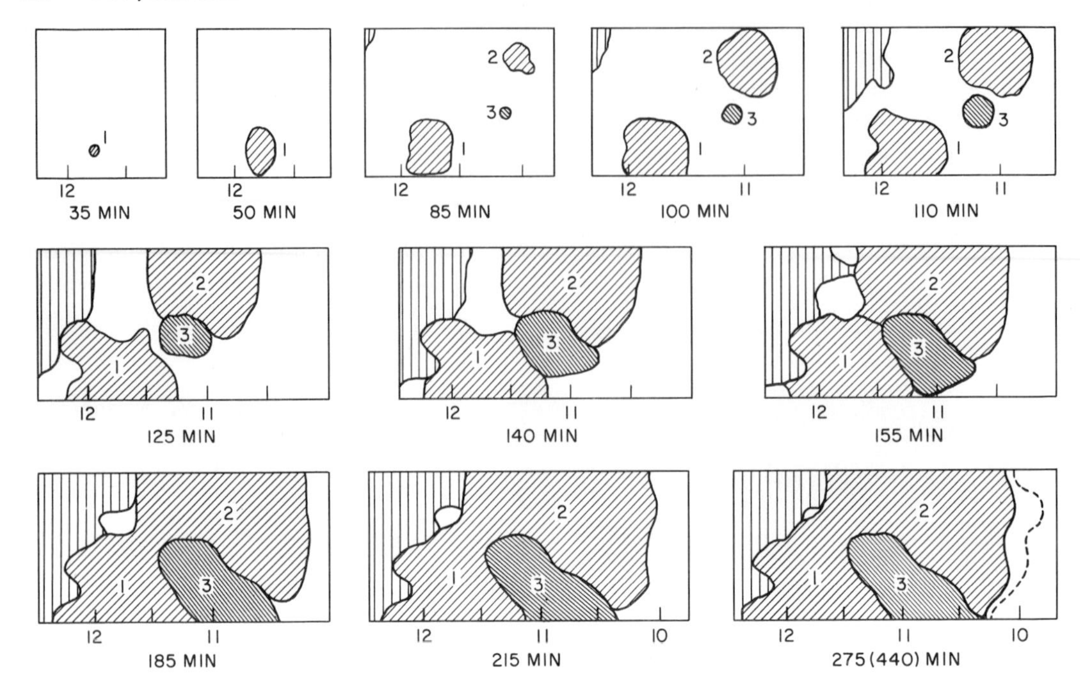

Fig. 7. Growth of new crystallites during recrystallization of a fine-grained aluminum plate, extended by a few per cent. Drawings were made after definite heating periods. It follows that the new grains did not form all at the same moment. (After Karnop and Sachs, 1930.)

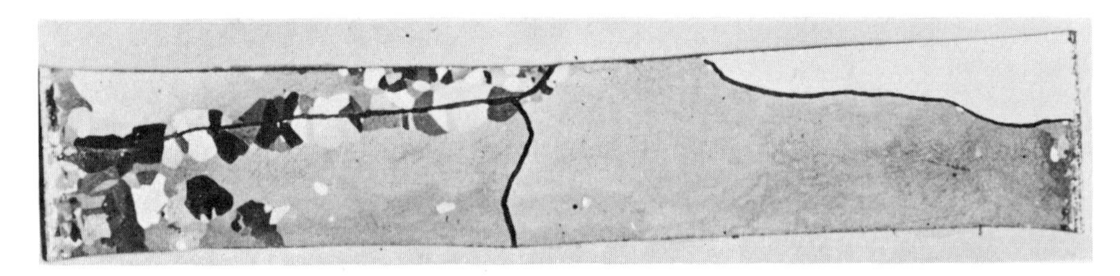

Fig. 8. Preferential formation of new crystallites at the boundary of two original crystals during recrystallization of a coarse-grained aluminum specimen, strained by extension (approximately natural size). (After van Arkel, 1932.)

Fig. 9. Fine-grained aluminum specimen, with incisions at upper and lower edge. After 3% extension, followed by recrystallization at 600°C for 4 min, new crystallites formed preferentially at the incisions, where the deformation was inhomogeneous (approximately natural size). (After van Arkel and Ploos van Amstel, 1930.)

tions of the boundaries have been made visible by a special oxidation technique. From an atomic point of view, one might say (see Fig. 12) that atoms on the concave side (*B*) are surrounded by more atoms belonging to the convex side (*A*) than by (*B*)

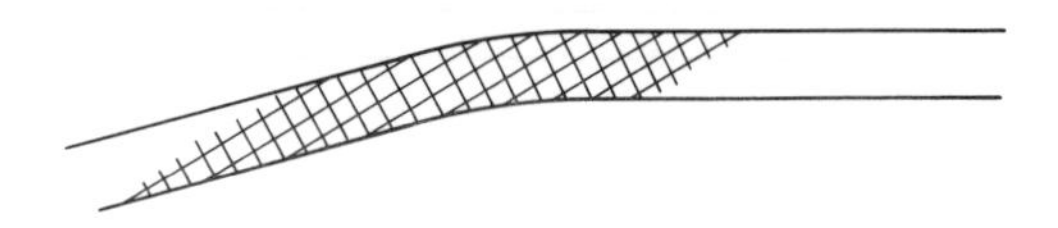

Fig. 10. Change in relative orientation of testpiece and growing crystal in a bent wire. (After Tiedema, 1949.)

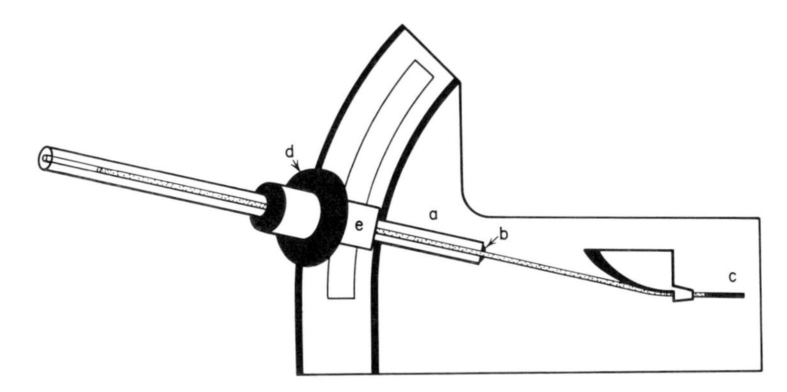

Fig. 11. Bending apparatus for aluminum wire. The crystal wire (c) of several centimeters length is fixed in the glass tube (a) by a drop of molten wax (b). The wire can be rotated about its axis by means of the disk (d) and bent by means of the slide (e). (After Tiedema, 1949.)

atoms, and thus tend to become allied with the A-grouping.*

Another point of view is to attribute the shortening to the action of boundary tensions. The random angles at which boundaries meet make it most improbable that an equilibrium of their tensions exists even if we assume that the tensions are independent of the particular types of boundaries (see p. 426, however). To approach equilibrium requires definite movements of the atoms which make new curvatures, these causing in their turn further movements (Fig. 15). A general coarsening results, in some ways comparable to the general coarsening of a soap froth—a point of view presented by several authors, in particular, C. S. Smith (1948, 1951). The actual attainment of equilibrium angles between adjoining crystallites is shown in Fig. 16 for a silicon iron plate containing three crystals (Dunn, Daniels, and Bolton, 1950).

By this general reasoning, it is understandable that a crystallite sufficiently large in comparison with its neighbors will tend to grow at the expense of them. If such a crystallite is considered to be a polygon, the average apex angle between two sides increases with the number of sides. Thus the chance that the interfacial tension between two of its neighbors will cause the sides of the larger crystal to become concave increases (see Fig. 17; grain I tends to grow at the cost of grains II to VI).

The critical size for grain growth depends on the ratio of the tension acting between the disappearing small grains (σ_{dd} in Fig. 17) and the tension between the large grain and its neighbors (σ_{Dd}). A high value of the ratio σ_{dd}/σ_{Dd} is favorable and requires a smaller surplus in size for the growing grain. For equiaxial grains the critical size would be given by

$$D \gtreqless \frac{2\sigma_{Dd}}{\sigma_{dd}} d$$

* This is only a general remark. We shall not discuss here theories concerning the atomic mechanism of grain boundary movements (compare, however, p. 436).

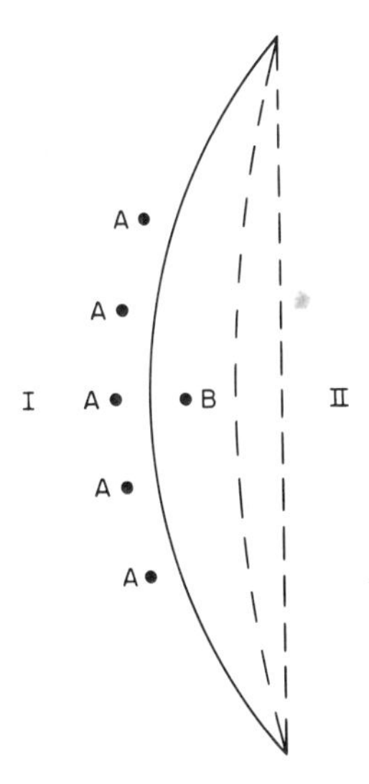

Fig. 12. Schematic movement of a curved boundary between two crystallites A and B.

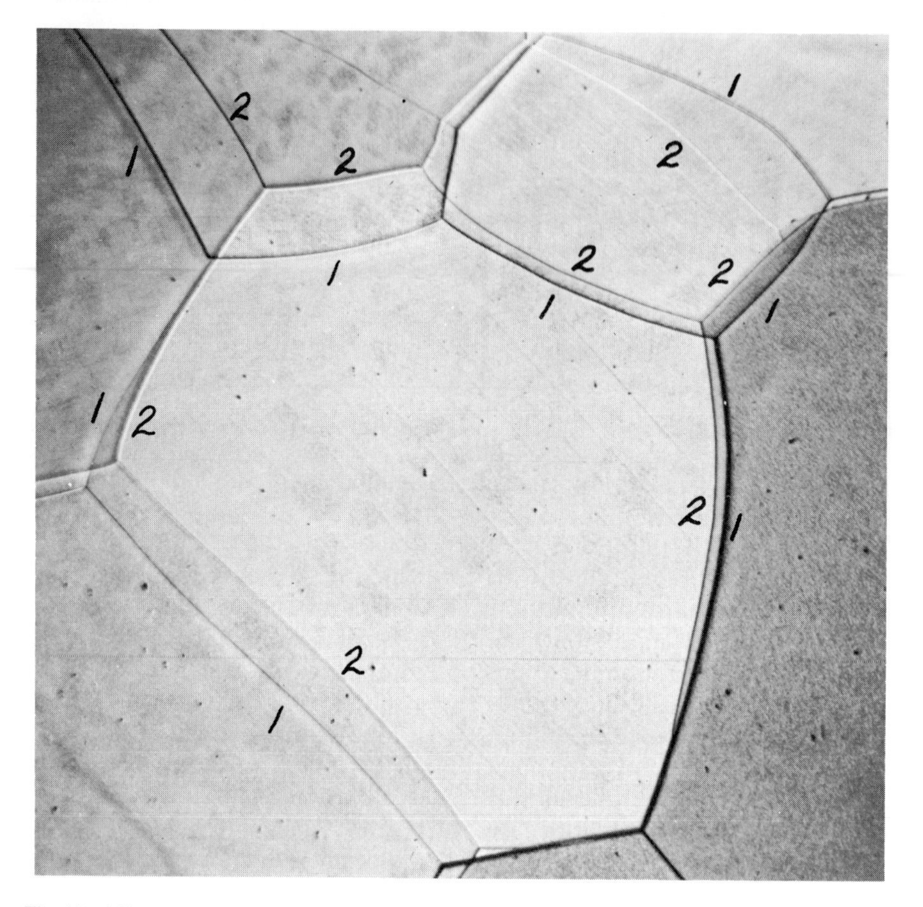

Fig. 13. Migration of grain boundaries toward their centers of curvature in high purity aluminum, annealed at 600°C (approximately 80 X). (After Beck and Sperry, 1950.)

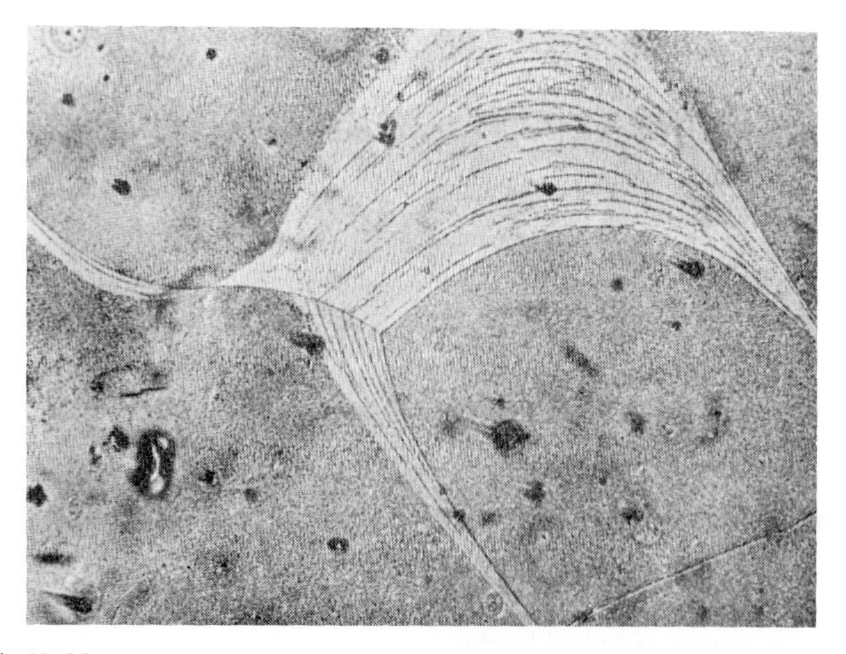

Fig. 14. Movement causing shortening of grain boundary in silicon iron annealed at 1100°C for a few minutes (20 X). (After Ferro, Sari, and Venturello, 1959.)

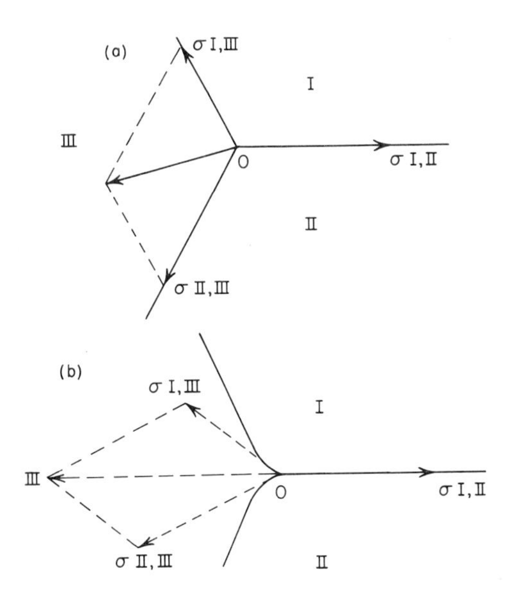

Fig. 15. The formation of curved boundaries due to the tendency of the interfacial tensions to reach equilibrium. (*a*) Original position of the boundaries. (*b*) After attainment of equilibrium.

in which D and d are the diameters of the large and the small grains respectively.*

A fine example of a growing large grain with concave boundaries towards the small surrounding grains is given in Fig. 18 due to Lacombe and Berghézan (1951).

* This can be easily seen as follows (see C. S. Smith, 1951). The large grain with circumference $\sim \pi D$ meets $\pi D/d$ small grains. An increase in size ΔD increases its circumference by $\pi\ \Delta D$ while diminishing the length of the boundaries between the small grains by $(\pi D/d)$ $(\Delta D/2)$. Multiplying with the corresponding interfacial energies and equating gives the above relation.

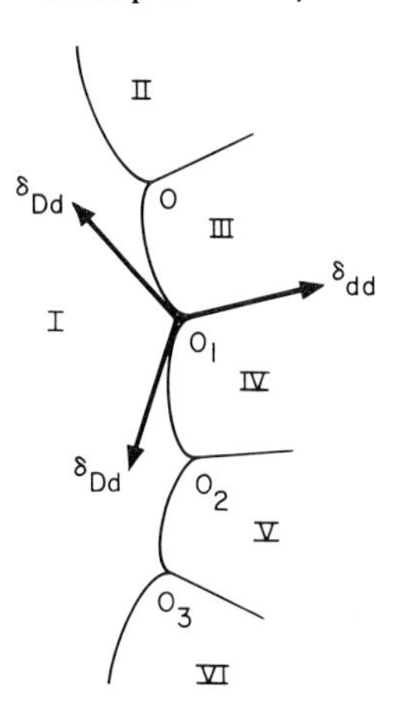

Fig. 17. The path of the boundary between a large crystallite and many surrounding ones of much smaller size.

Grain Boundary Energy

In the preceding discussion, boundary movement is ascribed to the action of the interfacial energy σ. For an individual boundary this action will be larger, the higher the energy and the smaller the radius of curvature R. As a measure of the driving force, we may take the ratio σ/R (see Burke 1948, 1949). On the other hand, the movement requires the passage of potential barriers, so by analogy with other processes we may ascribe to it a mobility M. Then the rate of movement is proportional to $(\sigma/R)M$. For obtaining insight into the conditions required for the preparation of large crystals, it is necessary to consider these factors more

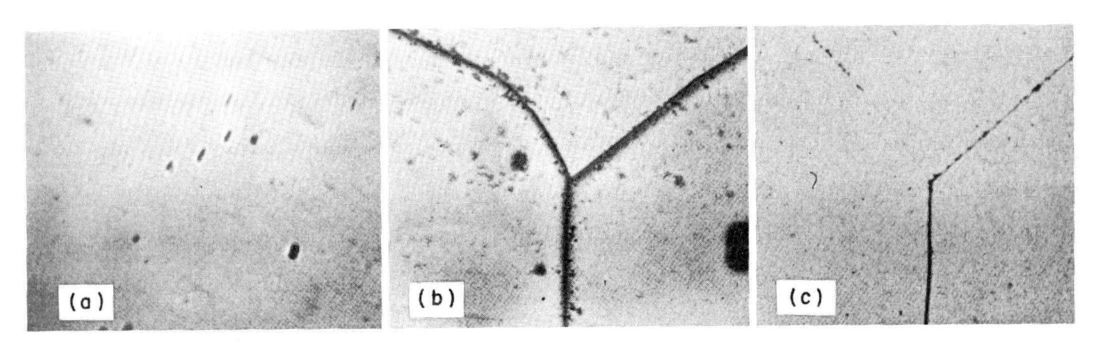

Fig. 16. Attainment of equilibrium positions by the boundaries between three silicon iron crystallites, meeting in one line (perpendicular to the flat testpiece). (*a*) Original state. (*b*) After heating at 1200°C for 24 hr. (*c*) After heating at 1400°C for 66 hr (approximately 500 X). (After Dunn, Daniels, and Bolton, 1950.)

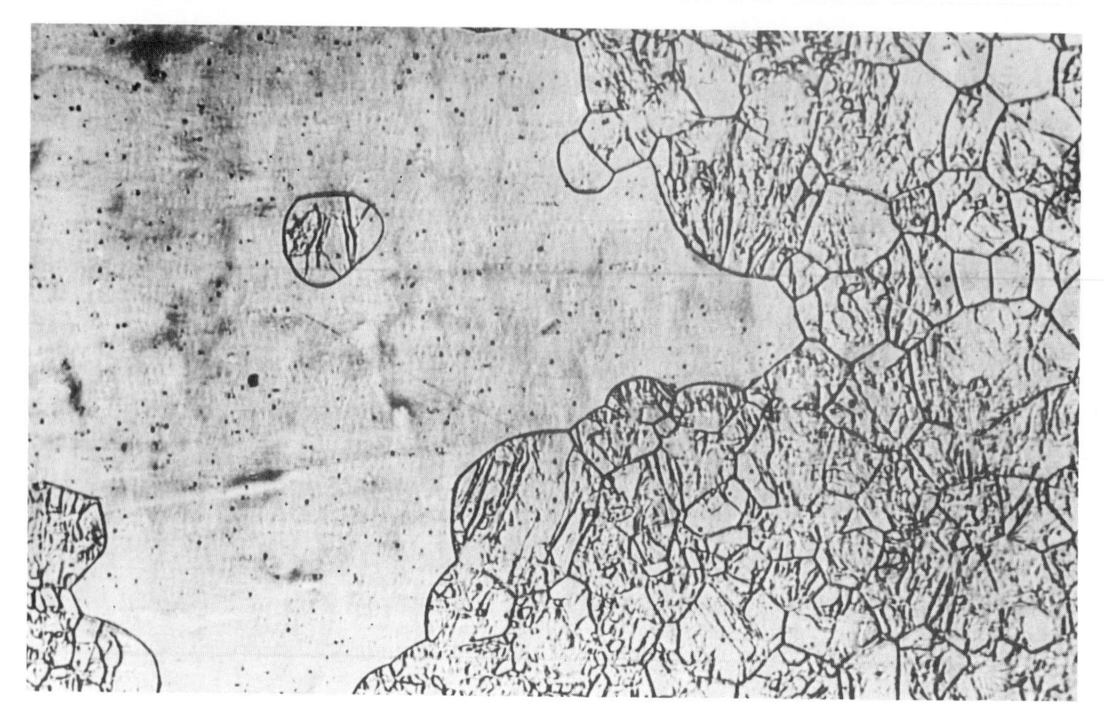

Fig. 18. Path of the boundary between a large crystallite and small grains being absorbed in an aluminum-zinc alloy, strained, and then annealed at 450°C (approximately 50 X). (After Lacombe and Berghezan, 1949, 1951.)

in detail. The most favorable combination of the parameters should enable one grain to grow at a faster rate than others.

For the interfacial energy, it is now well known that this quantity varies markedly with the orientation difference between two adjoining lattice regions, and also for a given difference with the orientation of the boundary plane between both regions. This is primarily due to the fact that grain boundaries are not "amorphous" transition regions but have a structure of their own. They form, broadly speaking, a region of best atomic fit between the regions they separate. The existence of differences among individual boundaries can be shown in various ways; for example, by a difference in etching attack (experiments by Lacombe and Yannaquis, 1948 with aluminum; Dunn and Lionetti, 1949 with silicon iron; and in particular by Vogel, Pfann, and co-workers, 1953, 1955 with germanium), or by the anisotropic character of intergranular diffusion in various metals (experiments by Leymonie and Lacombe, 1957, Smoluchowski, 1952, and Okkerse and co-workers, 1955). To illustrate this difference in yet another way, Fig. 19 shows variations in the amount of precipitation from a supersaturated Au-Pt solid-solution (Tiedema, Bouman and Burgers, 1957).

Quantitatively, the dependence of interfacial energy on orientation difference has been experimentally determined by various authors (see for a survey, Chalmers, 1952). Dunn (1949); Dunn, Daniels, and Bolton (1950), for example, prepared by the method mentioned on p. 420 plate-shaped specimens containing three oriented crystals meeting at a line. After long annealing, equilibrium among the boundary tensions was obtained and the angles between the three boundaries measured (see Fig. 16). From these, and by varying the angles in different specimens systematically, the dependence of energy on angle could be found. Similar measurements with lead specimens prepared from the melt were made by Aust and Chalmers (1950). In yet another method, one measures the groove formed on the surface of a flat specimen at the boundary between two crystallites of known orientation.

Boundary Structure

Boundaries between lattice regions with only slightly different orientation (they are often called *sub-boundaries*) are described on the basis of dislocation theory. One discriminates between a tilt-boundary, separating two regions slightly rotated about an axis lying in the plane of the boundary and a twist boundary in which the axis of rotation is perpendicular to the boundary plane. In the former case (Fig. 20) the boundary can be considered as a set of edge-dislocations of equal sign (J. M. Burgers, 1940; W. L. Bragg, 1940); in the latter case (Fig. 21), the transition consists of intercrossing grids of screw dislocations as first set forth by J. M. Burgers (1940). Both figures refer to cases where the boundary is assumed to be nearly parallel to a simple lattice plane. If it is not parallel to a simple plane, the boundary has to be built up of at least two sets of nonparallel dislocations as in Fig. 22 (J. M. Burgers, 1940; Shockley and Read, 1950). All these boundaries have the property that serious defects in crystal structure are, so to say, accumulated in the immediate neighborhood of the dislocation lines. Outside these places, the transition from one crystallite into the other involves only a slight elastic deformation.

Generally speaking, the complexity of the boundary increases with increasing orienta-

Fig. 19. Difference in amount of precipitated (platinum rich) phase at various crystal boundaries on decomposition of a 60 gold-40 platinum alloy at about 510°C for 24 hr (approximately 1000 X). (After Tiedema, Bouman, and Burgers, 1957.)

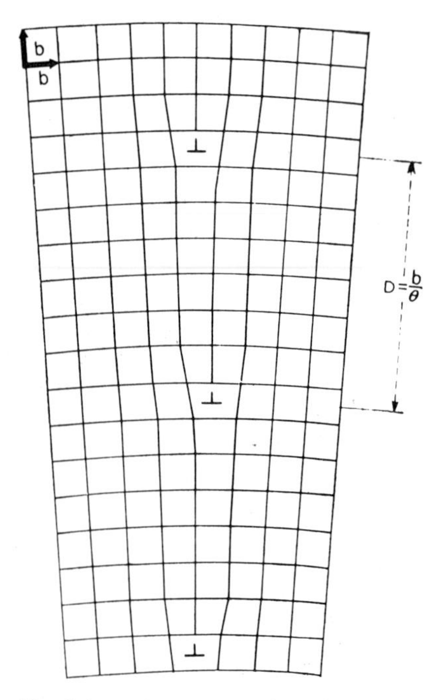

Fig. 20. Schematic representation of a tilt-boundary built-up of a series of equidistant edge dislocations of the same sign. The boundary separates two lattice regions rotated with respect to each other about an axis lying in the plane of the boundary. (From Read, 1953.)

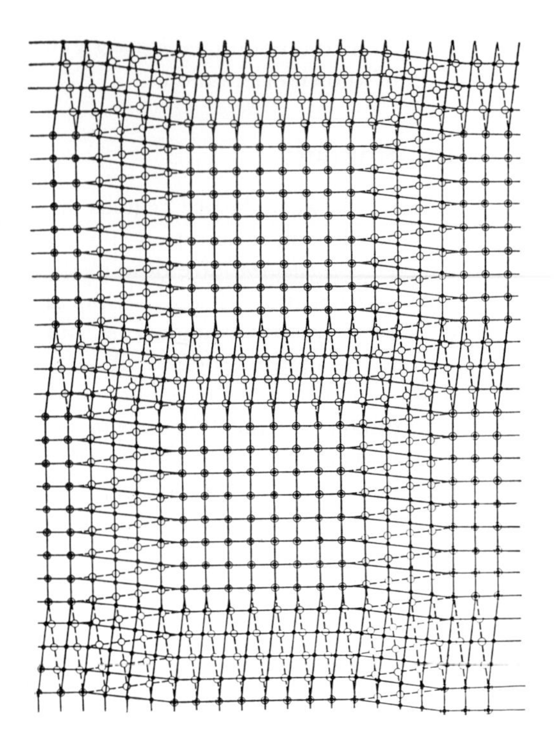

Fig. 21. Schematic representation of a twist boundary, built-up of two sets of intercrossing screw dislocations of the same sense. The boundary separates two lattice regions, rotated with respect to each other about an axis perpendicular to the plane of the boundary. (From Read, 1953.)

tion difference. For large angles it is not suitable to analyze the boundary structure in terms of separate dislocations as they come too near to each other.* Following Mott (1948), such boundaries may be conceived as consisting of alternating regions of good and bad fit, the latter perhaps viewed as regions equivalent to the undercooled molten state. Recently Li (1961) has presented a modified dislocation model for high-angle tilt-boundaries which takes into account the stress fields of the dislocations cores. Because of the volume expansion at each dislocation, the cores begin touching each other at higher angles, finally forming a "continuous slab."

There are, however, special lattice orientations with large angular differences that can have boundaries with a more simple structure. First are twin orientations, in which the two lattices have mirror symmetry with respect to the boundary. The plane between the partners of a twin can be a perfect lattice plane, having a fault in stacking only with respect to the directly adjacent

* However, Hornstra (1958, 1959, 1960) has shown that in the diamond lattice, large-angle boundaries can be described by well-defined dislocation arrays (see van Bueren, 1960).

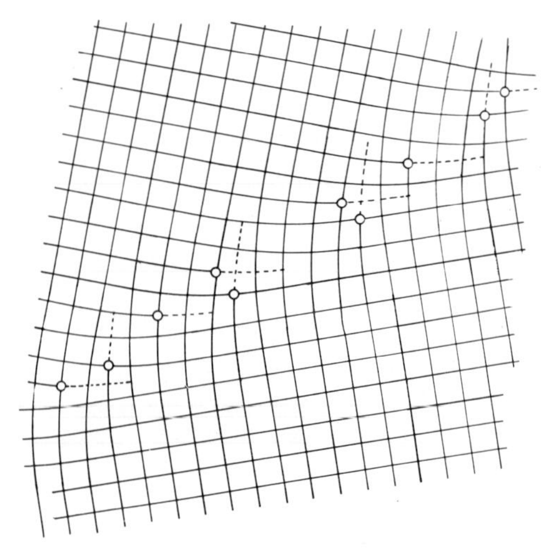

Fig. 22. Transition layer (grain boundary) built-up of *two* sets of dislocation lines of edge type. (After J. M. Burgers; see W. G. Burgers, 1947.)

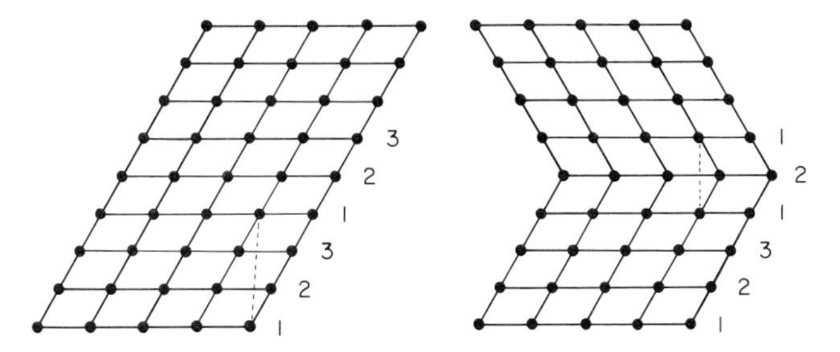

Fig. 23. Twinning in the face cubic-centered lattice: succession of close-packed {111} planes viewed along a ⟨110⟩ direction. Coherent twin plane.

planes. This holds, for example, for twins in the cubic close packed lattice (see Fig. 23). Other cases are given by "coincidence boundaries," first considered by Kronberg and Wilson (1949). In such boundaries, certain points of both lattice regions coincide. An example is formed by two cubic lattices rotated about a ⟨111⟩ axis by 22 or 38° on the plane lying perpendicular to this axis. Then, one-seventh of the atoms in two adjoining {111} planes occupy coincident positions.

Differences in boundary structure correspond to differences in boundary energy. Calculations of this, based on dislocation models in a simple cubic lattice, have been made sometime ago by Shockley and Read (1949, 1950). For a small-angle tilt boundary (the partners rotated with respect to each other about a common cubic axis by a small angle θ) the resulting expression is

$$E = E_0\theta(A - \ln\theta)$$

where E_0 and A are constants depending on the macroscopic elastic constants and on the orientation of the grain boundary. For high-angle boundaries, with "merging" dislocations, the dependence of energy on misfit is more complicated. Even for constant position of the boundary plane, the energy versus θ curve will show low energy cusps for certain angles between the lattices, for example, at twin boundaries or possibly at coincidence boundaries. Moreover, the energies for these orientations are very sensitive to changes in the orientation of the boundary plane.

Boundary Mobility

One may also expect a considerable dependence of mobility on boundary structure.

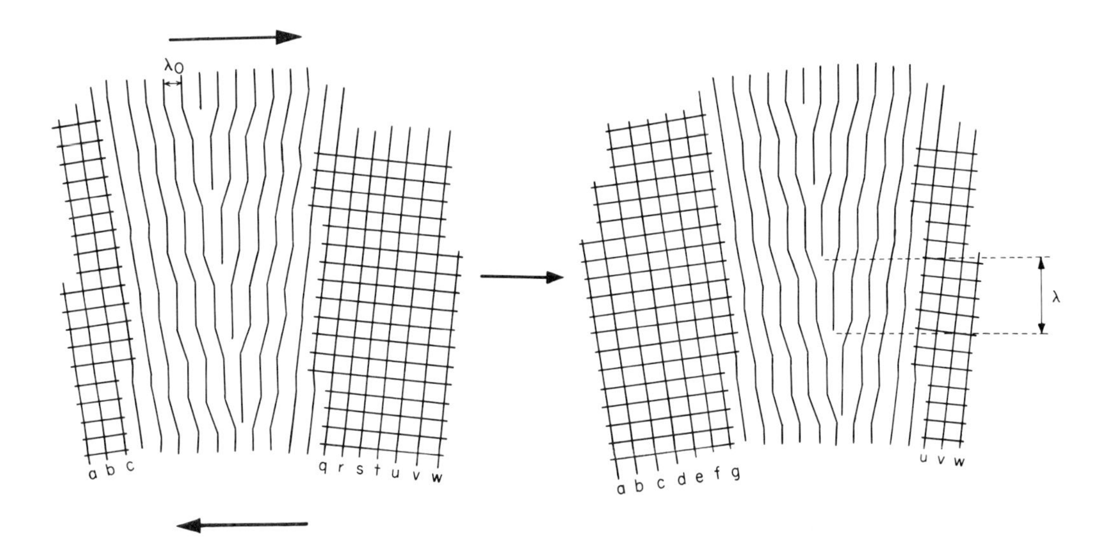

Fig. 24. Displacement of a low angle tilt-boundary due to a shear stress.

For a simple tilt-boundary, consisting of parallel dislocations at relatively large distances from each other, movement may be brought about by a cooperative, simultaneous gliding of all the dislocations in a manner shown schematically in Fig. 24. Such movements have been directly observed for zinc by Washburn, Parker, and co-workers (1952, 1953, 1954) and recently again by Vreeland (1961), as in Fig. 25. A cooperative movement requires a definite boundary structure and cannot be expected (and is not found) for an arbitrary boundary, even between very slightly misoriented regions (see Weissmann, Hirabayashi, and Fujita, 1961). If such a boundary should correspond to Fig. 22 and contain two sets of differently directed dislocations, a similar movement would require motions of dislocations by both gliding and climbing* as first pointed out by Shockley and Read (1949). This latter process involves diffusion of individual atoms, a slow process at low temperatures.

If, however, the misfit is bad enough and the boundary energy is relatively high, a general increase in atomic mobility may be expected. This enhances the possibility of boundary movements by individual atomic jumps. Direct interchanges with vacant lattice sites at boundaries will then play an important role (see Lücke, 1961). Thus a relatively high mobility may be foreseen for both low angle and high angle boundaries, whereas medium angle boundaries ($\theta \sim 1°$) may be relatively immobile (see Dunn and Daniels, 1951).†

Coincidence boundaries that show a high mobility (Aust and Rutter, 1959a) occupy an exceptional position. Theoretically, it is not easy to predict whether on account of the coincidence sites they are stable configurations or, on the contrary, movement is easy for them (see Lücke, 1961, also Stüwe 1961).

* Climbing of an edge dislocation involves shortening or lengthening of the "extra" half-plane of atoms.

† May and Erdmann-Jesnitzer (1959) conclude from measurements on fine-grained aluminum specimens that boundary mobility *decreases* for misfit angles larger than 20°. They applied a statistical method, however, and their conclusion cannot be transferred as such to direct measurements as discussed above.

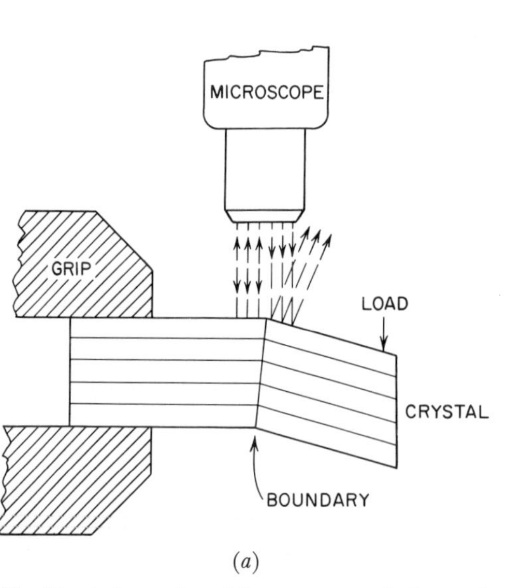

(*a*)

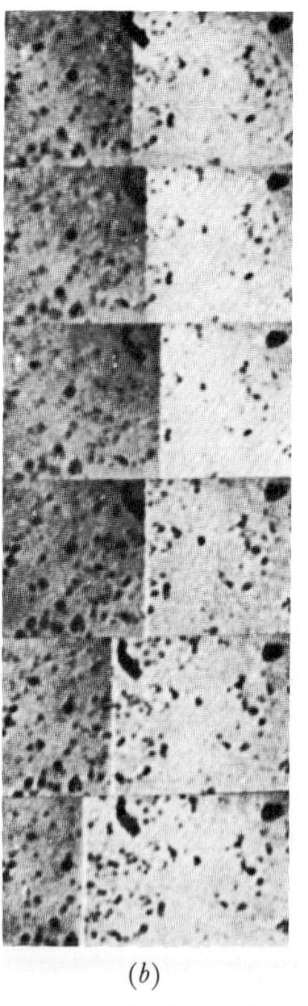

(*b*)

Fig. 25. Direct observation of the movement of a low angle tilt-boundary in a zinc bicrystal due to shear stress. (*a*) Experimental arrangement. (*b*) Microphotographs of the movement. By reversing the direction of the stress indicated in (*a*), the movement reverses (approximately 100 X). (After Li, Edwards, Washburn, and Parker, 1953.)

Fig. 26. Large aluminum crystal with inclusions (approximately natural size). (After Tiedema, May, and Burgers, 1949.)

The relatively low mobility of medium angle boundaries, as compared with high angle ones, finds experimental confirmation in the fact that large crystals, growing at the cost of a fine-grained matrix, often contain small included domains with deviating orientation, obviously unabsorbed grains of the original matrix (Fig. 26). Such inclusions persist after prolonged heat treatment. X-ray analysis and etch figures show that they have orientations corresponding within about 5° to twins of the growing crystal (Tiedema, May, and Burgers 1949; Lacombe and Berghézan, 1949a).

The dependence of interfacial energy and boundary mobility on relative orientation differences leads to the expectation that in a matrix of given grain size and state of deformation the rate of growth of a crystal depends on the relative orientation of the growing crystal and the matrix texture. In a qualitative way this could be deduced from the fact that while circular crystallites occur in a quasi-isotropic matrix (Fig. 5), a less isodimensional shape grows in a matrix that has a pronounced texture, for example, in drawn wires crystallites become elongated in the direction of the wire axis.*

Orientation Dependence of Boundary Mobility

Most clearly and quantitatively, the orientation dependence of boundary migration has been demonstrated in aluminum by Beck and co-workers (Beck, Sperry, and Hsun Hu, 1950; Beck, 1952, 1954, 1961). These authors found that growth is fastest if the orientation of the growing crystal is related to that of the matrix by a rotation around a $\langle 111 \rangle$ axis of about 40°. In one of their experiments, a slightly deformed large-grained aluminum specimen is locally deformed by a scratch on the surface and then recrystallized. Within the one original crystal (a matrix possessing a very narrow

* Some qualification of this statement is necessary because unequal growth rates can also be caused by the presence of insoluble impurities which form inpenetrable walls (see the experiments of Rieck, 1958 on the growth of tungsten crystals in drawn wires as mentioned on p. 436).

It is also possible that the deformation process has not acted uniformly over the whole specimen, causing so-called deformation bands. Then the state of deformation of the matrix varies with direction and so will the rate of growth of a new crystallite (see Czjzek and Haeszner, 1960).

range of lattice orientations) some of the new crystallites growing from the scratch are much larger than others (Fig. 27). The faster growing ones have the mentioned orientation with respect to the matrix.

In another striking experiment (Kohara, Parthasarathi, and Beck, 1958) a fine-grained aluminum matrix with a sharp texture was produced by rolling a monocrystal of a certain orientation and subjecting it to recrystallization. The flat specimen was then rubbed with emery paper on one side and subjected briefly again to recrystallization. In this way, a large number of randomly orientated nuclei are produced on one side of the matrix. Upon further heating, mainly those crystallites which obey to the mentioned orientation relationship grow through the specimen.

In order to obtain information regarding growth selectivity, Tiedema (1951) reversed this method of investigation. He prepared a sharp fine-grained texture by rolling a monocrystal, and then grew new grains of prechosen orientation into the matrix by growing a seed crystal "round the corner" (p. 420). According to his experiments, done

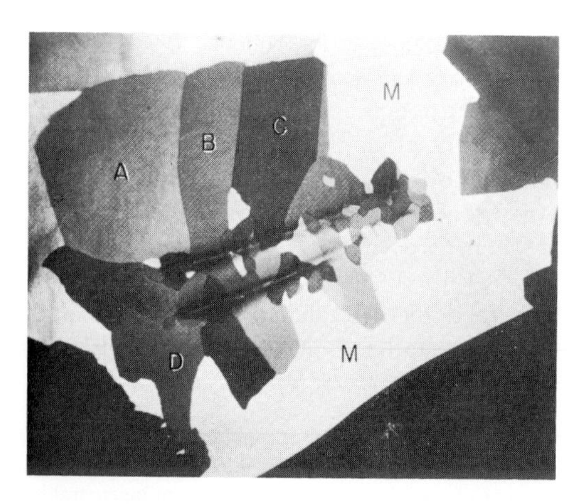

Fig. 27. Orientation dependence of the rate of boundary migration. An aluminum crystal (*M*) was locally deformed by scratching. Many small recrystallized grains of various orientations grew around the scratch. The grains *A, B, C,* and *D* have grown to a large size due to their suitable relative orientation (rotation about ⟨111⟩ over approximately 40°) with respect to the matrix crystal (approximately 20 X). (After Beck, Sperry, and Hsun Hu, 1950.)

with aluminum of commercial purity, considerable variation in the relative orientation of growing crystal and matrix texture could be realized without causing a marked change in rate of growth (not including, however, a ⟨111⟩-rotation). Dunn (1949b), in similar experiments with silicon iron, found differences up to 30%.

Because of the importance of the question, several groups of investigators have since occupied themselves with this problem in aluminum, but with partly contradictory results. Liebmann, Lücke, and Masing (1956) confirmed the existence of a high mobility for the ⟨111⟩-40° rotated orientation relationship, but Graham and Cahn (1956) found growth rates to be insensitive to the orientation difference between growing grain and matrix, at least for large-angle grain boundaries. The discrepancy, however, was partly resolved by further experiments of Green, Liebmann, and Yoshida (1958) which showed that impurities may play an important role with regard in orientation-dependent growth. The authors used a commercial aluminum with an iron content (0.09%) above the solubility limit. Depending on whether (because of heat treatment) the iron was present in supersaturated solid solution, or precipitated, a preference for ⟨111⟩-rotated growth was absent or present. It was thought that foreign atoms in solid solution could retard the boundary (see Lücke and Detert, 1957, mentioned on p. 436), limiting its speed to the speed of the foreign atoms diffusing behind it, and thus obscuring any true orientation dependence.

More recently, careful experiments have been made by Aust and Rutter (1959a, b; 1960a; Rutter and Aust, in press, A) with high purity lead with small additions of tin, silver, and gold. These authors conclude that the development of preferred orientations depends upon the specific impurity present and on its content; tin, for example, is active only in the range of 5 to 40 ppm by weight (under specific conditions of deformation and annealing).

Considering the available evidence, it seems well established that the presence of foreign atoms can overwhelm the influence of orientation difference on the rate of movement.*

Returning again to the orientation dependence itself, one more detail emerges from the work of Lücke, and Beck, and coworkers, namely, that even for two adjoining lattice regions which are rotated about a ⟨111⟩ direction, the rate of boundary motion depends on the orientation of the boundary with respect to the rotation axis. Mobility is much larger for a boundary parallel to the axis (that is, tilt-boundary) than for a boundary perpendicular to it (a twist boundary)† (Yoshida, Liebmann, and Lücke, 1959). Moreover, for a tilt-boundary, mobility seems to depend on the azimuthal position of the boundary plane with respect to the crystallographic directions in the {111}-plane that stands perpendicular to the boundary (Parthasarathi and Beck, 1961).

Formation of Recrystallization Twins

A few words should be said concerning the occurrence of twins in crystals growing in a solid. Recrystallization twins occur frequently in various face centered cubic metals, such as copper, silver, and gold. They appear mostly in the form of narrow or broad bands with parallel straight boundaries, these latter being the traces of the twin plane {111} with the surface of the specimen. It is now generally assumed (see Burke, 1950) that recrystallization twins form during growth as a consequence of a faulty deposition of atoms parallel to the twin plane. One factor favoring such an occurrence is a low value of the stacking fault energy. Hence, under analogous circumstances, a metal with a high stacking fault energy such as aluminum, shows less twinning than metals with lower stacking fault energies like copper or gold.

A second factor that controls twinning is the orientation relationship between the growing crystal and the polycrystalline matrix.‡ This is evident from experiments with copper by Burgers, Meys, and Tiedema (1953), in which a large crystal was made to grow in a bent strip at the expense of a sharply textured fine-grained matrix. In this procedure, the relative orientation of growing crystal and disappearing texture changed gradually, and a change in the concentration of annealing twins was observed. From an analysis of these experiments, the tendency for twinning, that is, for a faulty deposition of a layer of atoms, seems to be increased if the average orientation of the matrix is closer to the twin lattice orientation of the growing crystal than to its own lattice orientation.

Recently the foregoing statement has received some confirmation in a study of the growth of large crystals in lead by Aust and Rutter (1960b). These authors found that whenever twinning occurred, a random large-angle orientation between growing crystal and matrix (in their work a striated melt-grown crystal) was replaced by a coincidence-type boundary or by a near-twin boundary. The first twinning may be followed by a second and eventually a third

* In a recent paper, however, Parthasarathi and Beck (1961) point out that in the experiments as performed by Green et al. and by Aust and Rutter, the crystals grow into a matrix that is only weakly deformed or imperfect, so that the driving energy for growth is low. Parthasarathi and Beck found that for a crystal growing into a strongly deformed matrix, the presence of a considerable amount of impurity, even in solid solution, could not suppress preferred growth of ⟨111⟩ rotated crystallites.

In a forthcoming paper, Aust and Rutter (1962) stress the importance of the temperature dependence of grain boundary mobility on the development of preferred orientations on recrystallization. These authors found earlier (Rutter and Aust, in press, B) that the temperature dependence of grain boundary mobility in zone-refined lead is smaller for large-angle, coincidence-type boundaries than for large-angle, noncoincidence boundaries, leading to equal mobility at about 300°C. In accordance with this fact, new grains grown into a matrix with sharp texture showed random orientations when recrystallization took place at 300° whereas preferred "coincidence orientations" formed at a lower recrystallization temperature. The possibility that this is caused by a difference in interaction among boundaries and residual impurities in the lead is left open.

† For the stability of a twist boundary composed of intersecting screw dislocations, see an article by Holt (1959).

‡ We consider here twin-formation in a large crystal growing into a fine-grained matrix. For discussion of the formation of twin lamellae during normal grain-growth, we refer to a paper by Fullman and Fisher (1951), in which an energy criterion for twinning, based on the difference in interfacial energies, is given.

twinning.* Similar results were obtained by Aust (1961) for zone-refined aluminum.

Exaggerated Grain Growth

It was mentioned in Section 1 that in certain cases, after prolonged annealing of a fine-grained matrix, instead of a further increase of average grain size, only one (or a few) large grain suddenly grows at the expense of the whole specimen.

It is known from early investigations of aluminum by van Arkel, Carpenter, and Elam, and Feitknecht (see Burgers, 1941) that this phenomenon occurs preferentially in a matrix with a very pronounced texture as can be obtained, for example, by rolling and recrystallizing a monocrystal (see Fig. 28). Feitknecht (1926) and also van Arkel and van Bruggen (1928) established that the large crystals develop preferentially for a medium amount of rolling deformation. This was confirmed by other investigations, for example, for nickel-iron by Dahl and Pawlek (1936) and Pawlek (1935).

The fact that a pronounced primary texture is favorable for secondary recrystallization has been confirmed many times in later work; for aluminum (Beck and Hu, 1949; Burgers, 1941), copper (Bowles and Boas, 1948), nickel-iron (Dahl and Pawlek, 1936),

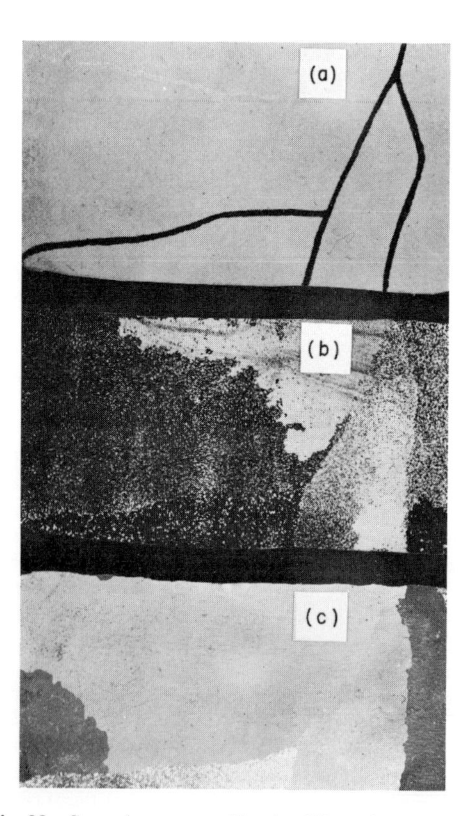

Fig. 28. Secondary recrystallization (discontinuous grain growth) in rolled aluminum crystals. (*a*) Drawing of an aluminum plate consisting of four crystals after rolling to about one-fourth of its original thickness. (*b*) The same plate after (primary) recrystallization at 600°C for a short time; within each original crystallite a fine-grained texture is formed (see Fig. 2). (*c*) On prolonged heating at 600°C large crystals have been formed by "secondary" recrystallization in the fine-grained domains, their boundaries coinciding approximately with those of the original crystals in the rolled matrix (*a*). (After van Arkel, 1932; see Burgers, 1941.)

* Inspection of Aust and Rutter's paper will show that several twin crystals observed in their experiments have a "pointed" shape with curved boundaries (see, in particular, Fig. 3 of their paper). This calls for the following remark. As said at the beginning of this section, annealing twins are mostly recognized by their appearance as bands with straight boundaries parallel to the twin plane. The straight boundaries are due to the fact that mother crystal and twin grow with equal rate parallel to the twin plane. For nickel iron strips this could be shown directly in the electron emission microscope (Burgers, 1938, 1941). In this case the large (secondary) crystals grew at the expense of a pronounced texture in the matrix (the so-called cube texture in which all the crystallites lie within 5 to 10° with their ⟨100⟩ directions parallel to the rolling and to the normal direction of the strip). This circumstance favors the formation of straight lamellae, as the relative orientation of growing crystal and matrix remains constant throughout. In a less preferentially oriented matrix, the formation of curved twin boundaries is quite possible. Several years ago it was shown by Sandee (1942) (and by Burgers and May, 1945; Burgers, 1947) that such curved twin boundaries are formed if the twin grows faster into the matrix than the mother crystal (Aust and Rutter, 1960 b), could not find different mobilities within a factor of two at 300°C; distinct curved boundaries arise, however, for rate differences of a few per cent (see May, 1950). The curved shape was analyzed mathematically by Sandee and by May for crystal growth in aluminum. Here an annealing twin, starting at a definite point of the growing mother crystal, grows with curved boundaries in a V-shape. Such "pointed" crystals, called *stimulated crystals* by Burgers and May, are annealing twins not immediately recognized as such; the fact that the frequency of their occurrence in recrystallized aluminum varies with the matrix material confirms again that abundancy of twinning is dependent on the relative orientation of growing crystal and matrix texture.

Finally it is of interest to remark that the same geometric shape may be produced in a growing colony of molds if by a sudden mutation a new species is formed which multiplies itself at a faster rate than the mother colony. (Waddell, 1945, 1946). An illustration of this phenomenon is given by Pontecorvo and Gemmell (1944).

silicon iron (Dunn, 1953). In view of the orientation dependence of grain growth, this behavior seems understandable. Suppose that the matrix contains, in addition to the textured grains a crystallite that is so much larger than its neighbors that the size-effect on growth capacity predominates. Then if its orientation with respect to the texture is favorable for growth (for example, in Al a 30 to 40° rotation about a ⟨111⟩ axis), it is favorably oriented with respect to most of the crystallites in the matrix, and it can thus continue to grow rapidly and so attain a large size.

In accordance with this view, Dunn (1953) (see also Gow, 1954) formulates the conditions for secondary recrystallization (exaggerated grain growth) as follows: primary recrystallization must produce a strong fine-grained texture containing some crystallites that have deviating orientations and are two to three times larger than the average of the textured grains. Perhaps a certain amount of prior deformation is favorable for reaching such a state.*

The relatively low grain boundary energies between the "primary" grains, due to their nearly parallel orientations, are unfavorable for their being consumed. However, once the larger crystal has obtained a sufficient size (see page 423), energy will be gained by its further growth.

Quite recently, Walter (1959) (Walter and Dunn, 1960) showed that an additional driving force may be derived from surface energy and its dependence on orientation. This was concluded from the fact that large secondary crystals growing in a special silicon-iron sheet had certain orientations with respect to the *surface* of the sheet; if a new surface was ground at a certain angle to the old one, secondary crystals formed in the new sample had changed orientations so as to retain the same orientation relative to the free surface.†

Usually, however, the secondary crystals take certain orientations with respect to the primary texture. This was already "qualitatively" established by Burgers and Basart (1929) (see Burgers, 1941) for rolled aluminum crystals, and emphasized by Beck and Hu, (1949; see Beck, 1954). These authors found an orientation relationship between primary texture and secondary crystals consisting again of a rotation of 30 to 40° about a ⟨111⟩-axis. Similar relationships, including also other orientations, have been found for crystals grown by secondary recrystallization in cube-texture material (obtained after primary recrystallization of severely cold-rolled strip); for example, in copper by Bowles and Boas (1948) and Kronberg and Wilson (1949). Another example of importance in practice is the secondary recrystallization of silicon-iron sheet (Assmus, Detert, and Ibe, 1957; Walter and Dunn, 1960).

The cause of such preferential orientations has been ascribed either to "oriented-growth" (Beck and co-workers) or to "oriented-nucleation" of the growing secondaries. Dunn (1954) has proposed a combination of these two viewpoints.

Beck, Holzworth, and Sperry (1949) showed that one other factor is favorable for exaggerated grain growth, the presence of a small quantity of a second, insoluble phase. This was clear from an investigation of grain growth in an aluminum-manganese alloy. It was found that the tendency for exaggerated grain growth was largest at temperatures at which the segregated phase nearly dissolved again. We return to this in the next section.

Finally, we remark that Verbraak (1960) has recently presented an entirely different mechanism for the formation of oriented secondary crystals based on the mechanism of martensite-type nucleation processes (see p. 441).

Influence of Foreign Atoms on Grain Growth‡

On general grounds one may anticipate that foreign atoms will exert an influence on boundary mobility. In experiments, the

* Gow (1954), in a paper on secondary recrystallization in extruded aluminum rods, discusses possible mechanisms for the occurrence of larger primary grains.

† From the observation that secondary nuclei are situated *inside* the free specimen surface, Philips and Lenhart (1961) conclude on the contrary that grain boundary energy, and not surface energy, is the driving force for secondary recrystallization.

‡ For some special cases the influence of foreign atoms is also discussed in other sections (pp. 432–436).

influence of even very small percentages of foreign atoms (of the order of 0.01 to 0.001% sometimes) on the rate of recrystallization is well established by many investigators (see for older literature Burgers, 1941). Often these results refer to recrystallization as a whole and do not allow conclusions as to whether the observed effects concern the rate of nucleation or the rate of growth. There are, however, a number of experiments that refer to boundary movements; those already mentioned in a previous section by Aust and Rutter on lead with additions of tin, gold, and silver, by Dimitrow (1959), and by Lücke, Masing, and Nölting (1956) on high purity aluminum. According to the latter authors, for example, 0.01 weight per cent manganese can lower the boundary mobility to 10^{-11} of its value in the pure metal. Significant in this connection are experiments of Vreeland (1961) on the motion of tilt-boundaries in zone-refined zinc crystals. Vreeland found that the stress required to move a tilt boundary in 99.999% zinc was more than ten times as large as that required in zone-refined metal.

The size of the foreign atoms compared to the matrix atoms, whether they are present in solid solution or as a second phase (to mention only two factors) will certainly have an influence. It seems natural to expect a retarding influence, either due to direct obstruction of the movements of the innate atoms of the metal or, if the boundary consists of dislocations, due to the hindering effects of "atmospheres" of solute atoms near the dislocations. A theory along such lines has been given by Lücke and Detert (1957). If the atmospheres lag behind, they exert a retarding force on the moving dislocations. Then, the rate of (volume) diffusion of the solute atoms determines the mobility of the boundary. The theory finds partial agreement in the measurements of Aust and Rutter with lead, but in the main their measurements do not support its quantitative predictions; neither do those of Holmes and Winegard (1959, 1960) on grain growth of tin with additions of lead, bismuth, or silver. The influence of impurities is it seems too complex to be understood from only one viewpoint.*

When impurities form a second phase, a direct obstruction of boundary movements seems natural. Quite early it was found by Jeffreys and Archer (1922) that the presence of insoluble thorium oxide in tungsten impedes the formation of large crystals during sintering. However, the same authors showed that for a low thorium oxide-content the inverse holds, and growth of large crystals is favored. In the latter circumstance growth is probably only locally hindered, which may result in an increase in size-contrast, a factor that favors growth. The way the second phase is distributed in the specimen may enhance this effects. For tungsten containing a second phase, according to Rieck and Meyering (see Rieck, 1958) the impurity forms tubelike walls which stop the growth of a crystal. The walls have, however, "leaks" at random places through which a crystal can enter the next tube and continue its growth.

The observation of Beck, Holzworth, and Sperry (1949) on aluminum-manganese alloys (see p. 435) is consistent with the same viewpoint. Here also the gradual removal of a second phase by heating near the solubility limit allowed the first few grains to grow to a size sufficient to consume most of the smaller ones.

3. NUCLEATION

General

Whereas crystal growth can be followed with a microscope, although its elementary steps occur on an atomic scale, nucleation is an event involving the movement of a small number of atoms and therefore escapes direct observation. The recent development of electron microscopic methods for thin metal foils has reduced the limit of direct observation to distances of 50 A and less, but even this resolution has not yet been sufficient to allow a precise state-

* It seems obvious to expect that vacancy diffusion plays an important role in boundary mobility. This point is discussed by Lücke (1961); in der Schmitten, Haasen, and Haeszner (1960); Oriani (1959); and, with reference to the dislocation model of grain boundaries, by Hornstra (1961). In this connection, one sees also Li's paper (1961) already mentioned on p. 428.

ment, in terms of atomic movements, of what actually happens at the birth of a new crystal.* Data concerning nucleation must therefore be deduced from indirect observations of a later stage, at which the process has progressed into the range of visibility. Such data refer to what Cohen (1958) calls *operational nucleation.* We start this section by discussing some of the laws of nucleation in this sense.

By following the growth of new crystals as a function of time, and extrapolating to zero size (see Fig. 7), one finds that not all nuclei have formed at the same moment. Thus one can introduce the concept of a "time of nucleation" (or incubation). If the matrix has been uniformly strained, the new grains grow at an approximately constant rate.

These features allow the recrystallization process to be described in terms of a rate of nucleation $\underline{N}$ (number of nuclei formed in unit volume in unit time) and a rate of growth $\underline{G}$, quite apart from questions about the underlying atomic processes. On this basis several authors (Johnson and Mehl, 1939; Anderson, and Mehl, 1945; Avrami, 1939, 1940; see Lücke, 1950; Burke and Turnbull, 1952) have developed formulas giving the recrystallized fraction as function of time for constant temperature (so called isothermal recrystallization curve). Inversely, such formulas have been used to analyze severely deformed specimens which develop large numbers of grains on recrystallization. By counting grains and making planimetric measurements of the size of the largest grains, the dependence of both the rate of nucleation and growth on the variables, temperature, degree of deformation, and time of heating, was deduced. Difficulties arise in this type of work regarding the exact way in which such a factor as the mutual impingement of growing crystals is to be put into the formulas.

It follows from such experiments that both $\underline{N}$ and $\underline{G}$ depend exponentially on the temperature, and can thus be characterized by an activation energy for nucleation (E_N) and for growth (E_g). Although of the same order of magnitude, the former is generally larger than the latter. For example, Dmitrov (1959) gives for zone-refined aluminum, E_N and E_g values of 18 and 13 calories/gram-atom, respectively. Such experiments, carried out with pure metals and with added foreign atoms, show that both quantities depend on the impurity content.† The same impurity, however, can have a very different effect on $\underline{G}$ than on $\underline{N}$. Dmitrov, in experiments just mentioned, found that the presence of 30 ppm of copper in aluminum reduced $\underline{N}$ to approximately 1/5000 of its original value, whereas $\underline{G}$ decreased only a little. Such results are an indication that the fundamental atomic processes underlying nucleation and growth are different.

We mention two other important observations that are essential for any understanding of nucleation in recrystallization, both already mentioned in Section 1 (Figs. 9 and 2): the first is the fact that newly formed crystals are preferentially formed at places where the deformation has been highly inhomogeneous and the second is the fact that the orientation of the new crystallites is almost never completely at random, but is related in some way to the texture of the matrix.

We shall not discuss in detail the various theories and conceptions of the atomic mechanism by which nucleation is brought about. A recent survey is given by Burgers (1961). For the purpose of this book it should suffice to present those salient points that may help in understanding the measures required for the preparation of large crystals by recrystallization of a fine-grained matrix.

Nucleation theories

From the available evidence it seems improbable that a nucleus is formed in a way directly comparable to what happens during crystal growth in a vapor. In the latter case a nucleus is gradually built up

* Perhaps the field-ion microscope will be able to fill this gap in time to come.

† For crystal growth this has been found in a more direct way by experiments of the kind discussed on p. 432.

from individual atoms subjected to thermal fluctuations. In a crystalline material, however, even in the most severely deformed state, the great majority of atoms are still arranged in crystallographic array. This is known because the Debye-Scherrer pattern differs from that of the annealed state by only a slight broadening or a decreased intensity of the diffraction lines, these changes being caused by elastic straining of the lattice, the presence of dislocations, stacking faults, and the like.

Dislocations are present both as networks within the lattice blocks and in transition layers between slightly misoriented blocks. Therefore lattice domains that are fit to serve as recrystallization nuclei need not be formed but are already plentiful. What is required is to make a particular domain capable of growing at the expense of its surroundings. As we have seen in Section 2 on boundary migration this happens if the domain is sufficiently larger than its neighbors and if it is separated from these by mobile boundaries. For a domain growing in a deformed matrix, not only does the grain boundary energy act as driving force but also the difference in internal energy between the growing domain and its surroundings. Since surplus free internal energy is associated with the network of dislocations mentioned, one may expect that for similar grain boundary states a domain with a lower density of dislocations will grow at the expense of one with a high density.

The dislocation density effect may be observed in a polycrystalline specimen that is given such a weak deformation that on subsequent annealing nucleation of new crystals does not occur. The deformation may, however, put different densities of dislocations into various original grains, causing them to have more or less dense networks of "sub-boundaries." Under such conditions a boundary tends to move in the direction of highest dislocation density, independently of its curvature. Examples of this phenomenon, called *strain-induced boundary-migration,* are given by Beck and Sperry (1950) for aluminum, and by Aust and Dunn (1957) for silicon-iron (Fig. 29).

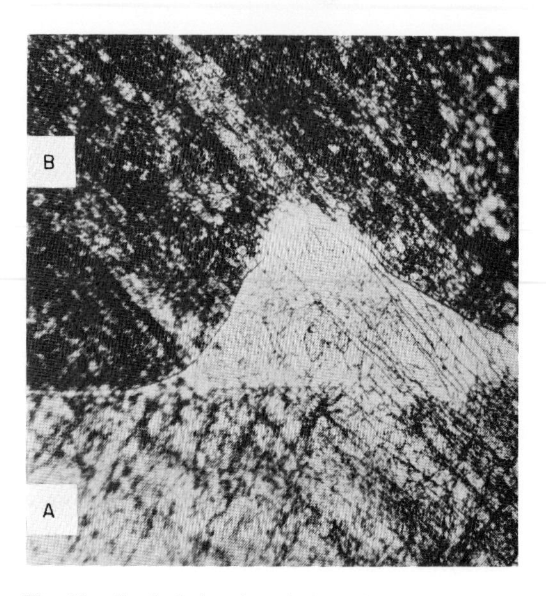

Fig. 29. Strain-induced grain-boundary migration in silicon-iron. The photograph shows part of a silicon-iron bicrystal ($3\frac{1}{4}$% Si-Fe), deformed by cold-rolling (12% reduction) and heat-treated at 1000°C for 15 min. Grain *A*, with lower dislocation density, has grown into grain *B* over a maximum distance of 0.09 mm (approximately 300 X). (After Aust and Dunn, 1957.)

Reasoning along these lines, a lattice domain in the cold-worked state might be capable of growth if it has sufficient size, possesses a lower dislocation density than surrounding domains, and differs sufficiently in orientation from its surroundings. As postulated by Beck (1949) and set forth in more detail by Cahn (1950), this state can be realized in "curved" lattice regions (as may be formed at places along the glide-planes where gliding is hindered in some way, for example, at crystal boundaries) by means of a process known as polygonization. This process is illustrated in Fig. 30 and consists of a rearrangement of dislocations both by glide and climb. The existence of this process has been confirmed by X-ray and microscopic investigations in numerous instances.

Polygonization transforms a continuously bent lattice, in which the dislocations are distributed irregularly, into a discrete set of "polygons," each keeping the orientation of its part of the bent lattice, but free from elastic strain. Figure 31, taken from Cahn's paper, illustrates this. It is suggested that

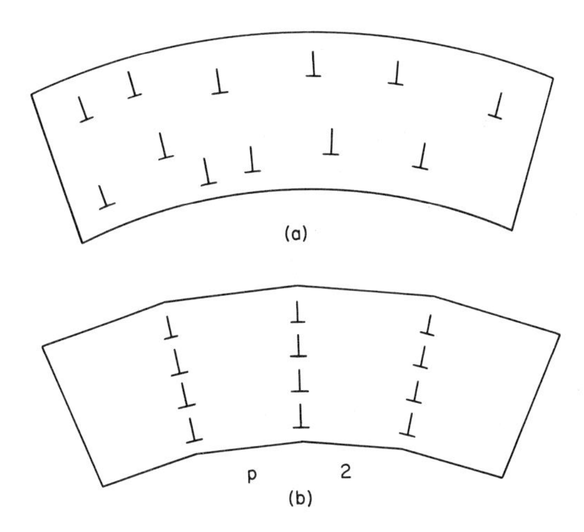

Fig. 30. Polygonization of a bent lattice. (*a*) Edge dislocations before polygonization. (*b*) After polygonization.

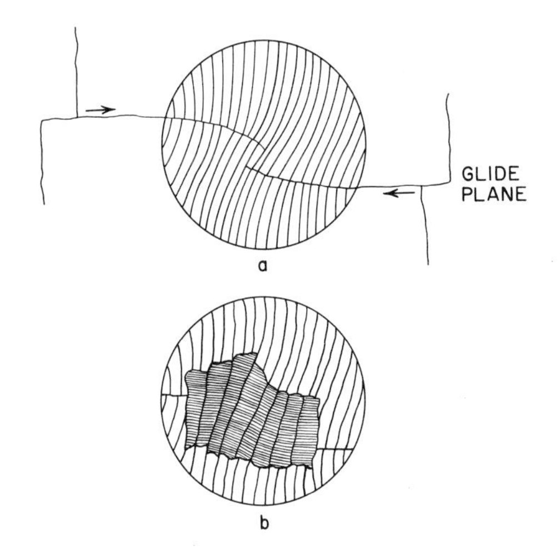

Fig. 31. Recrystallization nucleus formed by polygonization. (After Cahn, 1950.)

such a strain-free region is able to grow at the expense of its surroundings and is thus a potential nucleus. The incubation time for nucleation, according to this model, is the time required for polygonization. This will depend on the state of the curved region and may be different for various nuclei.

Investigations by means of X-rays and by electron microscopy support this mechanism, at least for various metals. By applying Guinier and Tennevin's (1949) X-ray technique, Tiedema (1950) showed that the central part of a new crystal formed by recrystallization in aluminum may consist of a small number of slightly misoriented lattice regions, the boundaries of which can be traced through the final macroscopic crystal. This is consistent with the polygonization process, for it is possible or even probable that not one single polygon, but a few adjacent polygons will grow out side by side to form the new crystal.

The transformation of a state with a high density of dislocations to one consisting of more or less dislocation-free regions separated by dislocation walls is evident from thin-foil electron microscopy: aluminum (Heidenreich, 1949; Hirsch et al., 1956,* Granzer and Haase, 1961); iron (McLean, 1959); nickel (Bollmann, 1959); silver (Bailey, 1960). As an example we reproduce Fig. 32 due to Granzer and Haase. It shows the formation of recrystallization nuclei in an aluminum foil. The foil was deformed by local compression during heating in the electron microscope. Recent electron diffraction work by these latter authors confirmed that the lattice orientation in the polygonized regions differed only slightly from the initial orientation in the same regions.

A necessary condition for the growth of a nucleus is the presence of a sufficient contrast in size and orientation between the growing polygon and its surroundings. For a metal of extreme purity the mobility of the dislocations can be so large that their rearrangement by polygonization leads to a general equilibrium between them without the necessary contrast. Then, growth of one special domain to macroscopic dimensions may not occur. Perhaps this happens in zone-melted iron and aluminum where, according to Talbot (1959) and Montuelle (1959), true recrystallization after cold-work and subsequent annealing does not take place. The recovery process stops after polygonization is complete. However, the addition of impurities may restore the capacity for recrystallization. The foreign atoms probably hinder the mobility of the dislocations locally, and contrast in size among the polygons is produced (see in this connection p. 436 concerning the influence of

* See also X-ray diffraction photographs of Hirsch and Kellar (1952).

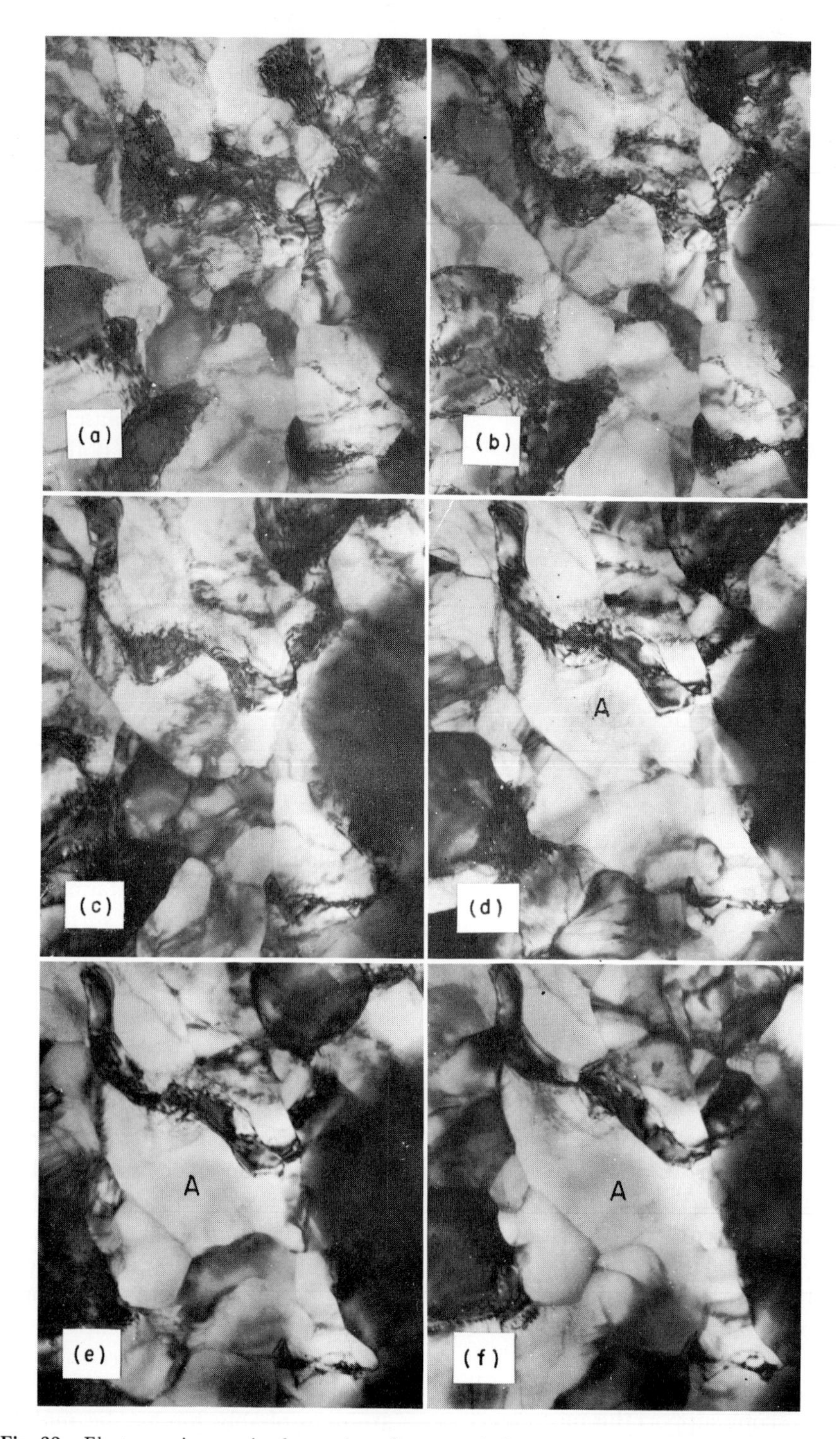

Fig. 32. Electron-microscopic observation of structural changes in a plastically deformed aluminum crystal. The dispersion of the dislocation clouds, and the formation of a dislocation free region (*A*) can be clearly seen (approximately 40,000 X). (After Granzer and Haase, 1961.)

a second phase on grain growth). Figure 33, due to Talbot (1959), shows an iron rod that was carburized in a temperature gradient, so the carbon content increases from left to right. After annealing, no recrystallization had taken place in the left-hand side where the carbon content is minimal (below 400 ppm). In contrast, in the part at the right large crystals have formed.

The model of nucleation as a process of rearrangement of dislocations is different from the model of growth, which involves movement of a boundary between two domains of different orientation. This agrees with the experimental fact (mentioned on p. 437) that the two processes have different activation energies and that they react differently to impurities.*

One further point must be mentioned. It seems plausible to expect that the capacity for polygonization and the rate at which it takes place depends on the deformation state of the bent region. We may assume that the state of deformation itself depends on the orientation of the principal deformation directions (direction of extension, drawing, or rolling) with respect to the original crystal lattice. This makes one anticipate that in a cold-worked specimen, lattice regions with a special orientation with respect to the deformation directions might polygonize and thus form nuclei sooner than other regions. This would cause a recrystallization texture, whose crystallographic character would depend on the relation between the original matrix and the directions of deformation. This is in agreement with observations of such preferred orientations as illustrated in Fig. 2.

* In three papers (Hu and Szirmae, 1961; Hu, 1962, 1963) on the annealing of a cold-worked silicon iron crystal, the authors, although confirming the occurrence of polygonization, did not observe migration of sub-boundaries but a "fading-away" of these boundaries, leading to the suggestion of nucleus formation by "coalescence" of subgrains, somewhat akin to a mechanism proposed by Nielsen (1954). A similar observation was made by Rzepski and Montuelle (1962) with aluminum-magnesium and aluminum-zinc alloys. A thermodynamic analysis of the possibility of subgrain rotation leading to coalescence is given by Li (1962).

It should be realized, however, that preferred orientations in recrystallized metals can be formed in a completely different way, vis., by a process of selective growth. As discussed previously, boundaries separating two lattice regions that have a special orientation difference have greater mobility than other boundaries. Beck and coworkers (1949) hold the view that such selective growth is the fundamental cause of recrystallization textures. Certainly the preferred orientation produced by rubbing an aluminum crystal on one side (p. 432) originated in this way. In other cases, both phenomena may play a part.

Martensitic Type of Nucleation

A final remark must be made. Both the processes described are only able to produce the growth of lattice domains already present in the cold-worked state. This means that the textures that are finally formed consist of orientations present before annealing. If these orientations encompass only a minor fraction of the cold-worked

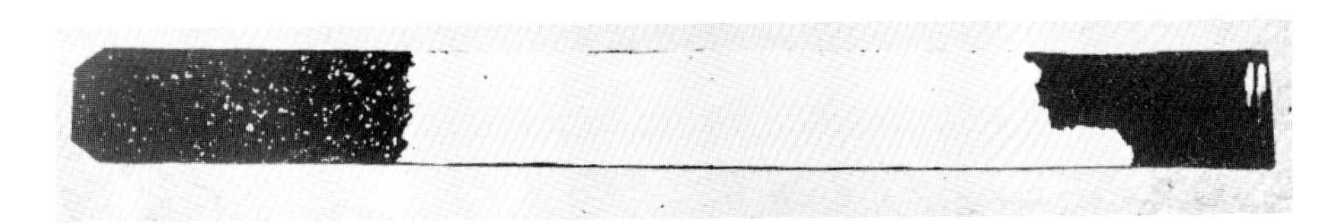

Fig. 33. Specimen of high purity iron, with carbon content varying along its length from 0.010% at the left-hand side to 0.050% at the right-hand side. The specimen was deformed by a 3% extension, and annealed in a hydrogen atmosphere at 880°C for 3 days. Although the region poorest in carbon (at the left) has remained fine-grained, large crystals have grown in the parts richer in carbon content (approximately natural size). (After Talbot, 1959.)

[A microphotograph shows, however, that the small grains have a polygonized structure (see Talbot, *loc. cit.*).]

matrix, the recrystallization texture may differ considerably from the deformation texture.

There are cases, however, where the recrystallization texture can not be traced experimentally to the deformed matrix and may not be unambiguously correlated to it by rotations of the type active in selective growth. This has raised the question of whether a nucleation process is conceivable that forms crystals with new orientations that are present in the deformed matrix. Verbraak and Burgers, 1957; Verbraak, 1958, 1960; see also Cook and Richards, 1940; Richards, 1955 have suggested that this could occur by a martensitic type of process, involving cooperative shear movements of complete lattice regions into new orientations. As with true martensitic transformations, such a process would probably involve distinct dislocation reactions, and hence could occur only in regions where suitable dislocations were present.

This concept is as yet not developed far enough or experimentally founded to justify a more detailed discussion here. It may suffice to refer the reader to the relevant literature. The idea is mentioned because the possibility can not be eliminated (Verbraak, 1960) that the remarkable coincidence orientations, as found and described by Kronberg and Wilson in recrystallized copper (see p. 429), may be the ultimate result of orientation changes produced by shear. Then they might not be produced by a process of selective growth in the sense used before.

4. PREPARATION OF LARGE CRYSTALS

In the foregoing sections the general principles concerning nucleation and growth during recrystallization have been outlined. Based on these we are now in a position to consider what experimental circumstances are favorable for the preparation of large crystals by this method.

Strain Anneal Method

In this method one generally starts from a fine-grained matrix that is subjected to a minimal critical strain just sufficient to cause the formation of one or at most a few new grains during annealing. To determine the critical strain for a given material, one usually subjects a set of specimens to gradually increasing strains, ranging from about 1 to 10%. They are then annealed at a suitable temperature for recrystallization. Such a recrystallized series of aluminum plates was shown in Fig. 6. In that case 1 to 2% would be a suitable critical strain to convert the whole specimen (a plate or a wire) into one crystal.

For crystal growth one applies a temperature gradient, as in most growth methods. The specimen is made to pass through a furnace at a certain speed. The furnace may be a long one in which the specimen enters at one end and leaves at the other, or it may be short. A temperature gradient during growth is necessary to avoid simultaneous nucleation in different parts of the specimen. It is most favorable if, on entering the hot zone of the furnace, one nucleus develops and grows fast enough to consume subsequent fine-grained material that enters the furnace before a second nucleus can be formed. The best gradient and velocity depend on the temperature dependences of the rates of nucleation and growth. Many times the first is more pronounced than the latter (see p. 437). This may be deduced from the fact that the number of grains formed in a specimen often increases with the rate of heating. For high purity aluminum, extremely pronounced temperature gradients have been used (Yoshida, Liebmann, and Lücke, 1957, Lommel, 1960). The tendency for grain growth in the pure aluminum is so large that the recrystallizing material must be prevented from entering the hot zone before it comes into contact with the growing grain. Otherwise, large grains can be formed by normal grain growth in front of the growing crystal and are left unconsumed.

In order to promote the formation of new crystallites, "localized nucleation" at one end of a specimen may be caused by severely compressing or pinching the specimen. This adds locally to the over-all critical deformation given to the specimen. When the severely deformed end enters the hot zone, several crystals form which on further growth compete until only one of them survives due to a faster rate of growth.

Other points to be considered are: the initial grain size must be sufficiently small and uniform to enable a nucleus, once it has grown larger than its immediate neighborhood, to consume all the rest. The presence in the matrix of grains much larger than the average may cause "inclusions" in the growing crystal that remain as unconsumed islands. On the other hand, the matrix must not be so fine-grained that even in the undeformed state annealing tends to increase its grain size. This again tends to cause "inclusions." A grain size of about 0.1 mm seems to be favorable in several instances (see, however, p. 445 for uranium).

Another factor to be considered is the texture of the matrix. The matrix is generally the product of one or more recrystallizations so its crystallites may show pronounced preferential orientations. On the one hand this may be favorable for producing a large crystal. If the orientation of a crystal is favorable for growth with respect to the main components of the texture, it will consume most of the original grains.

If, however, the orientation of the growing crystal is close to a twin orientation with respect to the main texture component, it may be considerably or even completely hampered in its growth. This results from the fact, discussed on p. 431, that a growing grain cannot consume a lattice region of its approximate twin orientation. Therefore in a matrix with a very pronounced texture (such as heavily drawn wire or a matrix obtained itself by cold-rolling and subsequent recrystallization of a monocrystal) one cannot grow crystals in any desired orientation. In particular, the orientation of the matrix itself and its twin orientation are excluded.

As an example of the effect of such restrictions, we mention that it is not possible to transform directly a drawn aluminum wire with a pronounced (111) texture parallel to the wire axis into a crystal with the same orientation. A growing crystal of this orientation would have its growth-plane perpendicular to the wire axis. It would meet many grains in an approximate twin or parallel orientation. As its growth-plane would be nearly perpendicular to the wire axis, even the grains that are rotated around the wire axis by the favorable angle of 40° would still be separated from it by a twist boundary that would have a low mobility (see p. 432).*

Another point concerns the formation of annealing twins (see Section 2). These form in several metals, often in the shape of thin lamellae with parallel boundaries. This makes it difficult to prepare large crystals of such metals by recrystallization. Since the tendency for twinning is correlated with a stacking-fault energy at the twin boundary, it seems inherent to the metal. The twinning tendency could probably be lowered if this energy could be increased by additions of certain foreign atoms. However, data on this point are scarce (Smallman and Westmacott, 1957). Moreover this remedy does not apply for a pure metal. We have seen, however, that a "lattice factor" is also involved in twin formation, the relative orientation of growing crystal and the matrix texture. According to Burgers, Meys, and Tiedema (1953), the tendency for twinning is particularly large if the grain to be consumed lies closer to the twin orientation of the growing grain than to its own orientation or, according to Fullman and Fisher (1951), if twinning results in a decrease in boundary energy between growing crystal and disappearing matrix grains (see also Aust and Rutter's (1960b) and Aust's (1961) observations on twinning in lead and aluminum mentioned on p. 433). Therefore twinning may sometimes be avoided by starting with a matrix having a sharp texture and making a crystal grow into

* An aluminum crystal with a ⟨111⟩-axis parallel to the wire axis could be obtained, however, in the following way: First the drawn wire is transformed into a crystal of arbitrary orientation by the usual method of annealing followed by critical strain. The crystal thus obtained is again subjected to a critical extension and a second crystal is made to grow at one end of the wire. This crystal will also have a chance orientation. However, by applying the method of "growing around the corner" (see p. 421) this crystal can be made to grow at the expense of the deformed matrix crystal in the desired ⟨111⟩ orientation.

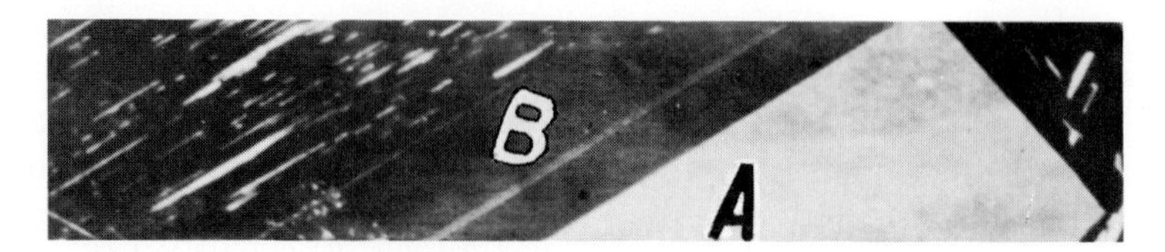

Fig. 34. Copper strip with two large crystals grown by recrystallization, one (*A*) without twins, the other (*B*) with many twin lamellae (approximately natural size). (After Burgers, Meys, and Tiedema, 1953.)

it with a lattice orientation that is neither so different from the matrix that twinning is favored nor so close to it that many grains of the matrix remain unabsorbed as inclusions. An example is given in Fig. 34* (Burgers, Meys, and Tiedema, 1953).

In certain cases (Takeyama, 1930) a decrease in the number of twins in recrystallized crystals can be obtained by deforming the crystal containing twins a second time and then recrystallizing it a second time.

Exaggerated Growth

Crystal formation by exaggerated growth was mentioned in Section 1, and further discussed in Section 2. In special cases continued annealing of a fine-grained specimen may result in very large crystals. A favorable condition for this to happen is the presence in the matrix of a sharp texture and a small quantity of an insoluble second phase. By impeding growth at certain points, a second phase allows boundary movements at other points so as to create a contrast in size. Excessive grain growth in an aluminum-manganese alloy at a temperature where the segregated phase dissolves (Beck, Holzworth, and Sperry, 1949) was mentioned as an example of this case.**

It has also been mentioned that surface energy as the driving force for exaggerated grain growth of a texture matrix has produced large crystals. The orientation of these depends on the difference in surface energy for different lattice planes at the gas metal interface. Therefore the orientation of the crystals formed in a certain matrix may be varied by changing the gas that surrounds the specimen. For example (Walter and Dunn, 1960), annealing of rolled high-purity silicon-iron of low oxygen content at 1200°C produces new crystals with a {110} plane parallel to the plane of rolling if the annealing is done in a nonoxidizing atmosphere such as vacuum. If, however, a slightly oxidizing atmosphere is used, {100} oriented grains result.

In secondary recrystallization crystal growth may be induced at a chosen spot of the specimen by a local deformation, such as a pin prick, (Rathenau and Custers, 1949) or even by direct contact with a second specimen. This latter method was used in remarkable experiments by Baer (1960) with silicon-iron and copper. Two flat specimens of the same metal, one containing large crystals and the other a fine primary texture, were pressed together and heated; some of the large crystals then grew into the fine-grained specimen at the places of contact.

The growth of large secondary crystals can sometimes be promoted by subjecting a matrix to a succession of two kinds of deformation, for example, rolling, and cutting, or hammering. This was observed by Müller (1940), and by Rathenau and Custers (1949) with nickel iron. If rolled sheet was cut with scissors (before being annealed), then primary recrystallization (to the cube texture) was followed by the excessive growth of large crystals at the cut edges. With rolling followed by hammering, the secondary crystals grew from the boundary between the hammered and the cubic region into the latter.

* The second crystal in the strip (*B*) shows many twin lamellae. They have all the same orientation as *A* and, in fact, *A* may be considered as a twin of *B*. *B*'s orientation with respect to the cube-oriented matrix is such that twinning is favored (see further, *loc. cit.*).

** See also Philip and Lenhart (1961) on the influence of manganese sulfide on the secondary recrystallization of silicon-iron.

To the same category of phenomena belongs the well-known fact that the formation of large crystals in tungsten lamp filaments (necessary for preventing "sagging" of the hot wires) can be enhanced by coiling the cold drawn wire (see Smithells, 1926). An interpretation of this curious behavior was given by van Liempt (1931). He thought that because of "interference" between the two strain-fields of different character (in van Liempt's trend of thought different "wavelength"), deformation gradients might be created at certain spots. If new crystals started to grow at such spots, they might attain sufficient size to create the necessary "contrast" for continued growth.

Allotropic Transition

In metals and alloys that have more than one allotropic modification, special measures are required to prepare monocrystals. Iron, for example, has a transition temperature at about 900°C, so crystals of the α-phase can only be prepared by recrystallization at a lower temperature. If crystals of the higher temperature modification are produced first (for example, for zirconium in the iodide process by thermal dissociation of ZrI_4 on a hot filament), they generally transform on cooling to more than one crystal of the lower temperature modification. These crystals show definite orientation relationships (Burgers, 1934), because of the fact that the transformation process may be of the "martensitic" type, taking place by shearing movements of the atoms parallel to definite crystallographic directions. See articles in *Progress in Metal Physics* by Bowles and Barrett (1952), and by Kaufman and Cohen (1958).

In some cases, however, a phase transformation may take place in such a way that a single crystal results. This has been achieved for iron by McKeehan (1927, 1928), and by Andrade and Greenland (see Andrade, 1937) by passing a proper temperature gradient along a wire at such a rate that the growing crystal of the lower temperature modification could keep up with the transformation. If not one crystal, then an array of stemlike crystals may be obtained (Wassermann 1935).

By applying a similar method (passing through a temperature gradient furnace) to uranium, making use of the $\beta \rightarrow \alpha$ phase change at ~670°C, Mercier, Calais, and Lacombe (1958) obtained imperfect monocrystals of the low temperature α-phase. If suitably oriented, such crystals could be recrystallized after a critical strain (at a temperature below the transition point) and transformed into crystals of high perfection.*

Other Methods

We conclude by mentioning two somewhat different methods where large crystals have been produced by recrystallization. Rutter and Aust (1958) were able to transform lead crystals, grown from the melt, with a pronounced "striated" structure, into stri-

* An exceptional case of the preparation of crystals involving a phase change is met with tin. Groen (1954, 1956) found that crystals of the cubic α-modification ("gray tin," stable below 13.2°C) can be obtained from the tetragonal β-modification ("white tin") if this latter contains a small amount of mercury. This transformation, however, is not a recrystallization process. As shown by van Lent (1961), due to the characteristics of the mercury-tin phase diagram, mercury can be liberated at the transformation front, forming a thin liquid layer between both phases through which tin diffuses as it were from the "white" to the "gray" phase. (See for details van Lent, 1962; see also Ewald and Tufte's (1958) method of preparing gray tin single crystals from a solution of tin in mercury.)

Fig. 35. Growth of striation-free high purity lead crystal into a striated crystal ($1\frac{1}{2}$ X). (After Rutter and Aust, 1958.)

ation-free monocrystals. Large-angle grain boundaries were locally introduced by plastic deformation followed by recrystallization. Then, on further annealing, a nucleated grain could grow into the undeformed striated crystal, the driving force being provided by the sub-boundary energy of this latter. Figure 35 is taken from this paper.

Leighly and Perkins (1960) prepared crystals of high purity aluminum by strain-induced boundary migration. A matrix with fairly large grains was critically strained so that no "new" nuclei were formed on subsequent annealing, but because of different degrees of straining in the matrix grains, some of them grew at the expense of others. By repeating this process, large crystals were finally obtained.

5. RECRYSTALLIZATION OF NONMETALLIC SUBSTANCES

Research carried out in this field has been far less than for metals. In part this is because of the enormous importance of metals as construction materials. Also, because of their plastic deformability, recrystallization in metals is not restricted to grain-growth in specimens that already consist of more or less perfect small crystallites. Rather, it encompasses the whole range of phenomena occurring during the transition from the cold-worked to the annealed state.

With nonmetals, deformability is often nearly nonexistent or very restricted in comparison with metals, and recrystallization is therefore restricted to the phenomenon of grain-growth. Exceptions include such substances as silver chloride and thallium chloride which, perhaps because of the high polarizibilities of their cations, can be plastically deformed like metals; for example, they can be rolled into plates. In such substances new crystallites can form during annealing of the cold-worked material and can enlarge in size by subsequent growth. The courses of these processes show considerable analogy with those observed in metals. As an example, Fig. 36 shows a set of recrystallized silver chloride plates obtained by subjecting fine-grained plates to a slight rolling deformation followed by annealing at 400°C. (Burgers and Tan Koen Hiok, 1946; Nye, 1949).

Extensive experiments on the recrystallization of rocksalt have been carried out by Müller (1934, 1935) and to some extent also by Przibram (1931, 1936). In Müller's experiments natural and synthetic crystals (the latter prepared from the melt) were plastically deformed at about 400°C by compression between flat plates.* Both the formation of new crystals and their growth

* Plastic deformation of rocksalt is also possible at room temperature if a fresh surface is obtained by wetting with water or if the crystal is immersed in water. Much research has been carried out, and several partly controversial theories have been given about the cause of this plasticity; see for older literature Schmid and Boas (1935), W. G. Burgers and J. M. Burgers (1935), and more recently, Stokes, Johnston, and Li (1960).

Fig. 36. Rolled and recrystallized plates of silverchloride with large crystals, after etching with sodiumthiosulfonate solution (approximately 2 X mag.). (After Burgers and Tan Koen Hiok, 1946.)

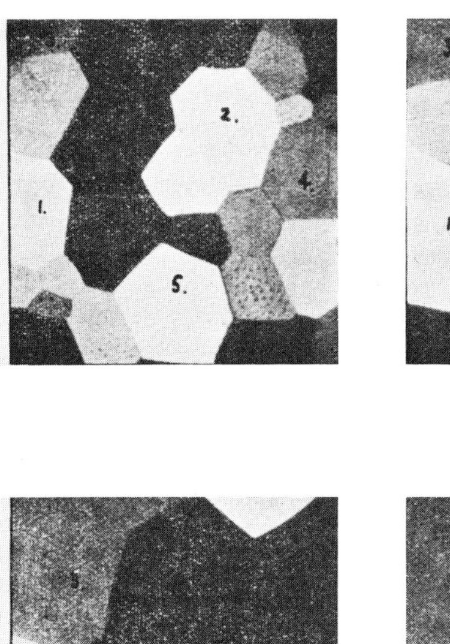

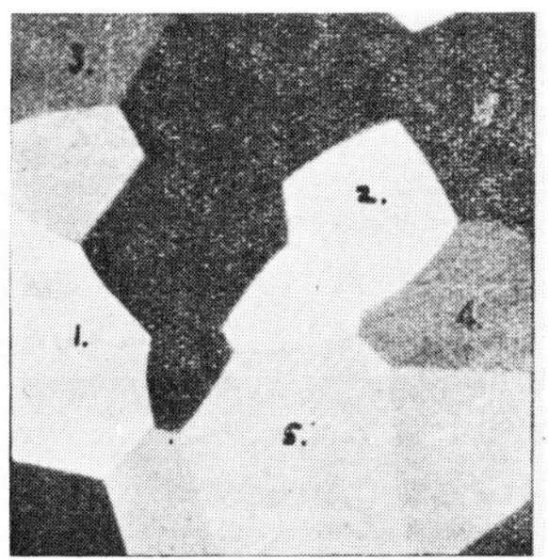

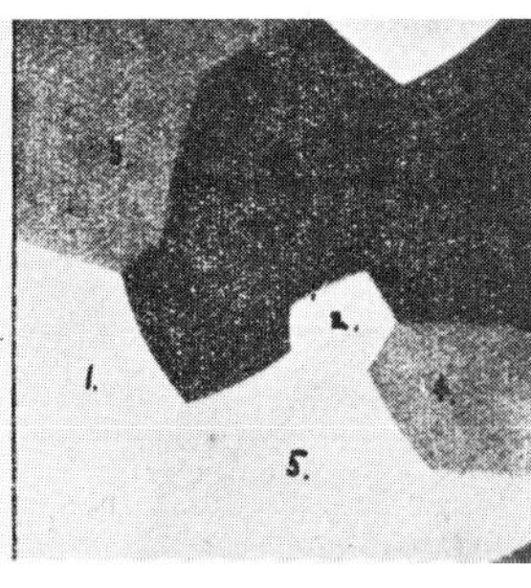

Fig. 37. Grain growth in an organic compound, octachloropropane heated at about 130°C. Crossed nicols (approximately 180 X). (After McCrone, 1949.)

into the deformed rocksalt matrix were studied by Müller. The laws governing these phenomena are quite analogous with the results obtained for metals.

Crystal growth phenomena have also been observed in solid organic substances. Figure 37, due to McCrone (1949), shows grain growth in thin plates of octachloropropane C_3Cl_8 at about 130°C. The sequence of photographs resembles closely the normal grain-growth process in metallic specimens. This seems remarkable here since complete molecules, each consisting of many atoms, must change their places.

6. RECRYSTALLIZATION DURING SINTERING

Sintering is a very complicated process, encompassing a variety of phenomena that take place when a more or less compressed powder, either dry or as a plasticized mass, is transformed by firing into a compact and coherent material. The phenomena include changes in porosity, density, shapes of grains and pores, recrystallization, and crystal growth (these all apart from possible chemical processes in case of mixtures of various compounds). Theoretical approaches to this process deal with questions of how far and to what extent surface, and/or volume diffusion, macroscopic flow, evaporation and condensation, and recrystallization play a part. For details we must refer to relevant textbooks and articles, of which some are Jones (1960); Gray (1957); Shaler and Wulff (1948).

Recent considerations are concerned especially with the role of vacancy diffusion for the motion of grain boundaries and the closing of pores (Hornstra, 1961; van Bueren and Hornstra, 1961).

An example of large crystals formed by sintering is shown by tungsten. When pressed ingots of tungsten powder are heated at high temperatures, grain growth is not restricted to the original powder particles but passes from one particle to another. Both the application of a temperature gradient and the addition of a certain amount of an insoluble phase (thorium-oxide) in order to increase size contrast (see p. 436) favor the formation of large crystals during the sintering process.

References

Anderson, W. A. and R. F. Mehl (1945), *Trans. AIME Inst. Met. Div.* **161,** 140.
da C. Andrade, E. N. (1937), *Proc. Roy. Soc. London* **A163,** 16.
van Arkel, A. E. (1932), *Polytechnisch Weekblad* **26,** 397ff.
van Arkel, A. E. and M. G. van Bruggen (1928), *Z. Physik* **51,** 520.
van Arkel, A. E. and J. J. A. Ploos van Amstel (1930), *Z. Physik* **62,** 43.
Assmus, F. K. Detert, and G. Ibe (1957), *Z. Metallk.* **48,** 344.
Aust, K. T. (1961), *Trans. Met. Soc. AIME,* **221,** 758.
Aust, K. T. and B. Chalmers (1950), *Proc. Roy. Soc.* **A201,** 210.
Aust, K. T. and C. G. Dunn (1957), *Trans. AIME, J. Metals* **9,** 472.
Aust, K. T. and J. W. Rutter (1959a), *Trans. Met. Soc. AIME* **215,** 119; (1959b), **215,** 820; (1960a), **218,** 50; (1960b), **218,** 1023; (1962), **224,** 111.
Avrami, M. (1939), *J. Chem. Physics* **7,** 1103; (1940), **8,** 212.
Baer, H. G. (1960), *Z. Metallk.* **51,** 650.
Bailey, J. E. (1960), *Phil. Mag.* **5,** 833.
Bainbridge, D. W., C. H. Li, and E. H. Edwards (1954), *Acta Met* **2,** 322.
Barten, P. G. J. (1954), *Ned. Tijdschrift v. Natuurkunde* **20,** 25.
Beck, P. A. (1949), *J. Appl. Phys.* **20,** 633; (1952), *Interface Migration in Recrystallisation,* Am. Soc. for Metals, Cleveland Symposium, 208; (1954), *Advances in Physics* **3,** 245; (1961), *Z. Metallk.* **52,** 13.
Beck, P. A. and Hsun Hu (1949), *Trans. AIME* **185,** 627.
Beck, P. A. and Ph. R. Sperry (1950), *J. Appl. Phys.* **21,** 150.
Beck, P. A., M. L. Holzworth, and Ph. R. Sperry (1949), *Trans. AIME* **180,** 163.
Beck, P. A., Ph. R. Sperry, and Hsun Hu (1950), *J. Appl. Phys.* **21,** 420.
Bollmann, W. (1959), *J. Inst. Metals* **87,** 439.
Bowles, J. S., and C. S. Barrett (1952), *Progress in Metal Physics,* **3,** 1.
Bowles, J. S. and W. Boas (1948), *J. Inst. Metals* **74,** 501.
Bragg, W. L. (1940), *Proc. Phys. Soc. London* **52,** 54.
Brinson, G. and A. J. W. Moore (1951), *J. Inst. Metals* **79,** 429.
Brugman, F. W. and T. J. Tiedema (1952), *Appl. Scien. Res.* **A3,** 250, 391.
van Bueren, H. G. (1960), *Imperfections in Crystals,* North-Holland Publishing Co. Amsterdam, p. 599.
van Bueren, H. G. and J. Hornstra (1961), *Reactivity of Solids,* Edited by J. H. de Boer et al., Elsevier, Amsterdam, 112.
Burgers, J. M. (1940), *Proc. Phys. Soc. London* **52,** 23.
Burgers, W. G. (1934), *Physica* **1,** 561; (1938), *Metallwirtschaft* **17,** 648; (1941), *Handb. d. Metall. physik* **III,** 2; (1947), *Proc. Roy. Acad. Sci.* Amsterdam **50,** 858; (1961), *Z. Metallk.* **52,** 19.
Burgers, W. G. and J. C. M. Basart (1929), *Z. Physik* **54,** 74.
Burgers, W. G. and J. M. Burgers (1935), *Verh. Roy. Acad. Sci. Amsterdam* **15,** 173; *Nature* **135,** 960.
Burgers, W. G. and W. May (1945), *Rec. Trav. Chim. d. Pays-Bas,* **64,** 5.
Burgers, W. G. and J. J. A. Ploos van Amstel (1937), *Physica* **4,** 15; (1938), *Physica* **5,** 305.
Burgers, W. G. and Tan Koen Hiok (1946), *Physica* **11,** 353.
Burgers, W. G. and T. J. Tiedema (1950), *Proc. Roy. Acad. Sci. Amsterdam* **53,** 1525.
Burgers, W. G. and C. A. Verbraak (1957), *Acta Met.* **5,** 765.
Burgers, W. G., J. C. Meys, and T. J. Tiedema (1953), *Acta Met.* **1,** 75.
Burke, J. E. (1948), *Met. Tech. T. P.* 2472; (1949), *Trans. AIME* **180,** 73; (1950), **188,** 1324.
Burke, J. E. and D. Turnbull (1952), *Progress in Metal Physics* **3,** 220.
Cahn, R. W. (1950), *Proc. Phys. Soc. London* **63,** 323.
Chalmers, B. (1952), *Progress in Metal Phys.* **3,** 293.
Cohen, M. (1958), *Trans. Met. Soc. AIME* **212,** 171.
Cook, M. and T. L. Richards (1940), *J. Inst. Met.* **66,** 1.
Czjzek, G. and F. Haeszner (1960), *Z. Metallk.* **51,** 567.
Dahl, O. and F. Pawlek (1936), *Z. Metallk.* **28,** 266.
Despujols, J. (1952), *C. R. Acad. Sci. Paris* **235,** 716.
Dmitrov, O. (1959), *C. R. Acad. Sci. Paris* **249,** 265; *Nouvelles propriétés physiques et chimiques des métaux de très haute pureté,* Colloque Internat. C.N.R.S., Paris, p. 79.
Dunn, C. G. (1949a), *Trans. AIME* **185,** 72; (1949b); *Symp. Cold Working of Metals,* Cleveland, Ohio, p. 113; (1953), *Acta Met.* **1,** 163; (1954), *Acta Met.* **2,** 386.
Dunn, C. G. and F. W. Daniels (1951), *Trans. AIME* **191,** 147.
Dunn, C. G. and F. Lionetti (1949), *Trans. AIME, J. Metals* **185,** 125.
Dunn, C. G., F. W. Daniels, and M. J. Bolton (1950), *Trans. AIME, J. of Metals* **188,** 1245.
Ewald, A. W. and O. N. Tufte (1958), *J. Appl. Phys.* **29,** 107.
Feitknecht, W. (1926), *J. Inst. Metals* **35,** 131.
Ferro, A., C. Sari, and G. Venturello (1959), *Acta Met.* **7,** 429.
Fujiwara, T. (1939), *J. Sci. Hiroshima Univ.* **A9,** 227; (1941), *J. Sci. Hiroshima Univ.* **A11,** 89.
Fullman, R. L. and J. C. Fisher (1951), *J. Appl. Phys.* **22,** 1350.
Gow, K. V. (1954), *Acta Met.* **2,** 396.
Graham, C. D. Jr. and R. W. Cahn (1956), *Trans. AIME, J. of Metals* pp. 504, 517.
Granzer, F., and G. Haase (1961), *Z. Physik.* **162,** 504, 517.
Gray, T. J. (1957), *The Defect Solid State,* Interscience, New York.
Green, R. E., B. G. Liebmann, and H. Yoshida (1958), *Tech. Rept. Met. Research Lab.,* Brown University.
Groen, L. J. (1954), *Nature* **174,** 1836; (1954), *Proc. Roy. Acad. Sci.: Amsterdam* **B57,** 122; (1956), Thesis, Delft.
Guinier, A. and J. Tennevin (1949), *Acta Cryst.* **2,** 133.
Hägg, G. and N. Karlsson (1952), *Acta Cryst.* **5,** 728.
Heidenreich, R. D. (1949), *J. Appl. Phys.* **20,** 993.

Hirsch, P. B. and J. N. Kellar (1952), *Acta Cryst.* **5,** 162, 168, and 172.

Hirsch, P. B., R. W. Horne, and M. J. Whelan (1956), *Dislocations and Mechanical Properties of Crystals,* John Wiley and Sons, New York. p. 92.

Holmes, E. L. and W. C. Winegard, 1959, *Can. J. Phys.* **37,** 496; 1960, *J. Inst. Metals* **88,** p. 468.

Holt, D. B. (1959), *Acta Met.* **7,** 446.

Hornstra, J. (1958), *Phys. Chem. Solids* **5,** 129; (1959), *Physica* **25,** 409; (1960), **26,** 198; (1961), **27,** 342.

Hu, Hsun (1962), *Trans. Met. Soc. AIME* **224,** 75; (1963), "Recrystallization by Subgrain Coalescence" in *Electron Microscopy and Strength of Crystals,* Interscience, New York.

Hu, Hsun, and A. Szirmae (1961), *Trans. Met. Soc. AIME* **221,** 839.

Jeffreys, Z. and R. S. Archer (1922), *Chem. Met. Eng.* **26,** 343, 402, and 449.

Johnson, W. A. and R. F. Mehl (1939), *Trans. AIME Iron Steel Div.* **135,** 416.

Jones, W. D. (1960), "Fundamental Principles of Powder Metallurgy (London).

Karnop, R. and G. Sachs (1930), *Z. Physik* **60,** 464.

Kaufman, L. and M. Cohen (1958), *Progress in Metal Physics* **7,** 165.

Kohara, S., M. N. Parthasarathi, and P. A. Beck (1958), *J. Appl. Phys.* **29,** 1125; *Trans. Met. Soc. AIME* **212,** 875.

Kronberg, M. L. and F. H. Wilson (1949), *Trans. AIME* **185,** 501.

Lacombe, P. and A. Berghézan (1949a), *Métaux et Corrosion* **25,** 1; (1949b), *Alluminio,* **7;** (1951), "Interface migration in recrystallisation," *Metal Interfaces,* Am. Soc. f. Metals, 211.

Lacombe, P. and D. Calais (1958), *Proc. Second U.N. Geneva Conference,* p. 1258.

Lacombe, P. and N. Yannaquis (1948), *Rev. Mét.* **45,** 68.

Leighly, H. P., Jr., and F. C. Perkins (1960), *Trans. Met. Soc. AIME* **218,** 379.

van Lent, P. H. (1961), *Acta Met.* **9,** 125; (1962), *Acta Met.* **10,** 1089; Thesis, Delft.

Leymonie, C. and P. Lacombe (1957), *Rev. Mét.* **54,** 653.

Li, J. C. M. (1961), *J. Appl. Phys.* **32,** 525; (1962) *ibid.,* **33,** 2958.

Li, C. H., E. H. Edwards, J. Washburn, and E. R. Parker, 1953, *Acta Met.* **1,** 223.

Liebmann, B., K. Lücke, and G. Masing (1956), *Z. Metallk.* **47,** 57.

van Liempt, J. A. M. (1931), *Z. anorg. allg. Chem.* **195,** 366.

Lommel, J. M. (1960), *Trans. Met. Soc. AIME* **218,** 374.

Lücke, K. (1950), *Z. Metallk.* **41,** 114; (1961), **52,** 1.

Lücke, K. and K. Detert (1957), *Acta Met.* **5,** 628.

Lücke, K., G. Masing, and P. Nölting (1956), *Z. Metallk.* **47,** 64.

May, M. and F. Erdmann-Jesnitzer (1959), *Z. Metallk.* **50,** 434.

May, W. (1950), Thesis, Delft.

McCrone, W. C. (1949), *Discussion Faraday Soc.* **5,** 158.

McKeehan, L. W. (1927), *Nature* **119,** 705; (1928), *Proc. AIME Inst. Met. Div.* 453.

McLean, D. (1959), *Nouvelles propriétés physiques et chimiques des métaux de très haute pureté,* Colloque Internat. C.N.R.S., Paris, 181.

Mercier, J., D. Calais, and P. Lacombe (1958), *Comptes. Rendus Acad. Sci. Paris* **246,** 110.

Montuelle, J. (1959), *Nouvelles propriétés physiques et chimiques des métaux de très haute pureté,* Colloque Internat. C.N.R.S., Paris, 185.

Mott, N. F. (1948), *Proc. Phys. Soc. London* **60,** 391.

Müller, H. G. (1934), *Physik. Z.* **35,** 646; (1935), *Z. Physik* **96,** 279; (1940), *Metallwirtschaft* **19,** 509.

Nielsen, J. P. (1954), *Trans. AIME* **200,** 1084.

Nye, J. P. (1949), *Proc. Roy. Soc. London* **A198,** 190; **A200,** 47.

Okkerse, B., T. J. Tiedema, and W. G. Burgers (1955), *Acta Met.* **3,** 300.

Oriani, R. A. (1959), *Acta Met.* **7,** 62.

Parthasarathi, M. N. and P. A. Beck (1961), *Trans. Met. Soc. AIME* **221,** 831.

Pawlek, F. (1935), *Z. Metallk.* **27,** 160.

Pfann, W. G. and L. C. Lovell, 1955, *Acta Met.* **3,** 512.

Philips, T. V. and R. E. Lenhart (1961), *Trans. Met. Soc. AIME* **221,** 439.

Pontecorvo, G. and A. R. Gemmell (1944), *Nature* **154,** 532.

Przibram, K. (1931), *Z. Physik* **67,** 89; (1936), **102,** 331.

Rathenau, G. W. and G. Baas (1951), *Physica* **17,** 117.

Rathenau, G. W. and J. F. H. Custers (1949), *Philips Research Rept.* **4,** 241.

Read, W. T. (1953), *Dislocations in Crystals,* McGraw-Hill Co., New York.

Richards, T. L. (1955), *X-Ray Diffraction by Polycrystalline Materials,* Inst. of Physics, London, p. 472.

Rieck, G. D. (1958), *Acta Met.* **6,** 360.

Rutter, J. W. and K. T. Aust (1958), *Acta Met.* **6,** 375; (1961), *Trans. Met. Soc. AIME* **221,** 641; (in press, A), "Kinetics of grain boundary migration in high-purity lead containing very small additions of silver and of gold"; (in press, B), "Mobility of ⟨100⟩ tilt grain boundaries in high-purity lead."

Rzepski, C. and J. Montelle (1962), *Comptes Rendus Acad. d. Sciences,* Paris, **254,** 1633.

Sachs, G. (1934), *Praktische Metallkunde,* **II,** Berlin.

Sandee, J. (1942), *Physica* **9,** 741.

Schmid, E. and W. Boas (1935), *Kristallplastizität* Springer, Berlin.

in der Schmitten, W., P. Haasen and F. Haeszner (1960), *Z. Metallk.* **51,** 101.

Shaler, A. J. and J. Wulff (1948), *Ind. Eng. Chem.* **40,** 838.

Shockley, W. and W. T. Read (1949), *Phys. Rev.* **75,** 692; (1950), **78,** 275.

Smallman, R. E. and K. H. Westmacott (1957), *Phil. Mag.* **2,** 669.

Smith, C. S. (1948), *Trans. AIME* **175,** 15; (1951), "Grain Shapes and other metallurgical applications of topology," *Metal Interfaces,* Am. Soc. f. Metals, p. 65.

Smithells, C. J. (1926), *Tungsten,* Chapman and Hall, London.

Smoluchowski, R. (1952), *Phys. Rev.* **87,** 483.

Stokes, R. J., T. L. Johnston, and C. H. Li (1960), *Trans. Met. Soc. AIME* **218,** 655.

Stüwe, H. P. (1961), *Z. Metallk.* **52,** 34.

Takeyama, S. (1930), *Mem. Coll. Sc., Kyoto Imp. Univ.* **A13,** 353.

Talbot, J. (1959), *Nouvelles propriétés physiques et chimiques des métaux de très haute pureté,* Colloque International C.N.R.S., Paris, 161.

Tiedema, T. J. (1949), *Acta Cryst.* **2,** 261; (1950), *Proc. Roy. Acad. Sci. Amsterdam* **53,** 1422; (1951), Thesis, Delft.

Tiedema, T. J. and C. L. D. Kooy (1953), *Ned. Tijdschrift v. Natuurkunde* **19,** 170.

Tiedema, T. J., J. Bouman, and W. G. Burgers (1957), *Acta Met.* **5,** 310.

Tiedema, T. J., W. May, and W. G. Burgers (1949), *Acta Cryst.* **2,** 151.

Verbraak, C. A. (1958), *Acta Met.* **6,** 580; (1960), *Acta Met.* **8,** 65; *Z. Metallk.* **51,** 646.

Vogel, F. L., W. G. Pfann, H. Corey, and E. Thomas (1953), *Phys. Rev.* **90,** 489; W. G. Pfann and L. C. Lovell (1955), *Acta Met.* **3,** 512.

Vreeland, T., Jr. (1961), *Acta Met.* **9,** 112.

Waddell, A. H. (1945), *Edinburgh Math. Notes,* 14; (1946), *Nature* **158,** 29.

Walter, J. L. (1959), *Acta Met.* **7,** 424.

Walter, J. L. and C. G. Dunn (1960), *Trans. Met. Soc. AIME* **218,** 914, 1033.

Washburn, J. and E. R. Parker (1952), *Trans. AIME* **194,** 1076; C. H. Li, E. H. Edwards, J. Washburn, and E. R. Parker (1953), *Acta Met.* **1,** 223; D. W. Bainbridge, C. H. Li, and E. H. Edwards (1954), *Acta Met.* **2,** 322.

Wassermann, G. (1935), *Mitt. K. W. Inst. Eisenf. Düsseldorf* **17,** 203.

Weik, H. (1953), *Z. angew. Physik* **5,** 119.

Weissmann, S., M. Hirabayashi, and H. Fujita (1961), *J. Appl. Phys.* **32,** 1156.

Yoshida, H., B. Liebmann, and K. Lücke (1959), *Acta Met.* **7,** 51.

Specific Substances

23

Large Crystals Grown by Recrystallization

K. T. Aust

Although many of the principles involved in producing single crystals by recrystallization are known, it is sometimes difficult to apply them to a particular material. This difficulty is due largely to the fact that the optimum conditions of prior treatment, straining and annealing, depend on the purity and state of the material. This chapter is concerned with recrystallization or solid state methods for preparing large crystals. The methods are described in Part I, and their application to several specific materials is discussed in Part II. Typical operating data are given wherever possible for the preparation of crystals having specified dimensions, purity, orientation, and perfection. It is hoped that this discussion of crystal preparation will enable the experimentalist to, first, decide whether a particular recrystallization method is suitable for the type of crystal that is required, and second, to successfully apply the method with a minimum amount of trial and error.

Several disadvantages are associated with the use of recrystallization methods. The control of orientation can be troublesome and the thickness of the specimen is sometimes limited. In some materials, such as germanium and silicon, it is difficult to obtain recrystallization, that is, the rearrangement of the atoms of the solid into an entirely new set of crystals. In other materials, especially many of the face-centered cubic metals, the high frequency with which annealing twins result during recrystallization makes it difficult to grow untwinned crystals.

There are, however, several reasons why solid state methods are often preferred for the preparation of crystals. One is that they may be used to obtain crystals in materials having a phase change between the melting point and room temperature. Another is that the shape of the specimen is under more precise control than, say, for solidification methods. As a result, crystals in the form of thin sheets or small diameter wires can be prepared readily in some cases. The methods have the additional advantage that there is, apparently, very little redistribution of solute during growth in the solid state. Therefore, it is simple to prepare homogeneous crystals of many single-phase alloys by recrystallization methods. Crystals of low melting point materials are generally more readily grown from the melt, although recrystallization methods could be applied in order to avoid gross imperfections such as lineage structure. It is sometimes advantageous, especially with very high purity materials, to combine solidification and solid state techniques in order to prepare large crystals with a minimum variation in orientation.

I. RECRYSTALLIZATION METHODS

1. STRAIN-ANNEAL

The strain-anneal method for producing monocrystals depends on the fact that the recrystallized grain size increases with decreasing amount of prior strain, as depicted in curve *A* of Fig. 1. A monocrystal can be obtained by annealing a polycrystalline specimen that has been given a critical amount of strain, that is, E_c in Fig. 1. A critical strain is required just sufficient to obtain recrystallization in order to allow as few nuclei as possible to grow into the deformed material at the annealing temperature. If less than the critical strain is applied, no recrystallization will occur on subsequent annealing (region *B* in Fig. 1); for larger strains, several crystals in the same specimen will result.

The starting material must be in a strain-free and uniform, fine-grained state, which is usually attained by fairly heavy cold work followed by a recrystallization anneal. A small and uniform grain size is needed to obtain homogeneous deformation during the critical straining so as to reduce the probability of nucleation of stray crystals. The fine-grained specimen, in the form of a strip or bar, is then given a slight deformation, for example, by a tensile strain of several per cent, and annealed by slowly heating to a temperature well above that necessary to obtain recrystallization. A few new crystal nuclei form at some temperature which cannot be exactly predicted, hence the need for a slowly rising temperature. These nuclei then invade the strained matrix and consume it entirely before the temperature has risen high enough to generate more nuclei.

A more reliable procedure is to pass the critically strained specimen through a temperature gradient. In this way, only a small portion at one end of the specimen will be at the proper temperature for recrystallization, as illustrated in Fig. 2. This allows nucleation to begin at one end of the specimen, and the temperature gradient insures that nucleation does not occur ahead of the main growth front. The temperature, in effect, travels along the specimen, thereby increasing the probability that a single grain will travel along with it, consuming the deformed matrix before the latter is able to nucleate new grains. The temperature gradient should be as large as possible in order to decrease the probability of forming many new grains by recrystallization. Williamson and Smallman (1953) theoretically treat the probability of nucleation in a specimen in terms of such factors as the velocity of travel, temperature gradient, and the activation energies for nucleation and growth.

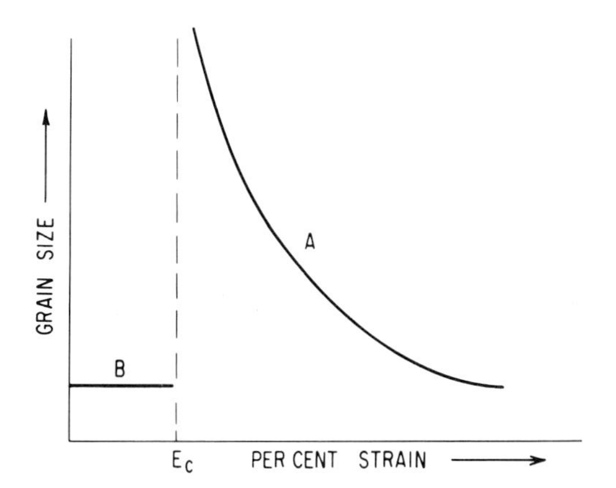

Fig. 1. Recrystallized grain size versus amount of prior strain.

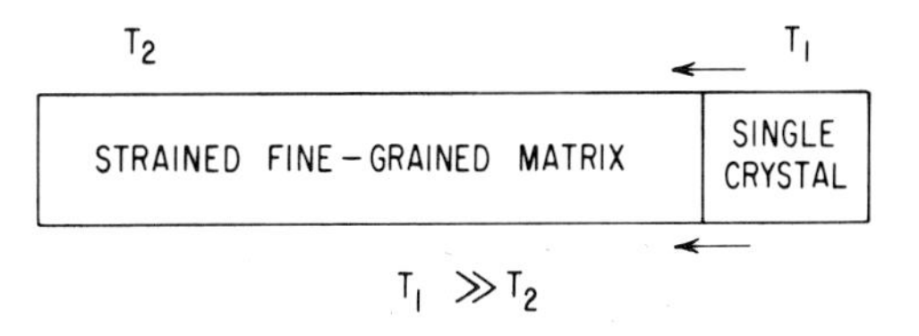

Fig. 2. Growth of a crystal by a critical strain followed by an anneal in a large temperature gradient.

A useful feature of the technique of critical strain followed by a temperature-gradient anneal is that the orientation of the crystal can be controlled (Fujiwara, 1939). The strip or wire is first critically strained and annealed at one end in a steep temperature gradient. The growth is allowed to proceed for a short distance, generally obtaining several grains with uncontrolled orientations. The orientation of each grain is determined, and the grains are cut in such a manner that only one grain (for example, *B* in Fig. 3) having the orientation closest to the desired orientation is connected to the main part of the specimen. The specimen is then bent and/or twisted, if necessary, so that the seed crystal "*B*" is correctly oriented with respect to the major part of the specimen. On further annealing, the seed will grow past the bend and thereby change its orientation relative to the specimen axis. The angle of

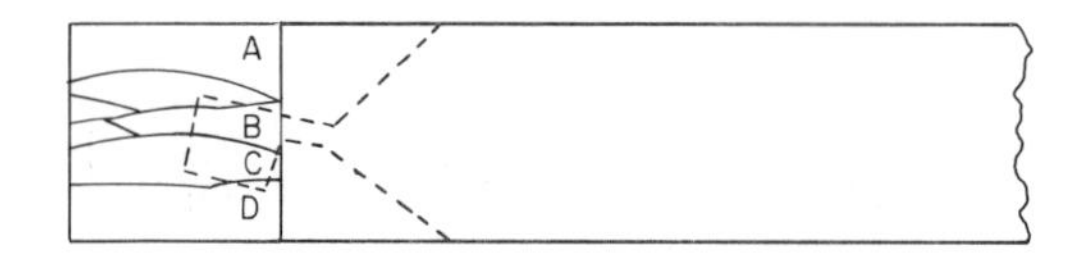

Fig. 3. Technique for producing an oriented crystal, showing method of selecting a seed crystal. (Courtesy of C. G. Dunn.)

bend and/or twist can be calculated to give crystals of any required orientation. This technique was extended by Dunn (1949) to a specimen containing two or three seed crystals with correctly adjusted orientations. Thus, it is possible to produce bicrystal or tricrystal specimens having controlled orientation.

Difficulties are generally encountered in attempts to grow crystals of high purity materials by the strain-anneal method. For example, a fine-grained, stable matrix structure prior to the critical strain and growth anneal is more difficult to attain because grain growth can occur more readily in a purer material. However, the introduction of a strong preferred orientation into a matrix structure may stabilize the matrix since the resulting low and medium angle grain boundaries have sufficiently low mobility to prevent normal grain growth. This texture-inhibited growth, which is the basis of the secondary recrystallization method, is helpful in the strain-anneal method. Another difficulty involves recovery effects such as polygonization which can occur more readily in the high purity strained matrix and thereby inhibit the recrystallization to new crystal orientations. However, polygonization may be retarded by the introduction of certain impurities, which can later be removed after the single crystal is obtained. This problem is considered further in the sections on preparing high purity crystals of aluminum and iron.

The application of the strain-anneal method to the growth of crystals of the following materials is discussed in Part II: Al and its alloys, Fe and its alloys, Ta, Th, Ti, and α-U.

2. GRAIN GROWTH

Another solid state method for obtaining single crystals is based on principles associated with grain growth. A very simple approach is to anneal a polycrystalline, cast material, and thereby obtain large grains by grain growth. However, a large increase in grain size is generally not observed in usual commercial-purity materials when annealed in the "as-cast" condition. In material of very high purity, it is possible to obtain large crystals from a polycrystalline casting by grain growth. For example, Fig. 4*a* shows a macrophotograph of a polycrystalline ingot of zone-refined lead, and Fig. 4*b* shows this same specimen after a 1 hr anneal at 300°C (Rutter and Aust, unpublished). Several large crystals are present after annealing which occupy the entire cross section of the ingot. The driving force for the growth shown in Fig. 4 is believed to be due largely to the reduction of interfacial area, that is, interfacial energy of the specimen. This simple method may be applied to other very high purity materials which have high grain-boundary mobility.

In some cases it is possible to allow the process of grain growth to occur so that only one grain grows. Andrade (1937) achieved this in molybdenum and tungsten by movement of localized hot zone along an electrically heated, polycrystalline wire, as shown schematically in Fig. 5. The number of possible growth centers is limited here by the use of a temperature gradient. In the original technique of Andrade, it is necessary to have wires of small enough size that grains extend across their section, or else to have wires in such a condition that heating them results in grain growth across the section. The gradient furnace is then moved at a suitable rate for crystal growth to occur.

It is apparently not necessary to deform intentionally the small diameter wire prior to heating, as is done in the strain-anneal method, in order to obtain a monocrystal. However, it is possible that strain energy introduced from the wire-drawing or accidentally (for example, during the mechanical polishing of the 1 mm or less diameter wire prior to the anneal) may supply the initial driving energy for crystal growth. For example, in order to grow crystals from larger diameter wire of molybdenum, niobium, or tantalum by the Andrade method, it is necessary to apply a small tensile strain

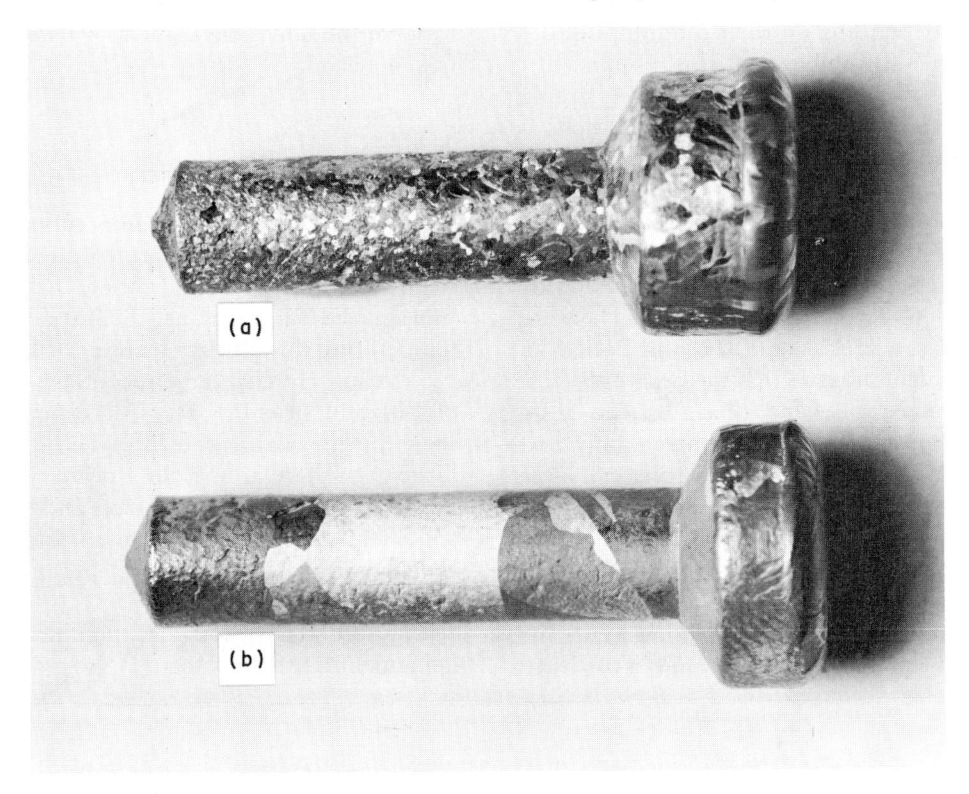

Fig. 4. Grain growth in a cast ingot of zone-refined lead. (*a*) "As-cast." (*b*) "As-cast" and annealed for 1 hr at 300°C. (J. W. Rutter and K. T. Aust, unpublished.)

of about 1% during the growth anneal (Maddin, 1958).

The Andrade method has several limitations. Small wires must be used so that they can be heated effectively to elevated temperatures by resistance heating. It is difficult to grow crystals in materials that are good thermal conductors because of the large heat losses to the grips holding the specimen. In addition, thermal expansion of the wire must be taken into account in the design of the equipment or the crystal will buckle and hence be deformed or even break. Oxidation must also be avoided to prevent the formation of local hot spots. It is evident that this method is limited to materials in wire form having high melting point and high thermal and electrical resistivity. The Andrade method, together with several modifications of it, are discussed in Part II for the following materials: Fe, Mo, Nb, and W. The grain growth method may also be applicable to Ta and Rh (Honeycombe, 1959).

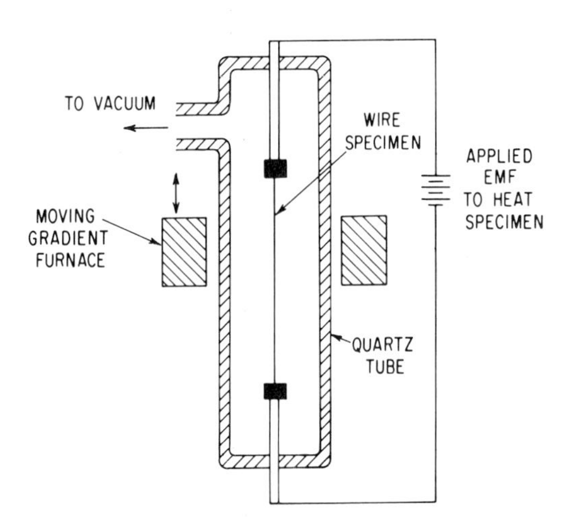

Fig. 5. Andrade's method for growing monocrystalline wires.

3. EXAGGERATED GRAIN GROWTH

Large grains are sometimes obtained by an exaggerated or discontinuous grain growth technique in which the growth of most of the grains in the matrix is inhibited by dispersed particles of a second phase. Inclusions, depending on their amount and dispersion, may inhibit grain growth sufficiently to stop the growth of recrystallized grains. However, if the amount of inclusions is initially small and their dispersion changes by solution or coalescence at the annealing temperature, very large grains are possible by a procedure of discontinuous grain growth in which the inhibitor breaks down locally and allows certain grains to grow. The work of Jeffries and Archer (1924) and Beck et al. (1949) demonstrates that the progressive dissolution of a second phase at the grain boundaries can, by permitting only few boundaries to move at first, develop a large disparity in the grain size.

This technique involving inclusion-inhibited growth is described for tungsten and α-uranium as well as for certain ceramic crystals obtained during sintering where the porosity present in the ceramic is believed to serve the same role as the growth-inhibiting inclusion.

4. SECONDARY RECRYSTALLIZATION

Secondary recrystallization occurs on annealing many heavily deformed metals. A strong primary recrystallization texture is first formed on annealing, then at a higher temperature this primary texture is removed by a relatively few, large secondary grains of different orientations. An example of secondary growth is shown in Fig. 6. The large grain here grew from a primary, fine-grained recrystallized matrix and shows the classical morphology with indented boundaries moving towards their centers of curvature.

The secondary recrystallization method is dependent on the fact that (1) the growth of

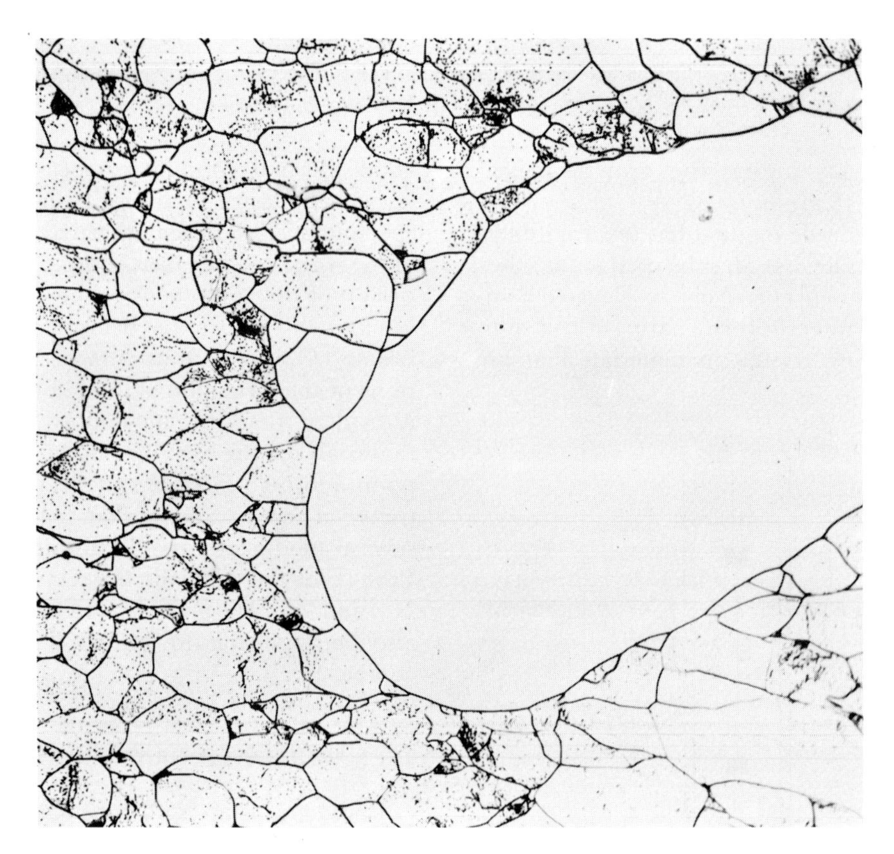

Fig. 6. Example of the growth of a large secondary grain into the primary recrystallized texture of silicon-iron (500 X). (Courtesy of C. G. Dunn.)

a few crystals is promoted by the presence of a strong preferred orientation and (2) the resulting crystals have a preference for certain orientation relationships to the matrix. The number of possible growth centers is limited by the fact that the primary recrystallization texture is so predominant that other orientations are generally quite scarce. As a consequence, this process is sometimes referred to as "texture-inhibited" growth. It should be noted that texture-inhibited growth and inclusion-inhibited growth, which was previously discussed, may be operative independently of each other. However, in many commercially pure materials they may both be active at the same time. Thus, it is not always easy to distinguish whether the large grains are produced by inclusions or by texture. In some cases the dispersed particles of a second phase are so small that they cannot be detected microscopically. For example, large grains can be obtained from a primary recrystallized matrix of 2S aluminum (commercial grade, approximately 99.2% aluminum). However, the growth here is believed to be due to the fact that the primary matrix is stabilized by inclusions instead of by a strong preferred orientation (for example, see Parthasarathi and Beck, 1958).

Large crystals, many square centimeters in area, can be grown in suitable materials in the form of thin sheets by secondary recrystallization. The application of this technique is discussed in Part II for Cu, and several other fcc materials, and also for α-uranium.

5. PHASE TRANSFORMATION

When a metal is cooled through a phase transformation, the grains of the high-temperature phase are replaced by a completely new set of grains of the low-temperature phase. For metals in which the phase change takes place by continuous growth from nuclei, it is possible to produce monocrystals by allowing the specimen to cool to the transformation temperature progressively along the specimen. This may be carried out by moving the specimen through a furnace with a suitable temperature gradient.

There is often, however, a substantial volume change which may lead to stresses, and subsequently to substructure formation in the crystal. For example, substructure is obtained in α-iron and α-uranium crystals produced by a phase transformation. Good crystals of α-uranium may be produced by giving a critical strain to the specimen after cooling through the transformation temperature to room temperature, followed by annealing just below the transformation temperature (Mercier et al., 1958). Alternative procedures may also be applied to materials in which the phase transformation is martensitic in character, that is, in which the growth of a crystallographically uniform transformed region is suppressed by strains which result in the transformation. The nucleation of such a transformation must occur at many independent points, thereby preventing the formation of a single crystal (Chalmers, 1953). This appears to be true for Zr and Ti. Zr crystals may be prepared however, by cooling through the phase transformation followed by prolonged annealing just below the transformation temperature (Langeron and Lehr, 1956, 1958). Large grains of Ti have also been obtained by cyclic heating and cooling through the phase change (Anderson et al, 1953).

The production of large crystals of Fe, Tl, Ti, α- and β-U, and Zr by phase transformation is discussed in Part II. Small crystals of grey Sn, approximately 1 sq mm in area and less than 2 mm long, have also been prepared by this technique (Becker, 1958).

6. SURFACE ENERGY

One of the ways a matrix structure can be stabilized in order to promote growth of a large single grain is by thermal grooving at the surface of the sample. When grains have diameters larger than about one to two times the sheet thickness, normal grain growth can cease as a result of a thickness effect (Beck et al., 1948). An example of this "specimen thickness effect" is given in Fig. 7; the maximum grain size attainable by gradual grain growth in zone-refined lead is shown as a function of the specimen

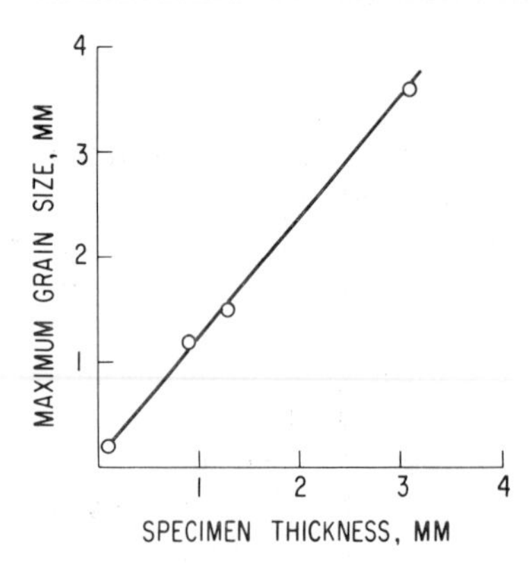

Fig. 7. Maximum grain size obtained by gradual grain growth in zone-refined lead as a function of specimen thickness, showing the "specimen thickness effect." (K. T. Aust and J. W. Rutter, unpublished.)

thickness. The thickness effect depends on the development of thermal grooves where grain boundaries meet the surface of an annealed specimen (Mullins, 1958). It is then possible to promote the growth of large grains by an additional driving force derived from a surface-energy dependence on grain orientation. A surface energy driving force may be considered in the form $2(\gamma_A - \gamma_B)/t$, where γ_A and γ_B are the specific surface energies of adjacent grains A and B respectively and t is the specimen thickness. This additional driving force provided by the lower surface energy can overcome restraining forces such as those caused by thermal grooving at the boundary between grains. A difference of surface energy of about 3% or more is believed to be sufficient to result in grain boundary migration without the boundaries becoming stuck in the grooves (Mullins, 1958).

Large grains in which the surface orientation is approximately parallel to high density or low energy planes have been obtained by this method in thin sheets of high purity silicon-iron (Walter and Dunn, 1959). A typical example is shown in Fig. 8. Impurities, either in the annealing atmosphere or in the metal itself, may change the surface energy such that crystals with either (100) or (110) in the plane of the sheet are formed. The surface energy of the (100) grain is believed to be the lowest in the presence of certain impurities, for example, oxygen, and the (110) grains have the planes of lowest energy in the absence of these impurities (Walter and Dunn, 1960). It is also possible to produce (100) [001] or (110) [001] grains by this technique, where the alignment of the [001] direction in the rolling direction of the sheet crystal depends on the prior rolling and annealing treatments.

Large crystals of Ag, about 1 to 5 mm in diameter, have been prepared by J. M. Lommel (unpublished) using the surface-energy driving force. The large crystals having (111) surfaces were obtained by annealing thin recrystallized sheets in air. When the sheets were annealed in vacuum or in purified argon, no large grains developed.

7. LINEAGE-INDUCED GROWTH

Substructure-free monocrystals may be prepared from very high purity materials

Fig. 8. Growth of large (110) grains in a matrix of (100) grains in high purity silicon-iron (1.5 X). (Courtesy of J. L. Walter.)

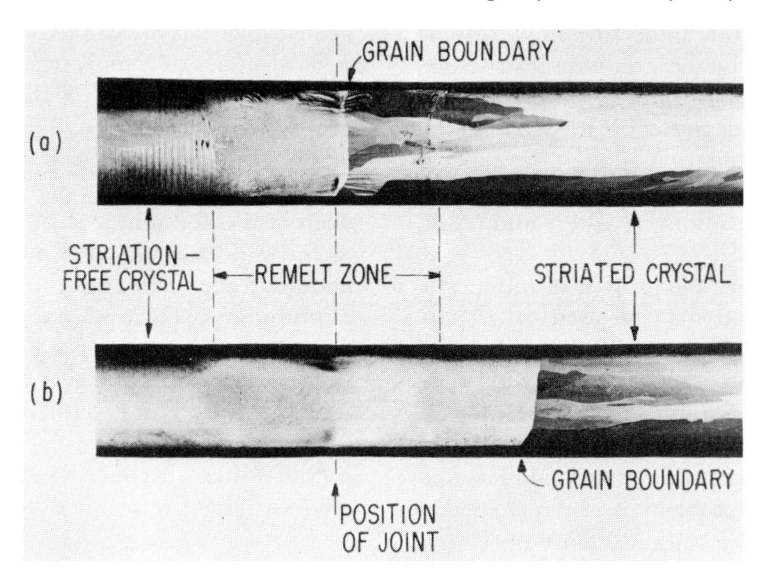

Fig. 9. Bicrystal of zone-refined lead having controlled orientations, prepared by a combined solidification and grain boundary migration technique. (*a*) After solidification. (*b*) After annealing at 300°C (1.5 X).

using a combination of solidification and solid state techniques (Rutter and Aust, 1958). The method is based on the fact that lineage or striation substructure, which is obtained in a melt-grown crystal, may be used as a source of driving energy for the growth in the solid of lineage-free grains.

A crystal of the high purity material, for example, zone-refined Pb, is first grown from the melt in a horizontal graphite boat using a modified Bridgman technique. Such a crystal is usually quite imperfect since an array of low-angle dislocation boundaries are formed during solidification. Since this striation or lineage structure is quite stable, it provides a suitable driving force for the growth of a crystal that contains no lineage. Recrystallized, lineage-free grains are next introduced into the striated crystal by plastic deformation of a small, localized region at one end of the specimen, followed by a recrystallization anneal. This latter step is accomplished in zone-refined lead simply by compressing one end of the striated crystal at room temperature. On suitable annealing, one or two of the recrystallized grains, produced in this manner, grow into the undeformed, striated crystal. The grain boundary moves through the striated crystal of zone-refined lead at speeds as high as 1 cm/min at temperatures near the melting point, thereby eliminating the striations.

The previous procedure does not provide control of the orientation of the crystal obtained. The following method can be used to prepare accurately oriented, substructure-free single crystals (Rutter and Aust, 1961). A striated crystal and a striation-free seed crystal of known orientations are placed end-to-end in a graphite trough with the seed crystal set at the desired orientation. The junction of the two crystals is then welded by melting and allowed to freeze to form a grain boundary. Upon subsequent annealing below the melting-point, the grain boundary moves through the striated crystal, producing an oriented nonstriated crystal. A typical specimen of zone-refined lead, produced by this method, is shown in Fig. 9*a* after solidification, and in Fig. 9*b* after grain boundary motion in the solid state.

The use of lineage structure as a driving force for grain boundary migration has been found to be satisfactory for preparing zone-refined crystals of Pb, Al, Cu, and Sn. The principal limitation appears to be the requirement of high purity material in order to maintain a sufficiently high grain boundary migration rate under the small driving force provided by the striation substructure. The technique provides a simple way of evaluating the degree of purity of a material since grain boundary motion at the expense of the lineage is greatly reduced by small solute concentrations (Rutter and Aust, 1960).

In addition to the growth of monocrystals, the method may be used to prepare oriented bicrystals. A difficulty that has been encountered in materials with relatively low twin boundary energies, for example, Pb and Cu, is the formation of annealing twins during grain boundary migration. This problem can be remedied to some extent by a suitable choice of relative orientations of the striated and seed crystals such that a lower energy grain boundary is formed which has sufficiently good mobility (Burgers et al., 1953; Aust and Rutter, 1960). In this way, the probability of obtaining an annealing twin that persists at the growth front is reduced in accord with the energy conditions proposed by Fullman and Fisher (1951). The principles underlying the avoidance of twins during the preparation of crystals are discussed more fully by Burgers in the previous chapter.

II. APPLICATION TO SPECIFIC MATERIALS

1. ALUMINUM AND ITS ALLOYS

Strain-Anneal Method

The strain-anneal method is frequently used to grow crystals of commercial purity Al. Since many of the experiments on preparing large crystals by the strain-anneal method have been performed on Al, the experimental conditions and results are considered in some detail. A small and uniform recrystallized grain size, approximately 0.1 mm diameter, is required in the starting material prior to the critical strain and growth anneal. Carpenter and Elam (1921), who first used the strain-anneal method to produce large single crystals, annealed their heavily worked Al at 550°C for several hours to remove the effects of previous strain and to provide the desired initial grain size. The purity of their Al was 99.6%, the principal impurities being Si and Fe, and the starting material was in the form of sheets 3 mm thick and 2.5 cm wide or in round test bars having a diameter of about 2.5 cm. The strain-free, fine-grained aluminum was next given a weak deformation, 1 to 3% strain, commonly by extension. The appropriate critical strain, which is usually not critical to more than 0.25% extension, may be determined by the use of taper specimens which permits a strain gradient to be introduced in a single specimen.

The surface of a deformed metal frequently shows a greater tendency to nucleate on annealing than does the interior. Graham and Maddin (1954) in their work on Al suggested that the nuclei originate at dislocation pileups against the surface oxide film. This surface nucleation is sometimes a difficulty in crystals grown by the strain-anneal method. It is therefore important after the critical straining to remove a thin surface layer of about 75 μ thick by electropolishing or etching.

In the earlier experiments on growing crystals by the strain-anneal method, the critically strained, commercially pure Al was heated slowly to a predetermined maximum temperature, 450 to 550°C, at a rate of 15 to 20°C per day. The time required to prepare a crystal by this method may be shortened by using the concept that a recovery anneal after deformation substantially reduces the number of grains growing in a metal on subsequent recrystallization. For example, annealed Al wire, of 99.5% purity, is given a critical strain of 4% and placed in a furnace at 320°C for 4 hr to obtain recovery. It is then heated in 1 hr to 450°C for a 2-hr anneal, resulting in monocrystalline lengths up to 15 cm in 1 mm diameter wire (Bagaryatskii and Kolonstava, 1959).

However, the usual procedure is to anneal the critically strained Al in a steep temperature gradient instead of the slow heating to temperature. The specimen shape must be such that a substantial temperature gradient can be imposed upon it, for example, rod up to about 6 mm diameter or square or strip up to 2 cm wide and several millimeters thick. Williamson and Smallman (1953) also recommend a radial temperature gradient in the furnace to eliminate nucleation of stray grains which tend to occur at the edges of the strip. They describe a furnace in which strip 1.2 cm $\times$ 1 mm and up to 5 cm long can be converted to a monocrystal by moving the specimen through the furnace at speeds up to 10 cm/min. However, rod specimens need no special furnace design. Honeycombe (1959) grew crystals in 3 mm square or round rod up to 5 cm in length in which a radial temperature gradient is not necessary; a longitudinal gradient of 20°C/cm is sufficient. A steep temperature gradient may be obtained by insertion at the end of the furnace of a water-cooled annulus through which the growing crystal passes.

High Purity Aluminum

The usual strain-anneal procedure is more difficult to apply successfully to the growth of crystals of purer Al that is 99.99% or greater. One of the difficulties is apparently due to the marked tendency for the purer material to recover by polygonization, or substructure formation, rather than to recrystallize and form new orientations. However, if the final growth anneal after critical straining is conducted in a uniform-temperature furnace, instead of a temperature gradient furnace, recrystallization to form large crystals results which is independent of the purity of the Al (Montuelle, 1959). It is suggested that a temperature-gradient anneal introduces stresses at elevated temperatures in high-purity Al which facilitate the motion of dislocations into a polygonized structure (Montuelle, 1959).

The tendency for polygonization may also be reduced by putting certain soluble impurities into the starting material. These can later be removed from the monocrystal. For example, if 0.04 weight per cent of Li is added to Al of 99.99% purity, recrystallization to very large crystals is obtained instead of small crystals having a polygonized structure (Montuelle and Chaudron, 1955). Al containing 0.035 weight per cent iron produces the same effect.

A strong preferred orientation in the annealed material prior to the critical strain is also helpful in preparing crystals from high purity Al. The low mobility of the boundaries associated with the strong texture is sufficient to stabilize the matrix grains and thereby promote the growth of a crystal with a deviating orientation. Lommel (1960) describes a strain-anneal technique for the growth of single crystals of Al of 99.993% purity in which the starting matrix must contain a strong preferred orientation. The material is heavily cold-rolled at $-196°C$ by cooling in liquid nitrogen before each rolling pass and annealed in a salt bath for 10 sec at 640°C followed by water quenching. The matrix that results from this treatment has a grain size of about 1 mm and contains a strong preferred orientation. The strip is then critically strained and pulled out of a water bath at 4 cm/hr through a temperature gradient of 100°C/cm and heated to 640°C, producing crystals 1 meter long.

Monocrystals can be prepared from coarse-grained (approximately 5 mm diameter) 99.99% Al by the application of straining treatments that are not severe enough to produce nucleation of new grains (Leighly and Perkins, 1960). The specimen is alternately strained and annealed until suitable single crystal sizes are obtained. The growth of grains during the annealing (at 640°C) occurs by strain-induced grain boundary migration in which a grain of lower strain energy enlarges itself at the expense of a grain of higher strain energy (Beck and Sperry, 1950; Aust and Dunn, 1957). This strain-induced technique is not successful with sheets wider than 2.5 cm or in lower purity Al since apparently nucleation of new crystals is obtained.

Orientation and Perfection of Crystals

If the original grains of the polycrystalline material, prior to the critical strain, are randomly oriented, then the orientation of the resulting monocrystal should be a matter of chance and the range of crystal orientations should be quite large. However, if a strong preferred orientation is present prior to the critical strain, the resulting crystals may be somewhat more restricted in orientation. The rate of growth in aluminum appears to be high (under certain conditions) when the orientation relationship between the growing crystal and the starting matrix is approximately 40° about a ⟨111⟩ axis (Beck et al., 1950). There is some indication that this favorable growth relationship is obtained when certain solutes, for example, Cu in very small amounts, are present in Al (Frois and Dimitrov, 1961), and is not obtained when an impurity such as Fe is present in solid solution (Green et al., 1959). A possible reason for these observations concerning the influence of solute on preferred orientation may be found in the work of Aust and Rutter (1960). They observed a selective action of solute impurity atoms resulting in differences in migration rates among various large-angle grain boundaries. On this basis, the development of a preferred orientation relationship in a dilute solid solution is the result of an interaction between solute and the grain boundaries, which is dependent on boundary structure and on the type and amount of solute.

The growth of Al crystals having controlled orientations may be obtained by critical strain plus an anneal in a steep temperature gradient using the method of Fujiwara (1939) as described in Part I. It is also possible to produce thin monocrystalline foils (0.06 mm thick) of Al having any desired orientation (Fujiwara and Ichiki, 1955). Oriented crystals of Al have also been prepared by Tiedema (1949) and Hägg and Karlsson (1952).

Within a crystal prepared by the strain-anneal method, smaller crystals that have not been consumed during growth sometimes appear on the surface. Such crystals tend to be of two classes of orientation relative to the matrix crystal: very nearly the same orientation, and twin orientations (Lacombe and Berghezan, 1949, Tiedema et al., 1949). Interfaces between such crystals and the matrix would be expected to have much lower rates of motion than high angle grain boundaries. Their occurrence usually indicates that the preannealing treatment was carried out at too high a temperature, that is, coarse initial grain size prior to the critical strain, or that the final growth anneal was at too low a temperature. These surface grains can often be absorbed by annealing at a temperature near the melting point or they can be removed by electropolishing or etching.

The perfection of strain-anneal crystals grown in commercial purity Al (99.2 to 99.6% pure) is usually quite good. They contain little substructure and also localized misorientations of less than 1 min of arc. Guinier and Tennevin (1950) using a very sensitive X-ray diffraction method found single crystals of Al, 99.99% purity, produced by the strain-anneal method, to possess a high degree of perfection with no detectable misorientations. However, strain-anneal crystals grown in higher purity Al (99.997%) often show localized misorientations in X-ray micrographs (Parthasarathi and Beck, 1958). In addition, low angle boundaries having misorientations from a few minutes of arc (Parthasarathi and Beck, 1958) to several degrees (Lacombe, 1948) are sometimes detected in the high purity crystals. These observations may be related to the fact that polygonization can occur more readily as the purity is increased. For example, Montuelle (1959) sometimes observed polygonized structure in single crystals of zone-refined Al if the specimen is not carefully supported or if even a small temperature gradient exists in the annealing furnace.

Aluminum Alloy Crystals

Crystals of single-phase Al alloys may also be prepared by the strain-anneal method and the chemical composition can be carefully controlled. Segregation is almost entirely eliminated, and alloy losses due to volatilization are kept to a minimum.

The experiments of Williamson and Smallman (1953) have shown that an important requisite for the growth of crystals of Al alloys by the strain-anneal method is a steep temperature gradient in the specimen. As alloying elements are introduced in increasing amounts, it becomes more difficult to grow crystals unless the temperature gradient is also increased. For example, Gane (1958) could not grow crystals of Al-Cu alloys containing more than about 3% Cu using a gradient furnace made with a resistance winding over a silica tube plus water-cooling to increase the gradient. However, increasing the temperature gradient by lowering specimens into a molten salt bath enabled crystals to be grown without any difficulty. Table 1 gives the operating conditions for the preparation of crystals of several Al alloys from heavily worked 3 mm sq rod. The best results are obtained by growing the crystals into a fine-grained and completely recrystallized matrix. The speed of growth should not exceed a critical value, although the speeds may be reduced somewhat below this value without detriment to successful growth. The influence of increased alloying is usually to decrease the speed that can be tolerated. Crystal lengths up to 45 cm in 3 mm sq rods have been obtained in some of these alloys, although shorter crystal lengths (1 to 5 cm) were common for the Al-Mg and Al-Ag alloys listed in Table 1 (Gane, 1958).

It is noteworthy that Montuelle (1959) was not able to prepare large crystals of *high purity* alloys of Al containing 6 to 15% Zn by the strain-anneal method. Instead, small polygonized crystals were obtained. However, the addition of Fe to the alloys, for example, 0.15 weight per cent Fe to Al-6% Zn, resulted in the formation of large crystals after critical straining and annealing. The amount of Fe necessary for high purity Al-Zn alloys is much larger than for high purity Al.

2. COPPER AND SEVERAL OTHER FCC METALS

The strain-anneal method is generally not useful for preparing Cu crystals because deviating orientations in the form of numerous twins are produced during the process (Carpenter and Tamura, 1927). The high frequency of twinning (associated with a low stacking fault energy and a low twin boundary energy) is a basic difficulty. A similar difficulty would be expected to occur in other metals that have low stacking fault and twin boundary energies, for example, Pb and Ag, but not in metals in which these energies appear to be high, for example, Al.

Copper Crystals Grown by Secondary Recrystallization

Large, thin crystals of Cu may be conveniently prepared by secondary recrystallization. The success of this technique is dependent on the fact that (1) a strong pre-

TABLE 1

ALUMINUM ALLOY CRYSTALS: STRAIN-ANNEAL METHOD

Alloying Addition	Prestraining Treatment	Critical Strain (%)	Growth Conditions	Growth Speed, cm/hr	Temperature Gradient, °C/cm
0.5% Ag	15 min 500°C	2	560°C	6.5	90
2% Ag	5 min 500°C	2	560°C	6.5	90
3%Cu	5 min 500°C	1.5	560°C	0.65	100
7%Zn	5 min 500°C	1.75	560°C	0.65	100
4–5%Cu 1.3% Si 4.2% Ge 4.5% Mg 8.6% Ag	Flash-anneal 10° below solidus	1.25	10–30° below solidus	0.5	Steep temperature gradient (Gane, 1958)

ferred orientation, that is, the cube texture, may be obtained by primary recrystallization and (2) the crystals obtained (termed the secondary grains) have a preference for certain orientation relationships toward the primary recrystallized matrix. For Cu, the secondary grains are related to the cube texture by Kronberg-Wilson rotations of 38° or 22° about a ⟨111⟩ axis (Kronberg and Wilson, 1949; Verbraak, 1960).

The general procedure of this method is the following. Annealed Cu strip is given a final straight-rolling reduction of more than 90% at room temperature. Suitably shaped specimens from the cold-worked sheet are then heated slowly in vacuo to 1000 to 1040°C and maintained at that temperature for several hours. During the heating, primary recrystallization to the cube texture occurs, to be replaced at higher temperatures by a few crystals or very often only one crystal. Very slow heating to temperature results in nearly twin-free material, whereas rapid heating yields fewer single crystals and much twinning. An alternative procedure that appears to be quite successful in reducing the incidence of twinning is described by Parthasarathi and Beck (1958) for electrolytic tough-pitch Cu. After the final cold-rolling reduction of 95%, the material is annealed for 10 hr at 600°C, resulting in a relatively strong cube texture. At this stage deviating orientations at the edges and near the rolled surfaces of the specimens can be eliminated by etching. If this is done, a final anneal near the melting point (for example, 72 hr at 1030°C) will result in large, secondary crystals that are free of annealing twins.

The perfection of the secondary crystals as revealed by the Schulz X-ray technique is quite high. No observable sub-boundaries or local misorientations are present if the specimens are handled with care and mechanically supported (such as by packing in a fine graphite powder) during the final anneal. There are also no unabsorbed matrix grains in the crystals produced by the texture-dependent coarsening. Seeding to grow specifically oriented crystals may be carried out as was previously described for the strain-anneal method.

Application of the Secondary Recrystallization Method to Other FCC Metals

Many fcc metals develop a strong cube texture during annealing after large cold-rolling reductions (80 to 95%). Barrett (1952) lists the following materials, besides Cu, in which this happens: Au, Al, Cu-Zn containing up to 1% Zn, Cu-0.2% Al, Cu-0.1% Cd, Cu-0.1% O_2, Ni-1% Mn, Fe-Ni alloys containing 30–100% Ni, and some ternary alloys of Fe, Ni, and Cu. Further heating after primary recrystallization may result in a few large, secondary grains that have special orientation relationships to the cube texture. With Fe-Ni alloys, the orientation of the secondary crystals with respect to the primary cube texture, (100) [001], may be described by a 30° rotation about the ⟨001⟩ axis. This rotation gives a (120) [001] secondary texture.

Large crystals of Ni, Cu-Al, Ag, Fe-Ni alloys have been prepared by this technique. It should also be useful for sheet specimens of Au and Pt or wherever secondary recrystallization results from a stabilized primary texture. However, the method is not widely used for preparing crystals from higher alloy concentrations. One of the difficulties is that in many of the alloys a single orientation texture is not obtained after primary recrystallization. Consequently, secondary recrystallization is not so pronounced. Another difficulty may be associated with a lower grain-boundary mobility as a result of alloying (Lommel, 1961).

3. IRON AND ITS ALLOYS

Crystals of Fe and its alloys may be prepared by the strain-anneal method. As with Al, the growth conditions are very dependent on the purity of the starting material. Several procedures are outlined in Table 2, and detailed growth conditions are considered in the following sections.

Iron with Purity $\lesssim$99.99%

If the starting material is Fe containing about 0.1% carbon (that is, mild steel), a decarburizing anneal in wet H_2 is used to reduce the carbon content to <0.01%. Apparently the presence of a relatively large amount of carbon in the impure Fe prevents the growth of large crystals. The grain size after decarburization and prior to critical straining should be quite small, ~0.1 mm. Starting materials such as Armco iron, or "Ferrovac E" vacuum-melted iron, in which the initial carbon content is already below about 0.1%, do not require the decarburization anneal but simply an anneal designed to produce uniform, fine-grained material. A preferred orientation in the matrix material of "Ferrovac E" iron may also be required, since an optimum reduction in area by rolling of about 50% is needed prior to growth. Attempts to grow crystals from strip material that had been reduced less than 40% were unsuccessful (Stein and Low, 1961).

The decarburized Fe is given a critical strain of ~3% in tension. An advantage is sometimes gained by shaping the specimens so that recrystallization during subsequent annealing is limited to a small volume of critically strained material (Holden and Hollomon, 1949). A very slow strain rate of 0.01/hr during straining of Ferrovac E prevents the formation of Luders bands that would produce inhomogeneous deformation. Cold-worked rods of 99.96% Fe may be heated to produce the necessary uniform, fine-grained structure and then quenched in water from 700°C to produce material having no yield point. Small, uniform, critical strains cannot be given when a marked yield point is present (Allen et al., 1956).

After the critical straining, the rods or strips are electropolished or etched to remove the surface layers. The specimens are then inserted into a furnace held at as high a temperature as possible without reaching the AC_3 point (at 910°C), for example, 880°C in pure, dry H_2 for 72 hr. Stein and Low (1961) conducted their final anneals in a temperature gradient furnace as developed by Dunn and Nonken (1953) for growing Fe-Si crystals. This furnace uses a water-cooled copper slot to obtain a sharp temperature gradient. A diagram of the furnace is shown in Fig. 10. After the final growth anneal, it is usually necessary to etch away a surface layer of fine-grained material to reveal the large crystals.

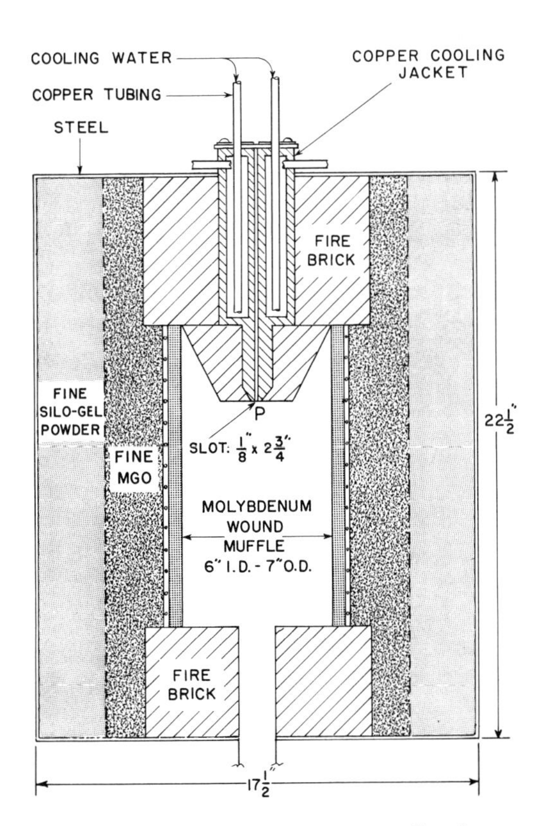

Fig. 10. Cross section of temperature-gradient furnace used to grow crystals of iron and silicon-iron. (Courtesy of C. G. Dunn.)

Pure Iron (99.99+%)

The usual procedures for growing crystals from relatively impure Fe by the strain-anneal method are not successful when applied to Fe of greater purity, that is, >99.99% purity (Talbot 1956, 1960a). Apparently, as with pure Al, the critically deformed matrix undergoes recovery effects such as polygonization instead of recrystallization. Talbot (1956) solved this problem by adding about 0.035% carbon in order to make polygonization more difficult. He also found that in less pure Fe (Armco iron) if the prestraining anneal were extended to 8 days at 950°C in H_2 instead of the 1 day anneal given in Table 2, large crystals were not obtained. It was suggested that some residual carbon is required in the less-pure Fe to

TABLE 2

PROCEDURES FOR GROWING CRYSTALS FROM DIFFERENT GRADES OF IRON BY THE STRAIN-ANNEAL METHOD

Starting Material	Prestraining Treatment	Critical Strain (%)	Growth Conditions	References
Mild steel hot-rolled $\frac{1}{8}$ in. thick plate or $\frac{1}{2}$ in. diameter bars	Decarburize at 950°C, 48 hr in H_2, slow cool; or at 750°C for 2 weeks in wet H_2	~3	880°C for 72 hr	Edwards and Pfeil (1924), also Gensamer and Mehl (1938)
Al-killed sheet 80 mil; machined into slightly tapered specimens	Decarburize at 730°C for 16 hr in wet H_2	3	880°C for 72 hr in pure dry H_2	Holden and Hollomon (1949)
Armco iron	Anneal $\gtrsim$ 24 hr in H_2 at 950°C	3	850°C for 72 hr in H_2	Talbot (1956); also see Stone (1948)
Ferrovac "E" Fe ~ 99.9% Vacuum-melted	Cold-rolled 50% to 0.080 in., and annealed 825°C, 3 hr in H_2, furnace cool	2.2–2.6	Annealed in temperature gradient maximum temperature 890°C	Stein and Low (1961)
α-Fe, 99.99+%	Carburize to 0.030–0.040% C by gaseous cementation at 920°C	3	Anneal in H_2; temperature raised to 880°C in 4 hr and hold 2 days	Talbot (1956, 1960)

obtain large crystals, and if this carbon is completely removed, no large crystals will result. It was concluded that recrystallization of Fe is inhibited by polygonization and the presence of carbon favors recrystallization. It should be noted that smaller amounts of carbon, 0.01% instead of 0.035%, will result in large crystals if the Fe is less pure (Talbot, 1960b). This indicates that elements other than carbon can play the same role. For example, about 0.030% silicon added to Fe has the same effect, but then the added impurity cannot be removed during the final growth anneal.

Carburizing can be done by placing the Fe specimens in an evacuated silica tube, and then introducing a fixed volume of acetylene gas calculated to give the desired carbon content. Carburization occurs during 4 hr at 920°C followed by a homogenization anneal in vacuum at the same temperature. The result is specimens having a uniform carbon content along their length. The critical value of strain depends on the carbon content; it increases from 2 to 5% as the carbon content decreases from 0.060 to 0.020%. For a carbon content of 0.035%, a critical strain of 3% is satisfactory.

The method of carrying out the final growth anneal on the pure iron containing the carbon addition is quite critical according to the findings of Talbot. The annealing treatment prescribed in Table 2 for pure Fe should be followed for two reasons: (1) the added carbon must be kept in the specimen long enough to prevent premature polygonization of the critically strained matrix; if the specimen is placed immediately into a furnace at 880°C in H_2, thereby rapidly removing the carbon instead of the recommended 4-hr heating time from room temperature to 880°C, only small polygonized crystals are obtained. (2) The carbon must eventually be eliminated so that growth is not hindered by carbon at the grain boundaries. Consequently, the growth anneal should be done in an atmosphere of H_2, not in a vacuum or an inert gas, in order to eliminate the carbon.

Iron Alloys

Ferrous alloys may also be made into large crystals by the strain-anneal method. The various procedures which have led to the successful growth of crystals from different iron alloys are outlined in Table 3.

Orientation and Perfection of Crystals

Similarities in the orientations of strain-anneal Fe crystals have been noted by Edwards and Pfeil (1924), and by Allen et al. (1956). For example, Allen and co-workers found that in 56 crystals, the directions of the specimen axes were mostly within 20° of ⟨011⟩; some were obtained near ⟨001⟩, but none near ⟨111⟩. This tendency for a preferred orientation is probably caused by the heavy cold work usually given to obtain a fine-grained material prior to the critical strain treatment. Crystals having controlled orientations have been prepared by the technique of Fujiwara (see Part I) in Fe-Si (Dunn and Nonken, 1953), and in Fe (Dohi and Yamoshita, 1957; Stein and Low, 1961).

Small, unabsorbed crystals are sometimes obtained in Fe and Fe alloys. The origin of these stray crystals is believed to be the same as for the included grains in Al crystals prepared by the strain-anneal technique. Laue X-ray photographs usually show no evidence of substructure in the Fe crystals (Gensamer and Mehl, 1938; Allen et al., 1956). Talbot (1956, 1960a) obtained crystals of high purity Fe having misorientations of only 1 to 2 min of arc, as revealed by the Guinier-Tennevin X-ray method.

Other Solid State Techniques for Growing Iron Crystals

The grain growth method of Andrade, as described in Part I, was first applied to 1 mm diameter wires of α-Fe, using a stationary temperature gradient and progressive cooling of the wire (Andrade, 1937). The Fe wires were polished with emery papers before insertion in the heating apparatus, and then heated well above 900°C by maintaining a temperature gradient. Finally, they were cooled through the phase transformation to obtain monocrystals of the α-phase. The method did not appear to depend for its success on a phase transformation since Mo crystals were prepared in the same way by Tsien and Chow (1937). The Andrade method was later modified by heating the wire specimen by the passage of a current

TABLE 3

IRON ALLOY CRYSTALS

Starting Alloy	Prestraining Treatment	Critical Strain (%)	Growth Conditions	References
Fe–3% Si plate 0.020–0.050 in. thick	20 hr in H_2 at 870°C	2.5	Temperature gradient furnace maximum temperature 1150°C, speed of growth 1 cm/hr temperature gradient 1000°C/cm	Dunn and Nonken (1953)
Fe–0.15% P tensile specimens chemically machined after critical strain	Annealed to produce 700 grains per mm²	3	Maximum temperature 880°C speed of growth 1 cm/hr. Very steep temperature gradient	Honeycombe (1959)
Fe–up to 6% Al 3 mm diameter rod	Decarburize at 900–970°C	2.4	1–2 days in range 880–1000°C	Yamamoto and Miyasawa (1954)
Fe–18 Cr–8 Ni Fe–20 Cr–20Ni Fe–20 Cr	Cold reduction 70% machine test specimens and recrystallize at 1000°C, 1 hr	2.5	Rapidly heat to 1350°C for 100 hr (1200°C for Fe–20 Cr)	Leggett, Read, and Paxton (1959)

and using a subsidiary moving furnace 2 cm long and 2 cm in diameter (Andrade and Chow, 1940). Monocrystals, 1 cm or so long, were regularly obtained in 0.5 to 1 mm diameter wires.

The fact that the Andrade method is successful for Fe indicates that a monocrystal of α-phase can result from transformation of the γ-phase. Crystals of α-Fe have also been obtained by progressive cooling through the $\gamma \rightarrow \alpha$ transformation temperature (McKeehan, 1927), or by slowly moving the specimen through a temperature gradient (Lehr, 1958). High purity of the Fe apparently favors the formation of large crystals during the phase transformation. However, the quality of the crystals is often not good because the crystals are usually polygonized.

Finally, a matrix structure stabilized by texture, inclusions, or thermal grooving (see Part I) should be suitable for crystal growth in thin sheets of Fe and Fe-base alloys. For example, large grains having the (110) [001] orientation are obtained in sheets of commercial Fe-Si from a primary recrystallized matrix stabilized by inclusions (Dunn, 1949b).

4. MOLYBDENUM

The Andrade method has been utilized to make crystals of Mo (Tsien and Chow, 1937). Mechanically polished wires of 1 mm diameter are heated in an evacuated quartz tube to about 1700°C by the passage of an electric current. A small furnace surrounds the tube for a short length, about 1 cm, and is maintained at about 1000°C. (See Fig. 5.) The temperature of the small furnace is considerably lower than that of the wire; its function is to raise the wire's temperature by diminishing radiation losses. Temperature gradients will result in the wire near the ends of the furnace, which is gradually moved along the vertical tube. Mo crystals of several centimeters in length can be made.

Larger crystals, up to about 1.2 cm diameter rod, have been grown using a modified Andrade furnace (Chen, Maddin, and Pond, 1951). A stationary temperature gradient and progressive cooling, without resort to an external furnace, produced crystals about 4 cm long after 7 hr at about 2000°C. However, to grow crystals in the larger diameters, 0.6 to 1.2 cm, it was necessary to preanneal, that is, recrystallize, the "as-received" rod and give it a small tensile strain of about 1% during the growth anneal.

5. NIOBIUM

The technique described by Chen et al. (1951) for Mo may be applied in principle to any refractory metal. For example, the same general procedure was used to grow Nb crystals (Maddin and Chen, 1953). An axial tension was applied to 3 mm diameter, 18 cm long rods in vacuum at temperatures in the vicinity of 2000°C for 2 to 4 hr. This resulted in crystals about 1 cm long which occupied the entire cross section of the rod. There is apparently no control of crystal orientation for Nb (and Mo) when grown by the Andrade or modified Andrade techniques. The crystals give undistorted Laue X-ray reflections with no evidence of splitting, provided the axial strain during growth does not exceed 2%.

6. TANTALUM

Crystals of Ta have been prepared from wire by the strain-anneal method. (Nichols, 1954; Seraphim et al., 1960). A suitable starting material consists of wire 5 to 40 mils diameter, having a purity of 99.9%. The polycrystalline wires are first annealed at 1800°C for several minutes in vacuum. A 2 to 3% uniform tensile strain is then applied, followed by annealing in vacuum at 2200°C. A steep temperature gradient and growth speeds from 0.3 to 1 cm hr are used. Crystals, in which the length/diameter ratio is greater than 100, are obtained under optimum conditions. The crystals can then be purified by heating in high vacuum (Seraphim et al., 1960).

7. THALLIUM

Sherby (1958) has utilized the phase-change method for obtaining single crystals of α-Tl having a purity of about 99.98%. The starting material, in the form of coarse-grained as-cast rods about 1 cm diameter and 30 cm long, is extruded at room temperature to about 0.6 cm diameter rod and annealed 1 hr at 215°C in a silicone oil bath to give an equiaxed grain structure of 1 mm average diameter. The Tl is then heated to the beta-range, above 230°C, and cooled slowly (10°C/min or less) through the transformation temperature. A single crystal invariably results with an orientation that is random. However, if the rate of cooling from beta to alpha is rapid, as in an oil quench, for example, a polycrystalline sample of the α-phase is usually obtained.

8. THORIUM

The use of solidification techniques for the production of Th crystals is difficult and complicated since molten Th shows great affinity for most metals and ceramics and since a high temperature (1400°C) phase change occurs in the solid state. A strain-anneal technique for obtaining large grains in Th metal prepared by the iodide process appears to be a suitable solid state method (Armstrong et al., 1959). An arc-melted Th sample, in the form of a bar 2.5 cm wide, 10 cm long, and 0.75 cm thick, is cold compressed by 5%. The Th is then uniformly heated in vacuo to 50°C above the phase transformation temperature, and subsequently lowered through a steep temperature gradient of about 225°C/cm at a rate of 1.5 cm/hr. After the thermal treatment, the bar should consist of comparatively few grains. The largest grain obtained by Armstrong et al. (1959) extended completely through the 0.75 cm dimension of the bar with cross-sectional dimensions of about 2.5 × 1.3 cm.

9. TITANIUM

Strain-Anneal Method

It is usually very difficult to make large crystals of hcp metals by the strain-anneal method because many twins form during critical straining. As a result of the twinning, too many recrystallization nuclei apparently are formed during subsequent annealing for successful operation of the method (Cahn, 1953). Churchman (1954) overcame this difficulty in Ti to some extent by obtaining a controlled surface orientation in the starting strip. However, it is often not desirable to develop the full preferred orientation in the starting material as the orientation of the crystals produced would then be restricted to a small range.

Churchman's starting material was Ti produced by the van Arkel process, having a purity of 99.85 to 99.95%. Specimens were cut from a starting strip having a small degree of preferred orientation such that the basal plane tended to be aligned with the plane of the strip. For this orientation, subsequent critical tensile straining (or compressive straining perpendicular to the strip) cannot produce $(10\bar{1}2)$ deformation twinning. Suitably shaped specimens were subjected either to tensile elongations of 1.5 to 2.5% or to transverse compressive strains of 0.25 to 1.0%, depending on the initial grain size. Precautions were taken to exclude O_2 and N_2 at all stages in the preparation of the crystals. The Ti was annealed in purified argon contained in silica tubes; Mo foil prevented contact between the silica and the Ti. The sealed tubes were "gettered" by heating Ti turnings to about 900°C at one end of the tube. The annealing furnaces were evacuated to 10^{-2} mm Hg pressure to insure that any porosity of the silica resulted in diffusion of gases out of the tube. After an annealing treatment of up to 200 hr at 860°C, approximately one-third of the specimens contained large, randomly oriented crystals.

Phase Change Method

Anderson et al. (1953) found that cyclic heating and cooling in vacuo through the transformation temperature results in large grains of α-Ti. Specimens were first heated to 1200°C for 4 hr and then transferred to an 850°C furnace where they were held for 3 to 5 days. This cycle was repeated at least three times before the specimens were cooled and removed from the sealed tubes.

Useful crystals, ranging in size from 0.5 to 5.0 cm long, with a fairly wide range of orientations, were obtained in about one-fourth of the specimens cycled. There was no evidence of imperfections in Laue X-ray reflections from the crystals. Cycling in vacuo had the disadvantage, however, that the specimen surfaces became roughened and the section thicknesses reduced by sublimation. When sublimation was suppressed by admitting one-fifth atmosphere of pure argon to the tubes, no large grains were obtained. Thus, proper surface conditions are required for the success of this method.

Titanium—13% Molybdenum

Single β-phase crystals, known to contain a small amount of unsuppressible ω-transition-phase, have been prepared by annealing 0.1 cm diameter rods of the alloy at high temperatures, thereby causing a large amount of grain growth (Spachner and Rostoker, 1959). The polycrystalline samples were heated in a Mo-element vacuum furnace to 1600°C, held at temperature for $\frac{1}{2}$ hr, and quenched by dropping the rod through the hot zone onto a cold plate in the bottom of the furnace. To transform any α-phase resulting from this slow quench, the rod is reheated in a helium-filled tube-furnace to 1000°C and quenched by flooding the tube with water. The crystals obtained by this method extend across the rod diameter.

10. TUNGSTEN

Large crystals of W and several other refractory metals are probably best prepared by special solidification techniques, for example, floating zone-melting in vacuum. Solid state techniques may be applied to small diameter W wires, which are of special use to workers in electron emission and surface physics.

One of the earliest methods for obtaining monocrystals of W was the commercial Pintsch process (Smithells, 1952). A filamentary wire is first obtained from a starting material consisting of very fine W powder made by reducing in H_2 tungstic oxide containing 2% thoria. The wire is then drawn slowly through a zone of high temperature (2000 to 2200°C) in dry H_2 under a very steep temperature gradient. This yields crystals meters in length which occupy the entire cross section of the wire. The method depends on the grain-coarsening that takes place in a suitable temperature gradient together with the presence of thoria in the metal to inhibit the growth of most of the grains. However, the resulting crystals contain a relatively large amount of thoria, making them unsuitable for many experimental purposes.

Use of Additives in High Purity Tungsten

When high purity tungsten, approximately 99.99% pure, is doped with suitable impurities, grain-coarsening will occur during the annealing of drawn wire, leading to crystal growth. In fine wires $\gtrsim$ 10 mil) crystals up to 1 cm long form readily. The impurities (for example a few per cent by weight of potassium silicate and aluminum chloride added to the tungstic oxide) largely disappear during subsequent processing involving powder metallurgy and wire-drawing techniques. The form in which the additives are finally present and their influence upon the growth process are still not satisfactorily explained, although numerous hypotheses have been suggested. It is often suggested that the additives are present in the form of dispersed particles of a second phase which inhibits grain growth (Rieck, 1958). However, because of the small amount of additives actually present in the wire ($<$100 ppm), there is considerable difficulty in establishing their form, whether they exist as discrete particles or in solid solution. Studies with the electron microscope, using replica techniques, have revealed no inclusions (or porosity) in the wire (Swalin and Geisler, 1957–58). It has also been suggested that the effect is caused by a very small quantity of impurities dissolved in the tungsten (Millner, 1957). This suggestion is supported by observations on zone-refined materials that very small quantities of solute have (1) a very strong retarding effect on crystal growth (Rutter and Aust, 1960) and (2) a selective effect that permits favorably oriented grains to grow preferentially (Aust and Rutter, 1960; Frois and Dimitrov, 1961).

Annealing Procedures

Several annealing techniques can be used to obtain large crystals in tungsten wire. One is Andrade's (1937) method as was described previously for molybdenum and iron crystals. Nichols (1940) and Hughes et al. (1959) also made use of a moving gradient method where the wire is first heated slowly to about 2300°C in dry H_2. Then a steep temperature gradient traverses the wire at speeds of 0.4 to 4.0 cm/hr. The wire is finally heated for several hours in H_2 at about 2700°C to absorb small surface crystals. Robinson (1942) grew long crystals in a commercial No. 218 and similar W wires by vacuum heating at a constant temperature (for example 18 hr at 1700°C). The growth process may be followed by observing the thermionic emission pattern with a cylindrical electron projection tube with a fluorescent screen. X-ray studies have indicated the presence of substructures in these crystals having misorientation angles of a few degrees.

Use of Additional Strain

Gifford and Coomes (1960) were not able to grow long crystals in certain commercial tungsten filaments either by the moving gradient method of Nichols (1940) or the grain growth method of Robinson (1942). Instead they found that long crystals could be reliably produced in 3 to 10 mil tungsten filaments (and also in molybdenum filaments) if a prescribed amount of plastic twist is first given and a favorable annealing temperature is used. The role of the torsional strain in the growth process is not understood, but it is known that the tendency for exaggerated grain growth is enormously increased when tungsten wire is coiled prior to annealing (Smithells, 1952). This problem is discussed by Burgers in the preceding chapter.

11. URANIUM

Since U undergoes two allotropic changes during cooling from the melting point, it is quite difficult to grow large crystals from the melt. Therefore, many solid state techniques have been used to make α-U crystals. The selection of one of the following methods depends on the size, shape, orientation, perfection, and purity of the crystal desired.

Phase Change Method

Crystals of α-U may be prepared by a phase change method similar to that described by Cahn (1953). A steep temperature gradient is used to reduce the probability that a new α-grain will be nucleated in the β-phase ahead of an advancing α-grain. In addition, the end of the specimen where the transformation begins may be tapered to a point to exclude copious initial nucleation. Large grains are formed consistently with the slowest speeds, 0.5 to 4.0 mm/hr. The volume change accompanying the transformation causes internal stresses and considerable substructures in the crystals. A scatter of 10° in orientation across a grain as a result of accumulated misorientations across subboundaries has been noted. However, if the imperfect crystal is subsequently critically strained at room temperature and then given a gradient anneal just below the transformation temperature, good crystals are obtained (Mercier, Calais, and Lacombe, 1958).

Secondary Recrystallization Method

Lacombe et al. (1959) have prepared α-phase crystals by secondary recrystallization of a strong primary recrystallization texture. The material is high purity U containing $\sim$35 ppm of carbon and smaller quantities of other impurities. Secondary recrystallization is obtained if the uranium sheet, heavily rolled at room temperature, is annealed in the α-phase region.

An essential requirement for the rapid growth of secondary crystals is a rapid heating of the cold-worked metal to a high temperature (650°C in the α-region). The incubation period prior to the appearance of the first secondary crystals varies from 10 hr with rapid heating (100°C/min) to 50 hr with slow heating (100°C/hr). The subsequent growth of secondary crystals is also affected by the rate of heating. An annealing period of 300 hr at 650°C is necessary after slow heating to produce a grain size comparable with that obtained after 48 hr at 650°C subsequent to rapid heating.

The secondary grains may reach a size of several millimeters and are free from substructure as revealed by Laue photographs. However, in no case do the secondary crystals completely absorb the primary grains. In addition, secondary recrystallization is not observed when the thickness of the specimen is less than a minimum value because of boundary anchoring by thermal grooving. The orientations of the secondary crystals in relation to the sheet surface and the rolling direction are of the type (203) [010], and (110) [110].

Strain-Anneal Method

The growth of large α-phase crystals in this same high purity uranium is also possible if a tensile deformation (2 to 6%) is applied to the primary texture before secondary growth starts at 650°C (Lacombe et al., 1959). The presence of a preferred orientation in the primary matrix prior to the critical straining may be important as it is for Al (Lommel, 1960) and Ti (Churchman, 1954). The mechanism does not appear to be deformation-accelerated secondary growth but a true recrystallization caused by critical straining. The value of critical strain increases with increase in size of the primary grains: from 2% for primary grain size of 10 μ diameter to 6% for grains of 35 μ diameter. The crystals developed extend throughout the thickness of the sheet and frequently attain a surface area of several square centimeters. These large crystals have a preferential texture in which the axis [010] remains parallel to the rolling direction, but the [100] and [001] axes assume all possible orientations relative to the secondary crystals.

Large crystals could not be obtained by the strain-anneal method in uranium of the usual nuclear purity, that is, an impurity content of ~200 to 300 ppm. (Cahn, 1953). Cahn proposed that the failure of the method is caused by the many twins that form during slight straining of α-U, resulting in too many recrystallization nuclei. However, attempts to produce large crystals by imparting the critical deformation at a temperature high enough to obviate twinning were not successful (Cahn, 1953). Lacombe (1960 to 1961) suggested that the difficulty with relatively impure uranium is that impurities have a strong influence on recrystallization because of their very small solubility. The very large effect of low solubility impurities such as Ag or Au in Pb on boundary migration is clearly evident (Rutter and Aust, 1960).

Grain Coarsening: Inclusion-Inhibited Growth

Fisher (1957) has prepared crystals of high purity α-U by a method that depends on a grain-coarsening phenomenon induced by dispersed particles in a fine-grained, recrystallized matrix. A small quantity of Si, approximately 10 to 20 ppm, is diffused into uranium rods to provide fine inclusions. This is done by reacting Si vapors with the surfaces of the rods at 1000°C in evacuated silica capsules. The rods are then quenched from the γ-phase, cold-worked, and annealed to obtain a fine α-grain size by recrystallization. On subsequent reannealing at 640 to 660°C, one or several large grains grow into the inclusion-stabilized, fine-grained matrix. The optimum conditions for growing the largest grains are described in detail by Fisher (1957).

Crystals up to 4 mm in diameter and ~10 mm long have been prepared by this method. Subgrain boundaries could not be detected in the crystals, either by polarized light or by the X-ray back-reflection method. However, the crystals were seldom completely free of twin bands. The crystal orientations are usually within ~15° of the [010] direction that is nearly parallel to the rod axis.

Beta-Uranium

Crystals of the tetragonal β-phase of uranium can be grown in material containing 0.5 to 1.5 atomic per cent Cr by a $\gamma \rightarrow \beta$ phase-transformation method (Holden, 1952). This method is successful apparently because the γ-phase is very soft, so the new β-phase crystals are not distorted. The polycrystalline wire is sealed off in an evacuated quartz tube which is suspended in a vertical tube furnace. The lower zone of the furnace is at ~850°C, well into the γ-range and the upper zone at 700°C in the β-range. The tube is initially positioned in the γ-zone and slowly raised into the β-zone. Finally, the tube is quenched into water at room temperature. Ductile β-crystals several centimeters long result. The crystals produced from the 1 atomic per cent U-Cr alloy contain a precipitate, whereas those grown in the 0.5 atomic per cent U-Cr alloy are free of precipitate. However, the β-phase is stable in the high-Cr alloy, whereas in the low-Cr alloy it will transform to the α-phase within a few hours at room temperature.

Crystals of β-phase uranium have also been obtained by progressive transformation from the α-phase. However, these crystals contain considerable substructure as in the case of α-phase crystals produced from the β-phase (Cahn, 1953).

12. ZIRCONIUM

Rapperport (1955) describes a grain growth method for obtaining large, stress-free grains of α-Zr from iodide crystal-bar Zr. Electrolytically polished samples are wrapped in tantalum foil and placed in a quartz tube sealed at one end which is then evacuated to 10^{-6} mm Hg. The samples are heated at a rate of about 100°C/hr, with the vacuum system constantly removing the H_2 evolved from the decomposition of zirconium hydride in the samples. After one week at 840°C, the samples are cooled to room temperature at a rate of about 100°C/hr. Crystals up to 0.5 cm sq and 1 cm long are obtained which give sharp Laue X-ray reflections.

Crystals of α-Zr can also be made by cooling from the β-phase region (150°C/hr), followed by prolonged annealing in the α-range (90 hr at 820°C). If this procedure is repeated several times, crystals greater than 1 cm in diameter are obtained (Langeron and Lehr, 1956, 1958). The crystal growth here is apparently stimulated by stresses arising from the transformation, and may involve strain-induced grain-boundary motion. The starting material consisted of Zr of van Arkel purity containing a total of about 500 ppm of H, O, N, and Fe. Large crystals could not be obtained by this method with less pure Zr. The large grains gave good Laue X-ray reflections, although electropolishing revealed lamellae of a second phase precipitated in well-defined directions. This phase appears to be zirconium hydride. Attempts to produce Zr crystals by the strain-anneal method resulted in grains only a few millimeters in diameter.

13. CERAMICS

The application of the recrystallization methods to nonmetallic materials is more restricted. A basic difficulty with materials such as refractory oxides appears to be that sufficient imperfections to support the nucleation and growth of new grains cannot be introduced. However, some minerals (anhydrite, fluorite, periclase, and corundum) when compressed and heated to temperatures well below the melting point display boundary migration similar to that observed in metals (Buerger and Washken, 1947). In the previous chapter, Burgers discusses the recrystallization and grain growth behavior of AgCl, TlCl, and NaCl in which he points out the analogy of these processes with those observed in metals.

Sintering Methods

There are instances in which grain growth during sintering is a practical and attractive method of obtaining large crystals. Pronounced grain growth at high temperatures in polycrystalline ferrites has been observed. Grains up to 5 mm have been seen in some polycrystalline *yttrium-iron-garnet* samples fired above 1450°C (Van Uitert et al., 1959). Marked grain growth has also been noted in this material and also in *copper-manganese-ferrite* sintered at 1400°C (Harrison, 1959).

Burke (1959) has utilized the principles of grain growth in metals to interpret observations of grain growth in ceramics. In metals, a large grain may grow readily at the expense of an array of small grains when the concentration of growth inhibiting inclusions is gradually reduced below the critical volume fraction just necessary to prevent uniform grain growth. In porous ceramics, the pores play the role of growth-inhibiting inclusions, and as sintering proceeds they gradually disappear. Thus conditions for exaggerated grain growth or grain-coarsening may be established in all single-phase ceramic systems at some stage during the sintering process. Indeed, this type of grain growth (large grains growing at the expense of small ones) is frequently observed in ceramic systems; for example, BeO (Duwez et al., 1949), Al_2O_3 (Wilder and Fitzsimmons 1955). The very largest, final grain sizes were obtained from very fine starting powders ~1 μ or less. When the starting powder was coarse, greater than 15 to 20 μ, the grain sizes in the sintered compacts were about the same as in the starting powder. Exaggerated grain growth usually occurs more readily in a fine-grained material than in a coarse-grained material (Burke, 1959).

It should be noted that exaggerated grain growth does not occur in Al_2O_3 doped with 0.25 weight per cent MgO. This has been attributed to an enhancement of the sintering rate by the presence of MgO (Coble, 1961). However, it is also possible that the addition of MgO, which is apparently mostly in solid solution, simply decreases the

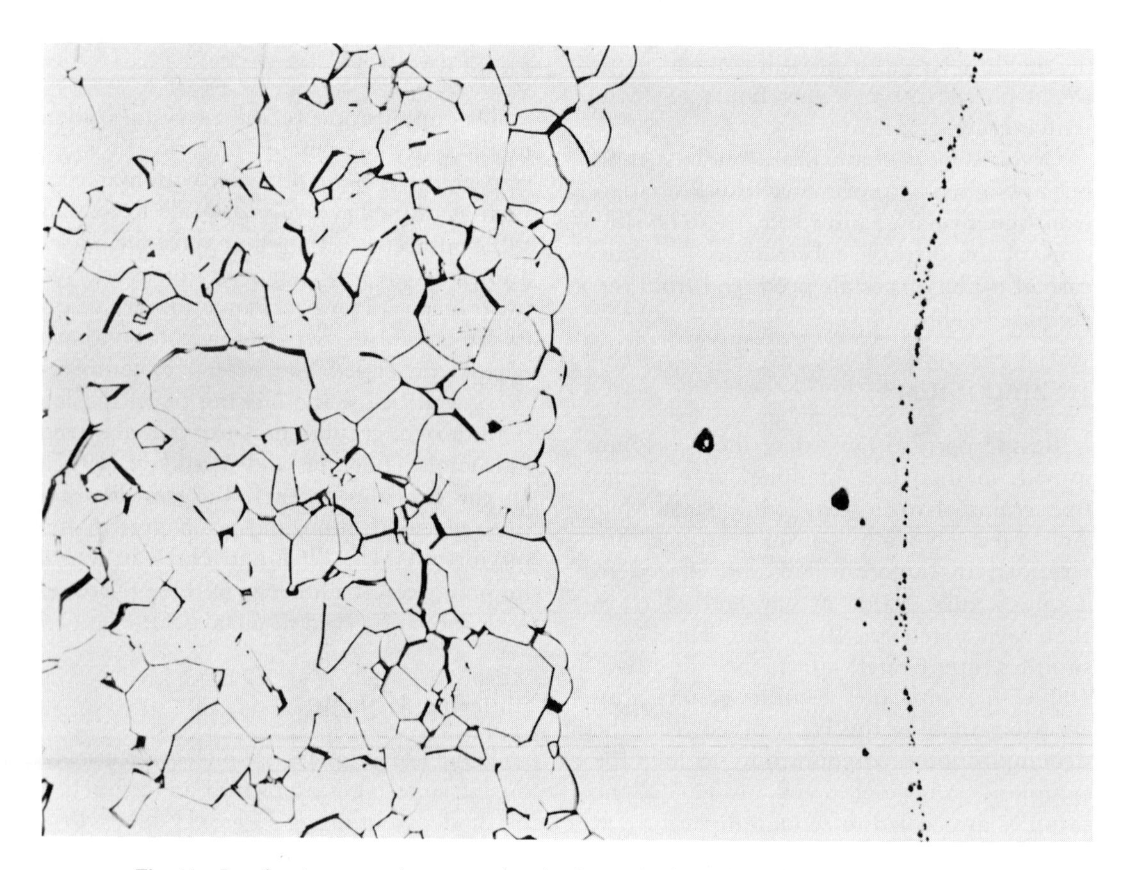

Fig. 11. Interface between a large crystal and a fine-grained matrix in aluminum oxide (470 X). (Courtesy of R. L. Coble and J. E. Burke.)

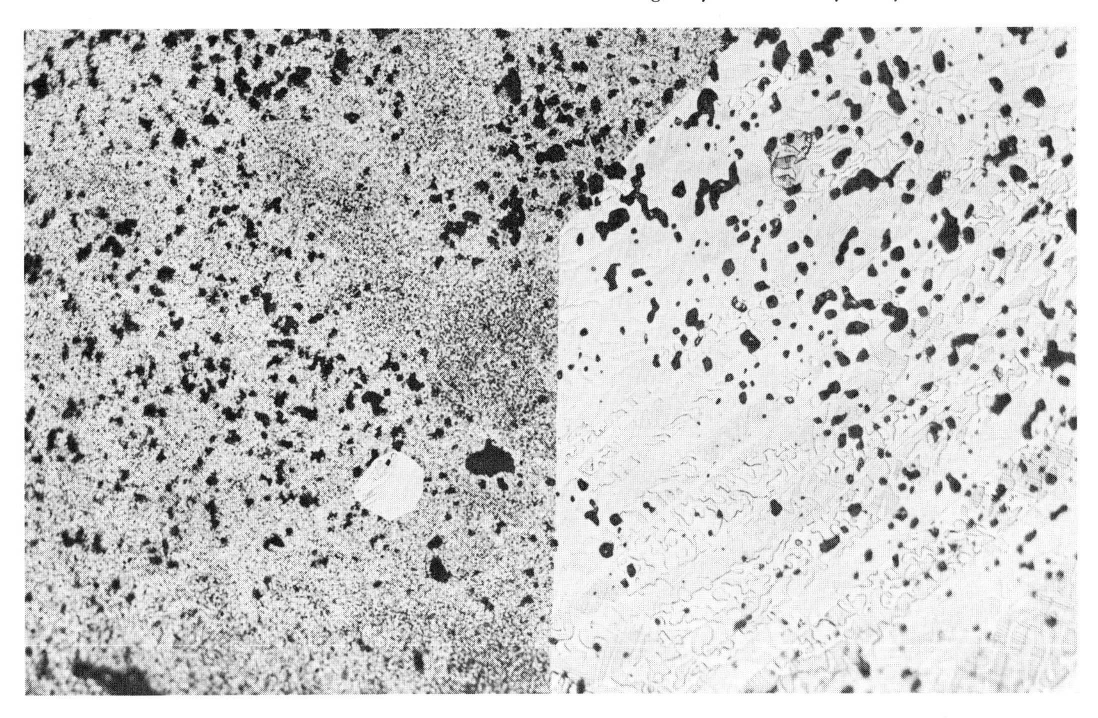

Fig. 12. Growth of a large grain (on the right) into a polycrystalline matrix of barium titanate (240 X). (Courtesy of R. C. DeVries.)

contrast between the rates of movement of the boundaries restrained by porosity and those that are not restrained. This would decrease the tendency for exaggerated grain growth (Coble and Burke, 1961). Burke (1961) has pointed out that this latter explanation may also apply in the case of metals. For example, Zn displays exaggerated grain growth upon annealing at 200°C, which may be due either to small particles of ZnO (Burke, 1951) or to small amounts of other impurities. However, an addition of 0.6 weight per cent Ag to the Zn prevents exaggerated grain growth completely, allowing only continuous grain growth even at higher annealing temperatures (Burke, 1951).

A novel method for producing a large crystal of Al_2O_3 by exaggerated grain growth is shown in Fig. 11. The specimen here was prepared by Coble by embedding an Al_2O_3 sphere in fine-grained Al_2O_3 powder and heating to obtain sintering and grain growth (Burke, 1959). In the illustration the boundary between the coarse grain and the fine grains is moving toward the centers of curvature of the fine grains. The initial position of the boundary is shown by a line of small pores. The curvature of the interface between the large Al_2O_3 crystal and the fine-grained matrix is stronger than that of the other boundaries. The morphology is similar to that during secondary recrystallization of metals. Continued encroachment of the small crystals results in the desired large crystal.

Special Growth Techniques

R. C. DeVries (unpublished) was able to produce a large grain in a polycrystalline rod of *barium titanate* by lowering the rod through a steep temperature gradient. The temperature at the center of the furnace coil was 1300°C and the rate of lowering of the specimen was 0.6 mm/hr. Figure 12 shows part of a grain about 3 mm long, 1 mm wide, and 0.5 mm thick. The straight boundaries are approximately parallel to {110} planes. It is not known whether this remarkably straight boundary (Fig. 12) is associated with a strong anisotropy of grain boundary energy or the presence of a liquid phase.

Navias (1961) describes a technique whereby an Al_2O_3 (sapphire) rod is completely converted to a spinel by exposure to MgO (periclase) vapor at 1900°C in a hydrogen atmosphere. Large crystals (1.5 × 2 mm) of spinel are developed as a result of crystal growth in the solid state.

Toth et al. (1960) describe a method involving exaggerated grain growth for making large crystals of *cuprous oxide.* The procedure is to oxidize copper plate (0.015 to 0.030 in.) completely by heating in air at 1020 to 1040°C and then to anneal the resulting polycrystalline Cu_2O at a preselected higher temperature (1085 to 1130°C for 5 to 150 hr). The temperature and time depend markedly on the plate thickness, thick plates requiring lower temperatures and longer times.

The growth behavior in Cu_2O always involved the growth of one grain at the expense of the fine-grained matrix. This, together with the observation that large grains were only formed in a restricted temperature range, suggests that the mechanism is similar to that of exaggerated grain growth in metals. The growth of the fine-grained matrix appears to be inhibited by porosity. In fact, a myriad of small pinholes was found in the centers of the large Cu_2O crystals. Crystals having surface areas of 1 in.^{2} or larger and thicknesses varying from 0.010 in. to 0.060 in. were consistently produced. X-ray analysis indicated that the crystals were free from gross substructure and had a preferred orientation on the surface with (211) and (311) planes predominating. The crystals were suitable for measurements of the semiconducting properties of Cu_2O.

Crystals of *silver chloride,* with a low residual density of dislocations, can be prepared by a combination of solidification and boundary migration (Mitchell, 1958). Thin sheet crystals, approximately 0.3 mm thick and 25 mm diameter, were first prepared by crystallizing disks of molten silver chloride between optically polished Pyrex plates. After separation from the Pyrex plates, the disks were cut with a razor blade to produce a region of intense local deformation along a chord near what had been the leading edge during crystallization. The sheets were thoroughly washed with chlorine water and then annealed in a sealed tube in chlorine at a pressure of 1 atmosphere at 420°C for periods between 1 and 12 hr. It was found that recrystallization began at the intensely deformed edge and produced a crystal in which there was a very low density of dislocations. During further annealing, the boundary between the crystal and the remaining polygonized crystal advanced steadily. When the crystals were annealed for 12 hr, quite large areas were produced in which the only dislocations present were aligned as small-angle tilt boundaries with the axis of tilt normal to the surfacc of thc crystal and an angle of tilt between 5 and 15 secs. This technique for obtaining good, single crystals is unsuitable in the case of crystals of sodium chloride and other alkali halides since dislocations in these substances can form during the cooling from the high annealing temperature to lower temperatures (Mitchell, 1958).

14. ORGANIC COMPOUNDS

Tammann and Dreyer (1929) published data showing that certain compounds (such as camphor, pinene hydrochloride, and ice) show boundary migration very similar to that observed in metals. McCrone (1949, 1957) found that boundary motion in many organic compounds can be motivated either by cold working or by a temperature change. Compounds such as octachloropropane and the previous ones studied by Tammann showed grain growth where grain shape, and not the orientations of the neighboring crystals, is the controlling factor.

Boundary motion in compounds such as TNT, DDT, Vitamin K is, however, dependent on orientation of the crystals and not on grain shape (McCrone , 1949). The crystals grow in a direction that can be predicted for a given compound from the known relative orientations. For example, the

(001) face will penetrate either the (100) or (010) planes of adjacent crystals in DDT, whereas the (010) face will always grow into the (001) and (100) faces in TNT. This orientation effect is attributed to the fact that the crystals are structurally anisotropic and must be elastically anisotropic. In compounds such as octachloropropane, etc., the crystal lattices are either cubic or, at least, almost plastically isotropic. The orientation effect leading to boundary motion appears to be similar to that observed in noncubic metals where the strains set up by anisotropic expansions during thermal cycling cause grain growth.

It is evident that solid state techniques involving boundary migration may be useful in some cases where crystals of organic materials are required.

ACKNOWLEDGMENTS

The author would like to thank the following for their many helpful discussions and useful criticisms: W. G. Burgers, J. E. Burke, R. C. DeVries, C. G. Dunn, C. D. Graham, J. M. Lommel, L. Navias, J. W. Rutter, and D. F. Stein.

References

Allen, N. P., B. E. Hopkins, and J. E. McLennan (1956), *Proc. Roy. Soc. (A)* **234,** 221.

Anderson, E. A., D. C. Jillson, and S. R. Dunbar (1953), *Trans. AIME* **197,** 1191.

Andrade, E. N. da C. (1937), *Proc. Roy. Soc. (A)* **163,** 16.

Andrade, E. N. da C. and Y. S. Chow (1940), *Proc. Roy. Soc.* **175,** 290.

Armstrong, P. E., O. N. Carlson, and J. F. Smith (1959), *J. Appl. Phys.* **30,** 36.

Aust, K. T. and C. G. Dunn (1957), *Trans. AIME* **209,** 472.

Aust, K. T. and J. W. Rutter (1960), *Trans. AIME* **218,** 50, 1023.

Bagaryatskii, Y. A. and E. V. Kolonstava (1959), *Soviet Physics—Crystallography* **4,** 935.

Barrett, C. S. (1952), *Structure of Metals,* McGraw-Hill Co. New York.

Beaulieu, C. de (1955), Thesis, Paris; (April 1956), *Publ. IRSID,* No. 132.

Beck, P. A., J. C. Kremer, J. L. Demer, and M. L. Holzworth (1948), *Trans. AIME* **175,** 372.

Beck, P. A., M. L. Holzworth, and P. R. Sperry (1949), *Trans. AIME* **180,** 163.

Beck, P. A. and P. R. Sperry (1950), *J. Appl. Phys.* **21,** 150.

Beck, P. A., P. R. Sperry, and H. Hu (1950), *J. Appl. Phys.* **21,** 420.

Becker, J. H. (1958), *J. Appl. Phys.* **29,** 1110.

Buerger, M. J. and E. Washken (1947), *Am. Miner.* **32,** 296.

Burgers, W. G., J. C. Meijs, and T. J. Tiedema (1953), *Acta Met.* **1,** 75.

Burke, J. E. (1959), *Conference on Kinetics of High Temp. Processes,* M.I.T. Technical Press and J. Wiley and Sons, New York, 109; (1951), *Atom Movements,* A.S.M. Cleveland, Ohio, 209. Private Communication (1961), from Burke, J. E.

Cahn, R. W. (1953), *Acta Met.* **1,** 176.

Carpenter, H. C. H. and C. F. Elam (1921), *Proc. Roy. Soc.* **100,** 329.

Carpenter, H. C. H. and S. Tamura (1927), *Proc. Roy. Soc.* **113,** 28.

Chalmers, B. (1953), *Modern Research Techniques in Physical Metallurgy,* A.S.M., Cleveland, Ohio, 170.

Chen, N. K., R. Maddin, and R. Pond (1951), *Trans. AIME* **191,** 461.

Churchman, A. T. (1954), *Proc. Roy. Soc.* **226,** 216.

Coble, R. L. (1961), *J. Appl. Phys.* **32,** 793.

Coble, R. L. and J. E. Burke (1961), *Sintering in Ceramics, Progress in Ceramic Science,* Vol. 3, to be published.

Dohi, S. and T. Yamoshita (1957), *Mem. Def. Acad., Yokosuka, Japan* **2,** No. II, July.

Dunn, C. G. (1949a), *Trans. AIME* **185,** 72; (1949b), *Cold Working of Metals,* A.S.M. Cleveland, Ohio, 121.

Dunn, C. G. and G. C. Nonken (1953), *Met. Progress* 71.

Duwez, P., F. Odell, and J. L. Taylor (1949), *J. Am. Ceram. Soc.* **32,** 1.

Edwards, C. A. and L. B. Pfeil (1924), *J. Fe and Steel Inst.* **109,** 129.

Fisher, E. S. (1957), *Trans. AIME* **209,** 882.

Frois, C. and O. Dimitrov (1961), *Compt. rend.* **252,** 1465.

Fujiwara, T. (1939), *J. Sci. Hiroshima Univ.* **9,** 227.

Fujiwara, H. and T. Ichiki (1955), *J. Phys. Soc., Japan* **10,** 468.

Fullman, R. L. and J. C. Fisher (1951), *J. Appl. Phys.* **22,** 1350.

Gane, N. (1958), *Bull. Inst. Met.* **4,** 94.

Gensamer, M. and R. F. Mehl (1938), *Trans. AIME* **131,** 372.

Gifford, F. E. and E. A. Coomes (1960), *J. Appl. Phys.* **31,** 235.

Graham, C. D. and R. Maddin (1954–1955), *J. Inst. Met.* **83,** 169.

Green, R. E. Jr., B. G. Liebmann, and H. Yoshida (1959), *Trans. AIME* **215,** 610.

Guinier, A. and J. Tennevin (1950), *Prog. Met. Phys.* **2,** 177.

Hägg, G. and N. Karlsson (1952), *Acta Cryst.* **5,** 728.

Harrison, F. W. (1959), *Research* **12,** 395.

Holden, A. N. (1952), *Acta Cryst.* **5,** 182.

Holden, A. N. and J. H. Hollomon (1949), *Trans. AIME* **185,** 179.

Honeycombe, R. W. K. (1959), *Metallurgical Rev.* **4,** 1.

Hughes, F. L., H. Levinstein, and R. Kaplan (1959), *Phys. Rev.* **113,** 1023.

Jeffries, Z. and R. S. Archer (1924), *The Science of Metals,* McGraw-Hill, New York.

Kronberg, M. L. and F. H. Wilson (1949), *Trans. AIME* **185,** 501.

Lacombe, P. (1960–1961), *J. Inst. Met.,* **89,** 358.
Lacombe, P. (1948), *Report Conf. Strength of Solids,* Bristol, Lond. Phys. Soc., 91.
Lacombe, P. and A. Berghezan (1949), *Metaux et Corrosion* **24,** 1.
Lacombe, P., N. Ambrosis de Libamati, and D. Calais, (1959), *Compt. rend.* **249,** 2769.
Langeron, J. P. and P. Lehr (1956), *Compt. rend.* **243,** 151; (1958), *Rev. Met.* **55,** 901.
Legett, R. D., R. E. Read, and H. W. Paxton (1959), *Trans. AIME* **215,** 679.
Lehr, P. (1958), *Comm. energie atom. (France) Rappt.* **800,** 68.
Leighly, H. P. Jr., and F. C. Perkins (1960), *Trans. AIME* **218,** 379.
Lommel, J. M. (1960), *Trans. AIME* **218,** 374; (1961), Private Communication.
Maddin, R. (1958), *The Metal Molybdenum,* A.S.M. Cleveland, Ohio, edited by J. J. Harwood, 214.
Maddin, R. and N. K. Chen (1953), *Trans. AIME* **197,** 1131.
McCrone, W. C. (1949), *Discussion Faraday Soc.* **5,** 158; (1957), *Fusion Methods in Chemical Microscopy,* Interscience, 192.
McKeehan, L. W. (1927), *Nature* **119,** 705.
Mercier, J., D. Calais, and P. Lacombe (1958), *Compt. rend.* **246,** 110.
Millner, T. (1957), *Acta Tech. Acad. Sci. Hung.* **17,** 67.
Mitchell, J. W. (1958), *Growth and Perfection of Crystals,* John Wiley and Sons, New York, 386.
Montuelle, J. (1959), *Compt. rend.* **248,** 1174.
Montuelle, J. and G. Chaudron (1955), *Compt. rend.* **240,** 1167.
Mullins, W. W. (1958), *Acta Met.* **6,** 414.
Navias, L. (1961), *J. Am. Ceram. Soc.* **44,** 434.
Nichols, M. H. (1940), *Phys. Rev.* **57,** 297; (1954) **94,** 309.
Parthasarathi, M. N. and P. A. Beck (1958), *Trans. AIME* **212,** 821.
Rapperport, E. J. (1955), *Acta Met.,* **3,** 208.
Rieck, G. D. (1958), *Acta Met.* **6,** 360.
Robinson, J. (1942), *J. Appl. Phys.* **13,** 647.
Rutter, J. W. and K. T. Aust (1958), *Acta Met.* **6,** 375; (1960), *Trans. AIME* **218,** 682; (1961), **221,** 641.
Seraphim, D. P., J. I. Budnick, and N. B. Ittner, III (1960), *Trans. AIME* **218,** 527.
Sherby, O. D. (1958), *Trans. AIME* **212,** 708.
Smithells, C. J. (1952), *Tungsten,* Chapman and Hall, London.
Spachner, S. A. and W. Rostoker (1959), *Trans. AIME* **215,** 463.
Stein, D. F. and J. R. Low, Jr. (1961), *Trans. AIME* **221,** 774.
Stone, F. G. (1948), *Trans. AIME* **175,** 908.
Swalin, R. A. and A. H. Geisler (1957–1958), *J. Inst. Met.* **86,** 129.
Talbot, J. (1956), *IRSID Rept. Series A,* No. 137; (1960a), *Compt. rend.* **251,** 243; (1960b), *Internat. Meeting, New Physical and Chemical Properties of Very High-Purity Metals,* Paris, October 1959, 161, discussion 170.
Tammann, G., and K. L. Dreyer (1929), *Z. Anorg. Chem.* **182,** 289.
Tiedema, T. J. (1949), *Acta Cryst.* **2,** 261.
Tiedema, T. J., W. May, and W. G. Burgers (1949), *Acta Cryst.* **2,** 151.
Toth, R. S., R. Kilkson, and D. Trivich (1960), *J. Appl. Phys.* **31,** 1117.
Tsien, L. C. and Y. S. Chow (1937), *Proc. Roy. Soc. A* **163,** 19.
Van Uitert, L. G., F. W. Swanekamp, and S. E. Haszko (1959), *J. Appl. Phys.* **30,** 363.
Verbraak, C. A. (1960), *Acta Met.* **8,** 65.
Walter, J. L. and C. G. Dunn (1959), *Trans. AIME* **215,** 465; (1960), *Acta Met.* **8,** 497.
Wilder, D. R. and E. S. Fitzsimmons (1955), *J. Am. Ceram. Soc.* **38,** 66.
Williamson, G. K. and R. E. Smallman (1953), *Acta Met.* **1,** 487.
Yamamoto, M. and R. Miyasawa (1954), *Sci. Rept. Res. Inst. Tohoku Univ. (A)* **6,** 333.

Substance Index

Subject Index